A user's guide to spectral sequences, second edition

Spectral sequences are among the most elegant, powerful, and complicated methods of computation in mathematics. This book describes some of the most important examples of spectral sequences and some of their most spectacular applications.

The first third of the book treats the algebraic foundations for this sort of homological algebra, starting from informal calculations, to give the novice a familiarity with the range of applications possible. The heart of the book is an exposition of the classical examples from homotopy theory, with chapters on the Leray-Serre spectral sequence, the Eilenberg-Moore spectral sequence, the Adams spectral sequence, and, in this new edition, the Bockstein spectral sequence. The last part of the book treats applications throughout mathematics, including the theory of knots and links, algebraic geometry, differential geometry, and algebra.

This is an excellent reference for students and researchers in geometry, topology, and algebra.

John McCleary is Professor of Mathematics at Vassar College.

D1322271

CAMBRIDGE STUDIES IN ADVANCED MATHEMATICS

Editorial Board:

B. Bollobas, W. Fulton, A. Katok, F. Kirwan, P. Sarnak

A USER'S GUIDE TO
SPECTRAL SEQUENCES

Second Edition

JOHN McCLEARY

Vassar College

CAMBRIDGE
UNIVERSITY PRESS

PUBLISHED BY THE PRESS SYNDICATE OF THE UNIVERSITY OF CAMBRIDGE
The Pitt Building, Trumpington Street, Cambridge, United Kingdom

CAMBRIDGE UNIVERSITY PRESS
The Edinburgh Building, Cambridge CB2 2RU, UK
40 West 20th Street, New York, NY 10011-4211, USA
10 Stamford Road, Oakleigh, Melbourne 3166, Australia
Ruiz de Alarcón 13, 28014 Madrid, Spain
Dock House, The Waterfront, Cape Town 8001, South Africa

http://www.cambridge.org

First published 1985
Second edition 2001

Printed in the United States of America

Typeface Times Roman 10/13 pt.   *System* TeXtures 1.7 [au]

A catalog record for this book is available from the British Library.

Library of Congress Cataloging in Publication data
McCleary, John, 1952–
A user's guide to spectral sequences/John McCleary.
p.   cm.– (Cambridge studies in advanced mathematics ; 58)
Includes bibliographical references and index.
ISBN 0-521-56141-8 – ISBN 0-521-56759-9 (paperback)
1. Spectral sequences (Mathematics)   I. Title.   II. Series.
QA612.8.M33   2000
514'.2–dc21   00-059866

ISBN 0 521 56141 8 hardback
ISBN 0 521 56759 9 paperback

To my family:

      to my Dad and to the memory of my Mom,

      to Carlie,

      and to my boys, John and Anthony.

Preface

to the second edition

"For I know my transgressions, . . . "

Psalm 51

The first edition of this book served as my introduction to the mysteries of spectral sequences. Since writing it, I have learned a little more in trying to do some algebraic topology using these tools. The sense that I had misrepresented some topics, misled the reader, even written down mistaken notions, grew over the years. When the first edition came to the end of its run, and was going out of print, I was encouraged by some generous souls to consider a second edition with the goal of eliminating many of the errors that had been found and bringing it somewhat up to date.

The most conspicuous change to the first edition is the addition of new chapters—Chapter 8bis on nontrivial fundamental groups and Chapter 10 on the Bockstein spectral sequence. In Chapter 8bis (an *address*, added on after Chapters 5 through 8, but certainly belonging in that neighborhood), I have found a natural place to discuss the Cartan-Leray and the Lyndon-Hochschild-Serre spectral sequences, as well as the important class of nilpotent spaces. This chapter is an odd mixture of topics, but I believe they hang together well and add details to earlier discussions that depended on the fundamental group. Chapter 10 acknowledges the fundamental role that the Bockstein spectral sequence plays in homotopy theory, especially in the study of H-spaces. It is as much a basic tool as the other spectral sequences of Part II.

Less conspicuously, I have changed the order of topics in Chapters 2 and 3 in order to focus better on convergence in Chapter 3, which includes an exposition of the important paper of [Boardman99]. I have reordered the topics in Chapter 8 to make it more parallel to Chapter 6. The proof of the existence and structure of the Leray-Serre spectral sequence is also significantly changed. I have followed the nice paper of [Brown, E94]. With this change, I have added a proof of the multiplicative structure that was not in the first edition. Chapter 9 now sports a discussion of the role of the Adams spectral sequence in the computation of cobordism rings.

Many of the intended improvements in this edition are small details that are mentioned in the acknowledgments. Details that are noticeable throughout include a change in the convention for citation. (*What was I thinking in the first edition?*) In this edition the reader is invited to read the citations as an integral part of the text. In the case of multiple papers in a given year, I have added a prime to the year to distinguish papers. The other little global change is an end of proof marker □ (suggested by Michele Intermont).

I had once thought that writing a book would be easy and the first edition cured me of that misconception. I have discovered that writing a second edition isn't easy either. I will thank others at the end of the Introduction, but I wish to thank certain folks whose encouragement, kind words, and steadfastness made the completion of the second edition possible. First are my teachers in the use of spectral sequences, and in the writing of books, Jim Stasheff, Bob Bruner, and Larry Smith. They have all given more than one could expect of a friend. To you, I owe many thanks for so many kindnesses. In an effort to avoid a second edition full of little errors that frustrate even the most diligent reader, Hal Sadofsky organized an army of folks who read the penultimate version of most chapters. This act of organization was most welcome, helpful and generous. Though I may have added new typos in an effort to fix found errors, I am sure that the book is much better for Hal's efforts. At Vassar, Diane Winkler gave some of her valuable time to help in the preparation of the bibliography and index. Ben Lotto solved all my computer problems, and Flora Grabowska hunted down reference material I always seemed to need yesterday. Much of the work on this edition was done during *une année sabbatique* à Strasbourg. My thanks to Christian Kassel and Jean-Louis Loday for their hospitality during that stay. In the department of steadfastness, many thanks go to my editor at Cambridge University Press, Lauren Cowles, whose patience is extraordinary. Finally, my thanks to my family—Carlie, John, and Anthony—for tolerating my projects and for their love through what seemed like a never ending story.

John McCleary   *July 17, 2000*
Poughkeepsie, NY

Introduction

"It is now abundantly clear that the spectral sequence is one of the fundamental algebraic structures needed for dealing with topological problems."

W.S. Massey

Topologists are fond of their machinery. As the title of this book indicates, my intention is to provide a user's manual for the class of complicated algebraic gadgets known as spectral sequences.

A 'good' user's manual for any apparatus should satisfy certain expectations. It should provide the beginner with sufficient details in exposition and examples to feel comfortable in starting to apply the new apparatus to his or her problems. The manual should also include enough details about the inner workings of the apparatus to allow a user to determine what is going on if it fails while in operation. Finally, a user's manual should include plenty of information for the expert who is looking for new ways to use the device. In writing this book, I have kept these goals in mind.

There are several classes of readers for whom this book is written. There is the student of algebraic topology who is interested in learning how to apply spectral sequences to questions in topology. This reader is expected to have seen a basic course in topology at the level of the texts by [Massey91] and [May99] on singular homology theory and including the definition of homotopy groups and their basic properties. This beginner also needs an acquaintance with the basic topics of the homological algebra of rings and modules, at the level of the first three chapters of the book of [Weibel94].

The next class of reader is principally interested in algebra and he or she wants an exposition of the basic notions about spectral sequences, hopefully without too much topology as prerequisite. Part I and Chapter 12 are intended for these readers, along with §7.1, §8bis.2, and §9.2.

Some sections of the book are intended for the experienced user and would offer an unenlightening detour for the novice. I have marked these sections with the symbol

for 'not for the novice.' As with other users' manuals, these sections will become useful when the reader becomes familiar with spectral sequences and has a need for particular results.

The material in the book is organized into three parts. Part I is called **Algebra** and consists of Chapters 1, 2, and 3. The intention in Part I is to lay the algebraic foundations on which the construction and manipulation of all subsequent examples will stand. Chapter 1 is a gentle introduction to the manipulation of first quadrant spectral sequences; the problem of how to construct a spectral sequence is set aside and some of the formal aspects of the algebra of these objects are developed. In Chapter 2, the algebraic origins of spectral sequences are treated in three classic cases—filtered differential graded modules, exact couples, and double complexes—along with examples of these ideas in homological algebra. The subtle notion of convergence is the focus of Chapter 3. Comparison theorems are introduced here and the underlying theory of limits and colimits is presented.

Part II is called **Topology**; it is the heart of the book and consists of Chapters 4 through 10. Part II treats the four classical examples of spectral sequences that are found in homotopy theory. The introduction to each chapter gives a detailed summary of its contents. We describe the chapters briefly here. Chapter 4 is a thumbnail sketch of the topics in basic homotopy theory that will be encountered in the development of the classical spectral sequences. Chapters 5 and 6 treat the Leray-Serre spectral sequence, and Chapters 7 and 8 treat the Eilenberg-Moore spectral sequence. Chapters 5 and 7, labeled as I, contain a construction of each spectral sequence and develop their basic properties and applications. Chapters 6 and 8, labeled as II, go into the deeper structures of the spectral sequences and apply these structures to less elementary problems. Alternate constructions of each spectral sequence appear in Chapters 6 and 8. Chapter 8bis gives an account of the effect of a nontrivial group on the Leray-Serre spectral sequence and the Eilenberg-Moore spectral sequence. Important topics, including nilpotent spaces, the homology of groups, and the Cartan-Leray and Lyndon-Hochschild-Serre spectral sequences, are developed. Chapter 9 treats the classical Adams spectral sequence (as constructed in the days before spectra). Chapter 10 treats the Bockstein spectral sequence, especially as a tool in the study of H-spaces. Throughout the book, I have followed an historical development of the topics in order to maintain a sense of the motivation for each development. In some of the proofs found in the book, however, I have strayed from the original papers and found other (hopefully more direct) proofs, especially based on the results of Part I.

Part III is called **Sins of Omission** and consists of Chapters 11 and 12. My first intention was to provide a catalogue of everyone's favorite spectral sequence, if it doesn't happen to be in Chapters 4 through 9. This has become too large an assignment as spectral sequences have become almost commonplace in many branches of mathematics. I have chosen some of the major examples and a few exotica to demonstrate the breadth of applications of spectral sequences. Chapter 11 consists of spectral sequences of use in topology. Chapter 12 includes examples from commutative algebra, algebraic geometry, algebraic K-theory, and analysis, even mathematical physics.

There are exercises at the end of all of the chapters in Parts I and II. They offer further applications, missing details, and alternate points of view. The novice should find these exercises helpful.

The bibliography consists of papers and books cited in the text. At the end of each bibliographic entry is a list of the pages where the paper has been cited. The idea of a comprehensive bibliography on spectral sequences is unnecessary with access to **MathSciNet** or the **Zentralblatt MATH Database**. These databases allow easy searches of titles and reviews of most of the publications written after the introduction of spectral sequences.

How to use this book

These instructions are intended for the novice who is seeking the shortest path to some of the significant applications of spectral sequences in homotopy theory. The following program should take the least amount of time, incur the least amount of pain, and provide a good working knowledge of spectral sequences.

(1) All of Chapter 1.
(2) §2.1, §2.2 (but skip the proof of Theorem 2.6), §2.3.
(3) §3.1 and §3.3.
(4) Chapter 4, as needed.
(5) §5.1 and §5.2.
(6) §6.1, §6.2, and §6.3.

From this grounding, the Bockstein (Chapter 10), the Cartan-Leray, and the Lyndon-Hochschild-Serre spectral sequences ($\S 8^{\mathrm{bis}}.2$) are accessible. The novice who is interested in the Eilenberg-Moore spectral sequence should include §2.4 with the above and then go on to Chapter 7 as desired. The novice who is interested in the Adams spectral sequence should also read §2.4 as well as §7.1 for the relevant homological algebra before embarking on Chapter 9.

Details and Acknowledgments

In the writing of both editions of this book, many people have offered their time, expertise, and support to whom I acknowledge a great debt. Along with a thanksgiving, I will say a little about the sources of each chapter.

This project began in Philadelphia, in the car with Bruce Conrad, between Germantown and Temple University. It was going to be a handy pamphlet, listing E_2-terms and convergence results, but it has since run amok. At the beginning, chats with Jim Stasheff, Lee Riddle, and Alan Coppola were encouraging. Chapter 1 is modeled on the second graduate course in algebraic topology I took from Jim and on unpublished notes for such a course written by David Kraines. David Lyons spotted a crucial misstatement in this chapter in the first edition. Michele Intermont gave it a good close reading for the second edition. Chapter 2 is classical in outline and owes much to my reading of the classics by [Cartan-Eilenberg56], [Eckmann-Hilton66], and [Mac Lane63]. Chapter 3 is my account of the foundational paper of [Boardman99], the 1981 version of which I count among the classics in algebraic topology. Alan Hatcher's close reading of the first edition version of the Zeeman comparison theorem resulted in some significant changes. Brooke Shipley read my first version of the new Chapter 3 and gave many helpful remarks. Chapter 4 was a suggestion of Lee Riddle whose notes on beginning homotopy theory were helpful.

Chapter 5 is based on the thesis of [Serre51]—it is a remarkable paper and I have added more of it to the second edition. The paper of [Brown, E94] is the backbone of §5.3. Special thanks to Ed Brown and Don Davis for close readings of part of this chapter. Chapter 6 is based on further work of [Serre51, 53], the thesis of [Borel53], the lovely paper of [Dress67], and subsequent developments. Chapter 7 is based on the Yale thesis of [Smith, L70]. I have learned a lot from the papers of and conversations with Larry Smith. His remarks on an early version of Chapters 7, 8, and 8bis made a big impact on the final version of these chapters. Jim Stasheff first taught me many of the ideas in Chapters 7 and 8, and deeply influenced their formulation. I wrote these chapters first in the first edition and I believe they bear the stamp of Jim's teaching.

Chapter 8bis was an idea based on a remark of Serre in a list of errata for the first edition he kindly sent me in 1985. He suggested that the class of nilpotent spaces should be presented in a discussion of spectral sequences. This chapter is the result. I have learned a lot from Emmanuel Dror-Farjoun and Bill Dwyer in the writing of this chapter. Coffee hours with Richard Goldstone have been very helpful as well. Chapter 9 is a result of my reading of the papers of Adams and Liulevicius and the beautiful book of [Mosher-Tangora68]. Bob Bruner and Norihiko Minami added much to the chapter in their readings of early versions of it. Chapter 10 is based on the papers of Bill Browder, Jim Lin, and Richard Kane. An outline of Chapter 10 appeared as the introduction to my paper, [McCleary87].

Many people have offered suggestions that enhanced my presentation and my understanding; they include Claude Schochet, Bill Massey, Nathan Habegger, Dan Grayson, Bob Thomason, John Moore, Andrew Ranicki, Frank Adams, Ian Leary, Alan Durfee, Carl-Friedrich Bödigheimer, Johannes Huebschmann, Lars Hasselholt, Jerry Lodder, Guido Mislin, and Jason Cantarella.

The army of readers organized by Hal Sadofsky are owed a mighty thanks for the degree of care and commitment they gave in making the book a better effort. They are: Zoran Petrovic (Chapter 2), Don Davis and Martin Crossley (Chapter 4), Martin Cadek and Dan Christensen (Chapter 5), Chris French (Chapter 6), Jim Stasheff (Chapters 7, 8, and 12), Tom Hunter and Kathryn Hess (Chapter 8), Richard Goldstone (Chapter 8bis), Christian Nassau (Chapter 9), Ethan Berkove and Kathryn Lesh (Chapter 10), and Frank Neumann (Chapters 11 and 12).

My thanks to Michael Spivak for his support of the first edition and for making the old TEX-files available for the second edition.

Finally, I wish to acknowledge the collective intellectual debt owed to the mathematicians whose work makes up this book—in particular, to Jean Leray (1906–1998), Henri Cartan, Jean-Louis Koszul, Jean-Pierre Serre, Armand Borel, Samuel Eilenberg (1913–1998), John Moore, and Frank Adams (1930–1989).

Table of Contents

Part I
Algebra

1

An Informal Introduction

"Sauter à pieds joints sur ces calculs; . . . telle est, suiv-
ant moi, la mission des géometres futurs; . . . "

E. Galois

In the chapters that follow, we will consider, in detail, the algebra of spectral sequences and furthermore, how this formalism can be applied to a topological problem. The user, however, needs to get acquainted with the manipulation of these gadgets without the formidable issue of their origins. This chapter is something of a tool kit, filled with computation techniques that may be employed by the user in the application of spectral sequences to algebraic and topological problems. We take a loosely axiomatic stance and argue from definitions, with most spectral sequences in mind. As in the case of long exact sequences or homology theory, this viewpoint still makes for a substantial enterprise. The techniques developed in this chapter, though elementary, will appear again and again in what follows. The user, facing a computation in later chapters, will profit by returning to this collection of tools and tricks.

1.1 "There is a spectral sequence . . . "

Let us begin with a **basic goal**: We want to compute H^* where H^* is a graded R-module or a graded k-vector space or a graded k-algebra or . . . . This H^* may be the homology or cohomology of some space or some other graded algebraic invariant associated to a space or perhaps an invariant of some algebraic object like a group, a ring or a module; in any case, H^* is often difficult to obtain. In order to proceed, we introduce some helpful conditions. Suppose further that H^* is **filtered**, that is, H^* comes equipped with a sequence of subobjects,

$$H^* \supset \cdots \supset F^n H^* \supset F^{n+1} H^* \supset \cdots \supset \{0\}.$$

For the sake of clarity, let's assume for this chapter that H^* is a graded vector space over a field k, unless otherwise specified and further, that $H^* = F^0 H^*$, that is, our filtration is **bounded below** by H^* in the 0th filtration. For example,

if H^* is any graded vector space with $H^n = \{0\}$ for $n < 0$, then there is an obvious filtration, induced by the grading, and given by $F^p H^* = \bigoplus_{n \geq p} H^n$.
A less trivial filtration is induced on the cohomology of a CW-complex, X, (see §4.1 for definitions) by filtering the space itself by successive skeleta:

$$X \supset \cdots \supset X^{(n)} \supset X^{(n-1)} \supset X^{(n-2)} \supset \cdots \supset X^{(0)} \supset \{*\},$$

and defining $F^p H^*(X) = $ the kernel of the map $H^*(X) \to H^*(X^{(p-1)})$, induced by the $(p-1)^{\text{st}}$ inclusion map.

A filtration of H^*, say F^*, can be collapsed into another graded vector space called the **associated graded vector space** and defined by $E_0^p(H^*) = F^p H^* / F^{p+1} H^*$. In the case of a locally finite graded vector space (that is, H^n is finite dimensional for each n), H^* can be recovered up to isomorphism from the associated graded vector space by taking direct sums, that is,

$$H^* \cong \bigoplus_{p=0}^{\infty} E_0^p(H^*).$$

If H^* is an arbitrary graded module over some ring R, however, there may be extension problems that prevent one from reconstructing H^* from the associated graded module (these problems will be discussed in §3.1). As a simple example of an associated graded vector space, consider the filtration induced by the grading on H^*; here we have $E_0^p(H^*) = H^p$.

Because H^* may not be easily computed, we can take as a first approximation to H^* the associated graded vector space to some filtration of H^*. *This is the target of a spectral sequence!* We then hope that H^* can be reassembled from $E_0^*(H^*)$. Before we give the definition, observe a simple property of the associated graded vector space to a filtered graded vector space H^*: $E_0^*(H^*)$ is **bigraded**. Using the degree in H^*, we define $F^p H^r = F^p H^* \cap H^r$ and

$$E_0^{p,q} = F^p H^{p+q} / F^{p+1} H^{p+q}.$$

The index q is called the **complementary degree** and the index p the **filtration degree**. The associated graded vector space $E_0^p(H^*)$ can be recovered by taking the direct sum of the spaces $E_0^{p,q}$ over the index q. To recover H^r directly, as a vector space, take the direct sum $\bigoplus_{p+q=r} E_0^{p,q}$.

We are now in position to introduce the objects of interest.

Definition 1.1 (First Definition). *A (first quadrant, cohomological)* **spectral sequence** *is a sequence of* **differential bigraded vector spaces**, *that is, for $r = 1, 2, 3, \ldots$, and for p and $q \geq 0$, we have a vector space $E_r^{p,q}$. Furthermore, each bigraded vector space, $E_r^{*,*}$, is equipped with a linear mapping $d_r : E_r^{*,*} \longrightarrow E_r^{*,*}$, which is a differential, $d_r \circ d_r = 0$, of bidegree $(r, 1-r)$,*

$$d_r : E_r^{p,q} \longrightarrow E_r^{p+r,q-r+1}.$$

Finally, for all $r \geq 1$, $E_{r+1}^{*,*} \cong H(E_r^{*,*}, d_r)$, that is,

$$E_{r+1}^{p,q} = \ker d_r \colon E_r^{p,q} \to E_r^{p+r,q-r+1} \Big/ \operatorname{im} d_r \colon E_r^{p-r,q+r-1} \to E_r^{p,q} .$$

We call the rth stage of such an object its E_r-**term** (or its E_r-*page*) and it may be pictured as a lattice in the first quadrant with each lattice point a vector space and the differentials as arrows (here for $r = 3$).

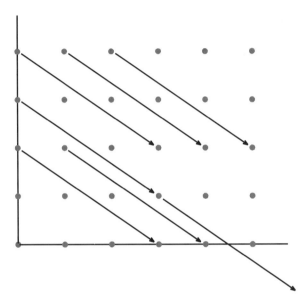

With this definition, consider $E_r^{p,q}$ for $r > \max(p, q+1)$; here the differentials, d_r, become trivial. Since $q+1-r < 0$, we have $E_r^{p+r,q-r+1} = \{0\}$ and so $\ker d_r = E_r^{p,q}$. Also $p - r < 0$ implies $E_r^{p-r,q+r-1} = \{0\}$ and $\operatorname{im} d_r = \{0\}$. Thus $E_{r+1}^{p,q} = E_r^{p,q}$ and, continuing in a similar fashion, $E_{r+k}^{p,q} = E_r^{p,q}$, for $k \geq 0$. We denote this common vector space by $E_\infty^{p,q}$. We can now complete the preliminary definitions.

Definition 1.2. *A spectral sequence $\{E_r^{*,*}, d_r\}$ of vector spaces is said to* **converge** *to a graded vector space H^* if H^* has a filtration F^* and*

$$E_\infty^{p,q} \cong F^p H^{p+q} / F^{p+1} H^{p+q} = E_0^{p,q}(H^*).$$

Though the notion of convergence requires conditions on the filtration in order to determine that the target is uniquely H^*, this definition will serve for this first exposure (see Chapter 3 for more sophisticated issues). It follows that our basic goal H^* is approximated, if we can find a spectral sequence converging to H^*. The "generic theorem" in this enterprise takes the following form.

"*Theorem I.*" *There is a spectral sequence with*

$$E_2^{*,*} \cong \text{"something computable"}$$

and converging to H^*, something desirable.

The important observation to make about the statement of the theorem is that *it gives an E_2-term of the spectral sequence but says nothing about the successive differentials d_r*. Though $E_r^{*,*}$ may be known, without d_r or some further structure, it may be impossible to proceed.

There is hope, however, that even without knowledge of the differentials, we can proceed as a formal consequence of the algebraic structure of some E_r-term. (An analogous situation arises in a long exact sequence where every third term is $\{0\}$; this implies a family of isomorphisms.) Since the differentials in a spectral sequence may carry otherwise inaccessible geometric information, such algebraic situations may lead to significant results. Our first example shows this feature of the existence and form of a spectral sequence. (Compare this example with Example 5.D.)

Example 1.A. *Suppose that there is a first quadrant spectral sequence of co-homological type with initial term $(E_2^{*,*}, d_2)$, converging to the graded vector space H^*, and satisfying $F^{p+k}H^p = \{0\}$ for all $k > 0$. Then $H^0 = E_2^{0,0}$, and there is an exact sequence,*

$$0 \to E_2^{1,0} \to H^1 \to E_2^{0,1} \xrightarrow{d_2} E_2^{2,0} \to E_\infty^{2,0} \to 0,$$

with $E_\infty^{2,0}$ a submodule of H^2.

Since there is a filtration on H^* whose associated graded vector space is given by the E_∞-term of this spectral sequence, then we can relate the final term to its target in the cases of H^0, H^1, and H^2. Now $F^0 H^0 = H^0$ and since $F^1 H^0 = \{0\}$, $E_\infty^{0,0} = H^0$. No nonzero differentials can involve $E_2^{0,0}$, so $E_2^{0,0} = E_\infty^{0,0}$.

For H^1 and H^2, the filtrations may be presented by

$$\{0\} \subset F^1 H^1 \subset F^0 H^1 = H^1, \qquad \{0\} \subset F^2 H^2 \subset F^1 H^2 \subset F^0 H^2 = H^2.$$

This leads to short exact sequences,

$$0 \to E_2^{1,0} \to H^1 \to E_2^{0,1} \to 0, \qquad 0 \to E_\infty^{2,0} \to H^2.$$

The only differential that appears in this lowest degree corner of the spectral sequence is $d_2 \colon E_2^{0,1} \to E_2^{2,0}$. This leads to the short exact sequence:

$$0 \to E_\infty^{0,1} \to E_2^{0,1} \xrightarrow{d_2} E_2^{2,0} \to E_\infty^{2,0} \to 0.$$

Putting this short exact sequence together with the ones that follow from the filtration, gives the exact sequence in the example.

For the rest of the chapter, we will consider various algebraic situations where the formal presentation of the data or some aspect of an extra structure allows us to compute effectively with a spectral sequence. More general spectral sequences (for example, not first quadrant or not restricted to vector spaces) will be defined and considered in Chapter 2.

1.2 Lacunary phenomena

Suppose that our generic 'Theorem I' holds and we have a first quadrant spectral sequence with H^* as the target of convergence. The simplest case of calculation occurs when a finite number of steps completes the computation.

Definition 1.3. *A spectral sequence,* $\{E_r^{*,*}, d_r\}$ *is said to* **collapse at the** N^{th} **term** *if* $d_r = 0$ *for* $r \geq N$.

Of course the immediate consequence of collapse at the Nth term is that $E_N^{*,*} \cong E_{N+1}^{*,*} \cong \cdots \cong E_\infty^{*,*}$ and we have recovered the graded vector space H^* up to isomorphism.

Example 1.B. *Suppose* n_1 *and* n_2 *are natural numbers and* $E_2^{p,q} = \{0\}$ *for* $p > n_1$ *or* $q > n_2$. *Then the spectral sequence collapses at the* N^{th} *term where* $N = \min(n_1 + 1, n_2 + 2)$.

We can picture the E_2-term in the two possible extreme cases as in the diagram:

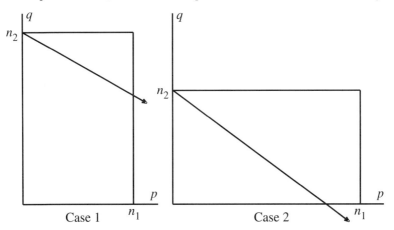

In case 1, $n_1 \leq n_2 + 1$ and so $N = n_1 + 1$. For $r \geq N, d_r \colon E_r^{p,q} \to E_r^{p+r,q-r+1}$ and so $p + r \geq p + N > n_1$ and $d_r = 0$. In case 2, $n_2 + 1 < n_1$ and $N = n_2 + 2$. Since $q - r + 1 \leq q - n_2 - 1 < 0$, we have $d_r = 0$. Thus the spectral sequence collapses at E_N and $E_N^{*,*} \cong E_{N+1}^{*,*} \cong \cdots \cong E_\infty^{*,*}$.

Example 1.C. *Suppose $E_2^{p,q} = \{0\}$ whenever p is even or q is odd. Then the spectral sequence collapses at E_2.*

This example may seem artificial but it can occur when the E_2-term is given by a tensor product $E_2^{p,q} = V^p \otimes_k W^q$ and V^* and W^* are graded vector spaces over k such that V^* is concentrated in odd dimensions and W^* is concentrated in even dimensions.

To prove the assertion that the spectral sequence collapses, observe, for any $r \geq 2$, $E_r^{p,q} \neq \{0\}$ implies $p + q \equiv 1 \pmod 2$. But $d_r \colon E_r^{p,q} \to E_r^{p+r,q-r+1}$ changes total degree, $p+q$, by one; $p+r+q-r+1 = p+q+1 \equiv 0 \pmod 2$ so $E_r^{p+r,q-r+1} = \{0\}$. Thus $d_r = 0$ for $r \geq 2$ and the spectral sequence collapses at the E_2-term.

In the previous examples the placement of "holes" (trivial vector spaces) in the spectral sequence forced a collapse. In the next example, the holes do not defeat the differentials but restrict their possible action.

Example 1.D. *Suppose $E_2^{p,q} = \{0\}$ unless $q = 0$ or $q = n$, for some $n \geq 2$. Then there is a long exact sequence*

$$\cdots \longrightarrow H^{p+n} \longrightarrow E_2^{p,n} \xrightarrow{d_{n+1}} E_2^{p+n+1,0} \longrightarrow$$
$$H^{p+n+1} \longrightarrow E_2^{p+1,n} \xrightarrow{d_{n+1}} E_2^{p+n+2,0} \longrightarrow \cdots .$$

To concoct some instances of these conditions, let $W^* = H^*(S^n; k)$ and V^* be any graded vector space with $E_2^{p,q} = V^p \otimes_k W^q$. In the study of the Leray-Serre spectral sequence, this example will be significant.

Pictorially, our conditions obtain two stripes in which the nontrivial data lie, as in the following diagram.

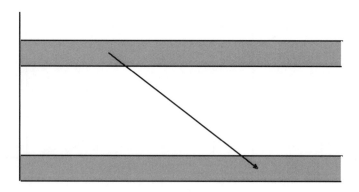

By the placement of trivial vector spaces in this E_2-term, the only possible nonzero differential is $d_{n+1} \colon E_2^{p,n} \to E_2^{p+n+1,0}$. Therefore we have $E_2 \cong \cdots \cong E_{n+1}$ and $E_{n+2} = H(E_{n+1}, d_{n+1}) \cong E_\infty$. Now d_{n+1} is the zero homomorphism on the bottom row and so $E_\infty^{*,0} \cong E_2^{*,0}/\operatorname{im} d_{n+1}$. Since everything above $E_2^{*,n}$ is trivial, $E_\infty^{*,n} \cong \ker d_{n+1}$. This yields the short exact sequence for each p,

$$0 \to E_\infty^{p,n} \to E_2^{p,n} \xrightarrow{d_{n+1}} E_2^{p+n+1,0} \to E_\infty^{p+n+1,0} \to 0.$$

Since $E_\infty^{*,*}$ is nontrivial only in these stripes, the filtration on H^* takes the form

$$H^{p+n} = F^0 H^{p+n} = \cdots = F^n H^{p+n} \supset F^{n+1} H^{p+n}$$
$$= F^{n+2} H^{p+n} = \cdots = F^{p+n} H^{p+n} \supset \{0\}.$$

Furthermore, $E_\infty^{p,n} = H^{p+n}/F^{p+1}H^{p+n} \cong H^{p+n}/E_\infty^{p+n,0}$ and so we have the short exact sequence for each p

$$0 \longrightarrow E_\infty^{p+n,0} \longrightarrow H^{p+n} \longrightarrow E_\infty^{p,n} \longrightarrow 0.$$

To obtain the long exact sequence of the example, we splice these short exact sequences together as in the diagram.

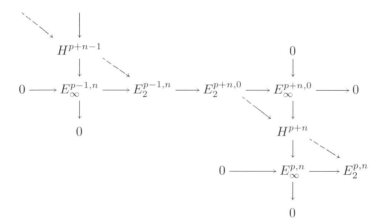

The long exact sequence given in Example 1.C is called the **Gysin sequence**. The analogous assertion holds when p and q are interchanged to give vertical stripes (see the exercises at the end of the chapter).

1.3 Exploiting further structure

It is often the case that H^*, the target of a spectral sequence, carries more structure than simply that of a graded vector space. For example, when H^* is the cohomology of a topological space, it is a graded algebra via the cup-product structure (see §4.4). Furthermore, if p is a prime number and $H^* = H^*(X; \mathbb{F}_p)$, for a space X, then H^* is a graded module over the mod p Steenrod algebra (see §4.4). In §1.1 we discussed how one might determine H^* as a graded vector space through a spectral sequence. If H^* is a graded algebra, we can ask how a spectral sequence might determine this structure as well. To be precise, we first endow the spectral sequence with an algebra structure and determine what it means for the algebra structure to converge to the algebra structure on H^*. This will turn out to be a subtle matter, though it has rich rewards. In §1.5, we take up the more difficult problem of reconstructing H^* from the bigraded algebra $E_\infty^{*,*}$.

Definition 1.4. *If H^* is a graded vector space over a field k, then let $H^* \otimes_k H^*$ be the graded vector space given by*

$$(H^* \otimes_k H^*)^n = \bigoplus_{p+q=n} H^p \otimes_k H^q.$$

H^* is a **graded algebra** *if there is a mapping of graded vector spaces, called the* **product**, *$\varphi \colon H^* \otimes_k H^* \to H^*$, that is, a linear mapping $\varphi \colon H^p \otimes_k H^q \to H^{p+q}$ for all p and q. The mapping φ must satisfy the following commutative diagram that expresses the fact that the multiplication on H^* is associative:*

$$
\begin{array}{ccc}
H^* \otimes H^* \otimes H^* & \xrightarrow{\ \varphi \otimes 1\ } & H^* \otimes H^* \\
{\scriptstyle 1 \otimes \varphi}\big\downarrow & & \big\downarrow{\scriptstyle \varphi} \\
H^* \otimes H^* & \xrightarrow{\quad \varphi \quad} & H^*.
\end{array}
$$

An algebra may also contain a unit element. This is expressed by a mapping $\eta \colon k \to H^*$ where k is the graded algebra determined by the field k in degree 0 and by $\{0\}$ in higher degrees. The mapping η must satisfy the following commutative diagram:

$$
\begin{array}{ccccc}
k \otimes H^* & \xrightarrow{\ \eta \otimes \mathrm{id}\ } & H^* \otimes H^* & \xleftarrow{\ \mathrm{id} \otimes \eta\ } & H^* \otimes k \\
{\scriptstyle \cong}\big\downarrow & & {\scriptstyle \varphi}\big\downarrow & & \big\downarrow{\scriptstyle \cong} \\
H^* & =\!\!=\!\!=\!\!= & H^* & =\!\!=\!\!=\!\!= & H^*.
\end{array}
$$

One can consider $\mathbb{Q}[x]$ as a graded vector space where the degree is given by the degree of a polynomial. With polynomial multiplication, $\mathbb{Q}[x]$ enjoys a graded algebra structure.

Definition 1.5. *Suppose $E^{*,*}$ is a bigraded vector space. Let $E^{*,*} \otimes_k E^{*,*}$ be the bigraded vector space given by*

$$(E^{*,*} \otimes_k E^{*,*})^{p,q} = \bigoplus_{\substack{m+n=p \\ r+s=q}} E^{m,r} \otimes_k E^{n,s}.$$

$E^{*,*}$ is a **bigraded algebra** *if there is a mapping of bigraded vector spaces, called the* **product**,

$$\varphi \colon E^{*,*} \otimes_k E^{*,*} \longrightarrow E^{*,*},$$

that is, φ has components $\varphi \colon E^{m,n} \otimes_k E^{r,s} \to E^{m+r,n+s}$. We assume further that φ satisfies the analogous conditions of associativity and having a unit.

As an example of a bigraded algebra, suppose A^* and B^* are graded algebras with φ and ψ their respective products. If we let $E^{p,q} = A^p \otimes_k B^q$, we define an algebra structure on $E^{*,*}$ by the composite:

$$\Phi \colon E^{p,q} \otimes_k E^{r,s} = A^p \otimes B^q \otimes A^r \otimes B^s \xrightarrow{\;1 \otimes T \otimes 1\;}$$

$$A^p \otimes A^r \otimes B^q \otimes B^s \xrightarrow{\;\varphi \otimes \psi\;} A^{p+r} \otimes B^{q+s} = E^{p+r,q+s}$$

where $T(b \otimes a) = (-1)^{(\deg a)(\deg b)} a \otimes b$.

The next ingredient in our definition is the differential. We give the graded and bigraded definitions. Let $a \cdot a' = \varphi(a, a')$.

Definition 1.6. *A* **differential graded algebra***, (A^*, d), is a graded algebra with a degree 1 linear mapping, $d \colon A^* \to A^*$, such that d is a* **derivation***, that is, satisfies the* **Leibniz rule**

$$d(a \cdot a') = d(a) \cdot a' + (-1)^{\deg a} a \cdot d(a').$$

A **differential bigraded algebra***, $(E^{*,*}, d)$, is a bigraded algebra with a total degree one mapping*

$$d \colon \bigoplus_{p+q=n} E^{p,q} \longrightarrow \bigoplus_{r+s=n+1} E^{r,s}$$

that satisfies the Leibniz rule $d(e \cdot e') = d(e) \cdot e' + (-1)^{p+q} e \cdot d(e')$, when e is in $E^{p,q}$ and e' is in $E^{r,s}$.

A simple example of a differential bigraded algebra can be constructed from two differential graded algebras (A^*, d) and (B^*, d') by letting $E^{*,*}$ be $A^* \otimes B^*$ and defining the differential $d_\otimes$ on $E^{*,*}$ by the formula

$$d_\otimes(a \otimes b) = d(a) \otimes b + (-1)^{\deg a} a \otimes d'(b).$$

Notice that the differentials in a spectral sequence are all mappings of total degree ± 1. We can now assemble these notions and give the expected definition.

Definition 1.7. $\{E_r^{*,*}, d_r\}$ *is a* **spectral sequence of algebras** *if, for each* r, $(E_r^{*,*}, d_r)$ *is a differential bigraded algebra and furthermore, the product* φ_{r+1} *on* $E_{r+1}^{*,*}$ *is induced by the product* φ_r *of* $E_r^{*,*}$ *on homology. That is, the product on* $E_{r+1}^{*,*}$ *may be expressed as the composite*

$$H(E_r^{*,*}, d_r) \otimes H(E_r^{*,*}, d_r) \xrightarrow{\cong} H(E_r^{*,*} \otimes E_r^{*,*}, d_r \otimes 1 \pm 1 \otimes d_r)$$
$$\xrightarrow[H(\varphi_r)]{} H(E_r^{*,*}, d_r).$$

To complete this series of definitions, we need to make precise how a spectral sequence can be used to recover a graded algebra H^*; to do this, we restrict how a filtration may behave with respect to a product.

Definition 1.8. *Suppose* F^* *is a filtration of* H^*, *a graded algebra with product* φ. *The filtration is said to be* **stable** *with respect to the product if*

$$\varphi(F^r H^* \otimes F^s H^*) \subset F^{r+s} H^*.$$

The reader will want to check that a filtration F^* on H^* that is stable with respect to a product on H^* induces a bigraded algebra structure on the associated bigraded module $E_0^{*,*}(H^*)$. We can now say what it means for a spectral sequence of algebras to **converge to** H^* **as a graded algebra**, that is, there is a spectral sequence $\{E_r^{*,*}, d_r\}$ of algebras and a stable filtration on H^* with the E_∞-term of the spectral sequence isomorphic *as a bigraded algebra* to the associated bigraded algebra, $E_0^{*,*}(H^*, F^*)$. The generic theorem in this case is given by the following statement.

"Theorem II." *There is a spectral sequence of algebras with*

$$E_2^{*,*} \cong \text{something computable},$$

and converging to H^*, *something desirable, as a graded algebra.*

Example 1.E. *Suppose* $E_2^{*,*}$ *is given as an algebra by*

$$E_2^{*,*} \cong \mathbb{Q}[x, y, z] \Big/ \big(x^2 = y^4 = z^2 = 0 \big),$$

where the bidegree of each generator is given by $\operatorname{bideg} x = (7, 1)$, $\operatorname{bideg} y = (3, 0)$ *and* $\operatorname{bideg} z = (0, 2)$. *Furthermore, suppose* $d_2(x) = y^3$ *and* $d_3(z) = y$. *In this case, the spectral sequence collapses at* E_4 *and, though* x *and* y *do not survive to* E_∞, *the product* xy *does.*

First observe that we only need to know the action of the differentials on the generators of the algebra since the Leibniz rule tells us how the differentials act on decomposable elements. With the given information we can list the entire E_2-term of the spectral sequence in the diagram.

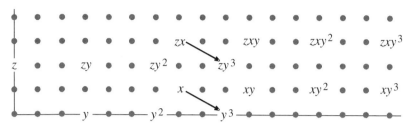

The E_2-term

Since the differential d_2, as shown, takes basis elements to basis elements, it is clearly an isomorphism of vector spaces when nonzero. We take the homology to obtain E_3, as in the diagram.

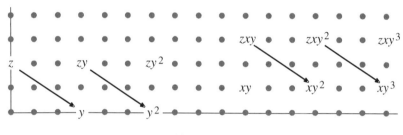

The E_3-term

Finally, by the placement of trivial vector spaces, $E_4 = E_\infty$, as presented in the diagram.

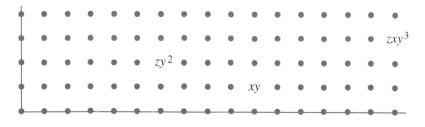

The E_∞-term

Notice that the product xy has lived to E_∞, even though the generators which gave rise to xy did not. Furthermore, we can read from the spectral sequence that the element represented by xy times the element represented by zy^2 is a nontrivial product zxy^3 in H^*. Though the example points out how a spectral sequence simplifies in the presence of a product structure, it is particularly simple. With regard to the problem of reconstructing the algebra structure on H^*, the reader should contrast this example with Example 1.J in which the E_∞-term of a spectral sequence is shown to be inadequate for the determination of the algebra structure of its target.

    In the next example, we introduce the Euler characteristic a useful arithmetic invariant of a graded vector space.

Definition 1.9. *Let A^* be a **locally finite** graded vector space over k, that is, $\dim_k A^n$ is finite for all n. The **Poincaré series** for A^* is the formal power series given by*

$$P(A^*, t) = \sum_{n=0}^{\infty} (\dim_k A^n) t^n.$$

On the face of it, the Poincaré series seems to carry very little information, especially if we treat A^* simply as a graded vector space. However, some immediate observations can be made:

(1) A^* is globally finite dimensional if and only if $P(A^*, t)$ is a polynomial.
(2) $P(A^* \otimes B^*, t) = P(A^*, t) \times P(B^*, t)$ where multiplication on the right is the Cauchy product of power series, that is,

$$\text{if } \sum_{i=0}^{\infty} a_i t^i \times \sum_{i=0}^{\infty} b_i t^i = \sum_{j=0}^{\infty} c_j t^j, \text{ then } c_j = \sum_{i=0}^{j} a_i b_{j-i}.$$

(3) We can define an **Euler characteristic** for A^* by $\chi(A^*) = P(A^*, -1)$, when this expression makes sense. If $A^* = H^*(M; \mathbb{Q})$ for M, a finite-dimensional manifold, $\chi(A^*)$ is the classical Euler-Poincaré characteristic, $\chi(M)$, of the manifold.

When A^* is a graded algebra, aspects of the algebra structure are reflected in the Poincaré series. For example, consider $A^* = \mathbb{Q}[x]$ where x is of degree n. Since A^* is a free commutative algebra on one generator, all of the powers of x persist in A^* and generate it as a graded vector space. It is easy to see then that

$$P(A^*, t) = \frac{1}{1 - t^n} = 1 + t^n + t^{2n} + t^{3n} + \cdots.$$

More generally, if $B^* = \mathbb{Q}[x_1, x_2, \ldots, x_r]$ with degree $x_i = n_i$, then

$$P(B^*, t) = \prod_{i=1}^{r} \frac{1}{1 - t^{n_i}}.$$

Whether the Poincaré series turns out to be a rational function of t expresses a measure of the complexity of the algebra structure. Some powerful numerical invariants, derived from the Poincaré series, have been useful in the study of local rings (see the work of [Assmus59], [Golod62] and [Avramov94]).

Suppose $E^{*,*}$ is a locally finite bigraded vector space (that is, $E^{p,q}$ is finite-dimensional over k for all p and q). We can define a Poincaré series for $E^{*,*}$ by

$$P(E^{*,*}, t) = \sum_{n=0}^{\infty} \dim_k \left(\bigoplus_{p+q=n} E^{p,q} \right) t^n.$$

For example, if $E_2^{*,*}$ is the bigraded vector space in Example 1.E, then

$$P(E_2^{*,*}, t) = (1 + t^{11})(1 + t^4 + t^8 + t^{12})(1 + t^3).$$

Because $H^r \cong \bigoplus_{p+q=r} E_0^{p,q}(H^*)$, it is immediate that a filtered graded vector space H^* has the same Poincaré series as its associated bigraded vector space, that is, $P(H^*, t) = P(E_0^{*,*}(H^*), t)$.

We can compare two Poincaré series at each degree. We say that $P(A^*, t) \geq P(B^*, t)$ if the power series $p(t) = P(A^*, t) - P(B^*, t)$ has all of its coefficients nonnegative. For example, whenever there is an epimorphism of locally finite graded vector spaces, $T: A^* \to B^*$, then $P(A^*, t) \geq P(B^*, t)$.

Example 1.F. *Suppose* $\{E_r^{*,*}, d_r\}$ *is a spectral sequence converging to* H^* *and that* $E_2^{*,*}$ *is locally finite. Then* H^* *is locally finite and*

$$P(E_2^{*,*}, t) \geq P(E_3^{*,*}, t) \geq \cdots \geq P(E_\infty^{*,*}, t) = P(H^*, t).$$

Furthermore, $P(H^*, t) = P(E_i^{*,*}, t)$ *for some finite* i *if and only if the spectral sequence collapses at the* i^{th} *term. Finally, if we let* $\chi(E_r^{*,*}) = P(E_r^{*,*}, -1)$, *whenever these expressions are meaningful, then* $\chi(E_r^{*,*}) = \chi(H^*)$, *for all* $r \geq 2$.

For the sake of accounting, let $d_i^{p,q}$ denote the ith differential,

$$d_i^{p,q}: E_i^{p,q} \longrightarrow E_i^{p+i,q-i+1}.$$

Let n be a fixed natural number and $p + q = n$. From elementary linear algebra we have $\dim_k(E_{i+1}^{p,q}) = \dim_k(\ker d_i^{p,q}) - \dim_k(\operatorname{im} d_i^{p-i,q+i-1})$. Since $\ker d_i^{p,q}$ is a subspace of $E_i^{p,q}$, we have for all $p + q = n$,

$$\dim_k(E_i^{p,q}) \geq \dim_k(\ker d_i^{p,q})$$
$$\geq \dim_k(\ker d_i^{p,q}) - \dim_k(\operatorname{im} d_i^{p-i,q+i-1}) = \dim_k(E_{i+1}^{p,q}).$$

By local finiteness, we have, for each n,

$$\dim_k \left(\bigoplus_{p+q=n} E_i^{p,q} \right) \geq \dim_k \left(\bigoplus_{p+q=n} E_{i+1}^{p,q} \right)$$

and so $P(E_i^{*,*}, t) \geq P(E_{i+1}^{*,*}, t)$.

If the spectral sequence collapses at the ith term, $E_i^{*,*} = E_{i+1}^{*,*} = \cdots = E_\infty^{*,*}$ and so $P(E_i^{*,*}, t) = P(E_\infty^{*,*}, t) = P(E_0^{*,*}(H^*), t) = P(H^*, t)$. Suppose $P(E_i^{*,*}, t) = P(H^*, t)$. Our first assertion implies that $P(E_i^{*,*}, t) = P(E_{i+j}^{*,*}, t)$ for all $j \geq 0$. To establish the collapse of the spectral sequence we can show that $\dim_k E_i^{p,q} = \dim_k E_{i+j}^{p,q}$ for all p, q and j. For all n, $\dim_k \left(\bigoplus_{p+q=n} E_i^{p,q} \right) = \dim_k \left(\bigoplus_{p+q=n} E_{i+j}^{p,q} \right)$. Since $E_2^{*,*}$ is locally finite, it follows that $E_r^{*,*}$ is locally finite for all $r \geq 2$ and

$$\dim_k \left(\bigoplus_{p+q=n} E_r^{p,q} \right) = \sum_{p+q=n} \dim_k(E_r^{p,q}).$$

Thus $\sum_{p+q=n} \dim_k(E_i^{p,q}) = \sum_{p+q=n} \dim_k(E_{i+j}^{p,q})$. Since dimension is always nonnegative and $\dim_k(E_i^{p,q}) \geq \dim_k(E_{i+j}^{p,q})$, the previous equation cannot hold unless $\dim_k E_i^{p,q} = \dim_k E_{i+j}^{p,q}$ for each p, q, $p+q=n$ and for all $j \geq 1$.

To establish the assertion about Euler characteristics, we introduce the partial sums for $\ell > 0$,

$$\chi_\ell(E_i^{*,*}) = \sum_{n=0}^{\ell} (-1)^n \dim_k \left(\bigoplus_{p+q=n} E_i^{p,q} \right).$$

Notice that $\lim_{\ell \to \infty} \chi_\ell(E_i^{*,*}) = \chi(E_i^{*,*})$. Since $\chi(E_\infty^{*,*}) = \chi(H^*)$, it suffices to show that $\chi(E_i^{*,*}) = \chi(E_{i+1}^{*,*})$ for all i. Recall that $\dim_k E_i^{p,q} = \dim_k(\ker d_i^{p,q}) + \dim_k(\operatorname{im} d_i^{p,q})$. We compute

$$\chi_\ell(E_i^{*,*}) = \sum_{n=0}^{\ell} (-1)^n \dim_k \left(\bigoplus_{p+q=n} E_i^{p,q} \right)$$

$$= \sum_{n=0}^{\ell} \sum_{p+q=n} \left[(-1)^n \dim_k(\ker d_i^{p,q}) + (-1)^n \dim_k(\operatorname{im} d_i^{p,q}) \right]$$

$$= \sum_{n=0}^{\ell} \left[\sum_{p+q=n} (-1)^n \dim_k(\ker d_i^{p,q}) + \sum_{r+s=n} (-1)^n \dim_k(\operatorname{im} d_i^{r,s}) \right]$$

$$= \dim_k(\ker d_i^{0,0}) + \sum_{n=1}^{\ell} \left[\sum_{p+q=n} (-1)^n \dim_k(\ker d_i^{p,q}) \right.$$

$$\left. - \sum_{r+s=n-1} (-1)^n \dim_k(\operatorname{im} d_i^{r,s}) \right] + \sum_{r+s=\ell} (-1)^\ell \dim_k(\operatorname{im} d_i^{r,s})$$

$$= \sum_{n=0}^{\ell} (-1)^n \dim_k \left(\bigoplus_{p+q=n} E_{i+1}^{p,q} \right) + \sum_{r+s=\ell} (-1)^\ell \dim_k(\operatorname{im} d_i^{r,s})$$

$$= \chi_\ell(E_{i+1}^{*,*}) + \sum_{r+s=\ell} (-1)^\ell \dim_k(\operatorname{im} d_i^{r,s}).$$

Let ℓ go to infinity; this incorporates the extra term into the limit and so $\chi(E_i^{*,*}) = \chi(E_{i+1}^{*,*})$.

As an application of the example, if one wants to compute the Euler characteristic for a manifold M, the job is simplified if there is a spectral sequence converging to $H^*(M; k)$. The E_2-term of the spectral sequence, which may be known, determines $\chi(M)$, even though $H^*(M; k)$ is not known.

Another piece of structure that can help in a calculation with a spectral sequence is the presence of an action of a graded algebra Γ^* on the target H^* that can be found on the spectral sequence. The motivating examples of such Γ^* are

(1) the Steenrod algebra, $\mathcal{A}_p$, which acts on $H^*(X; \mathbb{F}_p)$ and is present in many spectral sequences (see §4.4, §6.2, and §8.3 for more details);

(2) $H^*(G; k)$ that acts on the various equivariant topological invariants of a space X on which the group G acts (see [Greenlees88] and [Greenlees-May95] for examples).

Suppose Γ^* is a graded algebra over k. A graded vector space H^* is a Γ^*-**module** if there is a mapping of graded vector spaces

$$\psi: \Gamma^* \otimes_k H^* \longrightarrow H^*$$

satisfying the usual module axioms, that is, the following diagram commutes with m, the product on Γ^*:

$$
\begin{array}{ccc}
\Gamma^* \otimes \Gamma^* \otimes H^* & \xrightarrow{1 \otimes \psi} & \Gamma^* \otimes H^* \\
{\scriptstyle m \otimes 1}\downarrow & & \downarrow{\scriptstyle \psi} \\
\Gamma^* \otimes H^* & \xrightarrow{\psi} & H^*.
\end{array}
$$

As the reader might expect, we want to consider how a graded algebra, Γ^*, might act on a bigraded vector space $E^{*,*}$. In this instance, we consider only one case of such an action that represents the interesting case in later chapters; more general definitions are possible.

Definition 1.10. *Let $E^{*,*}$ be a bigraded vector space and Γ^* a graded algebra over a field k. We say that Γ^* **acts (vertically)** on $E^{*,*}$ if, for each $n \geq 0$, $E^{n,*}$ is a Γ^*-module; that is, there is a module structure $\psi_n: \Gamma^* \otimes_k E^{n,*} \to E^{n,*}$.*

The term "vertically" is appropriate since, componentwise, Γ^* acts on elements by moving them "up," $\psi_n: \Gamma^s \otimes_k E^{n,t} \to E^{n,s+t}$. Suppose H^* is a filtered graded vector space and a Γ^*-module such that $\Gamma^* \otimes F^p H^* \to F^p H^*$, that is, the Γ^*-action is **filtration-preserving**. If we descend to the associated bigraded vector space, $E_0^{*,*}(H^*)$, then Γ^* acts vertically on $E_0^{*,*}(H^*)$.

For a graded algebra Γ^* to **act on a spectral sequence**, $\{E_r^{*,*}, d_r\}$, we require that

(1) Γ^* acts on $E_r^{*,*}$ for each r,
(2) the differentials d_r are Γ^*-linear and
(3) the Γ^*-action on $E_{r+1}^{*,*}$ is induced through homology from the action of Γ^* on $E_r^{*,*}$.

Suppose H^* is a Γ^*-module. In applications, we want to determine H^* as a Γ^*-module, that is, to compute H^* along with its Γ^*-action. We say that a spectral sequence **converges to H^* as a Γ^*-module** if we have a spectral sequence, converging to H^*, on which Γ^* acts and the filtration on H^* induces a Γ^*-action on the associated bigraded space $E_0^{*,*}(H^*)$ that is isomorphic to the Γ^*-action on the E_∞-term of the spectral sequence. The generic setting for such applications is the following statement.

"Theorem III." *There is a spectral sequence on which Γ^* acts, with*

$$E_2^{*,*} \cong \text{something computable with a known } \Gamma^*\text{-action,}$$

and converging to H^*, something desirable, as a Γ^*-module.

In the next example we present a situation, albeit artificial, where the Γ^*-action plays an important role. In the natural examples, arguments similar to the one presented here can be employed with remarkable success (for example, the paper of [Serre53, §5]).

Example 1.G. *Suppose $\Gamma^* = \mathbb{Q}[a, b]$ with $\deg a = 2$ and $\deg b = 5$. Suppose we have a spectral sequence with $E_2^{*,*}$ a Γ^*-module. Suppose $E_2^{*,*}$ has Γ^*-module generators $\{x, y, z, w\}$ with $\operatorname{bideg} x = (8, 4)$, $\operatorname{bideg} y = (6, 0)$, $\operatorname{bideg} z = (0, 4)$ and $\operatorname{bideg} w = (10, 1)$ and Γ^* acts freely on this basis, except that $bx = 0$. Under these conditions the spectral sequence collapses at the E_2-term.*

First observe that, since the differentials, d_r, commute with the Γ^*-action, it suffices to show that $d_r = 0$ on the basis elements for all r. The generator z survives to E_∞ since it has total degree 4 and in total degree 5, $E_2^{*,*}$ is trivial—any differential originating on z lands in total degree 5 and so is zero. This implies that y also survives to E_∞ as no Γ^*-multiple of z hits y and y cannot bound any other element. Notice that a^2y could hit w by d_4. However, d_4 commutes with Γ^* and so $d_4(a^2y) = a^2 d_4(y) = 0$. Thus w survives since it cannot be hit by any Γ^*-multiple of x or y. The element x, however, could be mapped to aw by d_2. Since $bx = 0$, we can compute

$$0 = d_2(bx) = b d_2(x) = baw \neq 0.$$

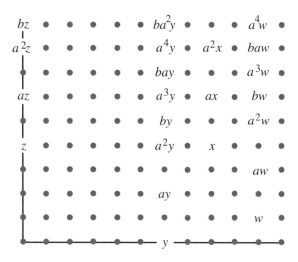

Thus $d_2(x) = aw$ leads to a contradiction and so $d_2(x) = 0$. For dimensional reasons, $d_r(x) = 0$ for $r > 2$ and so x survives to E_∞. The spectral sequence collapses at the E_2-term.

1.4 Working backwards

In this section we assume the following "theorem."

"Theorem IV." *There is a spectral sequence of algebras with*

$$E_2^{*,*} \cong V^* \otimes_k W^*, \text{ as bigraded algebras,}$$

where V^* and W^* are graded algebras, and converging to H^* as a graded algebra.

If H^* is an algebraic invariant of a topological space, we can think of V^* and W^* as similar invariants. (To rush ahead to Chapter 5 a moment, H^* may be $H^*(E; k)$, $V^* = H^*(B; k)$ and $W^* = H^*(F; k)$ where $F \hookrightarrow E \to B$ is a fibration.) Suppose that H^* is known as well as V^* or W^*. Does a spectral sequence have enough structure as an algebraic object to allow us to obtain W^* when V^* is known or vice versa? That is, can we work backward from the answer and part of the data to the rest of the data?

Before presenting some examples of this situation, we introduce an algebraic condition on our graded algebras that is characteristic of topological invariants.

Definition 1.11. *A graded algebra* H^* *is said to be* **graded commutative** *(also* **skew-commutative***) if, for* $x \in H^p$ *and* $y \in H^q$, $x \cdot y = (-1)^{pq} y \cdot x$.

Over the field of rational numbers, *free* graded commutative algebras come in two varieties. Suppose x_{2n} is the generator of A^* with x_{2n} of degree $2n$. Since powers of x_{2n} are all in even dimensions, they commute with each other and so $A^* \cong \mathbb{Q}[x_{2n}]$, that is, the polynomial algebra on one generator of dimension $2n$. Suppose B^* has one generator of odd degree, y_{2n+1}, of degree $2n + 1$. Since

$$y_{2n+1} \cdot y_{2n+1} = (-1)^{(2n+1)(2n+1)} y_{2n+1} \cdot y_{2n+1},$$

we deduce that $(y_{2n+1})^2 = 0$ and so any higher power of y_{2n+1} is zero. We denote B^* by $\Lambda(y_{2n+1})$, the **exterior algebra** on one generator of dimension $2n + 1$. (The Λ suggests the wedge product of differential 1-forms.) Any free graded commutative algebra over $\mathbb{Q}$ can be written as a tensor product of polynomial algebras and exterior algebras on the generators of appropriate dimensions.

Example 1.H. *Suppose "Theorem IV" holds and* $H^* \cong \mathbb{Q}$ *(that is,* H^* *is the graded algebra with* $H^0 \cong \mathbb{Q}$ *and* $H^i = \{0\}$ *for* $i \geq 1$*). If* $V^* \cong \mathbb{Q}[x_{2n}]$*, then* $W^* \cong \Lambda(y_{2n+1})$*. If* $V^* \cong \Lambda(x_{2n+1})$*, then* $W^* \cong \mathbb{Q}[y_{2n}]$*.*

This example is a simple case of a theorem of [Borel53] that we discuss in detail in later chapters. In particular, the reader can compare this example with Theorem 3.27 and Theorem 6.22.

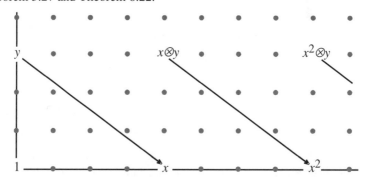

Recall that the differential d_i, applied to an element written as $w \otimes z$, satisfies the Leibniz rule

$$d_i(w \otimes z) = d_i(w) \otimes z + (-1)^{\deg w} w \otimes d_i(z).$$

Furthermore, the differential restricted to $V^* \cong E_2^{*,0}$ is null and restricted to $W^* \cong E_2^{0,*}$ it must have its image in $V^* \otimes W^*$. If $d_i(1 \otimes u) = \sum_j v_j \otimes w_j$, then

$$d_i(1 \otimes u^k) = k \left(\sum_j v_j \otimes (w_j u^{k-1}) \right).$$

In the first case, $V^* = \mathbb{Q}[x_{2n}]$ and we can display the E_2-term in the spectral sequence converging to $H^* = k$ as on the opposite page. Since x_{2n} does not survive to E_∞, there must be a y_{2n-1} in W^* so that

$$d_{2n}(1 \otimes y_{2n-1}) = x_{2n} \otimes 1.$$

Now, with y_{2n-1} in W^*, we have generated new elements in $E_2^{*,*}$, namely $(x_{2n})^m \otimes y_{2n-1}$. By the derivation property of differentials,

$$
\begin{aligned}
d_{2n-1}&((x_{2n})^m \otimes y_{2n-1}) \\
&= d_{2n-1}((x_{2n})^m) \otimes y_{2n-1} + (x_{2n})^m \otimes d_{2n-1}(y_{2n-1}) \\
&= m d_{2n-1}(x_{2n})(x_{2n})^{m-1} \otimes y_{2n-1} + (x_{2n})^{m+1} \otimes 1 \\
&= (x_{2n})^{m+1} \otimes 1.
\end{aligned}
$$

Thus we get the pattern of differentials in the picture. If W^* contains any other elements, they would give rise to classes that would persist to E_∞ and contribute to H^*. Thus $W^* \cong \Lambda(y_{2n-1})$.

To treat the other case, consider the following diagram.

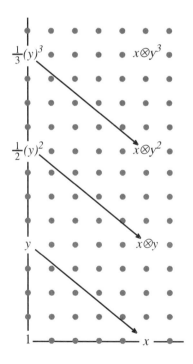

As in the previous example, the equations

$$d_{2n}(1 \otimes y_{2n}) = x_{2n+1} \otimes 1$$
$$d_{2n}\left(1 \otimes \left(\tfrac{1}{m}\right)(y_{2n})^m\right) = x_{2n+1} \otimes (y_{2n})^{m-1}$$

give the desired isomorphisms to obtain $H^* \cong \mathbb{Q}$. Notice that the characteristic of the field plays a role here; in $\mathbb{Q}$ there exist the denominators that allow d_{2n} to be one-one and onto. For fields of nonzero characteristic or for an arbitrary ring, this simple procedure can lead to elements that would persist to E_∞ unless W^* is different. For such cases, W^* would have the structure of a divided power algebra (see the exercises for the definition and discussion).

Example 1.I. *Suppose "Theorem IV" and* $H^* \cong \mathbb{Q}$*. If* $V^* \cong \mathbb{Q}[x_2]/(x_2)^3$*, then* $W^* \cong \Lambda(y_1) \otimes \mathbb{Q}[z_4]$*.*

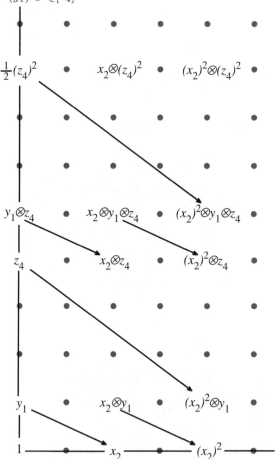

We argue by filling in the diagram. The pattern will be apparent. If we write V^* along the p-axis, then it is clear that we require $\Lambda(y_1)$ in W^* with $d_2(y_1) = x_2$. It follows that $d_2(x_2 \otimes y_1) = (x_2)^2$, leaving $(x_2)^2 \otimes y_1$ in need of a bounding element. Since $(x_2)^2 \otimes y_1$ has total degree 5, we want z_4 of degree 4 in W^*, with $d_4(z_4) = (x_2)^2 \otimes y_1$. Now $y_1 \otimes z_4$ bounds $x_2 \otimes z_4$ and $x_2 \otimes (y_1 \otimes z_4)$

takes care of $(x_2)^2 \otimes z_4$. Further $d_4((\frac{1}{2})(z_4)^2) = (x_2)^2 \otimes (y_1 \otimes z_4)$; this pattern continues to give the correct E_∞-term.

Arguments of this sort were introduced by [Borel53]. [Zeeman57] proved a very general result along these lines which is discussed in Chapter 3 (Theorem 3.26).

Working backward from a known answer can lead to invariants of interest. For example, in a paper on scissors congruence [Dupont82] has set up a certain spectral sequence, converging to the trivial vector space, with a known nonzero E_1-term. The differentials can be interpreted as important geometric invariants that generalize the classical Dehn invariant.

1.5 Interpreting the answer

After computing an entire spectral sequence to obtain $E_\infty^{*,*}$, we are finally in position to reconstruct H^*. As already noted, H^* is determined as a graded vector space. Suppose that $E_\infty^{*,*}$ is the associated bigraded algebra to some stable filtration of a graded algebra H^*. The problem is to determine the algebra structure on H^* from $E_\infty^{*,*}$. In Example 1.J, we show that this may be impossible without further information. That's the bad news. In Example 1.K there is some good news; in some reasonable cases, $E_\infty^{*,*}$ can completely determine H^* as an algebra.

If k is a field and $\{x, y, \dots\}$ a set, then we denote the vector space over k with basis $\{x, y, \dots\}$ by $k\{x, y, \dots\}$.

Example 1.J, *in which we present two graded algebras, H_1^* and H_2^*, with H_1^* not isomorphic to H_2^*, together with stable filtrations, F_1^* and F_2^*, such that $E_0^{*,*}(H_1^*) \cong E_0^{*,*}(H_2^*)$ as bigraded algebras.*

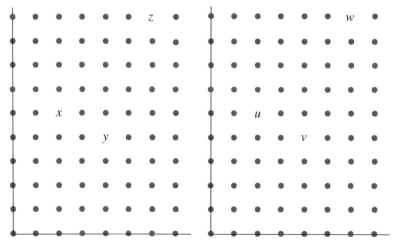

Let $H_1^* = \mathbb{Q}\{x, y, z\} \big/ \{\text{all products} = 0\}$, where $\deg x = 7$, $\deg y = 8$

and $\deg z = 15$. We filter H_1^* by

$$H_1^* = F_1^1 = F_1^2 \supset \mathbb{Q}\{y, z\} = F_1^4 \supset \mathbb{Q}\{z\} = F_1^6 = F_1^7 \supset \{0\},$$

to obtain an associated graded space $E_0^{*,*}(H_1^*)$, as pictured on the left in the diagram, with all products zero: Now let

$$H_2^* = \mathbb{Q}\{u, v, w\} \left/ \begin{array}{c} u^2 = v^2 = w^2 = 0 \\ uv = w \\ uw = vw = 0 \end{array} \right.$$

where $\deg u = 7$, $\deg v = 8$ and $\deg w = 15$. Filter H_2^* by

$$H_2^* = F_2^1 = F_2^2 \supset \mathbb{Q}\{v, w\} = F_2^4 \supset \mathbb{Q}\{w\} = F_2^6 = F_2^7 \supset \{0\}.$$

Notice that u, v are in F_2^2 and $u \cdot v = w$ is in F_2^4. Since this is the only nontrivial product, the filtration is stable. Taking the quotients for this filtration, we get the associated bigraded algebra, $E_0^{*,*}(H_2^*)$.

Since $\operatorname{bideg} u = (2, 5)$ and $\operatorname{bideg} v = (4, 4)$, the bidegree of $u \cdot v$ is $(6, 9)$ and so $u \cdot v = 0$ in $E_0^{*,*}(H_2^*)$. Since all products are zero, $E_0^{*,*}(H_1^*)$ is isomorphic to $E_0^{*,*}(H_2^*)$ as bigraded algebras.

It is clear from this example that the E_∞-term may not be enough to reconstruct H^* as an algebra. Some extra information may come from an earlier term in the spectral sequence (as in Example 1.E) or from the geometric or algebraic situation from which the spectral sequence arose (for example, when Massey products are involved; see §8.2).

In the topological applications, H^* is generally a graded commutative algebra. Therefore we introduce the analogous condition for a bigraded algebra. We say that $E^{*,*}$ is **graded-commutative** if, whenever x is in $E^{p,q}$ and y in $E^{r,s}$ then $x \cdot y = (-1)^{(r+s)(p+q)} y \cdot x$. To motivate this condition, consider the functor, total: **BigradedAlg** $\rightarrow$ **GradedAlg**, from the category of bigraded algebras with morphisms of bidegree (0,0) to the category of graded algebras with morphisms of degree 0, given by

$$(\text{total } E^{*,*})^n = \bigoplus_{p+q=n} E^{p,q}$$

and called the **total complex**. Since the multiplication on $E^{*,*}$ gives a map from $E^{p,r} \otimes E^{q,s}$ to $E^{p+q,r+s}$, if we treat x and y as lying in the total complex, the condition of graded commutativity of $E^{*,*}$ is that of graded commutativity for $\operatorname{total}(E^{*,*})$. Observe that $\operatorname{total} E_0^{*,*}(H^*) \cong H^*$ as graded vector spaces when H^* is filtered.

Example 1.K. *If there is a spectral sequence converging to H^* as an algebra and the E_∞-term is a free, graded-commutative, bigraded algebra, then H^* is a free, graded commutative algebra isomorphic to* total $E_\infty^{*,*}$.

Suppose that $E_\infty^{*,*}$ is a graded-commutative algebra on generators $x_1, \ldots, x_r$ where the bidegree of x_i is (p_i, q_i). Let A^* be the free graded commutative algebra on generators $y_1, \ldots, y_r$ where $\deg y_i = p_i + q_i$. Since we have free graded commutative algebras, $A^* \cong \text{total}(E_\infty^{*,*})$. We filter A^* by giving each element in A^* a weight and then use the weights to induce the filtration. Assign to each generator, y_i, the weight, p_i, and to each monomial, $w = \prod_{i=1}^r (y_i)^{s_i}$, the weight, $\sum_{i=1}^r s_i p_i$. If a homogeneous element in A^* is a sum of monomials, its weight is the minimum weight of the monomials in the sum. Filter A^* by defining

$$\bar{F}^p A^* = \{w \in A^* \mid \text{weight}(w) \geq p\}.$$

This is a decreasing stable filtration of A^* and the associated bigraded algebra, $\bar{E}_0^{*,*}(A^*, \bar{F})$, is free on generators $y_1, y_2, \ldots, y_r$ of bidegree (p_i, q_i), that is, $\bar{E}_0^{*,*}(A^*, \bar{F}) \cong E_\infty^{*,*}$.

We prove that A^* is isomorphic to H^* as an algebra. Notice that there is an obvious mapping of A^* to H^* since A^* is free. Next we show that A^* and H^* have isomorphic filtrations via a double induction: on i, the algebra degree, and $i - k$, the filtration degree. To begin, $H^0 \cong E_\infty^{0,0} \cong \bar{E}_0^{0,0}(A^*) \cong A^0$.

Suppose $A^j \cong H^j$ for $0 \leq j < i$ and consider the filtration of H^i

$$H^i \supseteq F^1 H^i \supseteq F^2 H^i \supseteq \cdots \supseteq F^i H^i \supseteq \{0\}.$$

Since $E_\infty^{i,0} = F^i H^i$, $\bar{E}_0^{i,0}(A^*) = \bar{F}^i A^i$, and $E_\infty^{i,0} \cong \bar{E}_0^{i,0}$, we get $F^i H^i \cong \bar{F}^i A^i$. Suppose $F^{i-k} H^i \cong \bar{F}^{i-k} A^i$ for $0 \leq k < j$. From the definition of E_∞ we get two short exact sequences with isomorphisms by the inductive hypotheses:

$$
\begin{array}{ccccccccc}
0 & \longrightarrow & F^{i-j+1} H^i & \longrightarrow & F^{i-j} H^i & \longrightarrow & E_\infty^{i-j,j} & \longrightarrow & 0 \\
 & & \cong \big\downarrow & & \big\downarrow & & \cong \big\downarrow & & \\
0 & \longrightarrow & \bar{F}^{i-j+1} A^i & \longrightarrow & \bar{F}^{i-j} A^i & \longrightarrow & \bar{E}_\infty^{i-j,j} & \longrightarrow & 0
\end{array}
$$

By the Five-lemma (see [Cartan-Eilenberg56, p. 5]), we get the missing isomorphism, that is, $F^{i-j} H^i \cong \bar{F}^{i-j} A^i$. By induction, we can let $j = i$ and obtain that $H^i = F^0 H^i \cong \bar{F}^0 A^i = A^i$. Thus $H^* \cong A^*$ and furthermore, $F^j H^* \cong \bar{F}^j A^*$ for all j. Since both filtrations are stable, no product in H^* can vanish unless it does so in A^*. Since A^* is free graded commutative and isomorphic to $\text{total}(E_\infty^{*,*})$, so is H^*.

This chapter has introduced the reader to some of the formal features of computations with spectral sequences. Chapter 2 focuses on the general algebraic foundations that give rise to spectral sequences and (finally) some classical examples that arise in purely algebraic settings. Chapter 3 contains a discussion of the manner in which a spectral sequence can or cannot determine its target uniquely. If this chapter provides a guide to the operation of a spectral sequence, then Chapter 2 goes deeper to give the blueprints for building one, and Chapter 3 tells one how to distinguish two of them.

Exercises

1.1. Prove the **Five-lemma**: Consider the commutative diagram of modules over a commutative ring with unit, R,

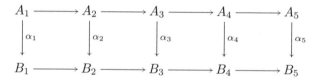

with rows exact. If α_1, α_2, α_4, and α_5 are all isomorphisms, then α_3 is an isomorphism. If α_1 is an epimorphism and α_2 and α_4 are monomorphisms, then α_3 is a monomorphism. If α_5 is a monomorphism and α_2 and α_4 are epimorphisms, then α_3 is an epimorphism.

1.2. Prove that a short exact sequence of differential graded modules with differentials of degree $+1$,

$$0 \to (K^*, \partial_K) \to (M^*, \partial_M) \to (Q^*, \partial) \to 0$$

gives rise to a long exact sequence of homology modules:

$$\to H^n(K^*, \partial_K) \to H^n(M^*, \partial_M) \to H^n(Q^*, \partial) \to H^{n+1}(K^*, \partial_K) \to$$

1.3. Suppose $\{E_r^{*,*}, d_r\}$ is a first quadrant spectral sequence, of cohomological type, converging to H^*. Suppose further that $E_2^{p,q} = \{0\}$ unless $p = 0$ or $p = n$ for some $n \geq 2$. Derive the **Wang sequence**:

$$\cdots \to H^k \to E_2^{0,k} \xrightarrow{d_n} E_2^{n,k-n+1} \to H^{k+1} \to E_2^{0,k+1} \to \cdots .$$

1.4. Show that any free graded commutative, locally finite, algebra over $\mathbb{Q}$ is isomorphic to a tensor product of polynomial algebras and exterior algebras.

1.5. Suppose that F^* is a stable filtration bounded below on a graded algebra H^*. Show that the associated graded module $E_0^{*,*}(H^*, F^*)$ is a bigraded algebra.

1.6. Suppose $\{E_r^{*,*}, d_r\}$ is a first quadrant spectral sequence, of cohomological type over $\mathbb{Z}$, associated to an bounded filtration, and converging to H^*. If the E_2-term is given by

$$E_2^{p,q} = \begin{cases} \mathbb{Z}/2\mathbb{Z}, & \text{if } (p,q) = (0,0), (0,4), (2,3), (3,2) \text{ or } (6,0), \\ \{0\}, & \text{elsewhere,} \end{cases}$$

then determine all possible candidates for H^*.

1.7. Suppose (A, d) is a differential graded vector space over k, a field. Let B^* be the graded vector space, $B^n = \mathrm{im}(d \colon A^{n-1} \to A^n)$. Show that

$$P(H(A^*, d), t) = P(A^*, t) - (1 + t)P(B^*, t).$$

1.8. Suppose we are working over $\mathbb{Z}$ and the "Theorem IV" of §1.4 holds. Suppose, as in Example 1.H, that E_r converges to a trivial E_∞-term (that is, $H^* \cong \mathbb{Q}$) and $V^* = \Lambda(x_{2n+1})$. Show that W^* is isomorphic to $\Gamma(y_{2n})$, that is, the **divided power algebra** on one generator of dimension $2n$, defined as having $\mathbb{Z}$-module generators $\gamma_i(y)$, for $i = 0, 1, 2, \ldots$ and

(1) $\gamma_0(y) = 1$, $\gamma_1(y) = y$, and $\deg \gamma_k(y) = 2nk$,
(2) $\gamma_k(y)\gamma_h(y) = (k, h)\gamma_{k+h}(y)$ where (k, h) is the binomial coefficient given by $(k + h)!/k!h!$.

The differentials in the spectral sequence are defined by

$$d(\gamma_k(y)) = (d(y)) \otimes \gamma_{k-1}(y).$$

2
What is a spectral sequence?

"A spectral sequence is an algebraic object, like an exact
sequence, but more complicated."

J. F. Adams

In Chapter 1 we restricted our examples of spectral sequences to the first
quadrant and to bigraded vector spaces over a field in order to focus on the
computational features of these objects. In this chapter we treat some deeper
structural features including the settings in which spectral sequences arise. In
order to establish a foundation of sufficient breadth, we remove the restrictions
of Chapter 1 and consider $(\mathbb{Z} \times \mathbb{Z})$-bigraded modules over R, a commutative
ring with unity. It is possible to treat spectral sequences in the more general
setting of abelian categories (the reader is referred to the thorough treatments
in [Eilenberg-Moore62], [Eckmann-Hilton66], [Lubkin80], and [Weibel96]).
The approach here supports most of the topological applications we want to
consider.

In this chapter we present two examples that arise in purely algebraic
contexts—the spectral sequence of a double complex and the Künneth spectral
sequence that generalizes the ordinary Künneth Theorem (Theorem 2.12). For
completeness we have included a discussion of basic homological algebra. This
provides a foundation for the generalizations that appear in later chapters.

2.1 Definitions and basic properties

We begin by generalizing our *First Definition* and identifying the basic
components of a spectral sequence.

Definition 2.1. *A **differential bigraded module** over a ring R, is a collection
of R-modules, $\{E^{p,q}\}$, where p and q are integers, together with an R-linear
mapping, $d\colon E^{*,*} \to E^{*,*}$, the **differential**, of bidegree $(s, 1-s)$ or $(-s, s-1)$,
for some integer s, and satisfying $d \circ d = 0$.*

The customary picture of a bigraded module is given on the next page and
may be displayed by imagining the R-module $E^{p,q}$ sitting at the integral lattice

point (p, q) in the Cartesian plane. The differential in this diagram has bidegree $(3, -2)$.

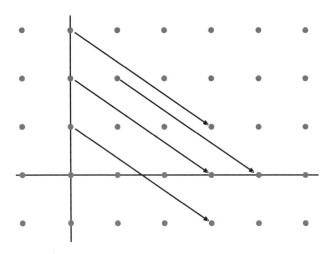

With the differential, we can take the **homology** of a differential bigraded module:

$$H^{p,q}(E^{*,*}, d) = \left. \ker d \colon E^{p,q} \to E^{p+s,q-s+1} \middle/ \operatorname{im} d \colon E^{p-s,q+s-1} \to E^{p,q}. \right.$$

Combining these notions we can give the definition.

Definition 2.2. *A **spectral sequence** is a collection of differential bigraded R-modules $\{E_r^{*,*}, d_r\}$, where $r = 1, 2, \dots$; the differentials are either all of bidegree $(-r, r - 1)$ (for a spectral sequence **of homological type**) or all of bidegree $(r, 1 - r)$ (for a spectral sequence **of cohomological type**) and for all p, q, r, $E_{r+1}^{p,q}$ is isomorphic to $H^{p,q}(E_r^{*,*}, d_r)$.*

It is worth repeating the caveat about differentials mentioned in Chapter 1: *knowledge of $E_r^{*,*}$ and d_r determines $E_{r+1}^{*,*}$ but not d_{r+1}*. If we think of a spectral sequence as a black box with input a differential bigraded module, usually $E_1^{*,*}$, then with each turn of the handle, the machine computes a successive homology according to a sequence of differentials. If some differential is unknown, then some other (*any other!*) principle is needed to proceed. From Chapter 1, the reader is acquainted with several algebraic tricks that allow further calculation. In the nontrivial cases, it is often a deep geometric idea that is caught up in the knowledge of a differential.

Although we have our spectral sequence indexed by $r = 1, 2, \dots$, it is clear that the indexing can begin at any integer and most often the sequence begins at $r = 2$, where $E_2^{*,*}$ is something familiar. In contrast with the first quadrant restriction of Chapter 1, the target of a general spectral sequence is less obvious to define. To identify this target, we present a spectral sequence as a tower of submodules of a given module. From this tower, it is clear where the algebraic information is converging and, as we saw already in Chapter 1, it may be possible to associate that information with some desired answer.

Let us begin with $E_2^{*,*}$. For the sake of clarity we suppress the bigrading (though the reader should keep track for a while to understand better the bidegrees of the differentials). Denote

$$Z_2 = \ker d_2 \quad \text{and} \quad B_2 = \operatorname{im} d_2.$$

The condition, $d_2 \circ d_2 = 0$, implies $B_2 \subset Z_2 \subset E_2$ and, by definition, $E_3 \cong Z_2/B_2$. Write $\bar{Z}_3$ for $\ker d_3 \colon E_3 \to E_3$. Since $\bar{Z}_3$ is a submodule of E_3, it can be written as Z_3/B_2 where Z_3 is a submodule of Z_2. Similarly $\bar{B}_3 = \operatorname{im} d_3$ is isomorphic to B_3/B_2 and so

$$E_4 \cong \bar{Z}_3/\bar{B}_3 \cong (Z_3/B_2)/(B_3/B_2) \cong Z_3/B_3.$$

These data can be presented as a tower of inclusions: $B_2 \subset B_3 \subset Z_3 \subset Z_2 \subset E_2$. Iterating this process, we present the spectral sequence as an infinite tower of submodules of E_2:

$$B_2 \subset B_3 \subset \cdots \subset B_n \subset \cdots \quad \cdots \subset Z_n \subset \cdots \subset Z_3 \subset Z_2 \subset E_2$$

with the property that $E_{n+1} \cong Z_n/B_n$ and the differential, d_{n+1}, can be taken as a mapping $Z_n/B_n \to Z_n/B_n$, which has kernel Z_{n+1}/B_n and image B_{n+1}/B_n. The short exact sequence induced by d_{n+1},

$$0 \longrightarrow Z_{n+1}/B_n \longrightarrow Z_n/B_n \xrightarrow{\ d_{n+1}\ } B_{n+1}/B_n \longrightarrow 0,$$

gives rise to isomorphisms $Z_n/Z_{n+1} \cong B_{n+1}/B_n$ for all n. Conversely, a tower of submodules of E_2, together with such a set of isomorphisms, determines a spectral sequence.

We say that an element in E_2 that lies in Z_r **survives to the rth stage**, having been in the kernel of the previous $r - 2$ differentials. The submodule B_r of E_2 is the set of elements that are **boundaries by the rth stage**. The bigraded module $E_r^{*,*}$ is called the E_r-**term** of the spectral sequence (or sometimes the E_r-**page**). Let $Z_\infty = \bigcap_n Z_n$ be the submodule of E_2 of elements that **survive forever**, that is, elements that are cycles at every stage. The submodule $B_\infty = \bigcup_n B_n$ consists of those elements that **eventually bound**. From the tower of inclusions it is clear that $B_\infty \subset Z_\infty$ and the fruit of our efforts appears: $E_\infty = Z_\infty/B_\infty$ is the bigraded module that remains after the computation of the infinite sequence of successive homologies. Anticipating the notion of convergence, it is the E_∞-term of a spectral sequence that is the general goal of a computation.

Under the best possible conditions, the computation ends at some finite stage; recall that a spectral sequence **collapses at the Nth term** if the differentials $d_r = 0$ for $r \geq N$. From the short exact sequence,

$$0 \longrightarrow Z_r/B_{r-1} \longrightarrow Z_{r-1}/B_{r-1} \xrightarrow{d_r} B_r/B_{r-1} \longrightarrow 0,$$

the condition $d_r = 0$ forces $Z_r = Z_{r-1}$ and $B_r = B_{r-1}$. The tower of submodules becomes

$$B_2 \subset B_3 \subset \cdots \subset B_{N-1}$$
$$= B_N = \cdots = B_\infty \subset Z_\infty = \cdots = Z_N$$
$$= Z_{N-1} \subset \cdots \subset Z_3 \subset Z_2 \subset E_2$$

and so $E_\infty = E_N$. The reader should try his or her hand at generating some examples of collapse analogous to those in §1.2.

2.2 How does a spectral sequence arise?

Now that we can describe a spectral sequence, how do we build one? In this section we present two general settings in which spectral sequences arise naturally: when one has a filtered differential module and when one has an exact couple. These approaches lay out the blueprints followed in the rest of the book.

Filtered differential modules

Definition 2.3. *A **filtration** F^* on an R-module A is a family of submodules $\{F^p A\}$ for p in $\mathbb{Z}$ so that*

$$\cdots \subset F^{p+1} A \subset F^p A \subset F^{p-1} A \subset \cdots \subset A \quad \textit{(decreasing filtration)}$$
$$\textit{or} \ \cdots \subset F^{p-1} A \subset F^p A \subset F^{p+1} A \subset \cdots \subset A \quad \textit{(increasing filtration)}.$$

An example of a filtered $\mathbb{Z}$-module is given by the integers, $\mathbb{Z}$, together with the decreasing filtration

$$F^p \mathbb{Z} = \begin{cases} \mathbb{Z}, & \text{if } p \leq 0, \\ 2^p \mathbb{Z}, & \text{if } p > 0. \end{cases}$$

$$\cdots \subset 16\mathbb{Z} \subset 8\mathbb{Z} \subset 4\mathbb{Z} \subset 2\mathbb{Z} \subset \mathbb{Z} \subset \mathbb{Z} \subset \cdots \subset \mathbb{Z}.$$

We can collapse a filtered module to its **associated graded module**, $E_0^*(A)$ given by

$$E_0^p(A) = \begin{cases} F^p A/F^{p+1} A, & \text{when } F \text{ is decreasing,} \\ F^p A/F^{p-1} A, & \text{when } F \text{ is increasing.} \end{cases}$$

In the example above, $E_0^p(\mathbb{Z}) = \{0\}$ if $p < 0$ and $E_0^p(\mathbb{Z}) \cong \mathbb{Z}/2\mathbb{Z}$ if $p \geq 0$.

The 2-adic integers, $\hat{\mathbb{Z}}_2 = \lim_{\leftarrow s} \mathbb{Z}/2^s\mathbb{Z}$ has a decreasing filtration given by

$$F^p\hat{\mathbb{Z}}_2 = \ker(\hat{\mathbb{Z}}_2 \to \mathbb{Z}/2^p\mathbb{Z})$$

for $p > 0$ and $F^p\hat{\mathbb{Z}}_2 = \hat{\mathbb{Z}}_2$ for $p \leq 0$. The projections $\phi_p \colon \hat{\mathbb{Z}}_2 \to \mathbb{Z}/2^p\mathbb{Z}$ give rise to short exact sequences

$$0 \to \hat{\mathbb{Z}}_2 \xrightarrow{\times 2^p} \hat{\mathbb{Z}}_2 \xrightarrow{\phi_p} \mathbb{Z}/2^p\mathbb{Z} \to 0$$

and so we obtain the same associated graded module, $E_0^p(\hat{\mathbb{Z}}_2) = \{0\}$ if $p < 0$ and $E_0^p(\hat{\mathbb{Z}}_2) \cong 2^p\hat{\mathbb{Z}}_2/2^{p+1}\hat{\mathbb{Z}}_2 \cong \mathbb{Z}/2\mathbb{Z}$ if $p \geq 0$.

Reconstruction of a filtered module from an associated graded module may be difficult. In Chapter 1, in the case of field coefficients and a first quadrant spectral sequence, dimension arguments allow the recovery of an isomorphic vector space from the associated graded one. For an arbitrary commutative ring R, however, extension problems may arise: Suppose A is a filtered R-module and the (decreasing) filtration is bounded above and below, that is, $F^k A = \{0\}$ if $k > n$. Further suppose that $F^k A = A$ for $k < 0$; we present the filtration

$$\{0\} \subset F^n A \subset F^{n-1} A \subset \cdots \subset F^1 A \subset F^0 A \subset F^{-1} A = A.$$

The associated graded module $E_0^*(A)$ is nontrivial only in degrees $-1 \leq k \leq n$, and we obtain the series of short exact sequences

$$0 \longrightarrow F^n A \xrightarrow{\;=\;} E_0^n(A) \longrightarrow 0$$

$$0 \longrightarrow F^n A \longrightarrow F^{n-1} A \longrightarrow E_0^{n-1}(A) \longrightarrow 0$$

$$\vdots \qquad\qquad \vdots \qquad\qquad \vdots$$

$$0 \longrightarrow F^k A \longrightarrow F^{k-1} A \longrightarrow E_0^{k-1}(A) \longrightarrow 0$$

$$\vdots \qquad\qquad \vdots \qquad\qquad \vdots$$

$$0 \longrightarrow F^1 A \longrightarrow F^0 A \longrightarrow E_0^0(A) \longrightarrow 0$$

$$0 \longrightarrow F^0 A \longrightarrow A \longrightarrow E_0^{-1}(A) \longrightarrow 0.$$

If one knows that the filtration satisfies such boundedness conditions, then $E_0^n(A)$ determines $F^n A$. However, $F^{n-1} A$ is only determined up to choice of extension of $F^n A$ by $E_0^{n-1}(A)$. Working downward, each $F^{k-1} A$ is determined by a choice of extension by $F^k A$ by $E_0^{k-1}(A)$ down to A itself, which is known only up to a series of choices. In general, we are left with some ambiguity about A unless some further structure guides our choices.

If H^* is a graded R-module and H^* is filtered, then we can examine the filtration on each degree by letting $F^p H^n = F^p H^* \cap H^n$. Thus the associated graded module is bigraded when we define

$$E_0^{p,q}(H^*, F) = \begin{cases} F^p H^{p+q}/F^{p+1} H^{p+q}, & \text{if } F^* \text{ is decreasing,} \\[2mm] F^p H^{p+q}/F^{p-1} H^{p+q}, & \text{if } F^* \text{ is increasing.} \end{cases}$$

We next combine the associated graded module with the definition of a spectral sequence.

Definition 2.4. *A spectral sequence* $\{E_r^{*,*}, d_r\}$ *is said to* **converge** *to* H^*, *a graded R-module, if there is a filtration F on* H^* *such that*

$$E_\infty^{p,q} \cong E_0^{p,q}(H^*, F),$$

where $E_\infty^{*,*}$ *is the limit term of the spectral sequence.*

Determination of a graded module H^* is generally the goal of a computation. If there is a spectral sequence converging to H^* *and* if it converges uniquely to H^* *and* if all of the extension problems can be settled, then H^* is determined (a lot of *ifs*).

With the fundamental definitions in place, we begin to describe a general setting in which spectral sequences arise.

Definition 2.5. *An R-module is a* **filtered differential graded module** *if*

(1) *A is a direct sum of submodules,* $A = \bigoplus\limits_{n=0}^{\infty} A^n$.

(2) *There is an R-linear mapping,* $d: A \to A$, *of degree* 1 $(d: A^n \to A^{n+1})$ *or degree* -1 $(d: A^n \to A^{n-1})$ *satisfying* $d \circ d = 0$.

(3) *A has a filtration F and the differential d respects the filtration, that is,* $d: F^p A \to F^p A$.

Since the differential respects the filtration, $H(A, d) = \ker d / \operatorname{im} d$ inherits a filtration

$$F^p H(A, d) = \text{ image } \left(H(F^p A, d) \xrightarrow{H(\text{inclusion})} H(A, d) \right).$$

It's time for the main theorem. For convenience, suppose that A is a filtered differential graded module with differential of degree $+1$ and a descending filtration. (This is often the case in cohomological examples. The case of a spectral sequence of homological type is treated in the exercises.)

Theorem 2.6. *Each filtered differential graded module* (A, d, F^*) *determines a spectral sequence,* $\{E_r^{*,*}, d_r\}$, $r = 1, 2, \ldots$ *with* d_r *of bidegree* $(r, 1 - r)$ *and*

$$E_1^{p,q} \cong H^{p+q}(F^p A / F^{p+1} A).$$

Suppose further that the filtration is **bounded**, *that is, for each dimension* n, *there are values* $s = s(n)$ *and* $t = t(n)$, *so that*

$$\{0\} \subset F^s A^n \subset F^{s-1} A^n \subset \cdots \subset F^{t+1} A^n \subset F^t A^n = A^n,$$

then the spectral sequence converges to $H(A, d)$, that is,

$$E_\infty^{p,q} \cong F^p H^{p+q}(A, d) \big/ F^{p+1} H^{p+q}(A, d).$$

Before giving a proof of the theorem, let's anticipate how it might be applied. Information about $H(A, d)$ is most readily obtained from A, but this module may be inaccessible, for example, when $A = C^*(X; R)$, the singular cochains on a space X with coefficients in a commutative ring R. If A can be filtered and some term of the associated spectral sequence identified as something calculable, then we can obtain $H(A, d)$ up to computation of the successive homologies and reconstruction from the associated graded module.

Theorem 2.6 first appeared in the work of [Koszul47] and [Cartan48] who had extracted the algebraic essence underlying the work of [Leray46] on sheaves, homogeneous spaces, and fibre spaces.

A word of warning: Though the guts of our black box are laid bare, such close examination may reveal details too fine to be enlightening. Thus we place the proof of the theorem in a separate section and recommend skipping it to the novice (Ⓝ, *not* for the novice) and to the weak of interest. The novice reader can take Theorem 2.6 on faith on the first reading of the book. Occasional reference may be made in later sections to details contained in the proof but the reader can easily identify the necessary details at that time.

After the proof of Theorem 2.6 we consider another setting in which spectral sequences arise—exact couples. Relations between these constructions are also determined.

The proof of Theorem 2.6 Ⓝ

In what follows, keep the decreasing filtration in mind:

$$\cdots \subset F^p A^{p+q} \subset F^{p-1} A^{p+q} \subset F^{p-2} A^{p+q} \subset \cdots ,$$

as well as the fact that the differential is stable, that is, $d(F^p A^{p+q}) \subset F^p A^{p+q+1}$. Consider the following definitions:

$$
\begin{aligned}
Z_r^{p,q} &= \text{elements in } F^p A^{p+q} \text{ that have boundaries in } F^{p+r} A^{p+q+1} \\
&= F^p A^{p+q} \cap d^{-1}(F^{p+r} A^{p+q+1}) \\
B_r^{p,q} &= \text{elements in } F^p A^{p+q} \text{ that form the image of } d \text{ from } F^{p-r} A^{p+q-1} \\
&= F^p A^{p+q} \cap d(F^{p-r} A^{p+q-1}) \\
Z_\infty^{p,q} &= \ker d \cap F^p A^{p+q} \\
B_\infty^{p,q} &= \operatorname{im} d \cap F^p A^{p+q}.
\end{aligned}
$$

The decreasing filtration and the stability of the differential give us the desired tower of submodules,

$$B_0^{p,q} \subset B_1^{p,q} \subset \cdots \subset B_\infty^{p,q} \subset Z_\infty^{p,q} \subset \cdots \subset Z_1^{p,q} \subset Z_0^{p,q}$$

as well as $d(Z_r^{p-r,q+r-1}) = d(F^{p-r}A^{p+q-1} \cap d^{-1}(F^p A^{p+q}))$
$$= F^p A^{p+q} \cap d(F^{p-r}A^{p+q-1})$$
$$= B_r^{p,q}.$$

The assumption that the filtration is bounded implies, for $r > s(p+q+1) - p$ and $r \geq p - t(p+q-1)$, that $Z_r^{p,q} = Z_\infty^{p,q}$ and $B_r^{p,q} = B_\infty^{p,q}$. This insures convergence.

Define, for all $0 \leq r \leq \infty$, $E_r^{p,q} = Z_r^{p,q}/(Z_{r-1}^{p+1,q-1} + B_{r-1}^{p,q})$ and define $\eta_r^{p,q} \colon Z_r^{p,q} \to E_r^{p,q}$ to be the canonical projection with $\ker \eta_r^{p,q} = (Z_{r-1}^{p+1,q-1} + B_{r-1}^{p,q})$. Observe that $d(Z_r^{p,q}) = B_r^{p+r,q-r+1} \subset Z_r^{p+r,q-r+1}$ and

$$d(Z_{r-1}^{p+1,q-1} + B_{r-1}^{p,q}) = d(Z_{r-1}^{p+1,q-1}) + d(B_{r-1}^{p,q})$$
$$\subset B_{r-1}^{p+r,q-r+1} + 0$$
$$\subset Z_{r-1}^{p+r+1,q-r} + B_{r-1}^{p+r,q-r+1}.$$

Thus the differential, as a mapping $d \colon Z_r^{p,q} \to Z_r^{p+r,q-r+1}$, induces a homomorphism, d_r, so that the following diagram commutes.

$$
\begin{array}{ccc}
Z_r^{p,q} & \xrightarrow{d} & Z_r^{p+r,q-r+1} \\
\eta \downarrow & & \downarrow \eta \\
E_r^{p,q} & \xrightarrow{d_r} & E_r^{p+r,q-r+1}
\end{array}
$$

Since $d \circ d = 0$, $d_r \circ d_r = 0$.

To complete the proof we must establish the following:

I. $H^*(E_r^{*,*}, d_r) \cong E_{r+1}^{*,*}$,
II. $E_1^{p,q} \cong H^{p+q}(F^p A/F^{p+1}A)$,
III. $E_\infty^{p,q} \cong F^p H^{p+q}(A,d)/F^{p+1}H^{p+q}(A,d)$.

Toward I, consider the diagram

$$
\begin{array}{ccccccc}
Z_r^{p+1,q-1} + B_r^{p,q} & \xrightarrow{\subset} & Z_{r+1}^{p,q} & \xrightarrow{\subset} & Z_r^{p,q} & \xrightarrow{d} & Z_r^{p+r,q-r+1} \\
\eta \downarrow & & \eta_r^{p,q} \downarrow & & & & \downarrow \eta_r^{p+r,q-r+1} \\
\ker d_r & & \longrightarrow & E_r^{p,q} & \xrightarrow{d_r} & & E_r^{p+r,q-r+1} \\
\downarrow & & & & & & \\
H^{p,q}(E_r^{*,*}, d_r) & & & & & & \\
\downarrow & & & & & & \\
0 & & & & & &
\end{array}
$$

First observe that $\eta_r^{p,q}(Z_{r+1}^{p,q}) = \ker d_r$. Consider $\eta^{-1}(\ker d_r)$. Since $d_r \circ \eta = \eta \circ d$, $d_r(\eta z) = 0$ if and only if dz is in $Z_{r-1}^{p+r+1,q-r} + B_{r-1}^{p+r,q-r+1}$ and, by the

definitions of $Z_*^{*,*}$ and $B_*^{*,*}$, this is so if and only if z is in $Z_{r+1}^{p,q} + Z_{r-1}^{p+1,q-1}$.
Thus $\eta^{-1}(\ker d_r) = Z_{r+1}^{p,q} + Z_{r-1}^{p+1,q-1}$ and so $\ker d_r = \eta(Z_{r+1}^{p,q} + Z_{r-1}^{p+1,q-1}) = \eta(Z_{r+1}^{p,q})$, since $Z_{r-1}^{p+1,q-1} \subset \ker \eta_r^{p,q}$.

Secondly observe that $Z_r^{p+1,q-1} + B_r^{p,q} = Z_{r-1}^{p,q} \cap ((\eta_r^{p,q})^{-1}(\operatorname{im} d_r))$. We
know that $\operatorname{im} d_r = \eta_r^{p,q}(d(Z_r^{p-r,q+r-1})) = \eta_r^{p,q}(B_r^{p,q})$ and so

$$
\begin{aligned}
(\eta_r^{p,q})^{-1}(\operatorname{im} d_r) &= B_r^{p,q} + \ker \eta_r^{p,q} \\
&= B_r^{p,q} + B_{r-1}^{p,q} + Z_{r-1}^{p+1,q-1} \\
&= B_r^{p,q} + Z_{r-1}^{p+1,q-1}.
\end{aligned}
$$

Since $Z_{r-1}^{p+1,q-1} \cap Z_{r+1}^{p,q} = F^{p+1}A^{p+q} \cap d^{-1}(F^{p+r}A^{p+q+1}) \cap F^p A^{p+q} \cap d^{-1}(F^{p+r+1}A^{p+q+1})$ and $F^{p+1}A^{p+q} \subset F^p A^{p+q}$, as well as $F^{p+r+1}A^{p+q} \subset F^{p+r}A^{p+q}$, we have $Z_{r-1}^{p+1,q-1} \cap Z_{r+1}^{p,q} = F^{p+1}A^{p+q} \cap d^{-1}(F^{p+r+1}A^{p+q+1}) = Z_r^{p+1,q-1}$. With the previous calculation we obtain

$$
Z_{r-1}^{p+1,q-1} \cap (\eta_r^{p,q})^{-1}(\operatorname{im} d_r) = Z_r^{p+1,q-1} + B_r^{p,q}.
$$

Finally, let $\gamma \colon Z_{r+1}^{p,q} \to H^{p,q}(E_r^{*,*}, d_r)$ be the composite mapping of $\eta_r^{p,q}$ with the canonical projection $\ker d_r \to H^{p,q}(E_r^{*,*}, d_r)$. The kernel of γ is given by $Z_{r+1}^{p,q} \cap (\eta_r^{p,q})^{-1}(\operatorname{im} d_r) = Z_r^{p+1,q-1} + B_r^{p,q}$. Since γ is an epimorphism, we have the isomorphism

$$
H^{p,q}(E_r^{*,*}, d_r) \cong Z_{r+1}^{p,q} \big/ (Z_r^{p+1,q-1} + B_r^{p,q}) = E_{r+1}^{p,q}.
$$

Thus γ induces the isomorphism making our tower a spectral sequence.

Toward II, observe $E_0^{p,q} = Z_0^{p,q}/(Z_{-1}^{p+1,q-1} + B_{-1}^{p,q})$ where we define

$$
Z_{-1}^{p+1,q-1} = F^{p+1}A^{p+q} \qquad \text{and} \qquad B_{-1}^{p,q} = d(F^{p+1}A^{p+q-1}).
$$

Since d respects the filtration

$$
\begin{aligned}
E_0^{p,q} &= F^p A^{p+q} \cap d^{-1}(F^p A^{p+q+1}) \big/ F^{p+1}A^{p+q} + d(F^{p+1}A^{p+q-1}) \\
&= F^p A^{p+q}/F^{p+1}A^{p+q}.
\end{aligned}
$$

The differential $d_0 \colon E_0^{p,q} \to E_0^{p,q+1}$ is induced by the differential $d \colon F^p A^{p+q} \to F^p A^{p+q+1}$ and so we have $E_1^{p,q} \cong H^{p+q}(F^p A/F^{p+1}A)$.

For III, let $\eta_\infty^{p,q} \colon Z_\infty^{p,q} \to E_\infty^{p,q}$ and $\pi \colon \ker d \to H(A,d)$ denote the canonical projections:

$$
F^p H^{p+q}(A,d) = H^{p+q}(\operatorname{im}(F^p A \to A), d) = \pi(F^p A^{p+q} \cap \ker d) = \pi(Z_\infty^{p,q}).
$$

Since $\pi(\ker \eta_\infty^{p,q}) = \pi(Z_\infty^{p+1,q-1} + B_\infty^{p,q}) = F^{p+1}H^{p+q}(A,d)$, we have that π induces a mapping $d_\infty \colon E_\infty^{p,q} \to F^p H^{p+q}(A,d)/F^{p+1}H^{p+q}(A,d)$. Observe further that

$$\begin{aligned}
\ker d_\infty &= \eta_\infty^{p,q}(\pi^{-1}(F^{p+1}H^{p+q}(A,d)) \cap Z_\infty^{p,q}) \\
&= \eta_\infty^{p,q}(Z_\infty^{p+1,q-1} \cap d(A) \cap Z_\infty^{p,q}) \\
&\subset \eta_\infty^{p,q}(Z_\infty^{p+1,q-1} + B_\infty^{p,q}) = \{0\}.
\end{aligned}$$

Thus d_∞ is an isomorphism. The boundedness of the filtration on (A,d) implies that the induced filtration on $H(A,d)$ is bounded and so a finite sequence of extension problems go from $E_\infty^{*,*}$ to $H(A,d)$. $\qquad\square$

In §3.1 the relation between a spectral sequence associated to a filtered differential graded module and the homology of that differential module is developed further. The role of assumptions like boundedness of the filtration in determining $H(A,d)$ uniquely is explored. Weaker conditions that guarantee convergence and uniqueness of the target are also discussed.

Exact couples

It can be the case that our objects of study are not explicitly filtered or do not come from a filtered differential object. In this section we present another general algebraic setting, exact couples, in which spectral sequences arise. The ease of definition of the spectral sequence and its applicability make this approach very attractive. Unlike the case of a filtered differential graded module, however, the target of the spectral sequence coming from an exact couple may be difficult to identify. Some results toward solving that problem are developed in Chapter 3. We also show how an exact couple results from a filtered differential graded module whose spectral sequence is the same as the one in Theorem 2.6. The idea of the exact couple was introduced by [Massey50].

Let D and E denote R-modules (which are bigraded in the relevant cases) and let $i\colon D \to D$, $j\colon D \to E$ and $k\colon E \to D$ be module homomorphisms. We present these data as in the diagram:

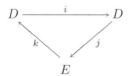

and call $\mathcal{C} = \{D, E, i, j, k\}$ an **exact couple** if this diagram is exact at each group, that is, $\operatorname{im} i = \ker j$, $\operatorname{im} j = \ker k$ and $\operatorname{im} k = \ker i$.

An important example of an exact couple comes from the long exact sequence in homology and a short exact sequence of coefficients. Let

$$0 \longrightarrow \mathbb{Z} \overset{\times p}{\longrightarrow} \mathbb{Z} \longrightarrow \mathbb{Z}/p\mathbb{Z} \longrightarrow 0$$

be the short exact sequence associated to the 'times p' map. Suppose (C^*, d) is a differential graded abelian group that is free in each degree. When we tensor C^* with the coefficients, the 'times p' map results in the short exact sequence

$$0 \longrightarrow C^* \xrightarrow{\times p} C^* \longrightarrow C^* \otimes \mathbb{Z}/p\mathbb{Z} \longrightarrow 0$$

and, on application of homology, an exact couple

The spectral sequence associated to this exact couple is known as the **Bockstein spectral sequence**, the topic of Chapter 10.

An immediate consequence of the exactness of a couple is that E becomes a differential R-module with $d\colon E \to E$ given by $d = j \circ k$. To see this, we compute: $d \circ d = (j \circ k) \circ (j \circ k) = j \circ (k \circ j) \circ k = 0$. The fundamental operation on exact couples is the formation of the **derived couple**: Let

$$E' = H(E, d) = \ker d / \operatorname{im} d = \ker(j \circ k) / \operatorname{im}(j \circ k), \qquad D' = i(D) = \ker j.$$

Also define

$$i' = i|_{iD} \colon D' \to D' \quad \text{and} \quad j' \colon D' \to E' \text{ by } j'(i(x)) = j(x) + dE \in E'$$

where $x \in D$. That j' is well-defined can be seen as follows: If $i(x) = i(x')$, then $x - x'$ is in $\ker i = \operatorname{im} k$ and there is a $y \in E$ with $k(y) = x - x'$. Thus $(j \circ k)(y) = d(y) = j(x) - j(x')$ and $j(x) = j(x') + d(y)$, that is, $j(x) + dE = j(x') + dE$ as cosets in E'. Finally, define

$$k' \colon E' \to D' \text{ by } k'(e + dE) = k(e).$$

If $e + dE = e' + dE$, then $e' = e + d(x)$ for some $x \in E$ and $k(e') = k(e) + k(d(x)) = k(e) + (k \circ j \circ k)(x) = k(e)$; thus k' is well-defined. Also, since $d(e) = 0$, we have that $k(e)$ is in $\ker j = \operatorname{im} i = D'$.

We call $C' = \{D', E', i', j', k'\}$ the *derived couple* of C. We prove the fundamental result.

Proposition 2.7. $C' = \{D', E', i', j', k'\}$, *the derived couple, is an exact couple.*

PROOF: We first consider exactness at the left D':

$$\begin{aligned} \ker i' &= \operatorname{im} i \cap \ker i = \ker j \cap \operatorname{im} k \\ &= k(k^{-1}(\ker j)) = k(\ker d) = k'(\ker d / \operatorname{im} d) \\ &= \operatorname{im} k'. \end{aligned}$$

Notice that $D' = iD = D/\ker i$. From this we can write

$$
\begin{aligned}
\ker j' &= {}^{j^{-1}(\operatorname{im} d)}\!/_{\ker i} = {}^{j^{-1}(j(\operatorname{im} k))}\!/_{\ker i} \\
&= {}^{(\operatorname{im} k + \ker j)}\!/_{\ker i} = {}^{(\ker i + \ker j)}\!/_{\ker i} \\
&= i(\ker j) = i(\operatorname{im} i) = \operatorname{im} i'.
\end{aligned}
$$

Finally consider $\ker k' = \ker k/\operatorname{im} d = \operatorname{im} j/\operatorname{im} d = jD/\operatorname{im} d = \operatorname{im} j'$ since $j \circ i = 0$. □

We can iterate this process to obtain the nth **derived couple** of $\mathcal{C}$,

$$
\mathcal{C}^{(n)} = \{D^{(n)}, E^{(n)}, i^{(n)}, j^{(n)}, k^{(n)}\} = (\mathcal{C}^{(n-1)})'.
$$

The connection with spectral sequences can be guessed by now since $E^{(n+1)} = H(E^{(n)}, d^{(n)})$. To solidify this connection we introduce a bigrading.

Theorem 2.8. *Suppose $D^{*,*} = \{D^{p,q}\}$ and $E^{*,*} = \{E^{p,q}\}$ are bigraded modules over R equipped with homomorphisms i of bidegree $(-1, 1)$, j of bidegree $(0, 0)$ and k of bidegree $(1, 0)$.*

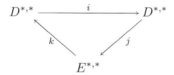

These data determine a spectral sequence $\{E_r, d_r\}$ for $r = 1, 2, \ldots$, of cohomological type, with $E_r = (E^{*,*})^{(r-1)}$, the $(r-1)$-st derived module of $E^{*,*}$ and $d_r = j^{(r)} \circ k^{(r)}$.

PROOF: It suffices to check that the differentials, d_r, have the correct bidegree, $(r, 1 - r)$. Let $E_1 = E^{*,*}$, $d_1 = j \circ k$ and so d_1 has bidegree $(1, 0) + (0, 0) = (1, 0)$. Now assume $j^{(r-1)}$ and $k^{(r-1)}$ have bidegrees $(r-2, 2-r)$ and $(1, 0)$, respectively. Since $j^{(r)}(i^{(r-1)}(x)) = j^{(r-1)}(x) + d^{(r-1)}E^{(r-1)}$, the image in $(E^{p,q})^{(r)}$ must come from $i^{(r-1)}(D^{p-r+2,q+r-2})^{(r-1)} = (D^{p-r+1,q+r-1})^{(r)}$ or $j^{(r)}$ has bidegree $(r-1, 1-r)$. Since $k^{(r)}(e + d^{(r-1)}E^{(r-1)}) = k^{(r-1)}(e)$ and $k^{(r-1)}$ has bidegree $(1, 0)$, so does $k^{(r)}$. Combining this with the inductive hypothesis gives us that $d^{(r)}$ has bidegree $(r, 1 - r)$ as required. □

A bigraded exact couple may be displayed as in the following diagram: Here the path made up of one vertical segment and two horizontal segments is exact.

$$
\begin{array}{ccccccc}
 & \downarrow i & & & & \downarrow i & \\
\xrightarrow{k} D^{p+2,q-1} & \xrightarrow{j} & E^{p+2,q-1} & \xrightarrow{k} & D^{p+3,q-1} & \xrightarrow{j} & \\
 & \downarrow i & & & & \downarrow i & \\
\xrightarrow{k} D^{p+1,q} & \xrightarrow{j} & E^{p+1,q} & \xrightarrow{k} & D^{p+2,q} & \xrightarrow{j} & \\
 & \downarrow i & & & & \downarrow i & \\
\xrightarrow{k} D^{p,q+1} & \xrightarrow{j} & E^{p,q+1} & \xrightarrow{k} & D^{p+1,q+1} & \xrightarrow{j} & \\
 & \downarrow i & & & & \downarrow i &
\end{array}
$$

Another useful presentation of exact couples is the **unrolled exact couple** (or *unraveled* exact couple; see the paper of [Boardman99]) where we suppress one of the bidegrees in the diagram:

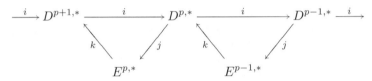

To an exact couple we associate a tower of submodules of E and an E_∞-term as we would do for any spectral sequence. There is an intrinsic expression (related only to the couple) for these objects that is useful when studying convergence. (See Chapter 3.) We use the notation of an unrolled exact couple.

Proposition 2.9. *Let* $Z_r^{p,*} = k^{-1}(\operatorname{im} i^{r-1} \colon D^{p+r,*} \longrightarrow D^{p+1,*})$ *and* $B_r^{p,*} = j(\ker i^{r-1} \colon D^{p,*} \longrightarrow D^{p-r+1,*})$ *designate submodules of* $E^{p,*}$. *Then these submodules determine the spectral sequence associated to the exact couple:*

$$
E_r^{p,*} = (E^{p,*})^{(r-1)} \cong Z_r^{p,*}/B_r^{p,*}.
$$

Furthermore, $E_\infty^{p,*} \cong \bigcap_r Z_r^{p,*}/\bigcup_r B_r^{p,*} \cong$

$$
\bigcap_r k^{-1}(\operatorname{im} i^{r-1} \colon D^{p+r,*} \to D^{p+1,*}) \Big/ \bigcup_r j(\ker i^{r-1} \colon D^{p,*} \to D^{p-r+1,*}).
$$

PROOF: For $r = 2$, $E_2^{*,*} = (E^{*,*})' = \ker d/\operatorname{im} d = \ker(j \circ k)/\operatorname{im}(j \circ k)$. Now $\operatorname{im}(j \circ k) = j(\operatorname{im} k) = j(\ker i)$. Also $\ker(j \circ k) = k^{-1}(\ker j) = k^{-1}(\operatorname{im} i)$. So

$$
E_2^{p,*} = k^{-1}(\operatorname{im} i \colon D^{p+2,*} \to D^{p+1,*}) \Big/ j(\ker i \colon D^{p,*} \to D^{p-1,*}).
$$

By induction, since $i^{(r)}$, $j^{(r)}$, and $k^{(r)}$ are induced by i, j, and k with the appropriate images in $D^{(r)} = \operatorname{im} i^{r-1}$, we obtain the E_r-term as described. It is clear that $Z_r^{p,*} \subset Z_{r-1}^{p,*}$ and $B_r^{p,*} \subset B_{r+1}^{p,*}$ where the inclusions follow by composition with another factor of the mapping i. We leave it to the reader to check that the differential

$$d_r : Z_{r-1}/B_{r-1} \to B_r/B_{r-1} \subset Z_{r-1}/B_{r-1}$$

is induced on our representation by $j \circ k$ with the proper kernel. This describes the tower of submodules. The description of the E_∞-term of the associated spectral sequence follows immediately. □

Proposition 2.9 implies the inclusion $j(D^{p,*}) \subset Z_r^{p,*}$ for all r. This follows because $k \circ j(D^{p,*}) = \{0\}$ and so $j(D^{p,*}) \subset \ker k \subset Z_r^{p,*}$. Thus elements in $E_r^{p,*}$ that come from the image of j are permanent cycles in the spectral sequence.

Another expression for the E_r-terms of the spectral sequence associated to an exact couple is given in the following corollary (first given by [Eckmann-Hilton66]).

Corollary 2.10. *For $r \geq 1$, there is an exact sequence:*

$$0 \longrightarrow D^{p,*} \Big/ \left(\ker i^r (D^{p,*} \to D^{p-r,*}) + iD^{p+1,*}\right) \xrightarrow{\ \bar{j}\ } E_{r+1}^{p,*}$$

$$\xrightarrow{\ \bar{k}\ } \operatorname{im} i^r (D^{p+r+1,*} \to D^{p+1,*}) \cap \ker i(D^{p+1,*} \to D^{p,*}) \longrightarrow 0.$$

PROOF: By Proposition 2.9 we get the following diagram with rows and columns exact:

$$j(\ker i^r : D^{p,*} \to D^{p-r,*})$$

$$\downarrow$$

$$k^{-1}(\operatorname{im} i^r : D^{p+r+1,*} \to D^{p+1,*}) \xrightarrow{\ k\ } \operatorname{im} i^r \cap \operatorname{im} k \longrightarrow 0$$

$$\downarrow \qquad\qquad \nearrow \scriptstyle{\bar{k}}$$

$$E_{r+1}^{p,*}$$

$$\downarrow$$

$$0$$

Let $\bar{k} : E_{r+1}^{p,*} \to \operatorname{im} i^r \cap \operatorname{im} k$ be induced by lifting an element and applying k. Since $k \circ j = 0$, this mapping is well-defined and since $\operatorname{im} i^r \cap \operatorname{im} k = \operatorname{im} i^r \cap \ker i$, we have the right half of the short exact sequence.

To construct the homomorphism $\bar{\jmath}$ we begin with the short exact sequences

$$0 \longrightarrow iD^{p+1,*} + \ker i^r \longrightarrow D^{p,*} \longrightarrow D^{p,*}\big/\left(iD^{p+1,*} + \ker i^r\right) \longrightarrow 0$$

$$\Big\downarrow j \qquad\qquad \Big\downarrow j \qquad\qquad \Big\downarrow \hat{\jmath}$$

$$0 \longrightarrow j(\ker i^r) \longrightarrow \operatorname{im} j \longrightarrow \operatorname{im} j\big/j(\ker i^r) \longrightarrow 0.$$

The mapping $\hat{\jmath}$ is an epimorphism by the Five-lemma. Also $\operatorname{im} j = \ker k = k^{-1}(0)$, so we have the homomorphism

$$\bar{\jmath}\colon D^{p,*}\big/\left(iD^{p+1,*} + \ker i^r\right) \to k^{-1}(0)\big/j(\ker i^r).$$

Consider the following diagram with $\bar{k}$ and both rows exact:

$$0 \longrightarrow k^{-1}(0) \longrightarrow k^{-1}(\operatorname{im} i^r) \xrightarrow{\ k\ } \operatorname{im} i^r \cap \operatorname{im} k \longrightarrow 0$$

$$\Big\downarrow \qquad\qquad\qquad \Big\downarrow \qquad\qquad\qquad \Big\|$$

$$0 \longrightarrow k^{-1}(0)\big/j(\ker i^r) \longrightarrow k^{-1}(\operatorname{im} i^r)\big/j(\ker i^r) \xrightarrow{\ \bar{k}\ } \operatorname{im} i^r \cap \operatorname{im} k \longrightarrow 0.$$

Since $E_{r+1}^{p,*} = k^{-1}(\operatorname{im} i^r)\big/j(\ker i^r)$, it suffices to show that $\bar{\jmath}$ is an isomorphism. We have that $\bar{\jmath}$ is an epimorphism already so we show that it is a monomorphism. Let $[a]$, $[b]$ be in $D^{p,*}\big/\left(iD^{p+1,*} + \ker i^r\right)$. If $\bar{\jmath}[a] = \bar{\jmath}[b]$, then $\bar{\jmath}[a - b] = 0$ and so $j(a - b)$ lies in $j(\ker i^r)$. Therefore, either $a - b$ is in $\ker i^r$ or $a - b$ is in $\ker j = \operatorname{im} i$. We conclude that $a - b$ is in $\operatorname{im} i + \ker i^r$ and so $[a] = [b]$.   □

The equivalence of the two approaches

A filtered differential graded R-module (A, d, F) leads to another example of an exact couple. For each filtration degree p, there is a short exact sequence of graded modules

$$0 \longrightarrow F^{p+1}A \longrightarrow F^pA \longrightarrow F^pA/F^{p+1}A \longrightarrow 0.$$

The fact that the differential respects the filtration gives us a short exact sequence of differential graded modules. When we apply the homology functor, we obtain, for each p, the long exact sequence

$$\cdots H^{p+q}(F^{p+1}A) \xrightarrow{i} H^{p+q}(F^pA) \xrightarrow{j} H^{p+q}(F^pA/F^{p+1}A)$$

$$\xrightarrow{k} H^{p+q+1}(F^{p+1}A) \xrightarrow{i} H^{p+q+1}(F^pA) \xrightarrow{j} \cdots$$

where k is the connecting homomorphism. Define the bigraded modules $E^{p,q} = H^{p+q}(F^p A/F^{p+1}A)$ and $D^{p,q} = H^{p+q}(F^p A)$. This gives an exact couple from the long exact sequences:

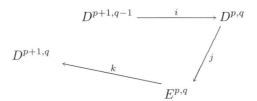

The bigradings agree with Theorem 2.8 to yield a spectral sequence. Furthermore, the E_1-terms of this spectral sequence and the one in Theorem 2.6 are the same.

Proposition 2.11. *For a filtered differential graded R-module (A, d, F), the spectral sequence associated to the (decreasing) filtration and the spectral sequence associated to the exact couple are the same.*

PROOF: It suffices to show that, in the spectral sequence for the exact couple, the E_r-term, as a subquotient of $F^p A/F^{p+1}A$, coincides with the subquotient given in the proof of Theorem 2.6. That is, in the notation of the proof of Theorem 2.6,

$$E_r^{p,q} = Z_r^{p,q} \Big/ Z_{r-1}^{p+1,q-1} + B_{r-1}^{p,q}.$$

Suppose z is in $H^{p+q}(F^p A/F^{p+1}A) = E_1^{p,q}$. Then z can be represented by $[x + F^{p+1}A]$ with x in $F^p A$ and $d(x)$ in $F^{p+1}A$. The boundary homomorphism k in the long exact sequence that is part of the exact couple can be described explicitly by

$$k([x + F^{p+1}A]) = [d(x)] \quad \text{in} \quad H^{p+q+1}(F^{p+1}A).$$

Thus $[x + F^{p+1}A]$ is in $k^{-1}(\operatorname{im} i^{r-1})$ if and only if $[d(x)]$ is in $\operatorname{im} i^{r-1}$ if and only if $d(x)$ is in $F^{p+r}A^{p+q+1}$. Since x is already in $F^p A^{p+q}$, then x lies in $F^p A^{p+q} \cap d^{-1}(F^{p+r}A^{p+q+1}) = Z_r^{p,q}$. Adding the indeterminacy, in the appropriate bigrading, we now have

$$k^{-1}(\operatorname{im} i^{r-1}) = Z_r^{p,q}/F^{p+1}A^{p+q}.$$

Consider $\ker i^{r-1} \subset H^{p+q}(F^p A)$. A class $[u]$ is in $\ker i^{r-1}$ if and only if u is in $F^p A^{p+q}$ and u is a boundary in $F^{p-r+1}A^{p+q}$. Then u lies in $F^p A^{p+q} \cap d(F^{p-r+1}A^{p+q-1}) = B_r^{p,q}$. Since j assigns to a class in $H^{p+q}(F^p A)$ its relative class modulo $F^{p+1}A$, we deduce that

$$j(\ker i^{r-1}) = B_{r-1}^{p,q}/F^{p+1}A^{p+q}.$$

By definition $Z^{p+1,q-1}_{r-1} \subset F^{p+1}A^{p+q}$ and so we have

$$E^{p,q}_r = k^{-1}(\operatorname{im} i^{r-1}) \Big/ j(\ker i^{r-1})$$

$$= Z^{p,q}_r / F^{p+1}A^{p+q} \Big/ B^{p,q}_r / F^{p+1}A^{p+q}$$

$$= Z^{p,q}_r / F^{p+1}A^{p+q} \Big/ (B^{p,q}_{r-1} + Z^{p+1,q-1}_{r-1}) / F^{p+1}A^{p+q}$$

$$\cong Z^{p,q}_r \Big/ (Z^{p+1,q-1}_{r-1} + B^{p,q}_{r-1}). \qquad \square$$

We next discuss the effect of extra structure on a filtered differential module.

2.3 Spectral sequences of algebras

Let (A, d_A) and (B, d_B) be differential graded modules over R. Recall the definition of the **tensor product of differential graded modules** over R; $(A \otimes_R B, d_\otimes)$ is given by

$$(A \otimes_R B)^n = \bigoplus_{p+q=n} A^p \otimes_R B^q$$

with $d_\otimes(a \otimes b) = d_A(a) \otimes b + (-1)^{\deg a} a \otimes d_B(b)$. Furthermore, a **differential graded algebra** over R is a differential graded module, (A, d) together with a morphism of differential graded modules,

$$\psi \colon (A \otimes_R A, d_\otimes) \longrightarrow (A, d)$$

for which the usual diagrams commute (expressing associativity and, if it exists, the property of a unit in A^0).

We next define the **tensor product of differential bigraded modules** over R. Given $(E^{*,*}, d_E)$ and $(\bar{E}^{*,*}, d_{\bar{E}})$ let

$$(E \otimes_R \bar{E})^{p,q} = \bigoplus_{\substack{r+t=p \\ s+u=q}} E^{r,s} \otimes_R \bar{E}^{t,u}$$

with $d_\otimes(e \otimes \bar{e}) = d_E(e) \otimes \bar{e} + (-1)^{r+s} e \otimes d_{\bar{E}}(\bar{e})$ when $e \in E^{r,s}$ and $\bar{e} \in \bar{E}^{t,u}$. Also we define a **differential bigraded algebra** over R to be a differential bigraded module $(E^{*,*}, d)$ together with a morphism of such, $\psi \colon (E \otimes E)^{*,*} \to E^{*,*}$ for which the usual diagrams commute.

Finally, we could try to define a "tensor product of spectral sequences" by forming the tensor product of differential bigraded modules at each term in the sequences. However, the defining isomorphism, $H(E^{*,*}_r \otimes_R \bar{E}^{*,*}_r, d_\otimes) \cong E^{*,*}_{r+1} \otimes_R \bar{E}^{*,*}_{r+1}$ may be too much to ask for; the Künneth theorem makes the difficulty precise. (The relevant definitions in this theorem are found in later sections of this chapter.) A proof of Theorem 2.12 may be found in [Weibel, §3.6].

Theorem 2.12 (the Künneth theorem). *If (A, d_A) and (B, d_B) are differential graded modules over R and, for each n, $Z^n(A) = \ker d_A \colon A^n \to A^{n+1}$ and $B^n(A) = \operatorname{im} d_A \colon A^{n-1} \to A^n$ are flat R-modules, then there is a short exact sequence*

$$0 \to \bigoplus_{r+s=n} H^r(A) \otimes_R H^s(B) \overset{p}{\to} H^n(A \otimes_R B)$$

$$\longrightarrow \bigoplus_{r+s=n-1} \operatorname{Tor}_1^R(H^r(A), H^s(B)) \to 0,$$

where the homomorphism p is given by $p([u] \otimes [v]) = [u \otimes v]$. In the case that $Z^n(A)$ and $H^n(A)$ are projective R-modules for all n, then the homomorphism p is an isomorphism.

The theorem indicates how $H(E_r \otimes E_r, d_r)$ need not give $E_{r+1} \otimes E_{r+1}$ except in special cases; for example, when R is a field. Since the notion of products on a spectral sequence remains desirable, we provide a workable definition as follows:

Definition 2.13. *A **spectral sequence of algebras** over R is a spectral sequence, $\{E_r^{*,*}, d_r\}$ together with algebra structures $\psi_r \colon E_r \otimes_R E_r \to E_r$ for each r, such that ψ_{r+1} can be written as the composite*

$$\psi_{r+1} \colon E_{r+1} \otimes_R E_{r+1} \overset{\cong}{\longrightarrow} H(E_r) \otimes_R H(E_r)$$

$$\underset{p}{\longrightarrow} H(E_r \otimes E_r) \xrightarrow{H(\psi_r)} H(E_r) \underset{\cong}{\longrightarrow} E_{r+1},$$

where the homomorphism p is given by $p([u] \otimes [v]) = [u \otimes v]$.

The first spectral sequences of [Leray46], [Koszul47], and [Cartan48] were spectral sequences of algebras—in fact, the term *spectral ring* was used until 1950 when *spectral sequence* was coined by [Serre50] to describe the more general case (see [McCleary99]).

Recall "Theorem II" of Chapter 1: A spectral sequence of algebras **converges** to H a graded algebra **as an algebra** when the algebra structure on $E_\infty^{*,*}$ is isomorphic to the induced algebra structure on the associated bigraded algebra $E_0^{*,*}(H, F)$.

Theorem 2.14. *Suppose (A, d, F) is a filtered differential graded algebra with product $\psi \colon A \otimes_R A \to A$. Suppose that the product satisfies the condition for all p, q,*

$$\psi(F^p A \otimes_R F^q A) \subset F^{p+q} A.$$

Then the spectral sequence associated to (A, d, F) is spectral sequence of algebras. If the filtration on A is bounded, then the spectral sequence converges to $H(A, d)$ as an algebra.

PROOF: Let $x \in E_r^{p,q}$ and $y \in E_r^{s,t}$. We represent x and y by classes

$$a \in Z_r^{p,q} = F^p A^{p+q} \cap d^{-1}(F^{p+r} A^{p+q+1}) \qquad \text{and}$$
$$b \in Z_r^{s,t} = F^s A^{s+t} \cap d^{-1}(F^{s+r} A^{s+t+1})$$

so that $x = a + B_r^{p,q}$ and $y = b + B_r^{s,t}$. By the properties of the product and the filtration, $a \cdot b \in F^{p+s} A^{p+q+s+t}$ and

$$d(a \cdot b) = (da) \cdot b + (-1)^{p+q} a \cdot (db) \in F^{p+s+r} A^{p+q+s+t+1}.$$

It follows that $a \cdot b \in F^{p+s} A^{p+s+q+t} \cap d^{-1}(F^{p+s+r} A^{p+s+q+t+1}) = Z_r^{p+s,q+t}$ and so $a \cdot b$ represents a class in $E_r^{p+s,q+t}$. Varying a and b by elements in $B_r^{*,*}$ changes neither the filtration degree nor the destination of the product on application of the differential. Hence the product ψ induces a product ψ_r on $E_r^{*,*}$ making it a bigraded algebra.

To prove that ψ_{r+1} is related to ψ_r, as the conditions for a spectral sequence of algebras require, it suffices to show that d_r is a derivation, that is, $d_r(x \cdot y) = (d_r x) \cdot y + (-1)^{p+q} x \cdot (d_r y)$. However, this follows from the Leibniz rule for (A, d, ψ).

Finally, by Theorem 2.6, we know that a bounded filtration implies the convergence of the spectral sequence to $H(A, d)$, that is,

$$E_\infty^{p,q} \cong F^p H^{p+q}(A, d) \big/ F^{p+1} H^{p+q}(A, d)$$
$$= \operatorname{im}(H^{p+q}(F^p A) \to H^{p+q}(A)) \big/ \operatorname{im}(H^{p+q}(F^{p+1} A) \to H^{p+q}(A)).$$

If we choose chain level representatives for products in $H(A, d)$, then the property $\psi(F^p A \otimes_R F^q A) \subset F^{p+q} A$ controls the products in the associated bigraded algebra $E_0^{*,*}(H(A, d), F)$. The isomorphism of $E_\infty^{*,*}$ with the filtration subquotients follows from the definition of the spectral sequence and is an isomorphism of bigraded algebras.   □

Product structures cannot be underestimated in their power to simplify. This theorem is a major tool in computations throughout the rest of the book.

2.4 Algebraic applications

In the previous section we defined the differential $d_\otimes$ on the tensor product of two differential graded modules, $(A, d_A) \otimes_R (B, d_B) = (A \otimes_R B, d_\otimes)$. Under simplifying assumptions the Künneth theorem allows us to determine $H(A \otimes_R B, d_\otimes)$ in terms of $H(A, d_A)$ and $H(B, d_B)$. The goal of the next two sections is a generalization of the Künneth theorem.

We first introduce double complexes (due to [Cartan48]) and devise two spectral sequences to calculate the homology of the total complex associated to a double complex. By taking $(A \otimes_R B, d_\otimes)$ as an example of a double complex, these spectral sequences are used to compute $H(A \otimes_R B, d_\otimes)$. In the second section we obtain the appropriate generalization of the Künneth theorem as a spectral sequence.

We remark that double complexes offer an example of the filtered differential graded module construction of a spectral sequence. Many spectral sequences are derived from double complexes. We also remark that the relationship between the short exact sequence of the Künneth theorem and the spectral sequence of its generalization is a paradigmatic example.

Double complexes

A **double complex**, $\{M^{*,*}, d', d''\}$, is a bigraded module over R, $M^{*,*}$, with two R-linear maps $d' : M^{*,*} \to M^{*,*}$ and $d'' : M^{*,*} \to M^{*,*}$ of bidegree $(1, 0)$, $d' : M^{n,m} \to M^{n+1,m}$ and bidegree $(0, 1)$, $d'' : M^{n,m} \to M^{n,m+1}$, which satisfy $d' \circ d' = 0$, $d'' \circ d'' = 0$ and $d' \circ d'' + d'' \circ d' = 0$.

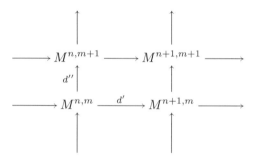

We associate to each double complex its **total complex**, $\mathrm{total}(M)$, which is the differential graded module over R defined by

$$\mathrm{total}(M)^n = \bigoplus_{p+q=n} M^{p,q}$$

with **total differential** $d = d' + d''$. The relations demanded of d' and d'' imply that $d \circ d = 0$.

An example of a double complex is given by two differential graded modules: If we let $K^{m,n} = A^m \otimes_R B^n$, $d' = d_A \otimes 1$ and $d'' = (-1)^m 1 \otimes d_B$, then we have a double complex such that $(\mathrm{total}(K), d) = (A \otimes_R B, d_\otimes)$.

How do we compute $H(\mathrm{total}(M), d)$? We construct two spectral sequences that exploit the fact that one can take the homology of $M^{*,*}$ in two directions. Let $H_I^{*,*}(M) = H(M^{*,*}, d')$, that is,

$$H_I^{n,m}(M) = \ker d' : M^{n,m} \to M^{n+1,m} \Big/ \mathrm{im}\, d' : M^{n-1,m} \to M^{n,m}.$$

Similarly, let $H_{II}^{*,*}(M) = H(M^{*,*}, d'')$. The condition $d' \circ d'' + d'' \circ d' = 0$ implies that $H_I^{*,*}(M)$ and $H_{II}^{*,*}(M)$ are each differential bigraded modules with differentials $\overline{d''}$ and $\overline{d'}$ induced by d'' and d', respectively. To be more precise, if x and $x + d'b$ represent $[x]$ in $H^{n,m}_I(M)$, then $d''x$ and $d''x + d''d'b$ are in the kernel of d' since $d'd''x = -d''d'x = 0$ and they differ by elements in the image of d'. Since $d' \circ d' = 0 = d'' \circ d''$, the induced morphisms, $\overline{d''}$ and $\overline{d'}$, are differentials. Let $H^{*,*}_I H_{II}(M) = H(H^{*,*}_{II}(M), \overline{d'})$ and $H^{*,*}_{II} H_I(M) = H(H^{*,*}_I(M), \overline{d''})$.

Theorem 2.15. *Given a double complex $\{M^{*,*}, d', d''\}$ there are two spectral sequences, $\{_I E^{*,*}_r, {}_I d_r\}$ and $\{_{II} E^{*,*}_r, {}_{II} d_r\}$ with*

$$_I E^{*,*}_2 \cong H^{*,*}_I H_{II}(M) \quad and \quad _{II} E^{*,*}_2 \cong H^{*,*}_{II} H_I(M).$$

If $M^{p,q} = \{0\}$ when $p < 0$ or $q < 0$, then both spectral sequences converge to $H^*(\text{total}(M), d)$.

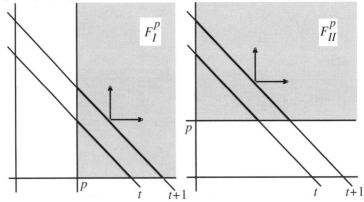

PROOF: We give a proof in the case of $\{_I E^{*,*}_r, {}_I d_r\}$; the other case follows by symmetry. The tool of choice is Theorem 2.6. Consider the following filtrations of $(\text{total}(M), d)$, as in the picture:

$$F^p_I(\text{total}(M))^t = \bigoplus_{r \geq p} M^{r,t-r} \quad and \quad F^p_{II}(\text{total}(M))^t = \bigoplus_{r \geq p} M^{t-r,r}.$$

We call F^*_I the *column-wise filtration* and F^*_{II} the *row-wise filtration*. Both are decreasing filtrations and d, the total differential, respects each filtration. Because $M^{p,q} = \{0\}$ when $p < 0$ or $q < 0$, this filtration is bounded and, by Theorem 2.6, we obtain two spectral sequences converging to $H(\text{total}(M), d)$. In the case of F_I we have

$$_I E^{p,q}_1 = H^{p+q}\left(F^p_I \text{total}(M) \middle/ F^{p+1}_I \text{total}(M), d \right).$$

It suffices to identify the E_2-term as described.

First, we claim that $_IE_1^{p,q} \cong H_{II}^{p,q}(M)$. Since the differential is given by $d = d' + d''$ and $d'(F_I^p\,\mathrm{total}(M)) \subset F_I^{p+1}\,\mathrm{total}(M)$, we get that

$$\left(F_I^p\,\mathrm{total}(M) \Big/ F_I^{p+1}\,\mathrm{total}(M) \right)^{p+q} \cong M^{p,q}$$

with the induced differential d''. Thus $_IE_1^{p,q} = H_{II}^{p,q}(M)$, as described.

Following Proposition 2.11, consider the diagram (where we write F^p for $F_I^p\,\mathrm{total}(M)$):

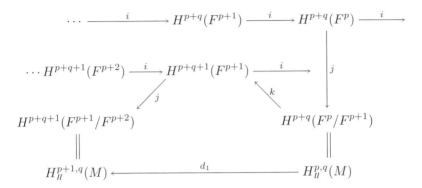

A class in $H^{p+q}(F^p/F^{p+1})$ can be written as $[x + F^{p+1}]$, where x is in F^p and dx is in F^{p+1} or it can be written as a class, $[z]$, in $H_{II}^{p,q}(M)$, z in $M^{p,q}$. Now k sends $[x + F^{p+1}]$ to $[dx]$ in $H^{p+q+1}(F^{p+1})$. Taking z as the representative, this determines $[d'z]$ which is in $H^{p+q+1}(F^{p+1})$ since $d''z = 0$. The morphism j assigns a class in $H^{p+q+1}(F^{p+1})$ to its representative mod F^{p+2}. Thus we can consider $d'z$ as an element of $M^{p+1,q}$. This gives $d_1 = j \circ k$ as the induced mapping of d' on $H_{II}^{p,q}(M)$ and so $d_1 = \overline{d'}$. Therefore, we have $_IE_2^{p,q} = H_I^{p,q}H_{II}(M)$.

To get the second spectral sequence from $F_{II}^*\,\mathrm{total}(M)$ reindex the double complex as its transpose: ${}^tM^{p,q} = M^{q,p}$, ${}^td' = d''$ and ${}^td'' = d'$. Then we have $\mathrm{total}({}^tM) = \mathrm{total}(M)$ and $F_{II}^*\,\mathrm{total}(M) = F_I^*\,\mathrm{total}({}^tM)$. The same proof goes over to obtain the result. □

The condition that $M^{p,q} = \{0\}$ for $p < 0$ or $q < 0$ guarantees that the associated spectral sequence is in the first quadrant. In this case, the spectral sequence converges to $H(\mathrm{total}(M^{*,*}), d)$. Convergence of the general case of a $\mathbb{Z} \times \mathbb{Z}$-graded double complex is susceptible to analysis using the tools of Chapter 3 (see Chapter 7 for an example where this is important).

The Künneth Spectral Sequence Ⓝ

In this section, we extend the Künneth theorem (Theorem 2.12) as an application of the spectral sequence associated to a double complex. We introduce the elementary parts of differential homological algebra here, not only for the sake of completeness, but also in anticipation of the generalizations that lie at the heart of Chapters 7, 8, and 9. The novice will find this material a distracting detour from the fastest route to the use of spectral sequences. On a second reading, especially with the Eilenberg-Moore spectral sequence as goal, the reader can find here motivation for the subsequent generalization of homological algebra.

Recall the familiar result that any abelian group, G, is the homomorphic image of a free abelian group. More descriptively, there is a short exact sequence of abelian groups:

$$0 \to F_1 \to F_0 \to G \to 0$$

with F_0 and F_1 free. The basis for F_0 maps onto a set of generators for G and the generators for F_1 give a record of the relations amid the products of generators of G.

If we wish to pass to different coefficients, we can tensor the short exact sequence with another abelian group A. Is it a simple matter to determine $G \otimes A$ from these data? Since $F_0 \otimes A$ and $F_1 \otimes A$ are easily described, the question reduces to asking if exactness is preserved after tensoring with the group A. The answer depends on the groups G and A and is given in terms of the functor $\mathrm{Tor}_1^{\mathbb{Z}}$; there is an exact sequence

$$0 \to \mathrm{Tor}_1^{\mathbb{Z}}(G, A) \to F_1 \otimes A \to F_0 \otimes A \to G \otimes A \to 0.$$

$\mathrm{Tor}_1^{\mathbb{Z}}(G, A)$ can be defined in terms of elements in G and A or simply as the kernel of the induced mapping $F_1 \otimes A \to F_0 \otimes A$ ([Mac Lane63, p. 150]).

If given a differential graded module, (K^*, d), taken as a differential abelian group, then this analysis is sufficient for determining how $H(K^* \otimes G)$ and $H(K^*) \otimes G$ compare; this is the **Universal Coefficient theorem** ([Mac Lane63, p. 171] and §12.1) that applies, for example, when we are studying singular cohomology with coefficients in a group as an abelian group.

Theorem 2.16. *If (K^*, d) is a differential graded abelian group with no elements of finite order and G is an abelian group, then, for each n, there is a short exact sequence*

$$0 \to H^n(K^*, d) \otimes G \xrightarrow{p} H^n(K^* \otimes G, d \otimes 1) \to \mathrm{Tor}_1^{\mathbb{Z}}(H^{n-1}(K^*, d), G) \to 0$$

which is split. The homomorphism p is given by $[u] \otimes g \mapsto [u \otimes g]$.

For coefficients of a more complicated nature, we need a generalization of these results to modules.

Let $0 \to M_1 \to M_0 \to N \to 0$ be a short exact sequence of modules over a ring R (here taken as *right* R-modules). What is the effect of tensoring this exact sequence with another R-module L (here taken as a *left* R-module)? The answer to this question comes in the form of a long exact sequence:

$$\cdots \to \mathrm{Tor}^R_n(M_1, L) \to \mathrm{Tor}^R_n(M_0, L) \to \mathrm{Tor}^R_n(N, L)$$
$$\to \mathrm{Tor}^R_{n-1}(M_1, L) \cdots \to \mathrm{Tor}^R_1(N, L) \to M_1 \otimes_R L$$
$$\to M_0 \otimes_R L \to N \otimes_R L \to 0.$$

To calculate the modules $\mathrm{Tor}^R_n(N, L)$ for any R-modules N and L, we generalize the short exact sequence of "generators and relators" for abelian groups. The key property of free abelian groups is the fact that they factor through epimorphisms, that is, if G is the image of a group A and $\rho\colon F \to G$ is any homomorphism, then ρ can be extended through A and there is a group homomorphism $\bar{\rho}\colon F \to A$ so that $\varphi \circ \bar{\rho} = \rho$ as in the diagram:

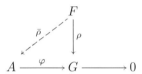

An R-module, P, that has the analogous property is called **projective**, that is, if $\varphi\colon M \to N$ is an epimorphism of modules over R and $\rho\colon P \to N$ is any R-linear homomorphism, then ρ can be extended through M, $\bar{\rho}\colon P \to M$, so that $\varphi \circ \bar{\rho} = \rho$.

$$
\begin{array}{ccc}
 & & P \\
 & \overset{\bar{\rho}}{\swarrow} & \downarrow{\rho} \\
M & \overset{\varphi}{\longrightarrow} N & \longrightarrow 0
\end{array}
$$

It is a standard result that a module is projective if and only if it is a direct summand of a free module ([Rotman79]). Results that are obtained using free modules can be improved sometimes using (often) smaller, more structured projective modules.

The analogue of the short exact sequence for an abelian group is the notion of a **projective resolution**. This is a long exact sequence,

$$\cdots \overset{d}{\to} P_2 \overset{d}{\to} P_1 \overset{d}{\to} P_0 \overset{\varepsilon}{\to} M \to 0$$

with each P_i a projective R-module. Since each free module is projective and free resolutions can be constructed, there is no question of the existence of projective resolutions. Furthermore, any two of them can be compared using the defining property for projectives:

$$
\begin{array}{ccccccccc}
\longrightarrow & P_2 & \overset{d_P}{\longrightarrow} & P_1 & \overset{d_P}{\longrightarrow} & P_0 & \overset{\varepsilon}{\longrightarrow} & M & \longrightarrow 0 \\
 & \downarrow{f_2} & & \downarrow{f_1} & & \downarrow{f_0} & & \| & \\
\longrightarrow & Q_2 & \underset{d_Q}{\longrightarrow} & Q_1 & \underset{d_Q}{\longrightarrow} & Q_0 & \underset{\varepsilon'}{\longrightarrow} & M & \longrightarrow 0.
\end{array}
$$

For example, f_2 exists and makes the square commute because P_2 is projective and fits into the diagram:

$$
\begin{array}{ccc}
& & P_2 \\
& \overset{f_2}{\dashleftarrow} & \Big\downarrow{\scriptstyle f_1 \circ d_P} \\
Q_2 \xrightarrow{\; d_Q \;} & \ker(d_Q : Q_1 \to Q_0) & \longrightarrow 0
\end{array}
$$

By analogy with the definition $\mathrm{Tor}_1^{\mathbb{Z}}(G, A) = \ker(F_1 \otimes A \to F_0 \otimes A)$, we define

$$\mathrm{Tor}_n^R(M, N) = H_n(P_* \otimes N, d_P \otimes 1).$$

The existence of maps between resolutions can be used to prove that $\mathrm{Tor}_n^R(M, N)$ does not depend on which projective resolution we use. In fact, we can use double complexes to prove a little more.

Proposition 2.17. *Let M be a right module and N a left module over R and*

$$
\begin{aligned}
\cdots &\to P_2 \xrightarrow{d_P} P_1 \xrightarrow{d_P} P_0 \xrightarrow{\varepsilon} M \to 0, \\
\cdots &\to Q_2 \xrightarrow{d_Q} Q_1 \xrightarrow{d_Q} Q_0 \xrightarrow{\varepsilon'} N \to 0
\end{aligned}
$$

be projective resolutions of M and N, respectively. Then

$$
\begin{aligned}
\mathrm{Tor}_n^R(M, N) &\cong H_n(P_* \otimes N, d_P \otimes 1) \\
&\cong H_n(M \otimes Q_*, 1 \otimes d_Q) \\
&\cong H_n(\mathrm{total}(P_* \otimes Q_*), D)
\end{aligned}
$$

where $D = d_P \otimes 1 + (-1)^i 1 \otimes d_Q$ on $P_i \otimes Q_j$.

PROOF: Let $K_{i,j} = P_i \otimes Q_j$, $d' = d_P \otimes 1$ and $d'' = (-1)^i 1 \otimes d_Q$. Then $\{K_{*,*}, d', d''\}$ is a double complex with differentials of bidegrees $(-1, 0)$ and $(0, -1)$, respectively. If we filter $\mathrm{total}(K_{*,*})$ by

$$
F_p^I(\mathrm{total}\, K)_t = \bigoplus_{r \le p} K_{r, t-r}, \qquad F_p^{II}(\mathrm{total}\, K)_t = \bigoplus_{r \le p} K_{t-r, r},
$$

we get increasing filtrations and D respects those filtrations. Using the dual versions of Theorems 2.6 and 2.15 we get two spectral sequences converging to $H(\mathrm{total}\, K, D)$.

In the first spectral sequence, $^I E_{*,*}^2 \cong H_{*,*}^I H^{II}(K)$. Since P_i and Q_j are projective, by the Künneth theorem and the exactness of the resolution $Q_* \xrightarrow{\varepsilon'} N \to 0$, we have

$$
\begin{aligned}
H^{II}(K) = H(K_{*,*}, d'') &= H(P_* \otimes Q_*, \pm 1 \otimes d_Q) \\
&= P_* \otimes H(Q_*, d_Q) = P_* \otimes N.
\end{aligned}
$$

Because $Q_* \to N \to 0$ is a projective resolution, H^{II} is concentrated in the strip of bidegrees $(n, 0)$ for $n \geq 0$, the horizontal bottom row. Therefore $H^I_{*,*} H^{II}(K)$ is simply $H(P_* \otimes N, d_P \otimes 1)$ or $\mathrm{Tor}^R_*(M, N)$ and the spectral sequence collapses. Since it lies entirely in one strip, there cannot be extension problems, so $H_*(\mathrm{total}\, K, D)$ is just $\mathrm{Tor}^R_*(M, N)$.

From the spectral sequence associated to the second filtration we get

$$H(M \otimes Q_*, 1 \otimes d_Q) \cong H(\mathrm{total}\, K, D) \cong \mathrm{Tor}^R_*(M, N). \qquad \square$$

In the classical sense (à la Hilbert) $\mathrm{Tor}^R_*(M, N)$ presents a sequence of essential invariants of the modules M and N that measure

(1) the deviation from M being projective (if M is projective, then $0 \leftarrow M \xleftarrow{=} M \leftarrow 0 \leftarrow 0 \cdots$ is a projective resolution of M and so $\mathrm{Tor}^R_i(M, N) = \{0\}$ for $i > 0$ and any N);
(2) the failure of the exactness of the functor $- \otimes_R N$.

This prompts the identification of the class of **flat modules**: A module N is **flat** if the functor $- \otimes_R N$ preserves exact sequences. It follows that if N is flat, $\mathrm{Tor}^R_i(M, N) = \{0\}$ for $i > 0$ and any module M.

We take the generalization one step further—from the category of modules over R to the category $\mathbf{DGMod}_R$ of differential graded modules over R and degree 0 module homomorphisms that commute with the differentials. Recall that an object in $\mathbf{DGMod}_R$ is a graded R-module, C^*, together with a degree 1 R-linear homomorphism for all $k \geq 0$, $d_C \colon C^k \to C^{k+1}$, such that $d_C \circ d_C = 0$. (Dually, we can have taken differentials of degree -1. In keeping with a preference for cohomology, degree +1 differentials are appropriate.)

A differential graded module (L^*, d_L) is said to be **flat** if $- \otimes_R L^*$, the functor that tensors a differential graded module with L^* over R, preserves exactness. A differential graded module (K^*, d_K) is **projective** if, whenever we have the diagram of differential graded modules,

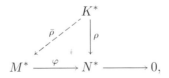

$\bar{\rho}$ always exists in $\mathbf{DGMod}_R$ with $\rho = \varphi \circ \bar{\rho}$. To measure the deviation from exactness and from being projective, we construct a differential graded version of Tor for $\mathbf{DGMod}_R$.

If we proceed dimensionwise by viewing (K^*, d_K) and (L^*, d_L) as the direct sum of their homogeneous parts, we can piece together $\mathrm{Tor}^R_n(K^*, L^*)$

from the ungraded version of Tor. Consider the double complex, for each n,

$$M_n^{p,q} = \operatorname{Tor}_n^R(K^p, L^q)$$
$$d' \colon \operatorname{Tor}_n^R(K^p, L^q) \to \operatorname{Tor}_n^R(K^{p+1}, L^q)$$
$$d'' \colon \operatorname{Tor}_n^R(K^p, L^q) \to \operatorname{Tor}_n^R(K^p, L^{q+1}),$$

where the differentials are induced by the differentials, d_K and d_L, and the functoriality of Tor_n^R in each variable. Define

$$\operatorname{Tor}_n^R(K^*, L^*) = \operatorname{total}(M_n^{*,*}).$$

For each n, this is a differential graded module.

In order to generalize constructions such as the long exact sequence that results from Tor applied to a short exact sequence to the category $\mathbf{DGMod}_R$, we adopt the following grading convention:

$$\operatorname{Tor}_R^{-n}(K^*, L^*) = \operatorname{Tor}_n^R(K^*, L^*).$$

Hence the differentials in the long exact sequence increase homological degree. With this convention, we write a **projective resolution** of a differential graded module (K^*, d_K) as a long exact sequence of negatively graded differential modules

$$\cdots \to P^{-2} \xrightarrow{d} P^{-1} \xrightarrow{d} P^0 \xrightarrow{\varepsilon} K \to 0.$$

A dimensionwise induction shows that such resolutions exist. We can display such a resolution as a double complex:

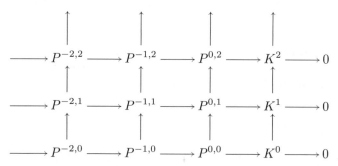

(If one is working in the dual category of differential graded modules with differentials of degree -1, the convention is to retain lower indices for Tor_n^R and the resulting projective resolutions give first quadrant double complexes.)

In these constructions, the differential is merely carried along as extra data. In order to involve this extra piece of structure more thoroughly, we identify a more restrictive type of projective resolution. We consider $H(K^*, d_K)$ as a differential graded module by giving it the zero differential.

Definition 2.18. *A **proper projective resolution** is an exact sequence in the category* **DGMod**$_R$: $\cdots \to P^{-2} \to P^{-1} \to P^0 \to K \to 0$ *where, for each n, the following are projective resolutions:*

P1. $\cdots \to P^{-2,n} \to P^{-1,n} \to P^{0,n} \to K^n \to 0$,

P2. $\cdots \to Z^n(P^{-2}) \to Z^n(P^{-1}) \to Z^n(P^0) \to Z^n(K) \to 0$ *(where* $Z^n(M) = \ker d_M : M^n \to M^{n+1}$*),*

P3. $\cdots \to H^n(P^{-2}) \to H^n(P^{-1}) \to H^n(P^0) \to H^n(K) \to 0.$

Lemma 2.19. *Every differential graded module, (K^*, d_K), has a proper projective resolution.*

PROOF: First notice that P1 comes for free from the definition of a projective resolution in **DGMod**$_R$. For each n, let $B^n(K) = \operatorname{im} d_K : K^{n-1} \to K^n$. We have short exact sequences $0 \to B^n(K) \to Z^n(K) \to H^n(K) \to 0$ and $0 \to Z^n(K) \to K^n \to B^{n+1}(K) \to 0$. Let $PB^{*,n} \xrightarrow{\varepsilon'} B^n(K) \to 0$ and $PH^{*,n} \xrightarrow{\varepsilon''} H^n(K) \to 0$ be projective resolutions of $B^n(K)$ and $H^n(K)$, respectively, and let $PZ^{k,n} = PB^{k,n} \oplus PH^{k,n}$. Clearly, $PZ^{*,*}$ is projective. This gives us short exact sequences, for all k and for all n, given by inclusion and projection,

$$0 \to PB^{k,n} \to PZ^{k,n} \to PH^{k,n} \to 0.$$

We next construct maps $\delta_Z^k : PZ^{k,n} \to PZ^{k+1,n}$ by induction to obtain a projective resolution of $Z^n(K)$. At the 0-level we have the diagram:

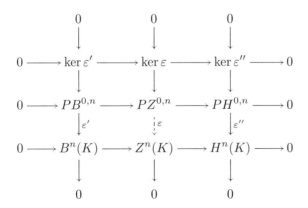

To fill in the homomorphism $\varepsilon : PZ^{0,n} \to Z^n(K)$, we can map the $PB^{0,n}$ factor of $PZ^{0,n}$ by ε' and the inclusion of $B^n(K)$ in $Z^n(K)$. The $PH^{0,n}$ factor is projective and, since ε'' is an epimorphism, we have a mapping of this factor to $Z^n(K)$. The sum of these two mappings is ε, which makes the bottom two squares commute. Two applications of the Five-lemma imply that the entire diagram is commutative and every row and every column exact.

The inductive step is exactly the same as applied to the diagram:

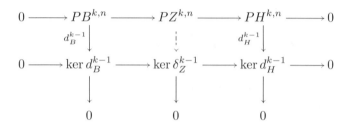

We now have a short exact sequence of projective resolutions

$$0 \to PB^{*,n} \to PZ^{*,n} \to PH^{*,n} \to 0.$$

Using the same argument this time with $PZ^{*,n} \to Z^n(K) \to 0$ and $PB^{*,n+1} \to B^{n+1}(K) \to 0$ we obtain a projective resolution of K^n, $P^{*,n} \to K^n \to 0$, that fits into a short exact sequence

$$0 \to PZ^{*,n} \to P^{*,n} \to PB^{*,n+1} \to 0.$$

Knitting the exact sequences together, we get the internal differential $d\colon P^{*,n} \to P^{*,n+1}$ given by the composition $d\colon P^{*,n} \to PB^{*,n+1} \to PZ^{*,n+1} \to P^{*,n+1}$. Since the right two maps are inclusions, $\ker d = PZ^{*,n}$ and, since $P^{*,n-1} \to PB^{*,n}$ is onto, $\operatorname{im} d = PB^{*,n}$. Thus $H^n(P^{*,*},d) = PH^{*,n}$ and $P^{*,*} \to K^* \to 0$ is a proper projective resolution.   □

We now tackle the problem of computing $H(K \otimes_R L, d_\otimes)$. The Künneth theorem requires that $Z(K)$ and $B(K)$ be flat. We can generalize a bit further and remove part of the flatness assumption. Because we have taken our differentials to be of degree $+1$, this leads to a second quadrant spectral sequence. The case for the dual category with differentials of degree -1 leads to a first quadrant spectral sequence; the arguments are exactly the same for the construction of both spectral sequences, and we give the less standard one.

Theorem 2.20 (the Künneth spectral sequence). *Let (K^*, d_K) and (L^*, d_L) be differential graded modules over R with K flat. Then there is a spectral sequence with $E_2^{p,q} = \bigoplus_{s+t=q} \operatorname{Tor}_R^p(H^s(K^*), H^t(L^*))$. If K^* and L^* have differentials of degree $+1$, this is a second quadrant spectral sequence, and if K^* and L^* have differentials of degree -1, a first quadrant spectral sequence. When $E_r^{*,*}$ converges, it does so to $H(K^* \otimes_R L^*, d_\otimes)$ in each case.*

PROOF: A proper projective resolution of L may be presented in the diagram:

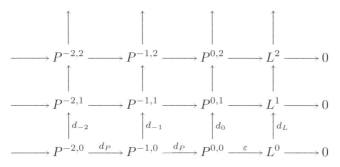

Consider the double complex $M^{p,q} = \bigoplus_{s+t=q} K^s \otimes_R P^{p,t}$ with differentials $d' \colon M^{p,q} \to M^{p+1,q}$, given by $d' = \sum(-1)^q 1 \otimes d_P$, and $d'' \colon M^{p,q} \to M^{p,q+1}$, given by $d'' = \sum d_K \otimes 1 + \sum(-1)^s 1 \otimes d_p$ on the summand $K^s \otimes_R P^{p,t}$. Here p is nonpositive and gives the homological degree while q is nonnegative and denotes a dimension. We proceed as in the case of a double complex.

In the second filtration, $F_{II}^p \mathrm{total}(M) = \bigoplus_{s \geq p} M^{r,s}$ and so F_{II}^q / F_{II}^{q+1} is the qth row with differential $\sum(-1)^q 1 \otimes d_P$. Thus

$$_{II}E_1^{p,q} = H\left(\bigoplus_{s+t=q} K^s \otimes P^{p,t}, \sum(-1)^q 1 \otimes d_P\right).$$

If we fix s and t and let p vary, then we have the projective resolution of L^t tensored with K^s and so the homology is $\mathrm{Tor}_R^*(K^s, L^t)$. Thus

$$_{II}E_1^{p,q} = \bigoplus_{s+t=q} \mathrm{Tor}_R^p(K^s, L^t).$$

But K^* is flat and so $\mathrm{Tor}_R^p(K^s, L^t) = \{0\}$ for $p \neq 0$ and $_{II}E_1^{*,*}$ degenerates to the column $_{II}E_1^{0,q} = \bigoplus_{s+t=q} K^s \otimes_R L^t = \mathrm{total}(K^* \otimes_R L^*)^q$ and so $_{II}E_2^{0,q} = H^q(K^* \otimes_R L^*)$. Furthermore, since our E_2-term is only a column, the spectral sequence collapses at $E_2 = E_\infty$. This establishes that the target module is $H(K^* \otimes_R L^*, d_\otimes)$.

Next consider the first filtration, $F_I^p \mathrm{total}(M) = \bigoplus_{r \geq p} M^{r,s}$. For this,

$$_IE_1^{p,*} = H(F_I^p / F_I^{p+1}) = H(p\text{th column}, d'').$$

The pth column of $M^{*,*}$, however, is $K^* \otimes_R P^{p,*}$ as the tensor product of two differential graded modules. Thus $_IE_1^{p,*} = H(K^* \otimes_R P^{p,*}, d'')$. By the Künneth theorem and because $P^{*,*} \to K^* \to 0$ is a proper projective resolution, $P^{p,*}$, $Z(P^{p,*})$ and $H(P^{p,*})$ are all projective, we have

$$_IE_1^{p,*} = H(K^* \otimes_R P^{p,*}) \cong H(K^*) \otimes_R H(P^{p,*}).$$

Continuing, we have $_IE_2^{p,*} = H(H(K^*) \otimes_R H(P^{p,*}), \overline{d'})$. Since

$$\cdots \to H(P^p) \to \cdots \to H(P^{-1}) \to H(P^0) \to H(L^*) \to 0$$

is a projective resolution of $H(L^*)$, $(H(K^*) \otimes_R H(P^*), \overline{d'})$ is the differential graded module from which we compute $\operatorname{Tor}_R^*(H(K^*), H(L^*))$. Thus

$$_IE_2^{p,q} = \bigoplus_{s+t=q} \operatorname{Tor}_R^p(H^s(K^*), H^t(L^*))$$

and we have the spectral sequence as described in the theorem.

When K^* and L^* have differentials of degree -1, the double complex $M^{*,*}$ lies in the first quadrant and the collapse of $_{II}E_r^{*,*}$ to $H(K^* \otimes_R L^*)$ and the convergence of $_IE_r^{*,*}$ imply that $_IE_r^{*,*}$ converges to $H(K^* \otimes_R L^*)$ as in Theorem 2.6. When K and L have differentials of degree $+1$, the convergence is a more delicate matter; in particular, our constructions do not give bounded filtrations automatically. Conditions may be found in Chapter 3 that guarantee that the spectral sequence converges to $H(K^* \otimes_R L^*)$. The reader should examine the filtrations carefully in the proof of the theorem to understand the possible difficulties in a second quadrant spectral sequence.   □

Exercises

2.1. Suppose (A_*, d, F) is a differential graded module with differential of degree -1 and an *increasing* filtration that respects the differential. Deduce the analogue of Theorem 2.1. In particular, show that this gives a spectral sequence of homological type.

2.2. Another manner in which a spectral sequence may arise is that of a **Cartan-Eilenberg system** that consists of a module $H(p, q)$ for each pair of integers, $-\infty \le p \le q \le \infty$ along with

(1) homomorphisms $\eta \colon H(p', q') \to H(p, q)$ whenever $p \le p', q \le q'$;
(2) for $-\infty \le p \le q \le r \le \infty$, we have a connecting homomorphism $\delta \colon H(p, q) \to H(q, r)$;
(3) $H(p, q) \to H(p, q)$ is the identity;
(4) if $p \le p' \le p''$ and $q \le q' \le q''$, then the following diagram commutes:

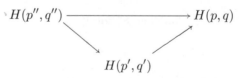

(5) if $p \le p', q \le q'$ and $r \le r'$, then the following diagram commutes:

$$\begin{array}{ccc} H(p', q') & \longrightarrow & H(q', r') \\ \downarrow & & \downarrow \\ H(p, q) & \longrightarrow & H(q, r) \end{array}$$

(6) for $-\infty \leq p \leq q \leq r \leq \infty$, the following sequence is exact:

$$\cdots \to H(q,r) \to H(p,r) \to H(p,q) \overset{\delta}{\to} H(q,r) \to \cdots$$

(7) $H(-\infty, q)$ is the direct limit of the system

$$H(q,q) \to H(q-1,q) \to H(q-2,q) \to \cdots .$$

With this definition we get a spectral sequence by letting

$$Z_r^p = \mathrm{im}\big(H(p, p+r) \to H(p, p+1)\big)$$
$$B_r^p = \mathrm{im}\big(H(p-r+1, p) \to H(p, p+1)\big)$$
$$E_r^p = Z_r^p / B_r^p.$$

Show that this is a spectral sequence and, if it converges, it does so to $H(-\infty, \infty)$. A filtered differential graded module, (A, d, F), gives rise to a Cartan-Eilenberg system given by $H(p,q) = H(F^p A / F^q A)$.

2.3. Suppose $\{ H(p,q), -\infty \leq p \leq q \leq \infty \}$ is a Cartan-Eilenberg system and for all $n, q, r \geq 0$ we have bilinear mappings

$$\varphi_r \colon H(n, n+r) \otimes H(q, q+r) \to H(n+q, n+q+r).$$

Suppose the following hold:

(1) if $n \geq n', q \geq q', n+r \geq n'+r'$ and $q+r \geq q'+r'$, then the following diagram commutes

$$
\begin{array}{ccc}
H(n, n+r) \otimes H(q, q+r) & \overset{\varphi_r}{\longrightarrow} & H(n+q, n+q+r) \\
\downarrow & & \downarrow \\
H(n', n'+r') \otimes H(q', q'+r') & \underset{\varphi_{r'}}{\longrightarrow} & H(n'+q', n'+q'+r').
\end{array}
$$

(2) For all $n, q \geq 0, r \geq 1$,

$$\delta(\varphi_r(a \otimes b)) = \varphi_1(\eta a \otimes \delta b) + (-1)^{\deg a}(\delta a \otimes \eta b),$$

where η and δ are the appropriate maps in the Cartan-Eilenberg system.

Show that these data give rise to a spectral sequence of algebras.

2.4. Suppose that $\{D, E, i, j, k\}$ is an exact couple of bigraded modules where $\mathrm{bideg}\, i = (1, -1)$, $\mathrm{bideg}\, j = (0, 0)$ and $\mathrm{bideg}\, k = (-1, 0)$. State and prove the analogue of Theorem 2.8 for such an exact couple leading to a spectral sequence of homological type.

2.5. Suppose that the conditions for Theorem 2.6 hold and we define

$$\bar{Z}_r^p = \text{im}\big(H(F^p A/F^{p+r} A) \to H(F^p A/F^{p+1} A)\big)$$
$$\bar{B}_r^p = \text{im}\big(H(F^{p-r+1} A/F^p A) \to H(F^p A/F^{p+1} A)\big)$$
$$\bar{E}_r^p = \bar{Z}_r^p/\bar{B}_r^p.$$

Show that this determines a spectral sequence converging to $H(A, d)$. (Hint: Look at the proof of Proposition 2.9.)

2.6. Suppose we are given a tower of submodules

$$B_2 \subset B_3 \subset \cdots \subset Z_N \subset \cdots \subset Z_3 \subset Z_2 \subset E_2$$

along with isomorphisms $Z_{p+1}/Z_p \cong B_p/B_{p+1}$, show that these data determine a spectral sequence whose tower of submodules is the one given. What kind of collapse results can you prove from this representation of a spectral sequence if you add the detail of keeping track of bidegrees? In the case of an exact couple, we defined $Z_r^p = k^{-1}(\text{im } i^{r-1} : D^{p+r} \to D^{p+1})$ and $B_r^p = j(\ker i^{r-1} : D^p \to D^{p-r+1})$. Show that $Z_r^p \subset Z_{r-1}^p$ and $B_r^p \subset B_{r+1}^p$ and that the differential $d_r : Z_{r-1}/B_{r-1} \to B_r/B_{r-1} \subset Z_{r-1}/B_{r-1}$ is induced by $j \circ k$ and has the appropriate kernel.

2.7. Ⓝ Determine conditions on an exact couple that guarantee that the associated spectral sequence is a spectral sequence of algebras. (Hint: Use a Cartan-Eilenberg system or consult the paper of [Massey54] if you get stuck).

2.8. Prove the Universal Coefficient theorem: Suppose that A is an abelian group and (C_*, ∂) is a complex of free abelian groups with differential ∂ of degree -1. Then for each $n > 0$ there is a short exact sequence with

$$0 \to H_n(C_*, \partial) \otimes A \to H_n(C_* \otimes A, \partial \otimes 1) \to \text{Tor}(H_{n-1}(C_*, \partial), A) \to 0.$$

Here $\text{Tor}(G, A)$ is the kernel of the homomorphism $1 \otimes \delta : G \otimes R_A \to G \otimes F_A$ where $0 \to R_A \xrightarrow{\delta} F_A \to A \to 0$ is a short exact sequence with F_A and R_A free abelian groups.

2.9. Prove the homological analogue of Theorem 2.15 holds for a double complex of the form $\{ M_{*,*}, d', d'' \}$ with d' of bidegree $(-1, 0)$ and d'' of bidegree $(0, -1)$, satisfying $d' \circ d' = d'' \circ d'' = d' \circ d'' + d'' \circ d' = 0$.

2.10. If p and q are relatively prime, show that $\mathbb{Z}/p\mathbb{Z}$ is a projective $\mathbb{Z}/pq\mathbb{Z}$-module.

2.11. Show that the definition of $\text{Tor}_R^*(K, L)$ for K and L in **DGMod**$_R$ does not depend on the choice of resolutions.

2.12. Consider the category of differential graded modules over a ring R with differentials of degree -1 and morphisms of degree 0. Develop the notions of projective and flat modules, projective resolutions, and proper projective resolutions for this category. Prove that there is a Künneth spectral sequence for this category.

2.13. Deduce the Künneth theorem from the Künneth spectral sequence.

3

Convergence of spectral sequences

"The machinery of spectral sequences, stemming from
the algebraic work of Lyndon and Koszul, seemed complicated and obscure to many topologists. Nevertheless,
it was successful ... "

G. W. Whitehead

In Chapter 2, we find recipes for the construction of spectral sequences. To develop these ideas further we need to clarify the relationship between a spectral sequence and its target; this is the goal of Chapter 3. To achieve this goal, it is necessary to introduce more refined ideas of convergence. These ideas require a discussion of limits and colimits of modules and the definition of a morphism between spectral sequences with which one can express the relevant theorems of comparison. In the case of a filtered differential graded module, conditions on the filtration guarantee that the associated spectral sequence converges uniquely to its target. The case of an exact couple is more subtle and we develop it after a discussion of some associated limits.

We express convergence results as comparison theorems that answer the questions: If two spectral sequences are isomorphic via a morphism of spectral sequences, then how do the targets of the spectral sequences compare? Need they be isomorphic? We end the chapter with some constructions and Zeeman's comparison theorem that reveals how special circumstances lead to powerful conclusions.

3.1 On convergence

Theorem 2.6 tells us that a filtered differential graded module, (A, d, F), determines a spectral sequence and, if the filtration is bounded, then the spectral sequence *determines* $H(A, d)$ (up to extension problems). We want to remove the restrictive hypothesis of a bounded filtration and still retain convergence to a *uniquely* determined target.

We begin with the case of a filtered differential module over a ring R. Let (A, d, F) denote a decreasing stable filtration on (A, d),

$$\cdots \subset F^{p+1}A \subset F^p A \subset F^{p-1}A \subset \cdots \subset A.$$

Notice that our first hurdle in understanding $H(A, d)$ is the fact that an inclusion $F^s A \subset F^t A$ need not induce an inclusion in homology, $H(\subset): H(F^s A) \to H(F^t A)$. We have defined (suppressing one of the bidegrees)

$$Z_r^p = F^p A \cap d^{-1}(F^{p+r} A) \qquad \text{and} \qquad B_r^p = F^p A \cap d(F^{p-r} A)$$

to get the tower of submodules

$$B_0^p \subset B_1^p \subset \cdots \subset B_r^p \subset \cdots \subset Z_r^p \subset \cdots \subset Z_1^p \subset Z_0^p.$$

The E_∞-term of the associated spectral sequence is given by this tower as

$$E_\infty^p = \bigcap_r Z_r^p \Big/ \bigcup_r B_r^p .$$

For an arbitrary filtered differential module, we ask, *how does this E_∞-term relate to the desired target, $H(A, d)$? Does it relate to some other filtered graded R-module as well?*

To obtain the induced filtration on $H(A, d)$ in Theorem 2.6, we defined $Z_\infty^p = F^p A \cap \ker d$ and $B_\infty^p = F^p A \cap \operatorname{im} d$ and showed that, for a bounded filtration,

$$Z_\infty^p / B_\infty^p \cong F^p H(A, d) / F^{p+1} H(A, d).$$

Observe, however, that these modules Z_∞^p and B_∞^p need *not* come from the tower. We extend the tower to include them:

$$B_0^p \subset B_1^p \subset \cdots \subset \bigcup_r B_r^p \subset B_\infty^p \subset Z_\infty^p \subset \bigcap_r Z_r^p \subset \cdots \subset Z_1^p \subset Z_0^p.$$

The equality, $B_\infty^p = \bigcup_r B_r^p$ or $F^p A \cap \operatorname{im} d = \bigcup_r (F^p A \cap d(F^{p-r} A))$ can fail if $\bigcup_s F^s A \neq A$, for example, when there is an x in $A - \bigcup_s F^s A$ with $d(x)$ in some $F^p A$. To avoid this pathology, we can require that the filtration be **exhaustive**, that is, $A = \bigcup_s F^s A$. In practice, this condition is satisfied by the reasonable examples.

Next we consider the submodule of infinite cycles.

Definition 3.1. *A filtration F of a differential graded module, (A, d), is said to be **weakly convergent** if, for all p, $Z_\infty^p = \bigcap_r Z_r^p$, that is, if $F^p A \cap \ker d = \bigcap_r (F^p A \cap d^{-1}(F^{p+r} A))$.*

Some simple conditions on the filtration imply that it is weakly convergent; for example, 1) the filtration is **bounded above**, that is, for each n, there is a value $s(n)$ with $F^{s(n)} A = \{0\}$ or, more generally, 2) $\bigcap_p F^p A = \{0\}$. The proof of Theorem 2.6 needs only slight modifications to prove the following result.

Theorem 3.2. *Let (A, d, F) be a filtered differential graded module such that the filtration is exhaustive and weakly convergent. Then the associated spectral sequence with $E_1^{p,q} \cong H^{p+q}(F^p A / F^{p+1} A)$ converges to $H(A, d)$, that is,*

$$E_\infty^{p,q} \cong F^p H^{p+q}(A, d) / F^{p+1} H^{p+q}(A, d).$$

The definition of weak convergence seems dependent on explicit knowledge of (A, d, F). It would be helpful to have some equivalent formulations of weak convergence in terms of cruder invariants such as the homology of subquotients of A or the homology of stages in the filtration.

Proposition 3.3. *The following conditions are equivalent on a filtration F of a differential graded module, (A, d):*

(1) *F is weakly convergent.*
(2) *$\bigcap_{r \geq 1} \operatorname{im}(H(F^p A / F^{p+r} A) \to H(F^{p+1} A)) = \{0\}$.*
(3) *For all p, the mappings induced by the filtration, $R^{p+1} \to R^p$, are monomorphisms, where $R^p = \bigcap_r \operatorname{im}(H(F^{p+r} A) \to H(F^p A))$.*

PROOF ([Cartan-Eilenberg56]): Consider the commutative diagram of R-modules where the bottom row is exact at B:

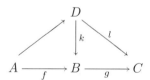

Since $\operatorname{im} k / \operatorname{im} f \cong \operatorname{im} k / \ker g$ and $\operatorname{im} k / \ker g \cong \operatorname{im}(g \circ k)$, g induces an isomorphism: $\operatorname{im} k / \operatorname{im} f \cong \operatorname{im} l$.

From a filtered differential module we obtain such a diagram

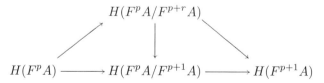

induced by the long exact sequences that arise when homology is applied to the short exact sequences $0 \to F^{p+1} A \to F^p A \to F^p A / F^{p+1} A \to 0$ and $0 \to F^{p+r} A \to F^p A \to F^p A / F^{p+r} A \to 0$ and to the inclusion $F^{p+r} A \subset F^{p+1} A$. To show that (1) is equivalent to (2), we begin with the observation that $\operatorname{im}\{H(F^p A / F^{p+r} A) \to H(F^{p+1} A)\}$ is isomorphic to the quotient of $\operatorname{im}\{H(F^p A / F^{p+r} A) \to H(F^p A / F^{p+1} A)\}$ by the subgroup $\operatorname{im}\{H(F^p A) \to H(F^p A / F^{p+1} A)\}$, which follows from the exactness of the bottom row. The cycles in $F^pA/F^{p+r}A$ are represented by those elements of F^pA with boundary in $F^{p+r}A$. It follows that $Z_r^p = F^pA \cap d^{-1}(F^{p+r}A)$ is isomorphic to the module of cycles in $F^pA/F^{p+r}A$ and similarly, Z_∞^p is the module of cycles in F^pA. From the canonical projections we obtain the composites $Z_r^p \longrightarrow H(F^pA/F^{p+r}A) \longrightarrow H(F^pA/F^{p+1}A)$ and $Z_\infty^p \longrightarrow H(F^pA)$ whose kernels are expressible as modules of lower filtration and lower degree in the tower. If (2) fails, we can identify some nonzero element in

$$\bigcap\nolimits_{r \geq 1} \operatorname{im}(H(F^pA/F^{p+r}A) \to H(F^{p+1}A))$$

and in turn some representative of it lies in $\bigcap_r Z_r^p$ but not in Z_∞^p and (1) fails. If (1) fails, then an element in $\bigcap_r Z_r^p$ not in Z_∞^p gives a class in $H(F^pA/F^{p+1}A)$ not in the image of $H(F^pA)$ and so (2) fails.

To establish the equivalence of (2) and (3) consider the commutative diagram induced by the filtration:

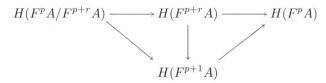

If $x \in R^{p+1} = \bigcap_r \operatorname{im}(H(F^{p+r}A) \to H(F^{p+1}A))$ then, by the exactness of the top row of the diagram, $x \in \ker(R^{p+1} \to R^p)$ if and only if x is in the intersection of the modules $\operatorname{im}\{H(F^pA/F^{p+r}A) \to H(F^{p+1}A)\}$ for all r. Therefore, for all p,

$$\bigcap\nolimits_{r \geq 1} \operatorname{im}\{H(F^pA/F^{p+r}A) \to H(F^{p+1}A)\} = \ker(R^{p+1} \to R^p).$$

and (2) is equivalent to (3). $\qquad\square$

A weakly convergent filtration guarantees that the spectral sequence from a filtered differential graded module, (A, d, F), converges to $H(A, d)$, in the sense that the E_∞-term is related directly to a filtration of $H(A, d)$. How do we know that the spectral sequence converges to $H(A, d)$ and not to something else, say $H(B, d)$? To illustrate, let K be a free graded R-module and (A, d, F) a filtered differential graded module. Consider $(A \oplus K, d \oplus 0, F')$ with the filtration F' given by $F'^p(A \oplus K) = F^pA \oplus K$. It is easily seen (check the associated tower of submodules) that the spectral sequence arising from (A, d, F) is the same as the spectral sequence coming from $(A \oplus K, d \oplus 0, F')$. Furthermore, if F is weakly convergent, so is F'. Thus the same spectral sequence converges to both $H(A, d)$ and $H(A, d) \oplus K$.

A condition on a filtration that prevents this trivial but malevolent example is to require that the filtration be **Hausdorff**, that is, it is weakly convergent and

$$\bigcap\nolimits_p F^pH(A, d) = \{0\}.$$

The terminology is meant to recall the function space situation where convergence of a sequence need not determine the limit. The Hausdorff condition will be developed in the later sections.

Morphisms of spectral sequences

Given two spectral sequences, $\{(E_r^{*,*}, d_r)\}$ and $\{(\bar{E}_r^{*,*}, \bar{d}_r)\}$, we define a **morphism of spectral sequences** to be a sequence of homomorphisms of bigraded modules, $f_r : (E_r^{*,*}, d_r) \longrightarrow (\bar{E}_r^{*,*}, \bar{d}_r)$, for all r, of bidegree $(0,0)$, such that f_r commutes with the differentials, that is, $f_r \circ d_r = \bar{d}_r \circ f_r$, and each f_{r+1} is induced by f_r on homology, that is, f_{r+1} is the composite

$$f_{r+1} : E_{r+1}^{*,*} \cong H(E_r^{*,*}, d_r) \xrightarrow{H(f_r)} H(\bar{E}_r^{*,*}, \bar{d}_r) \cong \bar{E}_{r+1}^{*,*}.$$

The class of spectral sequences, with morphisms so defined, constitutes a category, **SpecSeq**. It is useful at times to consider certain constructions as functors from a topological or algebraic category to **SpecSeq**.

Suppose we have a morphism, $\{f_r\} : \{(E_r, d_r)\} \to \{(\bar{E}_r, \bar{d}_r)\}$, of spectral sequences. Recall that each spectral sequence may be presented as a tower of submodules of its E_2-term. By restricting $f_2 : E_2 \to \bar{E}_2$, we get the diagram:

$$
\begin{array}{ccccccccccccc}
B_2 & \subset & B_3 & \subset & \cdots & \subset & \bigcap_r B_r & \subset & \bigcap_r Z_r & \subset & \cdots \subset Z_3 \subset Z_2 \subset E_2 \\
\downarrow & & \downarrow & & & & \downarrow & & \downarrow & & \downarrow \quad\quad \downarrow \quad\quad \downarrow \quad f_2 \\
\bar{B}_2 & \subset & \bar{B}_3 & \subset & \cdots & \subset & \bigcap_r \bar{B}_r & \subset & \bigcap_r \bar{Z}_r & \subset & \cdots \subset \bar{Z}_3 \subset \bar{Z}_2 \subset \bar{E}_2.
\end{array}
$$

The condition $f_r \circ d_r = \bar{d}_r \circ f_r$ allows us to identify f_{r+1} with the mapping induced by f_2:

$$f_{r+1} : E_{r+1} \cong Z_r / B_r \longrightarrow \bar{Z}_r / \bar{B}_r \cong \bar{E}_{r+1}.$$

Furthermore, such a morphism induces a mapping $f_\infty : E_\infty \to \bar{E}_\infty$.

The second condition, that f_{r+1} is induced by f_r on homology, can be expressed in the diagram

$$
\begin{array}{ccccccccc}
0 & \longrightarrow & b_r & \longrightarrow & c_r & \longrightarrow & E_{r+1} & \longrightarrow & 0 \\
& & \downarrow f_r & & \downarrow f_r & & \downarrow f_{r+1} & & \\
0 & \longrightarrow & \bar{b}_r & \longrightarrow & \bar{c}_r & \longrightarrow & \bar{E}_{r+1} & \longrightarrow & 0
\end{array}
$$

where $c_r = \ker d_r : E_r \to E_r$ and $b_r = \operatorname{im} d_r$. With these observations and the Five-lemma it is easy to prove the following result (exercise).

Theorem 3.4. *If* $\{f_r\}\colon \{(E_r^{*,*}, d_r)\} \to \{(\bar{E}_r^{*,*}, \bar{d}_r)\}$ *is a morphism of spectral sequences and, for some* n, $f_n\colon E_n \to \bar{E}_n$ *is an isomorphism of bigraded modules, then for all* r, $n \le r \le \infty$, $f_r\colon E_r \to \bar{E}_r$ *is an isomorphism.*

Thus an isomorphism at some stage of the spectral sequences gives an isomorphism of E_∞-terms.

Morphisms of spectral sequences arise in the case of filtered differential graded modules when there is a mapping $\phi\colon (A, d, F) \to (\bar{A}, \bar{d}, \bar{F})$ with $\phi\colon A \to \bar{A}$ a morphism of graded modules, such that $\phi \circ d = \bar{d} \circ \phi$, and ϕ *respects the filtration*, that is, $\phi(F^p A) \subset \bar{F}^p \bar{A}$. Such a mapping is a **morphism of filtered differential graded modules** and this notion leads to an appropriate category. It is immediate that ϕ induces a homomorphism of the associated towers of submodules and so a morphism of spectral sequences. Since the spectral sequences are taken to be approximations to $H(A, d)$ and $H(\bar{A}, \bar{d})$, it is natural to try to compare these modules through the associated spectral sequences as in the following result of [Moore53].

Theorem 3.5. *A morphism of filtered differential graded modules*

$$\phi\colon (A, d, F) \to (\bar{A}, \bar{d}, \bar{F}),$$

determines a morphism of the associated spectral sequences. If, for some n, $\phi_n\colon E_n \to \bar{E}_n$ *is an isomorphism of bigraded modules, then* $\phi_r\colon E_r \to \bar{E}_r$ *is an isomorphism for all* r, $n \le r \le \infty$. *If the filtrations are bounded, then* ϕ *induces an isomorphism* $H(\phi)\colon H(A, d) \to H(\bar{A}, \bar{d})$.

PROOF: It suffices, by Theorem 3.4, to prove the last part of the theorem. Since the filtration is bounded, there are functions $s = s(n)$ and $t = t(n)$ so that we can write

$$\{0\} = F^s H^n \subset F^{s-1} H^n \subset \cdots \subset F^{t+1} H^n \subset F^t H^n = H^n,$$

where H^n denotes $H^n(A, d)$ and $E_\infty^{p,q} \cong F^p H^{p+q}/F^{p+1} H^{p+q}$. Similar data hold for $H^n(\bar{A}, \bar{d})$ which we denote as $\bar{H}^n$. Since ϕ_∞ is an isomorphism, by the boundedness of the filtration, we have, for the same $s = s(n)$, $F^{s-1} H^n = F^{s-1} H^n/F^s H^n \cong E_\infty^{s-1,n-s+1} \cong \bar{E}_\infty^{s-1,n-s+1} \cong \bar{F}^{s-1} \bar{H}^n$. We can now apply induction downward to H^n and $\bar{H}^n$: Consider the commutative diagram with rows exact

$$\begin{array}{ccccccccc}
0 & \longrightarrow & F^p H^n & \longrightarrow & F^{p-1} H^n & \longrightarrow & E_\infty^{p-1,n-p+1} & \longrightarrow & 0 \\
& & \downarrow{\scriptstyle H(\phi)} & & \downarrow{\scriptstyle H(\phi)} & & \downarrow{\scriptstyle \phi_\infty} & & \\
0 & \longrightarrow & \bar{F}^p \bar{H}^n & \longrightarrow & \bar{F}^{p-1} \bar{H}^n & \longrightarrow & \bar{E}_\infty^{p-1,n-p+1} & \longrightarrow & 0.
\end{array}$$

We apply induction on i in the expression $p = s - 1 - i$ and so we assume that $H(\phi)$ induces an isomorphism $F^p H^n \cong \bar{F}^p \bar{H}^n$. Since ϕ_∞ is an isomorphism, by the Five-lemma, $H(\phi)\colon F^{p-1} H^n \to \bar{F}^{p-1} \bar{H}^n$ is an isomorphism. Thus, we arrive at $H^n = F^t H^n \cong \bar{F}^t \bar{H}^n = \bar{H}^n$. (Compare this proof with Example 1.K.) $\qquad\square$

3.2 Limits and colimits

In Theorem 3.5, we relied on the boundedness of the filtration to extract the comparison of the homology modules. In order to treat more general cases, we seek less restrictive conditions on the filtration. We first introduce some important constructions.

Definition 3.6. *Given a sequence of morphisms in a category* **C**,

$$\cdots \longrightarrow D^{s+1} \xrightarrow{g_{s+1}} D^s \xrightarrow{g_s} D^{s-1} \longrightarrow \cdots$$

an **limit** *(or* **inverse limit***) of this sequence is an object in* **C**, $D^\infty = \varprojlim_s \{D^s, g_s\}$, *together with morphisms* $h_s\colon D^\infty \to D^s$ *such that* $g_s \circ h_s = h_{s-1}$ *for all s, and, for any object E in* **C**, *together with morphisms* $j_s\colon E \to D^s$ *such that* $g_s \circ j_s = j_{s-1}$, *there is a unique morphism* $k\colon E \to D^\infty$ *so that, for all s,* $j_s = h_s \circ k$. *Dually, a* **colimit** *(or* **direct limit***) of the sequence is an object in* **C**, $D^{-\infty} = \varinjlim_s \{D^s, g_s\}$, *together with morphisms* $p_s\colon D^s \to D^{-\infty}$ *such that* $p_{s-1} \circ g_s = p_s$ *for all s, and, for any object E' in* **C**, *together with morphisms* $q_s\colon D^s \to E'$ *such that* $q_{s-1} \circ g_s = q_s$, *there is a unique morphism* $k'\colon D^{-\infty} \to E'$ *so that, for all s,* $q_s = k' \circ p_s$.

Defined by a universal property, limits and colimits are unique up to isomorphism. When the sequence $\{D^s, g_s\}$ consists of subobjects and inclusions, we have $\varprojlim_s \{D^s, g_s\} = \bigcap_s D^s$ and $\varinjlim_s \{D^s, g_s\} = \bigcup_s D^s$. Thus limits and colimits generalize intersections and unions.

We call a category **(sequentially) complete** if any sequence of morphisms has a limit and **(sequentially) cocomplete** if any sequence of morphisms has a colimit. In particular, the category of graded modules over a ring R is seen to be complete and cocomplete by letting

$$\varprojlim_s \{D^s, g_s\} = \left\{ (\dots, x_s, x_{s-1}, \dots) \in \prod_s D^s \;\middle|\; \text{for all } s,\, g_s(x_s) = x_{s-1} \right\}$$

with $h_s\colon \varprojlim_s \{D^s, g_s\} \to D^s$ the sth projection. The colimit is defined by

$$\varinjlim_s \{D^s, g_s\} = \bigcup_s D^s \big/ {\sim}$$

where $x \in D^s$ is related to $y \in D^t$ if there are integers n and m with

$$g_{s-n+1} \circ g_{s-n+2} \circ \cdots \circ g_s(x) = g_{t-m+1} \circ g_{t-m+2} \circ \cdots \circ g_t(y) \in D^{s-n} = D^{t-m}.$$

The morphisms $p_s : D^s \to D^{-\infty}$ are given by taking the inclusion into the union followed by the quotient.

Given two sequences of morphisms in **C**, say $\{D^s, g_s\}$ and $\{\bar{D}^s, \bar{g}_s\}$, then a **morphism of sequences in C** is a sequence of morphisms in **C**, $f_s : D^s \to \bar{D}^s$, for all s, satisfying $f_{s-1} \circ g_s = \bar{g}_s \circ f_s$. When **C** has notions of kernel and cokernel, then we can talk about a short exact sequence of sequences of morphisms in **C**. Among the important properties of limits is their behavior when applied to an exact sequence of sequences of morphisms in **C**. We restrict our attention to limits and colimits in the category of R-modules.

Lemma 3.7. *Suppose* $0 \to \{K^s, k_s\} \to \{D^s, g_s\} \to \{Q^s, q_s\} \to 0$ *is a short exact sequence of sequences of R-modules and module homomorphisms. Then on application of the colimit we obtain a short exact sequence of R-modules,*

$$0 \to \varinjlim_{\to s}\{K^s, k_s\} \to \varinjlim_{\to s}\{D^s, g_s\} \to \varinjlim_{\to s}\{Q^s, q_s\} \to 0.$$

Furthermore, on application of the limit we obtain a short exact sequence,

$$0 \to \varprojlim_{\leftarrow s}\{K^s, k_s\} \to \varprojlim_{\leftarrow s}\{D^s, g_s\} \to \varprojlim_{\leftarrow s}\{Q^s, q_s\}.$$

PROOF: We leave the proof of the exactness for the colimit to the reader. It is a simple exercise in definitions.

[Eilenberg-Moore62] compute the limit using the homomorphism

$$\Phi : \prod_s D^s \to \prod_s D^s \text{ determined by } p_n \circ \Phi = p_n - g_{n+1} \circ p_{n+1} \text{ for all } n.$$

By construction $\varprojlim_{\leftarrow s}\{D^s, g_s\} = \ker \Phi$. If we view $\Phi : \prod_s D^s \to \prod_s D^s$ as a chain complex, concentrated in degrees 0 and 1, then the exactness of the given sequence leads to the exactness of the short sequence of chain complexes

$$
\begin{array}{ccccccccc}
0 & \longrightarrow & \prod_s K^s & \longrightarrow & \prod_s D^s & \longrightarrow & \prod_s Q^s & \longrightarrow & 0 \\
& & \downarrow{\scriptstyle \Phi} & & \downarrow{\scriptstyle \Phi} & & \downarrow{\scriptstyle \Phi} & & \\
0 & \longrightarrow & \prod_s K^s & \longrightarrow & \prod_s D^s & \longrightarrow & \prod_s Q^s & \longrightarrow & 0
\end{array}
$$

and passing to homology, we get the short exact sequence associated to the limit. $\square$

The failure of the exactness of the limit is measured by continuing the exact sequence associated to the homology of chain complexes further. The **first derived functor** of the limit was introduced by [Milnor62] and is defined by

$$\lim_{\leftarrow s} {}^1 D^s = \text{cokernel of } \Phi.$$

This leads to the exact sequence

$$0 \to \lim_{\leftarrow s}\{K^s, k_s\} \to \lim_{\leftarrow s}\{D^s, g_s\} \to \lim_{\leftarrow s}\{Q^s, q_s\}$$

$$\xrightarrow{\delta} \lim_{\leftarrow s}{}^1\{K^s, k_s\} \to \lim_{\leftarrow s}{}^1\{D^s, g_s\} \to \lim_{\leftarrow s}{}^1\{Q^s, q_s\} \to 0.$$

Given a fixed R-module A we can form the sequence of homomorphisms all of which are the identity i_A on A. This is a constant sequence and it satisfies $\lim_{\leftarrow s}\{A, i_A\} = A = \lim_{\to s}\{A, i_A\}$ and $\lim_{\leftarrow s}{}^1\{A, i_A\} = \{0\}$. More generally, a sequence of homomorphisms that is eventually all isomorphisms has limit the eventual abstract R-module. If a sequence consists entirely of epimorphisms, then the limit maps onto each $D^{s,*}$ and the first derived functor vanishes. This has the amusing consequence that a sequence of epimorphisms with vanishing limit must be the sequence of trivial modules.

Given the filtered graded module, $(H(A, d), F)$, we obtain a sequence of graded modules together with the canonical homomorphisms

$$H(A)$$

$$\longrightarrow H(A)/F^{p+1}H(A) \longrightarrow H(A)/F^p H(A) \longrightarrow H(A)/F^{p-1}H(A) \longrightarrow$$

which induces a homomorphism $u\colon H(A) \to \lim_{\leftarrow p} H(A)/F^p H(A)$, unique up to isomorphism.

Definition 3.8. *A filtration on a differential graded module (A, d) is said to be **strongly convergent** ([Cartan-Eilenberg56]) or **complete** ([Eilenberg-Moore62]) if it is weakly convergent and the induced mapping $u\colon H(A) \to \lim_{\leftarrow p} H(A)/F^p H(A)$ is an isomorphism.*

Theorem 3.9. *Suppose $\phi\colon (A, d, F) \to (\bar{A}, \bar{d}, \bar{F})$, is a morphism of filtered differential graded modules so that for some n, $\phi_n\colon E_n^{*,*} \to \bar{E}_n^{*,*}$ is an isomorphism. If the filtrations are exhaustive and weakly convergent, then ϕ induces an isomorphism of associated bigraded modules $E_0^{*,*}(H(A, d), F) \cong E_0^{*,*}(H(\bar{A}, \bar{d}), \bar{F})$. If the filtrations are also complete, then ϕ induces an isomorphism on homology, $H(\phi)\colon H(A, d) \xrightarrow{\cong} H(\bar{A}, \bar{d})$.*

PROOF: By weak convergence, we know that $E_0^{*,*}(H(A,d),F) \cong E_\infty^{*,*}$ and $E_0^{*,*}(H(\bar{A},\bar{d}),\bar{F}) \cong \bar{E}_\infty^{*,*}$. Since ϕ induces an isomorphism of the E_∞-terms of the spectral sequences, the associated bigraded modules for $(H(A,d),F)$ and $(H(\bar{A},\bar{d}),\bar{F})$ are isomorphic.

We construct $H(\phi)$ with the E_∞-terms of the spectral sequences and the assumption of strong convergence. First we show that the isomorphism $\phi_\infty \colon E_\infty^{*,*} \to \bar{E}_\infty^{*,*}$, induces for all p and all $r \geq 0$,

$$F^{p-r}H(A)/F^p H(A) \xrightarrow[\phi_\infty]{\cong} \bar{F}^{p-r}H(\bar{A})/\bar{F}^p H(\bar{A}).$$

This follows by induction on r. Since

$$F^{p-1}H(A)/F^p H(A) \cong E_\infty^{p-1,*} \xrightarrow[\phi_\infty]{\cong} \bar{E}_\infty^{p-1,*} \cong \bar{F}^{p-1}H(\bar{A})/\bar{F}^p H(\bar{A})$$

the case of $r = 1$ is established. To complete the induction, apply the Five-lemma to the mappings between short exact sequences of the type

$$0 \to F^{p-r+1}H(A)/F^p H(A) \to$$
$$F^{p-r}H(A)/F^p H(A) \to E_\infty^{p-r+1,*} \to 0$$

where the first isomorphism is given by induction and the last by the isomorphism $E_\infty^{p-r+1,*} \to \bar{E}_\infty^{p-r+1,*}$. Since F and $\bar{F}$ are decreasing and exhaustive, ϕ_∞ induces an isomorphism, for all p,

$$\bigcup_r F^{p-r}H(A)/F^p H(A) = H(A)/F^p H(A)$$
$$\xrightarrow{\cong} H(\bar{A})/\bar{F}^p H(\bar{A}) = \bigcup_r \bar{F}^{p-r}H(\bar{A})/\bar{F}^p H(\bar{A}).$$

We can now pass this isomorphism to the limit to get the commutative diagram

$$
\begin{array}{ccc}
H(A,d) & \xrightarrow{\;u\;} & \varprojlim_p H(A)/F^p H(A) \\
\downarrow{\scriptstyle H(\phi)} & & \downarrow{\scriptstyle \cong} \\
H(\bar{A},\bar{d}) & \xrightarrow[\;\bar{u}\;]{} & \varprojlim_p H(\bar{A})/\bar{F}^p H(\bar{A})
\end{array}
$$

and by completeness, $H(\phi)$ is also an isomorphism.   □

This comparison theorem identifies the condition on a filtration of a differential graded module so that the associated spectral sequence determines the target module $H(A,d)$ (uniquely up to extensions). It also offers the motivating example for strong convergence of more general spectral sequences. The key property that makes strong convergence work for a filtered differential module is an inverse limit condition. We next consider more general situations.

Completion

The property of strong convergence is determined by comparison: If we know that a spectral sequence converges to a filtered R-module, (M, F), then $E_\infty^s \cong F^s M / F^{s+1} M$. Given $E_0(M, F)$, when can we reconstruct M? A filtered R-module (M, F) may be endowed with a topology: Take as basic open sets the collection of cosets of the submodules $F^s M$. A filtered module (M, F) is **complete** (in the topological sense) if any Cauchy sequence in the topological space M converges, that is, if a sequence of elements $\{x_n\}$ satisfies the condition, for all s, there is an integer $N(s)$ such that $x_n - x_m \in F^s M$ for any $n, m \geq N(s)$, then there is an element $y \in M$ such that, for all s, there is an integer $Q(s)$ such that $y - x_n \in F^s M$ for any $n \geq Q(s)$. We now prove the following properties relating the underlying algebra to this topology.

Lemma 3.10. *A filtered module (M, F) is Hausdorff in the the topology induced by the filtration if and only if $\bigcap_s F^s M = \{0\}$. The module is complete in this topology if and only if the mapping $M \to \varprojlim_s M/F^s M$ is an epimorphism. It follows that a filtered module (M, F) is Hausdorff if $\varprojlim_s F^s M = \{0\}$, and is complete if $\varprojlim_s^1 F^s M = \{0\}$. Dually, a filtration is exhaustive if and only if $\varinjlim_s M/F^s M = \{0\}$.*

PROOF: We leave the properties of M as a topological space to the reader. For the algebraic equivalents, consider the short exact sequence of sequences of modules where $q_s \colon M/F^s M \to M/F^{s-1} M$ is given by $m + F^s M \mapsto m + F^{s-1} M$:

$$0 \to \{F^s M, \subset\} \to \{M, \mathrm{id}_M\} \to \{M/F^s M, q_s\} \to 0.$$

When we apply the direct limit functor, then, by exactness, the filtration is exhaustive, $\varinjlim_s F^s M = \bigcup_s F^s M = M$, if and only if $\varinjlim_s M/F^s M = \{0\}$.

When we apply the limit functor, we obtain the short exact sequence:

$$0 \to \varprojlim_s F^s M \to M \to \varprojlim_s M/F^s M \to \varprojlim_s^1 F^s M \to 0.$$

Since the morphisms in the sequence of filtration submodules are all inclusions, the Hausdorff property is equivalent to the vanishing of first limit. With $\varprojlim_s F^s M = \{0\}$, vanishing of the first derived limit of the sequence $\{F^s M\}$ is equivalent to completeness. $\square$

Strong convergence of the spectral sequence associated to a filtered differential module (A, d, F) is a feature of the completeness of the filtration of $H(A, d)$. As in analysis, the failure of completeness is overcome by completing.

Definition 3.11. *To a module M with a decreasing filtration F we associate the* **completion** *of M with respect to F given by*

$$\hat{M} = \varprojlim_s M/F^s M.$$

The completion is equipped with a mapping $c \colon M \to \hat{M}$ induced by the quotients $M \to M/F^s M$ and called the **completion homomorphism**. *The filtration F on M induces a filtration $\hat{F}$ on $\hat{M}$ given by $\hat{F}^s \hat{M} = \varprojlim_t F^s M/F^t M$.*

Proposition 3.12. *The completion of a filtered module (M, F) satisfies the following properties:*

(1) *$(\hat{M}, \hat{F})$ is complete and Hausdorff.*
(2) *For all $t \geq s$, $\hat{F}^s \hat{M}/\hat{F}^t \hat{M} \cong F^s M/F^t M$.*
(3) *$\hat{F}$ is exhaustive if and only if F is exhaustive.*

PROOF: Consider the short exact sequence

$$0 \to F^s M/F^t M \to M/F^t M \to M/F^s M \to 0.$$

Varying t we get a short exact sequence of sequences of homomorphisms. Apply the inverse limit functor to get

$$0 \to \hat{F}^s \hat{M} \to \hat{M} \to M/F^s M \to \varprojlim_t{}^1 F^s M/F^t M \to \varprojlim_t{}^1 M/F^t M \to 0.$$

Since F is a decreasing filtration, the morphisms $F^s M/F^t M \to F^s M/F^{t-1} M$ are all epimorphisms and so $\varprojlim_t{}^1 F^s M/F^t M = \{0\}$. Thus the exact sequence reduces to

(3.13) $$0 \to \hat{F}^s \hat{M} \to \hat{M} \to M/F^s M \to 0.$$

Next consider the limit over s:

$$0 \to \varprojlim_s \hat{F}^s \hat{M} \to \hat{M} \to \varprojlim_s M/F^s M \to \varprojlim_s{}^1 \hat{F}^s \hat{M} \to 0.$$

Since $\varprojlim_s M/F^s M = \hat{M}$, the morphism in the middle is an isomorphism and so the modules on the ends vanish. It follows from Lemma 3.10 that $(\hat{M}, \hat{F})$ is Hausdorff and complete.

Notice that the short exact sequence (3.13) implies that $M/F^s M \cong \hat{M}/\hat{F}^s \hat{M}$ and so, by the Five-lemma and the diagram,

$$
\begin{array}{ccccccccc}
0 & \longrightarrow & F^s M/F^t M & \longrightarrow & M/F^t M & \longrightarrow & M/F^s M & \longrightarrow & 0 \\
 & & \downarrow & & \downarrow & & \downarrow & & \\
0 & \longrightarrow & \hat{F}^s \hat{M}/\hat{F}^t \hat{M} & \longrightarrow & \hat{M}/\hat{F}^t \hat{M} & \longrightarrow & \hat{M}/\hat{F}^s \hat{M} & \longrightarrow & 0,
\end{array}
$$

we get $\hat{F}^s \hat{M} / \hat{F}^t \hat{M} \cong F^s M / F^t M$. Furthermore, if we apply the direct limit functor to (3.13) we get

$$0 \to \bigcup_s \hat{F}^s \hat{M} \to \hat{M} \to \varinjlim_s M/F^s M \to 0.$$

By Lemma 3.10, $(\hat{M}, \hat{F})$ is exhaustive if and only if $\varinjlim_s M/F^s M = \{0\}$ if and only if (M, F) is exhaustive. $\qquad \square$

The dilemma that convergence of a spectral sequence presents is clarified by the completion. If we have only an isomorphism $E_\infty^{s,*} \cong E_0^{s,*}(M, F)$, is this enough to capture M up to extensions? The answer is NO in general, but it is the completion that emerges as the real target:

Proposition 3.14. *Let* $\phi \colon (M, F) \to (N, \mathcal{F})$ *be a morphism of filtered modules. If the filtrations* F *and* $\mathcal{F}$ *are exhaustive, then the following statements are equivalent.*

(1) $\hat{\phi} \colon (\hat{M}, \hat{F}) \to (\hat{N}, \hat{\mathcal{F}})$ *is an isomorphism of filtered modules, that is,* $\hat{\phi}| \colon \hat{F}^s \hat{M} \to \hat{\mathcal{F}}^s \hat{N}$ *is an isomorphism for all* s.
(2) $E_0(\phi) \colon E_0(M, F) \to E_0(N, \mathcal{F})$ *is an isomorphism.*

PROOF: If $\hat{\phi}$ induces an isomorphism of filtered modules, then it induces the isomorphism of associated graded modules $E_0(\phi) \colon E_0(M, F) \cong E_0(\hat{M}, \hat{F})$ $\xrightarrow{E_0(\hat{\phi})} E_0(\hat{N}, \hat{\mathcal{F}}) \cong E_0(N, \mathcal{F})$.

Suppose $E_0(\phi)$ is an isomorphism. This implies that, for all s, ϕ induces an isomorphism

$$E_0^s(\phi) \colon F^s M / F^{s+1} M \xrightarrow{\cong} \mathcal{F}^s N / \mathcal{F}^{s+1} N.$$

By induction and the Five-lemma it follows that ϕ induces an isomorphism $F^s M / F^{s+n} M \xrightarrow{\cong} \mathcal{F}^s N / \mathcal{F}^{s+n} N$ for all $n \geq 0$. Taking inverse limits over n gives us that $\hat{\phi}$ induces isomorphisms, for all s,

$$\hat{F}^s \hat{M} = \varprojlim_n F^s M / F^{s+n} M \xrightarrow[\phi]{\cong} \varprojlim_n \mathcal{F}^s N / \mathcal{F}^{s+n} N = \hat{\mathcal{F}}^s \hat{N}.$$

Finally, $\hat{\phi}$ induces an isomorphism of direct limits,

$$\hat{M} = \varinjlim_s \hat{F}^s \hat{M} \xrightarrow[\hat{\phi}]{\cong} \varinjlim_s \mathcal{F}^s \hat{N} = \hat{N}. \qquad \square$$

Corollary 3.15. *Suppose* $\{f_r\}\colon \{E_r^{*,*}, d_r\} \to \{\bar{E}_r^{*,*}, \bar{d}_r\}$ *is a morphism of spectral sequences and for some* n, $f_n\colon E_n \to \bar{E}_n$ *is an isomorphism. Suppose* E_r *converges to* (M, F) *and* $\bar{E}_r$ *converges to* $(N, \mathcal{F})$. *Then* f_∞ *induces an isomorphism of filtered modules,* $f\colon (\hat{M}, \hat{F}) \to (\hat{N}, \hat{\mathcal{F}})$.

Corollary 3.15 generalizes the successful comparison theorems (Theorems 3.5 and 3.9) for filtered differential modules. In these cases the target modules were already complete and so isomorphic to their completions.

We next turn to exact couples. A morphism between spectral sequence can arise via a **morphism of exact couples**,

$$\{\psi_1, \psi_2\}\colon \{D, E, i, j, k\} \to \{\bar{D}, \bar{E}, \bar{i}, \bar{j}, \bar{k}\}$$

by which we mean a pair of homomorphisms $\psi_1\colon D \to \bar{D}$ and $\psi_2\colon E \to \bar{E}$ compatible with the homomorphisms in the couple, that is, $\bar{i} \circ \psi_1 = \psi_1 \circ i$, $\bar{j} \circ \psi_1 = \psi_2 \circ j$, and $\bar{k} \circ \psi_2 = \psi_1 \circ k$. Under these conditions, ψ_1 and ψ_2 induce a morphism of the derived couples and hence a morphism of spectral sequences. In the topological applications morphisms of exact couples can arise from functorial constructions.

We saw in Chapter 2 (for example, Corollary 2.10), that the target of the spectral sequence associated to an exact couple can be more difficult to describe than the analogous problem for filtered differential graded modules. To an exact couple we associate two limit modules:

$$D^{\infty,*} = \lim_{\leftarrow s}\{D^{s,*}, i\} \text{ and } D^{-\infty,*} = \lim_{\to s}\{D^{s,*}, i\}.$$

Each module has a decreasing filtration given by

$$\mathcal{F}^s D^{\infty,*} = \ker(D^{\infty,*} \xrightarrow{h_s} D^{s,*}) \text{ and } F^s D^{-\infty,*} = \mathrm{im}(D^{s,*} \xrightarrow{p_s} D^{-\infty,*}).$$

In fact, these filtrations have very nice properties.

Proposition 3.16. *For an exact couple, the filtration F on the colimit $D^{-\infty,*}$ is exhaustive. The filtration $\mathcal{F}$ on the limit $D^{\infty,*}$ is Hausdorff and complete.*

PROOF: Recall that the colimit is a quotient of the union of the $D^{s,*}$ and so any element in $D^{-\infty,*}$ comes from some $D^{s,*}$ before the quotient and thus lies in some $F^s D^{-\infty,*}$. It follows that the filtration is exhaustive.

To prove that the filtration $\mathcal{F}$ on $D^{\infty,*}$ is Hausdorff, suppose that

$$(\ldots, x_s, x_{s-1}, \ldots) \in \bigcap_s \mathcal{F}^s D^{\infty,*}.$$

Since $h_s\colon D^{\infty,*} \to D^{s,*}$ is given by projection, to be in the kernel of h_s one has $x_s = 0$. Thus if an element of the limit is in all of the filtration submodules, then it must be the zero element and the filtration is Hausdorff.

To prove completeness we consider the isomorphism

$$D^{\infty,*}/\mathcal{F}^s D^{\infty,*} \cong \operatorname{im}(D^{\infty,*} \to D^{s,*}).$$

Applying the limit functor gives $\varprojlim_s D^{\infty,*}/\mathcal{F}^s D^{\infty,*} \cong \varprojlim_s \operatorname{im}(D^{\infty,*} \to D^{s,*})$. By the universal property of the inverse limit, the limit on the right can be identified with $D^{\infty,*}$ and so the filtration is complete. $\square$

We generalize Proposition 3.3 on convergence for filtered differential modules to exact couples.

Lemma 3.17. *Let* $R^p = \bigcap_r \operatorname{im}(i^r \colon D^{p+r,*} \to D^{p,*})$. *Then there is a short exact sequence for each* s:

$$0 \to \mathcal{F}^s D^{-\infty,*}/\mathcal{F}^{s+1} D^{-\infty,*} \to E_\infty^{s,*} \to R^{s+1} \xrightarrow{i} R^s.$$

PROOF ([Boardman99]): We apply Proposition 2.9 to rename the modules in the sequence. We begin with $\ker k \subset E^{s,*}$. Since

$$Z_\infty^{s,*} = \bigcap_r Z_r^{s,*} = \bigcap_r k^{-1}(\operatorname{im} i^{r-1} \colon D^{s+r,*} \to D^{s+1,*})$$
$$B_\infty^{s,*} = \bigcup_r B_r^{s,*} = \bigcup_r j(\ker i^{r-1} \colon D^{s,*} \to D^{s-r+1,*})$$

we see that $B_\infty^{s,*} \subset \ker k \subset Z_\infty^{s,*}$. Consider the short exact sequence

$$0 \to \ker k / B_\infty^{s,*} \to Z_\infty^{s,*}/B_\infty^{s,*} \to Z_\infty^{s,*}/\ker k \to 0.$$

Notice that $Z_\infty^{s,*} = k^{-1} R^{s+1}$ by the definition of R^{s+1}. Then k induces a mapping $\bar{k} \colon Z_\infty^{s,*}/B_\infty^{s,*} \to R^{s+1}$ by $\bar{k}(a + B_\infty^{s,*}) = k(a)$. The image of $\bar{k}$ is given by $R^{s+1} \cap \operatorname{im} k = R^{s+1} \cap \ker i$ and so the sequence

$$Z_\infty^{s,*}/B_\infty^{s,*} \xrightarrow{\bar{k}} R^{s+1} \xrightarrow{i} R^s$$

is exact at R^{s+1}. Furthermore, $\bar{k}$ may be taken as defined on $Z_\infty^{s,*}$ and so has kernel given by $\ker k$. Thus the mapping to R^{s+1} coincides with the mapping in the short exact sequence for $B_\infty^{s,*} \subset \ker k \subset Z_\infty^{s,*}$.

We next rewrite $B_\infty^{s,*}$ in terms of the colimit $D^{-\infty,*}$ to show

$$\bigcup_r (\ker i^{r-1} \colon D^{s,*} \to D^{s-r+1,*}) = \ker(p_s \colon D^{s,*} \to D^{-\infty,*}).$$

The union of kernels certainly lies in the kernel of p_s. If $p_s(a) = 0$, then $i^n(a) = 0 \in D^{s+n,*}$ for some n by the definition of the equivalence relation that determines the colimit. This implies $a \in \ker i^n$ and we get equality. It follows that $B_\infty^{s,*} = j(\ker p_s)$.

Using the homomorphism j we obtain an isomorphism

$$D^{s,*}/(\ker j + \ker p_s) \xrightarrow{\cong} j(D^{s,*})/j(\ker p_s) = \ker k/B_\infty^{s,*}.$$

Using the homomorphism $p_s \colon D^{s,*} \to D^{-\infty,*}$ we get a short exact sequence

$$0 \to p_s^{-1}(F^{s+1}D^{-\infty,*}) \to D^{s,*} \xrightarrow{p_s} F^s D^{-\infty,*}/F^{s+1}D^{-\infty,*} \to 0.$$

However, $p_s^{-1}(F^{s+1}D^{-\infty,*}) = p_s^{-1}(p_{s+1}(D^{s+1,*})) = \ker p_s + \operatorname{im} i = \ker p_s + \ker j$. It follows that F^s/F^{s+1} is isomorphic to $D^{s,*}/(\ker j + \ker p_s)$ and so to $\ker j/B_\infty^{s,*}$ and the lemma is proved.   □

It follows from the lemma that the modules R^s tell us if we can identify E_∞ in terms of the filtered module given by the colimit. Generalizing Proposition 3.3 we see that the spectral sequence associated to an exact couple converges to the colimit $D^{-\infty,*}$ whenever the morphisms $i \colon R^{s+1} \to R^s$ are monomorphisms for all s.

Conditionally convergent spectral sequences Ⓝ

[Boardman99] introduced a class of spectral sequences together with an invariant that aids in recognizing strong convergence *intrinsically*.

Definition 3.18. *The spectral sequence associated to an exact couple* $\{D^{s,*}, E^{s,*}, i, j, k\}$ *is said to be* **conditionally convergent** *to* $D^{-\infty,*} = \varinjlim_s \{D^{s,*}, i\}$ *if*

$$D^{\infty,*} = \varprojlim_s \{D^{s,*}, i\} = \{0\} \text{ and } \varprojlim_s{}^1 \{D^{s,*}, i\} = \{0\}.$$

We associate to the tower of submodules $\cdots \subset Z_r^{s,*} \subset Z_{r-1}^{s,*} \subset \cdots$ *the first derived module of the limit of this sequence of inclusions denoted by*

$$RE_\infty^{s,*} = \varprojlim_s{}^1 Z_r^{s,*}.$$

The notation hides a nice property of $\varprojlim^1$: If (M, F) is a filtered module with N a submodule that satisfies $N \subset F^s M$ for all s, then

$$\varprojlim_s \left(F^s M/N\right) \cong \left(\varprojlim_s F^s M\right)/N \text{ and } \varprojlim_s{}^1 F^s M \cong \varprojlim_s{}^1 \left(F^s M/N\right).$$

This follows from the short exact sequences $0 \to N \to F^s M \to F^s M/N \to 0$. Since $B_\infty^{s,*} \subset Z_r^{s,*}$ for all r, and $\varprojlim_r Z_r^{s,*} = Z_\infty^{s,*}$, we can write $RE_\infty^{s,*} = \varprojlim_r{}^1 Z_r^{s,*}/B_\infty^{s,*}$. The strength of these notions is born out by the following result. For the motivating topological application of this result see the discussion of [Adams74, Theorem 8.2].

Theorem 3.19. *Suppose $\{D^{s,*}, E^{s,*}, i, j, k\}$ is an exact couple satisfying $E^{s,*} = \{0\}$ for all $s < 0$. Suppose further that the associated spectral sequence converges conditionally to $D^{-\infty,*}$. Then the spectral sequence converges strongly to $D^{-\infty,*}$ if and only if $RE_\infty^{s,*} = \{0\}$ for all s.*

PROOF: We add a little more detail to the objects involved. Let

$$\operatorname{Im}^r D^{s,*} = \operatorname{im} i^r : D^{s+r,*} \to D^{s,*}.$$

These submodules are organized in two directions: $\operatorname{Im}^{r+1} D^{s,*} \subset \operatorname{Im}^r D^{s,*}$ for all r, and, fixing r, the homomorphisms i map $\operatorname{Im}^r D^{s,*}$ to $\operatorname{Im}^r D^{s-1,*}$. In fact, $i(\operatorname{Im}^r D^{s,*}) = \operatorname{Im}^{r+1} D^{s-1,*}$. This leads to two limits

$$\varprojlim_{r}\{\operatorname{Im}^r D^{s,*}, \subset\} \qquad \text{and} \qquad \varprojlim_{s}\{\operatorname{Im}^r D^{s,*}, i\}.$$

Lemma 3.20. *For all r and s we have $\varprojlim_{r} \operatorname{Im}^r D^{s,*} = \bigcap_r \operatorname{Im}^r D^{s,*} = R^s$,*

$$\varprojlim_{s}\{\operatorname{Im}^r D^{s,*}, i\} = \varprojlim_{s}\{D^{s,*}, i\}, \text{ and } \varprojlim_{s}{}^1\{\operatorname{Im}^r D^{s,*}, i\} = \varprojlim_{s}{}^1\{D^{s,*}, i\}.$$

PROOF: The first assertion follows from the sequence of inclusions and the definition of R^s. For the second assertion we work by induction. For $r = 1$ we have the short exact sequence

$$0 \to \operatorname{Im}^1 D^{s,*} \to D^{s,*} \to D^{s,*}/\operatorname{Im}^1 D^{s,*} \to 0.$$

These fit together into a short exact sequence of sequences where the homomorphism between the last modules are all the zero homomorphism. Since $\varprojlim_{s}\{M_s, 0\} = \{0\} = \varprojlim_{s}{}^1\{M_s, 0\}$ for any sequence of zero homomorphisms, the exact sequence relating $\varprojlim_{s}$ and $\varprojlim_{s}{}^1$ shows that the limit and first derived limit of $\{D^{s,*}, i\}$ and $\{\operatorname{Im}^1 D^{s,*}, i\}$ are isomorphic. Since $\operatorname{Im}^1(\operatorname{Im}^r D^{s,*}, i) = \operatorname{Im}^{r+1} D^{s,*}$, we prove the second assertion by induction. $\square$

Next we consider the sequence $\cdots \xrightarrow{i} R^{s+1} \xrightarrow{i} R^s \xrightarrow{i} \cdots$. The inverse limit may be written

$$\varprojlim_{s}\{R^s, i\} = \varprojlim_{s}\left(\varprojlim_{r}\{\operatorname{Im}^r D^{s,*}, \subset\}, i\right).$$

(It *would* be nice to simply commute the indices, but ...) We prove directly that $\varprojlim_{s}\{R^s, i\} = \varprojlim_{s}\{D^{s,*}, i\} = D^{\infty,*}$.

Consider the projections $h_s: D^{\infty,*} \to D^{s,*}$. Notice that the image lies in R^s: An element of $D^{\infty,*}$ may be written as $(\ldots, x_{s+1}, x_s, x_{s-1}, \ldots)$ with $x_s = i(x_{s+1}) = i^2(x_{s+2}) = \cdots$. Thus $x_s \in \operatorname{Im}^r D^{s,*}$ for all r. This fact induces a mapping $D^{\infty,*} \to \varprojlim_{s}\{R^s, i\}$. Furthermore, the inclusions

$R^s \subset D^{s,*}$ give a mapping of sequences and hence a homomorphism $\lim_{\leftarrow s} R^s \to \lim_{\leftarrow s} D^{s,*}$. By the universal property of limits we factor the identity $D^{\infty,*} \to \lim_{\leftarrow s} R^s \to D^{\infty,*}$ and so the mapping $\lim_{\leftarrow s} R^s \to D^{\infty,*}$ is onto. Checking at the level of elements, we see that it is also one-one having been induced by inclusions. Thus $\lim_{\leftarrow s} R^s = \lim_{\leftarrow s} D^{s,*}$.

From the definition of $\operatorname{Im}^r D^{s,*}$, we have that $Z_r^{s,*} = k^{-1} \operatorname{Im}^{r-1} D^{s+1,*}$ and there is a short exact sequence:

$$0 \to Z_r^{s,*}/\ker k \to \operatorname{Im}^{r-1} D^{s+1,*} \xrightarrow{i} \operatorname{Im}^r D^{s,*} \to 0.$$

If we apply the functor $\lim_{\leftarrow r}$, we get the exact sequence

$$0 \to Z_\infty^{s,*}/\ker k \to R^{s+1} \xrightarrow{i} R^s$$
$$\xrightarrow{\delta} RE_\infty^{s,*} \to \lim_{\leftarrow r}{}^1 \operatorname{Im}^{r-1} D^{s+1,*} \xrightarrow{i} \lim_{\leftarrow r}{}^1 \operatorname{Im}^r D^{s,*} \to 0.$$

We splice in the short exact sequence from Lemma 3.17 to get

$$0 \to F^s D^{-\infty,*}/F^{s+1} D^{-\infty,*} \to E_\infty^{s,*} \to R^{s+1} \xrightarrow{i} R^s$$
$$\xrightarrow{\delta} RE_\infty^{s,*} \to \lim_{\leftarrow r}{}^1 \operatorname{Im}^{r-1} D^{s+1,*} \xrightarrow{i} \lim_{\leftarrow r}{}^1 \operatorname{Im}^r D^{s,*} \to 0.$$

Suppose $RE_\infty^{s,*} = \{0\}$ for all s: From the exact sequence we find that $i: R^{s+1} \to R^s$ is an epimorphism for all s. By the definition of conditional convergence $\{0\} = D^{\infty,*} = \lim_{\leftarrow s} R^s$ and so we conclude that $R^s = \{0\}$ for all s. Thus $F^s D^{-\infty,*}/F^{s+1} D^{-\infty,*} \cong E_\infty^{s,*}$ and the spectral sequence converges to $(D^{-\infty,*}, F)$.

We next show that a right-half-plane spectral sequence with $RE_\infty^{s,*} = \{0\}$ for all s and conditionally convergent to $D^{-\infty,*}$ is, in fact, strongly convergent, that is, $(D^{-\infty,*}, F)$ is complete and Hausdorff. This is equivalent, in the language of limits, to $\lim_{\leftarrow s} F^s D^{-\infty,*} = \{0\} = \lim_{\leftarrow s}{}^1 F^s D^{-\infty,*}$.

A right-half-plane spectral sequence is given by the condition $D^{s,*} = D^{0,*}$ for all $s < 0$, which implies that $D^{-\infty,*} = D^{0,*}$ and $F^s D^{-\infty,*} = \operatorname{im} i^s: D^{s,*} \to D^{0,*} = \operatorname{Im}^s D^{0,*}$. It follows that

$$\lim_{\leftarrow s} F^s D^{-\infty,*} = \lim_{\leftarrow s} \operatorname{Im}^s D^{0,*} = R^0 = \{0\}.$$

Therefore we have the Hausdorff condition.

To study the derived limit consider the short exact sequence of sequences of homomorphisms:

$$
\begin{array}{ccccccccc}
0 & \longrightarrow & \ker i^r & \longrightarrow & D^{s+r,*} & \xrightarrow{i^r} & \operatorname{Im}^r D^{s,*} & \longrightarrow & 0 \\
& & \downarrow{i} & & \downarrow{i} & & \downarrow{\subset} & & \\
0 & \longrightarrow & \ker i^{r-1} & \longrightarrow & D^{s+r-1,*} & \xrightarrow{i^r} & \operatorname{Im}^{r-1} D^{s,*} & \longrightarrow & 0
\end{array}
$$

This leads to the exact sequence

$$0 \to \varprojlim_r \{\ker i^r, i\} \to \varprojlim_r \{D^{s+r,*}, i\} \to \varprojlim_r \{\operatorname{Im}^r D^{s,*}, \subset\}$$
$$\overset{\delta}{\to} \varprojlim_r{}^1 \{\ker i^r, i\} \to \varprojlim_r{}^1 \{D^{s+r,*}, i\} \to \varprojlim_r{}^1 \{\operatorname{Im}^r D^{s,*}, \subset\} \to 0.$$

Conditional convergence implies $\varprojlim_r{}^1 \{D^{s+r,*}, i\} = \{0\}$ and so the exact sequence forces $\varprojlim_r{}^1 \operatorname{Im}^r D^{s,*} = \{0\}$. In particular, $\varprojlim_r{}^1 \operatorname{Im}^r D^{0,*} = \{0\}$ and completeness follows.   □

[Boardman99] discusses more general situations and applications. The insight of conditional convergence is that the invariant $RE_\infty^{*,*}$ may vanish for reasons easily at hand; for example, when all of the modules $E_2^{*,*}$ are finitely generated, or there are only finitely many nonzero differentials at any given (s,t). Thus the raw data $E_2^{*,*}$ can hold the key to $RE_\infty^{*,*} = \{0\}$, an intrinsic condition with global consequences. Conditional convergence is a global condition that depends on the behavior of certain limits, and hence may be satisfied by virtue of certain structural features of the construction of a spectral sequence.

The Mittag-Leffler condition Ⓜ

Other global properties of the modules in an exact couple determine the behavior of limits and hence convergence of the associated spectral sequence. For example, if $R = \mathbb{Z}$ and, for every s, $D^{s,*}$ is a finite abelian group, then the groups

$$\operatorname{Im}^r D^{s,*} = \operatorname{im}(i^r : D^{s+r,*} \to D^{s,*})$$

must become constant for r large enough. It follows that $\varprojlim_r{}^1 \{\operatorname{Im}^r D^{s,*}, \subset\} = \{0\}$ because the sequence of inclusions stabilizes to equalities. Furthermore, since $Z_r^{s,*} = k^{-1}(\operatorname{Im}^{r-1} D^{s+1,*})$, we get a constant sequence after some finite index r, $Z_r^{s,*} = Z_{r+1}^{s,*} = Z_{r+2}^{s,*} = \cdots$ and it follows that $RE_\infty^{s,*} = \varprojlim_r{}^1 Z_r^{s,*} = \{0\}$. Putting these data into the exact sequence of the proof of Theorem 3.19, we get a short exact sequence

$$0 \to F^s D^{-\infty,*} / F^{s+1} D^{-\infty,*} \to E_\infty^{s,*} \to R^{s+1} \overset{i}{\to} R^s \to 0.$$

(Recall that $R^s = \varprojlim_r \operatorname{Im}^r D^s = \bigcap_s \operatorname{Im}^r D^s$.)

A generalization of these phenomena for R-modules is implied by the following condition (introduced by [Dieudonné-Grothendieck61]).

Definition 3.21.   *A sequence of homomorphisms* $\cdots \overset{i}{\to} D^{s+1,*} \overset{i}{\to} D^{s,*} \overset{i}{\to} D^{s-1,*} \overset{i}{\to} \cdots$ *satisfies the* **Mittag-Leffler condition** *if, for every* s, *there is an integer* $\gamma(s)$ *such that, whenever* $t \geq \gamma(s)$, *we have*

$$\operatorname{im}(D^{s+t,*} \to D^{s,*}) = \operatorname{im}(D^{s+\gamma(s),*} \to D^{s,*}).$$

We can define the Mittag-Leffler condition more generally and allow $\gamma(s)$ to depend on the codegree as well, that is, for every (s, t) there is an integer $\gamma(s, t)$ such that, whenever $n \geq \gamma(s, t)$, we have

$$\mathrm{im}(D^{s+n,t} \to D^{s,t}) = \mathrm{im}(D^{s+\gamma(s,t),t} \to D^{s,t}).$$

For ease of discussion we suppress the codegree.

The same argument for the case of finite abelian groups proves the following result.

Proposition 3.22. *Given an exact couple* $\{D^{s,*}, E^{s,*}, i, j, k\}$, *if the sequence of morphisms* $\cdots \xrightarrow{i} D^{s+1,*} \xrightarrow{i} D^{s,*} \xrightarrow{i} D^{s-1,*} \xrightarrow{i} \cdots$ *associated to the exact couple satisfies the Mittag-Leffler condition, then there is an exact sequence*

$$0 \to F^s D^{-\infty,*} / F^{s+1} D^{-\infty,*} \to E_\infty^{s,*} \to R^{s+1} \xrightarrow{i} R^s \to 0.$$

Corollary 3.23. *The Mittag-Leffler condition for a sequence* $\cdots \xrightarrow{i} D^{s,*} \xrightarrow{i} D^{s-1,*} \xrightarrow{i} \cdots$ *implies* $\varprojlim_s{}^1 D^{s,*} = \{0\}$.

PROOF: Following [Boardman99] we introduce the bigraded R-module

$$I^{s,t} = \begin{cases} \mathrm{im}(D^{t,*} \to D^{s,*}) = \mathrm{Im}^{t-s} D^s, & \text{for } t > s, \\ D^s, & \text{for } t \leq s. \end{cases}$$

Whenever $s \geq u$ and $t \geq v$, there is a homomorphism $I^{s,t} \to I^{u,v}$ determined by the commutative square

$$\begin{array}{ccc}
D^{t,*} & \xrightarrow{i^{t-s}} & D^{s,*} \\
{\scriptstyle i^{t-u}}\downarrow & & \downarrow{\scriptstyle i^{s-v}} \\
D^{u,*} & \xrightarrow{i^{u-v}} & D^{v,*}.
\end{array}$$

Fixing t, we get a sequence of homomorphisms:

$$\cdots D^{t+n,*} \xrightarrow{i} D^{t+n-1,*} \xrightarrow{i} \cdots \xrightarrow{i} D^{t,*} \to I^{t-1,t} \to I^{t-2,t} \to \cdots.$$

It follows that $\varprojlim_s I^{s,t} = \varprojlim_s D^{s,*}$ and $\varprojlim_s{}^1 I^{s,t} = \varprojlim_s{}^1 D^{s,*}$. To compute the inverse limit and derived functor we consider the exact sequence:

$$(3.24) \qquad 0 \to \varprojlim_s I^{s,t} \to \prod_s I^{s,t} \xrightarrow{\Phi} \prod_s I^{s,t} \to \varprojlim_s{}^1 I^{s,t} \to 0.$$

Since the first term in this exact sequence is independent of t, it follows that $\lim_{\leftarrow t}{}^1 \lim_{\leftarrow s} I^{s,t} = \{0\}$. We now apply two facts ([Eilenberg-Moore62]) left to the reader to prove:

(1) $\lim_{\leftarrow t} \prod_s I^{s,t} = \prod_s \lim_{\leftarrow t} I^{s,t}$ and $\lim_{\leftarrow t}{}^1 \prod_s I^{s,t} = \prod_s \lim_{\leftarrow t}{}^1 I^{s,t}$.

(2) If $0 \to \{A^s\} \to \{B^s\} \to \{C^s\} \to \{D^s\} \to 0$ is an exact sequence of sequences of morphisms with $\lim_{\leftarrow s}{}^1 A^s = \{0\}$, then there is an exact sequence

$$0 \to \lim_{\leftarrow s} A^s \to \lim_{\leftarrow s} B^s \to \lim_{\leftarrow s} C^s \to \lim_{\leftarrow s} D^s$$
$$\to \lim_{\leftarrow s}{}^1 B^s \to \lim_{\leftarrow s}{}^1 C^s \to \lim_{\leftarrow s}{}^1 D^s \to 0.$$

Apply the functor $\lim_{\leftarrow t}$ to the four-term exact sequence (3.24) and we get an exact sequence from facts (1) and (2),

$$0 \to \lim_{\leftarrow s} D^s \to \prod_s \lim_{\leftarrow t} I^{s,t} \to \prod_s \lim_{\leftarrow t} I^{s,t} \to \lim_{\leftarrow s}{}^1 D^s$$
$$\to \prod_s \lim_{\leftarrow t}{}^1 I^{s,t} \to \prod_s \lim_{\leftarrow t}{}^1 I^{s,t} \to 0.$$

Fixing s, the Mittag-Leffler condition gives the sequence of morphisms,

$$\cdots \xrightarrow{=} \operatorname{Im}^{\gamma(s)+n} D^{s,*} \xrightarrow{=} \operatorname{Im}^{\gamma(s)+n-1} D^{s,*} \xrightarrow{=} \cdots$$
$$\xrightarrow{=} \operatorname{Im}^{\gamma(s)} D^{s,*} \to \cdots \to \operatorname{Im}^r D^{s,*} \to \cdots$$

from which we conclude $\lim_{\leftarrow t} I^{s,t} = \operatorname{Im}^{\gamma(s)} D^{s,*} = R^s$ and $\lim_{\leftarrow t}{}^1 I^{s,t} = \{0\}$. This gives the exact sequence

$$0 \to \lim_{\leftarrow s} D^{s,*} \to \prod_s R^s \xrightarrow{\Phi} \prod_s R^s \to \lim_{\leftarrow s}{}^1 D^s \to 0.$$

Since $R^{s+1} \xrightarrow{i} R^s$ is an epimorphism for all s, coker $\Phi = \{0\}$ and so $\lim_{\leftarrow s}{}^1 D^{s,*} = \{0\}$. $\square$

Corollary 3.25. *Suppose $\{D^{s,*}, E^{s,*}, i, j, k\}$ is an exact couple such that the sequence of morphisms $\cdots \to D^{s,*} \xrightarrow{i} D^{s+1,*} \xrightarrow{i} \cdots$ satisfies the Mittag-Leffler condition, then the topology induced by the filtration on $\lim_{\to s} D^{s,*} = D^{-\infty,*}$ is complete.*

PROOF: The filtration on $D^{-\infty,*}$ is given by $F^s D^{-\infty,*} = \operatorname{im}(D^{s,*} \xrightarrow{p_s} D^{-\infty,*})$. This leads to the short exact sequence $0 \to \ker p_s \to D^{s,*} \to \operatorname{im} p_s \to 0$. Applying the inverse limit functor gives the exact sequence that ends with $\lim_{\leftarrow s}{}^1 D^{s,*} \to \lim{}^1 \operatorname{im} p_s \to 0$ and, since $\lim_{\leftarrow s}{}^1 D^{s,*} = \{0\}$, we have that $\lim_{\leftarrow s}{}^1 \operatorname{im} p_s = \lim_{\leftarrow s}{}^1 F^s D^{-\infty,*} = \{0\}$ and so the filtration is complete. $\square$

Corollary 3.25 tells us that the only thing standing in our way from identifying the target of strong convergence of a spectral sequence in the presence of the Mittag-Leffler condition is $\lim_{\leftarrow s} F^s D^{-\infty,*}$, that is, whether or not the filtration on the direct limit of $\{D^{s,*}, i\}$ is Hausdorff. Like the point-set interpretation of the Hausdorff condition, it determines if the spectral sequence converges to a *unique* R-module or not. Notice that $\lim_{\leftarrow s} F^s D^{-\infty,*}$ is also the kernel of the completion homomorphism when $\lim_{\leftarrow s}{}^1 F^s D^{-\infty,*}$ vanishes.

Recent studies of the convergence of spectral sequences have concentrated on the spectral sequence associated to a tower of fibrations introduced by [Bousfield-Kan72]. The associated towers of groups fit into the theory of pro-groups and pro-morphisms that emerged from the study of étale homotopy theory ([Artin-Mazur69], [Quillen69']). The basic reference for this theory as applied to homotopy theory is the book of [Bousfield-Kan72]. General convergence results for the homology spectral sequence associated to such towers are due to [Dwyer75'], [Bousfield87] and to [Shipley96].

3.3 Zeeman's comparison theorem

The next theorem shows how limited information can be turned into global information about a spectral sequence. Though the hypotheses seem specific in their algebraic formulation, the geometric interpretation of these results in Chapter 5 carries great generality. This theorem has become a standard tool in the application of spectral sequences.

Theorem 3.26 (Zeeman's comparison theorem). *Suppose $\{E_r^{*,*}, d_r\}$ and $\{\bar{E}_r^{*,*}, \bar{d}_r\}$ are first quadrant spectral sequences of cohomological type (that is, d_r and $\bar{d}_r$ of bidegree $(r, 1-r)$). Suppose $\{f_r\}$ is a morphism of spectral sequences and the following diagram commutes with rows exact:*

$$
\begin{array}{ccccccccc}
0 & \longrightarrow & E_2^{p,0} \otimes E_2^{0,q} & \longrightarrow & E_2^{p,q} & \longrightarrow & \mathrm{Tor}_1^R(E_2^{p+1,0}, E_2^{0,q}) & \longrightarrow & 0 \\
 & & \downarrow{\scriptstyle f_2 \otimes f_2} & & \downarrow{\scriptstyle f_2} & & \downarrow{\scriptstyle \mathrm{Tor}(f_2,f_2)} & & \\
0 & \longrightarrow & \bar{E}_2^{p,0} \otimes \bar{E}_2^{0,q} & \longrightarrow & \bar{E}_2^{p,q} & \longrightarrow & \mathrm{Tor}_1^R(\bar{E}_2^{p+1,0}, \bar{E}_2^{0,q}) & \longrightarrow & 0
\end{array}
$$

Then any two of the following conditions imply the third:

I. $f_2 \colon E_2^{p,0} \to \bar{E}_2^{p,0}$ *is an isomorphism for all p.*
II. $f_2 \colon E_2^{0,q} \to \bar{E}_2^{0,q}$ *is an isomorphism for all q.*
III. $f_\infty \colon E_\infty^{p,q} \to \bar{E}_\infty^{p,q}$ *is an isomorphism for all p, q.*

PROOF: That I and II imply III follows from the functoriality of Tor_1^R from which it follows that $\mathrm{Tor}(f_1, f_2)$ is an isomorphism and so, by the Five-Lemma, $f_2 \colon E_2^{p,q} \to \bar{E}_2^{p,q}$ is an isomorphism for all p,q. Then apply Theorem 3.4.

We establish that I and III imply II by sneaking up on II by induction. Let II_k designate the statement:

$$II_k. \quad f_2 \colon E_2^{0,q} \to \bar{E}_2^{0,q} \text{ is an isomorphism when } 0 \le q \le k.$$

Since $E_2^{0,0} = E_\infty^{0,0}$, III establishes II_0; we assume I, III and II_k. We first prove some facts.

Fact 1. $f_r \colon E_r^{p,q} \to \bar{E}_r^{p,q}$ is an epimorphism when $q \le k$ and an isomorphism when $q \le k - r + 2$.

Observe that Fact 1 holds when $r = 2$ by the Five-lemma, I and II_k. Suppose it is true for $r = m$. Consider the following diagram with the rows exact:

$$
\begin{array}{ccccccc}
0 & \longrightarrow & (\ker d_m)^{p,q} & \longrightarrow & E_m^{p,q} & \xrightarrow{d_m} & E_m^{p+m,q-m+1} \\
& & \downarrow{\tilde{f}_m} & & \downarrow{f_m} & & \downarrow{f_m} \\
0 & \longrightarrow & (\ker \bar{d}_m)^{p,q} & \longrightarrow & \bar{E}_m^{p,q} & \xrightarrow{\bar{d}_m} & \bar{E}_m^{p+m,q-m+1}
\end{array}
$$

If $q \le k$, then $q - m + 1 \le k - m + 1 < k - m + 2$ and so the rightmost morphism f_m is an isomorphism. If $q \le k$, then the preceding f_m is an epimorphism. Since f_m commutes with d_m and $\bar{d}_m$, it induces a homomorphism $\tilde{f}_m \colon \ker d_m \to \ker \bar{d}_m$.

Let $\bar{u} \in (\ker \bar{d}_m)^{p,q}$. This goes by inclusion to $\bar{u} \in \bar{E}_m^{p,q}$. Since f_m is onto, there is an element $u \in E_m^{p,q}$ with $f_m(u) = \bar{u}$. Since f_m is an isomorphism on the right, we have

$$d_m(u) = f_m^{-1} \circ \bar{d}_m \circ f_m(u) = f_m^{-1} \circ \bar{d}_m(\bar{u}) = 0,$$

so u is in $\ker d_m$ and $\tilde{f}_m(u) = \bar{u}$. Thus $\tilde{f}_m$ is onto. If $q \le k - m + 2$, then both morphisms f_m are isomorphisms. If $u, v \in (\ker d_m)^{p,q}$ satisfy $\tilde{f}_m(u - v) = 0$, then $f_m(u - v) = 0$ and so $u = v$ in $E_m^{p,q}$. Since $(\ker d_m)^{p,q}$ injects into $E_m^{p,q}$, $u = v$ in $(\ker d_m)^{p,q}$ and so $\tilde{f}_m$ is an isomorphism.

Consider the dual diagram:

$$
\begin{array}{ccccccc}
E_m^{p-m,q+m-1} & \xrightarrow{d_m} & E_m^{p,q} & \xrightarrow{Q} & E_m^{p,q}/\operatorname{im} d_m & \longrightarrow & 0 \\
\downarrow{f_m} & & \downarrow{f_m} & & \downarrow{\hat{f}_m} & & \\
\bar{E}_m^{p-m,q+m-1} & \xrightarrow{\bar{d}_m} & \bar{E}_m^{p,q} & \xrightarrow{\bar{Q}} & \bar{E}_m^{p,q}/\operatorname{im} \bar{d}_m & \longrightarrow & 0.
\end{array}
$$

Once again, since f_m commutes with the differentials, the homomorphism $\hat{f}_m \colon E_m^{p,q}/\operatorname{im} d_m \to \bar{E}_m^{p,q}/\operatorname{im} \bar{d}_m$ is induced by f_m.

Let $\bar{U} \in \bar{E}_m^{p,q}/\operatorname{im} \bar{d}_m$. Then there is an element $\bar{U}'$ with $Q(\bar{U}') = \bar{U}$ and $\bar{U}' \in \bar{E}_m^{p,q}$. When $q \leq k$, there is an element U' with $f_m(U') = \bar{U}'$, since f_m is an epimorphism. Let $U = Q(U')$. Then $\hat{f}_m(U) = \bar{U}$ and we see that $\hat{f}_m$ is onto. Suppose $q \leq k - m + 1$. Then $q + m - 1 \leq k$ and the leftmost f_m is an epimorphism, the next f_m an isomorphism. Suppose U, $V \in E_m^{p,q}/\operatorname{im} d_m$ satisfy $\hat{f}_m(U - V) = 0$. Suppose $U = Q(U')$, $V = Q(V')$. Then, $\bar{Q} \circ f_m(U' - V') = \hat{f}_m(U - V) = 0$ and so there is an element $\bar{W} \in \bar{E}_m^{p-m,q+m-1}$ with $\bar{d}_m(\bar{W}) = f_m(U' - V')$. Since $\bar{W} = f_m(W)$ for some $W \in E_m^{p-m,q+m-1}$, we have $d_m(W) = U' - V'$ and

$$U - V = Q(U' - V') = Q \circ d_m(W) = 0.$$

Thus $\hat{f}_m$ is an isomorphism.

Next consider the diagram

$$
\begin{array}{ccccccccc}
0 & \longrightarrow & (\operatorname{im} d_m)^{p,q} & \longrightarrow & E_m^{p,q} & \longrightarrow & E_m^{p,q}/\operatorname{im} d_m & \longrightarrow & 0 \\
& & \downarrow{f_m|} & & \downarrow{f_m} & & \downarrow{\hat{f}_m} & & \\
0 & \longrightarrow & (\operatorname{im} \bar{d}_m)^{p,q} & \longrightarrow & \bar{E}_m^{p,q} & \longrightarrow & \bar{E}_m^{p,q}/\operatorname{im} \bar{d}_m & \longrightarrow & 0.
\end{array}
$$

When $q \leq k$, f_m is an epimorphism, and we can conclude $f_m|$ is an epimorphism. When $q \leq k - m + 1$, both f_m and $\hat{f}_m$ are isomorphisms and so the argument for $\hat{f}_m$ works to show that $f_m|$ is an isomorphism.

Finally, let us consider the diagram:

$$
\begin{array}{ccccccccc}
0 & \longrightarrow & (\operatorname{im} d_m)^{p,q} & \longrightarrow & (\ker d_m)^{p,q} & \longrightarrow & E_{m+1}^{p,q} & \longrightarrow & 0 \\
& & \downarrow{f_m|} & & \downarrow{\tilde{f}_m} & & \downarrow{f_{m+1}} & & \\
0 & \longrightarrow & (\operatorname{im} \bar{d}_m)^{p,q} & \longrightarrow & (\ker \bar{d}_m)^{p,q} & \longrightarrow & \bar{E}_{m+1}^{p,q} & \longrightarrow & 0.
\end{array}
$$

If $q \leq k$, then $\tilde{f}_m$ is an epimorphism and hence so is f_{m+1}. If $q \leq k - m + 1$ then $f_m|$ and $\tilde{f}_m$ are both isomorphisms and so then is f_{m+1}. Since $k - m + 1 = k - (m + 1) + 2$, we have established Fact 1.

Fact 2. *When r is large enough, $E_{r+1}^{0,k+1} = E_\infty^{0,k+1}$ and so $f_{r+1}^{0,k+1} = f_\infty^{0,k+1}$ is an isomorphism.*

This follows because we have a first quadrant spectral sequence. From Fact 2 we argue that $f_r^{0,k+1}$ is an isomorphism and, descending to $r = 2$, establish II_{k+1}. By its placement in the first quadrant, $E_{r+1}^{0,k+1} = (\ker d_r)^{0,k+1}$. Consider the diagram:

$$
\begin{array}{ccccccccccc}
0 & \longrightarrow & (\ker d_r)^{0,k+1} & \longrightarrow & E_r^{0,k+1} & \xrightarrow{d_r} & E_r^{r,k-r+2} & \xrightarrow{Q} & E_r^{r,k-r+2}/\operatorname{im} d_r & \longrightarrow & 0 \\
& & \downarrow{\tilde{f}_r} & & \downarrow{f_r} & & \downarrow{f_r} & & \downarrow{\hat{f}_r} & & \\
0 & \longrightarrow & (\ker \bar{d}_r)^{0,k+1} & \longrightarrow & \bar{E}_r^{0,k+1} & \xrightarrow{\bar{d}_r} & \bar{E}_r^{r,k-r+2} & \xrightarrow{\bar{Q}} & \bar{E}_r^{r,k-r+2}/\operatorname{im} \bar{d}_r & \longrightarrow & 0.
\end{array}
$$

We have $\tilde{f}_r = f_{r+1}^{0,k+1} = f_\infty^{0,k+1}$, an isomorphism. By Fact 1, $\hat{f}_r$ is an isomorphism and so is the rightmost f_r.

If $\bar{u} \in \bar{E}_r^{0,k+1}$, then $\bar{u} = f_r(u) = \tilde{f}_r(u)$ when $d_r(\bar{u}) = 0$. Suppose $d_r(\bar{u}) = \bar{v} \in \bar{E}_r^{r,k-r+2}$, $\bar{v} \neq 0$. Then $\bar{v} = f_r(v)$ since f_r is an isomorphism. But $Q(v) = \hat{f}_r^{-1} \circ \bar{Q}(\bar{v}) = 0$ and so $v = d_r(u)$ for some $u \in E_r^{0,k+1}$. Thus $f_r \colon E_r^{0,k+1} \to \bar{E}_r^{0,k+1}$ is onto.

When $u, v \in E_r^{0,k+1}$ satisfy $f_r(u - v) = 0$, then

$$d_r(u - v) = f_r^{-1} \circ \bar{d}_r \circ f_r(u - v) = 0$$

so $u - v \in (\ker d_r)^{0,k+1}$. Now $f_r(u - v) = \tilde{f}_r(u - v) = 0$ implies $u = v$ and so $f_r \colon E_r^{0,k+1} \to \bar{E}_r^{0,k+1}$ is an isomorphism and II_{k+1} holds.

The proof that II and III imply I is similar. $\qquad\square$

We next give an application of Zeeman's theorem that generalizes the examples in §1.4. We remark once again, though the next theorem seems to be a quite specialized piece of algebra, the applications of it are spectacular.

Let $R = k$, a field, and suppose that the following hypotheses hold:

Hypotheses.

(1) $\{E_r^{*,*}, d_r\}$ *is a first quadrant spectral sequence of algebras with* $E_2^{*,*} \cong V^* \otimes_k W^*$, *that is,* $V^* \cong E_2^{*,0}$ *and* $W^* \cong E_2^{0,*}$, *both as algebras.*

(2) E_∞ *is the bigraded* k-*vector space* k, *that is,* $E_\infty^{0,0} \cong k$ *and* $E_\infty^{p,q} = \{0\}$ *for* $(p, q) \neq (0, 0)$.

(3) $W^* = \Lambda(x_1, x_2, \dots, x_m)$, *an exterior algebra on generators* x_1, x_2, $\dots, x_m$ *where each* x_i *is homogeneous of odd degree* $2r_i - 1$.

(4) *For each* i, $d_j(x_i) = 0$ *for* $2 \leq j < 2r_i$ *and* $d_{2r_i}(x_i) \neq 0$. *(Anticipating terminology, we say that* x_i *is* **transgressive**; *see* §6.1.)

(5) $\mathrm{char}\, k \neq 2$ *(a technical detail regarding exterior algebras).*

Theorem 3.27 ([Borel53])*. Under the hypotheses (1)–(5),* V^* *is isomorphic to* $k[y_1, y_2, \dots, y_m]$ *where* $y_i = d_{2r_i}(x_i)$.

PROOF ([Zeeman58]): Since we are working over a field, we have a definition of tensor products of spectral sequences. It is easy to establish the following:

a) if E and $\bar{E}$ have E_∞-terms isomorphic to k, so does $E \otimes \bar{E}$;

b) if $E^{*,*} \cong V^* \otimes W^*$ and $\bar{E}^{*,*} \cong \bar{V}^* \otimes \bar{W}^*$ then $(E \otimes \bar{E})^{*,*} \cong (V \otimes \bar{V}^*) \otimes (W \otimes \bar{W}^*)$:

c) an associative product for E induces a morphism of spectral sequences, for each n, $\underbrace{E \otimes E \otimes \cdots \otimes E}_{n \text{ copies}} \to E$.

Let s be an odd integer. We introduce $E(s)$, the **elementary spectral sequence** over k of degree s, given by $E(s)_2 = \Lambda(u_s) \otimes k[v_{s+1}]$ with $\deg u_s = s$, $\deg v_{s+1} = s+1$, $d_j = 0$ for $j \neq s+1$, and $d_{s+1}(u_s) = v_{s+1}$. Because the action of the differential d_{s+1} is determined by the Leibniz rule, $E(s)_\infty \cong k$, with the pattern of d_{s+1} given in the diagram.

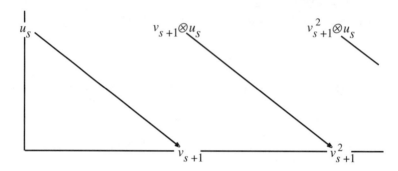

By hypothesis 4 we can construct a unique mapping, for each i, $1 \leq i \leq m$, $f_i \colon E(2r_i - 1)_2 \to E_2$, which extends to a morphism of spectral sequences, by $f_i(u_{2r_i-1}) = x_i$ and $f_i(y_{2r_i}) = d_{2r_i}(x_i)$. Since $E_2^{*,*} \cong V^* \otimes_k W^*$, everything else about x_i is determined by $d_{2r_i}(x_i) \neq 0$. Collecting these morphisms for $i = 1, 2, \ldots, m$, we obtain the composite

$$F \colon E(2r_1 - 1) \otimes E(2r_2 - 1) \otimes \cdots \otimes E(2r_m - 1) \xrightarrow{f_1 \otimes f_2 \otimes \cdots \otimes f_m}$$

$$E \otimes E \otimes \cdots \otimes E \xrightarrow{\text{multiplication}} E,$$

a morphism of spectral sequences. By definition

$$E(2r_1 - 1) \otimes \cdots \otimes E(2r_m - 1) \cong \Lambda(u_{2r_1 - 1}, \ldots, u_{2r_m - 1}) \otimes k[v_{2r_1}, \ldots, v_{2r_m}]$$

and F, at the E_2-term, maps $\Lambda(u_{2r_1 - 1}, \ldots, u_{2r_m - 1})$ isomorphically to $W^* = \Lambda(x_1, \ldots, x_m)$; thus we conclude that the tensor product of elementary spectral sequences satisfies part I of Theorem 3.26. Furthermore, since each spectral sequence has E_∞-term, k, F induces an isomorphism of the E_∞-terms and part III of Zeeman's comparison theorem is satisfied. Thus, F induces an isomorphism $k[v_{2r_1}, \ldots, v_{2r_m}] \cong V^*$ and the theorem is proved. $\square$

For a discussion of the case of char $k = 2$ see the paper of [Zeeman58]. The technique in this proof of comparing an algebraically constructed spectral sequence with an already existing one is standard and has wide applications. Important applications of Theorem 3.27 appear in the work of [Borel53] on the cohomology of Lie groups and their classifying spaces (see Chapters 5 and 6).

Exercises

3.1. Suppose (A, F) is a filtered module over R so that F is bounded and each $E_0^p(A, F)$ in the associated graded module is projective over R. Show that A is determined, up to isomorphism, by these conditions.

3.2. Show that a morphism of filtered differential graded modules leads to a morphism of the associated spectral sequences. Show that a morphism of exact couples determines a morphism of the associated spectral sequences.

3.3. Suppose (M, F) is a filtered module over R. Topologize M by taking all cosets of all submodules $F^p M$ as basic open sets. Show that the filtration is Hausdorff if and only if M is Hausdorff in this topology. Show that the filtration is strongly convergent if and only if every Cauchy sequence in M with this topology converges in M. This completes the proof of Lemma 3.10.

3.4. Suppose that $\{E_r^{*,*}, d_r\}$ is a first-quadrant spectral sequence, beginning with the E_1-page. Suppose further that the differential graded module, $(\text{total}(E_1^{*,*}), d_1)$ is filtered and d_1 respects this filtration. Show that there is a spectral sequence, derived from the filtration of $\text{total}(E_1^{*,*})$, converging to $E_2^{*,*}$ of the original spectral sequence. Show further that this new spectral sequence inherits a third grading from the bigrading of $E_1^{*,*}$. This spectral sequence is sometimes called the **algebraic May spectral sequence** ([May64]).

3.5. Prove Theorems 3.2 and 3.4.

3.6. Suppose $\{E_r^{*,*}, d_r\}$ is a first quadrant spectral sequence over a field, k, and E_r converges to (H^*, F). Let V be any graded vector space over k and filter $H^* \oplus V$ by $\hat{F}^p(H^* \oplus V) = F^p H^* \oplus V$. Show that E_r converges to $(H^* \oplus V, \hat{F})$. Suppose (H^*, F) is such that F is strongly convergent. Show that $(H^* \oplus V, \hat{F})$ is not strongly convergent. This shows that strongly convergent implies convergent but not vice versa.

3.7. Show that the definition of the inverse limit and direct limit for sequences of modules over a ring R satisfy the defining universal properties. Show that the direct limit functor is exact (Lemma 3.7). Finally, show that a sequence of epimorphisms of R-modules with vanishing inverse limit is a sequence of trivial modules.

3.8. Let f and g be morphisms of filtered differential graded modules,

$$f, g \colon (A, d, F) \to (\bar{A}, \bar{d}, \bar{F})$$

and suppose f is chain homotopic to g via $s \colon A \to \bar{A}$ of degree -1, that is, $ds + sd = f - g$. We say that s is a **homotopy of order** k if $s(F^p A) \subset \bar{F}^{p-k} \bar{A}$. Prove that, if a chain homotopy of order k exists between f and g, then for $r > k$, the induced maps on the spectral sequences satisfy $f_r = g_r$ and $f_\infty = g_\infty$.

3.9. A **Rees system** ([Eckmann-Hilton66]) or **spectral datum** ([Eilenberg-Moore]) is a diagram of modules over a ring R

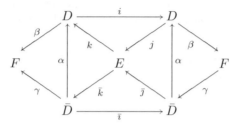

in which $\{D, E, i, j, k\}$ and $\{\bar{D}, E, \bar{i}, \bar{j}, \bar{k}\}$ are exact couples, all triangles commute and the sequence

$$\bar{D} \xrightarrow{\alpha} D \xrightarrow{\beta} F \xrightarrow{\gamma} \bar{D} \xrightarrow{\alpha}$$

is exact. Show that the derived couples of a Rees system give another Rees system in which F remains the same. Show that if two exact couples are part of a Rees system, then their associated spectral sequences are identical. The motivating example is given by (A, d, F), a filtered differential graded module: Derive the Rees system:

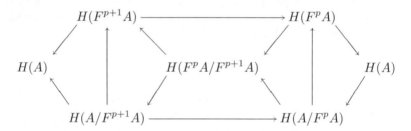

3.10. Suppose (M, F) is a filtered module with N a submodule that satisfies $N \subset F^s M$ for all s. Show that

$$\varprojlim_s \left(F^s M / N \right) \cong \left(\varprojlim_s F^s M \right) / N \quad \text{and} \quad \varprojlim_s {}^1 F^s M \cong \varprojlim_s {}^1 \left(F^s M / N \right)$$

3.11. Prove the assertions (1) and (2) of [Eilenberg-Moore62] used in the proof of Corollary 3.23.

3.12. Prove the assertions a), b), and c) about tensor products of spectral sequences made in the proof of Theorem 3.27.

3.13. Let $\{M^{*,*}, d', d''\}$ denote a double complex where there is no restriction on the bidegrees. Apply the notions of convergence of Chapter 3 to find conditions under which the associated (planar) spectral sequence converges (weakly, strongly, conditionally) to $H(\text{total}(M^{*,*}, D))$.

Part II
Topology

4

Topological Background

"Of course, one has to face the question, what is the good category of spaces in which to do homotopy theory?"

J. F. Adams

The classification of topological spaces up to homotopy equivalence is the central problem of algebraic topology. The method of attack on this problem is the application of various functors from the category of topological spaces (or a suitable subcategory) to certain algebraic categories. These functors do not distinguish between homotopy equivalent spaces, and the algebraic data they provide may be enough to distinguish nonequivalent spaces. In order to make the problem of classification more reasonable, it is necessary to identify a tractable category of spaces. A candidate for such a category should be large enough to contain all 'important' spaces (such as finite-dimensional manifolds) as well as contain the results of various constructions applied to these spaces (for example, it should be closed under suspension, loops, etc.). Furthermore, the classical homotopy functors, singular homology and cohomology and the homotopy groups, should be effective in distinguishing spaces in this category.

In this chapter, we present two categories of spaces that satisfy the desiderata of homotopy theory. In §4.1, CW-complexes are defined. The homotopy groups of CW-complexes are effective enough for a classification scheme, and the combinatorial structure allows the computation of their singular homology groups. In §4.2, the equivalent category of simplicial sets and mappings is presented. This category features a more rigid combinatorial structure in which the classical homotopy invariants can be defined and studied. Furthermore, the general notion of a simplicial object over an arbitrary category leads to many useful constructions in homotopy theory. Both a CW-complex and a simplicial set give rise to filtered spaces and so the algebraic technology developed in Chapters 1, 2, and 3 may be applied to the computation of the homotopy invariants of such spaces.

Another way in which the homotopy invariants of a space may be restricted enough to allow computation (or at least the construction of a spectral sequence) is to place the space in a system of spaces and maps with good exactness properties with respect to homotopy functors. Exactness properties can be translated into the problems of extension and lifting. Mappings with good extension properties are called cofibrations and they lead to sequences of mappings that give exact sequences when the functor $[-, Z]$, homotopy classes of maps to some space Z, is applied.

The dual notion of good lifting properties is enjoyed by fibrations (§4.3). Exact sequences result when the functor $[Z, -]$ is applied to a fibration and, in particular, if $Z = S^n$, these exact sequences consist of homotopy groups. By introducing a space 'dual' to an n-cell (with which one builds a CW-complex), one can construct a tower of fibrations that approximates a space one homotopy group at a time. The cell decomposition of a space (reflected in its homology groups) and its Postnikov tower (which carries the homotopy groups) are sources of homotopy invariants and so figure in the constructions found later in the book.

We finish this chapter (§4.4) with a discussion of some of the further structure that is available on singular homology and cohomology. With coefficients in a field, the dual of the cup product structure for cohomology is a coalgebra structure in homology. In the presence of a multiplication on a space (an H-space structure) such a coalgebra structure arises on cohomology and is intertwined with the cup product structure. For an H-space of finite type, the homology and cohomology are dual Hopf algebras, and the structure theory of Hopf algebras may be applied. The algebraic aspects of Hopf algebras play an important role in later chapters.

The cohomology of a space with mod p coefficients enjoys another significant piece of structure; $H^*(X; \mathbb{F}_p)$ is an algebra over the mod p Steenrod algebra, $\mathcal{A}_p$, of stable cohomology operations. The applications of this structure have led to some of the deepest results in homotopy theory. We record some of the basic properties of the Steenrod algebra in §4.4.

This chapter plays two roles in our exposition. First, it supports the constructions found in later chapters and so affords the reader a ready reference for topological facts that would take the reader too far afield to develop in the places where they are needed. Secondly, it provides the novice a thumbnail sketch of some of the fundamental constructions in homotopy theory. In both of these capacities, it is distracting to overburden the reader with details. Therefore, we omit most proofs and give copious references for the reader who needs details. The reader who is acquainted with these standard notions should skip to the next chapter and refer back when necessary.

4.1 CW-complexes

Let's begin by building a topological space. For building blocks, we take the n-**cell**, $e^n = \{\vec{x} \in \mathbb{R}^n \mid \|\vec{x}\| \leq 1\}$. Note that the boundary of the n-cell, ∂e^n, is the $(n - 1)$-sphere, S^{n-1}. We proceed a dimension at a time.

In dimension 0, let $X^{(0)}$ be a finite set of 0-cells (points). We next take a collection of mappings $f_{1j} \colon S^0 = \partial e^1 \to X^{(0)}$, $j = 1, \ldots, m$, called the **attaching maps**, and form the adjunction spaces, $X_1^{(0)} = X^{(0)} \cup_{f_{11}} e^1$,

$X_2^{(0)} = X_1^{(0)} \cup_{f_{12}} e^1$, and so on to $X^{(1)} = X_{m-1}^{(0)} \cup_{f_{1m}} e^1$. Recall that a mapping $g \colon S^n \to Y$ gives rise to the space $Y \cup_g e^{n+1}$ determined by the quotient space

$$Y \bigsqcup e^{n+1} \Big/ \{x \sim g(x) \text{ for } x \in S^n = \partial e^{n+1}\},$$

where $\bigsqcup$ denotes the disjoint union.

In dimension n, assume $X^{(n-1)}$ has been built and take a collection of mappings, $f_{nj} \colon S^{n-1} = \partial e^n \to X^{(n-1)}$, for $j = 1, \ldots, r$. Form the adjunction spaces, $X_1^{(n-1)} = X^{(n-1)} \cup_{f_{n1}} e^n$, $X_2^{(n-1)} = X_1^{(n-1)} \cup_{f_{n2}} e^n$, and so on, up to $X^{(n)} = X_{r-1}^{(n-1)} \cup_{f_{nr}} e^n$. If we continue to add higher dimensional cells, the space X that results in the direct limit (union) of this process is called a **cell-complex**. This notion was introduced by [Whitehead, JHC49] as a natural generalization of polyhedra. Spaces that can be built in this fashion are characterized by the following point-set description:

Definition 4.1. *A* **CW-complex** *is a topological space* X, *together with a* **cell decomposition** $\{e_\alpha^n \mid \alpha \in A\}$, *so that, for each* n *and* α, $e_\alpha^n \subset X$, *and*

(1) *X is the disjoint union of the collection* $\{\text{int } e_\alpha^n\}$. *(Here* int *denotes the interior of the cell and* int $e^0 = e^0$.)

(2) *e_α^n is closed in X.*

(3) *Let* $X^{(n)} = \bigcup_{\alpha \in A, m \le n} e_\alpha^m$; *the space* $X^{(n)}$ *is called the* n-**skeleton** *of X. The successive skeleta filter X; that is, there is a sequence of subspaces,* $X^{(0)} \subset X^{(1)} \subset \cdots \subset X^{(n)} \subset \cdots X$.

(4) *For each* $\alpha \in A$, *there is a mapping,* $f_\alpha \colon (e^n, S^{n-1}) \to (e_\alpha^n, \partial e_\alpha^n)$, *such that* $f_\alpha| \colon$ int $e^n \to$ int e_α^n *is a homeomorphism;* f_α *is called the* **characteristic mapping** *for the n-cell,* e_α^n.

(5) $\mathrm{cls}(e_\alpha^n)$ *is contained in finitely many cells (we say X is* **closure finite***).*

(6) *$K \subset X$ is closed in X if and only if, for all α, $K \cap \mathrm{cls}(e_\alpha^n)$ is closed in* int(e_α^n). *This is called the* **weak topology** *on the cell complex X.*

The cell-complexes built by adding cells a dimension at a time are CW-complexes. Most familiar spaces in homotopy theory are CW-complexes. For example, S^n can be given the structure of a CW-complex in at least two ways; $S^n = e^0 \cup_c e^n$, where $c \colon S^{n-1} \to e^0$ is the constant map, and

$$S^n = (e^0 \cup e^0) \cup_{i_1} (e^1 \cup e^1) \cup_{i_2} \cdots \cup_{i_n} (e^n \cup e^n),$$

where $i_j \colon S^{j-1} \hookrightarrow S^j$ is the equator and the two j-cells are the upper and lower hemispheres. These decompositions appear in later constructions.

If X is a CW-complex, the **dimension** of X, is defined

$$\dim X = \sup\{n \colon e_\alpha^n \text{ is in the decomposition of } X\};$$

thus $\dim X \leq \infty$. We speak of a **finite CW-complex**, which is built from finitely many cells, of a **countable CW-complex**, using countably many cells, and of a **locally finite CW-complex**, which is built of finitely many cells in each dimension.

If X and Y are two CW-complexes, then a mapping $f \colon X \to Y$ is **cellular** if $f(X^{(n)}) \subset Y^{(n)}$ for all n. We denote by **CW**, the category of CW-complexes with cellular maps. A first step in establishing this class of spaces as a convenient category for homotopy theory is to show that most of the relevant constructions are functors from this category to itself. We begin by identifying subobjects. A subspace A of a CW-complex X is a **subcomplex** if A is a CW-complex made up of a subset of cells of X. A CW-complex X with a subcomplex A form a **CW-pair** (X, A). By taking the cells in $X - A$ together with an appropriate 0-cell for A, one can show that the quotient, X/A, is a CW-complex for a CW-pair (X, A) and furthermore, that $X \to X/A$, the canonical projection, is cellular (see the book of [Fritsch-Piccinini90, pp. 62-63]). An example of a CW-complex built via quotients is **real projective n-space**, $\mathbb{R}P^n$ that is obtained from S^n by identifying antipodal cells in the decomposition $S^n = (e^0 \cup e^0) \cup_{i_1} (e^1 \cup e^1) \cup_{i_2} \cdots \cup_{i_n} (e^n \cup e^n)$; thus $\mathbb{R}P^n = e^0 \cup_{q_1} e^1 \cup_{q_2} \cdots \cup_{q_n} e^n$.

For two CW-complexes, X and Y, there is an evident cell-decomposition for $X \times Y$ and, if the topology is correct, we have a CW-complex. This works for countable CW-complexes or when one factor is locally finite. A counterexample, due to [Dowker52], shows that it does not work in general: Let X denote the wedge of uncountably many copies of the unit interval with basepoint 0 and form $X \times X$.

From the product and quotient constructions we can form the **cone** on a CW-complex, $CX = X \times I/X \times \{0\}$, and the **suspension**, $\Sigma X = CX/X \times \{1\}$. Both constructions give CW-complexes.

For a given cell-decomposition of a CW-complex, it is evident that many continuous functions to and from the complex may not be cellular. As in the classical case of polyhedra, cellular maps are sufficient, up to homotopy.

Theorem 4.2 (the cellular approximation theorem). *Let X and Y be CW-complexes and $f \colon X \to Y$ a continuous function. Then there is a cellular map $g \colon X \to Y$ with f homotopic to g.*

A thorough proof of this theorem for the more general case of mappings of pairs can be found in the book of [Fritsch-Piccinini90, §2.4]. We introduce one more level of generality into the discussion and consider the collection of **spaces of the homotopy type of a CW-complex**. Let $\mathcal{W}$ denote the category of such spaces with continuous functions as morphisms. [Milnor59] studied the operation of forming function spaces out of spaces in $\mathcal{W}$.

Theorem 4.3. *Suppose X is a space of the homotopy type of a (countable) CW-complex and $x_0 \in X$. Suppose K is a compact metric space. Then the* *space*

$$\mathrm{map}(K, X) = \{f \colon K \to X \mid f \text{ is continuous}\},$$

endowed with the compact-open topology, is of the homotopy type of a (countable) CW-complex. If $k_0 \in K$, then $\mathrm{map}((K, k_0), (X, x_0))$, the subspace of pointed maps, is of the homotopy type of a (countable) CW-complex.

An immediate corollary of the theorem is that, if X is in $\mathcal{W}$, then

$$\Omega X = \{\lambda \colon [0, 1] \to X \mid \lambda \text{ is continuous and } \lambda(0) = x_0 = \lambda(1)\},$$

the **space of based loops on** X, is also in $\mathcal{W}$. Restricting to $\mathcal{W}_0$, the category of spaces of the homotopy type of a *countable* CW-complex with continuous functions, we see that $\mathcal{W}_0$ is closed under products, subobjects, quotients, suspension and loops. [Milnor63] proved that every separable manifold is an object in $\mathcal{W}_0$ by a Morse theory argument.

Having introduced the category $\mathcal{W}$ of spaces of the homotopy type of a CW-complex, we now turn to the homotopy-theoretic properties of these spaces.

Theorem 4.4. *Suppose X is a space and $f, g \colon S^{n-1} \to X$ are two continuous mappings. If f is homotopic to g, then $X \cup_f e^n$ is homotopy-equivalent to $X \cup_g e^n$.*

PROOF: Suppose $H \colon S^{n-1} \times I \to X$ is a homotopy from f to g. Define a mapping $k \colon X \cup_g e^n \to X \cup_f e^n$ as follows:

$$k(x) = x \text{ for } x \in X,$$
$$k(t\vec{u}) = \begin{cases} 2t\vec{u} & \text{for } 0 \le t \le 1/2, \ \vec{u} \in S^{n-1}, \\ H(\vec{u}, 2t-1) & \text{for } 1/2 \le t \le 1, \ \vec{u} \in S^{n-1}. \end{cases}$$

Reversing the homotopy, one can construct the inverse to k and so k is a homotopy equivalence. □

This theorem points out the role of the combinatorial structure of a CW-complex in the homotopy classification problem. In the construction of a finite CW-complex, the n-skeleton, $X^{(n)}$, is built from the $(n-1)$-skeleton by a series of adjunctions. Suppose we have $X^{(n-1)}$ and the choice of mappings, $f_{n1}, f_{n2}, \ldots, f_{nr} \colon S^{n-1} \to X^{(n-1)}$. The classes $[f_{n1}], \ldots, [f_{nr}] \in \pi_{n-1}(X^{(n-1)})$ determine $X^{(n)}$ up to homotopy. It follows that the computation of $\pi_{n-1}(X^{(n-1)})$, a skeleton at a time, is sufficient to solve the homotopy classification problem for $\mathcal{W}_0$. The difficulty of this computation is a recurring theme in later chapters.

The next theorem, proved by [Whitehead, JHC49], shows how effective the homotopy groups of a space are in distinguishing the space from others.

Theorem 4.5 (the Whitehead theorem). *Suppose* $f\colon X \to Y$ *is a mapping of path-connected spaces of the homotopy type of locally finite CW-complexes. If* $\pi_i(f)\colon \pi_i(X) \to \pi_i(Y)$ *is an isomorphism of graded groups for* $i < n$ *and an epimorphism for* $i = n$, *then* $H_i(f)\colon H_i(X) \to H_i(Y)$ *is an isomorphism for* $i < n$ *and an epimorphism for* $i = n$. *If* $\pi_*(f)$ *is an isomorphism of graded groups for all* i, *then* f *is a homotopy equivalence. If* X *and* Y *are simply-connected and* $H_i(f)\colon H_i(X) \to H_i(Y)$ *is an isomorphism for* $i < n$ *and an epimorphism for* $i = n$, *then the same holds for* $\pi_i(f)$. *If* X *and* Y *are simply-connected and* $H_*(f)$ *is an isomorphism of graded abelian groups, then* f *is a homotopy equivalence.*

For arbitrary spaces, such a statement is false. For example, compare a space X with its plus construction X^+ with respect to a perfect normal subgroup $N \triangleleft \pi_1(X, x_0)$. The associated mapping $X \to X^+$ induces an isomorphism of homology groups, but $\pi_1(X^+, x_0) \cong \pi_1(X, x_0)/N$. (See chapter 2 of the book by [Srinivas96] for the relevant definitions and properties.) We note, furthermore, that it does not follow from an abstract isomorphism of graded groups, $\pi_*(X) \cong \pi_*(Y)$ or $H_*(X) \cong H_*(Y)$ that X and Y are homotopy equivalent. When X and Y are CW-complexes, the isomorphism must be induced by a continuous mapping. The machinery of obstruction theory may be employed to build continuous functions from algebraic data when possible. There is a thorough discussion of obstruction theory in the books of [Baues77] and [Whitehead, GW78].

In the next section we continue the study of the homotopy theory of CW-complexes in the more general setting of the homotopy properties of cofibrations.

Cofibrations

In constructions leading to spectral sequences, the exactness of certain functors implies relations between the invariants that are expressed in the spectral sequence. For homotopy theory, it is the exactness properties of the functors $[-, Z]$ and $[Z, -]$ that play the principal role. CW-complexes enjoy some of the most useful exactness properties by virtue of their construction.

We begin by identifying a topological property that leads to exactness.

Definition 4.6. *Suppose* X *is a topological space and* A *is a subspace. Then the inclusion mapping* $i\colon A \to X$ *is said to have the* **homotopy extension property** (HEP) *with respect to a space* Y, *if, for every mapping* $f\colon X \to Y$ *and homotopy* $G\colon A \times I \to Y$ *such that* $G(a, 0) = f(a)$ *for all* $a \in A$, *there is a homotopy* $F\colon X \times I \to Y$ *such that* $F(x, 0) = f(x)$ *and* $F(a, t) = G(a, t)$ *for* $x \in X, a \in A, t \in I$. *The inclusion* $i\colon A \to X$ *is called a* **cofibration** *if* i *has the* HEP *with respect to any space.*

Two examples of cofibrations are important. The first is given by the **mapping cylinder** of an arbitrary function. Let $f\colon X \to Y$ be a given continuous mapping and define

$$I_f = Y \bigsqcup X \times I \Big/ f(x) \sim (x, 0),$$

the quotient of the disjoint union of Y and $X \times I$ by the relation '$(x, 0)$ to be identified with $f(x)$.' The inclusion $i\colon X \hookrightarrow I_f$, $i(x) = (x, 1)$ is a cofibration (exercise). By construction, notice that Y, as a subset of I_f, is a deformation retract. In fact, the composite $X \hookrightarrow I_f \to Y$ is simply a factorization of f. Thus every mapping can be factored into a composite of a cofibration followed by a homotopy equivalence.

The second example is the **mapping cone** of a continuous function $f\colon X \to Y$. Define

$$M_f = Y \bigsqcup CX \Big/ f(x) \sim (x, 1).$$

By taking the quotient of I_f by the subspace X, we also obtain $M_f = I_f/X$. In this case, the inclusion $Y \hookrightarrow M_f$ is a cofibration (exercise). Recall that $CS^{n-1} \cong e^n$. If X is a CW-complex, then the adjunction spaces, $X_i^{(n-1)} = X_{i-1}^{(n-1)} \cup_{f_i} e^n = M_{f_i}$, can be written as mapping cones. Thus, the inclusions of subcomplexes $X^{(n)} \hookrightarrow X$ and $X^{(n)} \hookrightarrow X^{(n+k)}$ are examples of cofibrations.

Theorem 4.7. *Consider the sequence of mappings of pointed spaces* $(X, x_0) \xrightarrow{f} (Y, y_0) \xrightarrow{i} (M_f, y_0)$. *For any space* Z, *the sequence* $[X, Z] \xleftarrow{f^*} [Y, Z] \xleftarrow{i^*} [M_f, Z]$ *is an exact sequence of pointed sets.*

PROOF: First observe that $f^* \circ i^* = (i \circ f)^*$ carries a homotopy class of a mapping to the constant class since $i \circ f$ is the composite $X \to I_f \to I_f/X$. Thus $\operatorname{im} i^* \subset (f^*)^{-1}\{[c]\}$ where c denotes the constant map. Suppose $f^*[g] = [c]$, then $g \circ f \simeq c$, and so it can be extended over CX. Put this extension together with Y to get a mapping $g'\colon Y \cup_f CX \to Z$ with $i^*([g']) = [g]$. $\square$

Suppose we extend this sequence further by iterating the construction:

$$X \xrightarrow{f} Y \xrightarrow{i} Y \cup_f CX \xrightarrow{j} (Y \cup_f CX) \cup_i CY.$$

By identifying the cone on Y with the copy of Y in M_f, we have, up to homotopy, identified Y to a point. Making this line of argument precise, we find that $(Y \cup_f CX) \cup_i CY \simeq \Sigma X$. Continuing in this manner, we derive the **Barratt-Puppe sequence** ([Barratt55], [Puppe58])

$$X \xrightarrow{f} Y \xrightarrow{i} M_f \xrightarrow{j} \Sigma X \xrightarrow{\Sigma f} \Sigma Y \xrightarrow{\Sigma i} M_{\Sigma f} \xrightarrow{\Sigma j} \Sigma^2 X \xrightarrow{\Sigma^2 f} \Sigma^2 Y \to \cdots.$$

As a consequence of this sequence and Theorem 4.7, we have proved the exactness result:

Theorem 4.8. *Given a mapping* $f: (X, x_0) \to (Y, y_0)$ *and a space* Z, *there is a long exact sequence of pointed sets*

$$[X, Z] \xleftarrow{f^*} [Y, Z] \xleftarrow{i^*} [M_f, Z] \xleftarrow{j^*} [\Sigma X, Z] \xleftarrow{\Sigma f^*} [\Sigma Y, Z] \leftarrow \cdots .$$

Some further structure can be identified on this exact sequence of pointed sets. Let SX denote the **reduced suspension** of (X, x_0); that is,

$$SX = \Sigma X / \{x_0\} \times I = \left. X \times I \middle/ X \times \{0\} \cup X \times \{1\} \cup x_0 \times I \right. .$$

It follows that $SX \simeq \Sigma X$. A pointed mapping of SX to Z then takes the form $f(x, t) \in Z$ and so, by fixing x and varying t, we get a loop in Z based at $f(x_0, t) = z_0$. This determines a mapping $[(SX, *), (Z, z_0)] \to [(X, x_0), (\Omega Z, c_{x_0})]$, which is a bijection. Categorically speaking, the reduced suspension S and based loops Ω are adjoint functors on the category of spaces and homotopy classes of mappings.

Let us consider ΩZ more carefully. The multiplication of paths, given by

$$(\lambda * \mu)(t) = \begin{cases} \lambda(2t), & 0 \leq t \leq 1/2 \\ \mu(2t - 1), & 1/2 \leq t \leq 1 \end{cases}$$

for $\lambda, \mu \in \Omega Z$, determines a multiplication on ΩZ,

$$m: \Omega Z \times \Omega Z \longrightarrow \Omega Z.$$

This multiplication, in turn, induces a multiplication on $[X, \Omega Z]$,

$$m_*: [X, \Omega Z] \times [X, \Omega Z] \longrightarrow [X, \Omega Z].$$

The multiplication m, on the space level, enjoys associativity, an identity, and inverses up to homotopy and so these properties carry over to m_*. The next result is classical.

Proposition 4.9. *With the multiplication induced by multiplication of loops, the set* $[X, \Omega Z]$ *is a group. Furthermore,* $[X, \Omega^2 Z]$ *is an abelian group.*

A corollary of this proposition is that the long exact sequence of Theorem 4.8 is a long exact sequence of abelian groups after a certain point. To show this, it suffices to observe that the isomorphism, $[SX, Z] \cong [X, \Omega Z]$, commutes with the mappings in the sequence and that the induced mappings of pointed sets are in fact homomorphisms of groups (exercise).

Before leaving our discussion of these constructions, we record an important result about the suspension functor to be proved in later chapters. Given spaces X and Y, there is a mapping, $E: [X, Y] \to [SX, SY]$, given by $E([f]) = [Sf]$ ([Freudenthal37]).

Theorem 4.10 (the Freudenthal suspension theorem). *If Y is n-connected for $n \geq 2$ and X is a CW-complex with $\dim X = q$, then*

$$E\colon [X,Y] \longrightarrow [SX,SY]$$

is an isomorphism when $q \leq 2n$ and an epimorphism when $q = 2n + 1$.

We present a proof of Theorem 4.10 in Chapter 6 (Theorem 6.12). If we restrict our attention to finite CW-complexes, X, then for dimension reasons the sequence of mappings, $[X,Y] \underset{E}{\longrightarrow} [SX,SY] \underset{E}{\longrightarrow} [S^2X,S^2Y] \underset{E}{\longrightarrow} \cdots$, becomes stable after finitely many iterations. We denote this abelian group by $\{X,Y\}_0$, the set of **stable mappings** from X to Y. Taking this notion one step further we introduce the graded abelian group of mappings, $\{X,Y\}_*$, given by

$$\{X,Y\}_i = \lim_{\to E}[S^{i+n}X, S^nY].$$

These groups enjoy more structure than the unstable sets $[X,Y]$ and so may be studied as a first approximation to the unstable case. In particular, if $X = S^0 = Y$, then $\{S^0, S^0\}_* = \pi_*^S$, the **stable homotopy groups of spheres**, which is the subject of Chapter 9.

Cellular homology

We next take up the problem of computing the homology of CW-complexes. The combinatorial structure given by a cell decomposition simplifies the computation. This simplification should convince the reader that the category $\mathcal{W}$ of spaces of the homotopy type of a CW-complex and cellular maps is an acceptable place in which to do homotopy theory.

Suppose X is a CW-complex. We introduce a chain complex associated to X as follows: Let

$$\mathrm{Cell}_n(X) = H_n(X^{(n)}, X^{(n-1)})$$

and define $\partial^{\mathrm{cell}}\colon \mathrm{Cell}_n(X) \to \mathrm{Cell}_{n-1}(X)$ by taking the homomorphisms from the long exact sequences for the pairs $(X^{(n)}, X^{(n-1)})$ and $(X^{(n-1)}, X^{(n-2)})$, spliced together as in the diagram

$$
\begin{array}{ccc}
\mathrm{Cell}_n(X) & = & H_n(X^{(n)}, X^{(n-1)}) \\
& & \qquad\qquad\searrow{\scriptstyle \partial} \\
{\scriptstyle \partial^{\mathrm{cell}}}\downarrow & & \qquad H_{n-1}(X^{(n-1)}) \\
& & \qquad\qquad\nearrow{\scriptstyle i} \\
\mathrm{Cell}_{n-1}(X) & = & H_{n-1}(X^{(n-1)}, X^{(n-2)})
\end{array}
$$

Since $\partial \circ i = 0$ in the sequence $H_{n-1}(X^{(n-1)}) \xrightarrow{i} H_{n-1}(X^{(n-1)}, X^{(n-2)}) \xrightarrow{\partial} H_{n-2}(X^{(n-2)})$, it follows that $\partial^{\text{cell}} \circ \partial^{\text{cell}} = (i \circ \partial) \circ (i \circ \partial) = i \circ (\partial \circ i) \circ \partial = 0$. We call $(\text{Cell}_*(X), \partial^{\text{cell}})$ the **cellular chain complex** on X.

Because a space X may have many cellular decompositions, it is not immediate that $H(\text{Cell}_*(X), \partial^{\text{cell}})$ is a homotopy invariant of the space X. Furthermore, because $\text{Cell}_n(X)$ is a relative homology group, it is not clear what combinatorial data are carried by the complex. We address these questions by determining more of the structure of this complex.

Lemma 4.11. $\text{Cell}_n(X) = $ *the free abelian group generated by the n-cells of the cell decomposition of* X.

PROOF: Write $X^{(n)} = X^{(n-1)} \cup_F (\bigsqcup_\alpha e^n_\alpha) = A \cup B$ where F denotes the attaching maps taken together, A is the image of the open sets, $(1/2)e^n_\alpha = \{\vec{x} \in e^n \text{ such that } \|\vec{x}\| < 1/2\}$ and B is $X^{(n)} - \{\text{image of the centers of each } n\text{-cell}\}$. By the excision axiom for singular homology, we have

$$\text{Cell}_n(X) = H_n(X^{(n)}, X^{(n-1)}) \cong H_n(A, A \cap B)$$
$$\cong \bigoplus_\alpha H_n(e^n_\alpha, S^{n-1}_\alpha) \cong \bigoplus_\alpha \mathbb{Z}e^n_\alpha. \qquad \square$$

Write $\partial^{\text{cell}} : \bigoplus_\alpha \mathbb{Z}e^n_\alpha \to \bigoplus_\beta \mathbb{Z}e^{n-1}_\beta$ for the differential on the cellular chain complex. In this guise, ∂^{cell} is determined by its value on each basis cell and we can write

$$\partial^{\text{cell}}(e^n_\alpha) = \sum_\beta [e^n_\alpha, e^{n-1}_\beta] e^{n-1}_\beta .$$

The integer $[e^n_\alpha, e^{n-1}_\beta]$ is called the **incidence number** of e^n_α and e^{n-1}_β. For CW-complexes that have enough cells to fit together simply, these incidence numbers have a geometric interpretation. For a proof of the following result, we refer the reader to IX.§7 of the book by [Massey91].

Proposition 4.12. *Suppose X is a* **regular** *CW-complex, that is, the characteristic mappings* $\{f_\alpha : (e^n, S^{n-1}) \to (e^n_\alpha, \partial e^n_\alpha)\}$ *satisfy the condition that, for all α, $f_\alpha : e^n \to \text{cls}(e^n_\alpha)$ is a homeomorphism. Suppose further that the mappings f_α determine the isomorphisms* $\text{Cell}_n(X) \cong \bigoplus_\alpha \mathbb{Z}e^n_\alpha$. *Then*

$$[e^n_\alpha, e^{n-1}_\beta] = \begin{cases} 0, & \text{if } e^{n-1}_\beta \not\subset \text{cls}(e^n_\alpha), \\ \pm 1, & \text{if } e^{n-1}_\beta \subset \text{cls}(e^n_\alpha). \end{cases}$$

The isomorphisms determined by the f_α provide an orientation to each cell. We illustrate this proposition by computing $H(\text{Cell}_*(S^2), \partial^{\text{cell}})$ when S^2 has the regular cell decomposition $S^2 = (e^0 \cup e^0) \cup_{i_1} (e^1 \cup e^1) \cup_{i_2} (e^2 \cup e^2)$. Denote each cell with a $+$ or $-$ to determine upper or lower hemisphere, respectively. We find

$$\partial^{\mathrm{cell}}(e_+^1) = e_+^0 - e_-^0, \qquad \partial^{\mathrm{cell}}(e_-^1) = e_-^0 - e_+^0$$
$$\partial^{\mathrm{cell}}(e_+^2) = e_+^1 - e_-^1, \qquad \partial^{\mathrm{cell}}(e_-^2) = e_-^1 - e_+^1.$$

It follows that $H_2(\mathrm{Cell}_*(S^2), \partial^{\mathrm{cell}}) \cong \mathbb{Z}$, generated by the class $e_+^2 + e_-^2$; $H_1 \cong \{0\}$ and $H_0 \cong \mathbb{Z}$, generated by $e_+^0 + e_-^0$. Computations like this example can be done for such spaces as $\mathbb{R}P(n)$, $\mathbb{C}P(n)$ and the lens spaces (see exercises), where the geometric construction provides the cell structure and the incidence numbers.

It remains to show that cellular homology is a homotopy invariant of the CW-complex.

Theorem 4.13. *For X, a CW-complex, its integral singular homology $H_*(X)$ is isomorphic to $H(\mathrm{Cell}_*(X), \partial^{\mathrm{cell}})$.*

PROOF: Consider the bigraded exact couple arising from the long exact sequences for the collection of pairs $(X^{(p)}, X^{(p-1)})$ with $D_{p,q} = H_{p+q}(X^{(p)})$, $E_{p,q} = H_{p+q}(X^{(p)}, X^{(p-1)})$ and structure maps given by

$$H_{p+q}(X^{(p-1)}) \xrightarrow{\quad i \quad} H_{p+q}(X^{(p)})$$
$$\Big\downarrow{\scriptstyle j}$$
$$H_{p+q}(X^{(p)}, X^{(p-1)}) \xrightarrow{\quad \partial \quad} H_{p+q-1}(X^{(p-1)}).$$

By the proof of Lemma 4.11, it follows that, if $q \neq 0$, $E_{p,q}^1 = \{0\}$ and $d^1 \colon E_{p,0}^1 \to E_{p-1,0}^1$ is given by $\partial^{\mathrm{cell}} \colon \mathrm{Cell}_p(X) \to \mathrm{Cell}_{p-1}(X)$. Therefore,

$$E_{*,0}^2 = E_{*,0}^\infty = H(\mathrm{Cell}_*(X), \partial^{\mathrm{cell}}).$$

It suffices to show that $E_{*,0}^2 \cong H_*(X)$. This follows from a diagram chase based on the following

Fact. *For $r \geq 1$, $H_{p+r}(X^{(p)}) = \{0\}$ and $H_p(X^{(p+r)}) \cong H_p(X)$.*

To prove the first half of the fact, we proceed by induction on p. The case $p = 0$ certainly holds. Consider the exact sequence for the pair $(X^{9p}, X^{(p-1)})$:

$$\{0\} = H_{p+r+1}(X^{(p)}, X^{(p-1)}) \to H_{p+r}(X^{(p-1)}) \to$$
$$H_{p+r}(X^{(p)}) \to H_{p+r}(X^{(p)}, X^{(p-1)}) = \{0\}.$$

By induction, $H_{p+r}(X^{(p-1)}) = \{0\}$ when $r \geq 1$ and so, by exactness, $H_{p+r}(X^{(p)}) = \{0\}$.

To prove the second half of the fact, we reindex the same sequence:

$$\{0\} = H_{p+1}(X^{(p+r+1)}, X^{(p+r)}) \to H_p(X^{(p+r)}) \to$$
$$H_p(X^{(p+r+1)}) \to H_p(X^{(p+r+1)}, X^{(p+r)}) = \{0\}.$$

This implies that $H_p(X^{(p+r)})$ is isomorphic to $H_p(X^{(p+r+1)})$ for $r \geq 1$, and the isomorphism is induced by the inclusion. Thus we have the sequence of isomorphisms

$$H_p(X^{(p+1)}) \to H_p(X^{(p+2)}) \to \cdots \to H_p(X^{(p+r)}) \to \cdots .$$

induced by the inclusions of each skeleton in the next. Any p-chain in X, by the assumption that X is closure finite, must lie in some finite skeleton. Thus the direct limit of these isomorphisms is $H_p(X)$.

With this fact, we have collapsed the exact couple into a series of short exact sequences. Splicing these exact sequences together, we get the diagram:

Define the homomorphism $\rho\colon (\ker d^1 \subset H_p(X^{(p)}, X^{(p-1)})) \to H_p(X)$ by observing that $\ker d^1 = \ker \partial = \operatorname{im} j \cong H_p(X^{(p)})$ and $\rho(u) = i(v)$ if $j(v) = u$. Since i is onto, so is ρ and so it is sufficient to prove that $\ker \rho = \operatorname{im} d^1$. But, by the diagram, $\ker \rho \cong \ker i = \operatorname{im} \partial \cong j(\operatorname{im} \partial) = \operatorname{im} d^1$. It follows that

$$E_{p,0}^2 = \ker d^1 / \operatorname{im} d^1 \cong \ker d^1 / \ker \rho \cong H_p(X)$$

and so $H(\operatorname{Cell}_*(X), \partial^{\text{cell}}) \cong H_*(X)$. $\square$

For more details on CW-complexes and their properties, we refer the reader to the books of [Massey91], [Lundell-Weingram69], [Steenrod-Cooke-Finney67], [Fritsch-Piccinini90] and [May99].

4.2 Simplicial sets

The decomposition of a CW-complex into cells allows us to compute the homology of such a space more easily. Homotopy groups, however, remain difficult to compute. In this section we introduce simplicial sets and their relation to topological spaces. The combinatorial structure available in a simplicial set leads to another definition of homotopy groups (due to [Kan58]). Though the problem of effective computation is not solved by simplicial sets, this different view of a space (as a simplicial set) leads to new constructions.

Simplicial sets are the direct generalization of the structure of a triangulated polyhedron and so have their origins in the work of Euler and the inception of topology. A polyhedron may be described completely by giving its set of vertices and then the collections of vertices determined by the faces in each dimension. The description of a polyhedron doesn't suffer if we simply forget that we are talking about faces.

Definition 4.14. *A **simplicial set**, $K_\bullet$, is a sequence of sets, K_n, for $n = 0, 1, 2, \ldots$ together with functions*

$$d_i \colon K_n \longrightarrow K_{n-1}, \quad i = 0, 1, \ldots, n,$$
$$s_j \colon K_n \longrightarrow K_{n+1}, \quad j = 0, 1, \ldots, n,$$

*the **face** and **degeneracy** maps, respectively, which satisfy the **simplicial identities**

$$d_i \circ d_j = d_{j-1} \circ d_i, \qquad \text{for } i < j,$$
$$d_i \circ s_j = \begin{cases} s_{j-1} \circ d_i, & \text{for } i < j, \\ \text{identity}, & \text{for } i = j, j+1, \\ s_j \circ d_{i-1}, & \text{for } i > j+1, \end{cases}$$
$$s_i \circ s_j = s_{j+1} \circ s_i, \qquad \text{for } i \leq j.$$

The schema of ordering of vertices that describe a polyhedron immediately give the data for a simplicial set; a typical n-simplex can be written $\langle v_0, v_1, \ldots, v_n \rangle$ with $v_0 \leq v_1 \leq \cdots \leq v_n$. The face maps are given by

$$d_i \langle v_0, \ldots, v_n \rangle = \langle v_0, \ldots, \widehat{v_i}, \ldots, v_n \rangle,$$

where we omit the ith vertex, and the degeneracy maps are given by

$$s_j \langle v_0, \ldots, v_n \rangle = \langle v_0, \ldots, v_j, v_j, \ldots, v_n \rangle,$$

where the jth vertex is repeated. With this description, the face maps are dual to the inclusion of the ith face into a standard simplex and the degeneracy maps dual to the collapse of the standard simplex onto its jth face.

The canonical simplicial set is given by the opposite category of the small category, $\mathcal{O}$, with objects, $[n] = \{0, 1, \ldots, n\}$ and morphisms, $[n] \to [m]$, that are order-preserving. Let $\Delta = \mathcal{O}^{\mathrm{op}}$; then $d_i \colon [n] \to [n-1]$ is the map dual to the map $\{0, 1, \ldots, n-1\} \to \{0, 1, \ldots, n\}$ that skips i in the range and $s_j \colon [n] \to [n+1]$ is dual to the map that repeats j. The simplicial identities follow directly. The category Δ allows a different description of simplicial sets and a generalization to other categories.

Definition 4.15. *A **simplicial object** in a category **C** is a covariant functor, $K \colon \Delta \to \mathbf{C}$. A **morphism of simplicial objects** in **C** is a natural transformation, $f \colon K \to L$. Let **SimpC** denote the functor category, $\mathbf{C}^{\Delta}$, of simplicial objects in **C**.*

We write $K_n = K([n])$ and $d_i = K(d_i)$, $s_j = K(s_j)$ to describe the simplicial object as a sequence of objects and morphisms in **C**, and a morphism in **SimpC** can be described as a sequence of maps, $f_n \colon K_n \to L_n$, that commute with the face and degeneracy maps.

Let **Ens** denote the category of sets and $\mathcal{F}\!A \colon \mathbf{Ens} \to \mathbf{Ab}$, the functor from sets to abelian groups that assigns to each set the free abelian group generated by the set. This can be extended one dimension at a time to a functor, $\mathcal{F}\!A \colon \mathbf{SimpEns} \to \mathbf{SimpAb}$. We define the **homology groups** of a simplicial set $K_{\bullet}$ using $\mathcal{F}\!A$: For each n, let

$$C_n(K_{\bullet}) = \mathcal{F}\!A(K_n) \quad \text{and} \quad \partial = \sum_{i=0}^{n} (-1)^i \mathcal{F}\!A(d_i).$$

The simplicial identities imply that $\partial \circ \partial = 0$ and so we define $H_*(K_{\bullet}) = H(C_*(K_{\bullet}), \partial)$.

Before we introduce a notion of homotopy for the category **SimpEns**, we connect some of these ideas to the category of topological spaces. The classical construction, motivated by the structure of polyhedra, is the **singular complex** of a space X, (introduced by [Lefschetz33] and developed by [Eilenberg44])

$$\mathrm{Sing}_n(X) = \{f \colon \Delta^n \to X \mid f \text{ continuous}\},$$

where Δ^n is the standard n-simplex,

$$\Delta^n = \{(x_0, \ldots, x_n) \in \mathbb{R}^{n+1} \mid x_i \geq 0, \text{ for all } i, \text{ and } \sum x_i = 1\}.$$

We denote the **inclusion of the ith face** by $\varepsilon_i \colon \Delta^{n-1} \hookrightarrow \Delta^n$, defined by $\varepsilon_i(x_0, \ldots, x_{n-1}) = (x_0, \ldots, x_{i-1}, 0, x_i, \ldots, x_{n-1})$. The face maps for the singular complex are given by precomposition with these inclusions, that is, $d_i \colon \mathrm{Sing}_n(X) \to \mathrm{Sing}_{n-1}(X)$ is defined $d_i(f) = f \circ \varepsilon_i$. We denote the **collapse** of a simplex onto its jth face by $\eta_j \colon \Delta^{n+1} \to \Delta^n$, defined by

$\eta_j(x_0, \ldots, x_{n+1}) = (x_0, \ldots, x_{j-1}, x_j + x_{j+1}, x_{j+2}, \ldots, x_{n+1})$. The degeneracy maps for $\mathrm{Sing}_\bullet(X)$ are given by precomposition with the collapse maps, $s_j(f) = f \circ \eta_j$.

The singular complex determines a functor $\mathrm{Sing}_\bullet \colon \mathbf{Top} \to \mathbf{SimpEns}$, where $\mathrm{Sing}_\bullet(g)$ on a continuous mapping g in $\mathbf{Top}$ is given by composition.

The singular homology of a space X is defined as a composite of functors

$$\mathbf{Top} \xrightarrow{\mathrm{Sing}} \mathbf{SimpEns} \xrightarrow{\mathcal{FA}} \mathbf{SimpAb} \to \mathbf{DGAb} \xrightarrow{H} \mathbf{GAb}$$

where $\mathbf{DGAb}$ denotes the category of differential graded abelian groups and $\mathbf{GAb}$ the category of graded abelian groups. We note that it is possible to normalize the chain complex $C_*(X)$ by setting the subgroup generated by degenerate simplices (those in the image of the functions s_i) to zero. This does not change the homology (Normalization Theorem in [May67]) while focusing attention on the nondegenerate chains. We will use this fact in simplicial arguments.

If given an arbitrary simplicial set, what distinguishes it as being in the image of the functor $\mathrm{Sing}_\bullet$? The condition that must be satisfied was identified by [Kan58] and this condition plays the key role in defining homotopy groups for simplicial sets.

Definition 4.16. *A simplicial set $K_\bullet$ is said to satisfy the* **extension condition**, *if, for every collection, $x_0, x_1, \ldots, x_{i-1}, x_{i+1}, \ldots, x_{n+1}$, in K_n, such that*

$$d_k(x_l) = d_{l-1}(x_k) \qquad \text{for } k < l, \ k \neq i \neq l,$$

there exists x in K_{n+1} with $d_j(x) = x_j$ for $j \neq i$. If $K_\bullet$ satisfies the extension condition, it is called a **Kan complex***.*

Examples of Kan complexes are polyhedra, where the ordering of the vertices is used to establish the extension property, and $\mathrm{Sing}_\bullet(X)$, for a topological space X, where the retraction of an n-simplex onto all but one of its faces can be used to construct the desired simplex in the extension condition.

In the category $\mathbf{SimpEns}$, there is a unit interval given by $\Delta[1]_\bullet$ that may be described as a polyhedron with

$$\Delta[1]_n = \{\langle v_0, \ldots, v_n \rangle \mid v_i = 0 \text{ or } 1 \text{ and } v_0 \leq v_1 \leq \cdots \leq v_n\}.$$

More generally, $\Delta[m]_\bullet$ has n-simplices $\langle v_0, v_1, \ldots, v_n \rangle$ with $v_i \in \{0, \ldots, m\}$ and $v_0 \leq v_1 \leq \cdots \leq v_n$. We define the n-**skeleton** of a simplicial set $K_\bullet$ to be the sub-simplicial set that contains all nondegenerate ($\neq s_j(x)$) simplices of degree $\leq n$. Observe that $\Delta[1]_\bullet$ is isomorphic to the 1-skeleton of Δ which is the category $\Delta = \mathcal{O}^{\mathrm{op}}$ treated as a simplicial set, that is, as the inclusion functor $\Delta \subset$ **Ens**. An element of a simplicial set $K_\bullet$ that lies in the n-skeleton can be equivalently described by a simplicial mapping $\Delta[n]_\bullet \to K_\bullet$.

Given two simplicial sets $K_\bullet$ and $L_\bullet$, their **Cartesian product** $K_\bullet \times L_\bullet$ is the simplicial set

$$(K_\bullet \times L_\bullet)_n = K_n \times L_n, d_i = d_i \times d_i \text{ and } s_j = s_j \times s_j.$$

Definition 4.17. *If $f, g \colon K_\bullet \to L_\bullet$ are simplicial maps, then f is* **homotopic** *to g if there is a simplicial map, $F \colon K_\bullet \times \Delta[1]_\bullet \to L_\bullet$, with*

$$F(x, \langle 0, 0, \dots, 0 \rangle) = f(x) \quad and \quad F(x, \langle 1, 1, \dots, 1 \rangle) = g(x)$$

for all x in $K_\bullet$. If f is homotopic to g, we write $f \simeq g$.

The homotopy relation on simplicial maps $K_\bullet \to L_\bullet$ need not be well-behaved; however, we have the following result.

Lemma 4.18. *The relation $\simeq$ on the set of simplicial maps from $K_\bullet$ to $L_\bullet$,* **SimpEns**$(K_\bullet, L_\bullet)$, *is an equivalence relation whenever $L_\bullet$ is a Kan complex.*

The extension condition is there to overcome a difficulty in establishing the transitivity of $\simeq$. We refer the reader to the book of [May67] for a proof of the lemma.

We next define the homotopy groups of a Kan complex. Let $K_\bullet$ be a Kan complex and $x_0 \in K_0$, a choice of basepoint. Propagate x_0 to each K_p by choosing as distinguished element

$$\underbrace{s_0 \circ s_0 \circ \cdots \circ s_0}_{p \text{ times}}(x_0) \in K_p.$$

The n-sphere S^n may be modeled as the quotient simplicial set $\Delta[n]/\Delta[n]^{(n-1)}$, that is, $\Delta[n]_\bullet$ modulo its $(n-1)$-skeleton. The set of mappings of pairs,

$$\textbf{SimpEns}((\Delta[n]_\bullet, \Delta[n]^{(n-1)}), (K_\bullet, x_0)),$$

is defined by requiring a simplicial map to take elements in $(\Delta[n]^{(n-1)})_p$ to $s_0 \circ \cdots \circ s_0(x_0)$ in K_p. The relation $\simeq$ is an equivalence relation on this set when $K_\bullet$ is a Kan complex and so we define

$$\pi_n(K_\bullet, x_0) = \textbf{SimpEns}((\Delta[n]_\bullet, \Delta[n]^{(n-1)}), (K_\bullet, x_0)) \Big/ {\simeq}$$

$$=: [(\Delta[n]_\bullet, \Delta[n]^{(n-1)}), (K_\bullet, x_0)].$$

To determine further structure on this set, we give an intrinsic definition of $\pi_n(K_\bullet, x_0)$. Suppose x and y are in K_n. We write $x \sim y$ if, for all i,

$d_i(x) = d_i(y)$, and, for some k, $0 \le k \le n$, there is a w in K_{n+1} such that $d_k(w) = x$, $d_{k+1}(w) = y$ and $d_i s_k(x) = d_i(w) = d_i s_k(y)$ for $k \ne i \ne k+1$. The simplicial identities imply that $\sim$ is an equivalence relation. If we let $\tilde{K}_n = \{z \in K_n \mid d_i(z) = s_0 \circ \cdots \circ s_0(x_0), \text{ for all } i\}$, then we can form $\tilde{K}_n/\sim$, the set of equivalence classes of $\tilde{K}_n$ under $\sim$. Since any n-simplex, $z \in \tilde{K}_n$, is representable as a mapping $(\Delta[n]_\bullet, \Delta[n]^{(n-1)}) \to (K_\bullet, x_0)$, we find that $\pi_n(K_\bullet, x_0) = \tilde{K}_n/\sim$.

In this formulation, there is a binary operation, $+$, on $\pi_n(K_\bullet, x_0)$, given as follows: For a, b in $\tilde{K}_n$, let $w \in \tilde{K}_{n+1}$ be such that $d_0(w) = b$, $d_2(w) = a$ and $d_i(w) = s_0 \circ s_0 \circ \cdots \circ s_0(x_0)$ for $i \ge 3$. Such a class w exists by the extension condition. Define $[a] + [b] = [d_1(w)]$. For a proof of the following proposition, see the book of [May67, p. 12].

Proposition 4.19. *With $+$ defined as above, $\pi_n(K_\bullet, x_0)$ is a group for $n \ge 1$ and an abelian group, when $n \ge 2$.*

There are now two ways to associate homotopy groups to a topological space; we have the classical definition $\pi_n(X, x_0) = [(S^n, *), (X, x_0)]$ and we have $\pi_n(\text{Sing}_\bullet(X), x_0) = [(\Delta[n]_\bullet, \Delta[n]^{(n-1)}), (\text{Sing}_\bullet(X), x_0))]$. If the extra structure of simplicial sets is going to be an effective tool to study homotopy theory, these groups ought to be related if not isomorphic. The relation between these groups is provided by a construction introduced by [Milnor57].

Definition 4.20. *Let $K_\bullet$ be a simplicial set and define the **geometric realization** of $K_\bullet$ to be*

$$RK_\bullet = \left. \coprod_{n \ge 0} \Delta^n \times K_n \middle/ \approx \right.,$$

the quotient space of the disjoint union of pairs, $(\vec{v}, x)$, for $x \in K_n$, $\vec{v} \in \Delta^n$, by the relation $(\varepsilon_i(\vec{v}), x) \approx (\vec{v}, d_i(x))$, and $(\eta_j(\vec{v}), x) \approx (\vec{v}, s_j(x))$. $RK_\bullet$ has the quotient topology. This construction provides a functor $R\colon \mathbf{SimpEns} \longrightarrow \mathbf{Top}$.

By building $RK_\bullet$ one dimension at a time we see that $RK_\bullet$ is a CW-complex. Like CW-complexes, when forming products, one must be careful. For two simplicial sets, $K_\bullet$ and $L_\bullet$, $R(K_\bullet \times L_\bullet)$ is homeomorphic to $RK_\bullet \times RL_\bullet$ if $K_\bullet$ and $L_\bullet$ are both countable or if one of the spaces, $RK_\bullet$ or $RL_\bullet$, is *locally finite*, that is, every point is in the interior of a finite subcomplex. In particular, $R\Delta[1]$ is homeomorphic to $I = [0, 1]$ and so is locally finite. It follows that a simplicial homotopy $F\colon K_\bullet \times \Delta[1]_\bullet \to L_\bullet$ gives a topological homotopy $RF\colon RK_\bullet \times I \to RL_\bullet$. Furthermore, R induces a mapping $[K_\bullet, L_\bullet] \to [RK_\bullet, RL_\bullet]$, for $L_\bullet$ a Kan complex. In the opposite direction, a topological homotopy, $G\colon X \times I \to Y$, gives a simplicial homotopy as the composite

$$\text{Sing}_\bullet(X) \times \Delta[1]_\bullet \to \text{Sing}_\bullet(X) \times \text{Sing}_\bullet(I) \to$$
$$\text{Sing}_\bullet(X \times I) \xrightarrow[\text{Sing}_\bullet(G)]{} \text{Sing}_\bullet(Y).$$

Thus, application of the singular complex functor, $\text{Sing}_\bullet$, induces a mapping $[X, Y] \rightarrow [\text{Sing}_\bullet(X), \text{Sing}_\bullet(Y)]$.

The relation between the functors $\text{Sing}_\bullet$ and R is described in the following theorem of [Kan58'].

Theorem 4.21. $\text{Sing}_\bullet$ *and* R *are adjoint functors, that is, there is a natural bijection of sets*

$$\textbf{SimpEns}(K_\bullet, \text{Sing}_\bullet(X)) \longleftrightarrow \textbf{Top}(RK_\bullet, X);$$

moreover, this bijection preserves homotopy.

PROOF: If $f \colon K_\bullet \rightarrow \text{Sing}_\bullet(X)$ is a simplicial mapping, then define $\Psi(f)$ in $\textbf{Top}(RK_\bullet, X)$ by $\Psi(f)(\vec{v}, x) = f(x)(\vec{v})$. If $g \colon RK_\bullet \rightarrow X$ is a continuous mapping, let $\Phi(g) \colon K_\bullet \rightarrow \text{Sing}_\bullet(X)$ be the mapping, $\Phi(g)(x)(\vec{v}) = g(\vec{v}, x)$. Clearly, $\Psi \circ \Phi = \text{id}$ and $\Phi \circ \Psi = \text{id}$, giving the one-one correspondence.

That these mappings preserve homotopies follows from an argument similar to the discussion before the theorem. □

An immediate consequence of this theorem is that our definitions for the homotopy groups of a space X coincide; $\pi_n(X, x_0) \cong \pi_n(\text{Sing}_\bullet(X), x_0)$. Another consequence is the classical theorem of [Hurewicz35/36].

Theorem 4.22 (the Hurewicz theorem). *If* $K_\bullet$ *is a Kan complex, then there is a homomorphism, for* $m \geq 1$,

$$h_m \colon \pi_m(K_\bullet) \longrightarrow H_m(K_\bullet).$$

If $K_\bullet$ *is* $(n-1)$*-connected for* $n \geq 2$, *then* h_n *is an isomorphism.*

SKETCH OF PROOF: Consider the mapping $K_\bullet \rightarrow \mathcal{F}\mathcal{A}(K_\bullet)$, taking each generator in $K_\bullet$ to itself in the free abelian group on those generators. This is a simplicial mapping and furthermore, the group property of $\mathcal{F}\mathcal{A}(K_\bullet)$ allows one to show that $\mathcal{F}\mathcal{A}(K_\bullet)$ is a Kan complex. Thus we get a mapping $\pi_*(K_\bullet) \rightarrow \pi_*(\mathcal{F}\mathcal{A}(K_\bullet))$. We follow an argument of [Moore56] that $\pi_*(\mathcal{F}\mathcal{A}(K_\bullet)) \cong H_*(K_\bullet)$ by using the commutative multiplication and the intrinsic definition of π_*.

The second half of the theorem follows by replacing $K_\bullet$ with a simplicial set having one simplex in each dimension i, $i < n$, and which is homotopy-equivalent to $K_\bullet$. For such a simplicial set, the result follows directly from the definitions.

The classical Hurewicz theorem follows by taking $K_\bullet = \text{Sing}_\bullet(X)$ for X a topological space. □

We close this section by observing how faithfully the homotopy theory of simplicial sets represents the homotopy theory of spaces of the homotopy type of a CW-complex. As in the case of CW-complexes, an approximation theorem is needed for mappings.

Theorem 4.23 (the simplicial approximation theorem). *Given a continuous function, $f: RK_\bullet \to RL_\bullet$, where $K_\bullet$ is a simplicial set and $L_\bullet$ is a Kan complex, there is a simplicial mapping $g: K_\bullet \to L_\bullet$ with $Rg \simeq f$.*

Corollary 4.24. *The functors R and $\mathrm{Sing}_\bullet$ establish an equivalence between the homotopy category of Kan complexes and the homotopy category of spaces of the homotopy type of a CW-complexes. In particular, R induces a one-one correspondence, $[K_\bullet, L_\bullet]_{\mathrm{SimpEns}} \cong [RK, RL]_{\mathrm{Top}}$ for $K_\bullet$ a simplicial set and $L_\bullet$ a Kan complex. $\mathrm{Sing}_\bullet$ induces a one-one correspondence, $[X, Y]_{\mathrm{Top}} \cong [\mathrm{Sing}_\bullet(X), \mathrm{Sing}_\bullet(Y)]_{\mathrm{SimpEns}}$ for X of the homotopy type of a CW-complex and Y a topological space.*

These results establish the equivalence of homotopy theory in the category **SimpEns** with homotopy theory in $\mathcal{W}$ the category of spaces of the homotopy type of a CW-complex. We exploit this fact in later sections when a simplicial construction may be done more transparently than a topological one. For proofs of these results and the deeper developments of the ideas of [Kan58], we refer the reader to the book-length treatments of [May67], [Quillen67], [André67], [Lamotke68], [Curtis71], [Bousfield-Kan72], and [Goerss-Jardine99].

4.3 Fibrations

The construction of a CW-complex involves many natural and desirable features. The building blocks are simple, the operation of adjunction has nice exactness properties, and the homology of the resulting space is calculable from these geometric data. To understand further the importance of the homotopy groups of a space, we give another way to construct topological spaces that enjoys these exactness properties with respect to the homotopy groups functor. We first consider the dual of the homotopy extension property. This property is dual in the sense that it implies exactness on the application of $[Z \ -]$. :

Definition 4.25. *A mapping, $p: E \to B$, has the **homotopy lifting property** (HLP), with respect to a space Y if, given a homotopy $G: Y \times I \to B$ and a mapping $g: Y \times \{0\} \to E$ such that $p \circ g(y, 0) = G(y, 0)$, then there is a homotopy $\tilde{G}: Y \times I \to E$ such that $\tilde{G}(y, 0) = g(y, 0)$ and $p \circ \tilde{G} = G$.*

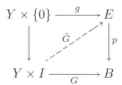

A mapping with the HLP *with respect to all spaces is called a **Hurewicz fibration**. A mapping with the* HLP *with respect to all n-cells is called a **Serre fibration**. By a **fibration** we mean a Hurewicz fibration.*

The property of being a Serre fibration is strictly weaker than that of being a Hurewicz fibration ([Brown, R66]). The properties of CW-complexes can be used to show that a Serre fibration has the HLP for all finite CW-complexes.

If $p\colon E \to B$ is a fibration, then we refer to the space B as the **base space** and E as the **total space** of the fibration. If b is a point in B, then we refer to $F_b = p^{-1}(b)$ as the **fibre** of p over b. Though F_b may vary for different choices of b, the homotopy lifting property restricts the homotopy type of F_b.

Proposition 4.26. *Suppose $p\colon E \to B$ is a fibration and that B is path-connected. Then, for b_0, $b_1 \in B$, F_{b_0} is homotopy-equivalent to F_{b_1}.*

PROOF: We prove a bit more here that can be applied later. Given a fibration, let WB denote the **free path space**, $\mathrm{map}([0,1], B) = \{\lambda\colon [0,1] \to B \mid \lambda$ is continuous$\}$, endowed with the compact-open topology. The evaluation mapping $\mathrm{ev}_0\colon WB \to B$, given by $\mathrm{ev}_0(\lambda) = \lambda(0)$, is continuous. Let

$$U_p = \{(\lambda, e) \in WB \times E \mid \lambda(0) = p(e)\}$$

denote the pullback of $p\colon E \to B$ over ev_0:

$$
\begin{array}{ccc}
U_p & \xrightarrow{\ pr_2\ } & E \\
{\scriptstyle pr_1}\downarrow & & \downarrow{\scriptstyle p} \\
WB & \xrightarrow[\ \mathrm{ev}_0\ ]{} & B.
\end{array}
$$

The homotopy $H\colon U_p \times I \to B$ given by $H((\lambda, e), t) = \lambda(t)$ poses the homotopy lifting problem:

$$
\begin{array}{ccc}
U_p & \xrightarrow{\ pr_2\ } & E \\
{\scriptstyle (-,0)}\downarrow & {\scriptstyle \tilde{H}}\nearrow & \downarrow{\scriptstyle p} \\
U_p \times I & \xrightarrow[\ H\ ]{} & B.
\end{array}
$$

When $p\colon E \to B$ is a fibration, we get a solution $\tilde{H}\colon U_p \times I \to E$. Let $\Lambda\colon U_p \to WE$ denote the **adjoint** of $\tilde{H}$ given by $\Lambda(\lambda, e)(t) = \tilde{H}((\lambda, e), t)$. This mapping satisfies the properties

$$p \circ \Lambda(\lambda, e) = \lambda, \quad \text{and} \quad \Lambda(\lambda, e)(0) = e.$$

Λ is called a **lifting function** for p. Lifting functions were introduced by [Hurewicz55] and developed by [Fadell-Hurewicz58] and [Brown, E59]. The universal properties of pullbacks lead to an equivalence between the HLP for a given fibration and the existence of a lifting function satisfying $\Lambda(c_{p(e)}, e) = c_e$ where c_e is the constant path at e and $\Lambda(\alpha * \beta, e) = \Lambda(\beta, \Lambda(\alpha, e)(1))$ ([Fadell60]).

Suppose $\alpha\colon I \to B$ is a path with $\alpha(0) = b_0$ and $\alpha(1) = b_1$. Consider the composite

$$\Phi_\alpha\colon F_{b_0} \to U_p \xrightarrow{\Lambda} WE \xrightarrow{\mathrm{ev}_1} E,$$

where the first mapping is $x \mapsto (\alpha, x)$. Since $p \circ \Lambda(\alpha, x) = \alpha$, $\Lambda(\alpha, x)(1) \in F_{b_1}$. Thus the composite $x \mapsto \Lambda(\alpha, x)(1)$ determines a continuous mapping $F_{b_0} \to F_{b_1}$. The adjoint of the composite $F_{b_0} \to WE$ gives a homotopy $h\colon F_{b_0} \times I \to E$ between the inclusion of F_{b_0} and $\Phi_\alpha\colon F_{b_0} \to F_{b_1} \hookrightarrow E$. Reverse the path to obtain the homotopy inverse of the mapping. $\qquad\square$

With Proposition 4.26, we can speak of *the* fibre of a fibration over a path-connected base space as a representative of the homotopy type of any fibre.

The lifting function provides some further structure. Let $b \in B$ and let $\Omega B = \Omega(B, b)$ denote the loops in B based at b and let $F = F_b$. Then $\Omega B \times F \subset U_p$ and the mapping $\mu = \mathrm{ev}_1 \circ \Lambda\colon \Omega B \times F \to E$ takes its image in F. This determines an action $\mu = \mathrm{ev}_1 \circ \Lambda\colon \Omega B \times F \to F$. Let $\alpha^{-1}(t) = \alpha(1-t)$.

Proposition 4.27. *To the action* $\mu = \mathrm{ev}_1 \circ \Lambda\colon \Omega B \times F \to F$ *and a loop* $\alpha \in \Omega B$ *associate the mapping* $h_\alpha = \mu(\alpha^{-1}, -)\colon F \to F$. *Then*

(1) *If* $\alpha \simeq \beta$, *then* $h_\alpha \simeq h_\beta$.
(2) *If* α *is homotopic to a constant map, then* $h_\alpha \simeq \mathrm{id}_F$.
(3) *If* $\alpha * \beta$ *denotes the loop multiplication of* α *and* β, *then* $h_{\alpha*\beta} \simeq h_\alpha \circ h_\beta$.

PROOF: (1) Suppose $K\colon I \times I \to B$ is a homotopy with $K(s,0) = \alpha^{-1}(s)$, $K(s,1) = \beta^{-1}(s)$, and $K(0,t) = b = K(1,t)$. The adjoint of K, $\hat{K}\colon I \to WB$ lands in ΩB. Thus $(x,t) \mapsto \mu(\hat{K}(t), x)$ is a homotopy $F \times I \to F$ between h_α and h_β.
(2) First notice that if $c = $ a constant map, then $\Lambda(c, x)(1) = x$ since the lifting problem is solved by constant maps. Now apply (1).
(3) Since $(\alpha*\beta)^{-1} = \beta^{-1}*\alpha^{-1}$, the definitions give that $\Lambda((\alpha*\beta)^{-1}, x)(1) = \Lambda(\alpha^{-1}, \Lambda(\beta^{-1}, x)(1))(1)$. Thus $h_{\alpha*\beta} \simeq h_\alpha \circ h_\beta$. $\qquad\square$

Corollary 4.28. *Let G denote an abelian group. If $p\colon E \to B$ is a fibration, $b \in B$, a path-connected space, then there is an action of the fundamental group $\pi_1(B, b)$ on $H_*(F; G)$ and on $H^*(F; G)$ induced by $[\alpha] \mapsto h_{\alpha*}$ and h_α^*, respectively.*

Lifting functions and their associated fundamental group action will be developed in later chapters.

Examples: (a) Let X be a path-connected space and x_0 a basepoint in X. Define the **space of based paths** in X as the subspace of $\mathrm{map}([0,1], X)$

$$PX = \{\lambda\colon [0,1] \to X \mid \lambda \text{ is continuous and } \lambda(0) = x_0\}.$$

The evaluation mapping $p\colon PX \to X$, given by $p(\lambda) = \lambda(1)$, is continuous. We show this mapping is a fibration by giving an explicit lifting of a given homotopy: Suppose $g\colon Y \to PX$ and $G\colon Y \times I \to X$ are mappings with $G(y,0) = p(g(y))$, then define $\tilde{G}\colon Y \times I \to PX$ by

$$\tilde{G}(y,t)(s) = \begin{cases} g(y)(s(t+1)), & \text{for } 0 \leq s \leq 1/(t+1), \\ G(y, s(t+1) - 1), & \text{for } 1/(t+1) \leq s \leq 1. \end{cases}$$

This establishes the HLP for p. The fibre over x_0 is the set of mappings, $\lambda\colon [0,1] \to X$ with $\lambda(0) = x_0 = \lambda(1)$, that is, the space of based loops ΩX.

(b) For a pair of spaces, B and F, the projection mapping, $p\colon B \times F \to B$, is called the **trivial fibration** with base B and fibre F. A **morphism of fibrations** is a pair of mappings $(\tilde{f}, f)\colon (E', B') \to (E, B)$ such that the following diagram commutes:

$$\begin{array}{ccc} E' & \xrightarrow{\tilde{f}} & E \\ {\scriptstyle p'}\downarrow & & \downarrow{\scriptstyle p} \\ B' & \xrightarrow{f} & B \end{array}$$

with p and p' fibrations. By restricting $p\colon E \to B$ to the subspace $p^{-1}(A)$ for $A \subset B$ we get a morphism of fibrations

$$\begin{array}{ccc} p^{-1}(A) & \xrightarrow{\subset} & E \\ {\scriptstyle p}\downarrow & & \downarrow{\scriptstyle p} \\ A & \xrightarrow{\subset} & B \end{array}$$

A fibration is called **locally trivial** if there is a covering of B by open sets $\{V_\alpha\}_{\alpha \in J}$ and a set of homeomorphisms $\{\varphi_\alpha\colon V_\alpha \times F \to p^{-1}(V_\alpha)\}_{\alpha \in J}$, each of which induces a morphism of fibrations

$$\begin{array}{ccccc} V_\alpha \times F & \xrightarrow{\varphi_\alpha} & p^{-1}(V_\alpha) & \xrightarrow{\subset} & E \\ {\scriptstyle \mathrm{pr}_1}\downarrow & & \downarrow{\scriptstyle p} & & \downarrow{\scriptstyle p} \\ V_\alpha & = = = & V_\alpha & \xrightarrow{\subset} & B. \end{array}$$

It is a slight extension of the proof of Proposition 4.26 to show that a fibration with base space a CW-complex is locally trivial up to homotopy. The main piece of the argument is a theorem of [Feldbau39] that a fibration over a contractible base is trivial. We can then cover the base space by the interiors of n-cells which are contractible.

(c) An example of a locally trivial fibration is constructed from a manifold, M, and its tangent bundle, $p \colon TM \to M$. The atlas of coordinate charts provides the covering by subspaces, each of which has the trivial fibration over it with fibre $\mathbb{R}^m$ where $m = \dim M$. In this case, the fibration enjoys a great deal more structure than described here. For some of this further structure, see §6.2.

(d) From the definition of a covering space, $f \colon E \to X$, the covering map is a fibration. The definition of a fibration can be thought of as a generalization of the lifting properties of a covering map. For a history of the notion of fibration see the article of [Zisman99].

(e) Let G be a Lie group and H a closed subgroup of G. By taking small open sets in G and using the group multiplication, one can show that the canonical projection, $G \to G/H$, onto the coset space, is a fibration with fibre H. See the classic book of [Steenrod51] for details of a proof. These fibrations are studied in Chapters 5, 7 and 8.

(f) Consider the classical division algebras over the real numbers: $\mathbb{R}$, the reals; $\mathbb{C}$, the complexes; $\mathbb{H}$, the quaternions; and $\mathbb{O}$, the Cayley numbers. If we denote any of these algebras by $\mathbb{A}$ and $d = \dim_{\mathbb{R}} \mathbb{A}$, then we define the **projective line** over $\mathbb{A}$, $\mathbb{A}P(1)$, to be the set of lines through the origin in $\mathbb{A} \times \mathbb{A}$. There is a mapping $S^{2d-1} \longrightarrow \mathbb{A}P(1)$, given by sending a unit vector in $\mathbb{A} \times \mathbb{A}$ to the line it spans. Note that $\mathbb{A}P(1)$ is the one-point compactification of $\mathbb{A}$ and so is homeomorphic to S^d. These mappings are called the **Hopf fibrations** and are denoted by $\iota \colon S^1 \to S^1$, $\eta \colon S^3 \to S^2$, $\nu \colon S^7 \to S^4$, and $\sigma \colon S^{15} \to S^8$. Their fibres are S^0, S^1, S^3 and S^7, respectively. Properties of the Hopf fibrations are considered in Chapters 5 and 9.

(g) Suppose $A \hookrightarrow X \to Y$ is a cofibration sequence and Z is a connected finite CW-complex. Then $\mathrm{map}(X, Z) \to \mathrm{map}(A, Z)$ is a fibration with fibre $\mathrm{map}(Y, Z)$. To prove this we check the homotopy lifting property. Let W denote a space. Suppose we have a commutative diagram as on the left:

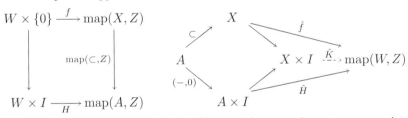

Using adjoints we can rewrite the lifting problem as a homotopy extension problem as in the diagram on the right. Here the maps are given by $\hat{f} \colon x \mapsto (w \mapsto f(w)(x))$ and $\hat{H} \colon (a, t) \mapsto (w \mapsto H(w, t)(a))$. Since $A \hookrightarrow X$ is a cofibration, $\hat{K}$ exists extending $\hat{f}$ and $\hat{H}$. Let $\tilde{H} \colon W \times I \to \mathrm{map}(X, Z)$ denote the adjoint of $\hat{K}$. Then $\tilde{H}$ solves the lifting problem for the data (f, H).

Given the cofibration sequence $A \hookrightarrow X \to Y$, we may take Y to be X/A. Fixing $g \colon A \to Z$, the subset of mappings $\hat{g} \colon X \to Z$ that extend g is given by extending g away from A, and meeting g at A. This describes a map in $\mathrm{map}(X/A, Z)$ and determines the fibre of the fibration $\mathrm{map}(\subset, Z)\colon \mathrm{map}(X, Z) \to \mathrm{map}(A, Z)$.

When a pointed space (X, x_0) satisfies the condition that the inclusion $\{x_0\} \hookrightarrow X$ is a cofibration, we say that X is **cofibrant**. It follows that the evaluation mapping

$$\mathrm{ev}_{x_0}\colon \mathrm{map}(X, Z) \to Z, \qquad f \mapsto f(x_0)$$

is a fibration. For example, the evaluation of free paths at any $t \in [0, 1]$ gives a fibration $\mathrm{ev}_t\colon WX \to X$.

In order to construct further examples of fibrations, we record some elementary constructions that can be applied to fibrations. The composition of two fibrations is a fibration (exercise). For (X, x_0) of the homotopy type of a countable CW-complex and $p\colon E \to B$ a fibration,

$$p \circ -\colon \mathrm{map}((X, x_0), (E, e_0)) \to \mathrm{map}((X, x_0), (B, p(e_0)))$$

is a fibration.

Suppose $p\colon E \to B$ is a fibration and $f\colon X \to B$ a continuous mapping. We can form the **pullback** of p over f by letting E_f denote the set $\{(x, e) \in X \times E \text{ such that } f(x) = p(e)\}$. The projection mappings on E_f give the diagram

$$
\begin{array}{ccc}
E_f & \longrightarrow & E \\
\downarrow{\scriptstyle p_f} & & \downarrow{\scriptstyle p} \\
X & \xrightarrow{\ f\ } & B.
\end{array}
$$

The universal property of a pullback has as input data mappings $u\colon Z \to X$ and $v\colon Z \to E$ such that $f \circ u = p \circ v$ and associates to them a unique mapping $w\colon Z \to E_f$ with all triangles and squares in the following diagram commutative:

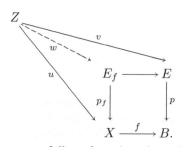

The homotopy lifting property follows from the universal property of pullbacks and the fact that p is a fibration. The fibre of $p_f\colon E_f \to X$ is the same space as the fibre of p.

We can use the pullback operation further to show that any mapping can be factored as a composite of a homotopy equivalence and a fibration (compare this with the dual situation for cofibrations). The evaluation mapping $\mathrm{ev}_0 \colon WY \to Y$ is a fibration with fibre ΩY. If we are given a mapping $f \colon X \to Y$, then form the pullback diagram

$$
\begin{array}{ccc}
U_f & \xrightarrow{\;pr_1\;} & WY \\
{\scriptstyle pr_2}\big\downarrow & & \big\downarrow{\scriptstyle \mathrm{ev}_0} \\
X & \xrightarrow{\quad f \quad} & Y.
\end{array}
$$

We show that the mapping $\pi = \mathrm{ev}_1 \circ pr_1 \colon U_f \to Y$ is a fibration, that $U_f \simeq X$ and finally, that the homotopy equivalence $X \simeq U_f$ followed by π is the mapping f. Consider the mapping $\xi \colon X \to U_f$ given by $\xi(x) = (c_{f(x)}, x)$, where $c_{f(x)}$ denotes the constant path at $f(x) \in Y$. The composite $pr_2 \circ \xi$ is the identity on X. Let $H \colon U_f \times I \to U_f$ denote the homotopy given by $H((\lambda, x), s) = ((t \mapsto \lambda(st)), x)$. At $s = 0$ we have $H((\lambda, x), 0) = (c_{\lambda(0)}, x) = (c_{f(x)}, x) = \xi(x)$ and at $s = 1$ we have the identity mapping on U_f. Thus $U_f \simeq X$. Furthermore, the composite $\pi \circ \xi$ equals f. It remains to show that π is a fibration.

Let Z be a space together with the lifting problem

$$
\begin{array}{ccc}
Z & \xrightarrow{\quad g \quad} & U_f \\
{\scriptstyle (-,0)}\big\downarrow & \nearrow{\scriptstyle \tilde{G}} & \big\downarrow{\scriptstyle \pi} \\
Z \times I & \xrightarrow{\quad G \quad} & Y
\end{array}
$$

Following [Whitehead, GW78], we introduce the mapping $\bar{G}$:

$$
\bar{G} \colon Z \times (\partial I \times I \cup I \times \{0\}) \to Y
$$

$$
\bar{G}(z, s, t) = \begin{cases} f((pr_2 \circ g)(z)) = (pr_1 \circ g)(z)(0) & s = 0, \\ G(z, t) & s = 1, \\ (pr_1 \circ g)(z)(s) & t = 0. \end{cases}
$$

Since $(\partial I \times I \cup I \times \{0\}) \subset I \times I$ is a cofibration (exercise), we can extend $\bar{G}$ to a mapping on $I \times I$. The adjoint of the extension $\bar{G} \colon Z \times I \times I \to Y$ with respect to the middle factor is a mapping $\widehat{G} \colon Z \times I \to WY$, given by $\widehat{G}(z, t) = (s \mapsto \bar{G}(z, s, t))$. The homotopy that solves the lifting problem is given by $\tilde{G}(z, t) = ((s \mapsto \widehat{G}(z, s, t)), pr_2 \circ g(z))$, and so π is a fibration.

Thus f can be factored as a composite of a homotopy equivalence and a fibration

$$
\begin{array}{ccc}
X & \xrightarrow[\simeq]{\;\xi\;} & U_f \\
& {\scriptstyle f}\searrow & \big\downarrow{\scriptstyle (\lambda, x) \mapsto \lambda(1)} \\
& & Y.
\end{array}
$$

We next consider the exactness properties of fibrations.

Theorem 4.29. *Let $E \xrightarrow{p} B$ be a fibration and b_0 a basepoint in B. If Z is a space, then for $F = p^{-1}(b_0)$*

$$[Z, F] \xrightarrow{i_*} [Z, E] \xrightarrow{p_*} [Z, B]$$

is an exact sequence of pointed sets. The same conclusion holds if p is a Serre fibration and Z has the homotopy type of a finite CW-complex.

PROOF: Suppose $g: Z \to E$ is a mapping with $p \circ g \simeq c_{b_0}$. Suppose $G: Z \times I \to B$ is a homotopy of $p \circ g$ to the constant mapping. Then we have the diagram

$$
\begin{array}{ccc}
Z \times \{0\} & \xrightarrow{\;g\;} & E \\
{\scriptstyle \mathrm{inc}}\downarrow & {\scriptstyle \tilde{G}}\nearrow & \downarrow{\scriptstyle p} \\
Z \times I & \xrightarrow[\;G\;]{} & B.
\end{array}
$$

By the homotopy lifting property, the homotopy $\tilde{G}$ exists. Let $f = \tilde{G}\big|_{Z \times \{1\}}$. Since G is a homotopy to the constant map, f determines a mapping into $F = p^{-1}(b_0)$. Furthermore, $\tilde{G}$ is a homotopy from $f: Z \to F \hookrightarrow E$ to g and so $[g]$ is in the image of i_*. $\square$

We can extend a fibration to a sequence of fibrations that is dual to the Barratt-Puppe sequence. The key to the construction is the fact that the space PX of based paths is contractible. We proceed as follows: Form the pullback of the path-loop fibration over B with respect to the fibration $p: E \to B$:

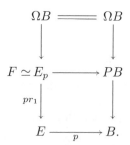

Because PB is contractible, E_p has the homotopy type of F, the fibre of p. This gives us a fibration up to homotopy $\Omega B \hookrightarrow F \to E$. By definition, $E_p = \{(\lambda, e,) \mid \lambda: (I, 0) \to (B, b_0), \lambda(1) = p(e)\}$. The mapping of ΩB to E_p is given by $\omega \mapsto (\omega, e_0)$ where e_0 is some choice of basepoint for E in the fibre over b_0. If $c = c_{b_0}$ is the constant loop at $b_0 \in B$, then the mapping $F \to E_p$ given by $x \mapsto (c, x)$ induces the homotopy equivalence. (Use the lifting function to construct the inverse.)

By a similar argument, we can analyze the pullback of the path-loop fibration over E with respect to the mapping $pr_2\colon E_p \to E$:

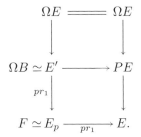

Here $E' = \{(\eta, \lambda, e) \mid \eta\colon (I, 0) \to (E, e_0), \lambda\colon (I, 0) \to (B, b_0), \eta(1) = e, \lambda(1) = p(e)\}$. The space E' is homotopy equivalent to ΩB by the mapping $(\eta, \lambda, e) \mapsto \lambda * (p \circ \eta)$, where $*$ is loop multiplication. The mapping of ΩE to E' is given by $\tilde{\omega} \mapsto (\tilde{\omega}, c, e_0)$ and continuing through to ΩB we get $\tilde{\omega} \mapsto c * (p \circ \tilde{\omega}) \simeq p \circ \tilde{\omega}$. Thus the fibration up to homotopy $\Omega E \to \Omega B \to F$ has Ωp as the 'inclusion' of the fibre ΩE in ΩB.

We iterate this procedure.

Theorem 4.30. *Given a fibration $p\colon E \to B$ with B path-connected and fibre F, there is a sequence of fibrations up to homotopy*

$$\cdots \to \Omega^n F \xrightarrow{\Omega^n i} \Omega^n E \xrightarrow{\Omega^n p} \Omega^n B \to \Omega^{n-1} F \to \cdots$$
$$\to \Omega B \to F \xrightarrow{i} E \xrightarrow{p} B.$$

Corollary 4.31. *For a Serre fibration, $F \hookrightarrow E \to B$ with B path-connected, there is a long exact sequence,*

$$\cdots \longrightarrow \pi_n(F) \xrightarrow{i_*} \pi_n(E) \xrightarrow{p_*} \pi_n(B) \longrightarrow \pi_{n-1}(F) \longrightarrow \cdots$$
$$\longrightarrow \pi_1(B) \longrightarrow \pi_0(F) \xrightarrow{i_*} \pi_0(E) \xrightarrow{p_*} \pi_0(B).$$

PROOF: Apply $[S^0, -]$ to the sequence in Theorem 4.30. Then Theorem 4.29 and the isomorphism, $[X, \Omega Y] \cong [SX, Y]$, imply the result. $\square$

Some immediate consequences of this corollary are well-known: If $\tilde{X}$ is the universal covering space of a space X with discrete fundamental group, then $\pi_n(\tilde{X})$ is isomorphic to $\pi_n(X)$ for $n \geq 2$. This corollary also implies that $\pi_n(X)$ is isomorphic to $\pi_{n-1}(\Omega X)$ since PX is contractible.

The long exact sequence of Corollary 4.31 shows how a fibration is a sort of exact sequence in **Top**, up to homotopy. In order to turn more algebra into homotopy theory, we use spaces whose homotopy groups are algebraically determined. These spaces and some resulting constructions are considered in the next section.

Eilenberg-Mac Lane spaces and Postnikov towers

The homotopy lifting property is dual to the homotopy extension property and developing the consequences of HLP leads to a sequence of fibrations that is dual to the Barratt-Puppe sequence of cofibrations. To extend this duality further, we recast one of the principal features of CW-complexes, namely, they are spaces that are built by successive cofibrations. The duality discussed in this chapter was recognized first by [Eckmann-Hilton58].

Suppose X is a CW-complex with a fixed cell-decomposition. The sequence of skeleta, $X^{(0)} \subset X^{(1)} \subset \cdots \subset X^{(n)} \subset \cdots \subset X$, has the following key properties:

(1) $X^{(i)} \subset X^{(j)}$ is a cofibration, for $i \leq j$.

(2) By the proof of Theorem 4.13,

$$H_*(X^{(n)}) \cong \begin{cases} H_*(X), & \text{if } * < n, \\ \{0\}, & \text{if } * > n. \end{cases}$$

(3) For a regular CW-complex, the quotient $X^{(n)}/X^{(n-1)}$ has the homotopy type of a bouquet of n-spheres, $\bigvee_\alpha S_\alpha^n$. Furthermore, this sequence of quotients determines $H_*(X)$ by the cellular chain complex given by the cell decomposition.

We can summarize these observations in the diagram of cofibrations and quotients:

$$X^{(0)} \xrightarrow{\subset} X^{(1)} \xrightarrow{\subset} \cdots \xrightarrow{\subset} X^{(n-1)} \xrightarrow{\subset} X^{(n)} \xrightarrow{\subset} \cdots \xrightarrow{\subset} X$$

$$\bigvee_{\alpha_1 \in I_1} S_{\alpha_1}^1 \qquad\qquad \bigvee_{\alpha_n \in I_n} S_{\alpha_n}^n$$

In this section, we develop a decomposition of a space X using fibrations, a dual of the cellular presentation. This new decomposition is based on the homotopy groups of a space rather than the homology.

We begin by dualizing the building blocks in a cell decomposition. The spheres that appear as quotients in the sequence of inclusions of skeleta are distinguished by homology as spaces with one nontrivial reduced homology group in the dimension determined by the sphere. A suitable analogue for homotopy is a space whose homotopy groups are concentrated in one dimension.

Definition 4.32. *Suppose G is an abelian group and $n \geq 0$ or G is a group and $n = 0$ or 1. If X is a space such that*

$$\pi_j(X) = \begin{cases} G, & \text{if } j = n, \\ \{0\}, & \text{if } j \neq n, \end{cases}$$

then X is called an **Eilenberg-Mac Lane space** *of type (G, n).*

Theorem 4.33. (1) *If G is an abelian group and $n \geq 0$ or G is a group and $n = 0$ or 1, then there exists an Eilenberg-Mac Lane space of type (G, n).* (2) *If two spaces are Eilenberg-Mac Lane spaces of type (G, n), then they have the same homotopy type. For any such space X, we write $X \simeq K(G, n)$.* (3) *If $\varphi \colon G \to H$ is a homomorphism of groups, then there is a continuous mapping $f \colon K(G, n) \to K(H, n)$ with $\pi_n(f) = \varphi$.*

SKETCH OF PROOF: For simplicity, let $n \geq 0$ and G be an abelian group. (The general case is similar.) Resolve G by free abelian groups in the short exact sequence

$$0 \to R \to F \to G \to 0.$$

If F has a basis $\{a_j\}_{j \in J}$ and R a basis $\{b_k\}_{k \in K}$, then these two groups and the homomorphism, $R \to F$, can be realized by a mapping $f \colon \bigvee_{k \in K} S^n \to \bigvee_{j \in J} S^n$. Let Y_n be the mapping cone of f. By the Hurewicz theorem, $\pi_n(Y_n) \cong G$. Proceed next by induction to attach cells in each dimension, $Y_{n+1} = Y_n \cup e^{n+2}_{\alpha_1} \cup \cdots \cup e^{n+2}_{\alpha_k}$, $Y_{n+2} = Y_{n+1} \cup e^{n+3}_{\beta_1} \cup \cdots \cup e^{n+3}_{\beta_l}$, and so forth, that kill the homotopy group in that dimension less one. The resulting complex, Y_∞ satisfies $\pi_n(Y_\infty) = G$ and $\pi_j(Y_\infty) = \{0\}$ for $j \neq n$. Thus $Y_\infty \simeq K(G, n)$.

To realize a mapping, $\varphi \colon G \to H$, build the mapping of resolutions

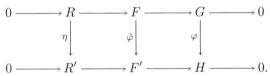

The mappings $\tilde{\varphi}$ and η can be realized as mappings of bouquets of spheres and so induce a mapping in the construction above. In particular, the identity mapping, $G \to G$, can be realized for two different complexes that are candidates for $K(G, n)$ by mapping appropriate resolutions. By the Whitehead theorem, these two complexes are homotopy-equivalent. □

Some examples of Eilenberg-Mac Lane spaces may be given from the theory of covering spaces. First, we have the fibration $\mathbb{Z} \hookrightarrow \mathbb{R} \to S^1$, the universal covering space of the circle. By the long exact sequence of homotopy groups for this fibration and the contractibility of $\mathbb{R}$, $S^1 \simeq K(\mathbb{Z}, 1)$.

For $G = \mathbb{Z}/2\mathbb{Z}$, there is the covering space for each n, given by $\mathbb{Z}/2\mathbb{Z} \hookrightarrow S^n \to \mathbb{R}P(n)$. These fibrations fit together in the system:

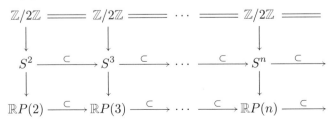

The direct limit is a fibration, $\mathbb{Z}/2\mathbb{Z} \hookrightarrow S^\infty \to \mathbb{R}P(\infty)$ with S^∞ **weakly contractible** (that is, $\pi_i(S^\infty) = \{0\}$ for all i). The long exact sequence of homotopy groups determines the homotopy groups of $\mathbb{R}P(\infty)$ and so $\mathbb{R}P(\infty) \simeq K(\mathbb{Z}/2\mathbb{Z}, 1)$.

For the complex projective spaces, we have a similar system of fibrations with fibre S^1; $S^1 \hookrightarrow S^{2n+1} \to \mathbb{C}P(n)$. The direct limit is a fibration

$$S^1 \hookrightarrow S^\infty \to \mathbb{C}P(\infty)$$

and, since $S^1 \simeq K(\mathbb{Z}, 1)$, the long exact sequence of homotopy groups implies that $\mathbb{C}P(\infty) \simeq K(\mathbb{Z}, 2)$.

Finally, if given a $K(G, n)$, the path-loop fibration

$$\Omega K(G, n) \hookrightarrow PK(G, n) \to K(G, n)$$

implies that $\Omega K(G, n) \simeq K(G, n - 1)$. We remark that the sequence of spaces, $K(G, 0)$, $K(G, 1)$, ... , $K(G, n)$, ... comprise an **Omega-spectrum** called the **Eilenberg-Mac Lane spectrum** for G. Spectra and generalized homology and cohomology theories are not developed in this book (see, for example, the recent book by [Kochman96]). It is useful to recall the following classical result. The proof of this theorem may be given by an obstruction theory argument ([Eilenberg47]).

Theorem 4.34. *Let G be an abelian group. Then there is a natural isomorphism, for each n and each CW-complex Y, $H^n(Y; G) \cong [Y, K(G, n)]$.*

This equivalence led to the definition of the generalized cohomology theories and plays a crucial role in many arguments in homotopy theory.

With the Eilenberg-Mac Lane spaces for building blocks, we next introduce the **Postnikov system** of a space X ([Postnikov51]), which provides a decomposition of X that is dual to the cellular decomposition.

Theorem 4.35. *To a simply-connected space X, there is a sequence of spaces, $\{ P_n X$ for $n = 0, 1, \dots \}$ and sequences of mappings,*

$$\{ p_i \colon P_i X \to P_{i-1} X \mid i = 1, 2, \dots \},$$
$$\{ f_j \colon X \to P_j X \mid j = 0, 1, \dots \}$$
$$\{ k^m \colon P_{m-1} X \to K(\pi_m(X), m + 1) \mid m = 1, 2, \dots \}$$

such that the following hold:

(1) *each mapping, p_i, is a fibration with fibre $K(\pi_i(X), i)$, pulled back over $P_{i-1} X$ from the path-loop fibration with respect to the mapping k^i:*

$$
\begin{array}{ccc}
P_i X & \longrightarrow & PK(\pi_i(X), i + 1) \\
{\scriptstyle p_i} \downarrow & & \downarrow \\
P_{i-1} X & \xrightarrow{\ k^i\ } & K(\pi_i(X), i + 1).
\end{array}
$$

(2) *For each space, $P_j X$,*

$$\pi_m(P_j X) \cong \begin{cases} \pi_m(X), & \text{for } m \leq j, \\ \{0\}, & \text{for } m > j, \end{cases}$$

and the mapping $f_j: X \to P_j X$ induces the isomorphism $f_{j*}: \pi_m(X) \to \pi_m(P_j X)$ for $m \leq j$.
(3) *For all j, $f_j = p_{j+1} \circ f_{j+1}$.*

Such a system of spaces and maps is called a Postnikov system for X, and the data contained in the system determine X up to homotopy.

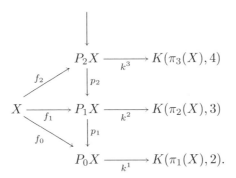

In the case that X is not simply-connected, the tower of fibrations $\{P_j X; p_j\}$ may be constructed with the fibre of p_j given by $K(\pi_j(X), j)$. However, the existence of the family of mappings $\{k^m\}$, called k-**invariants** of X, make each of the fibrations p_j a principal fibration. This would force the fundamental group $\pi_1(X)$ to act trivially on the higher homotopy groups of X, which is not true in general (see Chapter 8bis for more details).

SKETCH OF PROOF: Here is an instance where working simplicially allows us to make a very straightforward definition that would be cumbersome to make with spaces. Replace X with a Kan complex $L_\bullet$. Consider the following equivalence relation, defined for each n: Two q-simplices in $L_\bullet$, u and v, satisfy $u \sim_n v$ if each face of u of dimension $\leq n$ agrees with each face of v of dimension $\leq n$. Equivalently, if $\chi_u: \Delta[q]_\bullet \to L_\bullet$ and $\chi_v: \Delta[q]_\bullet \to L_\bullet$ are the characteristic mappings for u and v, then χ_u and χ_v agree on the n-skeleton $\Delta[q]_\bullet^{(n)}$ of $\Delta[q]_\bullet$. Let $(L_\bullet)_n = L_\bullet / \sim_n$.

It follows from the definitions in simplicial theory that each $(L_\bullet)_n$ is a Kan complex, that $\pi_i(L_\bullet) = \pi_i((L_\bullet)_n)$, for $i \leq n$ and $\pi_i((L_\bullet)_n) = \{0\}$ for $i > n$, that $(L_\bullet)_n \to (L_\bullet)_{n-1}$ is a simplicial fibration with fibre a $K(\pi_n(L_\bullet), n)$ (see the article of [Curtis71] for details).

To complete the proof for spaces, let $L_\bullet = \text{Sing}_\bullet(X)$ and then apply the realization functor to the subsequent construction. $\square$

The Postnikov system of a space and its generalization for mappings ([Moore58]) are powerful tools in the study of homotopy theory. Some of the computational potential is explored at the beginning of Chapter 6, and in Chapter 8bis; analogous constructions are considered in Chapter 9.

4.4 Hopf algebras and the Steenrod algebra

We close our consideration of topological preliminaries with a discussion of some of the properties of the singular homology and cohomology functors. The algebraic structures that we consider arise naturally from the functoriality of homology and cohomology and from a choice of basepoint. They have become basic structures in the study of algebraic topology and appear in many guises in the remainder of the book.

Fundamental to all subsequent structure is the following equivalence.

Theorem 4.36 (the Eilenberg-Zilber theorem). *For any pair of locally finite spaces, X and Y, and any field k, there are natural isomorphisms*

$$AW\colon H_*(X \times Y; k) \longleftrightarrow H_*(X; k) \otimes H_*(Y; k) : EZ,$$
$$AW\colon H^*(X \times Y; k) \longleftrightarrow H^*(X; k) \otimes H^*(Y; k) : EZ.$$

Furthermore, these equivalences are induced by chain equivalences on the singular chain and cochain complexes, respectively.

SKETCH OF PROOF: For complete details, the reader can consult the classic text of [Mac Lane63, Chapter VIII, §8]. We record the chain maps for later use:

$$AW\colon C_*(X \times Y; k) \to C_*(X; k) \otimes C_*(Y; k)$$

is known as the **Alexander-Whitney map** and it can be described simplicially by

$$AW(a \times b) = \sum_{i=0}^{n} \overrightarrow{d_{n-i}}a \otimes \overleftarrow{d_i}b,$$

the mapping that sends the product of simplices to the last $n - i$ faces of a ($\overrightarrow{d_{n-i}}a$) tensor the first i faces of b ($\overleftarrow{d_i}b$). We record that on chains AW is not a mapping of differential graded coalgebras since it is not cocommutative.

The inverse mapping (up to chain equivalence) is the **Eilenberg-Zilber map**

$$EZ\colon C_p(X; k) \otimes C_q(Y; k) \to C_{p+q}(X \times Y; k)$$

described by

$$EZ(a \otimes b) = \sum_{(p,q)-\text{shuffles } \sigma} (-1)^{\varepsilon(\sigma)}(s_{\sigma(1)} \cdots s_{\sigma(p)}a \times s_{\sigma(p+1)} \cdots s_{\sigma(p+q)}b),$$

where the sum is taken over all (p, q)-**shuffles**. A permutation of $p + q$ letters is a (p, q)-shuffle if

$$\sigma(1) < \cdots < \sigma(p) \quad \text{and} \quad \sigma(p+1) < \cdots < \sigma(p+q).$$

The mappings s_j are given by the degeneracy mappings for the underlying simplices and the sign is defined in various ways (see §7.2 for one of them). We note that the Eilenberg-Zilber map on cochains is an algebra mapping. (These ideas come up algebraically in Chapter 7.) □

Suppose (X, x_0) is a pointed space. Then X is immediately equipped with three continuous mappings: $\Delta \colon X \to X \times X$, the **diagonal mapping**, $\Delta(x) = (x, x)$; the **basepoint map**, $x_0 \colon * \to X$, $x_0(*) = x_0$, where $*$ is the one-point space and x_0 is both the basepoint and the mapping identifying it; finally, there is the unique mapping, $X \to *$, of X onto the terminal object in **Top**. We apply $H_*(\ ; k)$ to each of these mappings and obtain homomorphisms

$$\Delta = AW \circ \Delta_* \colon H_*(X; k) \to H_*(X; k) \otimes H_*(X; k),$$
$$\eta = (x_0)_* \colon k \longrightarrow H_*(X; k), \quad \text{and} \quad \varepsilon \colon H_*(X; k) \longrightarrow k.$$

We refer to Δ as the **comultiplication** on $H_*(X; k)$, η as a **unit** on $H_*(X; k)$ and ε as the **augmentation** of $H_*(X; k)$. We abstract the algebraic structures that result from these choices (following [Milnor-Moore65]).

Definition 4.37. *Let H_* denote a graded vector space over a field k. We say that H_* has a* **comultiplication** *if there is a mapping $\Delta \colon H_* \to H_* \otimes H_*$.*

(1) Δ *is* **coassociative** *if the following diagram commutes*

$$
\begin{array}{ccc}
H_* & \xrightarrow{\ \ \Delta\ \ } & H_* \otimes H_* \\
{\scriptstyle \Delta}\downarrow & & \downarrow{\scriptstyle \Delta \otimes 1} \\
H_* \otimes H_* & \xrightarrow[\ 1 \otimes \Delta\ ]{} & H_* \otimes H_* \otimes H_*.
\end{array}
$$

(2) Δ *has a* **counit** *if there is an* **augmentation** *of H_*, $\varepsilon \colon H_* \to k$ and the following diagram commutes*

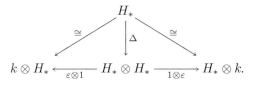

The triple $(H_*, \Delta, \varepsilon)$*, satisfying* (1) *and* (2)*, is called a* **coalgebra** *over* k.

    (3) *A homomorphism,* $\eta\colon k \to H_*$ *is a* **unit** *for the coalgebra* H_* *if the composite* $\varepsilon \circ \eta\colon k \to k$ *is the identity mapping on* k*. (A unit may also be called a* **supplementation** *of* H_**.)*

    (4) *A coalgebra is said to be* **cocommutative** *if the following diagram commutes,*

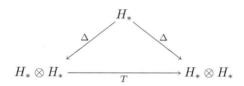

where $T(a \otimes b) = (-1)^{(\deg a)(\deg b)} b \otimes a$.

Proposition 4.38. *For a locally finite space* X *the singular homology of* X *with coefficients in a field* k *is a cocommutative coalgebra over* k*. For a choice of basepoint,* $H_*(X; k)$ *also has a unit.*

Some familiar algebraic objects enjoy the structure of a coalgebra. For example, let $\Lambda(x)$ denote the exterior algebra on a generator x with $\dim x$ odd. Equip $\Lambda(x)$ with the comultiplication determined by $\Delta(x) = 1 \otimes x + x \otimes 1$. This is the unique coalgebra structure on $\Lambda(x)$ up to multiplication by scalars. On $k[x]$, the polynomial algebra on x of even dimension, there is a comultiplication given by $\Delta(x) = 1 \otimes x + x \otimes 1$ and

$$\Delta(x^n) = 1 \otimes x^n + \binom{n}{1} x \otimes x^{n-1} + \cdots + \binom{n}{k} x^k \otimes x^{n-k} + \cdots + x^n \otimes 1.$$

These examples of coalgebras are realized by $H_*(S^{2n+1}; k)$ and $H_*(\mathbb{C}P(\infty); k)$.

An important subspace of a coalgebra is the subspace of **primitives**,

$$\mathrm{Prim}(H_*) = \{x \in H_* \mid \Delta(x) = 1 \otimes x + x \otimes 1\}.$$

The importance of the space of primitives becomes apparent when the coalgebra structure is combined with an algebra structure (as in a Hopf algebra). For our examples, $\mathrm{Prim}(\Lambda(x)) = k\{x\}$ and $\mathrm{Prim}(k[x]) = k\{x, x^p, x^{p^2}, \dots\}$, where $p = \mathrm{char}\, k$, and kS denotes the vector space with basis the set S.

If we apply cohomology with coefficients in a field to the diagonal, basepoint and terminal mappings, the Eilenberg-Zilber theorem gives a **multiplication**, the **cup product** on cohomology,

$$\smile\colon H^*(X; k) \otimes H^*(X; k) \longrightarrow H^*(X; k),$$

an **augmentation**, $\varepsilon\colon H^*(X;k) \to k$ and a **unit**, $\eta\colon k \to H^*(X;k)$. As defined in Chapter 1, a **graded algebra** over k is the appropriate abstraction of these notions. By applying the vector space dual to the definition of coalgebra, we can define an algebra as a triple, (H^*, φ, η), with φ, a multiplication that is associative and has a unit, η. Furthermore, the dual of Proposition 4.38 holds, namely, $H^*(X;k)$ is a graded commutative algebra over k and, for a choice of basepoint, $H^*(X;k)$ has an augmentation. The dual notion to the subspace of primitives of a coalgebra is the **space of indecomposables** of an algebra H^*: Let $I(H^*) = \ker \varepsilon$, then define

$$Q(H^*) = I(H^*) \Big/ \varphi\big(I(H^*) \otimes I(H^*)\big).$$

For the examples of $\Lambda(x)$ and $k[x]$, considered as algebras, $Q(\Lambda(x)) \cong k\{x\}$ and $Q(k[x]) \cong k\{x\}$.

We record some consequences of duality.

Theorem 4.39. *Let H_* denote a coalgebra over k with augmentation ε and unit η. Suppose H_* is of finite type, that is, $\dim_k H_n$ is finite for all n, then*

(1) *$(H_*^{\mathrm{dual}}, \Delta^{\mathrm{dual}}, \varepsilon^{\mathrm{dual}})$ is an algebra over k.*
(2) *If Δ is cocommutative, Δ^{dual} is commutative.*
(3) *η^{dual} is an augmentation of H_*^{dual}.*

Next we consider spaces with some further structure. Suppose that X is an **H-space**, that is, there is a continuous mapping, $m\colon X \times X \to X$, a **multiplication** and furthermore, for a choice of basepoint, $e \in X$, the mappings $m_l(x) = m(x, e)$ and $m_r(x) = m(e, x)$ are homotopic to the identity on X. Examples of H-spaces abound: If (G, μ, e) is a topological group, then it is an H-space. The loop multiplication $*$ on $\Omega(X, x_0)$ with basepoint c_{x_0} gives $\Omega(X, x_0)$ the structure of an H-space. Finally, notice that the group multiplication on an *abelian* group G is a group homomorphism and so induces a mapping, for each n, $K(G, n) \times K(G, n) \to K(G, n)$; this can be seen to be an H-space structure.

The homology of an H-space with field coefficients is endowed with a multiplication, $m_*\colon H_*(X;k) \otimes H_*(X;k) \to H_*(X;k)$, called the **Pontryagin product**. Dually, the cohomology algebra obtains a comultiplication, m^*. The properties that result are abstractly given in the following definition.

Definition 4.40. *Let H denote a graded vector space of finite type over a field k, equipped with a multiplication $\varphi\colon H \otimes H \to H$, a comultiplication, $\Delta\colon H \to H \otimes H$, a unit, $\eta\colon k \to H$, and an augmentation, $\varepsilon\colon H \to k$. Then H is a **Hopf algebra** over k if*

(1) *(H, φ, η) is an algebra over k with augmentation ε.*
(2) *(H, Δ, ε) is a coalgebra over k with unit η.*

(3) *The following diagram commutes*

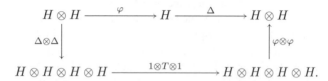

That is, Δ *is a morphism of algebras or, equivalently,* φ *is a morphism of coalgebras.*

We have assumed in the definition of algebra and coalgebra the conditions of associativity and coassociativity. Without these conditions (H, ϕ, Δ, η) is a **bialgebra**, a structure at the heart of the study of quantum groups ([Shnider-Sternberg93]).

Theorem 4.41. *If X is a locally finite H-space, then $H^*(X; k)$ is a commutative Hopf algebra over k and $H_*(X; k)$ is a cocommutative Hopf algebra over k. Furthermore, the Hopf algebras $H_*(X; k)$ and $H^*(X; k)$ are dual to each other.*

The dual of a Hopf algebra of finite type is also a Hopf algebra. The structure of Hopf algebras, as developed in the study of Lie groups and H-spaces, is well-known in the useful cases (see Theorems 6.37 and Theorem 10.1 for examples).

We next present some further structure on singular cohomology—the Steenrod operations. Taken together these operations have the structure of a Hopf algebra over the field of p elements $\mathbb{F}_p$ known as the Steenrod algebra.

When distinguishing two spaces by their cohomology with field coefficients, the first piece of structure available is that of a graded vector space. When two spaces have isomorphic graded vector spaces for their cohomology (for example, $\mathbb{C}P(2)$ and $S^2 \vee S^4$), then the cup product structure on the cohomology may distinguish the spaces by finding different graded algebras. However, this too may fail to distinguish the spaces (for example, $\Sigma\mathbb{C}P(2)$ and $S^3 \vee S^5$) and so further structure is desirable.

Definition 4.42. *A* **cohomology operation** *of type* (G, n, G', m) *is a natural transformation,* $\theta\colon H^n(\ ; G) \to H^m(\ ; G')$. *That is, for all spaces, X and Y, and mappings $f\colon X \to Y$, there are functions θ_X, θ_Y, such that the following diagram commutes*

$$
\begin{array}{ccc}
H^n(X; G) & \xrightarrow{\ \theta_X\ } & H^m(X; G') \\[4pt]
{\scriptstyle f^*}\big\uparrow & & \big\uparrow{\scriptstyle f^*} \\[4pt]
H^n(Y; G) & \xrightarrow[\ \theta_Y\ ]{} & H^m(Y; G').
\end{array}
$$

Over a field, k, the cup product on $H^*(X;k)$ allows us to define a squaring map $\theta(u) = u \smile u$. This operation θ is of type $(k, n, k, 2n)$, for each n.

Consider the short exact sequence of rings

$$0 \to \mathbb{Z}/p\mathbb{Z} \xrightarrow{\times p} \mathbb{Z}/p^2\mathbb{Z} \to \mathbb{Z}/p\mathbb{Z} \to 0.$$

If we tensor $C^*(X)$, the integral cochains, with this sequence of coefficients, we get a long exact sequence on homology

$$\cdots \to H^n(X; \mathbb{Z}/p^2\mathbb{Z}) \to H^n(X; \mathbb{Z}/p\mathbb{Z}) \xrightarrow{\beta} H^{n+1}(X; \mathbb{Z}/p\mathbb{Z}) \to \cdots.$$

For each $n \geq 1$, there is a connecting homomorphism, natural in X, that determines a sequence of operations, β, of type $(\mathbb{Z}/p\mathbb{Z}, n, \mathbb{Z}/p\mathbb{Z}, n+1)$. These operations are called the **Bockstein homomorphisms** (see Chapter 10).

More generally, one can use Theorem 4.34 to produce cohomology operations. Let θ be any class in $[K(G,n), K(G',m)]$ and u, a class in $H^n(X;G)$. Then $u = [f]$ for $f\colon X \to K(G,n)$ and so composition with θ determines a class in $[X, K(G',m)] = H^m(X;G')$ which is denoted by $\theta(u)$. By the same kind of obstruction theory argument that proves Theorem 4.34, one can prove the following characterization of operations.

Theorem 4.43. *Let* $\mathrm{Oper}(G, n, G', m)$ *denote the set of all cohomology operations of type* (G, n, G', m). *Then there is a one-one correspondence between* $\mathrm{Oper}(G, n, G', m)$ *and* $[K(G,n), K(G',m)]$.

Since $[K(G,n), K(G',m)] \cong H^m(K(G,n); G')$, it becomes necessary to compute the cohomology of Eilenberg-Mac Lane spaces in order to determine cohomology operations. This problem is taken up in Chapter 6. In Chapter 9, there is a detailed discussion of the application of these operations in homotopy calculations.

The collection of all cohomology operations, for a given field k, is sometimes a large, unmanageable set and so it is unsuitable for straightforward application. We reduce our attention to a more manageable set by considering sequences of operations that are independent of dimension and related by the suspension isomorphism.

Definition 4.44. *Let* $\{\,\theta_n \in \mathrm{Oper}(k, n, k, n+i)\,\}$ *denote a sequence of cohomology operations for i fixed. Then* $\{\theta_n\}$ *determines a **stable cohomology operation** of degree i if the following diagram commutes*

$$
\begin{array}{ccc}
\tilde{H}^n(X;k) & \xrightarrow{\ E\ } & \tilde{H}^{n+1}(SX;k) \\
{\scriptstyle \theta_n}\downarrow & & \downarrow{\scriptstyle \theta_{n+1}} \\
\tilde{H}^{n+i}(X;k) & \xrightarrow[\ E\ ]{} & \tilde{H}^{n+i+1}(SX;k)
\end{array}
$$

where $E \colon \tilde{H}^m(X;k) \to \tilde{H}^{m+1}(SX;k)$ *is the* **suspension isomorphism,**

$$E \colon \tilde{H}^m(X;k) \cong [X, K(k,m)] \cong [X, \Omega K(k,m+1)] \cong [SX, K(k,m+1)].$$

An example of such an operation is the Bockstein homomorphism: We can write the suspension isomorphism explicitly as the mapping determined by the diagram

$$
\begin{array}{ccc}
\tilde{H}^m(X;k) & \xrightarrow{\ \ E\ \ } & \tilde{H}^{m+1}(SX;k) \\
{\scriptstyle \delta}\downarrow{\scriptstyle \cong} & & \uparrow{\scriptstyle \cong} \\
\tilde{H}^{m+1}(C_+X, X;k) & \xleftarrow[\cong]{\text{excision}} & \tilde{H}^{m+1}(SX, C_-X;k)
\end{array}
$$

where $SX = C_+X \cup C_-X$, the union of two cones. The Bockstein homomorphism commutes with the mappings of pairs and so with the suspension isomorphism.

[Steenrod47] introduced a family of stable cohomology operations over $\mathbb{F}_2$. With composition as product, these operations form an algebra, denoted $\mathcal{A}_2$ and called the **mod 2 Steenrod algebra.** The analogous operations were constructed by [Steenrod52] over $\mathbb{F}_p$, for odd primes, p, to form the **mod p Steenrod algebra,** $\mathcal{A}_p$. In the early 1950's, [Cartan50], and [Adem52] established the structure of these algebras and then [Serre53] and [Cartan54] showed that Steenrod's constructions gave all possible stable cohomology operations over the prime characteristic fields.

Steenrod's constructions take place at the level of cochains and involve certain equivariant acyclic complexes and the cohomology of groups. We refer the reader to the readable accounts of [Steenrod-Epstein62] and [Mosher-Tangora68] for this construction. Alternative constructions of the Steenrod algebra may be found in the papers of [May70] and [Karoubi95]. The structure and action of the Steenrod algebra is described in the following theorem, which is stated twice—for the prime 2 and for the odd primes.

Theorem 4.45 (mod 2). *The mod 2 Steenrod algebra,* $\mathcal{A}_2$, *is generated by operations*

$$Sq^i \colon H^*(\ ;\mathbb{F}_2) \to H^{*+i}(\ ;\mathbb{F}_2),$$

for $i \geq 0$, satisfying

(1) $Sq^0 =$ *the identity homomorphism.*
(2) *If $x \in H^n(X;\mathbb{F}_2)$, then $Sq^n x = x^2$.*
(3) *If $x \in H^n(X;\mathbb{F}_2)$ and $i > n$, then $Sq^i x = 0$.*
(4) *For all $x, y \in H^*(X;\mathbb{F}_2)$,*

$$Sq^k(x \smile y) = \sum_{i=0}^{k} Sq^i x \smile Sq^{k-i} y, \qquad \text{the **Cartan formula.**}$$

(5) Sq^1 is the Bockstein homomorphism associated to the short exact sequence of coefficients,

$$0 \to \mathbb{Z}/2\mathbb{Z} \xrightarrow{\times 2} \mathbb{Z}/4\mathbb{Z} \to \mathbb{Z}/2\mathbb{Z} \to 0.$$

(6) *The following relations hold among the generators: if* $0 < a < 2b$

$$Sq^a Sq^b = \sum_{j=0}^{[a/2]} \binom{b-1-j}{a-2j} Sq^{a+b-j} Sq^j.$$

These relations are called the **Adem relations**.

Theorem 4.45 (for p an odd prime). *The mod p Steenrod algebra, $\mathcal{A}_p$, is generated by operations* $P^i \colon H^*(\ ; \mathbb{F}_p) \to H^{*+2i(p-1)}(\ ; \mathbb{F}_p)$, *for* $i \geq 0$, *along with the Bockstein homomorphism,* $\beta \colon H^*(\ ; \mathbb{F}_p) \to H^{*+1}(\ ; \mathbb{F}_p)$, *associated with the short exact sequence of coefficients,*

$$0 \to \mathbb{Z}/p\mathbb{Z} \xrightarrow{\times p} \mathbb{Z}/p^2\mathbb{Z} \to \mathbb{Z}/p\mathbb{Z} \to 0.$$

These operations satisfy the following

(1) $P^0 =$ *the identity homomorphism.*
(2) *If* $x \in H^{2n}(X; \mathbb{F}_p)$, *then* $P^n x = x^p$.
(3) *If* $x \in H^n(X; \mathbb{F}_p)$ *and* $k > 2n$, *then* $P^k x = 0$.
(4) *For all* $x, y \in H^*(X; \mathbb{F}_p)$, *the* **Cartan formula** *holds*

$$P^k(x \smile y) = \sum_{j=0}^{k} P^j x \smile P^{k-j} y.$$

Also, β is a derivation,

$$\beta(x \smile y) = \beta x \smile y + (-1)^{\dim x} x \smile \beta y.$$

(5) *The following relations hold among the generators:*

$$P^a P^b = \sum_{j=0}^{[a/p]} \binom{(p-1)(b-j)-1}{a-pj} P^{a+b-j} P^j$$

and if $a \leq pb$, *then*

$$P^a \beta P^b = \sum_{j=0}^{[a/p]} (-1)^{a+j} \binom{(p-1)(b-j)}{a-pj} \beta P^{a+b-j} P^j$$
$$+ \sum_{j=0}^{[(a-1)/p]} (-1)^{a+j-1} \binom{(p-1)(b-j)-1}{a-pj-1} P^{a+b-j} \beta P^j.$$

These relations are called the **Adem relations**.

In these theorems, all binomial coefficients are taken modulo the prime and [ ] denotes the greatest integer function. An elegant and useful reformulation of the Adem relations has been given by [Bullett-MacDonald82].

We can use the mod 2 Steenrod operations to distinguish $\Sigma\mathbb{C}P(2)$ and $S^3 \vee S^5$. The difference is detected by $Sq^2(H^3(\Sigma\mathbb{C}P(2);\mathbb{F}_2) \neq \{0\} = Sq^2(H^3(S^3 \vee S^5;\mathbb{F}_2)$. The applications of these operations pervade algebraic topology. For further examples we refer the reader to Chapters 6, 8, 9, and the classic book by [Steenrod-Epstein62].

We close this section by recording some useful facts about the Steenrod algebra that will be applied in later chapters.

FACT 1: The vector space of indecomposables of $\mathcal{A}_2$, is given by

$$Q(\mathcal{A}_2) = \mathbb{F}_2\{Sq^1, Sq^2, \dots, Sq^{2^i}, \dots\}.$$

For p, an odd prime,

$$Q(\mathcal{A}_p) = \mathbb{F}_p\{\beta, P^1, P^p, \dots, P^{p^k}, \dots\}.$$

The space of indecomposables of an algebra generates it by taking the span of all products of indecomposables. FACT 1 follows from the Adem relations. To determine a vector space basis for each Steenrod algebra, we push further with the Adem relations.

Theorem 4.46 (mod 2). *Let $I = (i_1, i_2, \dots, i_r)$ be a sequence of nonnegative integers. We say that I is* **admissible** *if $i_{s-1} \geq 2i_s$ for $r \geq s > 1$. Associate to I the product of generators $Sq^I = Sq^{i_1}Sq^{i_2} \cdots Sq^{i_r}$. We say that $Sq^I \in \mathcal{A}_2$ is admissible if the sequence I is admissible. The admissible products form a vector space basis for $\mathcal{A}_2$.*

Theorem 4.46 (for p an odd prime). *Let J be a sequence of nonnegative integers, $(\varepsilon_0, s_1, \varepsilon_1, \dots, s_k, \varepsilon_k)$, where $\varepsilon_i = 0$ or 1 for all i. We say that J is* **admissible** *if $s_i \geq ps_{i+1} + \varepsilon_i$ for $k > i \geq 1$. Associate to J the product of generators of $P^J = \beta^{\varepsilon_0}P^{s_1}\beta^{\varepsilon_1} \cdots P^{s_k}\beta^{\varepsilon_k}$ We say that $P^J \in \mathcal{A}_p$ is admissible if the sequence J is admissible. The admissible products form a vector space basis for $\mathcal{A}_p$.*

FACT 2: Let $\Delta: \mathcal{A}_p \to \mathcal{A}_p \otimes \mathcal{A}_p$ be the comultiplication that extends the Cartan formulas: for $p = 2$, $\Delta(Sq^k) = \sum_{i=0}^{k} Sq^i \otimes Sq^{k-i}$, and for p odd,

$$\Delta(\beta) = 1 \otimes \beta + \beta \otimes 1, \text{ and } \Delta(P^k) = \sum_{j=0}^{k} P^j \otimes P^{k-j}.$$

Then $\mathcal{A}_p$ is a cocommutative Hopf algebra over $\mathbb{F}_p$ for any prime, p.

This fact was first observed by [Milnor58]. It follows that $H^*(X; \mathbb{F}_p)$ is a module over the Hopf algebra $\mathcal{A}_p$ for any space X. Furthermore, the dual of $\mathcal{A}_p$, $\mathcal{A}_p^{\text{dual}}$, is a commutative Hopf algebra over $\mathbb{F}_p$. From the structure theorems for Hopf algebras proved by [Borel53] and [Milnor-Moore65] and the known action of $\mathcal{A}_p$ on certain test spaces, [Milnor58] computed the structure of $\mathcal{A}_p^{\text{dual}}$:

Theorem 4.47 (mod 2). *As an algebra, $\mathcal{A}_2^{\text{dual}}$ is isomorphic to the polynomial algebra*

$$\mathbb{F}_2[\xi_1, \xi_2, \xi_3, \dots]$$

where $\deg \xi_i = 2^i - 1$. *As a coalgebra, the comultiplication on $\mathcal{A}_2^{\text{dual}}$ is determined by* $\Delta(\xi_k) = \sum_{i=0}^{k} \xi_{k-i}^{2^i} \otimes \xi_i$.

Theorem 4.47 (for p an odd prime). *As an algebra, $\mathcal{A}_p^{\text{dual}}$ is isomorphic to the tensor product of an exterior and a polynomial algebra,*

$$\Lambda(\tau_0, \tau_1, \dots) \otimes \mathbb{F}_p[\xi_1, \xi_2, \dots],$$

where $\deg \tau_i = 2p^i - 1$ *and* $\deg \xi_j = 2(p^j - 1)$. *As a coalgebra, the comultiplication on $\mathcal{A}_p^{\text{dual}}$ is given by*

$$\Delta(\xi_k) = \sum_{i=0}^{k} \xi_{k-i}^{p^i} \otimes \xi_i, \quad \Delta(\tau_k) = \tau_k \otimes 1 + \sum_{i=0}^{k} \xi_{k-i}^{p^i} \otimes \tau_i.$$

The implications of these structure theorems become apparent in Chapter 9 where they figure prominently in the calculations associated with the Adams spectral sequence.

Exercises

4.1. Suppose that $f \colon X \to Y$ is a continuous function. Show that the inclusion $i \colon X \to I_f$, the mapping cylinder, is a cofibration. Show that Y is a deformation retraction of I_f.

4.2. Show that the composition of cofibrations is a cofibration. Show that the composition of fibrations is a fibration.

4.3. Suppose that $f \colon (X, x_0) \to (Y, y_0)$ is a map of pointed spaces. Show that the iterated mapping cone sequence

$$X \xrightarrow{f} Y \xrightarrow{i} M_f = Y \cup_f CX \xrightarrow{j} (Y \cup_f CX) \cup_i CY$$

gives a space $(Y \cup_f CX) \cup_i CY$ homotopy equivalent to ΣX.

4.4. Let SX denote the reduced suspension on X. Show that for (X, x_0) and (Y, y_0) pointed countable CW-complexes there is a isomorphism of groups

$$[(SX, sx_0), (Y, y_0)] \cong [(X, x_0), \Omega(Y, y_0)].$$

Show further that for a mapping of pointed spaces $f \colon (X, x_0) \rightarrow (Y, y_0)$, the mappings induced by the iterated mapping cone sequence give an exact sequence of groups:

$$[\Sigma X, Z] \xleftarrow{\Sigma f^*} [\Sigma Y, Z] \xleftarrow{\Sigma i^*} [\Sigma M_f, Z] \xleftarrow{\Sigma j^*} [\Sigma^2 X, Z] \leftarrow \cdots.$$

4.5. Prove that the suspension mapping $E \colon [S^k X, S^k Y] \rightarrow [S^{k+1} X, S^{k+1} Y]$ is a homomorphism for $k \geq 1$.

4.6. Extend the discussion of cellular homology to include coefficients in an abelian group G. Compute $H_*(\mathbb{R}P(n))$ and $H_*(\mathbb{R}P(n); \mathbb{F}_2)$ from an appropriate cell decomposition. What decomposition would be appropriate to compute $H_*(\mathbb{C}P(n))$ cellularly? Let $\mathbb{Z}/m\mathbb{Z}$ act on the circle as roots of unity (a generator of $\mathbb{Z}/m\mathbb{Z}$ corresponds to a primitive m^{th} root of unity in S^1). Since S^1 acts on S^{2n-1}, the unit vectors in $\mathbb{C}^n$, this gives an action of $\mathbb{Z}/m\mathbb{Z}$ on S^{2n-1}. Let $L^{2n-1}(m)$ denote the orbit space of this action; this is an example of a **lens space** ([Whitehead, GW78, p. 91]). Compute the cellular homology of $L^{2n-1}(m)$.

4.7. Determine explicitly the lifting function for the path-loop fibration over a space X. This leads to an action of ΩX on itself. Determine that action explicitly.

4.8. If $p \colon E \rightarrow B$ is a fibration and $(X, x_0) \in \mathcal{W}$, then show that precomposition with p gives a fibration:

$$p \circ - \colon \operatorname{map}((X, x_0), (E, e_0)) \rightarrow \operatorname{map}((X, x_0), (B, p(e_0))).$$

4.9. Show that $(\partial I \times I \cup I \times \{0\}) \subset I \times I$ is a cofibration.

4.10. Suppose that A is a connected Hopf algebra over a field k, that is, $A = k \oplus I(A)$, where $I(A)$ consists of elements in positive degrees. Show that the space of primitives is the kernel of the reduced comultiplication

$$\bar{\Delta} \colon I(A) \subset A \xrightarrow{\Delta} A \otimes A \rightarrow I(A) \otimes I(A)$$

where we can take $I(A)$ isomorphic to the cokernel of the unit $\eta \colon k \rightarrow A$. Show that there is a natural homomorphism $\operatorname{Prim}(A) \rightarrow Q(A)$.

4.11. A sequence $0 \rightarrow K \rightarrow H \rightarrow Q \rightarrow 0$ of Hopf algebras is a short exact sequence if K is a sub-Hopf algebra of H and $I(K) \cdot H = H \cdot I(K)$, that is, K is a *normal* sub-Hopf algebra, and $Q \cong k \otimes_K H \cong H/I(K) \cdot H$ as graded k-vector spaces. Show that the functor that associates to a Hopf algebra its subspace of primitives is left exact, that is, a short exact sequence of Hopf algebras, $0 \rightarrow K \rightarrow H \rightarrow Q \rightarrow 0$ gives a short exact sequence of vector spaces $0 \rightarrow \operatorname{Prim}(K) \rightarrow \operatorname{Prim}(H) \rightarrow \operatorname{Prim}(Q)$.

5

The Leray-Serre Spectral Sequence I

> " . . . the notion of a fibre space hides subordinate struc-
> tures, now revealed only in part by the array of differen-
> tials of the associated spectral sequence."
>
> S. Mac Lane

After 1930, new invariants of a topological space X were introduced: the higher homotopy groups $\pi_*(X)$ by [Hurewicz35/36] and [Čech32], and the cohomology ring by [Alexander35], [Kolmogoroff36], and [Whitney38]. The various homology theories (simplicial, de Rham, Čech, singular, etc.) were clarified through the axiomatization of [Eilenberg-Steenrod45].

Computations of these invariants proceeded slowly, especially for spaces of importance, such as manifolds, Lie groups, and their associated homogeneous spaces. The inclusion of a closed connected subgroup H of a compact Lie group G leads to a fibration (indeed, a fibre bundle) $H \hookrightarrow G \to G/H$. Considerable effort was directed at discovering the relations between the combinatorial invariants of H, G, and G/H. [Hurewicz35/36] already appreciated the relations among the new homotopy groups for $(H, G, G/H)$. Although he did not have the language of long exact sequences, he had shown what was equivalent to the exactness of the sequence of groups

$$\cdots \to \pi_n(H) \to \pi_n(G) \to \pi_n(G/H) \to \pi_{n-1}(H) \to \cdots$$

and he gave the isomorphism $\pi_j(S^3) \cong \pi_j(S^2)$ for $j > 2$ as an example.

Other geometric problems, like the existence of nonzero vector fields on manifolds, led to topological structures with properties analogous to the fibre bundle $(H, G, G/H)$. [Whitney35] introduced sphere bundles in this context and developed his theory of characteristic classes toward the classification of sphere bundles. Throughout the 30's and 40's a search for the most general and useful definition of a fibre space occupied many topologists (a history of this search has been written by [Zisman99]). The most influential efforts (recounted in the problem list of [Massey55]) were due to [Hurewicz-Steenrod41] who introduced slicing functions on metric spaces, to [Ehresmann-Feldbau41] who sought to generalize to topology the geometric notion of connection in a fibre bundle with smooth structural group (see [Steenrod52]), and to [Eckmann42] who focused on the homogeneous space case and its homotopy lifting properties.

During the Second World War, [Hopf41] introduced H-spaces as a generalization of Lie groups. He also showed that the cohomology of an H-space enjoyed the additional structure of a Hopf algebra. [Samelson41] developed Hopf's ideas further and introduced the notion of a closed subgroup H of a compact Lie group G being totally nonhomologous to zero in order to show that under these circumstances $H_*(H; \mathbb{Q}) \otimes H_*(G/H; \mathbb{Q})$ is isomorphic to $H_*(G; \mathbb{Q})$. [Hopf42] also studied the class of spaces dubbed *aspherical* by [Hurewicz35/36] for which $\pi_j(X) = \{0\}$ for $j > 1$. Hurewicz had shown that the homology groups of X were determined by the fundamental group. Hopf went much further when he gave an explicit formula in terms of the fundamental group, for any space Y, for $H_2(Y)/h(\pi_2(Y))$, where h denotes the Hurewicz map. When specialized to an aspherical space X, this gives the second homology of the group $\pi_1(X)$. This result launched the theory of the homology of groups (see Chapter 8bis). The notion of Eilenberg-Mac Lane spaces grew out of the generalization of Hopf's work to higher homotopy groups and higher orders of connectivity ([Eilenberg-Mac Lane45]).

[Leray46] solved the problem of relating the cohomology rings of spaces (F, E, π, B) making up a fibre space. He gave the first explicit example of a spectral sequence in the cadre of sheaves and a general cohomology theory that he had formulated while a prisoner-of-war (see [Leray45]). Leray's cohomology ring (based on his *couvertures*) specializes to de Rham cohomology, to Čech cohomology, to Alexander-Spanier cohomology, and to singular cohomology with the appropriate choice of complexes (see [Borel51, 98] and [Houzel90]). Refinements of Leray's algebraic apparatus were due to [Koszul47] and [Cartan48]. Their work adapted well to the study of homogeneous spaces. For the computations of homotopy groups, however, a tool like Leray's spectral sequence was needed for singular homology for which theorems like the Hurewicz theorem and the Whitehead theorem reveal a close computational relation.

Such a spectral sequence for singular homology appeared in a series of *Comptes Rendus* notes by [Serre50]. Complete details appeared in his classic thesis ([Serre51]) that introduced the very general notion of a (Serre) fibration making many computational examples possible and even computable. A short time later, [Borel53] published his Paris thesis in which the computational power of spectral sequences was used to extend the known results on the homotopy theory of Lie groups and homogeneous spaces much further, especially in the case of coefficients in a finite field. (For more details on the history of spectral sequences see [McCleary99].)

The main theorems of [Serre51] and of this chapter are the following:

Theorem 5.1 (the homology Leray-Serre spectral sequence). *Let G be an abelian group. Suppose $F \hookrightarrow E \xrightarrow{\pi} B$ is a fibration, where B is path-connected,* *and F connected. Then there is a first quadrant spectral sequence, $\{E^r_{*,*}, d^r\}$, converging to $H_*(E; G)$, with*

$$E^2_{p,q} \cong H_p(B; \mathcal{H}_q(F; G)),$$

the homology of the space B with local coefficients in the homology of the fibre of π. Furthermore, this spectral sequence is natural with respect to fibre-preserving maps of fibrations.

Theorem 5.2 (the cohomology Leray-Serre spectral sequence). *Let R be a commutative ring with unit. Suppose $F \hookrightarrow E \xrightarrow{\pi} B$ is a fibration, where B is path-connected and F is connected. Then there is a first quadrant spectral sequence of algebras, $\{E^{*,*}_r, d_r\}$, converging to $H^*(E; R)$ as an algebra, with*

$$E^{p,q}_2 \cong H^p(B; \mathcal{H}^q(F; R)),$$

the cohomology of the space B with local coefficients in the cohomology of the fibre of π. This spectral sequence is natural with respect to fibre-preserving maps of fibrations. Furthermore, the cup product $\smile$ on cohomology with local coefficients and the product $\cdot_2$ on $E^{*,*}_2$ are related by $u \cdot_2 v = (-1)^{p'q} u \smile v$ when $u \in E^{p,q}_2$ and $v \in E^{p',q'}_2$.

The E_2-terms of these spectral sequences are expressed in terms of local coefficient systems which are induced by the fibration over the space B. Such systems of coefficients arise naturally in the study of obstruction theory and in the study of sheaves. [Steenrod44] has given the definitive treatment on which we base much of our discussion.

In this chapter we derive the homology version of the spectral sequence (Theorem 5.1). It is possible to extend the statement of the main theorems to the case of fibrations of pairs and we will sometimes use the more general statement. The reader is invited to derive the relative cases in the exercises. In order to go quickly to the applications, we postpone the details of the proof to appendices (§5.3). In §5.1, the motivating ideas for the construction of the spectral sequence are given and the E^1-term and the differential d^1 are identified. We also describe the cohomology version of the spectral sequence and record its relevant multiplicative structure.

In §5.2, with machine in hand, we turn to applications. The algebraic examples from Chapter 1 can be applied to yield significant topological results. The classical theorems of Leray and Hirsch and of Borel and Serre are derived in short order with these tools. After such immediate applications, we turn to less general results and compute the cohomology of various Lie groups and homogeneous spaces. We also explore some of the implications in homotopy theory of computations for the based loop space made possible by the spectral sequence. Perhaps the most spectacular results are the generalization of a theorem of Marston Morse on the existence of infinitely many distinct geodesics joining two points in a Riemannian manifold and the proof that $\pi_i(S^{2n-1})$ is finite for $i > 2n - 1$. For the most part, the fibrations considered will have simple systems of local coefficients (that is, one can take them to be constant) and so experience with local coefficient systems is unnecessary (see Chapter 8bis for more subtle results).

In §5.3 the reader can find the details of a proof of the homology version of the Leray-Serre spectral sequence and the proof of the multiplicative properties of the cohomology spectral sequence. This requires developing the notion of homology and cohomology with coefficients in a bundle of abelian groups and then identifying $H(E^1, d^1)$ in terms of these notions. The exposition in §5.3 and its subsections owes much to the paper of [Brown, E94]. In the rest of the chapter, we have followed [Serre51] and [Borel53].

On first reading, the novice is encouraged to skip §5.3 and go on to further applications of the spectral sequence in Chapter 6. However, §5.1 may disguise the fact that it is the topological structure of a fibration that allows us to identify the E_2-term of the spectral sequence as something familiar. The appendices show that nontrivial results are needed in order to support the powerful machine we exercise in the applications.

5.1 Construction of the spectral sequence

We begin with a fibration: $F \hookrightarrow E \xrightarrow{\pi} B$. Recall that a fibration enjoys the homotopy lifting property (§4.3), that is, if K is a space and we have a mapping $g \colon K \to E$ and a homotopy $H \colon K \times I \to B$ so that $\pi \circ g(x) = H(x, 0)$, then the mapping H can be lifted to a homotopy $\tilde{H} \colon K \times I \to E$, so that $\pi \circ \tilde{H} = H$.

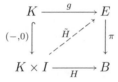

We take B to be in a "convenient category" of topological spaces, that is, B has the homotopy-type of a CW-complex. We replace B with an equivalent CW-complex in what follows. A CW-complex is equipped with a filtration by skeleta. We lift the filtration of B to a filtration on E by letting $J^s = \pi^{-1}(B^{(s)})$, the subspace of E that lies over the s-skeleton of B:

$$E \supset \cdots \supset J^s \supset J^{s-1} \supset \cdots \supset J^0 \supset \emptyset$$
$$B \supset \cdots \supset B^{(s)} \supset B^{(s-1)} \supset \cdots \supset B^{(0)} \supset \emptyset.$$

We use this filtration to obtain a spectral sequence. Putting the methods of Chapter 2 into action, we first obtain an exact couple giving the spectral sequence. Consider, for each s, the long exact sequence of homology groups for the pair (J^s, J^{s-1}) and assemble these sequences into an exact couple:

$$H_r(J^{s-1}; G) \xrightarrow{\quad i_* \quad} H_r(J^s; G)$$

$$H_{r-1}(J^{s-1}; G) \xleftarrow{\quad \partial \quad} H_r(J^s, J^{s-1}; G) \xrightarrow{\quad j_* \quad}$$

Thus $E^1_{p,q} = H_{p+q}(J^p, J^{p-1}; G)$ and the first differential d^1 is given by

$$j_* \circ \partial \colon H_r(J^s, J^{s-1}; G) \xrightarrow{\partial} H_{r-1}(J^{s-1}; G) \xrightarrow{j_*} H_{r-1}(J^{s-1}, J^{s-2}; G),$$

which is the boundary homomorphism in the long exact sequence of homology groups for the triple (J^s, J^{s-1}, J^{s-2}).

Alternatively, we can begin with $C_*(E; G)$, the singular chains on E with coefficients in G, and filter it by

$$F_s C_*(E; G) = \operatorname{im}(C_*(J^s; G) \to C_*(E; G)).$$

This is an increasing filtration and, by an easy argument (using the compactness and dimension of simplices), this filtration is exhaustive. The inclusions are monomorphisms, so that

$$E^1_{p,q} = H_{p+q}(F_p C_*(E; G)/F_{p-1} C_*(E; G)) = H_{p+q}(J^p, J^{p-1}; G).$$

The argument of Proposition 2.11 carries over and so we have that the two spectral sequences presented are the same. Since the filtration on E ends over the 0-skeleton of B, this is a first-quadrant spectral sequence. Adding the convergence of the spectral sequence we can summarize the discussion as follows:

Proposition 5.3. *Given a fibration $F \hookrightarrow E \xrightarrow{\pi} B$ with base space a CW-complex, there is a first-quadrant spectral sequence, $\{E^n_{*,*}, d^n\}$, with*

$$E^1_{s,r} \cong H_{r+s}(\pi^{-1}(B^{(s)}), \pi^{-1}(B^{(s-1)}); G)$$

and $d^1 = \Delta$, the boundary homomorphism in the exact sequence in homology for the triple $(\pi^{-1}(B^{(s)}), \pi^{-1}(B^{(s-1)}), \pi^{-1}(B^{(s-2)}))$. The spectral sequence converges to $H_*(E; G)$.

For cohomology with coefficients in a commutative ring R, we have the analogous long exact sequences for the pairs (J^s, J^{s-1}) and so the exact couple

$$H^r(J^s; R) \xrightarrow{\quad i^* \quad} H^r(J^{s-1}; R)$$

$$H^{r+1}(J^s; R) \xleftarrow{\quad j^* \quad} H^{r+1}(J^s, J^{s-1}; R) \xrightarrow{\quad \delta \quad}$$

and a spectral sequence with $d_1 = \Delta$, the boundary homomorphism in the long exact sequence in cohomology for the triple (J^s, J^{s-1}, J^{s-2}). The dual filtration of $C^*(E; R)$ is given by

$$F^s C^*(E; R) = \ker(C^*(E; R) \longrightarrow C^*(J^{s-1}; R)),$$

and the dual version of Proposition 5.3 holds for cohomology.

The next step in establishing Theorems 5.1 and 5.2 is the determination of the E^2-term. Consider the simplest case of a trivial fibration, $E = B \times F$. By induction over the skeleta of B, one can show that $(J^s, J^{s-1}) = (B^{(s)}, B^{(s-1)}) \times F$ in this case and so the Künneth theorem can be applied to show that

$$E^1_{p,q} \cong \text{Cell}_p(B) \otimes H_q(F; G)$$

with d^1 given by $d^1 = \partial^{\text{cell}} \otimes 1$. Thus $E^2_{p,q} \cong H_p(B; H_q(F; G))$. In the case of an arbitrary fibration, the 'twisting' of the fibre and base spaces in the total space is a global and possibly nontrivial phenomenon and so this prevents a simple expression for E^1 and hence for E^2.

In §5.3 we define the chains on a space with coefficients in a bundle of groups. The relevant bundle of (graded) groups associated to a fibration is given by

$$\mathcal{H}_*(F; G) = \{H_*(\pi^{-1}(b); G) \mid b \in B\},$$

equipped with the collection of isomorphisms,

$$\{h[\lambda] \colon H_*(\pi^{-1}(b_2); G) \to H_*(\pi^{-1}(b_1); G) \mid \lambda \in \Omega(B, b_1, b_2)\}$$

indexed over the homotopy classes of paths, λ, in B from b_1 to b_2. The main results of §5.3 are that

$$E^1_{p,q} \cong \text{Cell}_p(B; \mathcal{H}_q(F; G))$$

and that $d^1 = \partial^{\text{cell}}$ on these groups. This establishes Theorem 5.1.

In the examples of §5.2 a particular class of bundles of groups often occurs for which all of the isomorphisms $h[\lambda]$ may be taken to be the identity. This is called a **simple system of local coefficients** for which

$$E^1_{p,q} \cong \text{Cell}_p(B; H_q(F; G)), \quad d_1 = \partial^{\text{cell}},$$

that is, the contribution of the varying fibres gives the system of *constant* groups $H_q(F; G)$. If B is simply-connected, then any fibration over B leads to a simple system of local coefficients (Proposition 5.20); thus we can identify the E^1- and E^2-terms of this spectral sequence directly for a large class of examples. We state Theorem 5.1 for this case as follows:

Theorem 5.4. *Let G be an abelian group. Given a fibration, $F \hookrightarrow E \xrightarrow{\pi} B$, where B is path-connected, F is connected, and the system of local coefficients on B determined by the fibre is simple. Then there is a spectral sequence, $\{E^r_{*,*}, d^r\}$, with*

$$E^2_{p,q} \cong H_p(B; H_q(F; G)),$$

and converging to $H_*(E; G)$.

Multiplicative properties of the spectral sequence

Theorem 5.2 asserts that, for a commutative ring R and a fibration $F \hookrightarrow E \xrightarrow{\pi} B$ with B path-connected and F connected, there is a spectral sequence of algebras with

$$E^{p,q}_2 \cong H^p(B; \mathcal{H}^q(F; R))$$

and converging to $H^*(E; R)$ as an algebra. First a word about the proof: For the Čech or Alexander-Spanier cohomology theories, the multiplicative structure is carried along transparently in the construction of the spectral sequence and so we get a spectral sequence of algebras directly with convergence to $H^*(E; R)$ as an algebra ([Leray50], [Borel51]). The result for singular theory, however, is more difficult—it is one of the technical triumphs of Serre's celebrated thesis ([Serre51]).

We present a proof of the product structure for this spectral sequence in §5.3 following the exposition of [Brown, E94]. The interested reader should consult the thesis of [Serre51], not only for the first proofs of Theorems 5.1 and 5.2, but also for a model of exposition. Alternative presentations appear in Chapter 6, in exposé 9 of the 1950 Cartan Séminaire by [Eilenberg50], and in the books of [Hu59], [Hilton-Wylie60], and [Whitehead, GW78]. An exposition of the result for Čech theory appears in the book of [Bott-Tu82].

By introducing some simplifying hypotheses, the spectral sequence takes a manageable form.

Proposition 5.5. *Suppose that the system of local coefficients on B determined by the fibre is simple, that F is connected, and that F and B are of finite type; then, for a field k, we have*

$$E^{p,q}_2 \cong H^p(B; k) \otimes_k H^q(F; k).$$

We leave the proof to the reader with the hint that you should apply the Universal Coefficient theorem.

Suppose that the system of local coefficients is simple and that B and F are connected. Then

$$E^{p,0}_2 \cong H^p(B; H^0(F; R)) \cong H^p(B; R)$$

and $\qquad E^{0,q}_2 \cong H^0(B; H^q(F; R)) \cong H^q(F; R).$

For any bigraded algebra, like $E^{*,*}_2$, both $E^{*,0}_2$ and $E^{0,*}_2$ are subalgebras.

Proposition 5.6. *When restricted to the subalgebras $E_2^{*,0}$ and $E_2^{0,*}$, the product structure in the spectral sequence on $E_2^{*,*}$ coincides with the cup product structure on $H^*(B; R)$ and $H^*(F; R)$, respectively. Furthermore, if, for all p, q, $H^p(B; R)$ and $H^q(F; R)$ are free R-modules of finite type, and the system of local coefficients on B is simple, then $E_2^{*,*} \cong H^*(B; R) \otimes_R H^*(F; R)$ as a bigraded algebra.*

The proof of the first part of this proposition is postponed to Example 5.E where it follows from the naturality of the spectral sequence. The rest of the proposition is an application of the Universal Coefficient theorem.

As the reader begins the applications in §5.2, we remind her or him of the examples in Chapter 1; the techniques sketched there can be applied to the Leray-Serre spectral sequence with considerable success.

5.2 Immediate Applications

In this initial collection of applications we interpret topologically some of the formal consequences of having a spectral sequence. The second set of applications will be concerned with the problem of computing the cohomology of certain Lie groups by using the Leray-Serre spectral sequence as applied to the fibrations that result from quotients by subgroups. Finally, we investigate the path-loop fibration over a space.

We begin with some dimensional arguments.

Example 5.A. *Suppose $F \hookrightarrow E \xrightarrow{\pi} B$ is a fibration with B path-connected and the system of local coefficients on B induced by the fibre is simple. If R is Noetherian and two of the spaces F, E, or B have cohomology a finitely generated R-module in each dimension, then the other space also has cohomology finitely generated in each dimension.*

CASE 1: Suppose B and F have cohomology that is finitely generated in each dimension. The module $H^n(E; R)$ is associated to $\bigoplus_{p+q=n} E_\infty^{p,q}$ via the associated graded module to the filtration on $H^*(E; R)$ and so it suffices to show that each $E_\infty^{p,q}$ is finitely generated over R. We begin with an E_2-term given by $E_2^{p,q} \cong H^p(B; H^q(F; R))$ which is finitely generated by an argument using the Universal Coefficient theorem. Since $E_3^{p,q}$ is a subquotient of $E_2^{p,q}$, it is finitely generated because R is Noetherian. Similarly $E_4^{p,q}$, $E_5^{p,q}$, ... , and so finally $E_\infty^{p,q}$ are all finitely generated.

CASE 2: Suppose E and B have cohomology that is finitely generated in each dimension. We show by induction that $H^n(F; R)$ is also finitely generated. Since $H^0(F; R) = E_\infty^{0,0} = H^0(E; R)$, it is finitely generated. Let n be the least integer such that $H^n(F; R)$ is not finitely generated. Then $E_2^{0,n} = H^n(F; R)$ is not finitely generated and so neither is

$$E_3^{0,n} = \ker d_2 \colon E_2^{0,n} \to E_2^{2,n-1} \cong H^2(B; H^{n-1}(F; R))$$

because the target module is finitely generated by the induction hypothesis; $E_3^{3,n-2}$ is a subquotient of $H^3(B; H^{n-2}(F; R))$, so it is finitely generated and hence $E_4^{0,n} = \ker d_3 \colon E_3^{0,n} \to E_3^{3,n-2}$ cannot be finitely generated. Continuing this argument, we get that $E_\infty^{0,n}$ is not finitely generated but this is a contradiction, since $E_\infty^{0,n}$ is a subquotient of $H^n(E; R)$.

CASE 3: Suppose E and F have cohomology that is finitely generated in each dimension. By an argument similar to the one in case 2, we derive a contradiction if, for some n, $H^n(B; R) = E_2^{n,0}$ is not finitely generated. □

Suppose the assumptions of Example 5.A hold and that $R = k$, a field. Then we have $E_2^{p,q} \cong H^p(B; k) \otimes_k H^q(F; k)$ and we can speak of the Poincaré series and the Euler-Poincaré characteristic. We refer the reader to Example 1.F where we show that $P(E, t) \leq P(B, t) \times P(F, t)$ and $\chi(E) = \chi(B) \cdot \chi(F)$, whenever these expressions are meaningful.

Example 5.B. *Suppose* $F \hookrightarrow E \xrightarrow{\pi} B$ *is a fibration with* F *connected,* B *of finite type, path-connected, and the system of local coefficients on* B *induced by the fibre is simple. Suppose* k *is a field, that* $H^i(B; k) = \{0\}$ *for* $i > p$ *and* $H^j(F; k) = \{0\}$ *for* $j > q$. *Then* $H^i(E; k) = \{0\}$ *for* $i > p + q$ *and* $H^{p+q}(E; k) = H^p(B; k) \otimes H^q(F; k)$.

These assumptions imply that $E_2^{r,s} \cong H^r(B; k) \otimes_k H^s(F; k)$ and furthermore, that $E_2^{*,*}$ is nonzero only in the box pictured.

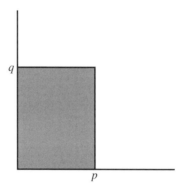

Since the spectral sequence converges to $H^*(E; k)$ and, k is a field, $H^n(E; k) \cong \bigoplus_{r+s=n} E_\infty^{r,s}$ as a vector space implies that $H^i(E; k) = \{0\}$ for $i > p + q$. Also from the picture, no differential can affect $E_2^{p,q}$, so it persists to $E_\infty^{p,q} \cong H^p(B; k) \otimes_k H^q(F; k)$. Since this is the only vector space contributing to $H^{p+q}(E; k)$, the result follows.

A corollary to this formal consequence of the data is the following seminal result of [Borel-Serre50].

Theorem 5.7. *Suppose $F \hookrightarrow \mathbb{R}^n \overset{\pi}{\to} B$ is a locally trivial fibration with B a polyhedron and F connected. Then B and F are acyclic spaces (that is, $\tilde{H}^*(B) \cong \{0\} \cong \tilde{H}^*(F)$).*

PROOF: By the tail end of the homotopy exact sequence,

$$\cdots \to \pi_1(\mathbb{R}^n) \to \pi_1(B) \to \pi_0(F) \to \cdots$$

we can see immediately that $\pi_1(B) = \{0\}$ and so the system of local coefficients induced by B on F is simple. Observe that $H^i(B)$ and $H^i(F)$ are trivial whenever $i > n$. To see this we make some simple observations from dimension theory:

(1) F is a subset of $\mathbb{R}^n$ and
(2) B has a system of neighborhoods, $\{U\}$, so that $\pi^{-1}(U)$ is homeomorphic to $U \times F$ for each U in the system.

Since $\pi^{-1}(U)$ is a subset of $\mathbb{R}^n$, U must have dimension $\leq n$. The argument of Example 5.B applies to show that

$$H^{p+q}(\mathbb{R}^n) \cong H^p(B) \otimes H^q(F) + \mathrm{Tor}_1^{\mathbb{Z}}(H^{p+1}(B), H^q(F)))$$
$$\cong H^p(B) \otimes H^q(F)$$

when p and q are the greatest nonzero dimensions in which B and F, respectively, have nontrivial integral homology. Since $\mathbb{R}^n$ is acyclic, $p + q = 0$ and we conclude that B and F are acyclic. (For the interested reader, we note that the argument of [Borel-Serre50] is given in terms of Alexander-Spanier cohomology with compact supports and original spectral sequence of [Leray50].) □

Another example of this kind of argument, using the spectral sequence as guide, is given by sphere bundles over spheres with spheres as total space. That is, we have a fibration $S^m \hookrightarrow S^n \overset{\pi}{\to} S^\ell$. In this case, using cohomology with coefficients in a field k, our E_2-term has the following form:

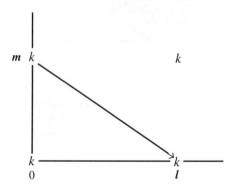

There is a nontrivial differential since the vector-spaces

$$E_2^{0,m} = H^0(S^\ell; H^m(S^m; k)) \cong H^m(S^m; k) = k$$

and $E_2^{\ell,0} \cong H^\ell(S^\ell; H^0(S^m; k)) = k$ must be mapped isomorphically to each other if we are to have the cohomology of a sphere at E_∞. Since differentials have bidegree $(r, 1 - r)$, we conclude that $m = \ell - 1$ and, following Example 5.B, $n = \ell + m = 2\ell - 1$. For $\ell = 2$, 4 and 8, the Hopf fibrations are examples of such fibrations.

Example 5.C (the Gysin sequence). *Suppose $F \hookrightarrow E \xrightarrow{\pi} B$ is a fibration with B path-connected and the system of local coefficients on B induced by the fibre is simple. Suppose further that F is a homology n-sphere, that is, $H_*(F) \cong H_*(S^n)$, for some $n \geq 1$. Then there is an exact sequence:*

$$\to H^k(B; R) \xrightarrow{\gamma} H^{n+k+1}(B; R) \xrightarrow{\pi^*} H^{n+k+1}(E; R) \xrightarrow{Q} H^{k+1}(B; R) \to$$

where $\gamma(u) = z \smile u$ for some z in $H^{n+1}(B; R)$ and, if n is even and $2 \neq 0$ in R, then $2z = 0$.

Since F is a homology n-sphere, $H^m(F; R) \cong R$ when $m = 0$ or n, and $H^m(F; R)$ is trivial in the other dimensions. With B path-connected, the E_2-term of the spectral sequence can be pictured:

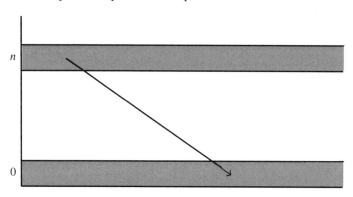

By the placement of holes, $E_2 \cong E_3 \cong \cdots \cong E_{n+1}$ and $H(E_{n+1}, d_{n+1}) \cong E_{n+2} \cong E_\infty$. By the argument of Example 1.D there is an exact sequence

$$\to E_2^{k,n} \xrightarrow{d_{n+1}} E_2^{k+n+1,0} \to H^{k+n+1}(E; R) \to E_2^{k+1,n} \xrightarrow{d_{n+1}} E_2^{k+n+2,0} \to$$

Suppose h is a generator of $H^n(F; R)$. Since we have a simple system of local coefficients and F is a homology n-sphere, we can write

$$E_{n+1}^{*,*} \cong (H^*(B; R) \otimes 1) \oplus (H^*(B; R) \otimes h).$$

Let z be the element in $H^{n+1}(B; R)$ such that $d_{n+1}(1 \otimes h) = z \otimes 1$. (Notice that $d_{n+1}(x \otimes 1) = 0$ for dimensional reasons.) Proposition 5.6 tells us what kind of spectral sequence of algebras we have, and so

$$(-1)^{n \deg x} d_{n+1}(x \otimes h) = d_{n+1}((1 \otimes h) \smile (x \otimes 1))$$
$$= d_{n+1}(1 \otimes h) \smile (x \otimes 1) + (-1)^n (1 \otimes h) \smile d_{n+1}(x \otimes 1)$$
$$= (z \otimes 1) \smile (x \otimes 1) = (z \smile x) \otimes 1.$$

Identifying $E_2^{k,n} = H^k(B; R) \otimes h$ and $E_2^{n+k+1,0} = H^{n+k+1}(B; R) \otimes 1$, we define $\gamma(x) = d_{n+1}(x \otimes h)$ and we get the exact sequence as described.

If n is even, then we use the Leibniz rule and the fact that $h \smile h = 0$ to compute:

$$0 = d_{n+1}(1 \otimes (h \smile h)) = d_{n+1}((1 \otimes h) \smile (1 \otimes h))$$
$$= (z \otimes 1) \smile (1 \otimes h) + (-1)^n (1 \otimes h) \smile (z \otimes 1)$$
$$= 2z \otimes h.$$

Since $2h \neq 0$, we must have $2z = 0$.

The long exact sequence in this example is called the **Gysin sequence** ([Gysin41]). With this sequence, we can compute $H^*(\mathbb{C}P(n); R)$ as an algebra. The n-dimensional complex projective space, $\mathbb{C}P(n)$, sits in a fibration with spherical fibre, $S^1 \hookrightarrow S^{2n+1} \to \mathbb{C}P(n)$. We have $\mathbb{C}P(1) \cong S^2$ and $\mathbb{C}P(n)$ is simply-connected for all n. Thus the system of local coefficients is always simple. We next show that

$$H^*(\mathbb{C}P(n); R) \cong R[x_2] \Big/ (x_2^{n+1}),$$

the **truncated polynomial algebra** of height $n+1$ on one generator of degree 2.

First notice that, $x_2 = \gamma(1)$ generates $H^2(\mathbb{C}P(n); R)$ from the initial part of the Gysin sequence:

$$H^0(\mathbb{C}P(n); R) \xrightarrow{\gamma} H^2(\mathbb{C}P(n); R) \to H^2(S^{2n+1}; R) \to \cdots$$

Moving on a little further, we find, for $n \geq 1$, the short exact sequence

$$\{0\} \to H^1(\mathbb{C}P(n); R) \xrightarrow{\gamma} H^3(\mathbb{C}P(n); R) \to \{0\}$$

where the trivial modules are $H^k(S^{2n+1}; R)$ for $k = 2, 3$. Since $\mathbb{C}P(n)$ is simply-connected, $H^1(\mathbb{C}P(n); R)$ is trivial and hence, so is $H^3(\mathbb{C}P(n); R)$.

By induction, suppose that $(x_2)^k$ generates $H^{2k}(\mathbb{C}P(n); R)$. Then consider the portion of the Gysin sequence,

$$\to H^{2k}(\mathbb{C}P(n); R) \xrightarrow{\gamma} H^{2k+2}(\mathbb{C}P(n); R) \to H^{2k+2}(S^{2n+1}; R) \to .$$

When $k < n$, we have that γ is an isomorphism, so $x_2 \smile (x_2)^k = (x_2)^{k+1}$ generates $H^{2k+2}(\mathbb{C}P(n); R)$. In odd dimensions, the pattern of trivial modules continues.

Finally, when $k = n$, we have the short exact sequence,

$$0 \to H^{2n+1}(\mathbb{C}P(n); R) \to H^{2n+1}(S^{2n+1}; R) \xrightarrow{Q}$$
$$H^{2n}(\mathbb{C}P(n); R) \xrightarrow{\gamma} H^{2n+2}(\mathbb{C}P(n); R) \to 0.$$

Since $\mathbb{C}P(n)$ is a $2n$-dimensional manifold, γ is the trivial homomorphism and Q is an isomorphism. Since $\gamma((x_2)^n) = 0$ and $\gamma((x_2)^n) = x_2^{n+1}$, we conclude $(x_2)^{n+1} = 0$.

We leave it for the reader to derive the **Wang sequence** ([Wang49]) for a fibration $F \hookrightarrow E \to B$ with B, a simply-connected homology n-sphere and F path-connected:

$$\cdots \to H^k(E; R) \to H^k(F; R) \xrightarrow{\theta} H^{k-n+1}(F; R) \to H^{k+1}(E; R) \to \cdots$$

In this case, if n is even, the mapping θ can be shown to be a derivation and, if n is odd, an antiderivation (that is, $\theta(x \smile y) = \theta(x) \smile y + x \smile \theta(y)$).

Example 5.D. *Suppose* $F \hookrightarrow E \xrightarrow{\pi} B$ *is a fibration with B path-connected and the system of local coefficients on B induced by the fibre is simple. Suppose further that* $H^i(B; R) = \{0\}$ *for $0 < i < p$ and $H^j(F; R) = \{0\}$ for $0 < j < q$. Then there is an exact sequence:*

$$0 \to H^1(B; R) \to H^1(E; R) \to H^1(F; R) \to H^2(B; R) \to \cdots$$
$$\to H^{p+q-2}(F; R) \to H^{p+q-1}(B; R) \to$$
$$H^{p+q-1}(E; R) \to H^{p+q-1}(F; R).$$

We first consider the diagram that pictures these conditions:

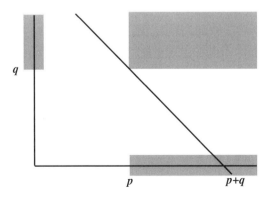

In total degrees less than $p + q$ the only possible nonzero differentials are of the form

$$d_j \colon H^{j-1}(F; R) = E_2^{0,j-1} \longrightarrow E_2^{j,0} = H^j(B; R).$$

Arguing as in the derivation of the Gysin sequence, we get a diagram of short exact sequences spliced together:

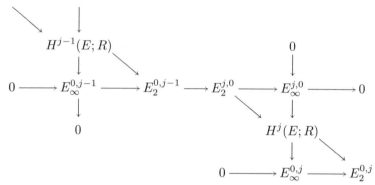

The sequence remains valid until $j = p + q$. Making the identifications $E_2^{0,j-1} \cong H^{j-1}(F; R)$ and $E_2^{j,0} \cong H^j(B; R)$ gives the desired sequence. As we will see in Example 5.E, the homomorphisms in the sequence are given by the differentials, d_j, plus π^* and i^* from the fibration. Furthermore, in §6.2, we will identify the d_j with the geometric homomorphism called the transgression. Without the connectivity assumptions, the resulting short exact sequence has five terms and is given in Example 1.A. The reader can establish the dual sequence in homology where the analogy with the long exact sequence in homotopy is more apparent. This sequence has become known as the **Serre exact sequence** ([Serre51, Chapitre III, §4]). We record some immediate corollaries.

Corollary 5.8. *For a fibration* $F \hookrightarrow E \xrightarrow{\pi} B$ *with* B *path-connected and for which the local coefficient system is simple,*

(a) *if* E *is acyclic and* $H^i(B; R) = \{0\}$ *for* $0 < i < p$, *then* $H^i(F; R) \cong H^{i+1}(B; R)$ *for* $i < 2p - 2$;

(b) *if* B *or* F *is acyclic then* $i \colon F \to E$ *or* $\pi \colon E \to B$, *respectively, is a weak homotopy equivalence.*

We return to these ideas later in this section where they apply to the path-loop fibration.

Example 5.E, *in which we exploit naturality.*

Consider the descending filtration on $H^q = H^q(E; R)$,

$$H^q = F^0 H^q \supset F^1 H^q \supset \cdots F^{q-1} H^q \supset F^q H^q \supset F^{q+1} H^q = \{0\},$$

and its associated graded module, $E_\infty^{n,q-n} \cong F^n H^q / F^{n+1} H^q$. Since $F^{q+1} H^q = \{0\}$, we have $E_\infty^{q,0} \subset H^q(E; R)$. Furthermore, there is the short exact sequence,

$$0 \longrightarrow F^1 H^q \longrightarrow H^q(E; R) \longrightarrow E_\infty^{0,q} \longrightarrow 0.$$

The modules along the bottom row of the E_2-term of a first quadrant spectral sequence are determined by quotients $E_2^{q,0}$ by the image of the incoming differential. Thus we have a series of quotients (here $\longrightarrow\!\!\!\rightarrow$ denotes an epimorphism):

$$E_2^{q,0} \longrightarrow\!\!\!\rightarrow E_3^{q,0} \longrightarrow\!\!\!\rightarrow \cdots \longrightarrow\!\!\!\rightarrow E_q^{q,0} \longrightarrow\!\!\!\rightarrow E_{q+1}^{q,0} = E_\infty^{q,0}.$$

The terms of the modules along the "y-axis" of the E_2-term of a first-quadrant spectral sequence are the kernels of successive differentials, and so we get a sequence of inclusions:

$$E_\infty^{0,q} = E_{q+1}^{0,q} \subset E_q^{0,q} \subset \cdots \subset E_3^{0,q} \subset E_2^{0,q}.$$

These facts are formal consequences of having a first-quadrant spectral sequence.

Theorem 5.9. *Given a fibration* $F \overset{i}{\hookrightarrow} E \overset{\pi}{\to} B$ *with B path-connected, F connected, and for which the system of local coefficients on B is simple; then the composites*

$$H^q(B; R) = E_2^{q,0} \longrightarrow\!\!\!\rightarrow E_3^{q,0} \longrightarrow\!\!\!\rightarrow \cdots$$
$$\longrightarrow\!\!\!\rightarrow E_q^{q,0} \longrightarrow\!\!\!\rightarrow E_{q+1}^{q,0} = E_\infty^{q,0} \subset H^q(E; R)$$

and $H^q(E; R) \longrightarrow\!\!\!\rightarrow E_\infty^{0,q} = E_{q+1}^{0,q} \subset E_q^{0,q} \subset \cdots \subset E_2^{0,q} = H^q(F; R)$ *are the homomorphisms*

$$\pi^*: H^q(B; R) \to H^q(E; R) \quad and \quad i^*: H^q(E; R) \to H^q(F; R).$$

PROOF: Consider the diagram of fibrations:

By the naturality of the Leray-Serre spectral sequence, we get induced mappings of the spectral sequences

$$E_r(B, B, *) \overset{\pi^*}{\longrightarrow} E_r(B, E, F) \overset{i^*}{\longrightarrow} E_r(*, F, F),$$

which converge to i^* and π^* at E_∞. However, $E_2(*, F, F)$ consists of the column $E_2^{0,*} = H^*(F; R)$ and so collapses at the E_2-term. Also $E_2(B, B, *)$ is the single row, $E_2^{*,0} = H^*(B; R)$ and the induced mappings of spectral sequences project $H^0(B; H^*(F; R))$ onto $H^*(F; R)$ and inject $H^*(B; R)$ into $H^*(B; H^0(F; R))$. In order to prove the theorem, it remains to consider how the mappings i^* and π^* are approached by the successive homologies in the spectral sequence and this is given in terms of the inclusions and quotients in the theorem.   □

Some immediate consequences of the theorem have already been mentioned. Since the induced maps are morphisms of algebras, we have established Proposition 5.6 on the product structure of $E_2^{0,*}$ and $E_2^{*,0}$. Furthermore, the corresponding mapping in the Gysin sequence is given by π^* and in the Serre exact sequence by π^* and i^*.

A topological invariant of the inclusion of F in E is the homomorphism i^*. We say that F is **totally nonhomologous to zero** in E with respect to the ring R if the homomorphism $i^* \colon H^*(E; R) \to H^*(F; R)$ is onto (introduced by [Samelson41]). In the case of a trivial fibration, this holds for obvious reasons. We can deduce something of a converse.

Theorem 5.10 (the Leray-Hirsch theorem). *Given a fibration* $F \overset{i}{\hookrightarrow} E \overset{\pi}{\to} B$ *with F connected, B of finite type, path-connected, and for which the system of local coefficients on B is simple; then, if F is totally nonhomologous to zero in E with respect to a field k, we have*

$$H^*(E; k) \cong H^*(B; k) \otimes_k H^*(F; k)$$

as vector spaces.

PROOF: The theorem follows immediately from the following stronger assertion.

Fact. F is totally nonhomologous to zero in E with respect to k if and only if the spectral sequence, $E_r(B, E, F)$, collapses at the E_2-term.

To establish this fact, examine the expression for i^* given by Theorem 5.9:

$$i^* \colon H^q(E; R) \longrightarrow\!\!\!\!\rightarrow E_\infty^{0,q} = E_{q+1}^{0,q} \hookrightarrow E_q^{0,q} \hookrightarrow \cdots$$
$$\hookrightarrow E_3^{0,q} \hookrightarrow E_2^{0,q} = H^q(F; R).$$

If the spectral sequence collapses at the E_2-term, then all of these inclusions are equalities and i^* is onto. Conversely, if i^* is onto, then all of these inclusions must be equalities and so d_r, restricted to the y-axis, must be zero. At E_2 however, $E_2^{p,q} = E_2^{p,0} \otimes E_2^{0,q}$ since k is a field and d_2 on $E_2^{p,0}$ is already zero. Since d_2 is a derivation with respect to this representation, $d_2 = 0$ and $E_3 = E_2$.

The same argument applies to d_3 and, continuing in this fashion, we establish that the spectral sequence collapses at E_2. □

By Theorems 5.9 and 5.10 it follows that if i^* is onto, then π^* is injective. It was conjectured that the converse also held. We sketch a counterexample due to G. Hirsch (see [Borel54]) showing that π^* can be injective without i^* onto. A systematic set of counterexamples was found by [Gottlieb77].

Hirsch's example begins with the Hopf fibration $\nu \colon S^7 \to S^4$ (see §4.3), which has fibre S^3. As mentioned in Example 5.B, the Leray-Serre spectral sequence for ν has a nonzero differential, d_4, taking the generator of $\tilde{H}^*(S^3; R)$ to the generator of $\tilde{H}^*(S^4; R)$. Consider the mapping of degree 1 from $S^2 \times S^2 \to S^4$ that can be constructed as the composite $S^2 \times S^2 \to S^2 \wedge S^2 = S^4$. Finally consider the projection $S^2 \times S^2 \xrightarrow{pr_1} S^2$ onto the first factor. Put the ingredients together in the diagram:

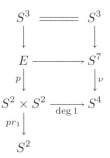

where E is the total space of the fibration gotten by pulling ν back over the degree 1 map. By naturality, the spectral sequence for $(S^3, E, S^2 \times S^2)$ does not collapse; $d_4 \neq 0$. Consider the fibration $S^3 \times S^2 \hookrightarrow E \xrightarrow{pr_1 \circ p} S^2$. Let $s \colon S^2 \to S^2 \times S^2$ split pr_1. Examine s^* on the generators of the E_2-term of the spectral sequence for $(S^3 \times S^2, E, S^2)$. For dimensional reasons, this splitting implies that $(pr_1 \circ p)^*$ is injective. However, the spectral sequence does not collapse since we know already that $H^*(E; R)$ is not isomorphic to $H^*(S^2 \times S^2 \times S^3; R)$. We leave it to the reader to determine the nonzero differential.

Particular fibrations

The importance of Lie groups in homotopy theory cannot be overemphasized. It is often necessary to compute the homotopy-theoretic invariants of various Lie groups and homogeneous spaces. Their structure naturally leads to some useful fibrations where the Leray-Serre spectral sequence can be applied. It would be impossible to sketch all of the applications in topology where Lie groups and homogeneous spaces play a crucial role. For a sampling of some of these results we refer the reader to the accounts of [Steenrod51], [Borel55], [Milnor-Stasheff74], [Mimura-Toda91], and [Dwyer-Wilkerson94]. In the next few examples, we compute the cohomology algebras of some of these spaces as much to illustrate some further techniques with spectral sequences as to obtain these important data. These computations are based on the thesis of [Borel53].

Example 5.F, in which we compute $H^*(\mathrm{SU}(n); R)$.

The classical Lie group $\mathrm{SU}(n)$ is the group of $(n \times n)$ complex-valued unitary $(A\bar{A}^t = I)$ matrices of determinant 1. If we fix a vector $\vec{x}_0 \in \mathbb{C}^n$, then we define $\rho \colon \mathrm{SU}(n) \to S^{2n-1}$ by sending a matrix A to $A\vec{x}_0$. This mapping can be proved to be a fibration (see §4.3) with fibre the subgroup of $\mathrm{SU}(n)$ that fixes $\vec{x}_0$, which is $\mathrm{SU}(n-1)$. The group $\mathrm{SU}(2)$ may be identified as the unit sphere in $\mathbb{R}^4$ or S^3 (by quaternionic representation). We next prove that $H^*(\mathrm{SU}(n); R) \cong \Lambda(x_3, x_5, \dots, x_{2n-1})$, the exterior algebra on generators, x_i, where $\deg x_i = i$.

We proceed by induction on n. For $n = 2$,

$$H^*(\mathrm{SU}(2); R) \cong H^*(S^3; R) = \Lambda(x_3).$$

Consider the fibration $\mathrm{SU}(n-1) \hookrightarrow \mathrm{SU}(n) \xrightarrow{\pi} S^{2n-1}$ and apply the Leray-Serre spectral sequence. Suppose y_{2n-1} generates $H^*(S^{2n-1}; R)$ as an exterior algebra. By induction, we have

$$H^*(\mathrm{SU}(n-1); R) \cong \Lambda(x_3, x_5, \dots, x_{2n-3}).$$

Since this is a spectral sequence of algebras, we need consider only the algebra generators in describing differentials. For $n \geq 2$, S^{2n-1} is simply-connected and so the system of local coefficients on the base space is simple. Furthermore, by an application of the Universal Coefficient theorem, we have

$$E_2^{*,*} \cong H^*(B; R) \otimes H^*(F; R) \cong \Lambda(y_{2n-1}) \otimes \Lambda(x_3, x_5, \dots, x_{2n-3}).$$

The algebra generators are found in bidegrees so that all of the differentials are zero and the spectral sequence collapses at the E_2-term. We argue as in Example 1.K to conclude that

$$H^*(\mathrm{SU}(n); R) \cong \mathrm{Tot}\, E_\infty^{*,*} \cong \Lambda(x_3, x_5, \dots, x_{2n-3}, y_{2n-1})$$

and the induction is completed.

We can extend the computation to $H^*(\mathrm{U}(n); R)$, where $\mathrm{U}(n)$ is the group of linear transformations of $\mathbb{C}^n$ which preserve the complex inner product. The group $\mathrm{U}(n)$ relates to $\mathrm{SU}(n)$ via the fibration:

$$\mathrm{SU}(n) \xrightarrow{\mathrm{inc}} \mathrm{U}(n) \xrightarrow{\det} S^1$$

and furthermore, $U(n) \simeq U(1) \times SU(n) \simeq S^1 \times SU(n)$. Thus, $H^*(U(n); R) \cong \Lambda(x_1, x_3, \ldots, x_{2n-1})$. These methods also apply to the symplectic group, $Sp(n)$, of linear transformations of $\mathbb{H}^n$ (quaternionic n-space) that preserve the inner product. It is left as an exercise to show that $H^*(Sp(n); R) \cong \Lambda(x_3, x_7, \ldots, x_{4n-1})$.

A space that is important in the study of cobordism (see Chapter 9) is the **infinite special unitary group**, SU, which is the direct limit (union) of the sequence of inclusions

$$SU(2) \subset SU(3) \subset \cdots \subset SU(n-1) \subset SU(n) \subset \cdots \subset SU.$$

By choosing a unit vector $\vec{x}_0$ in S^{2n-1}, we can realize the usual inclusion of $SU(n-1) \subset SU(n)$ as the inclusion of the fibre $SU(n-1) \hookrightarrow SU(n) \to S^{2n-1}$. By Theorem 5.9, we can describe

$$i^*: H^*(SU(n); R) \to H^*(SU(n-1); R)$$

as the mapping taking x_i to x_i except on x_{2n-1} which is sent to zero. Since each i^* is an epimorphism, $\lim\limits_{\leftarrow n, i^*}^1 H^*(SU(n); R) = \{0\}$ (§3.2) and so $H^*(SU; R) \cong \Lambda(\{x_{2i-1} \mid i = 2, 3, \ldots\})$. Similar results hold for the analogous groups U and Sp.

We take up the real case of $SO(n)$ in Example 5.H.

***Example 5.G**, in which we compute $H^*(V_k(\mathbb{C}^n); R)$.*

The Stiefel manifolds $V_k(\mathbb{R}^n)$, $V_k(\mathbb{C}^n)$, and $V_k(\mathbb{H}^n)$ consist of the orthonormal k-frames in each n-dimensional space over the given (skew) field. These spaces play an important role in the study of characteristic classes (see [Milnor- Stasheff74] and §6.2) and in the study of vector fields on manifolds; they are among the simplest examples of homogeneous spaces. In this example, we compute $H^*(V_k(\mathbb{C}^n); R)$ using the Leray-Serre spectral sequence (see §7.2 for another approach). We will take up $H^*(V_k(\mathbb{R}^n); R)$ in Example 5.H, and we leave the computation of $H^*(V_k(\mathbb{H}^n); R)$ to the reader.

Let $e_i \in \mathbb{C}^n$ denote the ith elementary vector,

$$e_i = (0, \ldots, 0, 1, 0, \ldots, 0),$$

with the 1 in the ith place. Consider the mapping (for $n \geq 2$), $SU(n) \to V_k(\mathbb{C}^n)$ given by applying the matrix A to each e_i in turn to get the orthonormal k-frame $(Ae_1, Ae_2, \ldots, Ae_k)$. This map is clearly onto and continuous and each given k-frame can be varied by the action of A on the $n - k$ remaining vectors in the standard n-frame for $\mathbb{C}^n$. Therefore, the fibre over a point in $V_k(\mathbb{C}^n)$ is a copy of $SU(n-k)$. By including a matrix in $SU(n-k)$ into $SU(n)$ as the bottom right hand $(n-k) \times (n-k)$ block of an otherwise $n \times n$ identity matrix, we see that $V_k(\mathbb{C}^n)$ is homeomorphic to $\mathrm{SU}(n)/\mathrm{SU}(n-k)$. Thus we have the fibration

$$\mathrm{SU}(n-k) \xrightarrow{\mathrm{inc}} \mathrm{SU}(n) \to V_k(\mathbb{C}^n)$$

It is immediate from the definitions that $V_1(\mathbb{C}^n) = S^{2n-1}$ and that $V_n(\mathbb{C}^n) = \mathrm{SU}(n)$. With the identification of $V_k(\mathbb{C}^n)$ as a homogeneous space, we can apply a little group theory to obtain some short exact sequences that give us new fibrations. The inclusion of subgroups $\mathrm{SU}(n-k-1) \subset \mathrm{SU}(n-k)$ provides us with the fibration

$$0 \to \mathrm{SU}(n-k)/\mathrm{SU}(n-k-1) \to \mathrm{SU}(n)/\mathrm{SU}(n-k-1)$$
$$\to \mathrm{SU}(n)/\mathrm{SU}(n-k) \to 0,$$

which may be identified as $S^{2(n-k)-1} \xrightarrow{\mathrm{inc}} V_{k+1}(\mathbb{C}^n) \to V_k(\mathbb{C}^n)$.

Proposition 5.11. $H^*(V_k(\mathbb{C}^n); R) \cong \Lambda(x_{2(n-k)+1}, x_{2(n-k)+3}, \cdots, x_{2n-1})$.

PROOF: We proceed by induction on k. For $k = 1$, $H^*(V_1(\mathbb{C}^n); R) = H^*(S^{2n-1}; R) = \Lambda(x_{2n-1})$ and the proposition holds. Now suppose the statement holds for the value k. The Leray-Serre spectral sequence for the fibration $(S^{2(n-k)-1}, V_{k+1}(\mathbb{C}^n), V_k(\mathbb{C}^n))$ has E_2-term given by

$$E_2^{*,*} = H^*(V_k(\mathbb{C}^n); H^*(S^{2(n-k)-1}; R))$$
$$\cong \Lambda(x_{2(n-k)+1}, x_{2(n-k)+3}, \cdots, x_{2n-1}) \otimes \Lambda(y_{2(n-k)-1})$$

by the Universal Coefficient theorem and the fact that the system of local coefficients is simple. We obtain the following diagram of the algebra generators in the E_2-term of the spectral sequence:

$y_{2(n-k)-1}$						
$\vdots$						
1	$\cdots$	$x_{2(n-k)+1}$		$x_{2(n-k)+3}$	$\cdots$	x_{2n-1}

For dimensional reasons, the spectral sequence collapses and so

$$E_\infty^{*,*} \cong \Lambda(y_{2(n-k)-1}, x_{2(n-k)+1}, \cdots, x_{2n-1}).$$

Once again, following Example 1.K, $H^n(V_{k+1}(\mathbb{C}^n); R) \cong \bigoplus_{p+q=n} E_\infty^{p,q}$ and this completes the induction. $\qquad \square$

Example 5.H. *We compute $H^*(V_k(\mathbb{R}^n); \mathbb{F}_2)$ and, by letting $k = n$, we compute $H^*(\mathrm{SO}(n); \mathbb{F}_2)$.*

The new feature about these computations is how naturality can determine certain differentials. We proceed a step at a time.

STEP 1: Compute $H^*(V_2(\mathbb{R}^n))$.

We start with the fibration, $S^{n-2} \hookrightarrow V_2(\mathbb{R}^n) \to V_1(\mathbb{R}^n)$ that leads to an E_2-term in the Leray-Serre spectral sequence as in the diagram

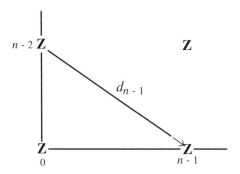

The only possible nontrivial differential may be written

$$d_{n-1} \colon H^{n-2}(F) \to H^{n-1}(B).$$

Consider the tangent bundle over S^{n-1}, $\mathbb{R}^{n-1} \hookrightarrow TS^{n-1} \to S^{n-1}$, and the associated sphere bundle of unit tangent vectors, $S^{n-2} \hookrightarrow T_0 S^{n-1} \to S^{n-1}$. By the geometry of the sphere elements in $T_0 S^{n-1}$ are pairs of vectors in $\mathbb{R}^n$, (u, v), with u in S^{n-1} and v orthogonal to u, that is, (u, v) is a 2-frame in $\mathbb{R}^n$. The mapping $T_0 S^{n-1} \to S^{n-1}$ is simply projection on the first factor so we get an isomorphism of fibrations

$$
\begin{array}{ccc}
S^{n-2} & = & S^{n-2} \\
\downarrow & & \downarrow \\
V_2(\mathbb{R}^n) & \longrightarrow & T_0 S^{n-1} \\
\downarrow & & \downarrow \\
V_1(\mathbb{R}^n) & \xrightarrow{\ =\ } & S^{n-1}.
\end{array}
$$

Thus to study the spectral sequence for $V_2(\mathbb{R}^n)$, we replace it with the spectral sequence for $S^{n-2} \hookrightarrow T_0 S^{n-1} \to S^{n-1}$. As a sphere bundle, we can apply the Gysin sequence for which the relevant portion is given by

$$H^0(S^{n-1}) \xrightarrow{z\smile} H^{n-1}(S^{n-1}) \to H^{n-1}(T_0 S^{n-1}) \to H^1(S^{n-1}) \to \cdots.$$

The unit $1 \in H^0(S^{n-1})$ goes to the cohomology class z following Example 5.C. If n is even, then the class z satisfies $2z = 0$. Since $H^{n-1}(S^{n-1}) \cong \mathbb{Z}$, this is only possible if $z = 0$.

When n is odd, we need a little more information about the class z. In fact, for a vector bundle like the tangent bundle, z is identified with the Euler class of the bundle. (See the accounts of [Milnor-Stasheff74] and [Husemoller66].) Since the class z is a multiple of the generator of $H^{n-1}(S^{n-1})$ and the cap product of z with the canonical generator of $H_{n-1}(S^{n-1})$ gives the Euler characteristic for S^{n-1}, we see that z is twice the generator and d_{n-1} is multiplication by two. It follows (for $n \geq 4$):

$$
H^*(V_2(\mathbb{R}^n)) = \begin{cases} \text{if } n \text{ is odd} \begin{cases} \mathbb{Z}, & \text{for } * = 0, 2n-3, \\ \mathbb{Z}/2\mathbb{Z}, & \text{for } * = n-1, \\ 0, & \text{elsewhere.} \end{cases} \\[2em] \text{if } n \text{ is even} \begin{cases} \mathbb{Z}, & \text{for } * = 0, n-2, n-1, 2n-3, \\ 0, & \text{elsewhere.} \end{cases} \end{cases}
$$

As a corollary, from the Universal Coefficient theorem, we have (for $n \geq 4$)

$$
H^*(V_2(\mathbb{R}^n); \mathbb{F}_2) = \begin{cases} \mathbb{F}_2, & \text{for } * = 0, n-2, n-1, 2n-3, \\ 0, & \text{elsewhere.} \end{cases}
$$

STEP 2: Some comments on simple systems of generators.

Definition 5.12. *Suppose H^* is a graded commutative algebra over R. A set of elements $\{x_i \mid i = 2, 3, \dots\}$ is called a **simple system of generators** if the elements 1 and $x_{i_1} \cdot x_{i_2} \cdots x_{i_n}$ with $i_1 < i_2 < \cdots < i_n$ form a basis over R for H^*.*

The notion of a simple system of generators was introduced by [Borel53]. If H^* is an R-algebra with a simple set of generators, then the homogeneous degree m part of H^* is a free R-module on the collection of monomials $x_{i_1} \cdot x_{i_2} \cdots x_{i_n}$ with $i_1 < i_2 < \cdots < i_n$ and $\sum_{j=1}^n \deg x_{i_j} = m$. Notice that this says less about the algebra structure of H^* than one would like; for example, these conditions do not determine $(x_j)^2$ for any j. The polynomial algebra $R[x]$ has a simple system of generators given by $\{x, x^2, x^4, x^8, \dots\}$. An exterior algebra, $\Lambda(x_1, x_2, \dots, x_n)$, has $\{x_1, \dots, x_n\}$ as a simple system of generators. The key theorem on simple systems of generators is the following analogue of Example 1.K. We leave the proof to the reader.

Theorem 5.13. *Suppose H^* is a filtered graded commutative algebra over R and $E_0^{*,*}$, the associated bigraded algebra, lies in the first quadrant. If $E_0^{*,*}$ has a simple system of generators $\{\bar{x}_i\}$ and the elements $\{x_i\}$ in H^* project to $\{\bar{x}_i\}$ in $E_0^{*,*}$, then $\{x_i\}$ is a simple system of generators for H^*.*

STEP 3: $H^*(\mathrm{SO}(n); \mathbb{F}_2)$ has a simple system of generators

$$\{x_1, x_2, \ldots, x_{n-1}\}, \qquad \text{where } \deg x_i = i.$$

We proceed by induction. For $n = 2$, $\mathrm{SO}(2) \cong S^1$ and the statement holds. Consider the fibration $\mathrm{SO}(n-1) \xrightarrow{\mathrm{inc}} \mathrm{SO}(n) \to S^{n-1}$. The spectral sequence, by the induction hypothesis, has a lacunary placement of algebra generators with d_{n-1} the only possibly nontrivial differential.

To show that d_{n-1} is the zero differential we consider the morphism of fibrations

$$
\begin{array}{ccccc}
\mathrm{SO}(n-1) & \longrightarrow & \mathrm{SO}(n) & \longrightarrow & S^{n-1} \\
\downarrow & & \downarrow & & \| \\
S^{n-2} & \longrightarrow & V_2(\mathbb{R}^n) & \longrightarrow & S^{n-1}.
\end{array}
$$

This morphism is a consequence of the following diagram of short exact sequences of groups:

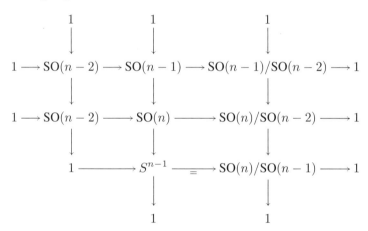

In the morphism of spectral sequences induced by the morphism of fibrations, the generator of $H^*(S^{n-1}; \mathbb{F}_2)$ is mapped to y_{n-1} and the generator of $H^*(S^{n-2}; \mathbb{F}_2)$ to x_{n-2}. In STEP 1, we established that d_{n-1} for the spectral sequence associated to $(S^{n-2}, V_2(\mathbb{R}^n), S^{n-1})$ is zero modulo 2. By naturality, d_{n-1} is zero in our original spectral sequence. Thus the spectral sequence collapses at the E_2-term. Furthermore, by Theorem 5.13, $E_2^{*,*}$ has a simple system of generators, $\{x_1, x_2, \ldots, x_{n-2}, y_{n-1}\}$ and the induction is completed.

STEP 4: $H^*(V_k(\mathbb{R}^n); \mathbb{F}_2)$ has a simple system of generators

$$\{x_{n-k}, x_{n-k+1}, \ldots, x_{n-1}\}, \qquad \text{where } \deg x_i = i.$$

As in STEP 3, we begin with the fibration

$$S^{n-k-1} \xrightarrow{\mathrm{inc}} V_{k+1}(\mathbb{R}^n) \to V_k(\mathbb{R}^n)$$

By induction and the spectral sequence associated to the fibration, we get a diagram of algebra generators, with, once again, one possibly nontrivial differential, d_{n-k}.

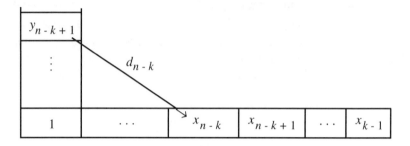

The assertion in STEP 4 is proved if we can show that d_{n-k} is the zero differential. To settle this, we consider the inclusion $V_{k-1}(\mathbb{R}^{n-1}) \subset V_k(\mathbb{R}^n)$ given by considering each $(n-1)$-vector as an n vector with 0 last entry and sending the $(k-1)$-frame $(v_1, v_2, \dots, v_{k-1})$ to $(v_1, v_2, \dots, v_{k-1}, e_n)$ (where e_n is the vector $(0, \dots, 0, 1)$). With this injection we get a sequence of maps of fibrations:

$$
\begin{array}{ccccccc}
S^{n-k-1} & = & S^{n-k-1} \cdots & = & S^{n-k-1} & = & S^{n-k-1} \\
\downarrow & & \downarrow & & \downarrow & & \downarrow \\
V_2(\mathbb{R}^{n-k+1}) & \longrightarrow & V_3(\mathbb{R}^{n-k+2}) \cdots & \longrightarrow & V_k(\mathbb{R}^{n-1}) & \longrightarrow & V_{k+1}(\mathbb{R}^n) \\
\downarrow & & \downarrow & & \downarrow & & \downarrow \\
S^{n-k} & \longrightarrow & V_2(\mathbb{R}^{n-k+2}) \cdots & \longrightarrow & V_{k-1}(\mathbb{R}^{n-1}) & \longrightarrow & V_k(\mathbb{R}^n).
\end{array}
$$

Each inclusion $V_{k-1}(\mathbb{R}^{n-1}) \subset V_k(\mathbb{R}^n)$ is the inclusion of the fibre in the fibration $V_{k-1}(\mathbb{R}^{n-1}) \hookrightarrow V_k(\mathbb{R}^n) \to S^{n-1}$ and so, by step 3, the class x_{n-k} is mapped to the corresponding x_{n-k} by i^*. By naturality, we have reduced the question of the collapse of the spectral sequences for all k to computing d_{n-k} in the spectral sequence for the particular fibration $(S^{n-k-1}, V_2(\mathbb{R}^{n-k+1}), S^{n-k})$ that was shown to be zero modulo 2 in STEP 1. As in STEP 3, the induction is completed with Theorem 5.13.   □

Loops on a space

Let X be a topological space. We associate two important spaces to X as follows: Suppose $*$ is a basepoint in X. Let $PX = \{\lambda \colon [0,1] \to X \mid \lambda$ is continuous and $\lambda(0) = *\}$ denote the **space of paths** in X based at $*$. Let $\Omega X = \{\lambda \colon [0,1] \to X \mid \lambda$ is continuous and $\lambda(0) = \lambda(1) = *\}$, the **space of based loops** in X at $*$ with the compact-open topology. The evaluation mapping $\mathrm{ev}_1 \colon PX \to X$ given by $\mathrm{ev}_1(\lambda) = \lambda(1)$ has fibre ΩX. In §4.3 we showed that $\text{ev}_1 \colon PX \to X$ has the homotopy lifting property and so we have a fibration

$$\Omega X \hookrightarrow PX \xrightarrow{\text{ev}_1} X.$$

The space PX is contractible with the immediate consequence that, for all $i > 0$, $\pi_i(\Omega X) \cong \pi_{i+1}(X)$ (Corollary 4.31).

An important feature of ΩX is that it is an H-space, that is, there is a multiplication, $\mu \colon \Omega X \times \Omega X \to \Omega X$,

$$\mu(\lambda_1, \lambda_2)(t) = \lambda_1 * \lambda_2(t) = \begin{cases} \lambda_1(2t), & \text{if } 0 \le t \le \frac{1}{2}, \\ \lambda_2(2t - 1), & \text{if } \frac{1}{2} \le t \le 1. \end{cases}$$

Up to homotopy, μ has an identity and is associative making the cohomology algebra of ΩX a Hopf algebra. We also recall that $\pi_1(\Omega X)$ is abelian as a consequence of the H-structure.

Example 5.I. *Suppose X is q-connected. Then, $H^i(\Omega X; R) \cong H^{i+1}(X; R)$ for $i < 2q - 2$. The analogous isomorphism, $H_i(\Omega X; R) \cong H_{i+1}(X; R)$ holds in homology for $i < 2q - 2$.*

This follows by the acyclicity of PX and Corollary 5.8, the Serre exact sequence. We discuss these isomorphisms in §6.2. We apply these ideas to prove the classical result.

Theorem 5.14 (the Hurewicz theorem). *If X is a connected space so that $\pi_i(X) = \{0\}$ for $1 \le i < q$, then $H_i(X) = \{0\}$ for $1 \le i < q$ and $\pi_q(X) \cong H_q(X)$.*

PROOF: We proceed by induction on the connectedness of X. If X is simply-connected, that is, $\pi_i(X) = \{0\}$ for $i < 2$, then

$$\pi_2(X) \cong \pi_1(\Omega X) \cong H_1(\Omega X) \cong H_2(X),$$

where the middle isomorphism is the classical theorem of Poincaré and the fact that the fundamental group of an H-space is abelian, and the last isomorphism follows from the Serre exact sequence.

Suppose the theorem is true for spaces that are $(q-1)$-connected and X is q-connected ($q \ge 2$). By the isomorphism, $\pi_{i-1}(\Omega X) \cong \pi_i(X)$, we have that ΩX is $(q-1)$-connected and so by induction, $H_i(\Omega X) = \{0\}$ for $0 < i < q-1$ and $\pi_{q-1}(\Omega X) \cong H_{q-1}(\Omega X)$. Finally we use the isomorphisms implied by the fibration in homology and homotopy

$$\pi_q(X) \cong \pi_{q-1}(\Omega X) \cong H_{q-1}(\Omega X) \cong H_q(X).$$

For $1 < i < q$, since $i < 2q - 2$, $H_i(X) \cong H_{i-1}(\Omega X) = \{0\}$, and $H_1(X) = \{0\}$ because $\pi_1(X) = \{0\}$. $\qquad\square$

Example 5.A for the case of the path-loop fibration implies that if R is Noetherian and $H^*(X; R)$ is finitely generated in each dimension, then so is $H^*(\Omega X; R)$.

Example 5.J, *in which we consider a remarkable theorem of* [Morse34].

Let $P(X, a, b)$ denote the space $\{\lambda \colon [0,1] \to X \mid \lambda(0) = a, \lambda(1) = b\}$ of paths joining a to b in X, a subspace of $\mathrm{map}([0,1], X)$. If we suppose that X is path-connected, then we can use a path joining a to b to construct a homotopy equivalence between $P(X, a, b)$ and $\Omega(X, a)$.

Theorem 5.15 *([Morse34]). Suppose M is a complete, connected Riemann-ian manifold, a and b are points in M and k is a field. Suppose further that $H_i(P(M, a, b); k) \neq \{0\}$ for infinitely many values of i. Then there are in-finitely many geodesics in M joining a to b.*

The question of the number of geodesics joining a to b in M is reduced by this theorem to a homological question. Since M is complete and connected, it is path-connected by the Hopf-Rinow theorem ([Hopf-Rinow31]) and so we study $H_*(\Omega M; k)$ and the following result of [Serre51].

Proposition 5.16. *Suppose k is a field and there is an integer $n \geq 2$ so that $H_i(X; k) = \{0\}$ for $i > n$ and $H_n(X; k) \neq \{0\}$. Then, for any $i \geq 0$, there is an integer j, $0 < j < n$, such that $H_{i+j}(\Omega X; k) \neq \{0\}$. That is, there does not exist a set of $(n-1)$ consecutive positive integers where the homology of ΩX with coefficients in k is trivial in those degrees.*

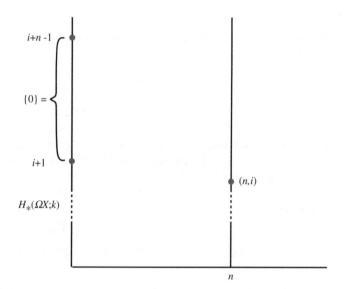

PROOF: In the spectral sequence associated with the path-loop fibration, no nonzero classes in $E^2_{*,*}$ persist to E^∞ since PX is contractible. Suppose i is the least integer for which the proposition fails and $H_i(\Omega X; k) \neq \{0\}$. Since k is a field,

$$E_{n,i}^2 = H_n(X; H_i(\Omega X; k)) \cong H_n(X; k) \otimes H_i(\Omega X; k) \neq \{0\}.$$

If the proposition fails, then we have the diagram on the previous page around (n, i) in the E^2-term. In particular, there are no nonzero vector spaces that can act as a target of a differential from $E_{n,i}^2$. By assumption, there are no differentials whose target is $E_{n,i}^2$. Thus $E_{n,i}^2 \cong E_{n,i}^\infty \neq \{0\}$, a contradiction. $\square$

Proposition 5.16 together with the theorem of Morse proves that, if M is a complete, connected, Riemannian manifold such that $H_i(M) \neq \{0\}$ for at least one value of $i > 1$, then any pair of points in M, a, b, $a \neq b$, are joined by infinitely many geodesics in M.

The reader can compute $H^*(\Omega S^{2n-1}; \mathbb{Q})$ to illustrate Proposition 5.16: This computation is the result of Example 1.H.

Example 5.K, *in which we prove that $\pi_i(S^{2n-1})$ is a finite group for $i > 2n - 1$.*

Since our methods all center on homology and cohomology, we need to turn problems about homotopy groups into problems in homology. To do this we exploit the classical isomorphism of Poincaré, $H_1(X) \cong \pi_1(X)/[\pi_1(X), \pi_1(X)]$. The proof of the assertion in this example is broken into steps (following [Serre51]).

STEP 1. Let X be a connected, simply-connected space. We associate to X a sequence of spaces as follows: $X_0 = X$, $T_1 = \tilde{X}_0$, the universal cover of X_0 and $X_1 = \Omega T_1$. Let $T_2 = \tilde{X}_1$ and $X_2 = \Omega T_2$. By recursion, we define $T_n = \tilde{X}_n$ and $X_{n+1} = \Omega T_n$.

The universal cover of a space Y is simply-connected and the covering $\tilde{Y} \to Y$ induces an isomorphism $\pi_i(\tilde{Y}) \cong \pi_i(Y)$ for $i > 1$. Among the homotopy groups of the spaces in the sequence, $\{X_j \mid j = 0, 1, \dots\}$, we have the following relations:

$$\pi_i(X_n) \cong \pi_{i+1}(X_{n-1}) \cong \cdots \cong \pi_{i+n}(X_0) = \pi_{i+n}(X).$$

Suppose $H_i(X)$ is finitely generated for all i, that is, X is **of finite type**. We proved in Example 5.A that X of finite type guarantees that ΩX is of finite type. We would like to extend this algebraic condition to homotopy groups. A naive argument can be made from the spaces $\{X_j\}$: If we knew that whenever X_i is of finite type, then so is X_{i+1}, we would have $H_1(X_i) \cong \pi_1(X_i) \cong \pi_{i+1}(X)$ and that $H_1(X_i)$ is finitely generated implies that $\pi_{i+1}(X)$ is finitely generated. However, the situation is more subtle than that. In order to make such an argument work, we need to study universal covers, which do not necessarily exist without some point set topological assumptions. [Massey67] shows that a connected, locally simply-connected space has a universal cover. To prove that universal covers exist for each of the spaces X_i, [Serre51] introduces the point set condition **ULC** (*uniformement localement contractile*): A space Y is ULC if there is a neighborhood U of the diagonal in $Y \times Y$ and a homotopy $F \colon U \times I \to Y$ such that $F(x, x, t) = x$ for all $x \in Y$ and $t \in I$; and $F(x, y, 0) = x$, $F(x, y, 1) = y$ for all $(x, y) \in U$. The relevant result is that if Y is ULC, then $\tilde{Y}$ exists, $\tilde{Y}$ is ULC and ΩY is ULC. Spaces that are ANRs (absolute neighborhood retracts, [Whitehead, GW78]) are ULC, and this includes spaces of the homotopy type of locally finite CW-complexes.

Proposition 5.17. *If X is ULC, of finite type, connected and simply-connected, then $\pi_i(X)$ is finitely-generated for all i.*

PROOF: To study the relationship between Y and $\tilde{Y}$, we do not have the Leray-Serre spectral sequence as a tool—the fibre is not connected. The tool of choice is the **Cartan-Leray spectral sequence** (see [Cartan48] and Theorem $8^{\text{bis}}.9$), converging to $H_*(X; A)$ for A, an abelian group, and for which

$$E^2_{p,q} \cong H_p(\pi_1(X), H_q(\tilde{X}; A)),$$

where we are using the homology of the group $\pi_1(X)$ with coefficients in the $\pi_1(X)$-module $H_*(\tilde{X}; A)$. (See Chapter 8bis for definitions.)

We proceed by induction. For $X_0 = X$, since X is simply-connected, $\tilde{X} = X = T_1$ and so $X_1 = \Omega T_1$ is of finite type by the argument of Example 5.A. By induction we suppose that X_{n-1} is of finite type and consider $T_n = \tilde{X}_{n-1}$. [Serre51] showed that the abelian group $\pi_1(X_{n-1})$ acts trivially on $H_*(T_n)$ (since X_{n-1} is an H-space—Corollary $8^{\text{bis}}.3$). The E^2-term of the Cartan-Leray spectral sequence for the covering $T_n \to X_{n-1}$ simplifies for a trivial action:

$$E^2_{p,q} \cong H_p(\pi_1(X_{n-1}), H_q(T_n))$$
$$\cong H_p(\pi_1(X_{n-1})) \otimes H_q(T_n) \oplus \text{Tor}^{\mathbb{Z}}_1(H_{p-1}(\pi_1(X_{n-1})), H_q(T_n)).$$

By induction, $\pi_1(X_{n-1}) \cong H_1(X_{n-1})$ is finitely generated, from which it follows that the homology groups of the group $\pi_1(X_{n-1})$ with coefficients in the trivial module $\mathbb{Z}$, $H_i(\pi_1(X_{n-1}))$, are finitely generated (they are subquotients of the bar construction). Since the target groups $H_{p+q}(X_{n-1})$ are finitely generated for all $p + q$ and the Tor terms are finitely generated, the argument of Example 5.A applies to prove that T_n is of finite type, and hence, so is $X_n = \Omega T_n$.

Using the Cartan-Leray spectral sequence establishes the necessary details for our naive argument, and so we have proved that $\pi_n(X)$ is finitely generated for all n, since $\pi_n(X) \cong \pi_1(X_{n-1}) \cong H_1(X_{n-1})$. $\quad\square$

If an abelian group is finitely generated, then by tensoring with $\mathbb{Q}$ we can determine whether the group is finite or not. In particular, if Π is finitely generated, $\Pi \otimes_{\mathbb{Z}} \mathbb{Q} = \{0\}$ if and only if the group Π is finite. We use the spaces X_i to study $\pi_*(X) \otimes_{\mathbb{Z}} \mathbb{Q}$. In fact, we do more and establish a generalization of the Hurewicz theorem with coefficients due to [Serre51]. We write $\otimes$ for $\otimes_{\mathbb{Z}}$ in what follows.

Theorem 5.18 (the rational Hurewicz theorem). *If X is ULC, connected, simply-connected, and of finite type, and $H_i(X; \mathbb{Q}) = \{0\}$ for $0 < i < n$, then $\pi_i(X) \otimes \mathbb{Q} = \{0\}$ for $0 < i < n$ and $\pi_n(X) \otimes \mathbb{Q}$ is isomorphic to $H_n(X; \mathbb{Q})$.*

PROOF: We establish the following fact for the sequence of spaces $\{X_i\}$.

Fact: *With X as in Theorem 5.18, $H_i(X_j; \mathbb{Q}) = \{0\}$ if $i > 0$ and $i + j < n$ and, $H_i(X_j; \mathbb{Q}) = H_n(X; \mathbb{Q})$, if $i + j = n$.*

PROOF OF FACT: If $j = 0$, the fact follows from the identification $X_0 = X$. Suppose it is true for $0 \le m \le j - 1$. For $j \ge 2$, X_{j-1} is a loop space and so the abelian group $\pi_1(X_{j-1})$ acts trivially on $H_*(T_j; \mathbb{Q})$. The Cartan-Leray spectral sequence takes the form

$$E_{p,q}^2 \cong H_p(\pi_1(X_{j-1})) \otimes H_q(T_j; \mathbb{Q}) \oplus \operatorname{Tor}_1^{\mathbb{Z}}(H_{p-1}(\pi_1(X_{j-1})), H_q(T_j; \mathbb{Q})).$$

The five-term exact sequence associated to the lower left hand corner of the E^2-term (Example 1.A for homology) takes the form

$$0 \to E_{2,0}^\infty \to E_{2,0}^2 \xrightarrow{d^2} E_{0,1}^2 \to H_1(X_{j-1}; \mathbb{Q}) \to E_{1,0}^2 \to 0.$$

By induction, $H_1(X_{j-1}; \mathbb{Q}) = \{0\}$ and this implies that

$$E_{0,1}^2 \cong H_1(\pi_1(X_{j-1})) \otimes H_0(T_j; \mathbb{Q}) \cong \{0\}.$$

Since $\pi_1(X_{j-1})$ is finitely generated—it isomorphic to $H_1(X_{j-1})$—it follows that $H_1(\pi_1(X_{j-1})) \cong \pi_1(X_{j-1})$ is finitely generated. Finally, $H_0(T_j; \mathbb{Q}) \cong \mathbb{Q}$ implies that $\pi_1(X_{j-1})$ is finite. In computing the homology of a group we can use explicit resolutions (see Chapter 8bis), and so we can deduce (from the bar construction) that $H_p(\pi_1(X_{j-1}))$ is finite for all $p > 0$ and so, when tensored with a rational vector space, gives $E_{p,q}^2 = \{0\}$ for $p > 0$. The Cartan-Leray spectral sequence collapses to the leftmost column and, by convergence, $H_i(T_j; \mathbb{Q}) \cong H_i(X_{j-1}; \mathbb{Q})$ for all $i > 0$.

Finally, by Example 5.I for $i < 2(n - j + 1) - 2$, we have $H_i(X_j; \mathbb{Q}) = H_i(\Omega T_j; \mathbb{Q}) \cong H_{i+1}(T_j; \mathbb{Q}) \cong H_{i+1}(X_{j-1}; \mathbb{Q})$. By the induction hypothesis, we have that $H_i(X_j; \mathbb{Q}) = \{0\}$ for $i > 0$ and $i + j < n$ and $H_{n-j}(X_j; \mathbb{Q}) \cong H_n(X; \mathbb{Q})$. $\square$

From the fact we prove the theorem. Consider $H_1(X_j) \cong \pi_1(X_j) \cong \pi_{j+1}(X)$. For $j < n-1, j+1 < n$ and so $H_1(X_j; \mathbb{Q}) = \{0\}$. By the Universal Coefficient theorem, it follows that $\{0\} = H_1(X_j) \otimes \mathbb{Q} \cong \pi_{j+1}(X) \otimes \mathbb{Q}$. For $j = n-1, \pi_n(X) \otimes \mathbb{Q} \cong \pi_1(X_{n-1}) \otimes \mathbb{Q} \cong H_1(X_{n-1}) \otimes \mathbb{Q} \cong H_1(X_{n-1}; \mathbb{Q}) \cong H_n(X; \mathbb{Q})$. $\qquad\square$

STEP 2. Consider the Eilenberg-Mac Lane spaces $K(\mathbb{Z}, n)$. We compute $H^*(K(\mathbb{Z}, n); \mathbb{Q})$ using the Leray-Serre spectral sequence as follows: We have the fibrations

$$K(\mathbb{Z}, n-1) = \Omega K(\mathbb{Z}, n) \hookrightarrow PK(\mathbb{Z}, n) \to K(\mathbb{Z}, n)$$

and $K(\mathbb{Z}, 1) = S^1$. To begin the induction, $H^*(K(\mathbb{Z}, 1); \mathbb{Q}) = \Lambda(x_1)$, an exterior algebra on one generator of dimension one. The reader may now apply the argument of Example 1.H to prove, for $\deg x_n = n$,

$$H^*(K(\mathbb{Z}, n); \mathbb{Q}) = \begin{cases} \Lambda(x_n), & \text{if } n \text{ is odd,} \\ \mathbb{Q}[x_n], & \text{if } n \text{ is even.} \end{cases}$$

STEP 3. There is a mapping $p \colon S^{2n-1} \to K(\mathbb{Z}, 2n-1)$ that classifies the generator of $H^{2n-1}(S^{2n-1})$. As we described in §4.3, we can turn p into a fibration with fibre F_{2n-1}. From the long exact sequence in homotopy, we know that $\pi_i(F_{2n-1}) \cong \pi_i(S^{2n-1})$ for $i > 2n-1$. We next show that $\pi_i(F_{2n-1}) \otimes \mathbb{Q} = \{0\}$ for $i > 2n-1$. Consider the mapping of fibrations:

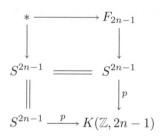

This mapping induces a morphism of spectral sequences in which

$$H^*(S^{2n-1}; \mathbb{Q}) = E_2^{*,0} \xleftarrow{p^*} \bar{E}_2^{*,0} = H^*(K(\mathbb{Z}, 2n-1); \mathbb{Q})$$

is an isomorphism and at E_∞ we have an isomorphism of bigraded modules because the class x_{2n-1} in the base is the only class of degree $2n-1$ that can survive (we know that F_{2n-1} is at least $(2n)$-connected). By the Zeeman comparison theorem (Theorem 3.26) the fibres must have isomorphic rational cohomology. Thus $\tilde{H}^*(F_{2n-1}; \mathbb{Q}) = \{0\}$. We apply Theorem 5.18 to conclude that $\pi_i(F_{2n-1}) \otimes \mathbb{Q} = \{0\}$ for $i > 2n-1$. Since S^{2n-1} is of finite type, for $i > 2n-1$, $\pi_i(S^{2n-1}) \otimes \mathbb{Q} = \{0\}$ implies that $\pi_i(S^{2n-1})$ is finite. $\qquad\square$

A slightly fancier argument (outlined in the exercises) shows that $\pi_i(S^{2n})$ is finite for $i > 2n$ except if $i = 4n - 1$.

Example 5.K first appeared in Serre's thesis. This paper firmly established the place of spectral sequences in homotopy theory. The reader who is interested in the further applications of the Leray-Serre spectral sequences should skip to Chapter 6 where more structure on the spectral sequence is developed and applications given. We close this chapter with a proof of Theorems 5.1 and 5.2, establishing the homology version of the spectral sequence and the multiplicative property for the cohomology version.

5.3 Appendices

Having sketched a construction of the spectral sequence in §5.1 and discussed many of its applications, we fill in the missing steps that establish Theorem 5.1. They are: 1) setting up the apparatus of singular chains in a system of local coefficients and recognizing the E^1-term of the spectral sequence in this context; 2) showing that d^1, the first differential in the spectral sequence, can be identified with the differential on these chain groups. Finally, we establish the multiplicative property of the cohomology spectral sequence (Theorem 5.2).

This exposition is based primarily on the article of [Brown, E94], which in turn is based on Serre's thesis ([Serre51]). In §6.3, there is a discussion of different constructions of the spectral sequence; most details of the proof due to [Dress67] are given. The reader can compare these different constructions to see what they reveal about the structure of fibrations. Finally, the reader should consult the proof of [Serre51] for the contrast between cubical singular theory and simplicial singular theory.

Systems of Local Coefficients Ⓝ

Instances in topology where local data are patched together into global data often require more sensitive systems of coefficients in homology or cohomology. For example, at a point in an n-dimensional manifold, $x \in M^n$, one can consider the group $H_n(M, M - \{x\})$, which is infinite cyclic by excision arguments. A coherent choice of generator for each of these groups as x varies over M determines an orientation of the manifold. Obstruction theory ([Steenrod43]) presents another example of the need for more sensitive coefficients when extending cellular maps a skeleton at a time. The obstructions to lifting maps through a fibration lie in cohomology groups where the coefficients are sensitive to which cell in the base you are lifting over. A more general system of coefficients will permit a precise expression of the geometric data in the exact couple of §5.1.

Let $\Omega(B, a, b)$ denote the set of paths in B joining a to b, $\{\lambda \colon I \to B \mid \lambda$ continuous, $\lambda(0) = a, \lambda(1) = b\}$.

Definition 5.19. *A* **bundle,** $\mathcal{G}$, **of groups over a space** B *is a collection of groups,* $\{G_b; b \in B\}$, *together with a collection of homomorphisms* $h[\lambda]\colon G_{b_1} \to G_{b_0}$, *one for each element in* $\Omega(B, b_0, b_1)$. *This collection of homomorphisms must satisfy:*

(1) *If* c_b *denotes constant path in* $\Omega(B, b, b)$, *then* $h[c_b] = \mathrm{id}\colon G_b \to G_b$.

(2) *If* λ *and* λ' *satisfy* $\lambda(0) = \lambda'(0)$, $\lambda(1) = \lambda'(1)$, *and* $\lambda \simeq \lambda'$ *relative to the endpoints, then* $h[\lambda] = h[\lambda']$.

(3) *If* $\lambda \in \Omega(B, b_0, b_1)$ *and* $\mu \in \Omega(B, b_1, b_2)$ *and* $\lambda * \mu$ *is the product path of* λ *and* μ, *then* $h[\lambda * \mu] = h[\lambda] \circ h[\mu]\colon G_{b_2} \to G_{b_0}$.

We also refer $\mathcal{G}$ *as a* **system of local coefficients** *on* B.

Notice that (1) and (2) together imply that each $h[\lambda]$ is an isomorphism: If we denote $\lambda^{-1}(t) = \lambda(1-t)$, then $\lambda * \lambda^{-1} \simeq c_{b_0} \in \Omega(B, b_0, b_0)$ and so $h[\lambda * \lambda^{-1}] = h[\lambda] \circ h[\lambda^{-1}] = \mathrm{id}$. Categorically speaking, a bundle of groups is a contravariant functor from the **fundamental groupoid of the space** B (the category with objects points in B and morphisms homotopy classes of paths between points) to the category of groups.

EXAMPLES: (a) Given any group G there is the **trivial bundle of groups** over B, also denoted by G, with $G_b = G$ for all b in B and $h[\lambda] = \mathrm{id}$ for all λ in $\Omega(B, b_0, b_1)$.

(b) Let Π denote the bundle of fundamental groups over B, a path-connected space. Here $G_b = \pi_1(B, b)$ for each b in B. If $\lambda \in \Omega(B, b_0, b_1)$, then λ determines an isomorphism $h[\lambda]\colon \pi_1(B, b_1) \to \pi_1(B, b_0)$ by $[\alpha] \mapsto [\lambda^{-1} * \alpha * \lambda]$. The properties of the $*$ operation imply that we have a bundle of groups.

(c) Suppose we have a fibration $F \hookrightarrow E \xrightarrow{\pi} B$ with B path-connected. For a ring R we can form a bundle of R-modules, $\mathcal{H}_n(F; R)$, as follows: Define $G_b = H_n(F_b; R)$ where $F_b = \pi^{-1}(b)$. Let $\Lambda\colon \Omega_p \to E^I$ be a lifting function for π (see §4.3) and suppose given $\lambda\colon I \to B$ with $\lambda(0) = b_0$ and $\lambda(1) = b_1$. Then define $\Phi_\lambda\colon F_{\lambda(1)} \to F_{\lambda(0)}$ by $\Phi_\lambda(x) = \Lambda(\lambda^{-1}, x)(1)$. We let

$$h[\lambda] = \Phi_{\lambda *}\colon H_n(F_{b_1}; R) \longrightarrow H_n(F_{b_0}; R).$$

By the properties of a lifting function, this is a bundle of R-modules. It is this bundle of groups that plays the key role in this chapter.

(d) If we fix a basepoint $*$ in B and let G be a group, then any representation, $\rho\colon \pi_1(B, *) \to \mathrm{Aut}(G)$, gives rise to a bundle of groups, G_ρ, over B. Let $G_b = G$, for each b in B; choose some $[\lambda_b] \in \pi_1(B, b, *)$ and let $h[\lambda_b] = \mathrm{id}$. Then, for any $[\mu] \in \pi_1(B, b_0, b_1)$ let $h[\mu] = \rho[\lambda_{b_0}^{-1} * \mu * \lambda_{b_1}]$. This construction yields a bundle of groups associated to the representation ρ. Conversely, a bundle of groups gives rise to a representation $\rho'\colon \pi_1(B, b) \to \mathrm{Aut}(G_b)$ for b in B.

(e) Given a path-connected n-dimensional manifold M, we can let $G_x = H_n(M, M - \{x\})$ for all $x \in M$. Via the topology of the manifold, we get a bundle of groups with which one can study the notion of orientation of the manifold.

(f) A sheaf of groups over a space B induces a bundle of groups on B, if B is a reasonable space (for example, a CW-complex). For the relevant definitions, the reader can consult the books of [Godement57] or [Warner71].

A **morphism of bundles of groups**, $\Xi \colon \mathcal{G}_1 \to \mathcal{G}_2$, is a collection of homomorphisms $\Xi_b \colon (G_1)_b \to (G_2)_b$, for each b in B, satisfying the property that the diagram

$$
\begin{array}{ccc}
(G_1)_{b_1} & \xrightarrow{\;h_1[\lambda]\;} & (G_1)_{b_0} \\
\Big\downarrow{\scriptstyle \Xi_{b_1}} & & \Big\downarrow{\scriptstyle \Xi_{b_0}} \\
(G_2)_{b_1} & \xrightarrow[\;h_2[\lambda]\;]{} & (G_2)_{b_0}
\end{array}
$$

commutes for all $[\lambda]$ in $\pi_1(B, b_0, b_1)$. A morphism of bundles of groups is a natural transformation of contravariant functors on the fundamental groupoid.

(g) Suppose we have a fibre-preserving map of fibrations, $\hat{f} \colon E \to E'$, that is a commutative diagram

$$
\begin{array}{ccc}
F & \xrightarrow{\;\tilde{f}|\;} & F' \\
\Big\uparrow & & \Big\uparrow \\
E & \xrightarrow{\;\tilde{f}\;} & E' \\
{\scriptstyle p}\Big\downarrow & & \Big\downarrow{\scriptstyle p'} \\
B & = \!\!= & B.
\end{array}
$$

Then $\tilde{f}_*$ induces a morphism of bundles $\mathcal{H}_n(F; R) \to \mathcal{H}_n(F'; R)$.

A morphism of bundles, $\Xi \colon \mathcal{G}_1 \to \mathcal{G}_2$ is an **isomorphism** if each Ξ_b is an isomorphism. We call a bundle of groups $\mathcal{G}$ **simple** if there is an isomorphism $\Xi \colon \mathcal{G} \to G$, the trivial bundle of groups for some group G. The topology of B can sometimes determine a bundle of groups.

Proposition 5.20. *If B is path-connected and simply-connected and $\mathcal{G}$ is a bundle of groups on B, then $\mathcal{G}$ is simple.*

PROOF: Let b be a point in B and $G = G_b$. By assumption $\pi_0(\Omega(B, b, x))$ contains only one homotopy class of paths. If we let $[\lambda]$ denote this class and $\Xi_x = h[\lambda]$, then $\Xi \colon \mathcal{G} \to G$ is an isomorphism. $\qquad\square$

Fix a bundle of abelian groups $\mathcal{G}$ over a space B. Let $C_*(B)$ denote the singular chains on B. If Δ^p denotes the standard p-simplex in $\mathbb{R}^{p+1}$, then let

v_0 denote the leading vertex of Δ^p, that is, the point $(1, 0, \dots, 0)$. Define the set of **singular p-chains with coefficients in the bundle** $\mathcal{G}$ by

$$C_p(B; \mathcal{G}) = \{\text{finite formal sums } \textstyle\sum_i g_i \otimes T_i \mid \text{where, for each } i,$$
$$T_i \colon \Delta^p \to B \text{ is continuous and } g_i \in G_{T(v_0)}\}.$$

The **singular p-cochains with coefficients in the bundle of groups** $\mathcal{G}$ are defined by

$$C^p(B; \mathcal{G}) = \{\text{functions } f \colon C_p(B) \to \textstyle\bigcup_{b \in B} G_b \mid f(T \colon \Delta^p \to B) \in G_{T(v_0)}\}.$$

We describe the differential on $C_*(B; \mathcal{G})$. The usual simplicial boundary operator behaves as follows with respect to the leading vertex (see §4.2 for a reminder of the formulas):

$$(\partial_i T)(v_0) = \begin{cases} T(v_0), & \text{if } i \neq 0, \\ T(v_1), & \text{if } i = 0. \end{cases}$$

Since the coefficients on a given simplex depend on the group in the bundle associated to the image of the leading vertex, a change of leading vertex must be entered into our local data. If $T \colon \Delta^p \to B$ is a p-simplex, let $TL_{v_1}^{v_0} \colon [0, 1] \to B$ be defined

$$TL_{v_1}^{v_0}(t) = T(tv_0 + (1 - t)v_1)$$

be the path joining $T(v_1)$ to $T(v_0)$. The boundary homomorphism ∂_h on $C_*(B; \mathcal{G})$ is defined on the basis by

$$\partial_h(g \otimes T) = h[TL_{v_1}^{v_0}](g) \otimes \partial_0 T + \sum_{j=1}^{p} (-1)^j g \otimes \partial_j T.$$

Lemma 5.21. $\partial_h \circ \partial_h = 0$.

PROOF: From the definition of the ∂_i we have $(\partial_i T)L_{v_1}^{v_0} = TL_{v_1}^{v_0}$ if $i > 1$ and $(\partial_1 T)L_{v_1}^{v_0} \simeq (\partial_0 T)L_{v_1}^{v_0} * TL_{v_1}^{v_0}$; now compute. □

We define the **homology of B with local coefficients in** $\mathcal{G}$ to be

$$H_*(B; \mathcal{G}) = H(C_*(B; \mathcal{G}), \partial_h).$$

Homology with local coefficients generalizes ordinary homology.

Proposition 5.22. *If the bundle of abelian groups $\mathcal{G}$ is simple, then the local coefficient system may be taken to be constant, that is, $H_*(B; \mathcal{G}) \cong H_*(B; G)$ where $G = G_x$ for any x in B.*

PROOF: Let $\Xi \colon \mathcal{G} \to G$ be an isomorphism of the bundles of groups. Since the isomorphism commutes with the homomorphisms $h[\lambda]$, Ξ induces an isomorphism of chain complexes, $\Xi \colon C_*(B; \mathcal{G}) \to C_*(B; G)$. In the case of a trivial bundle of groups, the expression given for ∂_h reduces to the ordinary boundary operator and so $H(C_*(B; \mathcal{G}), \partial_h)$ gives ordinary homology. □

Proof of the Main Theorem ⓃⓃ

Our strategy in proving Theorem 5.1 is to follow a proof of the Eilenberg-Zilber theorem, but for a twisted product. Recall that there are mappings

$$AW\colon C_*(X \times Y) \longleftrightarrow C_*(X) \otimes C_*(Y)\colon EZ,$$

satisfying (1) $AW \circ EZ = \mathrm{id}$ on $C_p(X) \otimes C_q(Y)$; (2) there is a chain homotopy $D_{EZ}\colon C_r(X \times Y) \to C_{r+1}(X \times Y)$ such that $EZ \circ AW - \mathrm{id} = \partial D_{EZ} + D_{EZ}\partial$. If we filter $C_r(X \times Y)$ by

$$F_p = \sum_{s \le p} \mathrm{im}(EZ\colon C_s(X) \otimes C_{r-s}(Y) \to C_r(X \times Y)),$$

the resulting spectral sequence leads to a proof of the Eilenberg-Zilber theorem, if we use homology with coefficients in a field, and to the Künneth theorem more generally.

Suppose $F \hookrightarrow E \xrightarrow{\pi} B$ is a fibration with F and B connected spaces. Recall that π comes equipped with a lifting function, $\Lambda\colon U_\pi \to WE$ where $U_\pi = \{(\lambda, e) \mid \lambda\colon [0,1] \to B, e \in E, \text{ and } \lambda(0) = \pi(e)\}$, $WE = \mathrm{map}([0,1], E)$, and Λ has the properties; $\pi \circ \Lambda(\lambda, e) = \lambda$ and $\Lambda(\lambda, e)(0) = e$ (§4.3).

Our goal is to analyze $C_*(E)$ in terms of $C_*(B)$ and $C_*(F)$ by introducing analogues of AW and EZ. We assume throughout that we have *normalized* singular chains, that is, degenerate singular simplices are set equal to zero.

We first introduce some combinatorial structure available in the setting of simplicial singular chains. Let $\Delta[p]_\bullet$ denote the combinatorial p-simplex,

$$\Delta[p]_n = \{(i_0, i_1, \dots, i_n) \mid i_j \in \{0, 1, \dots, p\} \text{ and } i_0 \le i_1 \le \cdots \le i_n\}.$$

Given $(i_0, i_1, \dots, i_n) \in \Delta[p]_n$ we can define a mapping $\Delta^n \to \Delta^p$ by requiring that the jth vertex of Δ^n map to the i_j^{th} vertex of Δ^p and extending linearly to the rest of Δ^n. We denote this mapping by $(i_0, i_1, \dots, i_n)\colon \Delta^n \to \Delta^p$. The construction leads to a pairing

$$C_p(X) \times \Delta[p]_n \to C_n(X)$$

given on the basis by

$$(T\colon \Delta^p \to X) \quad \mapsto \quad T(i_0, \dots, i_n)\colon \Delta^n \xrightarrow{(i_0, \dots, i_n)} \Delta^p \xrightarrow{T} X.$$

An example of the use of this pairing is the classical Alexander-Whitney mapping, $AW\colon C_*(X \times Y) \to C_*(X) \otimes C_*(Y)$. In this notation we have

$$AW(T) = \sum_{n=0}^{p+q} (pr_1 T)(0, 1, \dots, n) \otimes (pr_2 T)(n, n+1, \dots, p+q),$$

where pr_i denotes projection onto the ith factor.

We next introduce the analogue of $C_*(X) \otimes C_*(Y)$. By $C_p(B; C_q(F_b))$ we mean finite sums of tensor products, $U \otimes V$, where $V \colon \Delta^p \to B$ is a singular p-simplex on B and $U \colon \Delta^q \to F_{V(v_0)} = \pi^{-1}(V(v_0))$, a singular q-simplex on the fibre over $V(v_0)$, the image of the leading vertex of V in B. This is not exactly the group of p-chains with local coefficients since we have not introduced the homomorphisms required for such a system of groups. However, these groups will lead eventually to chains with local coefficients in $\mathcal{H}_q(F)$.

We construct the analogue of EZ based on an operation on $C_p(B; C_q(F_b))$. Suppose $U \otimes V$ is a generator where $V \colon \Delta^p \to B$ and $U \colon \Delta^q \to F_{V(v_0)}$. Let v and s denote points in Δ^p and denote the straight line path joining v to s by $L_v^s \colon [0,1] \to \Delta^p$, $L_v^s(t) = ts + (1-t)v$. We define $U \# V \colon \Delta^q \times \Delta^p \to E$ by

$$U \# V(s_1, s_2) = \Lambda(V \circ L_{v_0}^{s_2}, U(s_1))(1).$$

Since $V \circ L_{v_0}^{s_2}(0) = V(v_0) = \pi U(s_1)$, $U \# V$ is well-defined. Notice, from the properties of lifting functions, that $\pi(U \# V)(s_1, s_2) = V(s_2)$. Furthermore, if $\sigma \in \Delta[q]_{q'}$ and $\tau \in \Delta[p]_{p'}$, and $\tau(v_0) = v_0$, then

$$U \# V(\sigma \times \tau) = (U\sigma) \# (V\tau).$$

We can define the Eilenberg-Zilber map for π as

$$\psi \colon C_p(B; C_q(F_b)) \to C_{p+q}(E)$$
$$\psi(U \otimes V) = (U \# V)_*(EZ((0, 1, \dots, q) \otimes (0, 1, \dots, p))),$$

that is, ψ takes $U \otimes V$ to the image of the chain $(0, 1, \dots, q) \otimes (0, 1, \dots, p)$ under the composite

$$C_q(\Delta[q]_\bullet) \otimes C_p(\Delta[p]_\bullet) \xrightarrow{EZ} C_{p+q}(\Delta[q]_\bullet \times \Delta[p]_\bullet)$$
$$\to C_{p+q}(\Delta^q \times \Delta^p) \xrightarrow{(U \# V)_*} C_{p+q}(E).$$

The next ingredient in the proof is the filtration. Let $F_p C_r(E)$ be the subgroup of $C_r(E)$ generated by $T \colon \Delta^r \to E$, such that $\pi T = S(i_0, \dots, i_r)$, for $i_r \le p$ and $S \colon \Delta^p \to B$. This is an increasing filtration with $F_r C_r(E) = C_r(E)$.

Lemma 5.23. *The homomorphism $\psi \colon C_p(B; C_q(F_b)) \to C_{p+q}(E)$ satisfies the following properties:*

(1) $\psi(C_p(B; C_q(F_b)))$ *is contained in* $F_p C_{p+q}(E)$.
(2) $\partial \psi \equiv \psi(\partial_F \otimes 1) \bmod F_{p-1}$.

PROOF: From the properties of lifting functions we have the commutative diagram:

$$\begin{array}{ccc} \Delta^q \times \Delta^p & \xrightarrow{\ U\#V\ } & E \\ {\scriptstyle pr_2}\downarrow & & \downarrow{\scriptstyle \pi} \\ \Delta^p & \xrightarrow{\ \ V\ \ } & B. \end{array}$$

It follows that

$$\pi_*(U\#V)_*(EZ((0,1,\dots,q)\otimes(0,1,\dots,p)))$$
$$= V_*(EZ((0,1,\dots,q)\otimes(0,1,\dots,p))) = \sum V(i_0,\dots,i_{p+q})$$

where $i_{p+q}\le p$ for all summands.

To prove (2) we observe that the ordinary singular differential can be written as follows: For $W\in C_r(X)$,

$$\partial(W) = \sum_{i=0}^{r}(-1)^i\partial_i(W) = \sum_{i=0}^{r}(-1)^i W(0,1,\dots,\hat{i},\dots,r).$$

Since the singular differentials commute with all the maps that make up the composite that defines ψ, we have

$$\partial\psi(U\otimes V) = (U\#V)_*(EZ(\partial((0,\dots,q)\otimes(0,\dots,p))))$$
$$= (U\#V)_*\Big(EZ\Big(\sum_i(-1)^i(0,\dots,\hat{i},\dots,q)\otimes(0,\dots,p)$$
$$+ \sum_j(-1)^{j+q}(0,\dots,q)\otimes(0,\dots,\hat{j},\dots,p)\Big)\Big)$$
$$= ((\partial_F U)\#V)_*\Big(\sum(\sigma_i',\sigma_i'')\Big) + (-1)^q(U\#(\partial_B V))_*\Big(\sum(\tau_j',\tau_j'')\Big)$$
$$= \psi(\partial_F\otimes 1)(U\otimes V) \text{ modulo } F_{p-1}C_{p+q}(E),$$

where this last equality follows because $\partial_B V$ determines a chain in filtration $p-1$. □

Next up is the analogue of the Alexander-Whitney map: On generators T of $F_p C_{p+q}(E)$ let

$$\phi(T) = T(0,\dots,q)\otimes\pi T(q,\dots,p+q).$$

If $\phi(T)\ne 0$, then $\pi T(q,\dots,p+q)$ is a nondegenerate singular p-chain on B. Since $\pi T(q,\dots,p+q) = S(i_0,\dots,i_{p+q})(q,\dots,p+q)$ where $S\colon\Delta^p\to B$, we find that this is nondegenerate only when

$$(i_0,\dots,i_{p+q}) = (0,\dots,0,1,\dots,p).$$

It follows that $\pi T(0,\dots,q) = \pi T(0,\dots,0)$ and so the q-simplex $T(0,\dots,q)$ is a mapping $\Delta^q\to F_{\pi T(q,\dots,p+q)(v_0)}$. Thus $\phi\colon F_p C_{p+q}(E)\to C_p(B;C_q(F_b))$.

Lemma 5.24. *The homomorphism* $\phi\colon F_p C_{p+q}(E) \to C_p(B; C_q(F_b))$ *satisfies the following properties:*

(1) $\phi\psi = \mathrm{id}$.

(2) *There is a chain homotopy* $D\colon F_p C_{p+q}(E) \to F_p C_{p+q+1}(E)$ *with* $\partial D + D\partial = \psi\phi - \mathrm{id}$.

(3) $\phi(F_{p-1}C_{p+q}(E)) = 0$. *And so* ϕ *induces a homomorphism* $\phi^0\colon E^0_{p,q} \to C_p(B; C_q(F_b))$.

PROOF: To establish (1) we observe that we can write

$$EZ((0,\dots,q) \otimes (0,\dots,p))$$
$$= ((0,1,\dots,q,q,\dots,q) \times (0,0,\dots,0,1,\dots,p)) + \sum_i (\tau'_i, \tau''_i),$$

where $\tau''_i(0,\dots,p)$ is degenerate. If we write $\sigma' = (0,1,\dots,q,q,\dots,q)$ and $\sigma'' = (0,0,\dots,0,1,\dots,p)$, it follows that

$$\phi\psi(U \otimes V) = \phi((U\#V)_*(EZ((0,\dots,q) \otimes (0,\dots,p))))$$
$$= \phi((U\#V)_*(\sigma',\sigma''))$$
$$= (U\#V)_*(\sigma',\sigma'')(0,\dots,q) \otimes \pi(U\#V)_*(\sigma',\sigma'')(q,\dots,p+q).$$

Since $(0,\dots,0,1,\dots,p)(0,\dots,q) = (0,0,\dots,0)$, we get

$$= \Lambda(V(0,\dots,0)L^{(0,\dots,0)(-)}_{v_0}, U(0,\dots,q)(-))(1) \otimes V$$
$$= \Lambda(V \circ L^{v_0}_{v_0}, U(0,\dots,q)(-))(1) \otimes V$$
$$= U \otimes V.$$

Here we have used the property of the lifting function that constant paths lift to constant paths.

To establish the second assertion, we introduce another operation on singular simplices on E. If $T\colon \Delta^q \to E$ is a singular simplex on E, then define $\hat{T}\colon \Delta^q \times \Delta^q \to E$ by

$$\hat{T}(s_1, s_2) = \Lambda(\pi T \circ L^{s_2}_{s_1}, T(s_1))(1).$$

The relevant properties of $\hat{T}$ are that $\pi\hat{T}(s_1, s_2) = \pi T(s_2)$, that $\hat{T}(s_1, s_1) = T(s_1)$, and the following formula:

$$\hat{T}((0,\dots,q) \times (q,\dots,p+q)) = T(0,\dots,q)\#\pi T(q,\dots,p+q),$$

which can be seen by computing

$$T(0,\dots,q)\#\pi T(q,\dots,p+q)(s_1, s_2)$$
$$= \Lambda(\pi T \circ L(q,\dots,p+q)^{s_2}_{v_0}, T(0,\dots,q)(s_1))(1)$$
$$= \hat{T}((0,\dots,q) \times (q,\dots,p+q))(s_1, s_2).$$

Using this new operation we define the chain homotopy $D\colon C_{p+q}(E) \to C_{p+q+1}(E)$ by

$$D(T) = (\hat{T})_*(D_{EZ}((0,\dots,p+q) \times (0,\dots,p+q))).$$

Because $\partial D_{EZ} + D_{EZ}\partial = EZ \circ AW - \mathrm{id}$, we compute

$$\begin{aligned}
\partial D(T) &= \partial(\hat{T})_*(D_{EZ}((0,\dots,p+q) \times (0,\dots,p+q)))\\
&= (\hat{T})_*(\partial D_{EZ}((0,\dots,p+q) \times (0,\dots,p+q)))\\
&= (\hat{T})_*(-D_{EZ}\partial((0,\dots,p+q) \times (0,\dots,p+q))\\
&\quad + EZ \circ AW((0,\dots,p+q) \times (0,\dots,p+q))\\
&\quad - (0,\dots,p+q) \times (0,\dots,p+q))\\
&= -D(\partial T) + (\hat{T})_*(EZ \circ AW)((0,\dots,p+q) \times (0,\dots,p+q))\\
&\quad - (\hat{T})_*((0,\dots,p+q) \times (0,\dots,p+q)).
\end{aligned}$$

The last term in the list is T as it is $\hat{T}$ on diagonal elements. Thus the assertion comes down to showing that

$$(\hat{T})_*(EZ \circ AW)((0,\dots,p+q) \times (0,\dots,p+q)) = \psi\phi(T).$$

To establish this, chase through the commutative diagram:

The top triangle commutes because we have $\hat{T}$ on diagonal elements; the middle square is commutative by the definition of ϕ, and the bottom square commutes because $\hat{T}((0,\dots,q) \times (q,\dots,p+q)) = T(0,\dots,q)\#\pi T(q,\dots,p+q)$. Thus $\partial D + D\partial = \psi\phi - \mathrm{id}$.

We next show that ϕ induces a mapping on $E^0_{p,q} = F_p/F_{p-1}$. Suppose $U \in F_{p-1}C_{p+q}(E)$. Then $\phi(U) = U(0,\dots,q) \otimes \pi U(q,\dots,p+q)$ and $\pi U = S(i_0,\dots,i_{p+q})$ with $i_{p+q} \le p - 1$. But $(i_0,\dots,i_{p+q})(q,\dots,p+q)$ is always degenerate when $i_{p+q} \le p - 1$. Thus ϕ induces a mapping,

$$\phi^0\colon E^0_{p,q} \to C_p(B; C_q(F_b)). \qquad \square$$

Lemma 5.25. $\phi\partial(T) = (\partial_F \otimes 1)(\phi(T))$, that is, $\phi^0 \colon E^0_{p,q} \to C_p(B; C_q(F_b))$ is a chain mapping.

PROOF: We compute directly:

$$\phi(\partial T) = \sum_{i=0}^{p+q} (-1)^i \phi(T(0, \ldots, \hat{i}, \ldots, p+q))$$
$$= \sum_{i=0}^{p+q} (-1)^i [T(0, \ldots, \hat{i}, \ldots, p+q)(0, \ldots, q-1)$$
$$\otimes \pi T(0, \ldots, \hat{i}, \ldots, p+q)(q-1, \ldots, p+q-1)].$$

It follows from the definition of the pairing that

$$(0, \ldots, \hat{i}, \ldots, p+q)(q-1, \ldots, p+q-1)$$
$$= \begin{cases} (q, \ldots, p+q) & \text{for } i \leq q, \\ (q-1, q-1, q, \ldots, \hat{i}, \ldots, p+q) & \text{for } i > q. \end{cases}$$
$$(0, \ldots, \hat{i}, \ldots, p+q)(0, \ldots, q-1) = (0, \ldots, \hat{i}, \ldots, q), \text{for } i \leq q.$$

Since $\pi T = \pi T(0, \ldots, 0, 1, \ldots, p)$, we know that, for $j > 0$,

$$\pi T(q-1, q, \ldots, \widehat{q+j}, \ldots, p+q) = \pi T(0, 0, 1, \ldots, \hat{j}, \ldots, p),$$

which is degenerate. Our formula becomes

$$\phi(\partial T) = \sum_{i=0}^{q} (-1)^i [T(0, \ldots, \hat{i}, \ldots, q) \otimes \pi T(q, \ldots, p+q)]$$
$$= (\partial_F \otimes 1)(\phi(T)).$$

Thus we have the homomorphism $\phi^1 = H(\phi^0) \colon E^1_{p,q} \to C_p(B; \mathcal{H}_q(F))$.   □

So far we have proven that there is a chain equivalence between $E^0_{p,q}$ and $C_p(B; C_q(F))$. To finish the proof of Theorem 5.1, we need to show that we have a commutative diagram,

$$(5.26)$$

$$\begin{array}{ccc} E^1_{p,q} & \xrightarrow{\phi^1 = H(\phi^0)} & C_p(B; \mathcal{H}_q(F)) \\ d^1 \downarrow & & \downarrow (-1)^q \partial_h \\ E^1_{p-1,q} & \xrightarrow{\phi^1} & C_{p-1}(B; \mathcal{H}_q(F)). \end{array}$$

To do this we introduce another chain homotopy, this time

$$D_2 \colon C_p(B; \mathcal{H}_q(F)) \to F_{p-1}C_{p+q}(E).$$

To define the chain homotopy we require one more operation derived from the lifting function. Suppose given $V \colon \Delta^p \to B$ and $U \colon \Delta^q \to F_{V(v_0)}$. Define $U * V \colon \Delta^q \times \Delta^{p-1} \times \Delta^1 \to E$ by

$$U * V(s_1, s_2, t) = \Lambda(V \circ L^{(1, \ldots, p)(s_2)}_{(1-t)v_0+tv_1}, \Lambda(V \circ L^{v_1}_{v_0}, U(s_1))(t))(1).$$

Lemma 5.27. *For singular simplices* $V : \Delta^p \to B$ *and* $U : \Delta^q \to F_{V(v_0)}$ *we have the following*

(1) $U * V(s_1, s_2, 0) = U \# V(s_1, (1, \dots, p)(s_2))$.
(2) $U * V(s_1, s_2, 1) = \Phi_{V \circ L_{v_1}^{v_0}*}(U) \# (\partial_0 V)(s_1, s_2)$.
(3) $\pi(U * V)(s_1, s_2, t) = (\partial_0 V)(s_2)$.

PROOF: We leave (1) and (3) for the reader and establish (2). (Recall $\Phi_{\lambda*}$ is defined on p.164.) We observe first that

$$(\partial_0 V)(t_0, \dots, t_{p-1}) = V(0, t_0, \dots, t_{p-1}) = V(1, \dots, p)(t_0, \dots, t_{p-1}),$$

and so $(\partial_0 V) \circ L_{v_0}^{s_2} = V \circ L_{v_1}^{(1,\dots,p)(s_2)}$. It follows that

$$\begin{aligned}
U * V(s_1, s_2, 1) &= \Lambda(V \circ L_{v_1}^{(1,\dots,p)(s_2)}, \Lambda(V \circ L_{v_0}^{v_1}, U(s_1))(1))(1) \\
&= \Lambda((\partial_0 V) \circ L_{v_0}^{s_2}, \Phi_{V \circ L_{v_1}^{v_0}*}(U)(s_1))(1) \\
&= \Phi_{V \circ L_{v_1}^{v_0}*}(U) \# (\partial_0 V)(s_1, s_2). \qquad \square
\end{aligned}$$

Define the desired chain homotopy by

$$D_2(U \otimes V) = (-1)^{p+q}(U*V)_*(EZ(EZ((0, \dots, q) \otimes (0, \dots, p-1)) \otimes (0, 1))).$$

Lemma 5.28. $D_2(C_p(B; C_q(F_b))) \subset F_{p-1} C_{p+q}(E)$ *and*

$$\begin{aligned}
\partial \psi(U \otimes V) = {} & \psi(\partial_F U \otimes V) + (-1)^q \psi(\Phi_{V \circ L_{v_1}^{v_0}*}(U) \otimes (\partial_0 V)) \\
& + \sum_{i>0} (-1)^{i+q} \psi(U \otimes \partial_i V) + (-1)^q \partial D_2(U \otimes V) \\
& + (-1)^q D_2((\partial_F U) \otimes V) + W,
\end{aligned}$$

where W *is a sum of elements from* $F_{p-1} C_{p+q-1}(E)$.

PROOF: By Lemma 5.27(3), we get the commutative diagram

$$\begin{array}{ccc}
C_{p+q}(\Delta^q \times \Delta^{p-1} \times \Delta^1) & \xrightarrow{(U*V)_*} & C_{p+q}(E) \\
{\scriptstyle pr_2} \downarrow & & \downarrow {\scriptstyle \pi_*} \\
C_{p+q}(\Delta^{p-1}) & \xrightarrow[(\partial_0 V)_*]{} & C_{p+q}(B).
\end{array}$$

Since the simplices mapping into the second factor of $\pi D_2(U \otimes V)$ come from $EZ(0, \dots, p-1)$, D_2 takes its image in filtration $p-1$.

We compute directly

$$
\begin{aligned}
\partial\psi(U \otimes V) &= \partial(U\#V)_*(EZ((0,\dots,q) \otimes (0,\dots,p))) \\
&= (U\#V)_*(EZ(\partial((0,\dots,q) \otimes (0,\dots,p)))) \\
&= \psi((\partial_F U) \otimes V) + (-1)^q (U\#V)_*(EZ((0,\dots,q) \otimes (1,\dots,p))) \\
&\quad + \sum_{i=1}^{p} (-1)^{i+q}\psi(U \otimes (\partial_i V)).
\end{aligned}
$$

By examining the image of the Eilenberg-Zilber map EZ on nondegenerate simplices one establishes that

$$
\begin{aligned}
(U\#V)_*&(EZ((0,\dots,q) \otimes (1,\dots,p))) \\
&= (U * V)_*(EZ(EZ((0,\dots,q) \otimes (0,\dots,p-1)) \otimes (0))).
\end{aligned}
$$

Consider the chain homotopy:

$$
\begin{aligned}
\partial D_2&(U \otimes V) \\
&= (-1)^{p+q}\partial((U * V)_*(EZ(EZ((0,\dots,q) \otimes (0,\dots,p-1)) \otimes (0,1)))) \\
&= -D_2((\partial_F U) \otimes V) \\
&\quad + (-1)^{p+q+1}\sum_i (U * V)_*(EZ(EZ((0,..,q) \otimes (0,..,\hat{i},..,p-1)) \otimes (0,1))) \\
&\quad + (U * V)_*(EZ((0,\dots,q) \otimes (0,\dots,p-1),0)) \\
&\quad - (U * V)_*(EZ((0,\dots,q) \otimes (0,\dots,p-1),1)).
\end{aligned}
$$

Let W denote the big sum with terms coming from EZ applied to a product involving $(0,\dots,\hat{i},\dots,p-1)$. These classes are all of filtration $p-1$. By Lemma 5.27, the rest of the morass becomes

$$
\begin{aligned}
\partial D_2(U \otimes V) &+ D_2((\partial_F U) \otimes V) \\
&= W + (-1)^q \partial\psi(U \otimes V) - (-1)^q \psi((\partial_F U) \otimes V) \\
&\quad - \sum_{i=1}^{p} (-1)^i \psi(U \otimes \partial_i V) - \psi(\Phi_{V \circ L_{v_1}^{v_0}*}(U) \otimes (\partial_0 V)). \qquad \square
\end{aligned}
$$

For filtration reasons we have

$$
\phi(W) = \phi(\partial D_2(U \otimes V)) = \phi(D_2(\partial_F U \otimes V)) = 0.
$$

If U represents a class in $H_q(F_{V(v_0)})$, then we can apply ϕ to get

$$
(-1)^q \phi\partial\psi(U \otimes V) = \phi\psi(\partial_h(U \otimes V)).
$$

This establishes the commutativity of (5.26), and we have finished the proof of Theorem 5.1 for pairs $(B, \emptyset)$.

Multiplicative Matters Ⓝ

We close the chapter with a proof of Theorem 5.2. Duality allows us to identify the E_2-term as $H^*(B; \mathcal{H}^*(F))$. The dual filtration on cochains is given by

$$F^p C^{p+q}(E) = \{f \in C^{p+q}(E) \mid f(T) = 0 \text{ for all } T \in F_{p-1}C_{p+q}(E)\}.$$

Thus we only need to establish that we have a spectral sequence of algebras.

For any space X, suppose $f \in C^p(X)$ and $f' \in C^{p'}(X)$. We define the cup-product $f \smile f' \in C^{p+p'}(X)$ and the cup$_1$-product $f \smile_1 f' \in C^{p+p'-1}(X)$ as follows: If $T \in C_{p+p'}(X)$ and $U \in C_{p+p'-1}(X)$, then

$$(f \smile f')(T) = f(T(0, \dots, p))f'(T(p, \dots, p+p')),$$

and $(f \smile_1 f')(U) =$

$$\sum_{i=0}^{p+p'-1} \pm f(U(0, \dots, i, i+p', \dots, p+p'-1))f'(U(i, \dots, i+p')).$$

The cup$_1$-product plays a key role in proving the graded commutativity of the cup-product as the following formula shows:

$$\delta(f \smile_1 f') = f \smile f' - (-1)^{pp'}[f' \smile f] - [(\delta f) \smile_1 f'] - (-1)^p[f \smile_1 (\delta f')].$$

There is a subtlety to the duality when we use local coefficients: An element of $C^p(B; C^q(F_b))$ is a mapping of the form $f \colon C_p(B) \to \mathrm{Hom}(C_q(F_b), \mathbb{Z})$. If we employ the Hom-Tensor interchange, then we have a linear mapping $f \colon C_q(F_b) \otimes C_p(B) \to \mathbb{Z}$. However, the differential must satisfy

$$\langle \delta f, U \otimes V \rangle = \langle f, \partial(U \otimes V) \rangle.$$

Since ∂ is a derivation with respect to the tensor product, the pairing on the right is with $(\partial U) \otimes V + (-1)^q U \otimes (\partial V)$. To accommodate the cross term in a different bidegree we represent a class $u \in H^p(B; \mathcal{H}^q(F))$ as a pair $\{f, g\}$ with $f \colon C_p(B) \to \mathrm{Hom}(C_q(F_b), \mathbb{Z})$ and $g \colon C_{p+1}(B) \to \mathrm{Hom}(C_{q-1}(F_b), \mathbb{Z})$. This pair satisfies, for $T \in C_{p+1}(B)$, $U \in C_q(F_b)$,

$$\langle \delta f, T \rangle(U) = \langle f, \partial T \rangle(U) - \langle g, T \rangle(\partial U).$$

If $u = \{f, g\}$ is a cohomology class, then $\delta f(T) = 0$ and $f(\partial T) = g(T) \circ \partial$ as elements of $C^q(F_b)$.

Given $f \in C^p(B; C^q(F_b))$ and $f' \in C^{p'}(B; C^{q'}(F_b))$ we define a new cochain $D(f, f') \in C^{p+p'+q+q'-1}(E)$ by the following recipe: Let $T \in C_{p+p'+q+q'-1}(E)$. Form two new cochains, given by

$$U \in C_p(E) \mapsto f(\pi U)(T(0, \dots, q))$$
$$U' \in C_{q'}(E) \mapsto f'(\pi T(p+q+q'-1, \dots, p+p'+q+q'-1))(U').$$

Form their cup$_1$-product and apply it to $T(q, \dots, p+q+q'-1)$. This procedure defines $D(f, f')(T)$.

Lemma 5.29. $D(f, f')$ *has filtration* $p + p'$ *when* $f \in C^p(B; C^q(F_b))$ *and* $f' \in C^{p'}(B; C^{q'}(F_b))$. *Furthermore, we have*

$$\delta D(f, f') = \phi^*(f) \smile \phi^*(f') - (-1)^{pq'} \phi^*(f \smile f')$$
$$\pm D(\delta_1 f, f') \pm D(\delta_2 f, f') \pm D(f, \delta_1 f') \pm D(f, \delta_2 f'),$$

where $\delta_1 f(U)(S) = f(\partial U)(S)$, $\delta_2 f(U)(S) = f(U)(\partial S)$ *and* $\langle f, \phi_* T \rangle = \langle \phi^* f, T \rangle$.

PROOF: Let $U \in F_{p+p'-1} C_{p+p'+q+q'-1}(E)$. Then we can write

$$\pi U = \pi U(0, \ldots, 0, 1, \ldots, p + p' - 1).$$

To apply $D(f, f')$ to U we form a certain cup$_1$-product and evaluate on the chain $U(q, \ldots, p + q + q')$. However, $\pi U(q, \ldots, p + q + q')$ is degenerate and so $D(f, f')(U) = 0$. Therefore, $D(f, f') \in F^{p+p'} C^{p+p'+q+q'-1}(E)$.

   Suppose $V \in C_{p+p'+q+q'}(E)$. We compute $(\delta D(f, f'))(V)$ which is made up of cup products and cup$_1$-products evaluated on expressions involving $V(q, \ldots, p + q + q')$. Using the defining property of the cup$_1$-product, we can examine each summand in turn:

$$(f \smile f')(V(q, \ldots, p + q + q')) =$$
$$f(\pi V(q, \ldots, p + q + q')(0, \ldots, q))(V(0, \ldots, q))\cdot$$
$$f'(\pi V(p + p' + q', \ldots, p + p' + q + q'))(V(q, \ldots, p + q + q')(p, \ldots, p + q'))$$
$$= f(\pi V(q, \ldots, p + q))(V(0, \ldots, q))\cdot$$
$$f'(\pi V(p + q + q', \ldots, p + p' + q + q'))(V(p + q, \ldots, p + q + q'))$$
$$= \langle f, \phi_*(V(0, \ldots, p + q)) \rangle \cdot \langle f', \phi_*(V(p + q, \ldots, p + p' + q + q')) \rangle$$
$$= (\phi^*(f) \smile \phi^*(f'))(T).$$

The sign $(-1)^{pq'}$ in front of the next summand comes from the formula for the differential applied to a cup$_1$-product and the fact that we are using a p-cochain and a q'-cochain.

$$(f' \smile f)(V(q, \ldots, p + q + q')) =$$
$$f'(\pi V(p + q + q', \ldots, p + p' + q + q'))(V(q, \ldots, p + q + q')(0, \ldots, q'))\cdot$$
$$f(\pi V(q, \ldots, p + q + q')(q, \ldots, p + q'))(V(0, \ldots, q))$$
$$= f(\pi V(q + q', \ldots, p + q + q'))(V(0, \ldots, q))\cdot$$
$$f'(\pi V(p + q + q', \ldots, p + p' + q + q'))(V(q, \ldots, q + q'))$$
$$= \langle f, V(0, \ldots, p + q) \otimes \pi V(q + q', \ldots, p + q + q') \rangle\cdot$$
$$\langle f', V(q, \ldots, q + q') \otimes \pi V(p + q + q', \ldots, p + p' + q + q') \rangle$$
$$= \langle f \smile f', V(0, \ldots, q + q') \otimes \pi V(q + q', \ldots, p + p' + q + q') \rangle$$
$$= \langle f \smile f', \phi_*(T) \rangle = (\phi^*(f \smile f'))(T).$$

We are looking to establish a map on the E_2-term and so we may think of our class f as representing $u \in H^p(B; \mathcal{H}^q(F))$, that is, it is part of a pair $\{f, g\}$ with $\delta f(U)(S) = \delta_1 f(U)(S) + \delta_2 f(U)(S)$.

$$((\delta f) \smile_1 f')(V(q, \dots, p + q + q'))$$
$$= \sum_i \pm \langle \delta f, V(q, \dots, p + q + q')(0, \dots, i, i + q', \dots, p + q') \rangle \cdot$$
$$\langle f', V(q, \dots, p + q + q')(i, \dots, i + q') \rangle.$$

The cochain $\langle \delta f, V(q, \dots, p + q + q')(0, \dots, i, i + q', \dots, p + q') \rangle$ applied to a q-chain on F gives something of the form $\langle f, \partial U \rangle(S) + \langle g, U \rangle(\partial S)$. Since f is a cocycle, the first part of this sum vanishes and so $\langle \delta f, U \rangle(S) = \langle g, U \rangle(\partial S)$. Thus $(\delta f) \smile_1 f'$ has filtration $p + p' + 1$ and it vanishes when we take the associated graded module. At the E_2-term, $\langle g, U \rangle$ is a cocycle and so all terms involving δ_2 vanish. Since these remarks apply to all the terms coming from $(\delta f) \smile_1 f'$ and $f \smile_1 (\delta f')$, we get in the E_2-term

$$(\phi_2^* f) \smile (\phi_2^* f') = (-1)^{pq'} \phi_2^*(f \smile f')$$

and we have proved Theorem 5.2. $\qquad\qquad\qquad\qquad\qquad\qquad\square$

Exercises

5.1. Prove that the cellular filtration that gives rise to the Leray-Serre spectral sequence is exhaustive.

5.2. Derive the **Wang long exact sequence** for a fibration $F \hookrightarrow E \to B$ with B a simply-connected homology n-sphere, $n \geq 2$, and F, path-connected:

$$\cdots \to H^k(E; R) \xrightarrow{i^*} H^k(F; R) \xrightarrow{\theta} H^{k-n+1}(F; R) \xrightarrow{j^*} H^{k+1}(E; R) \to \cdots$$

Show further that θ is a derivation, if n is even and when n is odd, θ is an antiderivation $(\theta(x \smile y) = \theta(x) \smile y + x \smile \theta(y))$.

5.3. Establish the Serre exact sequence (Example 5.D) in the case of homology.

5.4. Prove Proposition 5.5.

5.5. Theorem 5.1 can be extended to fibrations over a CW-pair (B, A). Suppose $F \hookrightarrow E \xrightarrow{p} B$ is a fibration, $A \subset B$, $E_A \subset E$, and $p^{-1}(A) = E_A$, that is, we have a fibration of pairs,

$$F \hookrightarrow (E, E') \xrightarrow{p} (B, A).$$

Sketch a proof of the homology Leray-Serre spectral sequence for pairs for which the E^2-term takes the form

$$E_{p,q}^2 \cong H_p(B, A; \mathcal{H}_q(F; G)),$$

and the spectral sequence converges to $H_*(E, E_A; G)$.

5.6. There is another notion of fibration for pairs: Let $F \hookrightarrow E \xrightarrow{p} B$ is a fibration, $F_0 \subset F$, and $E_0 \subset E$, such that $p|_{E_0} : E_0 \to B$ is a fibration with fibre F_0. That is, we have a fibration of pairs

$$(F, F_0) \hookrightarrow (E, E_0) \xrightarrow{p} B.$$

Derive the homology Leray-Serre spectral sequence in this case for which $E_{p,q}^2 \cong H_p(B; \mathcal{H}_q(F, F_0; G))$, converging to $H_*(E, E_0; G)$.

5.7. Prove Corollary 5.8.

5.8. Show that F is totally homologous to zero with respect to k if and only if the following holds for the Poincaré series with respect to k:

$$P(E, t) = P(B, t) \times P(F, t).$$

5.9. Determine the nonzero differential in Hirsch's example of a fibration $F \hookrightarrow E \xrightarrow{p} B$ with p^* injective but for which the cohomology Leray-Serre spectral sequence does not collapse. (Another example is given by the **free loop space** associated to a space X: Let $\mathcal{L}X = \mathrm{map}(S^1, X)$, the space of unbased loops on X. Evaluation at $1 \in S^1$ gives a fibration $\Omega X \hookrightarrow \mathcal{L}X \xrightarrow{\mathrm{ev}_1} X$. Sending $x \in X$ to the constant loop at x gives a section of ev_1, and so ev_1^* is injective. However, the associated Leray-Serre spectral sequence need not collapse (see [Smith, L81] and [McCleary90] for explicit examples).

5.10. Show that $V_k(\mathbb{C}^n) \cong \mathrm{SU}(n)/\mathrm{SU}(n - k)$ where $\mathrm{SU}(n - k) \subset \mathrm{SU}(n)$ is included as the lower right hand corner of an otherwise $n \times n$ identity matrix.

5.11. Prove Theorem 5.13.

5.12. Show that $V_k(\mathbb{R}^n)$ is $(n - k)$-connected.

5.13. The symplectic groups, $\mathrm{Sp}(n)$, are the analogues of the special orthogonal or unitary groups over the quaternions. There are fibrations defined analogously, $\mathrm{Sp}(n - 1) \hookrightarrow \mathrm{Sp}(n) \to S^{4n-3}$. From $\mathrm{Sp}(1) \cong S^3$, compute $H^*(\mathrm{Sp}(n))$, for all n. The quaternionic Stiefel manifolds $V_k(\mathbb{H}^n)$ are defined similarly to the real and complex Stiefel manifolds. Compute $H^*(V_k(\mathbb{H}^n))$.

5.14. Compute $H^*(\mathrm{SO}(n); \mathbb{F}_p)$, for p, an odd prime.

5.15. The exceptional Lie group, G_2, is represented by the group of automorphisms of the Cayley numbers, $\mathcal{K}$. (See [Whitehead, GW78, Appendix A] for a discussion of this division algebra.) There is a fibration that results from this observation; $S^3 \hookrightarrow G_2 \to V_2(\mathbb{R}^7)$. Compute $H^*(G_2; \mathbb{F}_2)$.

5.16. Show that S^n is ULC.

5.17. Let $S^{n-1} \hookrightarrow V_2(\mathbb{R}^{n+1}) \to S^n$ be the fibration sending the orthonormal 2-frame (v_1, v_2) to v_1. Show that this fibration has no section if n is even. Modify the argument of Example 5.K, using this fibration, to show that $\pi_i(S^{2m})$ is finite, except when $i = 2m$ or $i = 4m - 1$. In particular, show that $\pi_{4m-1}(S^{2m}) \cong \mathbb{Z} \oplus$ finite. (Hint: Compute $H_*(V_2(\mathbb{R}^{n+1}); \mathbb{Q})$ and apply the rational Hurewicz theorem. Then study the long exact sequence of homotopy groups associated to the fibration.)

6

The Leray-Serre Spectral Sequence II

" ... the behavior of this spectral sequence ... is a bit
like an Elizabethan drama, full of action, in which the
business of each character is to kill at least one other
character, so that at the end of the play one has the stage
strewn with corpses and only one actor left alive (namely
the one who has to speak the last few lines)."

J. F. Adams

We begin with a classical computation that demonstrates further the potential of the Leray-Serre spectral sequence in homotopy theory; we also introduce several of the notions that will play as leitmotifs through this chapter. Our first goal is to prove the following result.

Theorem 6.1. $\pi_4(S^2) \cong \mathbb{Z}/2\mathbb{Z}$.

Though this fact may be obtained by more elementary means ([Whitehead, GW50]), the proof here is based on a general method (albeit limited) for investigating homotopy groups and which leads to such results as the proof of [Brown, E57] that the homotopy groups of finite complexes are finitely computable. The technique is to apply the spectral sequence to stages of the Postnikov tower of S^2 (§4.3). Each stage is a pullback of the path-loop fibration over an Eilenberg-Mac Lane space (S^2 is simply-connected). We compute $H^*(K(\pi, n))$ in some low dimensions in order to proceed. Further progress might be possible if more were known about $H^*(K(\pi, n))$. This problem was solved by [Cartan54]. The computation of $H^*(K(\pi, n); \mathbb{F}_p)$, however, is simplified significantly by the presence of the Steenrod algebra. The key to the mod p computation is the transgression and its properties. The development of this feature of the Leray-Serre spectral sequence occupies the first third of the chapter and culminates in the determination of $H^*(K(\mathbb{Z}/p\mathbb{Z}, n); \mathbb{F}_p)$ and $H^*(K(\mathbb{Z}, n); \mathbb{F}_p)$ due to [Serre53] for $p = 2$ and to [Cartan54] for p an odd prime. [Serre53] used these calculations to demonstrate the nonvanishing of the homotopy groups of finite complexes in infinitely many dimensions. His argument involves an ingenious use of Poincaré series and the method of 'killing homotopy groups'.

The middle third of the chapter turns the spectral sequence apparatus, together with the transgression, to computations of the cohomology of classifying spaces and characteristic classes of fibre bundles following [Borel53]. We begin with a sketch of the theory of classifying spaces and their construction. The computation of $H^*(BG; k)$ makes up a major portion of this section that closes with the relation of these computations to the notion of characteristic classes.

In the final third of the chapter, we discuss some different approaches to the construction of the Leray-Serre spectral sequence. We begin with a discussion of the original construction of [Leray50] and a structure result for differentials due to [Fadell-Hurewicz58]. The next construction considered is due to [Brown, E59] and has its roots in the classical theorem of [Hirsch, G54] on the cochains on the total space of a fibration. [Brown, E59] introduced the machinery of twisting cochains, a kind of generalization of the Eilenberg-Zilber theorem, to determine how the cochains on the total space are related to those on the base and fibre. A spectral sequence arises naturally from this construction that is identifiable with the Leray-Serre spectral sequence. The other construction considered in detail in §6.3 is due to [Dress67], where a bisimplicial set is associated with a fibration. The Leray-Serre spectral sequence is derived from the double complex associated to a bisimplicial set. Finally, some results on the uniqueness of the Leray-Serre spectral sequence, due to [Barnes85], are described to close out the chapter.

6.1 A Proof of Theorem 6.1

We begin by recalling the Postnikov tower for S^2:

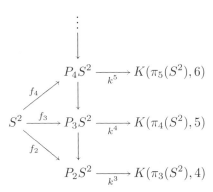

Each space, $P_n S^2$ satisfies the properties

(1) $$\pi_i(P_n S^2) \cong \begin{cases} \pi_i(S^2), & \text{for } i \leq n, \\ \{0\}, & \text{for } i > n. \end{cases}$$

(2) $f_{n*} \colon \pi_i(S^2) \to \pi_i(P_n S^2)$ induces the isomorphism for $i \leq n$.
(3) The mapping $P_n S^2 \to P_{n-1} S^2$ is the fibration pulled back from the path-loop fibration with base $K(\pi_n(S^2), n+1)$ over the mapping k^n.

By (1) and the fact that $\pi_2(S^2) \cong \mathbb{Z}$ (the classical theorem of Brouwer), P_2S^2 has the homotopy type of the Eilenberg-Mac Lane space $K(\mathbb{Z},2)$. Furthermore, we can identify $K(\mathbb{Z},2)$ with $\mathbb{C}P(\infty)$. Thus we know the integral cohomology of P_2S^2 completely; $H^*(P_2S^2) \cong \mathbb{Z}[x_2]$, where x_2 has degree 2.

Next observe that (2) and the Whitehead theorem imply that, for $i \leq n$, $H_i(f_n) \colon H_i(S^2) \to H_i(P_nS^2)$ is an isomorphism, and for $i = n+1$, an epimorphism. The Universal Coefficient theorem allows us to compute the low dimensional integral cohomology of P_nS^2:

$$H^i(P_nS^2) = \begin{cases} \mathbb{Z}, & \text{if } i = 0, 2, \\ \{0\}, & \text{if } i = 1, 3, 4, \dots, n+1. \end{cases}$$

As we work our way up the Postnikov tower, $H^*(P_nS^2)$ can be determined, in principle, in terms of $H^*(P_{n-1}S^2)$ and $H^*(K(\pi_n(S^2),n))$ via the Leray-Serre spectral sequence. We record some values, in low dimensions, of $H_*(K(\pi,n))$ and $H^*(K(\pi,n))$.

Lemma 6.2. *For π, a finite abelian group, and $n \geq 2$, $H_n(K(\pi,n)) \cong \pi \cong H^{n+1}(K(\pi,n))$ and $H^n(K(\pi,n)) = \{0\} = H_{n+1}(K(\pi,n))$.*

PROOF: By the Hurewicz theorem we know that $H_{n-1}(K(\pi,n)) = \{0\}$ and $H_n(K(\pi,n)) \cong \pi$. The Universal Coefficient theorem gives the short exact sequence

$$0 \to \text{Ext}(H_{n-1}(K(\pi,n)),\mathbb{Z}) \to H^n(K(\pi,n))$$
$$\to \text{Hom}(H_n(K(\pi,n)),\mathbb{Z}) \to 0.$$

Since $\text{Hom}(\pi,\mathbb{Z})$ is $\{0\}$ for π, a finite group, we find $H^n(K(\pi,n)) = \{0\}$.

We next consider the path-loop fibration

$$K(\pi,n-1) \to PK(\pi,n) \to K(\pi,n).$$

By the connectivity of $K(\pi,n)$ and repeated application of the Serre exact sequence (Example 5.D), we have

$$H_{n+1}(K(\pi,n)) \cong H_n(K(\pi,n-1)) \cong \cdots \cong H_3(K(\pi,2)).$$

By the fundamental theorem for finitely-generated abelian groups and the properties of Eilenberg-Mac Lane spaces, it suffices to prove the theorem for $\pi = \mathbb{Z}/m\mathbb{Z}$ and $n = 2$. The short exact sequence $0 \to \mathbb{Z} \to \mathbb{Z} \to \mathbb{Z}/m\mathbb{Z} \to 0$ gives a fibration $K(\mathbb{Z},2) \to K(\mathbb{Z},2) \xrightarrow{p} K(\mathbb{Z}/m\mathbb{Z},2)$ (up to homotopy) realizing the short exact sequence of groups on homotopy. In low dimensions there is a single differential in the associated spectral sequence to determine:

$$d_3 \colon H_3(K(\mathbb{Z}/m\mathbb{Z},2)) \to \mathbb{Z} = H_2(K(\mathbb{Z},2)).$$

If $H_3(K(\mathbb{Z}/m\mathbb{Z}, 2))$ were finite, then $d_3 \equiv 0$ and this finite group would persist to E^∞ to give a nontrivial contribution to $H_3(K(\mathbb{Z}, 2)) = \{0\}$.

The Serre exact sequence in homology gives the exact sequence

$$0 \to H_3(K(\mathbb{Z}/m\mathbb{Z}, 2)) \xrightarrow{d_3} H_2(K(\mathbb{Z}, 2))$$
$$\to H_2(K(\mathbb{Z}, 2)) \to H_2(K(\mathbb{Z}/m\mathbb{Z}, 2)) \to 0.$$

Thus $H_3(K(\mathbb{Z}/m\mathbb{Z}, 2))$ can be identified with a subgroup of $\mathbb{Z}$. It follows that $E_{0,2}^\infty \cong \mathbb{Z}/q\mathbb{Z}$ for some q. This presents an extension problem to reconstruct $H_2(K(\mathbb{Z}, 2)) \cong \mathbb{Z}$ with the data $E_{0,2}^\infty \cong \mathbb{Z}/q\mathbb{Z}$, $E_{1,1}^\infty = \{0\}$ and $E_{2,0}^\infty \cong \mathbb{Z}/m\mathbb{Z}$. Since this is impossible, $H_3(K(\mathbb{Z}/m\mathbb{Z}, 2)) = \{0\}$.

A second use of the Universal Coefficient theorem gives the exact sequence

$$0 \to \mathrm{Ext}(H_n(K(\pi, n)), \mathbb{Z}) \to H^{n+1}(K(\pi, n))$$
$$\to \mathrm{Hom}(H_{n+1}(K(\pi, n)), \mathbb{Z}) \to 0.$$

Since $\mathrm{Ext}(\pi, \mathbb{Z}) \cong \pi$, the short exact sequence gives $\pi \cong H^{n+1}(K(\pi, n))$. $\square$

From the properties of Hom and Ext, we deduce that, whenever π is a finitely generated abelian group, $H^n(K(\pi, n)) = \mathbb{Z}$ and $H^{n+1}(K(\pi, n)) = \{0\}$ imply that $\pi = \mathbb{Z}$.

Consider the next fibration in the Postnikov system:

$$K(\pi_3(S^2), 3) \xrightarrow{\mathrm{inc}} P_3 S^2 \to \mathbb{C}P(\infty).$$

The homology Leray-Serre spectral sequence for $P_3 S^2 \to P_2 S^2$ leads to the diagram.

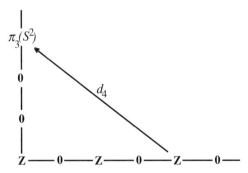

By the connectivity of the mapping $f_3 \colon S^2 \to P_3 S^2$ and the Whitehead theorem, $H_3(P_3 S^2) \cong H_4(P_3 S^2) = \{0\}$. But this can only occur if the differential $d_4 \colon E_{4,0}^4 \cong \mathbb{Z} \longrightarrow \pi_3(S^2) \cong E_{0,3}^4$ is an isomorphism. Thus $\pi_3(S^2) \cong \mathbb{Z}$.

To obtain further homology groups of P_3S^2, we must consider $H^*(K(\mathbb{Z}, 3))$ in a few more dimensions. The path-loop fibration over $K(\mathbb{Z}, 3)$ gives the following diagram for the cohomology spectral sequence in low dimensions:

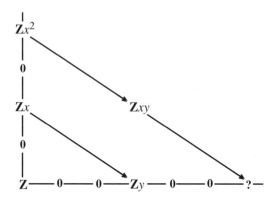

Since d_3 is a derivation, $d_3(x^2) = 2xy$. Furthermore, the group **?** in the diagram must be part of an exact sequence, $0 \to \mathbb{Z}x^2 \xrightarrow{d_3} \mathbb{Z}xy \xrightarrow{d_3} \; ? \; \to 0$. Thus **?** $\cong \mathbb{Z}/2\mathbb{Z} = H^6(K(\mathbb{Z}, 3))$. Putting this further data into the cohomology Leray-Serre spectral sequence for $K(\mathbb{Z}, 3) \to P_3S^2 \to P_2S^2 \simeq \mathbb{C}P(\infty)$, we see that the bottom degree class in $H^3(K(\mathbb{Z}, 3))$ must go to the generator of $H^4(\mathbb{C}P(\infty))$ in order that $H^3(P_3S^2) = \{0\}$. The algebra structure of the spectral sequence takes over; the next possible case of a differential must come from $H^6(K(\mathbb{Z}, 3))$ but such a class has nowhere to go and we conclude $H^6(P_3S^2) \cong \mathbb{Z}/2\mathbb{Z}$.

Finally, consider the fibration $K(\pi_4(S^2), 4) \to P_4S^2 \to P_3S^2$. In the cohomology spectral sequence, we have the data $E^{2,0} \cong \mathbb{Z}$, $E_2^{6,0} \cong \mathbb{Z}/2\mathbb{Z}$, and $E_2^{i,0} = \{0\}$ for $2 < i < 6$. This leaves $H^i(K(\pi_4(S^2), 4))$ unknown for $i = 4$ and 5.

Since $H^4(P_4S^2) \cong \{0\}$, it follows from the placement of holes that $H^4(K(\pi_4(S^2), 4)) = \{0\}$. (This is also a corollary of the argument of Example 5.K for even-dimensional spheres, that is, $\pi_4(S^2)$ is finite.) From the Universal Coefficient theorem we deduce that $H^5(P_4S^2) = \{0\}$. It follows that

$$d_6 \colon H^5(K(\pi_4(S^2), 4)) \to H^6(P_3S^2)$$

is an isomorphism: $H^5(K(\pi_4(S^2), 4)) \cong \mathbb{Z}/2\mathbb{Z}$ and, by Lemma 6.2, we have proved Theorem 6.1.   □

In the course of the proof of Theorem 6.1, we have computed some of the groups $H^*(K(\pi, n))$. The reader is encouraged to try his or her hand at extending the computation. The p-primary components of $H^*(K(\mathbb{Z}, 3))$ plan an important role and this reveals the difficulty of computing these groups in general. Employing the philosophy that problems, taken one prime at a time, may be more accessible, we turn to the computation of the algebras $H^*(K(\pi, n); \mathbb{F}_p)$, for p a prime and π a finitely generated group. The key tool in these computations is the transgression and its algebraic realization in the Leray-Serre spectral sequence.

6.2 The transgression

The Russian term for a fibration is literally a 'twisted product' (of the fibre and base). An algebraic measure of the nontriviality of the twisting is found in the long exact sequence of homotopy groups of the fibration:

$$\cdots \to \pi_q(F) \xrightarrow{i_*} \pi_q(E) \xrightarrow{p_*} \pi_q(B) \xrightarrow{\partial} \pi_{q-1}(F) \xrightarrow{i_*} \cdots.$$

The nontriviality of ∂, the connecting homomorphism, gives an indication of how the base and fibre are twisted together to form the total space. Though historically inaccurate, we follow [James61] and call ∂ the **transgression** of the fibration (F, E, p, B). (See [Chern48], [Hirsch48], and [Koszul50] for the first notions of the transgression.)

To define the transgression on homology groups we consider the following commutative diagram where $h \colon \pi_q(X) \to H_q(X)$ denotes the Hurewicz homomorphism:

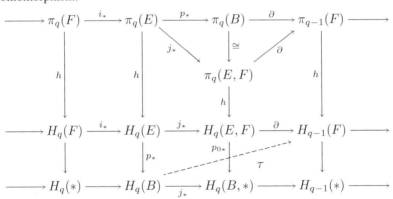

Here $p_0 \colon (E, F) \to (B, *)$ is the induced fibration of pairs. The exactness of the two bottom rows is used to define the transgression homomorphism.

Definition 6.3. *Let τ denote the homomorphism,*

$$\tau \colon j_*^{-1}(\operatorname{im} p_{0*}) \longrightarrow H_{q-1}(F)/\partial(\ker p_{0*})$$

*given by $\tau(z) = \partial r + \partial(\ker p_{0*})$ where $z \in j_*^{-1}(\operatorname{im} p_{0*})$, and $p_{0*}^{-1}(j_*(z)) = r + \ker p_{0*}$. The homomorphism τ is called the **transgression**.*

We leave it to the reader to prove that τ is well-defined. Notice that τ maps a subgroup of $H_*(B)$ to a quotient of $H_*(F)$. The definition of τ for homology with coefficients other than $\mathbb{Z}$ is clear from the diagram. The following is an immediate consequence of the definition.

Proposition 6.4. *If a class* $u \in H_q(B)$ *is spherical, that is,* u *is in the image of the Hurewicz homomorphism, then* $\tau(u)$ *is spherical in* $H_{q-1}(F)/\partial(\ker p_{0*})$, *that is,* $\tau(u)$ *lies in the image of the composite mapping*

$$\pi_{q-1}(F) \to H_{q-1}(F) \longrightarrow H_{q-1}(F)/\partial(\ker p_{0*}).$$

The dual notion may be defined for cohomology as follows: Consider the commutative diagram with exact rows:

$$
\begin{array}{ccccccccc}
\longrightarrow & H^{q-1}(*) & \xrightarrow{\;\delta\;} & H^q(B, *) & \xrightarrow{\;j^*\;} & H^q(B) & \longrightarrow & H^q(*) & \xrightarrow{\;\delta\;} \\
& \big\downarrow & & \Big\downarrow{\scriptstyle p_0^*} & {}^{\tau} & \big\downarrow{\scriptstyle p^*} & & \big\downarrow & \\
\longrightarrow & H^{q-1}(F) & \xrightarrow{\;\delta\;} & H^q(E, F) & \xrightarrow{\;j^*\;} & H^q(E) & \longrightarrow & H^q(F) & \xrightarrow{\;\delta\;}
\end{array}
$$

Define $\tau \colon (\delta^{-1}(\operatorname{im} p_0^*) \subset H_{q-1}(F)) \to H^q(B)/j^*(\ker p_0^*)$ by $\tau(z) = j^*(r) + j^*(\ker p_0^*)$ where $z \in \delta^{-1}(\operatorname{im} p_0^*)$ and $p_0^*(r + \ker p_0^*) = \delta z$. If we replace $H^*(\ )$ with $H^*(\ ; R)$, we get the definition of τ for cohomology with coefficients in a ring R. In particular, for $R = \mathbb{F}_p$, p a prime, the action of the Steenrod algebra commutes with all of the homomorphisms in the diagram.

Proposition 6.5. *If* θ *is in* $\mathcal{A}_p$, *the mod* p *Steenrod algebra, and* z, *a class in* $\delta^{-1}(\operatorname{im} p_0^*) \subset H^{q-1}(F; \mathbb{F}_p)$, *then* $\tau(\theta z) = \theta \tau(z)$.

Here equality means as cosets of $\ker p^*$ in $H^*(B; \mathbb{F}_p)$ or as elements of $\operatorname{im} p^* \cong H^*(B; \mathbb{F}_p)/\ker p^*$. It is this property of the transgression that is the key to later computations.

The main theorem of this section identifies τ, a feature of the homological data of the spaces in a fibration, with an algebraic feature in the Leray-Serre spectral sequence.

Theorem 6.6. *Given a fibration* $F \hookrightarrow E \xrightarrow{p} B$ *with base* B *and fibre* F *connected, the following hold for its associated Leray-Serre spectral sequence*

(1) $E_{n,0}^n \cong j_*^{-1}(\operatorname{im} p_0^*) \subseteq H_n(B)$,
(2) $E_{0,n-1}^n \cong H_{n-1}(F)/\partial(\ker p_{0*})$, *and*
(3) $d^n \colon E_{n,0}^n \to E_{0,n-1}^n$ *is the transgression.*

PROOF: We begin with a proof of part (1). Consider the map of fibrations

$$
\begin{array}{ccc}
E & \xrightarrow{\;j\;} & (E, F) \\
{\scriptstyle p}\big\downarrow & & \big\downarrow{\scriptstyle p_0} \\
B & \xrightarrow{\;j\;} & (B, *)
\end{array}
$$

By the naturality of the spectral sequence (Theorem 5.1), the inclusions induce a mapping $j_* \colon E^r_{m,q} \to \bar{E}^r_{m,q}$ for all r, m, q where $E^2_{m,*} \cong H_m(B; H_*(F))$ and $\bar{E}^2_{m,*} \cong H_m(B, *; H_*(F))$. Observe that $\bar{E}^2_{0,*} \cong H_0(B, *; H_*(F)) = \{0\}$, and $E^2_{m,q} \cong \bar{E}^2_{m,q}$ for $m > 0$. Since j_* commutes with the differentials, $j_* \colon E^k_{n,0} \cong \bar{E}^k_{n,0}$ for $k \leq n$. Furthermore, $\bar{E}^2_{0,*} = \{0\}$ implies $\bar{E}^{n+1}_{n,0} \cong \bar{E}^n_{n,0}$. By Theorem 5.9 $\bar{E}^{n+1}_{n,0}$ can be identified with $\mathrm{im}(p_{0*} \colon H_n(E, F) \to H_n(B, *))$ and so the isomorphism $j_* \colon E^n_{n,0} \to \bar{E}^n_{n,0}$ determines $E^n_{n,0} \cong j_*^{-1}(\mathrm{im}\, p_{0*})$. This proves assertion (1).

We next consider some important diagrams. The first displays a reformulation of the definition of the transgression:

$$
\begin{array}{ccccccccc}
0 & \longrightarrow & \ker p_{0*} & \longrightarrow & H_n(E, F) & \xrightarrow{p_{0*}} & \mathrm{im}\, p_{0*} & \longrightarrow & 0 \\
 & & \downarrow{\scriptstyle \partial} & & \downarrow{\scriptstyle \partial} & & \downarrow{\scriptstyle \tau} & & \\
0 & \longrightarrow & \partial(\ker p_{0*}) & \longrightarrow & H_{n-1}(F) & \longrightarrow & H_{n-1}(F)/\partial(\ker p_{0*}) & \longrightarrow & 0.
\end{array}
$$

The second diagram, with rows exact, is based on the naturality of the spectral sequence (Theorem 5.9).

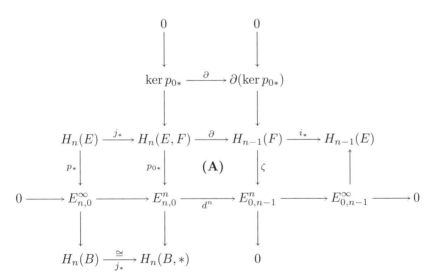

Most of the diagram commutes for obvious reasons; however, it remains to prove that the middlemost square commute. The homomorphism ζ denotes the surjection given by the composite

$$
H_{n-1}(F) \cong E^2_{0,n-1} \longrightarrow E^3_{0,n-1} \twoheadrightarrow \cdots \longrightarrow E^{n-1}_{0,n-1} \longrightarrow E^n_{0,n-1}.
$$

Lemma 6.7. **(A)** *is a commutative square.*

We postpone the proof of this lemma briefly and gather its corollaries. We first prove part (2) of Theorem 6.6: Suppose $u \neq 0$ is in $H_{n-1}(F)$ and $\zeta(u) = 0$. Then $i_*(u) = 0$ by the commutativity of the diagram and so u is in $\ker i_* = \operatorname{im} \partial$. Let w be in $H_n(E, F)$ with $\partial(w) = u$. By the commutativity of **(A)** $d^n(p_{0*}(w)) = 0$. If $p_{0*}(w) \neq 0$, then $p_{0*}(w)$ lies in $\ker d^n = E_{n,0}^\infty = \operatorname{im} p_*$. Thus there is a v in $H_n(E)$ with $p_{0*}(j_*(v)) = p_{0*}(w)$. Let $w' = w - j_*(v)$. Then $p_{0*}(w') = 0$ and $\partial(w') = \partial(w) - \partial j_*(v) = \partial(w) = u$. Thus $\ker \zeta = \partial(\ker p_{0*})$ and (2) follows.

To derive part (3) of Theorem 6.6, return to the first important diagram after the theorem and substitute the conclusions of parts (1) and (2) to obtain the commutative diagram

$$
\begin{array}{ccccccccc}
0 & \longrightarrow & \ker p_{0*} & \longrightarrow & H_n(E, F) & \longrightarrow & E_{n,0}^n & \longrightarrow & 0 \\
& & \downarrow{\scriptstyle \partial} & & \downarrow{\scriptstyle \partial} & & \downarrow{\scriptstyle d^n} & & \\
0 & \longrightarrow & \partial(\ker p_{0*}) & \longrightarrow & H_{n-1}(F) & \longrightarrow & E_{0,n-1}^n & \longrightarrow & 0.
\end{array}
$$

To complete the proof of Theorem 6.6 it suffices to establish Lemma 6.7.

PROOF OF LEMMA 6.7: We use the definition of the groups appearing in this part of the spectral sequence. In particular,

$$
E_{n,0}^n = Z_{n,0}^n / B_{n,0}^{n-1} + Z_{n-1,1}^{n-1}, \qquad E_{0,n-1}^n = Z_{0,n-1}^n / B_{0,n-1}^{n-1},
$$

where

$$
\begin{aligned}
Z_{n,0}^n &= \{x \in F_n C_n(E) \mid \partial(x) \in F_0 C_{n-1}(E) = C_{n-1}(F)\} \\
B_{n,0}^{n-1} &= \{x \in F_n C_n(E) \mid \text{there is } y \in F_{2n-1} C_{n+1}(E) \text{ with } \partial(y) = x\} \\
&= B_n(E) \\
Z_{n-1,1}^{n-1} &= \{x \in F_{n-1} C_n(E) \mid \partial(x) \in F_0 C_{n-1}(E) = C_{n-1}(F)\} \\
Z_{0,n-1}^n &= \{x \in F_0 C_{n-1}(E) \mid \partial(x) \in F_n C_{n-2}(E)\} = F_0 C_{n-1}(E) \\
B_{0,n-1}^{n-1} &= \{x \in F_0 C_{n-1}(E) \mid \text{there is } y \in F_{n-1} C_n(E) \text{ with } \partial(y) = x\}.
\end{aligned}
$$

At the level of the definitions, the differential d^n can be written as

$$
d^n(x + B_n(E) + Z_{n-1,1}^{n-1}) = \partial(x) + B_{0,n-1}^{n-1}.
$$

However, $H_n(E, F) = H_n(C_n(E)/F_0 C_n(E))$ and the mapping p_{0*} takes $u \in H_n(E, F)$ to the class $u + B_n(E) + Z_{n-1,1}^{n-1}$; since $\partial(u) \in F_0 C_{n-1}(E)$, and we have an increasing filtration, this is well-defined. Furthermore, $\partial(u) \in$

$H_{n-1}(F)$ is given by the class $\partial(u) + B_{n-1}(F)$. The mapping ζ takes $y + B_{n-1}(F) \in H_{n-1}(F)$ to the coset $y + B_{0,n-1}^{n-1}$. However, $B_{n-1}(F) \subset B_{0,n-1}^{n-1}$ so this is well-defined.

Finally, we chase around the square in each direction. A class $u + F_0 C_n(E)$ with $\partial(u) \in F_0 C_{n-1}(E)$ goes by $d^n \circ p_{0*}$ to $\partial(u) + B_{0,n-1}^{n-1}$. Likewise, the same class is the image of $u + F_0 C_n(E)$ under $\zeta \circ \partial$. $\qquad\square$

For completeness we include a statement of Theorem 6.6 for cohomology. The proof is left to the reader.

Theorem 6.8. *Given a fibration $F \hookrightarrow E \overset{p}{\longrightarrow} B$ with base B and fibre F connected, the following hold for its associated cohomology Leray-Serre spectral sequence*

(1) $E_n^{n,0} \cong H^n(B)/\ker p^*$,
(2) $E_n^{0,n-1} \cong \delta^{-1}(\operatorname{im} p_0^*) \subseteq H^{n-1}(F)$,
(3) $d_n \colon E_n^{0,n-1} \to E_n^{n,0}$ *is the transgression.*

Combining Theorem 6.6, Proposition 6.5 and Theorem 5.2 we obtain the following corollary for mod p cohomology.

Corollary 6.9. *Suppose u in $H^{n-1}(F; \mathbb{F}_p)$ survives to $E_n^{0,n-1}$ and θ is in $\mathcal{A}_p$, the mod p Steenrod algebra. If θ has degree k then*

$$d_{n+k}(\theta u) = \theta(d_n u).$$

We can present Corollary 6.9 pictorially in the diagram:

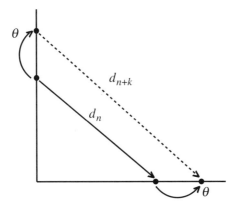

The homology transgression is an example of an *additive relation*, a homomorphism defined on a subquotient of a module to another subquotient. Consider the part of the long exact sequence in homology of the pair (E, F):

$$H_{q-1}(F) \overset{\partial}{\longleftarrow} H_q(E, F) \overset{p_{0*}}{\longrightarrow} H_q(B, *) = H_q(B).$$

The transgression is defined on the image of p_{0*} and it is determined up to the kernel of p_{0*}. So we have the equivalent expression

$$\tau \colon \operatorname{im} p_{0*} \cong H_q(E, F) / \ker p_{0*} \xrightarrow{\bar{\partial}} H_{q-1}(F) / \partial(\ker p_{0*}).$$

For a pointed space (X, x_0), we include $X \subset CX$ as the subspace of classes $[x, 1]$. For a cofibrant space, $CX/X \cong SX$ and the pair (CX, X) yields a similar pair of homomorphisms:

$$H_{q-1}(X) \xleftarrow{\partial} H_q(CX, X) \xrightarrow[\cong]{Q} H_q(SX, *).$$

Since the cone on X is contractible, ∂ is an isomorphism and the homomorphism $\Sigma = Q \circ \partial^{-1} \colon H_{q-1}(X) \to H_q(SX)$ is the classical suspension isomorphism.

Suppose (F, E, p, B) is a fibration and $H_q(E) = H_{q-1}(E) = \{0\}$. Then the homomorphism $H_{q-1}(F) \xleftarrow{\partial} H_q(E, F)$ is an isomorphism and we can form the composite, opposite to the transgression but in the order of the classical suspension, to produce the **homology suspension**

$$\sigma = p_{0*} \circ \partial^{-1} \colon H_{q-1}(F) \to H_q(B).$$

When $E = PY$ in the path-loop fibration, $(\Omega Y, PY, \mathrm{ev}_1, Y)$, the homology suspension is defined for all $q > 0$, $\Sigma_* \colon H_{q-1}(\Omega Y) \to H_q(Y)$. This homomorphism was introduced by [Eilenberg-Mac Lane50] in their study of the homology of $K(\pi, n)$'s, and further developed by [Serre51]. From the Serre exact sequence for the path-loop fibration, if Y is $(n - 1)$-connected, then $\Sigma_* \colon H_{q-1}(\Omega Y) \to H_q(Y)$ is the inverse of the transgression and an isomorphism for $q \leq 2n - 2$.

More generally, [Serre51] proved the following result.

Proposition 6.10. *Suppose $F \hookrightarrow E \xrightarrow{p} B$ is a fibration with B path-connected and F connected. Suppose that $H_q(E) \cong H_{q-1}(E) = \{0\}$. Then the homology suspension $\Sigma_* \colon H_{q-1}(F) \to H_q(B)$ is well-defined and makes the following diagram commute:*

$$
\begin{array}{ccc}
E_{q,0}^q & \xrightarrow[\cong]{d^q} & E_{0,q-1}^q \\
\Big\downarrow & & \Big\uparrow{\scriptstyle \zeta} \\
H_q(B) & \xleftarrow{\Sigma_*} & H_{q-1}(F)
\end{array}
$$

In particular, Σ_* has the same kernel as ζ and the same image as the inclusion $E_{q,0}^q \hookrightarrow H_q(B)$.

The proposition follows from Lemma 6.7.

We next relate the suspension isomorphism with the homology suspension.

Lemma 6.11. *Let* $f\colon X \to \Omega Y$ *be a pointed map and* $\mathrm{adj}(f)\colon SX \to Y$ *its adjoint. Then there is a mapping* $\tilde{f}\colon CX \to PY$ *making the following diagram commute*

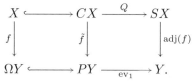

PROOF ([Adams60]): Recall the identity of all the maps in the diagram: If $f\colon X \to \Omega Y$ is written $f(x) = \lambda_x$, then $\mathrm{adj}(f)([x,t]) = \lambda_x(t) = f(x)(t)$. The quotient map $Q\colon CX \to SX$ takes $[x,t]$ to itself when $x \neq x_0$ and for $0 < t < 1$, and $[x,0]$, $[x,1]$, and $[x_0,t]$ all go to the same class, the basepoint of SX.

Since f is pointed, $f(x_0) = c_{y_0}$, the constant loop at $y_0 \in Y$, the basepoint of Y. Define $\tilde{f}\colon CX \to PY$ by $\tilde{f}([x,t])(s) = f(x)(st)$. Notice that $\tilde{f}([x,t])(0) = f(x)(0) = y_0$ and $\tilde{f}([x,1])(s) = f(x)(s)$ and so the left side of the diagram commutes. Since $\tilde{f}([x,t])(1) = f(x)(t) = \mathrm{adj}(f)([x,t])$, the right square commutes. □

On homology we get the following diagram relating Σ and Σ_*:

$$
\begin{array}{ccccc}
H_{q-1}(X) & \xleftarrow[\cong]{\partial} & H_q(CX,X) & \xrightarrow{Q_*} & H_q(SX) \\
\downarrow{\scriptstyle f_*} & & \downarrow{\scriptstyle \tilde{f}_*} & & \downarrow{\scriptstyle \mathrm{adj}(f)_*} \\
H_{q-1}(\Omega Y) & \xleftarrow[\cong]{\partial} & H_q(PY,\Omega Y) & \xrightarrow[\mathrm{ev}_1]{} & H_q(Y)
\end{array}
$$

with Σ spanning the top and Σ_* spanning the bottom.

When $Y = SX$ and $j\colon X \to \Omega SX$ is the mapping satisfying $\mathrm{adj}(j) = \mathrm{id}\colon SX \to SX$, we obtain the commutative diagram:

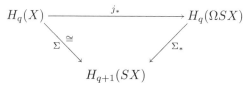

By the Serre exact sequence for $(\Omega SX, PSX, \mathrm{ev}_1, SX)$ we know that Σ_* is an isomorphism for $q \leq 2n$ where X is $(n-1)$-connected. This implies that j_* is an isomorphism in this range and the Whitehead theorem implies that $j_*\colon \pi_q(X, x_0) \to \pi_q(\Omega SX, Sx_0)$ is an isomorphism in this range and an epimorphism when $q = 2n + 1$. Since the adjoint of the identity mapping on SX induces the suspension mapping on homotopy groups, we have proved the classical theorem:

Theorem 6.12 (the Freudenthal suspension theorem). *If X is $(n-1)$-connected, then $\Sigma_*\colon \pi_q(X,x_0) \to \pi_{q+1}(SX,Sx_0)$ is an isomorphism for all $q \leq 2n$ and an epimorphism for $q = 2n+1$.*

We investigate the homology suspension and its dual further in §8.2. In the next section, we investigate another way in which the Steenrod algebra appears in the workings of the transgression.

Kudo's theorem

In this section we restrict our attention to mod p cohomology for p a prime. Many computations in which the transgression figures prominently involve fibrations with acyclic total space—a nontrivial transgression should be expected since $\partial\colon \pi_n(B) \to \pi_{n-1}(F)$ is an isomorphism in this case. Special attention must be given to every possibly nonzero class in the spectral sequence; in particular, the transgression can give rise to nonzero product elements. In light of this, we extend the definition of the transgression to include such classes.

Definition 6.13. *Suppose $u \in E_2^{s,t}$. We say that u is **transgressive** if $d_2(u) = d_3(u) = \cdots = d_t(u) = 0$ and $d_{t+1}(u) \neq 0$ in $E_{t+1}^{s+t+1,0}$. If v in $E_2^{s+t+1,0}$ survives to represent $d_{t+1}(u)$, then we say that u **transgresses** to v.*

Notice that it is not properly u that survives to $E_{t+1}^{s,t}$ but that u represents a class in $E_{t+1}^{s,t}$ with nonzero image under d_{t+1}. Elements in the subspace $\delta^{-1}(\operatorname{im} p_0^*)$ of $H^{n-1}(F;\mathbb{F}_p)$ (see §6.1) are transgressive according to this definition when their image under the cohomological transgression is nonzero.

Suppose a class $x \in H^{2k}(F;\mathbb{F}_p)$ is transgressive and $y \in H^{2k+1}(B;\mathbb{F}_p)$ represents $d_{2k+1}(x)$. Because d_{2k+1} is a derivation (Theorem 5.5), we have, for $1 \leq i < p$, $d_{2k+1}(x^i) = iy \otimes x^{i-1}$. However, two phenomena come into play when $i = p$. First, from the derivation property and the mod p coefficients, we have $d_{2k+1}(x^p) = py \otimes x^{p-1} = 0$. Alternatively, $x^p = P^k x$ and so, by Corollary 6.9, if $P^k x$ survives to $E_{2kp+1}^{0,2kp}$, then $d_{2kp+1}(P^k x) = P^k y$. This leaves $y \otimes x^{p-1}$ a nonzero class in $E_{2k+2}^{2k+1,2k(p-1)}$.

[Kudo56] determined the behavior of these elements in the spectral sequence in terms of the Steenrod algebra action on $H^*(B;\mathbb{F}_p)$.

Theorem 6.14 (the Kudo transgression theorem). *If a class x in $E_2^{0,2k} \cong H^{2k}(F;\mathbb{F}_p)$ is transgressive and x transgresses to the element represented by y in $E_2^{2k+1,0} \cong H^{2k+1}(B;\mathbb{F}_p)$, then $P^k x = x^p$ and $y \otimes x^{p-1}$ are also transgressive with $d_{2pk+1}(x^p) = P^k y$ and $d_{2(p-1)k+1}(y \otimes x^{p-1}) = -\beta P^k y$. (Here β denotes the mod p Bockstein in $\mathcal{A}_p$.)*

IDEA OF THE PROOF: A complete proof of Kudo's theorem requires careful consideration of a construction of the Steenrod algebra action at the cochain level on the cohomology of a space. Following [Steenrod57], [May70] gives such a construction in a very general setting. Steenrod operations are based on the $\mathbb{F}_p[\mathbb{Z}/p\mathbb{Z}]$-free, acyclic complex

$$W^0 \xrightarrow{\delta} W^1 \xrightarrow{\delta} \cdots \xrightarrow{\delta} W^n \xrightarrow{\delta} \cdots$$

with W^n the free $\mathbb{F}_p[\mathbb{Z}/p\mathbb{Z}]$-module on a single generator e_n. If $\mathbb{Z}/p\mathbb{Z}$ is generated by an element α (written multiplicatively), then

$$\delta(e_{2i}) = (1 + \alpha + \cdots + \alpha^{p-1})e_{2i-1}, \qquad \delta(e_{2i+1}) = (\alpha - 1)e_{2i}.$$

The complex (W^*, δ) plays a role in the cohomology of groups (see §8bis.2).

The additional structure needed to define Steenrod operations is a morphism Θ of $\mathbb{F}_p[\mathbb{Z}/p\mathbb{Z}]$-complexes, where $\mathbb{F}_p[\mathbb{Z}/p\mathbb{Z}]$ acts trivially on the target C^*, and permutes the factors of $(C^*)^{\otimes p}$,

$$\Theta \colon W^* \otimes (C^*)^{\otimes p} \to C^*.$$

The mappings $D_i(a) = \Theta_*(e_i \otimes a^{\otimes p})$ are identified with the cocycles representing the Steenrod operations applied to $[a] \in H(C^*)$. The structure map Θ extends the p-fold iterated product on $e_0 \otimes (C^*)^{\otimes p}$. [May70, pp. 166-7] presents an algorithm for constructing cochain representatives for these operations. We can apply the construction to $C^*(E; \mathbb{F}_p)$ and to representatives of transgressive elements. As we saw in the proof of Theorem 6.6, being transgressive is a condition on filtration degrees. [May70] shows that if $x \in E_{2k}^{0,2k}$ transgresses to $y \in E_{2k}^{2k+1,0}$, then $d_{2(p-1)k+1}(y \otimes x^{p-1})$ has a cochain representative $\partial\Theta(e_0 \otimes x^{p-1}y)$ that represents $-\beta P^k(y)$ modulo terms of higher filtration.

The key to the proof is the careful construction of cochain level operations along with the identification of the transgression with a condition on filtration degree. May's technical but elegant proof applies more generally to other spectral sequences. $\qquad\square$

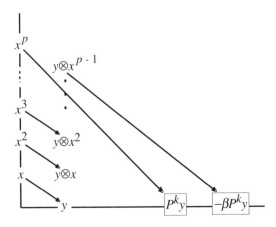

Kudo's theorem shows that the elements of the form $y \otimes x^{p-1}$ and the Steenrod algebra action play crucial roles in describing the transgression. We make two further comments along these lines.

In the calculation of $H_*(K(\pi, n); \mathbb{F}_p)$ and $H^*(K(\pi, n); \mathbb{F}_p)$ as algebras, [Cartan54] had already noticed the role of the classes $y \otimes x^{p-1}$. Cartan introduced the idea of an **acyclic construction** of algebras over $\mathbb{F}_p$: a triple, (A, N, M), with A, N, and M algebras where M possesses a differential and $H(M, d_M) \cong \mathbb{F}_p$; M is a twisted version of $A \otimes N$ with the differential on M a morphism $A \to N$ of degree ± 1 (depending on whether we are computing the homology or cohomology). The motivating idea is to obtain a version of the chains on the total space of a fibration from the chains on the base and fibre when the total space is acyclic. If one ignores the filtration in the Leray-Serre spectral sequence and considers the total complex,

$$(\text{total } E)_q = \bigoplus_{r+s=q} E^2_{r,s} \cong \bigoplus_{r+s=q} H_r(B; \mathbb{F}_p) \otimes H_s(F; \mathbb{F}_p)$$

along with all of the differentials simultaneously, then this gives a suggestive image for the idea of a construction. The technical details are given by [Cartan54] (also see [Moore76] and [Stasheff87]).

For (A, N, M), an acyclic, multiplicative, graded-commutative construction over $\mathbb{F}_p$, Cartan defined an additive function for $q > 0$,

$$\psi \colon {}_p(A_{2q}) \longrightarrow H_{2pq+2}(N)$$

where ${}_p(A_{2q}) = \{a \in A_{2q} \text{ such that } da = 0 \text{ and } a^p = 0\}$,. The function ψ determines the reduction in N of a class y in M with $dy = a^{p-1}x$ and $dx = a$. Under further conditions, satisfied in the problem of computing $H_*(K(\pi, n); \mathbb{F}_p)$, the mapping ψ determines a homomorphism

$$\varphi \colon H_{2q}(A) \longrightarrow H_{2pq+2}(N)$$

called the **transpotence**. In the case of Eilenberg-Mac Lane spaces, the transpotence is a function

$$\varphi_p \colon H_{2q}(K(\pi, n); \mathbb{F}_p) \longrightarrow H_{2pq+2}(K(\pi, n+1); \mathbb{F}_p)$$

that is dual to the mapping sending x in $H^{2q}(K(\pi, n); \mathbb{F}_p)$ to $-\beta P^q(\tau x)$ in $H^{2pq+2}(K(\pi, n+1); \mathbb{F}_p)$. The transpotence is the extra piece of structure needed to design constructions inductively and so compute $H_*(K(\pi, n); \mathbb{F}_p)$ for all n.

Returning to the action of the Steenrod algebra, we observe that $\mathcal{A}_p$ acts on $E^{*,0}_2$ and $E^{0,*}_2$ as isomorphic to $H^*(B; \mathbb{F}_p)$ and $H^*(F; \mathbb{F}_p)$, respectively. A natural question, posed by [Massey55], is whether there is an $\mathcal{A}_p$-action on all of $E^{*,*}_r$, for all r, that converges to the action of $\mathcal{A}_p$ on $H^*(E; \mathbb{F}_p)$. A solution to this problem was given independently by [Vázquez57] and [Araki57] who proved the following result.

Theorem 6.15. *On the mod p cohomology spectral sequence associated to a fibration* $F \hookrightarrow E \xrightarrow{p} B$, *there are operations*

$$\text{for } p, \text{ odd} \begin{cases} {}_F P^s \colon E_r^{a,b} \to E_r^{a,b+2s(p-1)}, & 1 \le r \le \infty, \\ {}_B P^s \colon E_r^{a,b} \to E_r^{a+(2s-b)(p-1),pb}, & 2 \le r \le \infty \end{cases}$$

$$\text{for } p = 2 \begin{cases} {}_F Sq^i \colon E_r^{a,b} \to E_r^{a,b+i}, & 1 \le r \le \infty \\ {}_B Sq^i \colon E_r^{a,b} \to E_r^{a+i-b,2b}, & 2 \le r \le \infty \end{cases}$$

that converge to the action of $\mathcal{A}_p$ on $H^*(E; \mathbb{F}_p)$, commute with the differentials in the spectral sequence, satisfy analogues of Cartan's formula and the Adem relations and reduce to the $\mathcal{A}_p$-action on $H^*(B; \mathbb{F}_p)$ and $H^*(F; \mathbb{F}_p)$ when $r = 2$ and $a = 0$ or $b = 0$ (that is, on $E_2^{*,0}$ and $E_2^{0,*}$).

These operations have been developed further by [Kristensen62]. [Kuo65] has given a general approach to the construction of families of operations on spectral sequences. [Singer73] has constructed Steenrod operations on certain first quadrant spectral sequences that include the Leray-Serre spectral sequence as an example.

On $H^*(K(\pi, n); \mathbb{F}_p)$; the theorems of Cartan and Serre

We can now turn to the major goal of this section—the computation of $H^*(K(\pi, n); \mathbb{F}_p)$ for a fixed prime p and π a finitely generated abelian group. By the fundamental theorem for abelian groups and the fact that $K(\pi_1 \oplus \pi_2, n) \cong K(\pi_1, n) \times K(\pi_2, n)$, the task reduces to the consideration of the cases $\pi = \mathbb{Z}$, $\mathbb{Z}/\ell\mathbb{Z}$ and $\mathbb{Z}/\ell^k\mathbb{Z}$ for some prime ℓ. We reduce the task further to the case $\ell = p$.

Lemma 6.16. *If π is a finitely generated abelian group, $n \ge 1$, and k is a field, then $\tilde{H}_*(K(\pi, n); k) = \{0\}$ if and only if $\pi \otimes k = \{0\}$.*

PROOF: If $\tilde{H}_*(K(\pi, n); k) = \{0\}$, then we have $H_n(K(\pi, n); k) = \{0\}$. By the Hurewicz theorem and the Universal Coefficient theorem, $H_n(K(\pi, n); k) \cong \pi \otimes k$ and so $\pi \otimes k = \{0\}$.

If $\pi \otimes k = \{0\}$, by the fundamental theorem for finitely generated abelian groups, we can fix $\pi = \mathbb{Z}/m\mathbb{Z}$ for an integer m relatively prime to the characteristic of the field k. The fibration $K(\pi, n-1) \to PK(\pi, n) \to K(\pi, n)$ and the Leray-Serre spectral sequence for homology with coefficients in the field k show that if the lemma holds for $n - 1$, then it holds for n. It suffices to prove the lemma for $\tilde{H}_q(K(\mathbb{Z}/m\mathbb{Z}, 1); k)$.

Consider the fibration $K(\mathbb{Z}/m\mathbb{Z}, 1) \to K(\mathbb{Z}, 2) \xrightarrow{-\times m} K(\mathbb{Z}, 2)$ where the mapping, $- \times m$, induces the '*times m*' map on the second homotopy groups and hence on the second homology groups of $K(\mathbb{Z}, 2) = \mathbb{C}P(\infty)$. Since $\mathbb{Z}/m\mathbb{Z} \otimes k = \{0\}$ implies that m is a nonzero unit in k, the homomorphism induced by $-\times m$ induces an isomorphism on $H(\ ;k)$ between the base and total spaces. This implies that the first nonzero vector space $H_q(K(\mathbb{Z}/m\mathbb{Z},1);k)$ persists to E^∞ and hence contributes to $H_q(\mathbb{C}P(\infty);k)$ which has already been accounted for by the base space. This proves the lemma. $\square$

We recall from Chapter 4 some facts about the Steenrod algebra. When p is an odd prime, a sequence of nonnegative integers $I = (\varepsilon_0, s_1, \varepsilon_1, \dots, s_m, \varepsilon_m)$, with $\varepsilon_j = 0$ or 1 for all j, represents the operation $St^I = \beta^{\varepsilon_0} P^{s_1} \beta^{\varepsilon_1} \cdots P^{s_m} \beta^{\varepsilon_m}$ in $\mathcal{A}_p$. Such a sequence represents an element in a basis for $\mathcal{A}_p$ over $\mathbb{F}_p$ whenever the sequence is **admissible**, that is, $s_i \geq ps_{i+1} + \varepsilon_i$ for $m > i \geq 1$. When $p = 2$, we write such a sequence as $I = (s_1, \dots, s_m)$ to which we associate the operation $St^I = Sq^{s_1} Sq^{s_2} \cdots Sq^{s_m} \in \mathcal{A}_2$. A sequence is admissible mod 2 when $s_i \geq 2s_{i+1}$ for $m > i \geq 1$.

Definition 6.17. *For p an odd prime let $I = (\varepsilon_0, s_1, \varepsilon_1, \dots, s_m, \varepsilon_m)$ be an admissible sequence. The **degree** of the associated operation is denoted by $|I|$ and is given by $|I| = 2(p-1)(s_1 + \cdots + s_m) + \varepsilon_0 + \cdots + \varepsilon_m$. If x is a class in $H^r(X;\mathbb{F}_p)$, then $St^I x$ is a class in $H^{r+|I|}(X;\mathbb{F}_p)$. The **excess** of I is given by*

$$e(I) = 2(s_1 - ps_2) + 2(s_2 - ps_3) + \cdots + 2(s_{m-1} - ps_m)$$
$$+ 2s_m + \varepsilon_0 - \varepsilon_1 - \cdots - \varepsilon_m$$
$$= 2s_1 p + 2\varepsilon_0 - |I|.$$

When $p = 2$, $I = (s_1, \dots, s_m)$, $|I| = s_1 + \cdots + s_m$ and $e(I) = 2s_1 - |I|$.

Before stating the main theorem, we state a rather technical lemma that describes how admissible sequences of a certain excess can be constructed from those of lower excess. The proof is an exercise in the definitions of admissibility and excess, with some careful bookkeeping, and is left to the reader. This lemma is crucial in the proof of the main result of this section.

Lemma 6.18. *Let p be an odd prime and define the admissible sequences*

$$J_k^t = (0, p^t k, 0, p^{t-1}k, \dots, 0, pk, 0, k) \qquad \bar{J}_k^t = (1, p^t k, J_k^{t-1}).$$

Suppose $I = (\varepsilon_0, s_1, \varepsilon_1, \dots, s_m, \varepsilon_m)$ is an admissible sequence with $e(I) \leq n$ for some n.

(1) *When $n + |I|$ is even and $t \geq 0$, let $k = (1/2)(n+|I|)$. Write $J \circ I = (J, I)$ for the concatenation of admissible sequences. Then $J_k^t \circ I$ is admissible and $e(J_k^t \circ I) = n$. Furthermore, $\bar{J}_k^t \circ I$ is admissible and $e(\bar{J}_k^t \circ I) = n+1$.*

(2) *If K is any admissible sequence with $e(K) = n+1$, then there is a $t \geq 0$ and a subsequence I of K with $e(I) \leq n$ such that $K = J_k^t \circ I$ and $2k = n+1+|I|$; or $K = \bar{J}_k^t \circ I$ with $2k = n+|I|$; or $K = J_{l_1}^t \circ \bar{J}_{l_2}^{t'} \circ I$ with $l_1 = l_2 p^{t'+1} + 1$ and $2l_2 = n+1+|I|$.*

The analogous lemma for the prime 2 is left to the reader.

Theorem 6.19 ([Cartan54])**.** *Let p be an odd prime.* (1) $H^*(K(\mathbb{Z}/p\mathbb{Z}, n); \mathbb{F}_p)$ *is a free, graded-commutative algebra on generators $St^I \iota_n$ where ι_n denotes the fundamental class in $H^n(K(\mathbb{Z}/p\mathbb{Z}, n); \mathbb{F}_p)$ and I is an admissible sequence with $e(I) < n$ or with $e(I) = n$ and $I = (1, s, I')$.* (2) $H^*(K(\mathbb{Z}, n); \mathbb{F}_p)$ *is a free, graded-commutative algebra on generators $St^I \iota_n$ where ι_n denotes the fundamental class in $H^n(K(\mathbb{Z}, n); \mathbb{F}_p)$ and I is an admissible sequence, $I = (\varepsilon_0, s_1, \varepsilon_1, \ldots, s_m, \varepsilon_m)$, with $\varepsilon_m = 0$ and $e(I) < n$, or $e(I) = n$ and $I = (1, s, I')$.* (3) *For $k > 1$, $H^*(K(\mathbb{Z}/p^k\mathbb{Z}, n); \mathbb{F}_p)$ is a free, graded-commutative algebra on generators $St_k^I \iota_n$ where $\iota_n \in H^n(K(\mathbb{Z}/p^k\mathbb{Z}, n); \mathbb{F}_p)$ and $I = (\varepsilon_0, s_1, \varepsilon_1, \ldots, s_m, \varepsilon_m)$ is admissible with $e(I) < n$ or $e(I) = n$ and $I = (1, s, I')$, and if $\varepsilon_m = 1$, then $St_k^I \iota_n = St^{(\varepsilon_0, s_1, \varepsilon_1, \ldots, s_m)}(\beta_k \iota_n)$, where β_k is the kth power Bockstein operation.*

Theorem 6.19 ([Serre53])**.** (1) $H^*(K(\mathbb{Z}/2\mathbb{Z}, n); \mathbb{F}_2)$ *is a polynomial algebra on elements $St^I \iota_n$ where $\iota_n \in H^n(K(\mathbb{Z}/2\mathbb{Z}, n); \mathbb{F}_2)$ is the fundamental class and $I = (s_1, s_2, \ldots, s_m)$ is an admissible sequence with $e(I) < n$.* (2) *If $n > 1$, then $H^*(K(\mathbb{Z}, n); \mathbb{F}_2)$ is a polynomial algebra on generators $St^I \iota_n$ where $\iota_n \in H^n(K(\mathbb{Z}, n); \mathbb{F}_2)$ is the fundamental class and $I = (s_1, s_2, \ldots, s_m)$ is an admissible sequence with $e(I) < n$ and $s_m > 1$.* (3) *For $k > 1$, $H^*(K(\mathbb{Z}/2^k\mathbb{Z}, n); \mathbb{F}_2)$ is a polynomial algebra on elements $St_k^I \iota_n$ where $\iota_n \in H^n(K(\mathbb{Z}/2^k\mathbb{Z}, n); \mathbb{F}_2)$ is the fundamental class and $I = (s_1, s_2, \ldots, s_m)$ is an admissible sequence with $e(I) < n$ and if $s_m = 1$, then $St_k^I \iota_n = St^{(s_1, \ldots, s_{m-1})} \beta_k \iota_n$, where β_k is the kth power Bockstein homomorphism.*

The kth power Bockstein operation is defined by examining the long exact sequence in cohomology that arises from the short exact sequence of coefficients,

$$0 \to \mathbb{Z}/p\mathbb{Z} \to \mathbb{Z}/p^{k+1}\mathbb{Z} \to \mathbb{Z}/p^k\mathbb{Z} \to 0.$$

The connecting homomorphism carries the fundamental class, ι_n, to $\beta_k \iota_n$;

$$\cdots \to H^n(K(\mathbb{Z}/p^k\mathbb{Z}, n); \mathbb{Z}/p^k\mathbb{Z}) \xrightarrow{\beta_k} H^{n+1}(K(\mathbb{Z}/p^k\mathbb{Z}, n); \mathbb{F}_p) \to \cdots.$$

See Chapter 10 for other details of such operations.

We present the proof of Theorem 6.19 in detail for p an odd prime and $\pi = \mathbb{Z}/p\mathbb{Z}$. This case was first computed by [Cartan54] using acyclic constructions. Our proof follows the argument of [Serre53] for the case $p = 2$ as outlined by [Postnikov66] for odd primes.

We proceed by induction. The key to the general inductive step is a theorem of [Borel53] for which we need an odd primary notion that substitutes for the mod 2 simple system of generators (Definition 5.12).

Definition 6.20. *A graded-commutative algebra over $\mathbb{F}_p$, W^*, is said to have a p-**simple system of generators** $\{x_i \mid i \in J\}$ for J some totally ordered set, if*

W^* is generated as a vector space over $\mathbb{F}_p$ by the monomials $x_{i_1}^{m_1} x_{i_2}^{m_2} \cdots x_{i_k}^{m_k}$ where $i_1 < i_2 < \cdots < i_k$ and, for $1 \leq j \leq k$, $0 \leq m_j \leq p-1$, if $\deg x_{i_j}$ is even; $0 \leq m_j \leq 1$, if $\deg x_{i_j}$ is odd.

A canonical example is $\mathbb{F}_p[u]$, the polynomial algebra, with u of even degree. The uniqueness of the base p expansion of a natural number implies that $\{u, u^p, u^{p^2}, \dots\}$ is a p-simple system of generators. This system of generators can be represented using Steenrod operations: Suppose $\deg u = 2k$, then the sequence becomes $\{u, P^k u, P^{kp}(P^k u), P^{kp^2}(P^{kp}(P^k u)), \dots\}$. The reader can compare the next theorem with Example 1.I and Theorem 3.27.

Theorem 6.21. *Suppose $\{E_r^{*,*}, d_r\}$ is a first-quadrant spectral sequence of algebras over $\mathbb{F}_p$ satisfying the following hypotheses:*

(1) $E_2^{*,*} \cong V^* \otimes W^*$ *with* $E_2^{*,0} \cong V^*$ *and* $E_2^{0,*} \cong W^*$ *as algebras,*
(2) $E_\infty^{*,*}$ *is trivial,*
(3) W^* *has a p-simple system of transgressive generators $\{x_i \mid i \in J\}$.*

Then V^* is a free, graded-commutative algebra on generators $\{y_i, z_j \mid i, j \in J\}$ where $y_i = \tau(x_i)$ and $z_j = \tau(y_j \otimes x_j^{p-1})$ when x_j has even degree. Here τ denotes the transgression, $\deg y_i = 1 + \deg x_i$ and $\deg z_j = 2 + p \deg x_j$.

PROOF ([Borel53, §§16, 17]): The proof follows the same outline as the proof of Theorem 3.27. We introduce two versions of elementary spectral sequences: For s odd, let $E(s)$ denote the spectral sequence with

$$E(s)_2^{*,*} \cong \mathbb{F}_p[y] \otimes \Lambda(x)$$

where $\mathrm{bideg}\, x = (0, s)$, $\mathrm{bideg}\, y = (s+1, 0)$, and $\tau(x) = y$. We can picture $E(s)$ in this case as in the diagram:

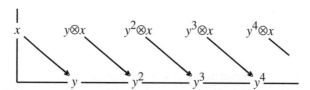

The spectral sequence $E(s)$ satisfies the hypotheses and conclusion of Theorem 6.21 and furthermore, $\mathbb{F}_p[y]$ has a p-simple system of generators.

For s even, let $E(s)$ denote the spectral sequence with

$$E(s)_2^{*,*} \cong \Lambda(y) \otimes \mathbb{F}_p[z] \otimes \mathbb{F}_p[x]\big/\langle x^p = 0\rangle$$

where $\mathrm{bideg}\, x = (0, s)$, $\mathrm{bideg}\, y = (s+1, 0)$, $\mathrm{bideg}\, z = (2+sp, 0)$, $\tau(x) = y$ and $\tau(y \otimes x^{p-1}) = z$. We picture $E(s)$ in this case as in the diagram on the next page. Once again, $E(s)$ satisfies the hypotheses and conclusion of the theorem.

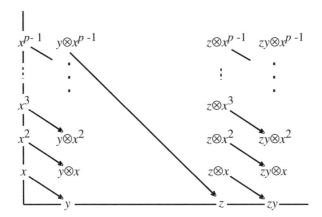

Suppose the hypotheses of Theorem 6.21 hold and $\deg x_i = s_i$ for the generators of W^*. Form the spectral sequence, $\bigotimes_{i \in J} E(s_i)$ together with the mapping of E_2-terms, $\bigotimes_{i \in J} E(s_i)_2^{*,*} \longrightarrow E_2^{*,*}$ that induces an isomorphism $\bigotimes_{i \in J} E(s_i)_2^{0,*} \to E_2^{0,*} \cong W^*$. By Zeeman's comparison theorem (Theorem 3.26) we deduce that $\bigotimes_{i \in J} E(s_i)_2^{*,0} \to E_2^{*,0} \cong V^*$ is an isomorphism of $\mathbb{F}_p$-vector spaces.

Examining $E(s_i)_2^{*,0}$ in each case, we find the expected free, graded-commutative algebras. It remains to show that V^* is free. The argument is similar to the one given in Example 1.K and so we leave it to the reader. □

To prove Theorem 6.19 we apply induction on the integer n and the Leray-Serre spectral sequence for the fibration

$$K(\mathbb{Z}/p\mathbb{Z}, n) \to PK(\mathbb{Z}/p\mathbb{Z}, n+1) \to K(\mathbb{Z}/p\mathbb{Z}, n+1).$$

When $n = 1$, $K(\mathbb{Z}/p\mathbb{Z}, 1)$ has the homotopy type of the lens space $L^\infty(p)$, which can be constructed as the direct limit of orbit spaces,

$$\cdots \hookrightarrow S^{2k+1}\big/\mathbb{Z}/p\mathbb{Z} \hookrightarrow S^{2k+3}\big/\mathbb{Z}/p\mathbb{Z} \hookrightarrow \cdots.$$

Here S^{2k+1} is taken to be the unit vectors in $\mathbb{C}^{k+1}$ and $\mathbb{Z}/p\mathbb{Z}$ acts diagonally as the group of pth roots of unity in $\mathbb{C}$. It is elementary to compute $H^*(L^\infty(p); \mathbb{F}_p)$ from the cell structure (see, for example, [Whitehead, GW78, p.92]):

$$H^*(K(\mathbb{Z}/p\mathbb{Z}, 1); \mathbb{F}_p) = \Lambda(\iota_1) \otimes \mathbb{F}_p[\beta\iota_1],$$

where $\iota_1 \in H^1(K(\mathbb{Z}/p\mathbb{Z}, 1); \mathbb{F}_p)$ is the fundamental class. To begin the induction, we establish that $H^*(K(\mathbb{Z}/p\mathbb{Z}, 1); \mathbb{F}_p)$ satisfies the conditions of Theorem

6.19. According to the definition of excess, $e(I) = 0$ can only be satisfied by the admissible sequence (0) giving the element ι_1. When $e(I) = 1$ we have the sequences (1) and $(0, p^t, 0, p^{t-1}, \dots, 0, p, 0, 1, 1)$, corresponding to $(\beta\iota_1)^{p^t}$ for $t \geq 0$. Only (1) produces a generator, and the sequences $J_1^t \circ (1)$ determine a p-simple system of generators for $H^*(K(\mathbb{Z}/p\mathbb{Z}, 1); \mathbb{F}_p)$.

We apply Borel's theorem as the inductive step. Suppose that the algebra $H^*(K(\mathbb{Z}/p\mathbb{Z}, n); \mathbb{F}_p)$ is given as described in the theorem. If $I = (\varepsilon_0, s_1, \varepsilon_1, \dots, s_m, \varepsilon_m)$ is an admissible sequence with $e(I) \leq n$, then the generator $St^I \iota_n$ transgresses to $St^I \iota_{n+1}$ by Corollary 6.9 and the fact that ι_n transgresses to ι_{n+1}. When $St^I \iota_n$ is of even degree, it contributes the p-simple system of generators $\{St^I \iota_n, (St^I \iota_n)^p, (St^I \iota_n)^{p^2}, \dots\}$ to W^*. We can write this system as

$$\{ St^I \iota_n, P^k St^I \iota_n, P^{kp}(P^k St^I \iota_n), \dots \}$$

where $2k = n + |I|$. These elements transgress to generators $St^{J_k^t \circ I} \iota_{n+1}$ where J_k^t is described in Lemma 6.18. Thus $H^*(K(\mathbb{Z}/p\mathbb{Z}, n); \mathbb{F}_p)$ has a p-simple system of transgressive elements.

The spectral sequence for the path-loop fibration over $K(\mathbb{Z}/p\mathbb{Z}, n + 1)$ provides the rest of the hypotheses and so $H^*(K(\mathbb{Z}/p\mathbb{Z}, n + 1); \mathbb{F}_p)$ is a free, graded-commutative algebra on certain generators. It remains to determine that the generators are the ones described in the theorem.

When $\deg(St^I \iota_n)$ is odd. Then $St^I \iota_{n+1}$ generates a polynomial subalgebra of $H^*(K(\mathbb{Z}/p\mathbb{Z}, n + 1); \mathbb{F}_p)$.

When $\deg St^I \iota_n$ is even, let $2k = n + |I|$ and we get elements $\{St^I \iota_{n+1}, St^{J_k^0 \circ I} \iota_{n+1}, St^{J_k^1 \circ I} \iota_{n+1}, \dots\}$ in $H^{\mathrm{odd}}(K(\mathbb{Z}/p\mathbb{Z}, n + 1), \mathbb{F}_p)$, all of the form $St^{I'} \iota_{n+1}$ where $e(I') \leq n$. The new generators take the form

$$\{\tau[St^I \iota_{n+1} \otimes (St^I \iota_n)^{p-1}], \tau[P^k(St^I \iota_{n+1}) \otimes (P^k St^I \iota_n)^{p-1}], \dots \}.$$

Here $2k = n + |I|$ with $e(I) \leq n$. The Kudo transgression theorem implies that

$$\tau[St^{J_k^t \circ I} \iota_{n+1} \otimes (St^{J_k^t \circ I} \iota_n)^{p-1}] = -\beta P^{kp^t} St^{J_k^t \circ I} \iota_{n+1} = -St^{\bar{J}_k^{p^{t+1}} \circ I} \iota_{n+1}$$

From Lemma 6.19 we conclude $e(\bar{J}_k^{p^{t+1}} \circ I) = n+1$. Since the converse part of Lemma 6.19 describes all admissible sequences of excess equal to $n+1$, we have proved that $H^*(K(\mathbb{Z}/p\mathbb{Z}, n+1); \mathbb{F}_p)$ is a free, graded-commutative algebra on generators $St^K \iota_{n+1}$ where K is an admissible sequence with $e(K) < n + 1$ or $e(K) = n + 1$ and $K = (1, s, I)$ for some sequence I with $e(I) \leq n$.

To prove the stated results for $K(\mathbb{Z}, n)$ and $K(\mathbb{Z}/p^k\mathbb{Z}, n)$ we simply observe that the inductions begin as described. For $K(\mathbb{Z}, 2)$ we take $\mathbb{C}P(\infty)$ as a representative of its homotopy type and $H^*(\mathbb{C}P(\infty); \mathbb{F}_p) \cong \mathbb{F}_p[\iota_2]$ satisfies the theorem. The same proof goes over *mutatis mutandis*. For higher powers of p, we can represent $K(\mathbb{Z}/p^k\mathbb{Z}, 1)$ as the lens space $L^\infty(p^k)$. By a careful accounting of the Bockstein homomorphism, it is elementary to prove that

$$H^*(L^\infty(p^k); \mathbb{F}_p) \cong \Lambda(\iota_1) \otimes \mathbb{F}_p[\beta_k\iota_1]$$

and the induction is off and running. Finally, we remark that the analogous argument works for the prime 2. □

Having completed these calculations, let us put them to work. For the rest of this section, we fix $p = 2$. (The case of odd primes is similar but the bookkeeping is more complicated. See the paper of [Umeda59].) In this case all the algebras involved are free and strictly commutative and have simple systems of generators as in Definition 5.12.

Recall from Example 1.F, the definition of the Poincaré series of a space, here given for coefficients in the field $\mathbb{F}_2$:

$$P(X,t) = \sum_{i=0}^{\infty} (\dim_{\mathbb{F}_2} H^i(X; \mathbb{F}_2))t^i.$$

Let $P(\pi, q, t) = P(K(\pi, q), t)$. Theorem 6.19 states

$$H^*(K(\mathbb{Z}/2\mathbb{Z}, q); \mathbb{F}_2) \cong \mathbb{F}_2[St^I \iota_q \mid I = (s_1, \dots, s_m), \text{ admissible}, e(I) \leq q].$$

It follows that $P(\mathbb{Z}/2\mathbb{Z}, q, t) = \prod_{I, \text{admissible}, e(I) \leq q} \dfrac{1}{1 - t^{q+|I|}}.$

In what follows, we make a closer analysis of these formal power series. To wit, consider an admissible sequence $I = (s_1, s_2, \dots, s_m)$. Since $s_i \geq 2s_{i+1}$, the sequence can be described by an associated one, $(\alpha_1, \alpha_2, \dots, \alpha_m)$ defined by $\alpha_m = s_m$ and, for $i < m$, $\alpha_i = s_i - 2s_{i+1}$. Notice that the excess (how much s_i exceeds $2s_{i+1}$) is equal to $\sum_i \alpha_i$. If we denote a typical factor of $P(\mathbb{Z}/2\mathbb{Z}, q, t)$ by $1/(1 - t^N)$, then $N = q + |I|$, and

$$N - q = |I| = \sum_{i=1}^{m} s_i = \sum_{i=1}^{m} \alpha_i(2^i - 1).$$

The last equation can be seen as follows;

$$s_m = \alpha_m \quad s_{m-1} = \alpha_{m-1} + 2s_m$$
$$= \alpha_{m-1} + 2\alpha_m$$
$$s_{m-2} = \alpha_{m-2} + 2s_{m-1}$$
$$= \alpha_{m-2} + 2\alpha_{m-1} + 4\alpha_m,$$

and one continues by induction. Introduce $\alpha_0 = q - 1 - \sum_{i=1}^{m} \alpha_i$. This allows us to write

$$|I| = N - q = \sum_{i=1}^{m} \alpha_i 2^i - \sum_{i=1}^{m} \alpha_i = \sum_{i=1}^{m} \alpha_i 2^i - (q - 1 - \alpha_0)$$
$$= \sum_{i=0}^{m} \alpha_i 2^i - q + 1.$$

Thus I is determined by a sequence $(\alpha_0, \alpha_1, \ldots, \alpha_m)$ whose sum is $q - 1$ and for which

$$
\begin{aligned}
n = q + |I| &= 1 + \sum_{i=0}^{m} \alpha_i 2^i \\
&= 1 + \underbrace{2^0 + \cdots + 2^0}_{\alpha_0} + \underbrace{2^1 + \cdots + 2^1}_{\alpha_1} + \cdots + \underbrace{2^m + \cdots + 2^m}_{\alpha_m} \\
&\qquad\qquad\qquad\qquad\qquad \underbrace{}_{q-1 \text{ powers of } 2} \\
&= 1 + 2^{h_1} + 2^{h_2} + \cdots + 2^{h_{q-1}} \text{ with } 0 \le h_1 \le h_2 \le \cdots \le h_{q-1}.
\end{aligned}
$$

This implies the following expression.

Proposition 6.22. $P(\mathbb{Z}/2\mathbb{Z}, q, t) = \displaystyle\prod_{0 \le h_1 \le \cdots \le h_{q-1}} \frac{1}{1 - t^{1 + 2^{h_1} + \cdots + 2^{h_{q-1}}}}.$

A similar development for $P(\mathbb{Z}, q, t)$ yields

$$
P(\mathbb{Z}, q, t) = \prod_{0 \le h_1 \le \cdots \le h_{q-3} < h_{q-2}} \frac{1}{1 - t^{1 + 2^{h_1} + \cdots + 2^{h_{q-2}}}}.
$$

In the case of higher order torsion, a higher order Bockstein takes the place of the ordinary Bockstein, and it follows that $P(\mathbb{Z}/2^k\mathbb{Z}, q, t) = P(\mathbb{Z}/2\mathbb{Z}, q, t)$.

We record some facts about these Poincaré series with regard to their analytic properties. For proofs, the interested reader should consult the proofs of [Serre53], [Ahlfors66, p. 191] or the local analytic number-theorist.

Fact 1. *For π, a finitely generated abelian group, $P(\pi, q, t)$ is a convergent series, on the open unit disk $\{ t \in \mathbb{C}; |t| < 1 \}$ with an evident essential singularity at $t = 1$.*

In order to study the behavior of these functions near $t = 1$, Serre introduced the functions, defined for x large,

$$
\phi(\pi, q, x) = \log_2 P(\pi, q, 1 - 2^{-x}).
$$

Fact 2. $\phi(\mathbb{Z}/2\mathbb{Z}, q, x) \sim \dfrac{x^q}{q!},$ *and* $\phi(\mathbb{Z}, q, x) \sim \dfrac{x^{q-1}}{(q-1)!},$ *where* $f(x) \sim g(x)$ *if* $\displaystyle\lim_{x \to \infty} f(x)/g(x) = 1.$

Since $P(\mathbb{Z}/2\mathbb{Z}, 1, t) = \dfrac{1}{1 - t}$ $(K(\mathbb{Z}/2\mathbb{Z}, 1) \cong \mathbb{R}P(\infty))$, $\phi(\mathbb{Z}/2\mathbb{Z}, 1, x) = x$. The proof of **Fact 2** follows by induction and a careful use of the product expansion in Proposition 6.22.

From these facts about the Poincaré series of Eilenberg-Mac Lane spaces, we prove a remarkable theorem of [Serre53].

Theorem 6.23. *If X is a finite CW-complex that is connected and simply-connected, and X is not contractible, then $\pi_i(X)$ contains a subgroup isomorphic to $\mathbb{Z}$ or $\mathbb{Z}/2\mathbb{Z}$ for infinitely many i.*

Toward the proof, we introduce another construction associated to a space X, a tower of fibrations with X at the bottom.

$$
\begin{array}{c}
\downarrow \\
K(\pi_2(X), 1) \longrightarrow X\langle 3\rangle \\
\downarrow \\
K(\pi_1(X), 0) \longrightarrow X\langle 2\rangle \\
\downarrow \\
X
\end{array}
$$

At each successive stage, we "kill" the bottom dimensional homotopy group of the previous stage by attaching the appropriate series of higher dimensional cells.

Definition 6.24. *The* **upside-down Postnikov tower** *of a space X is a tower of fibrations satisfying the properties:*

(1) $X\langle 2\rangle$ *is the universal cover of X,*
(2) *the composite map, $p\colon X\langle n\rangle \to X$ induces an isomorphism, if $j \geq n$,*

$$
\pi_j(p)\colon \pi_j(X\langle n\rangle) \to \pi_j(X).
$$

(3) $\pi_j(X\langle n\rangle) = \{0\}$ *for $j < n$.*

The homotopy long exact sequence for each fibration $X\langle n\rangle \to X\langle n-1\rangle$ shows that the fibre is a $K(\pi_{n-1}(X), n-2)$. Such towers were introduced independently by [Cartan-Serre52] and by [Whitehead, GW52]. The upside-down Postnikov tower can be related to the usual Postnikov tower by observing that the space $X\langle n\rangle$ can be constructed as the homotopy-theoretic fibre of the mapping $X \to P_n X$ where $P_n X$ is the nth space in the Postnikov tower for X. Up to homotopy we have the fibration $X\langle n\rangle \to X \to P_n X$.

PROOF OF THEOREM 6.23: Suppose N is the largest integer for which $\pi_N(X)$ contains a subgroup $\mathbb{Z}$ or $\mathbb{Z}/2\mathbb{Z}$. Consider the fibration

$$
K(\pi_N(X), N-1) \to X\langle N+1\rangle \to X\langle N\rangle.
$$

If $\tilde{H}^*(X\langle N+1\rangle; \mathbb{F}_2) = \{0\}$, then the Zeeman comparison theorem implies $H^*(X\langle N\rangle; \mathbb{F}_2) \cong H^*(K(\pi_N(X), N); \mathbb{F}_2)$. We prove the vanishing of

$\tilde{H}^*(X\langle N+1\rangle; \mathbb{F}_2)$: By Lemma 6.16, $\tilde{H}^*(K(\pi_{N+k}(X), N+k-2); \mathbb{F}_2) = \{0\}$ for all $k > 0$ and so the mod 2 cohomology Leray-Serre spectral sequence for the fibration $K(\pi_{N+k}(X), N+k-2) \to X\langle N+k\rangle \to X\langle N+k-1\rangle$ collapses to yield $H^*(X\langle N+k\rangle; \mathbb{F}_2) \cong H^*(X\langle N+k-1\rangle; \mathbb{F}_2)$. Stringing these isomorphisms together we have that $H^*(X\langle N+1\rangle; \mathbb{F}_2) \cong H^*(X\langle N+k\rangle; \mathbb{F}_2)$ for all $k > 0$. Since $X\langle N+k\rangle$ can be made to be arbitrarily highly connected, $\tilde{H}^*(X\langle N+1\rangle; \mathbb{F}_2) = \{0\}$.

Recall, from Chapter 1, some of the properties of Poincaré series:

(1) If $\{E_r^{*,*}, d_r\}$ is a first quadrant spectral sequence, converging to H^*, then $P(E_2^{*,*}, t) \geq P(H^*, t)$. (Recall that $P(A, t) \geq P(B, t)$ if the power series $p(t) = P(A, t) - P(B, t)$ has all nonnegative coefficients.) In particular, if $F \hookrightarrow E \xrightarrow{p} B$ is a fibration, then $P(B, t) \times P(F, t) \geq P(E, t)$.

(2) If $\pi = \pi_1 \oplus \pi_2$, all finitely generated abelian groups, then $P(\pi, q, t) = P(\pi_1, q, t) \times P(\pi_2, q, t)$. (Here $\times$ denotes the Cauchy product of power series.)

When we apply these observations to the upside-down Postnikov tower, we have

$$
\begin{aligned}
P(X\langle N\rangle, t) &= P(\pi_N(X), N, t) \\
&\leq P(\pi_{N-1}(X), N-2, t) \times P(X\langle N-1\rangle, t) \\
&\leq P(\pi_{N-1}(X), N-2, t) \times P(\pi_{N-2}(X), N-3, t) \times P(X\langle N-2\rangle, t) \\
&\vdots \\
&\leq P(X, t) \times \prod_{j=2}^{N-1} P(\pi_j(X), j-1, t)
\end{aligned}
$$

Because these power series have only positive coefficients, the inequality of power series gives an actual inequality on evaluation when t is in the interval $[0, 1)$. Also, $\log_2$ is an order-preserving function. Thus, for x large, we have

$$
\phi(\pi_N(X), N, x) \leq \log_2(P(X, 1 - 2^{-x})) + \sum_{j=2}^{N-1} \phi(\pi_j(X), j-1, x).
$$

If we express a finitely generated abelian group π as a direct sum

$$
\pi \cong \mathbb{Z} \oplus \mathbb{Z} \oplus \cdots \oplus \mathbb{Z} \oplus \mathbb{Z}/2^{k_1}\mathbb{Z} \oplus \cdots \oplus \mathbb{Z}/2^{k_r}\mathbb{Z} \oplus \text{ stuff of odd torsion,}
$$

then, at the prime 2, $P(\pi, q, t) = \left(\prod_1^s P(\mathbb{Z}, q, t)\right) \times \left(\prod_1^r P(\mathbb{Z}/2\mathbb{Z}, q, t)\right)$ and it follows that $\phi(\pi, q, x) = \sum_1^s \phi(\mathbb{Z}, q, x) + \sum_1^r \phi(\mathbb{Z}/2\mathbb{Z}, q, x)$. From this equation we have that $\phi(\pi, q, x) \sim rx^q/q!$, or, if $r = 0$, then $\phi(\pi, q, x) \sim sx^{q-1}/(q-1)!$ where s is the number of factors of $\mathbb{Z}$ appearing in the decomposition.

Since the space X is a finite CW-complex, $P(X,t)$ is a polynomial in t and so $\lim_{x\to\infty} P(X, 1 - 2^{-x}) = P(X, 1)$. Thus, $\log_2(P(X, 1 - 2^{-x})) \sim$ a constant. Putting this together with the inequality bounding $P(X\langle N\rangle, t)$, we see that the left hand side (lhs) $\phi(\pi_N(X), N, x)$ converges like $r_N x^N/N!$ or $s_N x^{N-1}/(N - 1)!$ as x grows unbounded, depending on the decomposition of $\pi_N(X)$. On the right hand side (rhs) of the inequality, we find the relation

$$\log_2(P(X, 1 - 2^{-x})) + \sum_{j=2}^{N-1} \phi(\pi_j(X), j - 1, x) \sim A \cdot \frac{x^{N-2}}{(N - 2)!}.$$

However, as x goes to infinity, the inequality must be violated at some x since

$$r_N \frac{x^N}{N!} \sim \text{lhs} \leq \text{rhs} \sim A\frac{x^{N-2}}{(N - 2)!}.$$

This contradiction to the existence of such an N establishes the theorem.   □

Around 1950, little was known about homotopy groups. These groups were known to be denumerable for finite CW-complexes. For spheres, $\pi_n(S^n)$, $\pi_3(S^2)$, $\pi_7(S^4)$ and $\pi_{15}(S^8)$ had been computed and shown to be isomorphic to $\mathbb{Z}\oplus$finite. Other cases were computed by [Freudenthal37] ($\pi_{n+1}(S^n) \cong \mathbb{Z}/2\mathbb{Z}$) and [Whitehead, GW50] ($\pi_{n+2}(S^n) \cong \mathbb{Z}/2\mathbb{Z}$). It was not known whether the groups were nontrivial after a certain range of dimensions (by analogy with homology groups). In his thesis [Serre51] proved that $\pi_i(X)$ is finitely generated when X is a simply-connected, finite CW-complex (Proposition 5.17). He also established the finiteness of $\pi_i(S^n)$ for $i \neq n$ (and $i \neq 2n - 1$ for n even) as in Example 5.K. With the further development of these methods (Theorem 6.23) [Serre53] proved the nontriviality of infinitely many homotopy groups of any noncontractible finite complex. To quote [Whitehead, GW83] from his history of homotopy theory, "It is no exaggeration to say that Serre's thesis revolutionized the subject."

The analogous theorem for odd primes given by Theorem 6.23 was proved by similar means by [Umeda59]. The theorem for all primes has been extended by [McGibbon-Neisendorfer84] to remove the finiteness assumption and replace it with the conditions that X be simply-connected, $H_n(X;\mathbb{F}_p) \neq \{0\}$ for some $n \geq 2$, and $H_n(X;\mathbb{F}_p) = \{0\}$ for n sufficiently large. They conclude further that there is p-torsion in infinitely many homotopy groups of X. Their proof relies on the study of [Miller, H84] of the space of mappings $\text{map}(B\mathbb{Z}/2\mathbb{Z}, X)$ for X a finite complex.

The upside-down Postnikov tower may be used to prove a generalization of the rational Hurewicz theorem (Theorem 5.18) to fields of arbitrary characteristic.

Theorem 6.25 (the Hurewicz-Serre theorem). *Suppose that k is a field and X a simply-connected space of finite type. If $H_i(X;k) = \{0\}$ for $1 \leq i < n$, then $\pi_i(X) \otimes k = \{0\}$ for $1 \leq i < n$ and $\pi_n(X) \otimes k \cong H_n(X;k)$.*

PROOF: We show by induction on j that, under the assumptions, $H_i(X\langle j\rangle; k) = \{0\}$ for $0 < i < n$ and $0 \leq j \leq n$. For $j = 1$, $X\langle 2\rangle = X$ because X is simply-connected. Furthermore, $\pi_1(X) \otimes k = \{0\}$ for trivial reasons.

The key to the induction step is the Leray-Serre spectral sequence for the fibrations $K(\pi_j(X), j - 1) \to X\langle j + 1\rangle \to X\langle j\rangle$. We assume that $H_i(X\langle j\rangle; k) = \{0\}$ for $0 < i < n$. The E^2-term for the fibration takes the form $E^2_{p,q} \cong H_p(X\langle j\rangle; k) \otimes_k H_q(K(\pi_j(X), j - 1); k)$. If $\pi_j(X) \otimes k \cong H_{j-1}(K(\pi_j(X), j-1); k)$ were nonzero, then $E^2_{0,j-1} \neq \{0\}$ and since $E^2_{i,0} = \{0\}$ for $0 < i < n$ and $j < n$, there is no nonzero differential to hit $E^2_{0,j-1}$ and so $H_{j-1}(X\langle j + 1\rangle; k) \neq \{0\}$, which contradicts the fact that $X\langle j + 1\rangle$ is j-connected. Thus $\pi_j(X) \otimes k = \{0\}$ for $j < n$ and, by Lemma 6.16, $\tilde{H}_*(K(\pi_j(X), j - 1); k) = \{0\}$. This implies that $H_*(X\langle j + 1\rangle; k) \cong H_*(X\langle j\rangle; k)$ and the induction is complete.

When $j = n - 1$, we have shown that $H_n(X; k) \cong H_n(X\langle n\rangle; k) \cong \pi_n(X) \otimes k$. □

In the special case of $X = S^{2n-1}$ we can be very explicit about the computation. The following result of [Serre51] extends our knowledge of the homotopy groups of spheres.

Proposition 6.26. For $n > 1$, $H_i(S^{2n-1}\langle 2n\rangle; \mathbb{F}_p) = \{0\}$ for $0 < i < 2n + 2p - 3$ and $H_{2n+2p-3}(S^{2n-1}\langle 2n\rangle; \mathbb{F}_p) \cong \mathbb{F}_p$.

PROOF: Since we know that $\pi_{2n-1}(S^{2n-1}) \cong \mathbb{Z}$, we have the fibration

$$K(\mathbb{Z}, 2n - 2) \to S^{2n-1}\langle 2n\rangle \to S^{2n-1}.$$

By Theorem 6.19, $H^*(K(\mathbb{Z}, 2n - 2); \mathbb{F}_p)$ is a free graded-commutative algebra on generators $St^I \iota_{2n-2}$ for admissible sequences $I = (\varepsilon_0, s_1, \varepsilon_1, \dots, s_m, \varepsilon_m)$ with $\varepsilon_m = 0$ and $e(I) \leq 2n - 2$. The transgression takes ι_{2n-2} to the generator $u \in H^{2n-1}(S^{2n-1}; \mathbb{F}_p)$ and the powers ι_{2n-2}^j map to $u \otimes \iota_{2n-2}^{j-1}$ for $j < p$. The class in least degree that persists to E^∞ and determines the connectivity of $H^*(S^{2n-1}\langle 2n\rangle; \mathbb{F}_p)$ is $P^1\iota_{2n-2} \in H^{2n-2+2(p-1)}(K(\mathbb{Z}, 2n - 2); \mathbb{F}_p)$. Since P^1 corresponds to the admissible sequence $(0, 1, 0)$ of excess 2, $P^1\iota_{2n-2}$ is found in $H^*(K(\mathbb{Z}, 2n - 2), \mathbb{F}_p)$ for all $n > 1$. □

By construction $\pi_j(S^{2n-1}\langle 2n\rangle) \cong \pi_j(S^{2n-1})$ for $j > 2n - 1$ and so the Hurewicz-Serre theorem implies the following result.

Corollary 6.27. $\pi_i(S^{2n-1}) \otimes \mathbb{F}_p = \{0\}$ for $2n - 1 < i < 2n + 2p - 3$ and $\pi_{2n+2p-3}(S^{2n-1}) \otimes \mathbb{F}_p \cong \mathbb{F}_p$.

Analogous results for even-dimensional spheres can be obtained by similar arguments. The argument for Corollary 6.27 and the Freudenthal suspension provide the basis for an induction that motivates the Adams spectral sequence of Chapter 9.

6.3 On classifying spaces and characteristic classes

Among the successful applications of homotopy theory, few are as spectacular as those found in the study of vector fields on manifolds. It was already known, in the "ancient" origins of the subject, that geometric problems of this sort could be tackled with topological tools. For example, the Poincaré index theorem ([Poincaré1881]) relates the index of a vector field to the Euler characteristic of the manifold.

Certain problems on differentiable manifolds can be seen to involve, if not reduce to, the problem of the existence of linearly independent vector fields on a manifold. An example is the problem of embedding an m-dimensional manifold M in a Euclidean space $\mathbb{R}^N$. Such an embedding leads to a bundle of $(N-m)$-dimensional vector spaces ν_M over M of normal vectors to the tangent space. A necessary condition that the manifold be embedded in a dimension lower in $\mathbb{R}^{N-1}$ is a nonzero vector field of vectors in ν_M. Another problem that can be formulated this way is the problem of the existence of a division algebra structure on $\mathbb{R}^n$ (see [Adams60] and Chapter 9 for this idea).

[Stiefel36] in Switzerland [Whitney35], then at Harvard, introduced the idea of *characteristic classes* toward solving the problem of determining the number of linearly independent vector fields on a manifold in a given vector bundle. Following [Steenrod51], these characteristic classes arise as follows: Let $G_k(\mathbb{R}^n)$ denote the **Grassmann manifold** of k-dimensional linear subspaces in $\mathbb{R}^n$. Triangulate the manifold M and suppose there is a bundle of n-dimensional real vector spaces over M. Assign k linearly independent vectors at each vertex of the triangulation. This gives a mapping, $M^{(0)} \to G_k(\mathbb{R}^n)$, sending each vertex in the 0-skeleton of M to the k-dimensional subspace spanned by the k vectors in an identification of the vector space at a point with $\mathbb{R}^n$. The problem is to extend this assignment from the 0-skeleton to the 1-simplices, and then to $M^{(1)}$, the 1-skeleton of M, and so on, proceeding a dimension at a time. The obstruction to extending the assignment at dimension $(i-1)$ to an assignment on the i-skeleton is a cocycle in $C^i(M, \pi_{i-1}(G_k(\mathbb{R}_M^n)))$, the cochains on M with coefficients in the system of local coefficients,

$$\pi_{i-1}(G_k(\text{the vector space at each point in } M)).$$

By reducing mod 2, these obstructions determine classes $w_i \in H^i(M; \mathbb{F}_2)$ now known as the **Stiefel-Whitney classes** of the bundle under consideration. Analogous constructions given by [Chern48] for bundles of complex vector spaces and by [Pontryagin47] are important tools in the study of manifolds.

Characteristic classes are useful in the classification of vector bundle structures on manifolds. One can reverse the process by building vector bundles on manifolds using the properties of specially constructed spaces called *classifying spaces*. The algebraic topology of classifying spaces is the subject to which we next turn the Leray-Serre spectral sequence. We first introduce some of the fundamental notions of vector bundles with structure and sketch the construction and properties of classifying spaces. We then study the cohomology of classifying spaces with the aid of the Leray-Serre spectral sequence. These computations were first done by [Borel53] in his celebrated thesis. Finally, we reconsider characteristic classes, having computed the cohomology of the classifying spaces and apply those results to some geometric problems.

The details of the theory of characteristic classes would easily double the length of this book and take us, not unhappily, far afield of our central concern. There are many good expositions of these topics, for example, the papers of [Borel53, 55], and the books of [Milnor-Stasheff74], [Dupont78], [Husemoller66], and [Bott-Tu82]. Our goal is to demonstrate to the reader the crucial role played by the Leray-Serre spectral sequence in this development.

On vector bundles and classifying spaces

We begin with the observation that the tangent bundle to a manifold is more than a fibration—it carries extra structure.

Definition 6.28. *A fibration* $F \hookrightarrow E \overset{p}{\to} B$ *is a* **fibre bundle with structure group** G *if the following hold*

(1) G *is a topological group and there is an action of* G *on* F, *the fibre,* $\mu\colon G \times F \to F$, *satisfying, for all* $g_1, g_2 \in G$ *and all* $x \in F$, $\mu(g_1, \mu(g_2, x)) = \mu(g_1 g_2, x)$ *and* $\mu(e, x) = x$. *Furthermore, the action is* **effective**, *that is, if* $\mu(g, x) = x$ *for all* x *in* F, *then* $g = e$.

(2) *There is a family,* $\{U_j \mid j \in J\}$ *of open sets of* B *that cover* B *and, for each* j *in* J, *a homeomorphism,* $\varphi_j\colon U_j \times F \to p^{-1}(U_j)$, *satisfying the equation* $p\varphi_j(x, y) = x$.

(3) *For each pair,* i, j *in* J, *and* x *in* $U_i \cap U_j$, *the mapping* $g_{ij}(x)$ *defined by*

$$g_{ij}(x) = \varphi_i(x, -)^{-1} \circ \varphi_j(x, -)\colon F \to p^{-1}(\{x\}) \to F$$

is continuous in x *and given by left multiplication by an element of* G, *that is, the map* g_{ij} *determines a continuous function* $U_i \cap U_j \to G$ *and* $g_{ij}(x)(y) = \mu(g_{ij}(x), y)$.

The principal example is the tangent bundle to a manifold. There the fibre is $\mathbb{R}^m$ for M, an m-dimensional manifold, and the structure group may be taken to be $Gl(m, \mathbb{R})$, the group of invertible $(m \times m)$-matrices. If we consider only the unit tangent vectors at each point in M, we have a fibre bundle with S^{m-1} as fibre and structure group $O(m)$, the group of invertible, real $(m \times m)$-matrices A with $A^t A = 1$. If the manifold M is orientable, then the structure group for the associated sphere bundle may be taken as $SO(m)$.

In order to classify the fibre bundles with given fibre F and structure group G over a fixed space B, we introduce a notion of equivalence of such bundles.

Definition 6.29. *A* **morphism** *of fibre bundles,*

$$(f, \tilde{f}) \colon (G, F, E, p, B) \longrightarrow (G, F', E', p', B')$$

is a pair of maps, $f \colon B \to B'$ and $\tilde{f} \colon E \to E'$ such that the following diagram commutes,

$$
\begin{array}{ccc}
E & \xrightarrow{\tilde{f}} & E' \\
\downarrow{\scriptstyle p} & & \downarrow{\scriptstyle p'} \\
B & \xrightarrow{f} & B',
\end{array}
$$

and $\tilde{f}\big|_F \colon F \to F'$ over each point is a G-equivariant map, that is, for all $g \in G$ and $x \in F$, $\tilde{f}(\mu(g, x)) = \mu'(g, \tilde{f}(x))$. For $B = B'$, we say that a morphism $(1, \tilde{f})$ is an **equivalence of bundles** *if $\tilde{f}\big|_F \colon F \to F'$, over each point, is a homeomorphism.*

A sufficiently differentiable mapping of manifolds, $f \colon M \to N$, provides a bundle morphism of the tangent bundles and a diffeomorphism of a manifold to itself provides an example of an equivalence of the tangent bundle with itself.

The first reduction of the problem of classifying all fibre bundles over a space B with a given fibre and structure group is to focus on bundles of a certain type.

Definition 6.30. *A fibre bundle is said to be a* **principal bundle** *if it has structure group G and fibre G with the action of G on the fibre given by left multiplication.*

Let $\xi = (G, F, E, p, B)$ denote a fibre bundle with structure group G. We associate a principal bundle to this bundle by taking the data $\varphi_j \colon U_j \times F \to p^{-1}(U_j)$ and $g_{ij}(x) \colon F \to F$ and replacing F with G everywhere. Property (3) of Definition 6.28 guarantees that another fibre bundle results. We call the new bundle constructed in this manner the **associated principal bundle** to ξ, and denote it by $\mathrm{Prin}(\xi) = (G, G, E, p, B)$.

Suppose F is a G-space with an effective action $\mu \colon G \times F \to F$. Suppose further that $\eta = (G, G, E, p, B)$ is a principal bundle. We construct a fibre bundle over B with fibre F, structure group G, and μ the given action as follows: Let $\{U_i\}$ be the open cover of B given in the principal bundle structure on E. Let $Z = \coprod U_i \times G \times F$ denote the disjoint union of the sets $U_i \times G \times F$. Define an equivalence relation on Z for which an element of $U_i \times G \times F$ is denoted by (x, i, g, f) and we require, for $x \in U_i \cap U_j$, $(x, i, g, f) \sim (x, j, g_{ij}(x)g, f)$ and, for all $h \in G$, $(x, i, g, f) \sim (x, i, gh, h^{-1}f)$. We denote by $E \times_G F$ the quotient space of Z by this relation together with the projection $\bar{p}([x, i, g, f]) = x$. The reader can check that $\eta[F] = (G, F, E \times_G F, \bar{p}, B)$ is a fibre bundle (see [Husemoller66, §5.3.2]). This procedure is seen to be effective for our purposes in the next proposition. For a proof the reader can consult the classic books of [Steenrod51] or [Husemoller66].

Proposition 6.31. *Let* $\xi = (G, F, E, p, B)$ *denote a fibre bundle with fibre* F *and structure group* G. *Let* $\eta = (G, G, E', p', B)$ *denote a principal bundle over* B *with structure group* G. *Then* ξ *is equivalent as a bundle to* $\mathrm{Prin}(\xi)[F]$ *and* η *is equivalent as a bundle to* $\mathrm{Prin}(\eta[F])$.

The proposition reduces the classification problem to determining principal G-bundles over B. If we view G as a group of transformations on the fibre G, then the motivating example of classification is the theory of covering spaces and the fundamental group; any covering space over a space can be constructed from the universal covering space. By analogy we seek a universal principal G-bundle over a space. The following proposition allows us to identify the defining property of such a universal principal bundle. It is also makes use of another ingredient in the construction of fibrations—the pullback of a fibration over a continuous mapping (§4.3).

Proposition 6.32. *Suppose* $\xi = (G, G, E, p, B)$ *is a principal* G-bundle over B, *a Hausdorff, paracompact space. Suppose* $f, g \colon X \to B$ *are homotopic maps, then* $f^*\xi = (G, G, f^*E, f^*p, X)$ *and* $g^*\xi = (G, G, g^*E, g^*p, X)$, *the induced principal bundles over* X, *are equivalent.*

SKETCH OF PROOF: Consider the diagram with H a homotopy from f to g.

$$
\begin{array}{ccc}
f^*E \times I & \longrightarrow & E \\
{\scriptstyle f^*p \times 1}\downarrow & & \downarrow{\scriptstyle p} \\
X \times I & \xrightarrow{\ \ H\ \ } & B
\end{array}
$$

This diagram is completed by the homotopy lifting property for fibrations and so this gives the data for a bundle homotopy from f^*E to g^*E. One can argue from the unique lifting of paths that the bundle homotopy provides an isomorphism of bundles. For all of the details, see [Husemoller66, §4.9]. □

Definition 6.33. *A principal* G-bundle $\eta = (G, G, E, p, B)$ *is* n-**universal** *if* $\pi_i(E) = \{0\}$ *for* $i \leq n$. *If* $\pi_i(E) = \{0\}$ *for all* i, *then we say that* η *is a* **universal bundle** *and we denote* E *by* EG, B *by* BG. *The space* BG *is called the* **classifying space** *for* G.

To establish the "universality" of these bundles one supposes given a principal G-bundle, (G, G, E, p, B). Over each finite skeleton of B, a mapping can be constructed, $B^{(k)} \to BG$, that pulls back to the given bundle. The obstructions to extending the mapping to the next skeleton lie in the cohomology of B with coefficients in the homotopy of EG and so they vanish. If we restrict B to be a CW-complex of dimension n, then a principal G-bundle over B is the pullback of an n-universal bundle over a continuous mapping. When we combine these ideas with Proposition 6.31, we have proved the following result.

Theorem 6.34. *If* $\chi(n) = (G, G, E(n, G), p, B(n, G))$ *is an* n*-universal bundle and* B *a CW-complex of dimension* $\leq n$, *then the set* { *equivalence classes of principal* G*-bundles over* B } *is in one-to-one correspondence with the set of homotopy classes of maps* $[B, B(n, G)]$.

Corollary 6.35. *If* $\chi = (G, G, EG, p, BG)$ *is a universal bundle for* G *and* B *of the homotopy type of a CW-complex, then the set* { *equivalence classes of principal* G*-bundles over* B } *is in one-to-one correspondence with* $[B, BG]$.

Some examples are in order:

(a) Suppose $G = \mathbb{Z}/2\mathbb{Z}$. Then, for each n, we have the bundle

$$\mathbb{Z}/2\mathbb{Z} \to S^n \to \mathbb{R}P(n)$$

where $\mathbb{R}P(n)$ denotes the real projective n-space. Thus $\mathbb{R}P(n)$ is a version of $B(n-1, \mathbb{Z}/2\mathbb{Z})$. The series of equatorial inclusions gives the system

$$
\begin{array}{ccccccc}
S^n & \longrightarrow & S^{n+1} & \longrightarrow & \cdots & \longrightarrow & S^\infty \\
\downarrow & & \downarrow & & & & \downarrow \\
\mathbb{R}P(n) & \longrightarrow & \mathbb{R}P(n+1) & \longrightarrow & \cdots & \longrightarrow & \mathbb{R}P(\infty).
\end{array}
$$

The direct limit is the $\mathbb{Z}/2\mathbb{Z}$-bundle $\mathbb{Z}/2\mathbb{Z} \hookrightarrow S^\infty \to \mathbb{R}P(\infty)$ with S^∞ weakly contractible, a universal bundle for $\mathbb{Z}/2\mathbb{Z}$. Thus $\mathbb{R}P(\infty)$ is a classifying space for $\mathbb{Z}/2\mathbb{Z}$.

(b) Recall from Chapter 5 the Stiefel manifolds of orthonormal k-frames in $\mathbb{R}^n$, $V_k(\mathbb{R}^n)$. Consider the mapping, $V_k(\mathbb{R}^n) \to G_k(\mathbb{R}^n)$ which sends a k-frame in $\mathbb{R}^n$ to the k-dimensional linear subspace it spans. This determines a fibre bundle with $O(k)$ as structure group. A careful argument with the integral homology Leray-Serre spectral sequence shows that $\pi_i(V_k(\mathbb{R}^n)) = \{0\}$ for $k \neq n$ and $i < n - k$. Thus $G_k(\mathbb{R}^n)$ is an $(n - k - 1)$-classifying space for $O(k)$, and the standard inclusions, $V_k(\mathbb{R}^n) \hookrightarrow V_k(\mathbb{R}^{n+1})$ and $G_k(\mathbb{R}^n) \hookrightarrow G_k(\mathbb{R}^{n+1})$, yield, in the direct limit, the space $G_k(\mathbb{R}^\infty)$ a classifying space, $BO(k)$.

These examples were known in the 1940's and, in the book of [Steenrod51] there are classifying spaces for closed subgroups of $O(k)$ constructed by using representation theory. A direct construction of BG for any topological group G was first given by [Milnor56] with later refinements by [Dold-Lashof59], [Stasheff63'] and [Milgram67]. We give a construction, following [Milgram67]: If G is a topological group, let BG denote the quotient space

$$BG = \bigcup_n \Delta^n \times G^n / \sim$$

where G^n is $G \times G \times \cdots \times G$ (n times) and $\sim$ denotes the relations

$$(t_0, \ldots, t_n, g_1, \ldots, g_n) \sim \begin{cases} (\ldots, \widehat{t_i}, \ldots, g_i g_{i+1}, \ldots), & \text{if } t_i = 0 \\ (\ldots, t_{i-1} + t_i, \ldots, \widehat{g_i}, \ldots), & \text{if } g_i = e. \end{cases}$$

The interested reader can refer to the cited papers for details. Here are some pertinent facts about this construction.

(1) This definition can be extended to define classifying spaces for associative H-spaces. The role of associativity and applications of these more general classifying spaces can be found in [Stasheff63, 63'], [May75] and [Madsen-Milgram79].

(2) The construction of [Milgram67] is functorial on the category of topological groups. An example of this functoriality at work is the case when one classifies the fibration associated to a closed subgroup $H \hookrightarrow G$.

$$H \hookrightarrow G \to G/H$$

The following diagram commutes:

$$
\begin{array}{ccccc}
H & =\!=\!= & H & \xrightarrow{\ i\ } & G \\
{\scriptstyle i}\downarrow & & \downarrow & & \downarrow \\
G & \longrightarrow & EH & \longrightarrow & EG \\
\downarrow & & \downarrow & & \downarrow \\
G/H & \longrightarrow & BH & \xrightarrow[Bi]{} & BG
\end{array}
$$

where the mapping $G/H \to BH$ is the classifying map. From functoriality and the long exact sequence of homotopy groups, we obtain relations among these spaces. First, the acyclicity of EG implies that $\pi_i(G) \cong \pi_{i+1}(BG)$, or more descriptively, that G has the same homotopy type as ΩBG. Indeed, if G is a simply-connected CW-complex, the Whitehead theorem implies that $G \simeq \Omega BG$. Also observe that the classifying map $G/H \to BH$ acts like the inclusion of the fibre of $Bi \colon BH \to BG$. Thus, for homotopy-theoretic purposes, we have the fibration

$$G/H \xrightarrow{\ \text{inc}\ } BH \xrightarrow{\ Bi\ } BG.$$

This plays an important role in Chapter 8 where we compute $H^*(G/H; k)$.

(3) Finally, suppose G is a discrete group. Then $\pi_0(G) \cong G$ and $\pi_i(G) = \{0\}$ for $i > 0$. Thus $\pi_1(BG) \cong G$ and $\pi_i(BG) = \{0\}$ for $i \neq 1$. It follows that $BG \simeq K(G, 1)$ and EG is its universal covering space. Thus the theory of classifying spaces is a generalization of the theory of covering spaces.

On the cohomology of classifying spaces

With Corollary 6.28, we have reduced the classification of principal G-bundles over a space B to the determination of $[B, BG]$, the set of homotopy classes of maps from B to the classifying space BG. A crude way to study

$[B, BG]$ is to consider the image of elements in this set under the cohomology functor, that is, $[f]: B \to BG$ is sent to

$$[f] \in [B, BG] \quad \mapsto \quad f^*: H^*(BG; R) \to H^*(B; R).$$

We call the image, $f^*(H^*(BG; R)) \subseteq H^*(B; R)$, the **characteristic ring** of the bundle whose classifying map is f; nonisomorphic characteristic rings imply nonequivalent bundles. To make use of the characteristic ring for classification, we need then to compute $H^*(BG; R)$ for various rings R.

We would like to apply the Leray-Serre spectral sequence to the fibration $G \hookrightarrow EG \to BG$. If we can work backward from knowledge of $H^*(G; R)$ and the fact that $H^*(EG; R) \cong R$, then we can compute $H^*(BG; R)$. This situation recalls the computation of $H^*(K(\pi, n); \mathbb{F}_p)$, where we knew the cohomology of the fibre as the inductive hypothesis and proceeded to compute the cohomology of the base space.

The structure of $H^*(G; R)$ is based on the fact that cohomology of an associative H-space of finite type is a graded-commutative Hopf algebra. The algebraic structure of such Hopf algebras was first studied by [Hopf41] for k, a field of characteristic zero and, for other fields, by [Borel53].

Theorem 6.36 (the Hopf-Borel theorem). *Let k denote a field of characteristic p where p may be zero or a prime. A connected Hopf algebra H over k is said to be* **monogenic** *if H is generated as an algebra by 1 and one homogeneous element x of degree > 0. If H is a monogenic Hopf algebra, then*

(1) *if $p \neq 2$ and degree x is odd, then $H \cong \Lambda(x)$,*
(2) *if $p \neq 2$ and degree x is even, then $H \cong k[x]/\langle x^s \rangle$ where s is a power of p or is infinite ($H \cong k[x]$),*
(3) *if $p = 2$, then $H \cong k[x]/\langle x^s \rangle$ where s is a power of 2 or is infinite.*

If k contains the pth root of each of its elements, k is said to be **perfect**. *A graded-commutative Hopf algebra H, of finite type over a perfect field k is isomorphic as an algebra to a tensor product of monogenic Hopf algebras.*

The algebra generators in the isomorphism given in the theorem constitute a simple or p-simple system of generators for H. Thus all Hopf algebras of the type described have p-simple systems of generators.

PROOF: We prove the first part of the theorem for the prime p and the field $\mathbb{F}_p$, which is perfect. The reader can provide the characteristic zero case for himself or herself from the argument here. Chapter 10 also contains the result. If x has odd degree in H, a monogenic graded-commutative Hopf algebra, then $x \cdot x = -x \cdot x$, and if p is odd, then $H \cong \Lambda(x)$. If H has finite dimension and the generator x has even degree (or $p = 2$), then, for some $h > 0$, $x^h = 0$ while $x^{h-1} \neq 0$. Such a positive integer h is called the **height** of the element

x. Since x is also the element of least nonzero degree in H, x is primitive, that is, $\Delta(x) = 1 \otimes x + x \otimes 1$. It follows that

$$
\begin{aligned}
0 = \Delta(x^h) &= (\Delta(x))^h \\
&= (1 \otimes x + x \otimes 1)^h \\
&= \sum_{i=0}^{h} \binom{h}{i} x^i \otimes x^{h-i}.
\end{aligned}
$$

Since $\binom{h}{i}$ is congruent to zero mod p for all $0 < i < h$ if and only if $h = p^s$ for some power s (Lucas's Lemma), the stated form for the monogenic Hopf algebra holds. When H is infinite-dimensional, no power of the generator vanishes and $H \cong k[x]$.

To prove the second half of the theorem, we order the algebra generators of H by degree: $\deg x_1 \leq \deg x_2 \leq \deg x_3 \leq \cdots$. Let A_n denote the subalgebra of H generated by the elements $x_1, \ldots, x_n$ and 1. The degreewise ordering of the generators implies that the coproduct Δ is closed on A_n and so A_n is a sub-Hopf algebra of H. Likewise, A_{n-1} is a sub-Hopf algebra of A_n.

We write $B_n = A_n // A_{n-1}$ for $A_n/(A_{n-1}^+ \cdot A_n)$ where A_{n-1}^+ is the degree positive part of A_{n-1} ($A_{n-1} = A_{n-1}^+ \oplus k$) and $A_{n-1}^+ \cdot A_n$ is the ideal of A_n consisting of products from A_{n-1}^+ and A_n. This quotient is also a Hopf algebra and monogenic with generator $\bar{x}_n$. We admit for the moment that we can choose $x_n \in A_n$ with $x_n \mapsto \bar{x}_n$ by the quotient $A_n \to B_n$ and, more delicately, that the height of x_n is the same as the height of $\bar{x}_n$. Let $\eta\colon B_n \to A_n$ denote the algebra monomorphism $\eta(\bar{x}_n) = x_n$.

Consider the composite:

$$
A_{n-1} \otimes B_n \xrightarrow{\text{inc} \otimes \eta} A_n \otimes A_n \xrightarrow{\mu} A_n \xrightarrow{\Delta} A_n \otimes A_n \xrightarrow{1 \otimes pr} A_n \otimes B_n.
$$

Applied to an element $a \otimes \bar{x}_n$ we get

$$
(1 \otimes pr) \circ \Delta \circ \mu(a \otimes \bar{x}_n) = (1 \otimes pr)(\Delta(a) \cdot \Delta(x_n)) = a \otimes \bar{x}_n + \sum_i y_i' \otimes y_i'',
$$

where the elements y_i' and y_i'' are determined by Δ on x_n and a. In particular, no $y_i' \otimes y_i''$ equals $-a \otimes \bar{x}_n$, and the composite is a monomorphism. Since A_{n-1} together with $x_n = \eta(\bar{x}_n)$ generate all of A_n, the first part of the composite, $A_{n-1} \otimes B_n \xrightarrow{\text{inc} \otimes \eta} A_n \otimes A_n \xrightarrow{\mu} A_n$ is an epimorphism, and hence an isomorphism. This is the inductive step. Writing $A_1 = B_1$, the induction can start and the theorem is proved.

It remains to show that we can choose $x_n \in A_n$ of the same height as $\bar{x}_n \in B_n$. Let's assume that char $k = p$ and either $p = 2$ or $\deg x$ is even. The case of interest is $B_n \cong k[\bar{x}_n]/\langle \bar{x}^{p^s} \rangle$ for some $s > 0$. The problem is that there could be some nontrivial relation—it could happen that $x_n^{p^s} \in A_{n-1}^+ \cdot A_n$

and $x_n^{p^s} \neq 0$ for all choices of x_n that map to $\bar{x}_n$. We show that we can choose a representative for x_n such that $x_n^{p^s} = 0$. Consider the **Frobenius homomorphism** $\zeta \colon H \to H$ given by $\zeta(a) = a^p$. This mapping preserves the coproduct structure, the algebra structure when H is graded-commutative, and takes primitives to primitives. Let $Z = \zeta^s(A_{n-1})$ and form the diagram of Hopf algebras (we have used the property that if $C \subset B \subset A$ is a sequence of normal inclusions of Hopf algebras, then $(A//C)//(B//C) \cong A//B$:

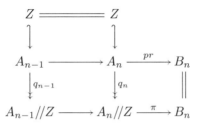

Since π is onto, there is a $y \in A_n//Z$ with $\pi(y) = \bar{x}_n$. We show that $y^{p^s} = 0$. Consider $\Delta(y) = y \otimes 1 + 1 \otimes y + \sum_j a_j' \otimes a_j''$. Since $\pi(a_j') = 0 = \pi(a_j'')$, the elements a_j', a_j'' lie in $A_{n-1}//Z$. Therefore $\Delta(y^{p^s}) = y^{p^s} \otimes 1 + 1 \otimes y^{p^s}$, that is, y^{p^s} is primitive. However, there are no primitives in B_n of degree $p^s \deg \bar{x}_n$ and there are no primitives in $A_{n-1}//Z$ in that dimension. The relations between the terms of the bottom row imply the exactness of the sequence of primitives:

$$0 \to \mathrm{Prim}(A_{n-1}//Z) \to \mathrm{Prim}(A_n//Z) \to \mathrm{Prim}(B_n).$$

Thus $y^{p^s} = 0$. If we let w be an element in A_n with $q_n(w) = y$, then $q_n(w^{p^s}) = y^{p^s} = 0$ and so $w^{p^s} \in Z$. Since $Z = \zeta^s(A_{n-1})$, there is a class $v \in A_{n-1}$ with $w^{p^s} = v^{p^s}$. But then $(w-v)^{p^s} = 0$ and $q_n(w-v) = y$. We take $w - v$ for our representative of x_n and it has the same height as $\bar{x}_n$. $\square$

We apply this theorem to $H^*(G; k)$, which has the structure of a connected Hopf algebra over k when G is a topological group (actually, as in §4.4, it suffices for G to be an H-space with homotopy associative product). If G has the homotopy type of a finite CW-complex, then $H^*(G; k)$, for k perfect, has a decomposition, as an algebra,

$$H^*(G; k) \cong \Lambda(x_1, \ldots, x_r) \otimes k[y_1]/\langle y_1^{p^{h_1}} \rangle \otimes \cdots \otimes k[y_s]/\langle y_s^{p^{h_s}} \rangle$$

with all $h_i < \infty$.

In the case that $\mathrm{char}\, k = 0$, the finite dimensionality of $H^*(G; k)$ implies that $H^*(G; k) \cong \Lambda(x_1, \ldots, x_r)$. This is Hopf's original theorem (and Theorem 10.2). When the integral cohomology, $H^*(G)$, has no torsion at the prime p, then its p-primary information is the reduction of the torsion-free part of $H^*(G)$ and this is calculable from $H^*(G; \mathbb{Q})$.

Proposition 6.37. *If G is a topological group of the homotopy type of a finite CW-complex, k a perfect field of characteristic p, and $H^*(G)$ has no torsion at the prime p, then $H^*(G; k) \cong \Lambda(x_1, \dots, x_r)$ as an algebra over k with each x_i of odd degree.*

In this case, Theorem 3.27 applies to compute $H^*(BG; k)$.

Theorem 6.38. *If G is a topological group of the homotopy type of a finite CW-complex, k a perfect field of characteristic p, and $H^*(G)$ has no torsion at the prime p, $H^*(BG; k) \cong k[y_1, \dots, y_r]$ where y_i corresponds to x_i under the transgression.*

The computations in Chapter 5 provide data for this theorem. In Example 5.F, we computed $H^*(U(n); R) \cong \Lambda(x_1, x_3, \dots, x_{2n-1})$ for any commutative ring of coefficients. For a perfect field k, then, Theorem 6.39 implies that $H^*(BU(n); k) \cong k[c_2, c_4, \dots, c_{2n}]$, where the polynomial generators c_i have degree $c_i = i$ and each c_i corresponds to x_{i-1} under the transgression. By the Universal Coefficient Theorem, we lift the generators to integer coefficients and so $H^*(BU(n)) \cong \mathbb{Z}[c_2, c_4, \dots, c_{2n}]$. The inclusions $U(n) \subset U(n+1) \subset \cdots \subset U$ determine the infinite unitary group. Computing the inverse limit of the homomorphisms induced by the inclusions we find no $\varprojlim^1$-term and so $H^*(BU) \cong \mathbb{Z}[c_2, c_4, c_6, \dots]$.

For the group $SO(n)$ as in Example 5.H (step 3), the Hopf-Borel theorem (and direct calculation) implies that $H^*(SO(n); \mathbb{F}_2)$ has a simple system of generators $\{x_1, \dots, x_{n-1}\}$. By Theorem 6.27, we obtain $H^*(BSO(n); \mathbb{F}_2) \cong \mathbb{F}_2[w_2, \dots, w_n]$. Taking the limit of inclusions $SO(2) \subset SO(3) \subset \cdots \subset SO$, we get $H^*(BSO; \mathbb{F}_2) \cong \mathbb{F}_2[w_2, w_3, w_4, \dots]$.

To compute $H^*(BO(n); \mathbb{F}_2)$ recall that the determinant homomorphism together with the inclusion, $SO(n) \subset O(n)$, gives a sequence of maps

$$SO(n) \hookrightarrow O(n) \xrightarrow{\det} \mathbb{Z}/2\mathbb{Z} \to BSO(n) \xrightarrow{Bi} BO(n) \xrightarrow{B\det} B\mathbb{Z}/2\mathbb{Z}$$

any three of which form a fibration. Thus $BSO(n)$ is a universal covering space for $BO(n)$ and $H^1(BO(n); \mathbb{F}_2) \cong \mathbb{F}_2$ with generator w_1. If we apply the Cartan-Leray spectral sequence for the universal covering $BSO(n) \to BO(n)$, the E_2-term is given by $E_{p,q}^2 \cong H^p(\mathbb{Z}/2\mathbb{Z}; H^q(BSO(n); \mathbb{F}_2))$. Since $\mathbb{Z}/2\mathbb{Z}$ acts trivially on $H^*(BSO(n); \mathbb{F}_2)$ and $H^*(\mathbb{Z}/2\mathbb{Z}, \mathbb{F}_2) \cong H^*(\mathbb{R}P(\infty); \mathbb{F}_2) \cong \mathbb{F}_2[w_1]$, we can rewrite the E_2-term as

$$E_2^{*,*} \cong \mathbb{F}_2[w_1] \otimes_{\mathbb{F}_2} \mathbb{F}_2[w_2, \dots, w_n].$$

By studying auxiliary quotient spaces related to this covering (using Stiefel manifolds), we find that the spectral sequence collapses and so $H^*(BO(n); \mathbb{F}_2) \cong \mathbb{F}_2[w_1, \dots, w_n]$. Furthermore, $H^*(BO; \mathbb{F}_2) \cong \mathbb{F}_2[w_1, w_2, \dots]$.

In these cases, we observe that a classifying map, $f: B \to BU(n)$, or $f: B \to BO(n)$ has its characteristic ring in $H^*(B)$ and $H^*(B; \mathbb{F}_2)$, generated as an algebra by the classes $f^*(c_2), \dots, f^*(c_{2n})$ or $f^*(w_1), \dots, f^*(w_n)$. In the next section, we relate these generators to the characteristic classes.

Applications

We turn our attention, once again, to vector bundles and the problem of computing the characteristic classes associated to a bundle. Using the vector space structure on each fibre, we define the following operation on vector bundles:

Definition 6.39. *Suppose $\xi = (\mathbb{R}^n, E, p, B)$ and $\eta = (\mathbb{R}^m, E', p', B)$ are two vector bundles over a space B. The **Whitney sum** of ξ and η, $\xi \oplus \eta = (\mathbb{R}^{n+m}, E \oplus E', p \oplus p', B)$ is the vector bundle given by the pullback diagram:*

$$
\begin{array}{ccc}
\mathbb{R}^{n+m} & \xrightarrow{\cong} & \mathbb{R}^n \times \mathbb{R}^m \\
\downarrow & & \downarrow \\
E \oplus E' & \longrightarrow & E \times E' \\
{\scriptstyle p \oplus p'} \downarrow & & \downarrow {\scriptstyle p \times p'} \\
B & \xrightarrow{\text{diag}} & B \times B
\end{array}
$$

The definition allows us to take the fibres of ξ and η over a point in B and form their direct sum over that point. The pullback operation provides the appropriate topological glue.

Stiefel-Whitney classes for a vector bundle ξ, $w_i(\xi)$ in $H^i(M; \mathbb{F}_2)$, can be described constructively in the obstruction theory framework and they can be constructed via the action of the Steenrod algebra in conjunction with the Thom isomorphism as proved by [Thom52] based on the work of [Wu50]. We describe these classes uniquely through a set of axioms first suggested by [Hirzebruch66].

Axioms for Stiefel-Whitney classes. *Let $\xi = (\mathbb{R}^n, E, p, M)$ be a vector bundle over a base space M.*

I. *There are classes $w_i(\xi) \in H^i(M; \mathbb{F}_2)$ and $w_0(\xi) = 1, w_j(\xi) = 0$ for $j > n$.*

II. *If $f: B_1 \to B_2$ is a continuous mapping that is covered by a bundle map $(f, \tilde{f}): \xi_1 \to \xi_2$ that is a linear isomorphism on the fibres, then $w_i(\xi_1) = f^*(w_i(\xi_2))$.*

III. *For two vector bundles, ξ and η, the **Whitney sum formula** holds;*

$$ w_k(\xi \oplus \eta) = \sum_{i+j=k} w_i(\xi) \smile w_j(\eta). $$

IV. *Let $\gamma_1^1 = (\mathbb{R}, E, p, \mathbb{R}P(1))$ be the canonical line bundle over $\mathbb{R}P(1)$, then $w_1(\gamma_1^1) \neq 0$.*

The proof that these axioms uniquely determine the Stiefel-Whitney classes can be found in the book of [Milnor-Stasheff74, p. 86]. We establish the existence of Stiefel-Whitney classes by constructing them from the cohomology of the appropriate classifying space. To wit, let $\xi = (\mathbb{R}^n, E, p, M)$ be a vector bundle over M and suppose ξ is classified by a map $f_\xi \colon M \to BO(n)$, that is, following Theorem 6.34, ξ is equivalent to $f_\xi^* \chi(n)[\mathbb{R}^n]$ where $\chi(n) \colon EO(n) \to BO(n)$ denotes the universal principal $O(n)$-bundle and $f_\xi^* \chi(n)$ is the pullback over f_ξ. We define classes in $H^*(M; \mathbb{F}_2)$ for ξ by the formula

$$w_i(\xi) = f_\xi^*(w_i)$$

where $w_i \in H^i(BO(n); \mathbb{F}_2)$ is one of the transgressive generators identified in Theorem 6.38.

Theorem 6.40. *The cohomology classes defined by the classifying map satisfy the axioms for the Stiefel-Whitney classes.*

SKETCH OF A PROOF: We provide a series of observations (with references) as a complete proof would take us too far afield. First notice that Axiom I holds trivially from our definition and the cohomology of $BO(n)$.

Axiom II, the naturality of the Stiefel-Whitney classes, follows from the universal properties of the pullback operation and the definition of a bundle map to replace the bundle ξ_1 with $f^* \xi_2$. Then the classifying map for ξ_1, $f_{\xi_1} \colon B_1 \to BO$, is homotopic to $f_{\xi_2} \circ f$.

$$
\begin{array}{ccccccc}
E(\xi_1) & \xrightarrow{\tilde{f}} & E(\xi_2) & \longrightarrow & EO & \longleftarrow & E(\xi_1) \\
\downarrow & & \downarrow & & \downarrow & & \downarrow \\
B_1 & \xrightarrow{f} & B_2 & \xrightarrow{f_{\xi_2}} & BO & \xleftarrow{f_{\xi_1}} & B_1
\end{array}
$$

The relations between the Stiefel-Whitney classes follow.

Axiom III, the Whitney sum formula, hangs on an extra piece of structure. Consider the mapping $\varphi_{n,m} \colon O(n) \times O(m) \to O(n+m)$ defined on matrices $A \in O(n)$ and $B \in O(m)$ by

$$\varphi_{n,m}(A, B) = \begin{pmatrix} A & 0 \\ 0 & B \end{pmatrix}.$$

This group homomorphism induces a mapping, $B\varphi_{n,m} \colon BO(n) \times BO(m) \to BO(n+m)$, whose key property is recorded in the following lemma.

Lemma 6.41. *The homomorphism,*

$$(B\varphi_{n,m})^* \colon H^*(BO(n+m); \mathbb{F}_2) \to H^*(BO(n); \mathbb{F}_2) \otimes H^*(BO(m); \mathbb{F}_2),$$

satisfies $(B\varphi_{n,m})^*(w_i) = \sum_{r+s=i} w_r \otimes w_s$.

A proof of this lemma may be found in [Borel53', §6]. The crucial feature of Borel's proof is the identification of an important abelian subgroup of $O(n)$, $Q(n)$, of diagonal matrices with entries ± 1. The subgroup $Q(n)$ is isomorphic to the product of n copies of $\mathbb{Z}/2\mathbb{Z}$ and we can examine the induced mapping B inc: $BQ(n) \to BO(n)$. Since $BQ(n)$ is the product of n copies of $B\mathbb{Z}/2\mathbb{Z} = \mathbb{R}P(\infty)$, we know its mod 2 cohomology: $H^*(BQ(n); \mathbb{F}_2) \cong \mathbb{F}_2[y_1, \ldots, y_n]$ where $\deg y_i = 1$ for all i. The representation theory of Lie groups may be pressed into service to identify the image of

$$(B \operatorname{inc})^* \colon H^*(BO(n); \mathbb{F}_2) \to H^*(BQ(n); \mathbb{F}_2)$$

as the symmetric functions in $\{y_1, \ldots, y_n\}$, $\operatorname{Sym}(y_1, \ldots, y_n)$. By a counting argument, $(B \operatorname{inc})^*$ is an isomorphism of $H^*(BO(n); \mathbb{F}_2)$ onto the algebra $\operatorname{Sym}(y_1, \ldots, y_n)$. From this rather deep structural result, the lemma follows.

The Whitney sum formula can now be proved by examining the pullback diagram that describes the associated principal bundle for $\xi \oplus \eta$:

$$
\begin{array}{ccccccc}
E(\operatorname{Prin}(\xi \oplus \eta)) \to f_\xi^* \chi(n) \times f_\eta^* \chi(m) & \to & EO(n) \times EO(m) & \longrightarrow & EO(n+m) \\
\downarrow & & \downarrow & & \downarrow & & \downarrow \\
M \xrightarrow{\ \text{diag}\ } M \times M & \xrightarrow{f_\xi \times f_\eta} & BO(n) \times BO(m) & \xrightarrow[B\varphi_{n,m}]{} & BO(n+m)
\end{array}
$$

Finally, we establish Axiom IV for the canonical line bundle over $\mathbb{R}P(1)$ by considering the geometry of the bundle γ_1^1 carefully. The canonical line bundle over $\mathbb{R}P(n)$ can be described by the total space,

$$E(\gamma_1^1) = \{\, (\{\pm \vec{x}\}, t\vec{x}), \text{ where } \vec{x} \in S^n \text{ and } t \in \mathbb{R} \,\};$$

the first projection provides the bundle map. For $n = 1$, this description identifies $E(\gamma_1^1)$ as the open Möbius band. It is an elementary fact that $E(\gamma_1^1)$ is not homotopy equivalent to $\mathbb{R}P(1) \times \mathbb{R}$, the total space of the trivial bundle. On the classifying space level, we can identify $w_1(\gamma_1^1)$ with $[f_{\gamma_1^1}]$ in $[\mathbb{R}P(1), BO(1)]$. Since $O(1) \cong \mathbb{Z}/2\mathbb{Z}$, we compute: $[\mathbb{R}P(1), BO(1)] = [\mathbb{R}P(1), B\mathbb{Z}/2\mathbb{Z}] = [\mathbb{R}P(1), K(\mathbb{Z}/2\mathbb{Z}, 1)] = H^1(\mathbb{R}P(1); \mathbb{F}_2)$. Finally, $\mathbb{R}P(1)$ is homeomorphic to S^1, and so $w_1(\gamma_1^1)$ is either the generator of H^1 or zero. If $w_1(\gamma_1^1) = 0$, then $f_{\gamma_1^1}$ would be null-homotopic and γ_1^1 would be equivalent to the trivial bundle. Since this would imply the equivalence of the total spaces, the observation shows that $w_1(\gamma_1^1) \neq 0$. $\qquad\square$

Similar axioms were given by [Hirzebruch66] for the **Chern classes** of a complex vector bundle over a space M. In the original paper of [Chern48], these classes are constructed using de Rham cohomology and specific differential forms, which have relevance to developments in mathematical physics (see also [Milnor-Stasheff74, appendix C], [Dupont78], or [Bott85]). In our formulation, the classifying space for n-dimensional complex bundles is $BU(n)$ and so we can take the construction of the Stiefel-Whitney classes and carry it over to define, for ξ a complex bundle,

$$c_{2i}(\xi) = f_\xi^*(c_{2i}) \in H^{2i}(M).$$

Notice that the Chern classes lie in the integral cohomology of M (a fact related to the quantization of magnetic charge; see the papers of [Dirac35], [Atiyah79], and, for an elementary exposition of these topics, [McCleary92]).

For a complex vector bundle, $\xi = (\mathbb{C}^n, E, p, M)$ we have its Chern classes $c_{2i}(\xi) \in H^{2i}(M)$. These classes reduce mod 2 to determine classes $\bar{c}_{2i}(\xi) \in H^{2i}(M; \mathbb{F}_2)$. Furthermore, by identifying $\mathbb{C}^n$ with $\mathbb{R}^{2n}$, we obtain Stiefel-Whitney classes for ξ, $w_i(\xi) \in H^i(M; \mathbb{F}_2)$. How do these reduced Chern classes, $\bar{c}_{2i}(\xi)$, relate to the Stiefel-Whitney classes?

Theorem 6.42. *For a complex vector bundle, $\xi = (\mathbb{C}^n, E, p, M)$, the Stiefel-Whitney classes of the associated $2n$-dimensional real bundle satisfy the relations $w_{2i+1}(\xi) = 0$ and $w_{2i}(\xi) = \bar{c}_{2i}(\xi)$, for $0 \le i \le n$.*

PROOF: The theorem follows if these relations hold amid the universal classes. Consider the orthogonal group $O(2m)$ of metric preserving linear transformations of $\mathbb{R}^{2m}$. The identification of $\mathbb{C}^m$ with $\mathbb{R}^{2m}$ allows us to treat $U(m)$ as a subgroup of $O(2m)$. Consider the subgroup $O(2m-1) \subset O(2m)$. If we take the intersection, $O(2m-1) \cap U(m)$, because $U(m)$ acts on coordinates "pairwise," we get $O(2m-1) \cap U(m) = U(m-1)$. Furthermore, this intersection behaves correctly with respect to the inclusion and so we get the diagram:

$$
\begin{array}{ccc}
U(m-1) & \xrightarrow{\ \subset\ } & O(2m-1) \\
\downarrow & & \downarrow \\
U(m) & \xrightarrow{\ \subset\ } & O(2m) \\
\downarrow & & \downarrow \\
\end{array}
$$

$$U(m)/U(m-1) = S^{2m-1} = O(2m)/O(2m-1).$$

This gives a mapping of fibrations

$$
\begin{array}{ccccc}
S^{2m-1} & \hookrightarrow & BU(m-1) & \xrightarrow{B\,\text{inc}} & BU(m) \\
\| & & B(\subset)\downarrow & & B(\subset)\downarrow \\
S^{2m-1} & \hookrightarrow & BO(2m-1) & \xrightarrow[B\,\text{inc}]{} & BO(2m).
\end{array}
$$

In the Leray-Serre spectral sequence for these fibrations the class $s_{2m-1} \in E_2^{0,2m-1}$ corresponding to the generator of $H^{2m-1}(S^{2m-1}; \mathbb{F}_2)$ transgresses to $w_{2m} \in H^{2m}(BO(2m); \mathbb{F}_2)$ and $\bar{c}_{2m} \in H^{2m}(BU(m); \mathbb{F}_2)$, respectively. If we identify $BU(1)$ with $BSO(2)$, then we can prove inductively that

$$H^*(BO(2m); \mathbb{F}_2) \xrightarrow{B(\subset)^*} H^*(BU(m); \mathbb{F}_2)$$

takes w_{2i+1} to zero and w_{2i} to $\bar{c}_{2i}$ by the naturality of the transgression modulo its indeterminacy.

The theorem follows by observing that the classifying map for ξ, as a $2n$-dimensional real vector bundle, factors through $BU(n) \to BO(2n)$. ☐

An immediate geometric corollary of these relations is the following result.

Corollary 6.43. *Suppose M is a real $2n$-dimensional manifold with tangent bundle τ_M. If M can be obtained from an n-dimensional complex manifold by the usual identifications, then $w_{2i+1}(\tau_M) = 0$ and $w_{2i}(\tau_M)$ is the reduction of an integral class for $0 \le i \le n$.*

Thus the Stiefel-Whitney classes provide obstructions to the existence of a complex manifold structure on an even-dimensional real manifold. The satisfaction of these conditions is not sufficient to guarantee a complex structure (see the paper of [Massey61]) but these classes provide a first set of obstructions.

In this short discussion of classifying spaces and characteristic classes we have touched on very few topics in a rich area of study. The reader who is interested in these matters will find a wealth of applications and deeper developments in the literature. The role of spectral sequences in this work has been firmly established.

6.4 Other constructions of the spectral sequence Ⓝ

In Chapter 5, we constructed the Leray-Serre spectral sequence for simplicial singular homology and identified the E^2-term. In this section we discuss some different settings that lead to the construction of a spectral sequence for fibrations. We supply some of the details of an elegant construction due to [Dress67]. The apparatus introduced there has other applications in topology; for example, it provided [Singer73] and [Turner98] with a useful framework in which to construct Steenrod operations in spectral sequences.

The motivating examples of fibre spaces are the homogeneous spaces, $H \hookrightarrow G \to G/H$, where G is a compact, connected, Lie group and H a closed subgroup. Investigations of the topological invariants of Lie groups had proceeded using analytic tools like de Rham cohomology and the associated Lie algebra cohomology ([Koszul50], [Cartan51]). The problem of relating the homology groups of the fibre, base and total space was undertaken by J. Leray in a series of boldly original papers that appeared from 1946 to 1950 ([Leray46,

50]). Leray based his investigations on a cohomology theory that generalized features of de Rham cohomology to more topological settings.

Suppose $f: E \to X$ is a surjective mapping where X is a compact polyhedron with a *good* covering, $\mathcal{U}$; *good* is taken here to mean that any finite intersection, $U_{\alpha_1} \cap U_{\alpha_2} \cap \cdots \cap U_{\alpha_n}$, of subsets in the cover is contractible. Because f is surjective, this covering lifts to a cover of E, $f^{-1}\mathcal{U}$. As an instance of Leray's theory, we can consider the Čech cochains determined by this cover with coefficients in the constant presheaf R, a commutative ring, $\check{C}^*(f^{-1}\mathcal{U}; R)$. (For definitions of Čech theory and presheaves see [Spanier66].) [Leray50] studied the cochains $\check{C}^*(f^{-1}\mathcal{U}; R)$ within a product structure (of the *couverture* with the sheaf) that compares with $\check{C}^*(\mathcal{U}; \check{C}^*(f^{-1}U_*; R))$, the cochains on X with coefficients in the nonconstant sheaf $\check{C}^*(f^{-1}U_*; R)$. The subsequent filtration induced by weights leads to the spectral sequence. When f is a fibration, $f^{-1}(U_\alpha) \simeq F$, the fibre, and so we get a system of local coefficients on X. The spectral sequence associated to the filtration has E_2-term identifiable with $\check{H}^*(X; \check{\mathcal{H}}^*(F; R))$, where F is the fibre of the mapping $f: E \to X$. The morass of algebra in [Leray46] was greatly simplified by [Koszul47] who introduced the standard construction of a spectral sequence. The desired applications to the structure of Lie groups and homogeneous spaces soon followed in work by [Leray50], [Koszul50], [Cartan51], and [Borel53]. A careful modern treatment of this construction may be found in [Bott-Tu82]. A discussion of the development of Leray's work is given by [Borel98].

The next development was a spectral sequence of this kind for simplicial singular homology and cohomology, which had been recognized as more versatile tools in homotopy theory. This was accomplished by [Serre51] in his celebrated thesis. A key technical feature of Serre's work is the determination of the homology spectral sequence. Serre established this result and the multiplicative structure of the cohomology spectral sequence using cubical cochains on a space rather than the classical simplicial cochains. That these multiplicative properties held for simplicial singular theory was proved by [Gugenheim-Moore57] in their thorough study of singular theory applied to fibrations via the method of acyclic models.

While we are discussing the various manifestations of classical cohomology, it is natural to ask about analogous results for generalized cohomology theories. Though beyond the intended scope of this book, we mention that the filtration by cellular skeleta of a CW-complex leads to a spectral sequence introduced by [Atiyah-Hirzebruch69] (though known to [Lima59] and [Whitehead, GW62]) for the computation of the generalized cohomology of the complex. Analogous results for a fibration lead to a generalization of this spectral sequence with coefficients in the generalized cohomology of the fibre (see [Dyer69], [Switzer75], or [Prieto79]). For more details we refer the reader to §11.2.

Further developments of the structure of fibrations revealed other features of the Leray-Serre spectral sequence. One result in this vein is due to [Fadell-Hurewicz58].

Theorem 6.44. *Suppose (F, E, p, B) is a fibration and B is n-connected for some $n \geq 2$. Then, in the associated homology spectral sequence, the differentials, d^i, are trivial for $2 \leq i \leq n$. Furthermore, for all $p \geq 0$, $q \geq 0$,*

$$H_p(B; H_q(F; G)) \cong E^{n+1}_{p,q} \xrightarrow{d^{n+1}} E^{n+1}_{p-n-1,q+n} \cong H_{p-n-1}(B; H_{q+n}(F; G))$$

is given by $d^{n+1}(x) = \gamma \cap x$ where the class γ is the fundamental class of B in $H^{n+1}(B; \pi_{n+1}(B))$ and the cap product is defined by a suitable pairing $\cap \colon \pi_{n+1}(B) \otimes H_q(F; G) \to H_{q+n}(F; G)$.

The differential in the theorem is similar in spirit to the transgression—in both cases, a geometric interpretation is found for a particular differential in the spectral sequence. These results were proved independently by [Shih62]. The proof of [Fadell-Hurewicz58] uses the lifting function for the fibration and the pairing takes the form $\cap \colon H_n(\Omega B; G) \otimes H_q(F; G) \to H_{q+n}(F; G)$, induced by evaluation at 1 (see §4.3).

Another basic viewpoint from which to study the algebraic topology of fibrations is the Eilenberg-Zilber theorem. In particular, does the local product structure of a fibre space mean we can compare $C_*(E)$ with $C_*(B)$ and $C_*(F)$? In the case of a trivial fibration, we are comparing $C_*(B \times F)$ with $C_*(B) \otimes C_*(F)$ and the Eilenberg-Zilber theorem provides a mapping at the chain level that induces an isomorphism $H_*(B \times F) \cong H_*(B) \otimes H_*(F)$. For an arbitrary fibration, [Hirsch54] generalized the Eilenberg-Zilber theorem by showing, when k is a field, that a differential can be constructed on $C^*(B; k) \otimes H^*(F; k)$ along with a homomorphism $C^*(B; k) \otimes H^*(F; k) \to C^*(E; k)$ inducing an isomorphism, $H(C^*(B; K) \otimes H^*(F; k)) \cong H^*(E; k)$. This result provides a smaller cochain complex for $H^*(E; k)$. We could also try to provide $C_*(B) \otimes C_*(F)$ with a new differential, D, so that $H(C_*(B) \otimes C_*(F), D) \cong H_*(E)$. Such a generalization was obtained by [Brown, E59]. Two pieces of the structure of a fibration come into play: The first is the *lifting function* for a fibration and the action of ΩB on the fibre F it provides, $\mu \colon \Omega B \times F \to F$, $\mu(\alpha, x) = \lambda(\alpha, x)(1)$. At the chain level, this induces an action of $C_*(\Omega B)$ on $C_*(F)$, $C_*(\Omega B) \otimes C_*(F) \to C_*(F)$. The second piece of structure introduced in [Brown, E59] is the notion of a **twisting cochain**, $\Phi \colon C_*(B) \to C_{*-1}(\Omega B)$. We give the definition in the more general setting of differential homological algebra.

Definition 6.45. *Let (C, d_C) denote a supplemented, differential graded coalgebra and (A, d_A) an augmented, differential graded algebra, $\eta \colon R \to C_0$ and $\varepsilon \colon A_0 \to R$ the supplementation and augmentation, respectively. An R-module homomorphism $\tau \colon C \to A$ of degree -1 is a **twisting cochain** if $\tau\eta = 0 = \varepsilon\tau$ and $d_A\tau = (-1)^n \tau d_C + \tau \smile \tau$ where $\tau \smile \tau$ is the composite*

$$C \xrightarrow{\text{coalgebra}} C \otimes C \xrightarrow{\tau \otimes \tau} A \otimes A \xrightarrow{\text{algebra}} A.$$

The diagonal and Alexander-Whitney maps provide $C_*(B)$ with a natural coalgebra structure and the multiplication of loops in ΩB provides $C_*(\Omega B)$ with an algebra structure. [Brown, E59] constructed a twisting cochain in this case using the simplicial structure of singular chains and acyclic models. Suppose $\Phi \colon C_*(B) \rightarrow C_{*-1}(\Omega B)$ is a twisting cochain; this gives a differential on $C_*(B) \otimes C_*(F)$ as the composite

$$d_\Phi \colon C_*(B) \otimes C_*(F) \xrightarrow{\text{coalgebra} \otimes 1} C_*(B) \otimes C_*(B) \otimes C_*(F)$$
$$\xrightarrow[1 \otimes \Phi \otimes 1]{} C_*(B) \otimes C_*(\Omega B) \otimes C_*(F) \xrightarrow[1 \otimes \mu_*]{} C_*(B) \otimes C_*(F).$$

The properties of the twisting cochain imply that $D = d_\otimes - d_\Phi$ is a differential on $C_*(B) \otimes C_*(F)$. The main theorem proved by [Brown, E59] is that the chain complex $(C_*(B) \otimes C_*(F), d_\otimes - d_\Phi)$ is chain equivalent to $C_*(E)$ and hence has the same homology.

There is a filtration of $(C_*(B) \otimes C_*(F), D)$ as a double complex

$$F_p = \bigoplus_{q \leq p} C_q(B) \otimes C_*(F).$$

The associated spectral sequence converges, by Brown's theorem, to $H_*(E)$. The properties of a twisting cochain, along with Theorem 2.15 imply that the the E^2-term is isomorphic to $H_*(B; \mathcal{H}_*(F; R))$. Brown carefully compared this spectral sequence with Serre's to prove that they are isomorphic. Furthermore, Theorem 6.44 is a corollary of this comparison of spectral sequences. The notion of a twisting cochain was introduced by [Cartan54] as the algebraic analogue of the simplicial twisting functions introduced by [Moore56]. The interested reader can consult the book of [May67] or the paper of [Husemoller-Moore-Stasheff74] for more details. An example of an explicit spectral sequence calculation using this level of control of the first differential can be found in [McCleary90].

An algebraic development founded on these ideas is known as *homological perturbation theory* [Gugenheim-Lambe-Stasheff91]. The main idea is to generalize the relation between $d_\otimes$ and $d_\otimes - d_\Phi$ on $C_*(F) \otimes C_*(B)$. Given a tensor product of differential graded modules, $(M, d_M) \otimes (N, d_n)$, there is the tensor differential on $M \otimes N$ given by $d_\otimes = d_M \otimes 1 + (\pm)1 \otimes d_N$. A *perturbation of $d_\otimes$ with respect to a decreasing filtration F_* of $M \otimes N$ is a sum

$$D = d_\otimes + d_2 + d_3 + \cdots,$$

with $d_s(F_k) \subset F_{k-s}$. The condition $D \circ D = 0$ places a strong restriction on the homomorphisms d_i. When a perturbation exists there is a spectral sequence with E_2-term given by $H(M \otimes N, d_\otimes)$ and converging to $H(M \otimes N, D)$. Furthermore, the differentials in the spectral sequence are closely related to the d_i in the sum. [Gugenheim60] gave this formulation of the results of [Brown,

E59] and [Hirsch54]. Since then a very elegant and satisfactory theory has been developed. See the papers of [Lambe92] and [Huebschmann-Kadeishvili91] for recent reviews of the subject, as well as the work of [Chen77] and [Gugenheim82].

Bisimplicial sets and Dress' construction ⓝ

The idea of a locally trivial product with a global twisting is realized algebraically in the twisted tensor product (the tensor product with the new differential). We next describe a more geometric construction that leads to the Leray-Serre spectral sequence by exploiting the local product structure. The construction is due to [Dress67]. The double complex involved was introduced in [Fadell-Hurewicz58] without the explicit simplicial structure. There is also a nice discussion of this work to be found in [Liulevicius67].

Our point of departure is an object that carries the local product data of a fibration in an fashion analogous to the way that the simplicial structure of a space carries its homology.

Definition 6.46. *A **bisimplicial set**, $X_{\bullet\bullet}$, is a bigraded collection of sets, $X_{p,q}$, for $p, q \geq 0$, together with vertical and horizontal face and degeneracy maps, $\partial_i^h \colon X_{p,q} \to X_{p-1,q}$, $s_j^h \colon X_{p,q} \to X_{p+1,q}$, $\partial_k^v \colon X_{p,q} \to X_{p,q-1}$, and $s_l^v \colon X_{p,q} \to X_{p,q+1}$, indexed over the usual sets, commuting with each other and satisfying the standard simplicial identities in each direction.*

A bisimplicial set may also be defined as a simplicial object over the category of simplicial sets (*a simplicial simplicial set*). The fundamental example is the bisimplicial set $\Delta_{\bullet\bullet}$ with $\Delta_{pq} = \Delta^p \times \Delta^q$ and $\partial_i^h, s_j^h, \partial_k^v, s_l^v$, the usual face and degeneracy maps.

To an epimorphism of spaces, $f \colon E \to B$, we associate a bisimplicial set $K_{\bullet\bullet}(f)$ as follows: Let $K_{pq}(f) = \{(u, v) \mid u \colon \Delta^p \times \Delta^q \to E, v \colon \Delta^p \to B$ continuous with $f \circ u = v \circ pr_1\}$; here pr_1 is the first projection mapping. Thus (u, v) is in $K_{pq}(f)$ if the following diagram commutes

$$
\begin{array}{ccc}
\Delta^p \times \Delta^q & \xrightarrow{\ u\ } & E \\
{\scriptstyle pr_1}\downarrow & & \downarrow{\scriptstyle f} \\
\Delta^p & \xrightarrow[\ v\ ]{} & B.
\end{array}
$$

Let (u, v) be in $K_{pq}(f)$. The face maps

$$\partial_i^h \colon K_{pq}(f) \to K_{p-1,q}(f) \quad \text{and} \quad \partial_k^v \colon K_{pq}(f) \to K_{p,q-1}(f)$$

are described by the diagrams for $\partial_i^h(u, v)$ and $\partial_k^v(u, v)$;

$$\partial_i^h(u, v): \quad \begin{array}{ccccc} \Delta^{p-1} \times \Delta^q & \xrightarrow{\varepsilon_i \times 1} & \Delta^p \times \Delta^q & \xrightarrow{u} & E \\ {\scriptstyle pr_1} \downarrow & & {\scriptstyle pr_1} \downarrow & & \downarrow {\scriptstyle f} \\ \Delta^{p-1} & \xrightarrow{\varepsilon_i} & \Delta^p & \xrightarrow{v} & B \end{array}$$

$$\partial_k^v(u, v): \quad \begin{array}{ccccc} \Delta^p \times \Delta^{q-1} & \xrightarrow{1 \times \varepsilon_k} & \Delta^p \times \Delta^q & \xrightarrow{u} & E \\ {\scriptstyle pr_1} \downarrow & & {\scriptstyle pr_1} \downarrow & & \downarrow {\scriptstyle f} \\ \Delta^p & = = = = = & \Delta^p & \xrightarrow{v} & B. \end{array}$$

Here ε_j denotes the inclusion of the jth-face of the image simplex. The degeneracies are defined similarly.

Apply the free abelian group functor $\mathcal{F}A$ to $K_{\bullet\bullet}(f)$ to get a bisimplicial group that we denote by $S_{pq}(f) = \mathcal{F}A(K_{pq}(f))$. This bisimplicial group determines a double complex by taking $d': S_{pq}(f) \to S_{p-1,q}(f)$ and $d'': S_{pq}(f) \to S_{p,q-1}(f)$ to be

$$d' = \sum_{j=0}^p (-1)^j \mathcal{F}A(\partial_j^h) \quad \text{and} \quad d'' = \sum_{k=0}^q (-1)^{q+k} \mathcal{F}A(\partial_k^v).$$

The signs and the simplicial identities conspire to give $d' \circ d' = d'' \circ d'' = d' \circ d'' + d'' \circ d' = 0$. When we apply the analogue of Theorem 2.15 for such double complexes, we obtain two spectral sequences.

Theorem 6.47. *For the double complex* $\{S_{*,*}(f), d', d''\}$, *the following hold;*

$$H_p(H_q(S_{*,*}(f), d''), \bar{d}') \cong \begin{cases} H_q(E), & \text{if } p = 0, \\ 0, & \text{if } p \neq 0. \end{cases}$$

If f is a fibration with fibre F, then

$$H_p(H_q(S_{*,*}(f), d'), \bar{d}'') \cong H_p(B; \mathcal{H}_q(F))$$

where $\mathcal{H}_*(F)$ is the local coefficient system given by the homology of the fibre over each point in B.

Thus the double complex $(S_{*,*}(f), d', d'')$ gives rise to the Leray-Serre spectral sequence. This approach provides considerable simplicity of the construction of the spectral sequence as well as a framework for studying the spectral sequence for various chain and cochain functors that support classical homology and cohomology, for example, de Rham cohomology (see [Grivel79] and [McCleary82]). The argument given here can also be generalized to obtain the spectral sequence for fibrations of pairs ([Liulevicius67]).

PROOF: We display the double complex $S_{*,*}(f)$ as follows

$$
\begin{array}{cccc}
\downarrow & \downarrow & \downarrow & \downarrow \\
S_{20}(f) \longleftarrow S_{21}(f) \longleftarrow S_{22}(f) \longleftarrow S_{23}(f) \longleftarrow \\
\downarrow & \downarrow & \downarrow & \downarrow \\
S_{10}(f) \longleftarrow S_{11}(f) \longleftarrow S_{12}(f) \longleftarrow S_{13}(f) \longleftarrow \\
d' \downarrow & \downarrow & \downarrow & \downarrow \\
S_{00}(f) \underset{d''}{\longleftarrow} S_{01}(f) \longleftarrow S_{02}(f) \longleftarrow S_{03}(f) \longleftarrow
\end{array}
$$

First consider $H_*(S_{i*}(f), d'')$. This is clearly the homology of the set $K_{i*}(f)$. Furthermore, $H_*(S_{0*}(f), d'') = H_*(E)$ follows from the definition of $K_{0q}(f)$ and that of the singular complex for E. To establish the first part of Theorem 6.47, we examine the complex of graded abelian groups:

$$
H_*(S_{0*}(f), d'') \xleftarrow{\partial_{0*} - \partial_{1*}} H_*(S_{1*}(f), d'') \xleftarrow{\partial_{0*} - \partial_{1*} + \partial_{2*}} H_*(S_{2*}(f), d') \leftarrow
$$

Lemma 6.48. *For all $j \geq 0$, $K_{0*}(f) \simeq K_{j*}(f)$.*

PROOF: In fact, we prove a bit more by showing that there are mappings $J_j \colon K_{0,*}(f) \to K_{j,*}(f)$ that are homotopy equivalences and that the following diagram commutes up to homotopy

$$
\begin{array}{ccc}
K_{0,*}(f) & = & K_{0,*}(f) \\
J_j \downarrow & & \downarrow J_{i-1} \\
K_{j,*}(f) & \xrightarrow{\partial_j} & K_{j-1,*}(f).
\end{array}
$$

The mappings, J_j, are induced by taking the collapse map $\Delta^j \to \Delta^0$, which is a homotopy equivalence with homotopy inverse $\Delta^0 \to \Delta^j$ defined by taking Δ^0 to the barycenter of Δ^j. The maps induced by the collapse and its inverse give a homotopy equivalence. Furthermore, this homotopy equivalence is chosen to be compatible with the face maps $\varepsilon \colon \Delta^j \to \Delta^{j+1}$ that induce the simplicial structure on $K_{\bullet\bullet}(f)$.

It follows that the homotopy equivalence induces a mapping of simplicial spaces:

$$
\begin{array}{ccccccc}
\xrightarrow{\;\text{id}\;} & & \xrightarrow{\;\text{id}\;} & & & & \\
\vdots\; & K_{0*} & \xrightarrow{\;\text{id}\;} & K_{0*} & \xrightarrow{\;\text{id}\;} & K_{0*} & \\
\xrightarrow{\;\;\;} & \;\Big\downarrow J_2 & \xrightarrow{\;\text{id}\;} & \;\Big\downarrow J_1 & \xrightarrow{\;\text{id}\;} & =\Big\| J_0 & \\
& & & & & & \\
\xrightarrow{\;\partial_0\;} & & \xrightarrow{\;\partial_0\;} & & \xrightarrow{\;\partial_0\;} & & \\
\vdots\; & K_{2*} & \xrightarrow{\;\partial_1\;} & K_{1*} & \xrightarrow{\;\partial_1\;} & K_{0*}(f). & \\
\xrightarrow{\;\;\;} & & \xrightarrow{\;\partial_2\;} & & & &
\end{array}
$$

The mappings J_j induce the identity on $H_*(E)$ for all i and so mappings induced by ∂_i may be taken to be the identity for all i. $\qquad\square$

Applying the result of the lemma, the graded complex that computes $H_*(S_{**}(f), d'')$ can be identified as the complex:

$$
H_*(E) \xleftarrow{\;0\;} H_*(E) \xleftarrow{\;\text{id}\;} H_*(E) \xleftarrow{\;0\;} H_*(E) \xleftarrow{\;\text{id}\;} \cdots
$$

and so part (i) of the theorem is established.

We sketch a proof of the second half of Theorem 6.47. Examine a (p,q)-simplex, (u,v).

$$
\begin{array}{ccc}
\Delta^p \times \Delta^q & \xrightarrow{\;\;u\;\;} & E \\
{\scriptstyle pr_1}\Big\downarrow & & \Big\downarrow{\scriptstyle f} \\
\Delta^p & \xrightarrow[\;\;v\;\;]{} & B.
\end{array}
$$

This identifies a p-simplex of B and a family of q-simplices of $f^{-1}(v(x_0))$, parametrized by points x_0 in Δ^p. Since Δ^p is contractible to its leading vertex, if f is a fibration, then $f^{-1}(v(x_0))$ is homotopy-equivalent to $f^{-1}(v(e_0)) = F_{v(e_0)}$, the fibre over $v(e_0)$, the image of the leading vertex of Δ^p. Fix $v \colon \Delta^p \to B$ and consider the set $\{\text{chains } \sum_j a_j(u_j, v) \in S_{pq}(f), \text{ for any } q\}$. Each (u_j, v) identifies a mapping, $\Delta^q \to f^{-1}(v(e_0 \in \Delta^p))$, and so the homology, with respect to d', of this set over v is $H_*(f^{-1}(v(e_0 \in \Delta^p))) \cong H_*(F_{v(e_0)})$. Each v and each (u_j, v) contribute an generating chain to $C_p(B; \mathcal{H}_*(F))$ (see §5.3 for the relevant definitions) and so we have established that $H(S_{*,*}(f), d') \cong C_*(B; \mathcal{H}_*(F))$.

To complete the proof, it suffices to show how $\bar{d}''$ induces the differential for chains on B with local coefficients. To do this, one must keep track of the leading vertex of an elementary chain on B through the identifications and boundary maps ∂_i^v. When the leading vertex does not map to a leading vertex, however, the identification of the coefficients with $H_*(F_{v(e_0)})$ is carried across the simplex along the path given in the definition of homology with local coefficients (see §5.3) and so $\bar{d}''$ determines the correct differential. This establishes the second half of the theorem. [Dress67] identified the notion of systems of local coefficients with certain functors on the category of simplicial objects over a fixed space. The details of his proof provide a combinatorial description of local coefficients. $\qquad\square$

We have discussed several constructions of the Leray-Serre spectral sequence. We mention one other method of construction of the Leray-Serre spectral sequence: [Eilenberg50] and [Damy96] gave constructions based on the use of a Cartan-Eilenberg system. The fact that these different constructions give rise to the same spectral sequence reflects the restrictions imposed by the structure of a fibration. These restrictions can be made very precise. [Barnes85] introduced axioms that a functor **F**, from the category of fibrations with all spaces locally finite CW-complexes to the category of filtered chain complexes, must satisfy in order to give a spectral sequence like the Leray-Serre spectral sequence. Such a functor is called a **fibration spectral sequence constructor**. He shows that the constructions we have discussed fit into this framework. Furthermore, there is a comparison theorem for natural transformations between constructors $t\colon \mathbf{F} \to \mathbf{F}'$ that implies when two such functors give rise to the same spectral sequence. There are notions of a free and cofree functor, for which there are always natural transformations and the construction of [Dress67] is a free functor. [Barnes85] proved the following uniqueness result via an acyclic models argument.

Theorem 6.49. *Let* **F** *and* **F**$'$ *be two fibration spectral sequence constructors. Then there is a canonical isomorphism* $t\colon E^r_{*,*}\mathbf{F} \to E^r_{*,*}\mathbf{F}'$, *for all* $r \geq 2$, *of the associated spectral sequences as functors on the category of CW-fibrations.*

The Leray-Serre spectral sequence takes information about the fibre and base of a fibration and proceeds to information about the total space. As we have already seen in Theorems 6.19 and 6.38, we may want to go from data about the fibre and total space to the base (for example, the cohomology of classifying spaces) or from data about the base and total space to the fibre (for example, the cohomology of ΩX). In the most successful examples (Eilenberg-Mac Lane spaces and classifying spaces) the answer depends heavily on the manageable form of the input—for example, when there is a p-simple system of generators. Then the output takes the form of a homological functor on the input. How systematic is this algebraic phenomenon? Can it be generalized? There **is** a unified approach to these computations, and once again, it comes in the form of a spectral sequence. The construction, due to [Eilenberg-Moore66], and its applications are the subject of the next two chapters.

Exercises

6.1. We used certain facts about Ext in the computation of $\pi_4(S^2)$. Prove that $\mathrm{Ext}(\pi, \mathbb{Z}) \cong \pi$ when π is a finite group.

6.2. Show that the transgression (Definition 6.3) is well-defined.

6.3. Suppose $F \hookrightarrow E \xrightarrow{p} B$ is a fibration and $\tilde{H}^*(E; R) \cong \{0\}$. Show that $H^i(F; R) = \{0\}$ for $0 < i < q$ if and only if $H^j(B; R) = \{0\}$ for $0 < j < q + 1$ and furthermore, the transgression, $\tau \colon H^t(F; R) \to H^{t+1}(B; R)$ is an isomorphism for $t < 2q$.

6.4. Show that the transgression annihilates products.

6.5. Prove the following theorem of [Borel50]: Suppose k is a field and $F \hookrightarrow E \to B$ is a fibration with $H^*(E; k) \cong H^*(S^n; k)$. Suppose further that $H^*(F; k) \cong H^*(S^{2m_1+1} \times S^{2m_2+1} \times \cdots \times S^{2m_s+1}; k)$, the cohomology of a product of odd spheres and B a finite complex. Then $s = 1$.

6.6. Prove Theorem 6.8.

6.7. Compute $\pi_5(S^2)$ by the method of killing homotopy groups, that is, using the associated upside-down Postnikov system.

6.8. Prove Lemma 6.18.

6.9. Complete the proof of Theorem 6.19 (theorems of Cartan and Serre) by computing $H^*(K(\Pi, n); \mathbb{F}_p)$ when $\Pi = \mathbb{Z}/p^k\mathbb{Z}$. Also prove the mod 2 version of Theorem 6.19.

6.10. Use Theorem 4.43 and the results of Cartan and Serre in §6.2 to show that the Steenrod algebra, as described in Chapter 4, contains all stable cohomology operations over $\mathbb{Z}/p\mathbb{Z}$.

6.11. Show that the transgression, $\tau \colon H^{n-1}(S^{n-1}) \to H^n(S^n)$, for the fibration, $S^{n-1} \hookrightarrow T_0 S^n \to S^n$, the sphere bundle associated to the tangent bundle, is given by multiplication by the Euler number of S^n.

6.12. Generalize the Hopf algebra argument of the proof of the Hopf-Borel theorem: Suppose $A \subset H$ is a sub-Hopf algebra of a given commutative Hopf algebra H. Let $M \cong H//A$ be a sub-vector space of M such that $q| \colon M \to H//A$, the restriction of the canonical quotient map, is an isomorphism. Show that $A \otimes M$ is isomorphic to H as left A-modules. (Hint: Consider the composite mapping

$$A \otimes M \xrightarrow{\text{inc} \otimes (q|)^{-1}} H \otimes H \xrightarrow{\mu} H \xrightarrow{\Delta} H \otimes H \xrightarrow{1 \otimes q} H \otimes (H//A).$$

Show that the composite is a monomorphism. Show also that A together with M generate H as an algebra, so the composite of the first two maps in the display is onto.)

6.13. Prove that the Whitney sum formula follows from Lemma 6.41.

6.14. The axioms for Chern classes are given for $\xi = (\mathbb{C}^n, E, p, M)$, a complex vector bundle over the base space M, as follows:

I. There are classes $c_{2i}(\xi) \in H^{2i}(M)$ and $c_0(\xi) = 1$, $c_{2i}(\xi) = 0$ if $i > n$.

II. If $f\colon M_1 \to M_2$ is a continuous mapping that is covered by a bundle map, $(f, \tilde{f})\colon \xi_1 \to \xi_2$, that is a linear isomorphism on the fibres, then $c_{2i}(\xi_1) = f^*(c_{2i}(\xi_2))$.

III. For two vector bundles, ξ and η, the Whitney sum formula holds:

$$c_{2k}(\xi \oplus \eta) = \sum_{i+j=k} c_{2i}(\xi) \smile c_{2j}(\eta).$$

IV. Let $\eta_1^1 = (\mathbb{C}^1, E, p, \mathbb{C}P(1))$ be the canonical complex line bundle over $\mathbb{C}P(1)$. Then $c_2(\eta_1^1) \neq 0$.

Prove the analogue of Theorem 6.22 for Chern classes and $H^*(BU(n))$. You can assume that the analogue of Lemma 6.23 holds for the mapping $BU(n) \times BU(m) \to BU(n+m)$.

6.15. To prove the analogue of Lemma 6.41 for $BU(n)$, one uses the maximal torus in $U(n)$, $T^n \cong S^1 \times \cdots \times S^1$ (n times). The inclusion of T^n in $U(n)$ induces an isomorphism,

$$H^*(BU(n)) \cong H^*(BT^n)^{\Sigma_n}$$
$$= \{\, \text{Symmetric polynomials in } y_1, \dots, y_n \mid \deg y_j = 2 \,\}.$$

The result still holds mod p and $H^*(BT^n; \mathbb{F}_p)$ has a particularly simple $\mathcal{A}_p$-structure. From this, deduce the $\mathcal{A}_p$-structure on $H^*(BU(n); \mathbb{F}_p)$. Do the analogous computation for $H^*(BSO(n); \mathbb{F}_2)$.

6.16. Compute $H^*\big(O(2m)/U(m); \mathbb{F}_2\big)$ and $H^*\big(U(n)/T^n\big)$.

7

The Eilenberg-Moore Spectral Sequence I

" ... , the application to topology of homological
algebra leads to somewhat different developments ... ,
which may be included under the heading of differential
homological algebra."

J. C. Moore

The Leray-Serre spectral sequence provides a method for computing the cohomology of the total space of a fibration from knowledge of the cohomology of the base space and the fibre. By arguing backward through the spectral sequence, the inverse problems of computing the cohomology of the fibre (as for the path-loop fibration) or the cohomology of the base space (as in the case of classifying spaces or Eilenberg-Mac Lane spaces) from the cohomology of the other two spaces in the fibration can sometimes be solved (Theorems 5.16, 6.20, and 6.39). In the particular case of the computation of $H^*(BG; k)$ from $H^*(G; k)$ when G is a compact Lie group, the algebraic relation between $H^*(G; k)$ and $H^*(BG; k)$ is often expressible in the language of homological algebra and derived functors. In pioneering work, [Cartan54], [Moore59], and [Eilenberg-Moore66] developed the correct algebraic setting to explain this relation. We present in this chapter the homological framework that leads to a general method of computation.

To begin, we extend the problem of computing the cohomology of the fibre from the cohomology of the base and total space to a more general question. Suppose $\pi\colon E \to B$ is a fibration with fibre F and $f\colon X \to B$ is a continuous function. A new fibration, $\pi'\colon E_f \to X$, can be constructed from these data, namely the pullback (§4.3) of $f\colon X \to B \leftarrow E\colon \pi$, as in the diagram

$$
\begin{array}{ccc}
F & = & F \\
\uparrow & & \uparrow \\
E_f & \longrightarrow & E \\
{\scriptstyle \pi'}\downarrow & & \downarrow{\scriptstyle \pi} \\
X & \xrightarrow{\ f\ } & B.
\end{array}
$$

Problem: *Compute $H^*(E_f; R)$ from knowledge of $H^*(E; R)$, $H^*(B; R)$, $H^*(X; R)$, f^* and π^*, where R is a commutative ring with unit.*

This problem includes the problem of computing $H^*(F; R)$ from $H^*(B; R)$ and $H^*(E; R)$: Let b_0 be a basepoint for B and $b_0 \colon * \to B$, the unique map in **Top** that selects the basepoint. If we take $f = b_0$ in the pullback diagram, then $E_f = F$ and the initial data are $H^*(E; R)$, $H^*(B; R)$ and π^*.

The solution to the problem comes in the form of a spectral sequence, introduced by [Moore59] and developed by [Eilenberg-Moore66].

Theorem 7.1. *Let $\pi \colon E \to B$ be a fibration with connected fibre for which the system of local coefficients on B determined by π is simple. Let $f \colon X \to B$ be a continuous function and E_f be the total space of the pullback fibration of π over f. Then there is a second quadrant spectral sequence, $\{ E_r^{*,*}, d_r \}$, converging to $H^*(E_f; R)$, with*

$$E_2^{*,*} \cong \mathrm{Tor}_{H^*(B;R)}^{*,*}(H^*(X; R), H^*(E; R)).$$

When B is simply-connected, the spectral sequence converges strongly.

New algebraic functors, $\mathrm{Tor}_{H^*(B;R)}^{*,*}(-,-)$ and $\mathrm{Tor}_{(C^*(B;R),\delta)}^{*,*}(-,-)$ were introduced by [Moore59] that generalize the derived functors $\mathrm{Tor}_R(-,-)$ of the tensor product to the category of differential graded modules over a differential graded algebra. The cup-product structures on $H^*(X; R)$ and $H^*(E; R)$ together with f^* and π^*, respectively, provide each with an $H^*(B; R)$-module structure. The bulk of this chapter (§7.1) treats the algebraic foundations that support the construction of the spectral sequence of Theorem 7.1 and make it calculable. In §7.2 the algebra is translated into topology. In §7.3 we further streamline the homological computations in special cases using the Koszul resolution. In this context we compute again the cohomology of the complex Stiefel manifolds (compare Example 5.G). The Stiefel manifolds are a special case of the computation of the cohomology of a homogeneous space: If H is a closed connected subgroup of the Lie group G, then G/H may be taken to be the fibre of the mapping $B(\mathrm{inc}) \colon BH \to BG$ induced by the inclusion. Theorem 6.38 provides $H^*(BG; k)$ in many cases and so Theorem 7.1 may be applied to compute $H^*(G/H; k)$ (see §8.1). We also consider the computation of the cohomology of the based loop space of an H-space.

In §7.4 we discuss the dual case of a pushout diagram and the Eilenberg-Moore spectral sequence for homology. This is especially useful in computing $H_*(BG; k)$. We postpone a complete discussion of the problem of strong convergence of the Eilenberg-Moore spectral sequence to Chapter 8bis. More sophisticated applications and the development of the deeper structure of the spectral sequence are taken up in Chapter 8.

The reader with little exposure to homological algebra can consult §2.4 for a primer on the homological algebra that is generalized in this chapter.

7.1 Differential homological algebra

Let R denote a commutative ring with unit and let $C^*(Y)$ denote $C^*(Y; R)$, the singular cochains on Y with coefficients in R. We begin our discussion of homological algebra with the motivating topological example.

The Alexander-Whitney map provides $C^*(Y)$ with a multiplication,

$$\smile : C^p(Y) \otimes_R C^q(Y) \longrightarrow C^{p+q}(Y)$$

for all p, q. The differential, $\delta: C^p(Y) \to C^{p+1}(Y)$, satisfies the Leibniz rule, $\delta(a \smile b) = \delta(a) \smile b + (-1)^{\deg a} a \smile \delta(b)$, and so $(C^*(Y), \delta)$ is a differential graded algebra over R. We can apply the cochain functor to a pullback diagram to get the commutative square:

$$
\begin{array}{ccc}
E_f \longrightarrow E & \quad & C^*(E_f) \longleftarrow C^*(E) \\
\pi' \downarrow \quad \downarrow \pi & \overset{\Longrightarrow}{} & \uparrow \qquad \qquad \uparrow \pi^* \\
X \underset{f}{\longrightarrow} B & \scriptstyle C^*(\ ;R) & C^*(X) \underset{f^*}{\longleftarrow} C^*(B).
\end{array}
$$

The morphisms of differential graded algebras, f^* and π^*, endow $C^*(E)$ and $C^*(X)$ with $C^*(B)$-module structures: The (right) module structure map for $C^*(X)$ is given by $\varphi: C^*(X) \otimes_R C^*(B) \longrightarrow C^*(X)$, which denotes the composite

$$C^*(X) \otimes_R C^*(B) \overset{1 \otimes f^*}{\longrightarrow} C^*(X) \otimes_R C^*(X) \overset{\smile}{\longrightarrow} C^*(X)$$

and similarly, $\psi = \smile \circ (\pi^* \otimes 1): C^*(B) \otimes_R C^*(E) \to C^*(E)$ denotes the (left) module structure map for $C^*(E)$. We take these module structures as the basic data. The space E_f associated to $f: X \to B \leftarrow E : \pi$ enjoys the universal properties of pullbacks. It is natural to ask whether or not $C^*(E_f)$ shares any dual universal properties in the category of $C^*(B)$-modules. A first guess for $C^*(E_f)$ might be $C^*(E_f) \cong C^*(X) \otimes_{C^*(B)} C^*(E)$ (with the right hand side properly defined to generalize the case of the tensor product of two R-modules). We will see later (Lemma 7.3) that this implies $H^*(E_f; R) \cong H^*(X; R) \otimes_{H^*(B;R)} H^*(E; R)$. If we apply this guess to the path-loop fibration, it fails to compute the cohomology of the based loop space.

Notice, however, that when $B = *$, the guess above is correct. A more sophisticated second guess can be made by observing that the first guess works for $B = *$ because $C^*(X)$ and $C^*(E)$ are *free* $C^*(*)$-modules. (This is the Künneth theorem.) Furthermore, if π and f are projections, the same argument works. The deviation from "freeness" of $C^*(X)$ and $C^*(E)$ as $C^*(B)$-modules reflects the nontrivial twisting of the base and fibre as a product in the fibration, $\pi: E \to B$, and further as pulled back over the mapping $f: X \to B$. In the case of modules over R and differential graded modules over R (see

§2.4), this information is expressed through the derived functors $\mathrm{Tor}_R(-, -)$ of the tensor product. The correct expression for $H^*(E_f; R)$ in terms of the data provided by $f\colon X \to B \leftarrow E\colon \pi$ is based on a generalization of the theory of differential graded modules over R to a theory of differential graded modules over a differential graded algebra over R. In our exposition, we follow [Moore59] and [Smith, L67].

Differential graded modules over a differential graded algebra

Let (Γ, d) denote a differential graded algebra over R. For completeness, we recall Definition 1.4:

(1) $\Gamma = \bigoplus_{n \geq 0} \Gamma^n$ is a differential graded module over R. We take d, the differential, to have degree $+1$.

(2) There is an R-module homomorphism, $m\colon \Gamma \otimes_R \Gamma \to \Gamma$, the multiplication, which is associative, that is, the following diagram commutes:

$$
\begin{array}{ccc}
\Gamma \otimes_R \Gamma \otimes_R \Gamma & \xrightarrow{\;m \otimes 1\;} & \Gamma \otimes_R \Gamma \\
{\scriptstyle 1 \otimes m}\big\downarrow & & \big\downarrow{\scriptstyle m} \\
\Gamma \otimes_R \Gamma & \xrightarrow{\quad m \quad} & \Gamma.
\end{array}
$$

(3) d is a derivation with respect to m, that is, if we write $m(a \otimes b) = a \cdot b$, then $d(a \cdot b) = d(a) \cdot b + (-1)^{\deg a} a \cdot d(b)$.

(4) Γ has a unit, that is, there is an injection $\eta\colon R \to \Gamma$ of rings, so that the following diagram commutes

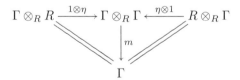

The ring R may be taken to be the differential graded algebra, $(R, d = 0)$, with R in degree zero, $\{0\}$ in higher degrees, and the zero differential.

Those differential graded module morphisms, $\varphi\colon (\Gamma, d) \to (\Gamma', d')$, of degree zero that commute with the multiplications and the units comprise the morphisms of differential graded R-algebras and so determine a category, denoted **DGAlg**$_R$. Singular cochains on a space with coefficients in R and the cohomology of a space (with zero differential), both with the cup product, are functors **Top** $\to$ **DGAlg**$_R$ from spaces to differential graded algebras.

A **left module**, N, **over** Γ is a differential graded module over R, together with an **action of** Γ, $\psi\colon \Gamma \otimes_R N \to N$, which is a morphism of differential graded R-modules such that the following diagrams commute:

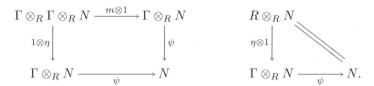

A **morphism of left Γ-modules**, $\rho \colon N \to N'$, is a homomorphism of differential graded R-modules that is compatible with the Γ-actions. The left (differential graded) Γ-modules constitute a category that we denote by $\mathbf{DG_\Gamma Mod}$. The category of **right** (differential graded) Γ-**modules**, $\mathbf{DGMod_\Gamma}$, is defined similarly.

Suppose M is a right Γ-module and N a left Γ-module with structure maps φ and ψ, respectively. We define $M \otimes_\Gamma N$, the **tensor product** of M and N over Γ, via the short exact sequence in $\mathbf{DGMod}_R$

$$M \otimes_R \Gamma \otimes_R N \xrightarrow{\varphi \otimes 1 - 1 \otimes \psi} M \otimes_R N \to M \otimes_\Gamma N \to 0.$$

Thus $M \otimes_\Gamma N$ can be described as $M \otimes_R N$ modulo the relations $(m \cdot \gamma) \otimes n = m \otimes (\gamma \cdot n)$, for all $\gamma \in \Gamma$, $m \in M$, and $n \in N$. For N in $\mathbf{DG_\Gamma Mod}$, the unit in Γ may be used to construct an isomorphism between N and $\Gamma \otimes_\Gamma N$ as left Γ-modules: Consider the R-module homomorphisms $f \colon N \to \Gamma \otimes_\Gamma N$ given by $f(n) = 1 \otimes_\Gamma n$, and $g \colon \Gamma \otimes_\Gamma N \to N$ given by $g(\gamma \otimes_\Gamma n) = \gamma \cdot n$. By the definition of $\otimes_\Gamma$, f and g are inverses. We consider $\Gamma \otimes_\Gamma N$ as a left Γ-module with the action $\gamma \otimes (\gamma' \otimes_\Gamma n) \mapsto (\gamma\gamma') \otimes_\Gamma n$. The identification $(\gamma\gamma') \cdot n = \gamma \cdot (\gamma' \cdot n)$ is equivalent to the fact that f and g are Γ-module homomorphisms. Thus $N \cong \Gamma \otimes_\Gamma N$ as left Γ-modules. By the symmetric argument for a right Γ-module M, $M \otimes_\Gamma \Gamma \cong M$ as right Γ-modules.

We next define the derived functors of $- \otimes_\Gamma -$, for which we need a notion of exact sequence. Let

$$\cdots \to (Q^n, d_n) \to (Q^{n+1}, d_{n+1}) \to (Q^{n+2}, d_{n+2}) \to \cdots$$

be a sequence of morphisms of left Γ-modules. Recall from the case of differential graded modules over R (§2.4) that the sequence is **proper exact** if it satisfies

(1) $\cdots \to (Q^n)^\# \to (Q^{n+1})^\# \to (Q^{n+2})^\# \to \cdots$ is exact in the category of graded $\Gamma^\#$-modules, where $(\ )^\# \colon \mathbf{DG_\Gamma Mod} \to \mathbf{G_\Gamma Mod}$ is the forgetful functor that ignores differentials.

(2) $\cdots \to Z(Q^n) \to Z(Q^{n+1}) \to Z(Q^{n+2}) \to \cdots$ is exact in the category of graded R-modules, where $Z(Q^n) = Z(Q^n, d_n) = \ker d_n$.

(3) $\cdots \to H(Q^n, d_n) \to H(Q^{n+1}, d_{n+1}) \to H(Q^{n+2}, d_{n+2}) \to \cdots$ is exact in the category of graded R-modules.

If N is a left Γ-module, we define a **proper resolution** of N to be a proper exact sequence in $\mathbf{DG_\Gamma Mod}$:

$$\cdots \xrightarrow{\delta} Q^{-n} \xrightarrow{\delta} Q^{-n+1} \xrightarrow{\delta} \cdots \xrightarrow{\delta} Q^{-1} \xrightarrow{\delta} Q^0 \xrightarrow{\varepsilon} N \to 0.$$

Having defined the notion of exactness, there is a corresponding notion of projective. We say that a Γ-module, P, is (proper) **projective** if, for any morphism, $\sigma \colon P \to N$ and *proper* epimorphism $\zeta \colon M \to N$ in $\mathbf{DG_\Gamma Mod}$, there is an extension, $\tilde{\sigma} \colon P \to M$ in $\mathbf{DG_\Gamma Mod}$ so that $\zeta \circ \tilde{\sigma} = \sigma$:

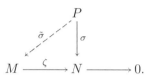

We observe that the category $\mathbf{DG_\Gamma Mod}$ has enough projective modules.

Proposition 7.2. *Suppose Γ and $H(\Gamma)$ are flat modules over R and N is a left Γ-module. Then there is a projective Γ-module, P, and a proper epimorphism, $\pi \colon P \to N$. That is, every left Γ-module is the image of a proper projective Γ-module.*

PROOF: We first introduce the notion of an **extended module**. Suppose V is a proper projective differential graded R-module; then V, $Z(V)$ and $H(V)$ are all projective modules over R (see §2.4). Consider $\tilde{V} = \Gamma \otimes_R V$ and give $\tilde{V}$ the left Γ-module structure, $c_V \colon \Gamma \otimes_R \tilde{V} \to \tilde{V}$ as in the diagram

$$
\begin{array}{ccc}
\Gamma \otimes_R \tilde{V} & \xrightarrow{\ c_V\ } & \tilde{V} \\
\| & & \| \\
\Gamma \otimes_R \Gamma \otimes_R V & \xrightarrow[m \otimes 1]{} & \Gamma \otimes_R V.
\end{array}
$$

We claim that $\tilde{V}$ is a projective Γ-module.

Suppose $\zeta \colon M \to N$ is a proper epimorphism and $\sigma \colon \tilde{V} \to N$ is a (left) Γ-module homomorphism. Observe that V sits in $\tilde{V}$ as $1 \otimes_R V$ since Γ has a unit.

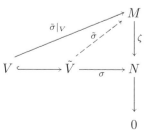

Because ζ is a proper epimorphism, $M \xrightarrow{\zeta} N \to 0$ is exact in $\mathbf{DGMod}_R$ and so we can extend $\sigma|_V : V \to N$ to an R-module homomorphism $\tilde{\sigma}|_V : V \to M$. Extend $\tilde{\sigma}|_V$ to all of $\tilde{V}$ by $\tilde{\sigma}(\gamma \otimes v) = \gamma \cdot \tilde{\sigma}|_V (v)$. Since σ is a Γ-module homomorphism, $\tilde{\sigma}$ is our desired extension.

To complete the proof of the proposition, we use the fact that there are enough projective differential graded R-modules and take $\pi_0 : P^0 \to N$, any proper epimorphism in $\mathbf{DGMod}_R$. Extend π_0 to the Γ-module homomorphism $\pi : \Gamma \otimes P^0 \to N$, which is the composite

$$\pi : \Gamma \otimes_R P^0 \xrightarrow{1 \otimes \pi_0} \Gamma \otimes_R N \xrightarrow{\varphi_N} N.$$

It remains to show that $\pi : \Gamma \otimes P^0 \to N$ is proper epic.

Lemma 7.3. *Suppose that Γ and $H(\Gamma)$ are flat modules over R, P an extended (hence projective) left Γ-module and B a right Γ-module. Then $H(B \otimes_\Gamma P) \cong H(B) \otimes_{H(\Gamma)} H(P)$.*

PROOF: With the assumption of flatness, the Künneth theorem allows us to give $H(B)$ an $H(\Gamma)$-module structure by

$$H(B) \otimes_R H(\Gamma) \xrightarrow{\cong} H(B \otimes_R \Gamma) \xrightarrow{H(\psi_B)} H(B).$$

Since P is an extended module over Γ, we can write $P = \Gamma \otimes_R V$ and compute, applying the Künneth theorem at the appropriate points,

$$\begin{aligned}
H(B \otimes_\Gamma P) = H(B \otimes_\Gamma \Gamma \otimes_R V) &\cong H(B \otimes_R V) \\
&\cong H(B) \otimes_R H(V) \cong H(B) \otimes_{H(\Gamma)} H(\Gamma) \otimes_R H(V) \\
&\cong H(B) \otimes_{H(\Gamma)} H(\Gamma \otimes_R V) \cong H(B) \otimes_{H(\Gamma)} H(P). \qquad \square
\end{aligned}$$

We define a **proper projective resolution** of a left Γ-module, N to be a proper exact sequence

$$\cdots \xrightarrow{\delta} P^{-i} \xrightarrow{\delta} P^{-i+1} \xrightarrow{\delta} \cdots \xrightarrow{\delta} P^{-1} \xrightarrow{\delta} P^0 \xrightarrow{\varepsilon} N \to 0$$

for which each P^{-i} is a proper projective Γ-module for $i \geq 0$. Lemma 2.19 guarantees that proper projective resolutions can always be constructed.

Corollary 7.4. *Suppose Γ and $H(\Gamma)$ are flat modules over R and*

$$\cdots \xrightarrow{\delta} P^{-i} \xrightarrow{\delta} P^{-i+1} \xrightarrow{\delta} \cdots \xrightarrow{\delta} P^{-1} \xrightarrow{\delta} P^0 \xrightarrow{\varepsilon} N \to 0$$

is a proper projective resolution of N, then

$$\xrightarrow{H(\delta)} H(P^{-i}) \xrightarrow{H(\delta)} \cdots \xrightarrow{H(\delta)} H(P^{-1}) \xrightarrow{H(\delta)} H(P^0) \xrightarrow{H(\varepsilon)} H(N) \to 0$$

is a projective resolution of $H(N)$ as an $H(\Gamma)$-module.

PROOF: By Lemma 7.3 we may take all of the P^{-i} to be extended modules over Γ. Since Γ is flat, $H(P^{-i}) \cong H(\Gamma) \otimes_R H(V^{-i})$ by the Künneth theorem. We take V^{-i} to be proper projective, so $H(V^{-i})$ is projective over R. Thus $H(P^{-i})$ is an extended module over $H(\Gamma)$. Since a proper projective resolution is proper exact, and $H(\Gamma)$ is flat, the sequence on homology is long exact. Hence we have an $H(\Gamma)$-projective resolution of $H(N)$. $\qquad\square$

As in Chapter 2, we have chosen the homological grading for the resolutions in such a way that, with the internal degree, we have a double complex:

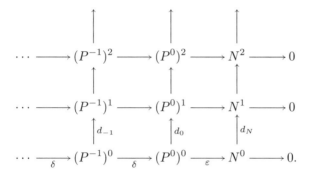

Suppose $(P^\bullet, \delta)$ is the complex

$$\cdots \to P^{-i} \xrightarrow{\delta} P^{-i+1} \xrightarrow{\delta} \cdots \xrightarrow{\delta} P^{-1} \xrightarrow{\delta} P^0.$$

This complex has homology $H(P^\bullet, \delta) = (N, d_N)$. Form the $\mathbb{Z}$-graded differential module, $(\mathrm{total}(P^\bullet), D)$ by letting

$$[\mathrm{total}(P^\bullet)]^j = \bigoplus_{m+n=j} (P^m)^n, \qquad D = \sum_m (\delta + (-1)^m d_m).$$

(We write the direct sum with the word of warning that, in the general case, this sum may be infinite and so the weak direct sum is intended. In certain situations of topological interest, such large sums do not occur.)

There is a Γ-module structure on $(\mathrm{total}(P^\bullet), D)$ given by restricting the Γ-action, $\Gamma \otimes_R P^{-i} \to P^{-i}$ to $(P^{-i})^{m+i}$. Since the action determines a mapping $\Gamma^k \otimes_R (P^{-i})^{m+i} \to (P^{-i})^{m+i+k}$, we get $\Gamma^k \otimes_R [\mathrm{total}(P^\bullet)]^m \to [\mathrm{total}(P^\bullet)]^{m+k}$. If M is a right Γ-module, we form the $\mathbb{Z}$-graded differential R-module

$$(M \otimes_\Gamma \mathrm{total}(P^\bullet), d_M \otimes 1 + (-1)^{\deg} \otimes D)$$

where the sign is given by $d_M \otimes 1 + (-1)^q 1 \otimes D$ on elements represented in $M^q \otimes \mathrm{total}(P^\bullet)$.

Definition 7.5. $\operatorname{Tor}_\Gamma(M, N) = H(M \otimes_\Gamma \operatorname{total}(P^\bullet), d_M \otimes 1 \pm 1 \otimes D)$.

We leave it as an exercise (compare Proposition 2.17) to verify that the definition is independent of the choice of proper projective resolution as well as independent of the choice to resolve N in $\mathbf{DG_\Gamma Mod}$ and tensor on the left with M or to resolve M in $\mathbf{DGMod_\Gamma}$ and tensor on the right with N or resolve both and tensor the resolutions together. (In the topological applications to come, we will resolve right modules.) Furthermore, the reader should verify that $\operatorname{Tor}_\Gamma$ is the correct generalization of Tor_R, in the sense that $\operatorname{Tor}_\Gamma(M, -)$ behaves as the left derived functors of $M \otimes_\Gamma -$.

We make the following useful observations:

(1) $\operatorname{Tor}_\Gamma(M, N)$ is bigraded.

This follows by writing

$$\operatorname{Tor}_\Gamma^{-i,*}(M, N) = H^{-i,*}(M \otimes_\Gamma \operatorname{total}(P^\bullet), d_M \otimes 1 + (-1)^{\deg} \otimes D)$$
$$= \text{a subquotient of } M \otimes_\Gamma P^{-i}.$$

Those elements coming from $M^m \otimes_\Gamma (P^{-i})^n$ have bidegree $(-i, m+n)$. The first degree is the homological degree and the second is the internal degree.

(2) $\bigoplus_i \operatorname{Tor}_\Gamma^{0,i}(M, N) = M \otimes_\Gamma N$.

This follows from the fact that $H(P^\bullet, \delta) = (N, d_N)$. Thus $\operatorname{Tor}_\Gamma(M, N)$ contains not only $M \otimes_\Gamma N$ but also the deviation from exactness of the functor $M \otimes_\Gamma -$ and the deviation from being projective of N. In the topological interpretation, $\operatorname{Tor}_{C^*(B)}(C^*(X), C^*(E))$ measures the twisting effect of f^* as the deviation from exactness of $C^*(X) \otimes_{C^*(B)} -$ and the nontriviality of π^* as the deviation from freeness of $C^*(E)$ as a $C^*(B)$-module.

(3) $\operatorname{total}(P^\bullet)$ is filtered.

An increasing filtration of $\operatorname{total}(P^\bullet)$ is a consequence of having a double complex and so we get the filtration:

$$\{0\} = F^1 \subset F^0 \subset F^{-1} \subset \cdots \subset F^{-n} \subset \cdots \subset \operatorname{total}(P^\bullet)$$

defined by $(F^{-n})^r = \bigoplus (P^i)^j$ where the sum is over $i + j = r$ and $i \geq -n$. We can picture the submodule F^{-n} as in the diagram:

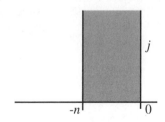

It is immediate that $D\colon F^{-n} \to F^{-n}$ and each F^{-n} is a $\mathbb{Z}$-graded Γ-module. We filter $M \otimes_\Gamma \mathrm{total}(P^\bullet)$ by $\mathcal{F}^{-n} = M \otimes_\Gamma F^{-n}$. By Theorem 2.6 this filtration leads to a spectral sequence.

Theorem 7.6 (the first Eilenberg-Moore theorem). *Suppose Γ and $H(\Gamma)$ are flat modules over R. Then there is a second quadrant spectral sequence with*

$$E_2^{*,*} \cong \mathrm{Tor}_{H(\Gamma)}^{*,*}(H(M), H(N))$$

and converging to $\mathrm{Tor}_\Gamma^{*,*}(M, N)$.

PROOF: Having set up the algebra carefully, we use Theorem 2.6 to get most of the theorem. It suffices to compute the E_2-term. The initial term of the spectral sequence is $E_1^{-n,*} \cong H_*(\mathcal{F}^{-n}/\mathcal{F}^{-n+1})$ and, as in the proof of Theorem 2.20 (the Künneth spectral sequence), we have

$$E_1^{-n,*} \cong H_*(\mathcal{F}^{-n}/\mathcal{F}^{-n+1}) \cong H_*(M \otimes_\Gamma P^{-n}).$$

By Lemma 7.3 we obtain $E_1^{-n,*} \cong H(M) \otimes_{H(\Gamma)} H(P^{-n})$ and $d_1 = 1 \otimes H(\delta)$. Corollary 7.4 and the definition of Tor give us the desired E_2-term. $\quad\square$

As in the discussion of Theorem 2.20, the strong convergence of this spectral sequence may be delicate—the stages in the filtration are $\mathbb{Z}$-graded. A judicious choice of resolution can sometimes be made that insures strong convergence (for example, in the discussion to follow on the bar construction). General convergence results are discussed in Chapter 8bis. An important corollary of the theorem is due to [Moore59] that motivates a number of results in Chapter 8.

Corollary 7.7. *Suppose Γ and Λ are differential graded algebras over R, M is a right Γ-module, N is a left Γ-module, M' is a right Λ-module, and N' is a left Λ-module. Suppose we have homomorphisms, $f\colon \Gamma \to \Lambda$, $g\colon M \to M'$ and $h\colon N \to N'$ satisfying*

(1) *$f\colon \Gamma \to \Lambda$ is a homomorphism of differential graded R-algebras.*

(2) *The following diagrams commute*

$$
\begin{array}{ccc}
M \otimes_R \Gamma & \xrightarrow{\varphi_M} & M \\
{\scriptstyle g \otimes f}\downarrow & & \downarrow{\scriptstyle g} \\
M' \otimes_R \Lambda & \xrightarrow{\varphi_{M'}} & M'
\end{array}
\qquad
\begin{array}{ccc}
\Gamma \otimes_R N & \xrightarrow{\psi_N} & N \\
{\scriptstyle f \otimes h}\downarrow & & \downarrow{\scriptstyle h} \\
\Lambda \otimes_R N' & \xrightarrow{\psi_{N'}} & N'.
\end{array}
$$

(3) *$H(f)$, $H(g)$ and $H(h)$ are all isomorphisms.*

Then, when the associated spectral sequences converge strongly,

$$\mathrm{Tor}_f(g,h)\colon \mathrm{Tor}_\Gamma(M,N) \to \mathrm{Tor}_\Lambda(M',N')$$

is an isomorphism.

PROOF: The corollary follows directly from Theorem 3.9 and the fact that $\mathrm{Tor}_{H(f)}(H(g),H(h))$ is an isomorphism of the E_2-terms of the associated spectral sequences.   □

This corollary indicates how Tor, under suitable conditions, "sees" only the homology of the objects involved. The subtleties of what is lost in this procedure figure prominently in the applications of this spectral sequence to the computation of the cohomology of homogeneous spaces.

At this point we should introduce the topology and interpret this algebraic spectral sequence for topologists. The impatient reader can look ahead to §7.2. However, before we leave the algebra, it is useful to recast our results in a (theoretically) computable rather than categorical context.

The bar construction

In order to compute $\mathrm{Tor}_\Gamma(M,N)$, we need a proper projective resolution of the left Γ-module N. Since any proper projective resolution will do, a constructive, functorial complex is desirable that perhaps carries some extra structure that can be exploited in calculation. One of the most useful explicit constructions in homological algebra is the **bar construction**, which was introduced by [Eilenberg-Mac Lane53] in their study of the cohomology of abelian groups. For the sake of simplicity, we assume that our commutative ring R is a field k. The interested reader can extend the construction (led by the Künneth theorem) to arbitrary rings (see the exercises).

Suppose that Γ is connected, that is, the unit, $\eta\colon k \to \Gamma$ is an isomorphism in degree zero. Denote the cokernel of η by $\bar{\Gamma}$. Since Γ is connected, we have that $\bar{\Gamma} = \Gamma^+ = \{\gamma \in \Gamma \mid \deg\gamma > 0\}$. Define

$$\mathrm{B}^{-n}(\Gamma,N) = \Gamma \otimes_k \overbrace{\bar{\Gamma} \otimes_k \cdots \otimes_k \bar{\Gamma}}^{n \text{ times}} \otimes_k N.$$

Notice that $\mathrm{B}^{-n}(\Gamma,N)$ is a left Γ-module with the extended module action. It is customary to write an element of $\mathrm{B}^{-n}(\Gamma,N)$ as $\gamma[\gamma_1 \mid \gamma_2 \mid \cdots \mid \gamma_n]a$ and to write $\gamma[\ ]a$ for elements of $\mathrm{B}^0(\Gamma,N)$. To simplify signs in many formulas, we adopt the following

Convention: If α is an element of a differential graded module, then

$$\bar{\alpha} = (-1)^{1+\deg\alpha}\alpha.$$

The internal differential d^{-n} defined on $\mathrm{B}^{-n}(\Gamma, N)$ takes the form

$$d^{-n}(\gamma[\gamma_1 \mid \cdots \mid \gamma_n]a) =$$
$$(d_\Gamma \gamma)[\gamma_1 \mid \cdots \mid \gamma_n]a + \sum_{i=1}^{n} \bar{\gamma}[\bar{\gamma}_1 \mid \cdots \mid \bar{\gamma}_{i-1} \mid d_\Gamma \gamma_i \mid \gamma_{i+1} \mid \cdots \mid \gamma_n]a$$
$$+ \bar{\gamma}[\bar{\gamma}_1 \mid \cdots \mid \bar{\gamma}_n](d_N a).$$

We leave it to the reader to check that d^{-n} is Γ-linear.

We assemble the differential graded Γ-modules $(\mathrm{B}^{-n}(\Gamma, N), d^{-n})$ into a resolution by introducing the homological differential $\delta \colon \mathrm{B}^{-n}(\Gamma, N) \to \mathrm{B}^{-n+1}(\Gamma, N)$.

$$(-1)^{\deg \gamma} \delta(\gamma[\gamma_1 \mid \cdots \mid \gamma_n]a) =$$
$$(\gamma \cdot \gamma_1)[\gamma_2 \mid \cdots \mid \gamma_n]a + \sum_{i=1}^{n-1} \gamma[\bar{\gamma}_1 \mid \cdots \mid \bar{\gamma}_{i-1} \mid \bar{\gamma}_i \cdot \gamma_{i+1} \mid \cdots \mid \gamma_n]a$$
$$+ \gamma[\bar{\gamma}_1 \mid \cdots \mid \bar{\gamma}_{n-1}](\gamma_n a).$$

(Our sign convention follows [May67] and agrees with [Smith, L67]. Though more complicated than [Mac Lane70], it simplifies some notation used in §8.2.) The reader can verify easily that $\delta\delta = 0$, δ is Γ-linear and $d^{-n+1}\delta + \delta d^{-n} = 0$. If we let $\varepsilon = \varphi_N \colon \mathrm{B}^0(\Gamma, N) = \Gamma \otimes_k N \to N$, then we have the complex of Γ-modules

$$\cdots \xrightarrow{\delta} \mathrm{B}^{-n}(\Gamma, N) \xrightarrow{\delta} \cdots \xrightarrow{\delta} \mathrm{B}^{-1}(\Gamma, N) \xrightarrow{\delta} \mathrm{B}^0(\Gamma, N) \xrightarrow{\varepsilon} N \to 0$$

(ε is onto, because Γ has a unit).

Proposition 7.8. $(\mathrm{B}^\bullet(\Gamma, N), \delta, d^\bullet, \varepsilon)$, *as a complex of differential graded Γ-modules, satisfies*

  (1) $H(\mathrm{B}^\bullet(\Gamma, N), \delta, \varepsilon) = \{0\}$, *that is, the sequence is exact.*
  (2) $H(\mathrm{B}^{-n}(\Gamma, N), d^{-n}) = \mathrm{B}^{-n}(H(\Gamma), H(N))$.
  (3) $(\mathrm{B}^\bullet(\Gamma, N), \delta, \varepsilon)$ *is a proper projective resolution of N over Γ.*

PROOF: Consider the contracting homotopy:

$$s(\gamma[\gamma_1 \mid \cdots \mid \gamma_n]a) = \begin{cases} 1[\gamma \mid \gamma_1 \mid \cdots \mid \gamma_n]a, & \text{if } \deg \gamma > 0, \\ 0, & \text{if } \deg \gamma = 0, \end{cases}$$

and $s(a) = 1[\ ]a$. We verify $\delta s + s\delta = \mathrm{id}$:

$$\delta s(\gamma[\gamma_1 \mid \cdots \mid \gamma_n]a) = \delta(1[\gamma \mid \gamma_1 \mid \cdots \mid \gamma_n]a)$$
$$= \gamma[\gamma_1 \mid \cdots \mid \gamma_n]a + \sum_{i=1}^{n-1} 1[\bar{\gamma} \mid \bar{\gamma}_1 \mid \cdots \mid \bar{\gamma}_i \gamma_{i+1} \mid \cdots \mid \gamma_n]a$$
$$+ 1[\bar{\gamma} \mid \bar{\gamma}_1 \mid \cdots \mid \bar{\gamma}_{n-1}](\gamma_n a) + 1[\bar{\gamma}\gamma_1 \mid \cdots \mid \gamma_n]a.$$

Next we compute $s\delta$:

$$s\delta(\gamma[\gamma_1 \mid \cdots \mid \gamma_n]a) =$$
$$s\left(-\bar{\gamma}\gamma_1[\gamma_2 \cdots \mid \gamma_n]a\right) - \bar{\gamma}[\bar{\gamma}_1 \mid \cdots \mid \bar{\gamma}_{n-1}](\gamma_n a))$$
$$- s\left(\sum_{i=1}^{n-1} \bar{\gamma}[\bar{\gamma}_1 \mid \cdots \mid \bar{\gamma}_i\gamma_{i+1} \mid \cdots \mid \gamma_n]a\right)$$
$$= -1[\bar{\gamma}\gamma_1 \mid \gamma_2 \mid \cdots \mid \gamma_n]a - 1[\bar{\gamma} \mid \bar{\gamma}_1 \mid \cdots \mid \bar{\gamma}_{n-1}](\gamma_n a)$$
$$- \sum_{i=1}^{n-1} 1[\bar{\gamma} \mid \bar{\gamma}_1 \mid \cdots \mid \bar{\gamma}_i\gamma_{i+1} \mid \cdots \mid \gamma_n]a.$$

Combining the two displays we find that $\delta s + s\delta = \text{id}$ when $n \geq 1$. For the bottom of the resolution, observe that $\delta s(a) = \delta(1[\ ]a) = a$ and $s\delta(a) = s(0) = 0$.

It follows immediately from $\delta s + s\delta = \text{id}$ that the complex is exact. Observe from $\delta d^{-n} + d^{-n+1}\delta = 0$ that

$$\delta(Z(\mathrm{B}^{-n}(\Gamma, N), d^{-n})) \subset Z(\mathrm{B}^{-n+1}(\Gamma, N), d^{-n+1}).$$

The reader can verify that $sd^{-n} + d^{-n-1}s = 0$ with the consequence that s carries $\ker d^{-n}$ into $\ker d^{-n-1}$. If we restrict s to $Z(\mathrm{B}^\bullet(\Gamma, N), d^\bullet)$, it remains a contracting homotopy and so the sequence of graded R-modules

$$\longrightarrow Z(\mathrm{B}^{-n}(\Gamma, N), d^{-n}) \xrightarrow{\delta} Z(\mathrm{B}^{-n+1}(\Gamma, N), d^{-n+1}) \xrightarrow{\delta} \cdots$$
$$\cdots \xrightarrow{\delta} Z(\mathrm{B}^{-1}(\Gamma, N), d^{-1}) \xrightarrow{\delta} Z(\mathrm{B}^0(\Gamma, N), d_0) \xrightarrow{\varepsilon} Z(N, d_N) \longrightarrow 0$$

is exact. Furthermore, when we ignore the internal differentials, the complex remains exact because $\delta s + s\delta = \text{id}$ does not involve the internal differential. This proves (1). To complete the proof of (3), we establish (2). From the Künneth theorem and the fact that k is a field, (2) follows from

$$H(\mathrm{B}^{-n}(\Gamma, N), d^{-n}) = H(\Gamma) \otimes_k \overbrace{H(\bar{\Gamma}) \otimes_k \cdots \otimes_k H(\bar{\Gamma})}^{n \text{ times}} \otimes_k H(N)$$
$$= \mathrm{B}^{-n}(H(\Gamma), H(N)).$$

Passing to homology, $\delta s + s\delta$ becomes $H(\delta)H(s) + H(s)H(\delta) = \text{id}$ and the sequence of homologies of the $\mathrm{B}^{-n}(\Gamma, N)$ is exact. $\square$

The proper projective resolution of N over Γ, $(\mathrm{B}^\bullet(\Gamma, N), \delta, d^\bullet, \varepsilon)$, is called the **bar resolution** of N over Γ. We define

$$\mathrm{B}^{-n}(M, \Gamma, N) = M \otimes_\Gamma \mathrm{B}^{-n}(\Gamma, N)$$
$$= M \otimes_k \underbrace{\bar{\Gamma} \otimes_k \cdots \otimes_k \bar{\Gamma}}_{n \text{ times}} \otimes_k N,$$

with internal differential, $d_M \otimes 1 + (-1)^{\deg} \otimes d^\bullet$, and external differential, $1 \otimes_\Gamma \delta$. The next result follows from the definitions.

Corollary 7.9. (1) $H(\mathrm{total}(\mathrm{B}^\bullet(M, \Gamma, N)), \mathrm{total}\, D) = \mathrm{Tor}_\Gamma(M, N)$.

(2) *If Γ is connected and simply-connected (that is, $\Gamma^1 = \{0\}$), then the spectral sequence of Theorem 7.6 converges strongly.*

PROOF: The first part of the corollary follows from Proposition 7.8 and the definition of Tor. To establish the second part, observe that a typical element $\gamma[\gamma_1 \mid \cdots \mid \gamma_n]a \in \mathrm{B}^{-n}(M, \Gamma, N)$ has internal degree

$$\deg \gamma + \deg a + \sum_{i=1}^n \deg \gamma_i \geq 2n.$$

Thus each $\mathrm{B}^{-n}(M, \Gamma, N)$ is at least $(2n - 1)$-connected and the resulting filtration on $\mathrm{total}(\mathrm{B}^\bullet(M, \Gamma, N))$ is bounded. By the proof of Theorem 2.6 and Definition 3.8, the spectral sequence converges strongly. □

Let us consider a simple example that illustrates a calculation with the bar construction. Let $\Gamma = \Lambda(x)$, the exterior algebra over k with $\deg x = m$ and trivial differential. To compute $\mathrm{Tor}_{\Lambda(x)}(k, k)$ consider $\mathrm{B}^\bullet(k, \Lambda(x), k)$. In homological degree $-n$, we find

$$\mathrm{B}^{-n}(k, \Lambda(x), k) \cong k \otimes_{\Lambda(x)} \Lambda(x) \otimes_k \overbrace{\bar\Lambda \otimes_k \cdots \otimes_k \bar\Lambda}^{n \text{ times}} \otimes_k k$$
$$\cong \bar\Lambda \otimes_k \cdots \otimes_k \bar\Lambda \quad (n \text{ times}).$$

Here $\bar\Lambda = \mathrm{coker}\, \eta$ is the vector space generated by x in dimension m and so $\mathrm{B}^{-n}(k, \Lambda(x), k)$ has dimension 1 over k generated by $[x \mid \cdots \mid x]$ of bidegree $(-n, mn)$. Since $x^2 = 0$, the formula for the external differential gives us that $\delta = 0$ and so

$$\mathrm{Tor}_{\Lambda(x)}(k, k) \cong \mathrm{B}^\bullet(k, \Lambda(x), k).$$

$\mathrm{Tor}^{*,*}_{\Lambda(x)}(k, k)$ can be displayed as a bigraded module giving the E_2-term of the spectral sequence of Theorem 7.6 in the diagram:

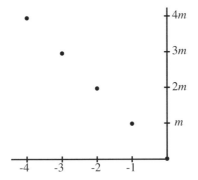

Suppose (Γ, d) is any connected differential graded algebra over k for which $H(\Gamma, d) \cong \Lambda(x)$. By the placement of the nontrivial entries of $\mathrm{Tor}_{\Lambda(x)}(k, k)$, no nontrivial differentials can arise in the spectral sequence and so we conclude

$$\mathrm{Tor}_{\Gamma}(k, k) \cong \mathrm{Tor}_{\Lambda(x)}(k, k).$$

In §7.2 we apply this calculation to compute $H^*(\Omega S^{2l+1}; k)$.

$\mathrm{Tor}_{\Gamma}(k, k)$ enjoys an additional useful piece of structure.

Proposition 7.10. *If Γ is a connected, differential graded algebra over k, then* $\mathrm{B}^{\bullet}(k, \Gamma, k)$ *is a differential coalgebra and this induces a coalgebra structure on* $\mathrm{Tor}_{\Gamma}(k, k)$ *as well as on the spectral sequence with* $E_2 \cong \mathrm{Tor}_{H(\Gamma)}(k, k)$, *converging to* $\mathrm{Tor}_{\Gamma}(k, k)$ *as a coalgebra.*

PROOF: The coproduct,

$$\Delta \colon \mathrm{B}^{-n}(k, \Gamma, k) \longrightarrow \bigoplus_{r+s=n} \mathrm{B}^{-r}(k, \Gamma, k) \otimes \mathrm{B}^{-s}(k, \Gamma, k),$$

is defined on a typical element $[\gamma_1 \mid \cdots \mid \gamma_n]$ by

$$\Delta([\gamma_1 \mid \cdots \mid \gamma_n]) = \sum_{j=0}^{n} [\gamma_1 \mid \cdots \mid \gamma_j] \otimes [\gamma_{j+1} \mid \cdots \mid \gamma_n],$$

where the empty bracket, $[\ ] = 1$, is taken when $j = 0$ or n. It then follows, by explicit computations, that Δ commutes with the external and internal differentials on $\mathrm{B}^{\bullet}(k, \Gamma, k)$ and so we have a differential coalgebra. The mapping induced on homology by Δ provides the coalgebra structure on $\mathrm{Tor}_{\Gamma}(k, k)$.

Observe that the comultiplication Δ respects the filtration of $\mathrm{B}^{\bullet}(k, \Gamma, k)$ that gives rise to the spectral sequence of Theorem 7.6, that is, $\Delta(F^{-n}) \subset \bigoplus_{r+s=n} F^{-r} \otimes F^{-s}$. This follows because the filtration in this case simply counts the "number of bars." Thus the spectral sequence is a spectral sequence of coalgebras and this coalgebra structure converges to the coalgebra structure on $\mathrm{Tor}_{\Gamma}(k, k)$. $\qquad\square$

This coproduct has been applied in the study of local rings to determine the structure of the Poincaré series associated to $\mathrm{Tor}_R(k, k)$ (see [Assmus59]).

There is a great deal of further structure that can be imposed on the Eilenberg-Moore spectral sequence via the bar construction and more generally from the structure of $\mathrm{Tor}_{\Gamma}(M, N)$. Furthermore, these structures have useful topological interpretations that we will present in the later sections of the chapter. Before leaving the bar construction, we consider some further structure in the special case of $\mathrm{B}^{\bullet}(k, \Gamma, k)$.

Lemma 7.11. *Let Γ and Λ be two differential graded algebras over k. Then, as coalgebras,*

$$\text{Tor}_\Gamma(k, k) \otimes \text{Tor}_\Lambda(k, k) \cong \text{Tor}_{\Gamma \otimes \Lambda}(k, k).$$

PROOF: We exploit another feature of the bar construction—it can be viewed as a simplicial object (§4.2). Let $\hat{B}_n(\Gamma) = \Gamma^{\otimes n}$ and endow $\hat{B}_\bullet(\Gamma)$ with face and degeneracy maps

$$\partial_i([\gamma_1 \mid \cdots \mid \gamma_n]) = \begin{cases} [\gamma_2 \mid \cdots \mid \gamma_n], & i = 0 \\ (-1)^i[\gamma_1 \mid \cdots \mid \bar{\gamma}_i \gamma_{i+1} \mid \cdots \mid \gamma_n], & 0 < i < n \\ (-1)^n[\bar{\gamma}_1 \mid \cdots \mid \bar{\gamma}_{n-1}], & i = n. \end{cases}$$

$$s_j([\gamma_1 \mid \cdots \mid \gamma_n]) = [\gamma_1 \mid \cdots \mid \gamma_j \mid 1 \mid \gamma_{j+1} \mid \cdots \mid \gamma_n], \qquad 0 \leq j \leq n.$$

The complex $(\hat{B}_\bullet, d)$ with differential $d = \sum_i (-1)^i \partial_i$ is called the **unnormalized bar construction** and it has the same homology as the bar construction (see [Mac Lane70]), that is, $H(\hat{B}_\bullet(\Gamma), d) \cong \text{Tor}_\Gamma(k, k)$. The simplicial nature of the unnormalized bar construction suggests the key idea of the proof.

Let $\hat{B}_\bullet(\Gamma) \times \hat{B}_\bullet(\Lambda)$ denote the product of simplicial sets (recall that means $(\hat{B}_\bullet(\Gamma) \times \hat{B}_\bullet(\Lambda))_n = \hat{B}_n(\Gamma) \times \hat{B}_n(\Lambda)$). There is a natural identification

$$\hat{B}_\bullet(\Gamma) \times \hat{B}_\bullet(\Lambda) \longleftrightarrow \hat{B}_\bullet(\Gamma \otimes \Lambda)$$

given by identifying $[\gamma_1 \mid \cdots \mid \gamma_n] \times [\lambda_1 \mid \cdots \mid \lambda_n]$ with $[\gamma_1 \otimes \lambda_1 \mid \cdots \mid \gamma_n \otimes \lambda_n]$ and this is an isomorphism of coalgebras.

To prove the lemma we compare $\hat{B}_\bullet(\Gamma) \otimes \hat{B}_\bullet(\Lambda)$ and $\hat{B}_\bullet(\Gamma) \times \hat{B}_\bullet(\Lambda)$ by taking the homomorphism at the heart of the Eilenberg-Zilber theorem (Theorem 4.36): There is a chain equivalence of coalgebras

$$\hat{B}_\bullet(\Gamma) \otimes \hat{B}_\bullet(\Lambda) \xrightarrow{EZ} \hat{B}_\bullet(\Gamma) \times \hat{B}_\bullet(\Lambda) \cong \hat{B}_\bullet(\Gamma \otimes \Lambda)$$

given by the shuffle map

$$EZ([\gamma_1 \mid \cdots \mid \gamma_p] \otimes [\lambda_1 \mid \cdots \mid \lambda_q])$$
$$= \sum_{(p,q)\text{-shuffles, } \sigma} (-1)^{s(\sigma)} [c_{\sigma(1)} \mid c_{\sigma(2)} \mid \cdots \mid c_{\sigma(p+q)}]$$

where $c_{\sigma(i)} = \gamma_{\sigma(i)} \otimes 1$ if $1 \leq \sigma(i) \leq p$ and $1 \otimes \lambda_{\sigma(i)-p}$ if $p+1 \leq \sigma(i) \leq p+q$. The sign is determined by $s(\sigma) = \sum (\deg c_i + 1)(\deg c_{j+p} + 1)$ summed over all pairs $(i, j+p)$ with $\sigma(i) > \sigma(p+j)$ (this is the sign convention that reflects a change of sign when elements are permuted past each other, according to their total degrees). The classical argument of [Eilenberg-Mac Lane53] completes the proof. $\square$

The bar construction $B^\bullet(k, \Gamma, k)$ is a functor from the category of differential graded algebras over k to the category of differential graded coalgebras. A differential graded algebra morphism $\phi \colon \Gamma \to \Lambda$ induces a mapping of coalgebras

$$\mathrm{Tor}_\phi(1, 1) \colon \mathrm{Tor}_\Gamma(k, k) \longrightarrow \mathrm{Tor}_\Lambda(k, k).$$

If (Γ, d) is a graded-commutative differential graded algebra, then the multiplication is a morphism of differential graded algebras. Putting the induced mapping $B^\bullet(k, \Gamma \otimes \Gamma, k) \to B^\bullet(k, \Gamma, k)$ together with Lemma 7.11 we get the following result.

Corollary 7.12. *If (Γ, d) is a graded-commutative, differential graded algebra over k, then $B^\bullet(k, \Gamma, k)$ is a differential graded Hopf algebra and $\mathrm{Tor}_\Gamma(k, k)$ is a commutative, cocommutative Hopf algebra with product induced by the shuffle product followed by multiplication on Γ. Furthermore, the spectral sequence converging to $\mathrm{Tor}_\Gamma(k, k)$ with E_2-term $\mathrm{Tor}_{H(\Gamma)}(k, k)$ is a spectral sequence of Hopf algebras.*

Finally, we record an algebraic consequence of the presence of a Hopf algebra structure on a spectral sequence that is most handy in topological applications. (See, especially, Chapter 10.)

Lemma 7.13. *If $\{E_r, d_r; r = 2, 3, \dots\}$ is a spectral sequence of Hopf algebras, then, for each r, in the lowest degree on which the differential d_r is nontrivial, it is defined on an indecomposable element and has as value a primitive element.*

PROOF: Since d_r is a derivation with respect to the multiplication it follows from the Leibniz formula that in the least degree in which it is nontrivial d_r is defined on an indecomposable element. Suppose that element is a. Then

$$\Delta(d_r(a)) = d_r(\Delta(a)) = d_r\left(a \otimes 1 + 1 \otimes a + \sum_i a_i' \otimes a_i''\right)$$
$$= d_r(a) \otimes 1 + 1 \otimes d_r(a) + \sum_i \left(d_r(a_i') \otimes a_i'' \pm a_i' \otimes d_r(a_i'')\right).$$

Since $\deg a_i' < \deg a$ and $\deg a_i'' < \deg a$, the assumption of least degree implies $d_r(a_i') = 0 = d_r(a_i'')$ and so $\Delta(d_r(a)) = d_r(a) \otimes 1 + 1 \otimes d_r(a)$. $\quad\square$

7.2 Bringing in the topology

We return to the basic problem of this chapter and the fibre square

$$\begin{array}{ccc}
E_f & \xrightarrow{\tilde{f}} & E \\
{\scriptstyle \pi'}\downarrow & & \downarrow{\scriptstyle \pi} \\
X & \xrightarrow{f} & B
\end{array}$$

with π a fibration with fibre F and π' the fibration induced by π over f. Let $C^*(\ ) = C^*(\ ;k)$ denote the cochain functor with coefficients in a field k. Applying $C^*(\ )$ to the fibre square yields $C^*(B)$-modules $C^*(X)$ and $C^*(E)$ via f^* and π^* and the cup-product. To make our "second guess" at $H^*(E_f;k)$ we compare $\mathrm{Tor}_{C^*(B)}(C^*(X), C^*(E))$ with $C^*(E_f)$ or $H^*(E_f;k)$.

Define the mapping $\alpha\colon C^*(X) \otimes_k C^*(E) \to C^*(E_f)$ by the composite

$$\alpha\colon C^*(X) \otimes_k C^*(E) \xrightarrow{(\pi')^* \otimes \tilde{f}^*} C^*(E_f) \otimes_k C^*(E_f) \overset{\smile}{\to} C^*(E_f).$$

In the definition of the tensor product over Γ, we also have the homomorphism

$$\xi = (1 \smile f^*) \otimes 1 - 1 \otimes (\pi^* \smile 1)\colon$$
$$C^*(X) \otimes_k C^*(B) \otimes_k C^*(E) \to C^*(X) \otimes_k C^*(E).$$

We show that $\alpha \circ \xi = 0$:

$$\begin{aligned}
\alpha((x &\smile f^*(b)) \otimes e - x \otimes (\pi^*(b) \smile e)) \\
&= [(\pi')^*(x \smile f^*(b)) \smile (\tilde{f})^*(e)] - [(\pi')^*(x) \smile (\tilde{f})^*(\pi^*(b) \smile e)] \\
&= [(\pi')^*(x) \smile (f \circ \pi')^*(b) \smile (\tilde{f})^*(e)] \\
&\qquad - [(\pi')^*(x) \smile (\pi \circ \tilde{f})^*(b) \smile (\tilde{f})^*(e)].
\end{aligned}$$

Since $\pi \circ \tilde{f} = f \circ \pi'$, we have $\alpha \circ \xi = 0$ and so α induces a mapping

$$\bar{\alpha}\colon \left[C^*(X) \otimes_k C^*(E) \Big/ \mathrm{im}\, \xi \right] = C^*(X) \otimes_{C^*(B)} C^*(E) \to C^*(E_f).$$

Suppose $Q^\bullet \xrightarrow{\varepsilon} C^*(X) \to 0$ is a proper projective resolution of $C^*(X)$ as a right $C^*(B)$-module. There is a homomorphism $\varepsilon\colon \mathrm{total}(Q^\bullet) \to C^*(X)$ given on $\mathrm{total}^n(Q^\bullet)$ by ε on $(Q^0)^n$ and zero on $(Q^{-i})^{n+i}$ when $i > 0$. Consider the composite θ as in the diagram

$$
\begin{array}{ccc}
\mathrm{total}(Q^\bullet) \otimes_{C^*(B)} C^*(E) & & \\
{\scriptstyle \varepsilon \otimes 1}\Big\downarrow & \searrow{\scriptstyle \theta} & \\
C^*(X) \otimes_{C^*(B)} C^*(E) & \xrightarrow[\bar{\alpha}]{} & C^*(E_f).
\end{array}
$$

The relationship between the topology of the pullback diagram and the resulting algebra of cochains is described in a remarkable theorem of [Eilenberg-Moore66].

Theorem 7.14 (the second Eilenberg-Moore theorem). *Suppose* $\pi\colon E \to B$ *is a fibration with connected fibre F for which the system of local coefficients induced by the fibre on B is simple. Suppose $f\colon X \to B$ is a continuous mapping and $\pi'\colon E_f \to X$ is the pullback fibration of π over f. Then the mapping $\theta = \bar{\alpha} \circ (\varepsilon \otimes 1)$ is a homology isomorphism, that is,*

$$\theta\colon \mathrm{total}(Q^\bullet) \otimes_{C^*(B)} C^*(E) \to C^*(E_f)$$

induces an isomorphism on homology,

$$\theta^*\colon \mathrm{Tor}_{C^*(B)}(C^*(X), C^*(E)) \to H^*(E_f; k).$$

PROOF: The strategy of proof is to (a) filter the domain and codomain of θ conveniently, (b) obtain spectral sequences that are isomorphic at E_2 via θ and (c) apply Theorem 3.5 to secure the desired isomorphism. Since $C^*(E_f)$ and $C^*(E)$ are each algebras of cochains of the total space of a fibration, they can be filtered as in the Leray-Serre spectral sequence (§5.1). Explicitly, we write

$$F^q C^*(E_f) = \ker(i^*\colon C^*(E_f) \to C^*(J_f^{q-1}))$$

where J_f^{q-1} is the subspace of E_f that lies over the $(q-1)$-skeleton of X. If we filter $C^*(E)$ similarly over the skeleta of B, and filter $C^*(B)$ by degree, $F^q C^*(B) = \bigoplus_{i \geq q} C^i(B)$, then we obtain a tensor product filtration on $C^*(B) \otimes_k C^*(E)$:

$$F^r(C^*(B) \otimes_k C^*(E)) = \bigoplus_{i=0}^{r} F^i C^*(B) \otimes_k F^{r-i} C^*(E).$$

The cochains in $F^q C^*(B)$ are supported on skeleta of dimension at least q and so $\pi^*(F^q C^*(B)) \subset F^q C^*(E)$. Also, the cup product on $C^*(E)$ respects the filtration by inverse images of skeleta. Thus, the $C^*(B)$-module structure on $C^*(E)$ is filtration-preserving:

$$F^q(C^*(B) \otimes_k C^*(E)) \xrightarrow{\pi^* \otimes 1} \bigoplus_{i=0}^{q} F^i C^*(E) \otimes_k F^{q-i} C^*(E) \xrightarrow{\smile} F^q C^*(E).$$

Filter $\mathrm{total}(Q^\bullet)$ by total degree, that is, $F^q(\mathrm{total}(Q^\bullet)) = \bigoplus_{i+j \geq q} (Q^i)^j$; this respects the $C^*(B)$-module structure. Filter $\mathrm{total}(Q^\bullet) \otimes_{C^*(B)} C^*(E)$ as a tensor product

$$F^q(\mathrm{total}(Q^\bullet) \otimes_{C^*(B)} C^*(E)) = \bigoplus_{i+j=q} F^i \,\mathrm{total}(Q^\bullet) \otimes_{C^*(B)} F^j C^*(E).$$

Such a tensor product on the right is possible because both filtrations preserve the $C^*(B)$-module structure. Observe that the map θ is filtration-preserving:

$$\theta\colon F^q(\mathrm{total}(Q^\bullet) \otimes_{C^*(B)} C^*(E)) \xrightarrow{\varepsilon \otimes 1} F^q(C^*(X) \otimes_{C^*(B)} C^*(E))$$

$$\xrightarrow{(\pi')^* \otimes (\tilde{f})^*} F^q(C^*(E_f) \otimes_k C^*(E_f)) \xrightarrow{\smile} F^q(C^*(E_f)).$$

The mapping $\varepsilon \otimes 1$ is filtration-preserving, since it only sees the total degree and $\smile$ is already known to be filtration-preserving; $(\pi')^*$ and $(\tilde{f})^*$ preserve the filtration as geometric maps that preserve skeleta. Thus θ induces a map of spectral sequences by Theorem 3.5.

Recall from the construction of the Leray-Serre spectral sequence and the assumptions of the theorem, that the filtration on $C^*(E_f)$ has E_1-term, $C^*(X; \mathcal{H}^*(F; k))$ and because k is a field and the system of local coefficients is simple, E_2-term given by $H^*(X; k) \otimes_k H^*(F; k)$. For the filtered differential graded module, $\text{total}(Q^\bullet) \otimes_{C^*(B)} C^*(E)$, we have the E_0-term, $F^q/F^{q+1} =$

$$\bigoplus_{i+j=q} F^i \text{total}(Q^\bullet) \otimes_{C^*(B)} F^j C^*(E) \bigg/ \bigoplus_{l+m=q+1} F^l \text{total}(Q^\bullet) \otimes_{C^*(B)} F^m C^*(E).$$

The filtration on $\text{total}(Q^\bullet)$ is given by total degree and $d^\bullet + \delta$ raises total degree. It follows that d_0, the zeroth differential in the Leray-Serre spectral sequence for E, is zero on the $\text{total}(Q^\bullet)$ part. Since the filtration is decreasing, we compute that $F^q/F^{q+1} \cong [\text{total}(Q^\bullet)]^q \otimes E_0^q C^*(E)$ and $d_0 = 1 \otimes \bar{d}_0$. Thus $E_1^{q,*} = H(F^q/F^{q+1}) \cong [\text{total}(Q^\bullet)]^q \otimes_{C^*(B)} C^*(B; \mathcal{H}^*(F; k))$. Since the local coefficient system induced by the fibre is simple, we can write $E_1^{q,*} \cong$

$$[\text{total}(Q^\bullet)]_q \otimes_{C^*(B)} \otimes_k C^*(B) \otimes_k H^*(F; k) \cong [\text{total}(Q^\bullet)]_q \otimes_k H^*(F; k)$$

with d_1 given by $(d^\bullet + \delta) \otimes 1$. To compute $H(\text{total}(Q^\bullet), d^\bullet + \delta)$, recall from the proof of Theorem 2.20 that, if we view $\text{total}(Q^\bullet)$ as a double complex, then there are two filtrations, one of which yields a spectral sequence that collapses to a column at the E_2-term. This gives

$$H(\text{total}(Q^\bullet), d^\bullet + \delta) = H(C^*(X), d_X) = H^*(X; k).$$

Therefore, the E_2-term derived from this filtration on $\text{total}(Q^\bullet) \otimes_{C^*(B)} C^*(E)$ is isomorphic to $H^*(X; k) \otimes_k H^*(F; k)$.

To complete the proof of the theorem, it suffices to show that θ induces an isomorphism of E_2-terms. First, at the E_1-term, the induced homomorphism

$$E_1\theta \colon \text{total}(Q^\bullet) \otimes_k H^*(F; k) \longrightarrow C^*(X) \otimes_k H^*(F; k),$$

is induced by $(\tilde{f})^*$ on the $H^*(F; k)$ factor and so, by the construction of a pullback, is an isomorphism. Furthermore, the mapping $E_1\theta$ on the $\text{total}(Q^\bullet)$ factor is induced by the map ε, which is the mapping in the proof of Theorem 2.20 (the Künneth spectral sequence) that yields $H(\text{total}(Q^\bullet), d^\bullet + \delta) \cong H(C^*(X), d_X)$. Thus, $E_2\theta$ is an isomorphism. By Theorems 3.5 and 5.2 (the Leray-Serre spectral sequence) we have $H(\text{total}(Q^\bullet) \otimes_{C^*(B)} C^*(E)) = \text{Tor}_{C^*(B)}(C^*(X), C^*(E))$ and $\text{Tor}_{C^*(B)}(C^*(X), C^*(E)) \xrightarrow[\theta^*]{\cong} H^*(E_f; k)$. $\square$

The interested reader may want to rework the proof with coefficients in a ring R. We remark that the entire algebraic setting and topological interpretation (for pushouts) can be developed dually in homology ([Eilenberg-Moore66]); this is described later in the chapter.

Suppose B is simply-connected. By normalizing the cochains on B, we can take $C^*(B)$ to be simply-connected as a differential graded algebra. Combining the first and second Eilenberg-Moore theorems with Corollary 7.9 (2), we obtain the solution to the problem posed in the introduction to the chapter.

Theorem 7.15. *Suppose B is simply-connected, $\pi: E \to B$, a fibration with connected fibre F, and $f: X \to B$, a continuous mapping. Let $\pi': E_f \to X$ be the induced fibration over f given as a pullback. Then there is a spectral sequence, lying in the second quadrant, with*

$$E_2 \cong \operatorname{Tor}_{H^*(B;k)}(H^*(X;k), H^*(E;k))$$

and converging strongly to $H^*(E_f; k)$.

An immediate corollary to Theorem 7.15 is a spectral sequence with E_2-term $\operatorname{Tor}_{H^*(X;k)}(k,k)$ converging strongly to $H_*(\Omega X; k)$ when X is simply-connected. More generally, the cohomology of the fibre of a fibration is recoverable from the base and total space.

Corollary 7.16. *If $\pi: E \to B$ is a fibration with base space B simply-connected and fibre F connected, then there is a spectral sequence with E_2-term given by $\operatorname{Tor}_{H^*(B;k)}(k, H^*(E;k))$ and converging strongly to $H^*(F; k)$.*

With this corollary, we can apply the calculation of $\operatorname{Tor}_{\Lambda(x)}(k,k)$ in §7.1 to a topological computation. Consider the fibre square

$$
\begin{array}{ccc}
\Omega S^{2n+1} & \longrightarrow & PS^{2n+1} \simeq * \\
\downarrow & & \downarrow \\
* & \longrightarrow & S^{2n+1}.
\end{array}
$$

By Corollary 7.16, $H^*(\Omega S^{2n+1}; k)$ is the target of the Eilenberg-Moore spectral sequence with E_2-term $E_2^{i,j} \cong \operatorname{Tor}_{\Lambda(x_{2n+1})}^{i,j}(k,k)$, where $\deg x_{2n+1} = 2n+1$. Our calculation shows that this spectral sequence collapses and we conclude the classical result—as vector spaces,

$$H^t(\Omega S^{2n+1}; k) \cong \operatorname{total}^t(\operatorname{Tor}_{\Lambda(x_{2n+1})}^{*,*}(k,k)) \cong \begin{cases} k, & \text{if } t = 2nr \text{ for } r \geq 0, \\ \{0\}, & \text{otherwise.} \end{cases}$$

In fact, we can do a little more here. Since ΩS^{2n+1} is an H-space, we can wonder about the Hopf algebra structure on $H^*(\Omega S^{2n+1}; k)$. Since

$H^*(\Omega S^{2n+1}; k) \cong \mathrm{Tor}_{\Lambda(x_{2n+1})}(k, k)$, there is a product structure already at the level of the bar resolution. Define

$$\gamma_m([x]) = [x \mid \cdots \mid x] \qquad (m \text{ times}),$$

where $[x]$ has bidegree $(-1, 2n + 1)$ or total degree $2n$. The shuffle product on these generators is given by

$$\gamma_p([x]) * \gamma_q([x]) = \sum_{(p,q)\text{-shuffles } \sigma} (-1)^{s(\sigma)} \gamma_{p+q}([x]).$$

Since $2n + 1$ is odd, $s(\sigma)$ is even for all (p, q)-shuffles and so

$$\gamma_p([x]) * \gamma_q([x]) = \binom{p + q}{p} \gamma_{p+q}([x]),$$

where $\binom{p + q}{p}$ is the binomial coefficient; $\binom{p + q}{p}$ is the number of (p, q)-shuffles. The reader can find this structure in the definition found in Exercise 1.7 of a **divided power algebra** on an element $[x]$ of even degree $2n$.

The comultiplication resulting from the bar construction is described by

$$\Delta(\gamma_s([x])) = \sum_{i=0}^{s} \gamma_i([x]) \otimes \gamma_{s-i}([x]),$$

where $\gamma_0([x]) = 1$. Thus, $\mathrm{Tor}_{\Lambda(x)}(k, k) \cong \Gamma([x])$, the divided power algebra on $[x]$, as a Hopf algebra over k.

It is tempting to conclude that we have determined the Hopf algebra structure on $H^*(\Omega S^{2n+1}; k)$. However, to do that we need to assume that $\mathrm{Tor}_{C^*(S^{2n+1};k)}(k, k)$ is isomorphic as a Hopf algebra to $\mathrm{Tor}_{H^*(S^{2n+1};k)}(k, k)$, and that the Hopf algebra structure coincides with the Hopf algebra structure on $H^*(\Omega S^{2n+1}; k)$. With a bit more effort, we will be able to conclude this. However, observe that $C^*(S^{2n+1}; k)$ is not a graded-commutative differential graded algebra like its homology! This subtlety needs to be considered next.

Multiplicative matters

Our goal in this section is to prove that, in fact, the Eilenberg-Moore spectral sequence carries information on the cup-product structure of $H^*(E_f; k)$.

Proposition 7.17. *On* $\mathrm{Tor}_{C^*(B)}(C^*(X), C^*(E))$, *there is a natural algebra structure and furthermore,*

$$\theta \colon \mathrm{Tor}_{C^*(B)}(C^*(X), C^*(E)) \longrightarrow H^*(E_f; k)$$

is an isomorphism of algebras.

To prove the proposition, we begin with a natural product structure on the $\mathrm{Tor}_\Gamma(-, -)$ functor of the form

$$\mathrm{Tor}_\Gamma(M_1, N_1) \otimes_k \mathrm{Tor}_\Gamma(M_2, N_2) \xrightarrow{\Phi} \mathrm{Tor}_{\Gamma \otimes \Gamma}(M_1 \otimes_k M_2, N_1 \otimes_k N_2).$$

Let N_1 and N_2 be left Γ-modules and let

$$\cdots \to P^{-i} \to P^{-i+1} \to \cdots \to P^0 \to N_1 \to 0,$$
$$\cdots \to Q^{-i} \to Q^{-i+1} \to \cdots \to Q^0 \to N_2 \to 0$$

be left proper projective resolutions of N_1 and N_2 over Γ. Let

$$R^{-i} = \bigoplus_{m+n=i} P^{-m} \otimes_k Q^{-n};$$

the Künneth theorem implies that

$$\cdots \to R^{-i} \to R^{-i+1} \to \cdots \to R^0 \to N_1 \otimes N_2 \to 0$$

is a left proper projective resolution of $N_1 \otimes N_2$ over $\Gamma \otimes \Gamma$ that acts on $N_1 \otimes N_2$ by the composite

$$\Gamma \otimes \Gamma \otimes N_1 \otimes N_2 \xrightarrow{1 \otimes T \otimes 1} \Gamma \otimes N_1 \otimes \Gamma \otimes N_2 \xrightarrow{\varphi_1 \otimes \varphi_2} N_1 \otimes N_2.$$

Let $\xi \colon M \otimes \Gamma \otimes N \to M \otimes N$ be the mapping $\xi(m \otimes \gamma \otimes n) = m\gamma \otimes n - m \otimes \gamma n$ for $M \in \mathbf{DGMod}_\Gamma$ and $N \in \mathbf{DG}_\Gamma\mathbf{Mod}$. Recall that $M \otimes_\Gamma N = \mathrm{coker}\,\xi$. If M_1 and M_2 are right Γ-modules, then $\mathrm{im}(\xi \otimes \xi) \subset (M_1 \otimes N_1) \otimes (M_2 \otimes N_2)$ is carried by the interchange $1 \otimes T \otimes 1$ to the image of ξ:

$$
\begin{array}{ccc}
(M_1 \otimes M_2) \otimes (\Gamma \otimes \Gamma) \otimes (N_1 \otimes N_2) & \xrightarrow{\xi} & (M_1 \otimes M_2) \otimes (N_1 \otimes N_2) \\
{\scriptstyle 1 \otimes T \otimes T \otimes 1}\downarrow & & \uparrow \\
(M_1 \otimes \Gamma \otimes M_2) \otimes (N_1 \otimes \Gamma \otimes N_2) & & {\scriptstyle 1 \otimes T \otimes 1} \\
{\scriptstyle 1 \otimes 1 \otimes T \otimes 1 \otimes 1}\downarrow & & \\
(M_1 \otimes \Gamma \otimes N_1) \otimes (M_2 \otimes \Gamma \otimes N_2) & \xrightarrow[\xi \otimes \xi]{} & (M_1 \otimes N_1) \otimes (M_2 \otimes N_2)
\end{array}
$$

and so the interchange induces a homomorphism, for all m and n,

$$(M_1 \otimes_\Gamma P^{-m}) \otimes (M_2 \otimes_\Gamma Q^{-n}) \to (M_1 \otimes M_2) \otimes_{\Gamma \otimes \Gamma} (P^{-m} \otimes Q^{-n}).$$

This homomorphism induces the pairing

$$\Phi \colon \mathrm{Tor}_\Gamma^m(M_1, N_1) \otimes_k \mathrm{Tor}_\Gamma^n(M_2, N_2) \to \mathrm{Tor}_{\Gamma \otimes \Gamma}^{m+n}(M_1 \otimes_k M_2, N_1 \otimes_k N_2).$$

Recall from §4.4 that the coproduct on the chains of a space is the compo-sition:

$$C_*(X) \xrightarrow{\Delta_*} C_*(X \times X) \xrightarrow{AW} C_*(X) \otimes C_*(X).$$

This factorization, however, is not one of coalgebra mappings. In particular, AW is not a coalgebra mapping. The inverse of AW up to homotopy is the Eilenberg-Zilber map EZ, based on shuffles; EZ *is* a mapping of coalgebras. Thus we get an honest coproduct structure on homology but only up to homotopy at the chain level. In cohomology, to factor the cup-product in terms of algebra mappings we meet a new problem: $(C_*(X) \otimes C_*(X))^{\text{dual}}$ is rarely $C^*(X) \otimes C^*(X)$—it is a completed tensor product. We can, however, consider the maps

$$C^*(X) \otimes C^*(X) \xrightarrow{i} (C_*(X) \otimes C_*(X))^{\text{dual}} \xleftarrow{EZ^*} C^*(X \times X) \xrightarrow{\Delta^*} C^*(X),$$

where i denotes the inclusion into the completion. Every mapping here is a differential graded algebra mapping and on homology we can fit the induced homomorphisms into a product because EZ is a homology equivalence. Its inverse on homology is also an algebra mapping.

To prove the proposition, we consider the diagram in which $A = C^*(B)$

$$
\begin{array}{ccc}
\text{Tor}_A(C^*(X),C^*(E)) \otimes_k \text{Tor}_A(C^*(X),C^*(E)) & \xrightarrow[\cong]{\theta^* \otimes \theta^*} & H^*(E_f) \otimes_k H^*(E_f) \\
\Big\downarrow{\Phi} & & \Big\uparrow \\
\text{Tor}_{A \otimes A}(C^*(X) \otimes_k C^*(X), C^*(E) \otimes_k C^*(E)) & & \\
\Big\downarrow{\text{Tor}_i(i,i)} & & \cong \Big\downarrow{EZ^*} \\
\text{Tor}_{(C_*(B) \otimes C_*(B))^*}((C_*(X) \otimes C_*(X))^*, (C_*(E) \otimes C_*(E))^*) & & \\
\Big\uparrow{\text{Tor}_{EZ^*}(EZ^*,EZ^*)} \cong & & \\
\text{Tor}_{C^*(B \times B)}(C^*(X \times X), C^*(E \times E)) & \xrightarrow[\cong]{\theta^*} & H^*(E_f \times E_f) \\
\Big\downarrow{\text{Tor}_\Delta(\Delta,\Delta)} & & \Big\downarrow{\Delta^*} \\
\text{Tor}_{C^*(B)}(C^*(X), C^*(E)) & \xrightarrow[\theta^*]{\cong} & H^*(E_f).
\end{array}
$$

We leave it to the reader to check that the Eilenberg-Zilber maps, EZ, commute with the actions of $C^*(B)$ and $C^*(B \times B)$ on the appropriate modules, so they induce the vertical isomorphism as shown; since they are homology isomorphisms, Corollary 7.7 implies that $\text{Tor}_{EZ^*}(EZ^*, EZ^*)$ is an isomorphism.

We claim that the entire diagram commutes. The bottom square com-mutes by the naturality of the homomorphism θ and the geometric mappings of

fibrations

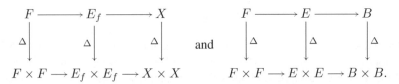

The top square commutes because the morphisms constructed to carry the mappings cover the composite $\Delta^* \circ (EZ^*)^{-1} \circ i^*$, which is the cup-product. Since the composite on the left side is the product induced by Δ on Tor and the product on the right is the cup-product, we have proved Proposition 7.17. □

In the proof of Proposition 7.17, the reader can check that all maps in sight preserve the filtration leading to the Eilenberg-Moore spectral sequence. This implies the following useful result.

Corollary 7.18. *The Eilenberg-Moore spectral sequence is a spectral sequence of algebras, converging to its target as an algebra.*

In the special case of the path-loop fibration, our factorization of the cup-product into differential graded algebra morphisms, albeit dizzying in direction, induces a Hopf algebra structure on Tor because the crucial (backward pointing) morphism is an equivalence.

Corollary 7.19. *The Eilenberg-Moore spectral sequence associated to the path-loop fibration on X, a simply-connected space, with $E_2 \cong \mathrm{Tor}_{H^*(X;k)}(k, k)$ is a spectral sequence of Hopf algebras, converging to $H^*(\Omega X; k)$ as a Hopf algebra.*

In §7.1 we computed $\mathrm{Tor}_{(A^*, \partial)}(k, k) \cong \Gamma([x])$ for any differential graded algebra (A^*, ∂) with $H(A^*, \partial) \cong \Lambda(x)$ and $\deg x$ odd. Corollary 7.19 implies that $\Gamma([x]) \cong E^0(H^*(\Omega S^{2n+1}; k))$ as Hopf algebras. To obtain the Hopf algebra $H^*(\Omega S^{2n+1}; k)$ exactly, we need to settle the extension problems. As a vector space, there are no problems as $\Gamma([x])$ is concentrated in even degrees. As an algebra the extension problem depends on the characteristic of k. If char $k = 0$, then $\Gamma([x])$ is isomorphic to a polynomial algebra on one generator and there are no vanishing products. As in Example 1.K, we have established the product structure completely. If the characteristic of k is finite, then we must turn to another argument. Here the structure of the Leray-Serre spectral sequence for the path-loop fibration settles the extension problems. To wit, the issue at hand is the vanishing of pth powers of the class $[x]$. They vanish in the structure of a divided power algebra (Proposition 7.26). In the cohomology of ΩS^{2n+1}, the appearance of a pth power implies a class on which the first (and only) nontrivial differential in the Leray-Serre spectral sequence vanishes. This leaves a nontrivial class in $H^+(PS^{2n+1}; k)$ and gives a contradiction. Thus the product structure on $H^*(\Omega S^{2n+1}; k)$ coincides with the product structure of $\Gamma([x])$ and we have recovered $H^*(\Omega S^{2n+1}; k)$ as a Hopf algebra.

The cup-product on cochains fails to be a differential graded algebra mapping. However, special circumstances (an explicit and nice homotopy inverse) allowed us to produce an algebra structure on the spectral sequence. If (A^*, d_A) and (B^*, d_B) are differential graded algebras, we can ask more generally for conditions on a k-module morphism $(A^*, d_A) \to (B^*, d_B)$ to define a morphism on Tor, even for the case of $\mathrm{Tor}_{(A^*, d_A)}(k, k) \to \mathrm{Tor}_{(B^*, d_B)}(k, k)$. This question with $A^* = C^*(X; k) \otimes C^*(X; k)$ and $B^* = C^*(X; k)$ includes the case of the cup-product. We discuss this question in the next chapter.

7.3 The Koszul complex

With the bar construction and Corollary 7.19 the Eilenberg-Moore spectral sequence leads to the computation of $H^*(\Omega S^{2n+1}; k)$ as a Hopf algebra. The E_2-term in this case is rather simple and the spectral sequence collapses. In contrast, the same computation with the Leray-Serre spectral sequence (as outlined in Example 1.H) involves a flurry of nontrivial differentials, some of which settle the extension problems for the E_∞-term of the Eilenberg-Moore spectral sequence.

Lest the reader come to believe that the Eilenberg-Moore spectral sequence is a more facile method for computation, we describe a symmetric situation where the Leray-Serre spectral sequence associated to a fibration is quite simple while the Eilenberg-Moore spectral sequence requires a panoply of differentials.

Consider the Hopf fibration, $S^3 \hookrightarrow S^7 \xrightarrow{\nu} S^4$, which is associated to the quaternionic multiplication. We can compute $H^*(S^3; k)$ as the cohomology of the fibre of ν by applying the Eilenberg-Moore spectral sequence. Here the E_2-term is $\mathrm{Tor}_{H^*(S^4; k)}(k, H^*(S^7; k))$ which we compute using the bar construction. In particular, as a vector space, $\mathrm{B}^{-n}(k, H^*(S^4; k), H^*(S^7; k))$ is generated by

$$\overbrace{[x_4 \mid \cdots \mid x_4]}^{n \text{ times}} \quad \text{and} \quad \overbrace{[x_4 \mid \cdots \mid x_4]}^{n \text{ times}} y_7.$$

On each generator, the exterior differential, δ, is zero and so

$$\mathrm{Tor}_{H^*(S^4; k)}(k, H^*(S^7; k)) \cong \mathrm{B}^\bullet(k, H^*(S^4; k), H^*(S^7; k)).$$

By a dimension argument, the only possibly nontrivial differential is d_2. If $d_2([x_4 \mid x_4]) = [\ ]y_7$, then the multiplicative structure implies the rest of the nontrivial differentials needed to obtain a finite-dimensional E_∞-term. Thus

$[x_4]$ remains to generate $H^*(S^3; k)$ in total degree $3 = 4 - 1$.

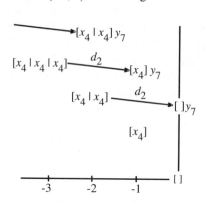

This computation is not at all efficient as a way to determine the cohomology of the fibre of ν. It does reveal, however, some of the algebraic topology associated to the Hopf fibration: In Chapter 8, the nature of the differentials in the Eilenberg-Moore spectral sequence is discussed and the interpretation of differentials in terms of higher cohomology operations may be applied in this case. The reader can contrast this computation with the discussion of the Leray-Serre spectral sequence for fibrations of spheres by spheres after Example 5.B.

In any application of the Eilenberg-Moore spectral sequence, it is necessary to compute $\mathrm{Tor}_{H(\Gamma)}(H(M), H(N))$ in order to begin with the E_2-term. In the examples we have considered so far, the bar construction has been manageable as a result of the sparseness of the algebraic input. For algebras with more than one generator and some relations, the bar construction can become quite large and complicated. One of the features of homological algebra is the invariance of the derived functors with regard to the choice of resolution and so the construction of smaller and more manageable resolutions is important. Though this cannot be achieved in general, there is a considerable reduction possible when $H(\Gamma)$ is a free graded commutative algebra over k.

Let J denote a graded set. The **free graded commutative algebra, $S(J)$, generated by** J is determined by a universal property: The set J includes in $S(J)$ as a generating set for the algebra. If A is any graded commutative algebra over k and $J \to A$ any mapping of sets, then there is a unique mapping of algebras $S(J) \to A$ extending $J \to A$.

We may write $J = J^{\mathrm{even}} \cup J^{\mathrm{odd}}$ where $J^{\mathrm{even}} = \{x \in J \mid \deg x \text{ is even}\}$ and $J^{\mathrm{odd}} = J - J^{\mathrm{even}}$ (If char $k = 2$, then let $J^{\mathrm{even}} = J$). If J is a singleton set $J = \{x\}$, then $S(\{x\}) \cong k[x]$ when x has even degree or char $k = 2$, and $S(\{x\}) \cong \Lambda(x)$ when x has odd degree. If J has finitely many elements in each degree, then an application of the universal property shows $S(J) \cong k[J^{\mathrm{even}}] \otimes_k \Lambda(J^{\mathrm{odd}})$, that is, the tensor product of the polynomial algebra on the elements in J^{even} and the exterior algebra on the elements in J^{odd}.

Definition 7.20. *Let J be a positively graded set (that is, $J^r = \emptyset$, for $r \leq 0$) which is finite in each degree. The* **Koszul complex** *associated to $S(J)$ is the differential graded algebra*

$$\mathcal{K}(J) = (G(s^{-1}J) \otimes_k S(J), d_{\mathcal{K}}),$$

where $s^{-1}J$ is the **desuspension** *of J, given by the graded set $(s^{-1}J)^m = J^{m+1}$ and*

$$G(s^{-1}J) = \Lambda(s^{-1}J^{\mathrm{even}}) \otimes_k \Gamma(s^{-1}J^{\mathrm{odd}}).$$

The differential $d_{\mathcal{K}}$ is given on generators by

$$d_{\mathcal{K}}(1 \otimes x) = 0, \text{ for } x \in S(J),$$
$$d_{\mathcal{K}}(s^{-1}x \otimes 1) = 1 \otimes x, \text{ for } x \in J^{\mathrm{even}},$$
$$d_{\mathcal{K}}(\gamma_n(s^{-1}x) \otimes 1) = \gamma_{n-1}(s^{-1}x) \otimes x, \text{ for } x \in J^{\mathrm{odd}},$$

and extended as a derivation.

Proposition 7.21. *For J a positively graded set that is finite in each degree, the homology of the associated Koszul complex is trivial, that is,*

$$H(\mathcal{K}(J), d_{\mathcal{K}}) \cong \{0\}.$$

PROOF: If $J = \{x\}$ is a singleton set and x has even degree, then the argument given in Example 1.H applies to prove this case. If $J = \{x\}$ and x has odd degree, then observe that $\Gamma(s^{-1}x) \cong B^\bullet(k, \Lambda(x), k)$ via the mapping,

$$\gamma_m(s^{-1}x) \mapsto [x \mid \cdots \mid x] \quad (m \text{ times}).$$

It follows that $\mathcal{K}(\{x\}) \cong \mathrm{total}\, B^\bullet(\Lambda(x), \Lambda(x))$, which is acyclic.

To complete the proof for an arbitrary generating set J, filter J by degree and apply the Künneth theorem and induction to the successive filtrations. $\square$

In order to construct a resolution from the Koszul complex, we introduce a bigrading: Because we can write

$$\mathcal{K}(J) = (\Lambda(s^{-1}J^{\mathrm{even}}) \otimes_k \Gamma(s^{-1}J^{\mathrm{odd}})) \otimes_k (k[J^{\mathrm{even}}] \otimes_k \Lambda(J^{\mathrm{odd}})),$$

we assign the bigradings as follows: If $x \in J$, $\mathrm{bideg}(1 \otimes x) = (0, \deg x)$ and $\mathrm{bideg}(s^{-1}x \otimes 1) = (-1, \deg x)$ and, if $y \in J^{\mathrm{odd}}$, $\mathrm{bideg}(\gamma_n(s^{-1}y) \otimes 1) = (-n, n \deg y)$. All other elements in $\mathcal{K}(J)$ are bigraded as in the tensor product of bigraded algebras. Taking $G(s^{-1}J)$ to be bigraded according to this scheme, we observe that $\mathcal{K}^{-p,*}(J) = G^{-p,*}(s^{-1}J) \otimes_k S(J)$, that is, $\mathcal{K}^{-p,*}(J)$ is an extended $S(J)$-module. Furthermore, the differential on the Koszul complex is defined so that $d_{\mathcal{K}}| : \mathcal{K}^{-p,*}(J) \to \mathcal{K}^{-p+1,*}(J)$.

Corollary 7.22. *The sequence,*

$$\cdots \xrightarrow{d_{\mathcal{K}}} \mathcal{K}^{-p,*}(J) \xrightarrow{d_{\mathcal{K}}} \cdots \xrightarrow{d_{\mathcal{K}}} \mathcal{K}^{-1,*}(J) \xrightarrow{d_{\mathcal{K}}} S(J) \xrightarrow{\varepsilon} k \to 0,$$

*is a resolution of k as a right S(J)-module, called the **Koszul resolution**.*

We apply the resolution immediately to compute $\mathrm{Tor}_{S(J)}(k,k)$. Following Definition 7.5,

$$\begin{aligned}
\mathrm{Tor}_{S(J)}(k,k) &= H(\mathcal{K}(J) \otimes_{S(J)} k, d_{\mathcal{K}} \otimes 1) \\
&= H(G(s^{-1}J) \otimes S(J) \otimes_{S(J)} \otimes k, d_{\mathcal{K}} \otimes_{S(J)} 1) \\
&= H(G(s^{-1}J), 0) = G(s^{-1}J).
\end{aligned}$$

More generally, suppose L is a left $S(J)$-module.

Corollary 7.23. *For L a left S(J)-module,*

$$\mathrm{Tor}_{S(J)}(k, L) \cong H(G(s^{-1}J) \otimes_k L, d_L),$$

where the complex $G(s^{-1}J) \otimes L$ has the differential, d_L, given by

$$d_L(1 \otimes l) = 0,$$
$$d_L(s^{-1}x \otimes l) = 1 \otimes xl, \text{ for } x \in J,$$
$$d_L(\gamma_n(s^{-1}x) \otimes l) = \gamma_{n-1}(s^{-1}x) \otimes xl, \text{ for } x \in J^{\mathrm{odd}}.$$

If L is a graded commutative algebra over S(J), then $\mathrm{Tor}_{S(J)}(k, L)$ is isomorphic to $H(G(s^{-1}J) \otimes L, d_L)$ as bigraded algebras over k.

The result follows from Corollary 7.22 and the definition of Tor:

$$\begin{aligned}
\mathrm{Tor}_{S(J)}(k, L) &\cong H(G(s^{-1}J) \otimes_k S(J) \otimes_{S(J)} L, d \otimes_{S(J)} 1) \\
&\cong H(G(s^{-1}J) \otimes_k L, d_L).
\end{aligned}$$

To recover the algebra structure on $\mathrm{Tor}_{S(J)}(k, L)$ check that the following diagram of graded algebras commutes:

$$\begin{array}{ccc}
G(s^{-1}J) \otimes L \otimes G(s^{-1}J) \otimes L & \longrightarrow & L \otimes L \\
\downarrow & & \downarrow \\
G(s^{-1}J) \otimes L & \longrightarrow & L.
\end{array}$$

We next apply this new tool to a familiar cohomology computation, Example 5.G. The complex Stiefel manifold, $V_l(\mathbb{C}^n)$, of l-frames in $\mathbb{C}^n$, is homeomorphic to $U(n)/U(n-l)$, the quotient of the unitary groups, $U(n-l) \subset U(n)$, by the usual inclusion. We have already remarked that a homogeneous space G/H can be taken to be the fibre of the mapping $B\mathrm{inc}\colon BH \to BG$. Thus we have the fibre square

$$
\begin{array}{ccc}
V_l(\mathbb{C}^n) & \longrightarrow & BU(n-l) \\
\downarrow & & \downarrow{\scriptstyle B\mathrm{inc}} \\
* & \longrightarrow & BU(n).
\end{array}
$$

The Eilenberg-Moore spectral sequence for this fibre square begins

$$
E_2^{*,*} \cong \mathrm{Tor}_{k[c_2,\dots,c_{2n}]}^{*,*}(k, k[c_2,\dots,c_{2(n-l)}]),
$$

therefore, we can apply Corollary 7.23 to compute the E_2-term. The action of $k[c_2,\dots,c_{2n}]$ on $k[c_2,\dots,c_{2(n-l)}]$ may be described by writing the algebra $k[c_2,\dots,c_{2(n-l)}]$ as the quotient $k[c_2,\dots,c_{2n}]/(c_{2(n-l)+2},\dots,c_{2n})$. Corollary 7.23 gives the complex

$$
\Lambda(x_1,\dots,x_{2n-1}) \otimes_k k[c_2,\dots,c_{2(n-l)}],
$$

which we can rewrite as

$$
\Lambda(x_{2(n-l)+1},\dots,x_{2n-1}) \otimes_k \Lambda(x_1,\dots,x_{2(n-l)-1}) \otimes_k k[c_2,\dots,c_{2(n-l)}].
$$

The explicit differential on the complex is fashioned out of the $k[c_2,\dots,c_{2n}]$-action on $k[c_2,\dots,c_{2(n-l)}]$. This complex splits as the acyclic Koszul complex $(\Lambda(x_1,\dots,x_{2(n-l)-1}) \otimes_k k[c_2,\dots,c_{2(n-l)}], d_{\mathcal{K}})$ for $k[c_2,\dots,c_{2(n-l)}]$, leaving $E_2 \cong \Lambda(x_{2(n-l)+1},\dots,x_{2n-1})$. Because all of the algebra generators for $E_2^{*,*}$ lie in $E_2^{-1,*}$, the spectral sequence collapses and

$$
E_\infty^{*,*} \cong \Lambda(x_{2(n-l)+1},\dots,x_{2n-1}).
$$

Since this is a free graded commutative algebra, there are no extension problems and so we conclude, as in Example 5.G,

$$
H^*(V_l(\mathbb{C}^n); k) \cong \Lambda(x_{2(n-l)+1},\dots,x_{2n-1}).
$$

The Koszul resolution plays an important role in the study of the cohomology of homogeneous spaces to be considered more deeply in Chapter 8.

The free graded commutative algebra $S(J)$ when J is locally finite enjoys some further structure—$S(J)$ is a graded commutative Hopf algebra. The Hopf-Borel structure theorem (Theorem 6.36) allows us to write a locally finite graded commutative Hopf algebra H over a perfect field as a tensor product of monogenic Hopf algebras. To compute $\mathrm{Tor}_H(k,k)$ in this case we can use the splitting of H into monogenic factors (Lemma 7.11). The Koszul complex handles the free graded commutative pieces leaving the case of a truncated polynomial algebra factor over a field of characteristic $p > 0$. This case admits a similar resolution described by [Tate57] in the ungraded case and extended to the graded case by [Jozefiak72].

Proposition 7.24. *Suppose that k is a field of characteristic $p > 0$. Then* $\mathrm{Tor}_{k[x_{2n}]/(x_{2n}^{p^s})}(k, k) \cong \Lambda(a_{2n-1}) \otimes \Gamma(b_{2np^s-2})$, *where a_{2n-1} corresponds to* $[x_{2n}]$ *and b_{2np^s-2} corresponds to* $[x_{2n}^{p^s-1} \mid x_{2n}]$ *in the bar construction.*

The element b_{2np^s-2} is called the *transpotence element* as it plays the role of the element $(\tau x)^{p^s-1} \otimes x$ in the Leray-Serre spectral sequence (see §6.2).

PROOF: To carry out the computation we proceed naively. We can form $(\Lambda(a_{2n-1}) \otimes k[x_{2n}]/(x_{2n}^{p^s}), d)$ with differential given by $d(a_{2n-1}) = x_{2n}$, $d(x_{2n}) = 0$ and extended as a derivation. This does not yield a resolution as the homology of this complex is nonzero. In particular, there is a nonzero homology class given by $a \otimes x^{p^s-1}$ and the homology is isomorphic to an exterior algebra generated by this class. Next form the differential graded algebra

$$(\Gamma(b_{2np^s-2}) \otimes \Lambda(a_{2n-1}) \otimes k[x_{2n}]/(x_{2n}^{p^s}), d')$$

where $d' = d$ on the subalgebra where d is defined, and $d'(b) = a \otimes x^{p^s-1}$. The extension of the differential to the other generators of the divided power algebra is given by

$$d'(\gamma_m(b)) = \gamma_{m-1}(b) \otimes a \otimes x^{p^s-1}.$$

The proposition follows by showing that this complex is acyclic. If we bigrade the complex according to the rule $\mathrm{bideg}(x) = (0, 2n)$, $\mathrm{bideg}(a) = (-1, 2n)$ and $\mathrm{bideg}(b) = (-2, 2np^s)$, then filtering by homological degree leads to a spectral sequence that separates out the $\Gamma(b)$ factor from $\Lambda(a) \otimes k[x_{2n}]/(x_{2n}^{p^s-1})$ in the first term and looks like $\Gamma(b) \otimes \Lambda(a \otimes x^{p^s-1})$ in the next term. The argument for the Koszul complex proves acyclicity and the bigrading provides the resolution. $\square$

When (X, μ) is a locally finite H-space, $H^*(X; k)$ is a graded commutative Hopf algebra and $\mathrm{Tor}_{H^*(X;k)}(k, k)$ is the E_2-term of the Eilenberg-Moore spectral sequence that computes $H^*(\Omega X; k)$. With such complete control of the E_2-term in this case, [Gitler62] and [Clark65] proved the following result.

Theorem 7.25. *Let (X, μ) be a simply-connected H-space that is of the homotopy type of a finite complex. Then, in the Eilenberg-Moore spectral sequence converging to $H^*(\Omega X; \mathbb{F}_p)$ for p a prime, the only possible nontrivial differentials d_r satisfy $r = p^s - 1$ or $r = 2p^s - 1$.*

PROOF: From the assumptions we know that $H^*(X; \mathbb{F}_p)$ is isomorphic as an algebra to a tensor product of the following form

$$H^*(X; \mathbb{F}_p) \cong \Lambda(x_1, \dots, x_m) \otimes \left(\mathbb{F}_p[y_1]/(y_1^{p^{s_1}})\right) \otimes \cdots \otimes \left(\mathbb{F}_p[y_n]/(y_n^{p^{s_n}})\right).$$

It follows that $\mathrm{Tor}_{H^*(X;\mathbb{F}_p)}(\mathbb{F}_p, \mathbb{F}_p) = E_2$ takes the form

$$\Gamma(a_1, \dots, a_m) \otimes \Lambda(b_1, \dots, b_n) \otimes \Gamma(c_1, \dots, c_n)$$

where the a_i have bidegree $(-1, |x_i|)$, the b_j have bidegree $(-1, |y_j|)$ and the c_j have bidegree $(-2, |y_j|)$. By Lemma 7.13 the lowest degree differential takes an indecomposable to a primitive. We first consider the primitives. According to the Hopf algebra structure on E_2, the primitives are spanned over $\mathbb{F}_p$ by the set $\{a_i, b_j, c_j\}$. Since the bidegree of the differentials is $(r, 1 - r)$ and $r \geq 2$, the primitives are all permanent cycles.

To determine the indecomposables of E_2, we need to know a little more about divided power algebras over $\mathbb{F}_p$. Since a divided power algebra is locally finite and commutative, it is described by the Hopf-Borel theorem as a tensor product of monogenic Hopf algebras.

Lemma 7.26. $\qquad \Gamma(u_{2l}) \cong \bigotimes_{i=0}^{\infty} \left(\mathbb{F}_p[\gamma_{p^i}(u_{2l})]/(\gamma_{p^i}(u_{2l})^p) \right).$

PROOF: Let $u = u_{2l}$. The Poincaré series for $\Gamma(u)$ is given by $P(\Gamma(u), t) = \dfrac{1}{1 - t^{2l}}$. This can be seen by identifying $\Gamma(u)$ with $\mathrm{B}^\bullet(\mathbb{F}_p, \Lambda(su), \mathbb{F}_p)$. The Poincaré series for the tensor product of truncated polynomial algebras is given by

$$\prod_{i=0}^{\infty} \frac{1 - t^{2lp^{i+1}}}{1 - t^{2lp^i}} = \frac{1}{1 - t^{2l}}.$$

Notice that the tensor product of truncated algebras has the property that every element has p^{th} power zero. We note that this is also true in $\Gamma(u)$: Given any generators, $\gamma_r(u)$ and $\gamma_s(u)$, the defining relation for $\Gamma(u)$ is

$$\gamma_r(u)\gamma_s(u) = \binom{r + s}{s} \gamma_{r+s}(u).$$

By induction $(\gamma_r(u))^p = \binom{pr}{r} \gamma_{pr}(u)$ and the binomial coefficient is zero mod p.

Since every p^{th} power vanishes in $\Gamma(u)$, there is a homomorphism of algebras $\phi \colon \bigotimes_{i=0}^{\infty} \mathbb{F}_p[\gamma_{p^i}(u)]/((\gamma_{p^i}(u))^p) \to \Gamma(u)$ taking $\gamma_{p^i}(u)$ to itself. From the equality of the Poincaré series it suffices to show that this homomorphism is an epimorphism.

Recall **Lucas's Lemma** ([Hardy-Wright38]): If A and B are positive integers and $A = \alpha_0 + \alpha_1 p + \cdots + \alpha_\nu p^\nu$ and $B = \beta_0 + \beta_1 p + \cdots + \beta_\nu p^\nu$ are the base p expressions of A and B (in particular, $0 \leq \alpha_i, \beta_j < p$), then

$$\binom{A}{B} \equiv \prod_{i=0}^{\nu} \binom{\alpha_i}{\beta_i} \bmod p.$$

We show that we can express a typical generator $\gamma_A(u)$ in the divided power algebra in terms of the $\gamma_{p^i}(u)$: Write $A = \alpha_0 + \alpha_1 p + \cdots + \alpha_\nu p^\nu$.

Claim: $\gamma_A(u) = C_A(\gamma_1(u))^{\alpha_0}(\gamma_p(u))^{\alpha_1} \cdots (\gamma_{p^\nu}(u))^{\alpha_\nu}$ with $C_A \not\equiv 0 \bmod p$.

We prove the claim by induction on ν. For $\nu = 0$, there are only the generators $\gamma_i(u)$ where $1 \le i < p$. We have the identity $\gamma_i(u) = \binom{i}{1}(\gamma_1(u))^i$ in this range and so the induction is begun.

Assuming the claim for sums out to $p^{\nu-1}$, we write

$$A = \alpha_0 + \alpha_1 p + \cdots + \alpha_\nu p^\nu$$
$$= \alpha_0 + p(\alpha_1 + \cdots + \alpha_\nu p^{\nu-1}) = \alpha_0 + pA'.$$

The defining relation for the divided power algebra gives

$$\binom{\alpha_0 + pA'}{\alpha_0}\gamma_A(u) = \gamma_{\alpha_0}(u)\gamma_{pA'}(u).$$

By Lucas's Lemma, $\binom{\alpha_0 + pA'}{\alpha_0} \equiv 1 \bmod p$ and so we have

$$\gamma_A(u) = C_{\alpha_0}(\gamma_1(u))^{\alpha_0}\gamma_{pA'}(u).$$

Consider the mapping $F\colon \gamma_i(u) \mapsto \gamma_{pi}(u)$ defined on $\Gamma(u)$. Though this is not a morphism in the category of graded vector spaces, it is nonetheless a multiplicative monomorphism by Lucas's Lemma. By induction we can express $\gamma_{A'}(u)$ as in the claim. Apply this Frobenius mapping F on $\Gamma(u)$ to obtain the expression for $\gamma_A(u)$ in the claim. It follows that the homomorphism ϕ is an epimorphism and the lemma is proved.   □

The decomposition in Lemma 7.26 implies that the indecomposables of $\operatorname{Tor}_{H^*(X;\mathbb{F}_p)}(\mathbb{F}_p, \mathbb{F}_p)$ lie in bidegrees $(-p^s, (\deg a_i)p^s)$ or $(-2p^s, (\deg c_j)p^s)$. These bidegrees have even total degree and so a differential cannot have an image among the primitives of even total degree. Thus a differential that is nontrivial must land in bidegree $(-1, \text{even})$. It follows that the first nontrivial differential in the spectral sequence must be defined on E_n where $n = p^s - 1$ or $n = 2p^s - 1$.

If we take such a differential and restrict our attention to its domain and image, we find a differential Hopf algebra of the form $(\Gamma(U) \otimes \Lambda(V), d)$, where $U = \gamma_{p^s}(u_{2l})$ and $d(U) = V$, a class among the odd degree primitives in $\operatorname{Tor}^{-1,*}$. The homology of this piece of the spectral sequence is given by

$$H(\Gamma(U) \otimes \Lambda(V), d) \cong \Gamma(U)/(\gamma_t(U) \mid t \ge p^s).$$

Though this homology produces a new primitive, it is in an even degree and so determines a permanent cycle. Furthermore, no new indecomposables are created. If we compare the initial differential Hopf algebra with the quotient by this differential sub-Hopf algebra, then we get a splitting of any E_n-term into differential Hopf algebras of this form and ones with trivial differentials. This proves the theorem. $\square$

More can be said about the differentials that appear in Theorem 7.25. We postpone the discussion to Chapter 8 where we develop a general setting for describing differentials in the Eilenberg-Moore spectral sequence. Theorem 7.25 plays an important role in the study of torsion in H-spaces—one of the main topics of Chapter 10 (see the paper of [Kane75]).

7.4 The homology of quotient spaces of group actions

In this section we consider another situation where the differential homological algebra of §7.1 may be applied and the spectral sequence of Theorem 7.6 interpreted topologically. We assume that the reader is acquainted with the following notions: G is a topological group, and G acts on spaces X and Y. Denote these actions by $\psi_X \colon X \times G \to X$ (a right action) and $\psi_Y \colon G \times Y \to Y$ (a left action). To such data, we associate a new space, $X \times_G Y$, the pushout of $\psi_X \times 1$ and $1 \times \psi_Y$, as in the diagram

$$
\begin{array}{ccc}
X \times G \times Y & \xrightarrow{\psi_X \times 1} & X \times Y \\
{\scriptstyle 1 \times \psi_Y}\Big\downarrow & & \Big\downarrow \\
X \times Y & \xrightarrow{} & X \times_G Y,
\end{array}
$$

that is, $X \times_G Y$ is the quotient of $X \times Y$ by the relation $(xg, y) \sim (x, gy)$ for all $x \in X$, $y \in Y$ and $g \in G$. Some examples of this situation are:

(1) Suppose G acts on Y and Y/G is its orbit space. Then, if $X = *$, a point endowed with the trivial action of G, we recover Y/G as $* \times_G Y$. Observe that whenever the action of G on Y is free, we get a principal fibration, $G \hookrightarrow Y \to Y/G$.

(2) An important example of the situation in (1) has been described in Chapter 6: When EG is an acyclic space on which G acts freely, then $EG/G = BG$ is the classifying space for principal G-bundles and we have the fibration $G \hookrightarrow EG \to BG$.

(3) Suppose $G \hookrightarrow E \to X$ is a principal G-bundle and F is a right G-space. Then we can form the fibre bundle over X with fibre F and structure group G. This is the associated fibration

$$
F \hookrightarrow F \times_G E \to X = E/G.
$$

This construction is discussed in §6.2.

From the description of $X \times_G Y$, we pose the problem dual to the problem of the introduction to the chapter.

Problem. *Compute* $H_*(X \times_G Y; k)$ *from knowledge of* $H_*(X; k)$, $H_*(Y; k)$, $H_*(G; k)$ *and the actions of* G *on* X *and* Y.

In order to place ourselves in the algebraic setup of §7.1, apply the singular chain functor, with coefficients in a field k, to the G-action maps. We write $C_*(\ )$ for $C_*(\ ; k)$. Then $C_*(G)$ is a differential graded algebra with the Pontryagin multiplication induced by the group multiplication on G. Further we have actions of $C_*(G)$,

$$C_*(X) \otimes C_*(G) \longrightarrow C_*(X) \quad \text{and} \quad C_*(G) \otimes C_*(Y) \longrightarrow C_*(Y)$$

induced by ψ_{X*} and ψ_{Y*}. Therefore, $C_*(X)$ and $C_*(Y)$ become right and left modules over $C_*(G)$. In this setting we can construct a proper projective resolution of $C_*(X)$ as a right $C_*(G)$-module, $Q_\bullet \xrightarrow{\varepsilon} C_*(X) \to 0$. Note that $Q_\bullet$ can be presented as a first quadrant double complex since the differential on $C_*(X)$ has degree -1. From the pushout diagram, there is a mapping of $C_*(G)$-modules,

$$\bar{\alpha} \colon C_*(X) \otimes_{C_*(G)} C_*(Y) \to C_*(X \times_G Y).$$

We introduce the composite θ, as in the diagram

$$\text{total}(Q_\bullet) \otimes_{C_*(G)} C_*(X)$$

$$\downarrow{\varepsilon \otimes 1} \qquad\qquad \searrow{\theta}$$

$$C_*(X) \otimes_{C_*(G)} C_*(Y) \xrightarrow[\bar{\alpha}]{} C_*(X \times_G Y).$$

The following theorem is due to [Moore59].

Theorem 7.27. *Suppose* G *is a connected topological group,* X *is a right* G-*space, and* Y *is the total space of a principal* G-*bundle where* G *acts on* Y *on the left. Then the mapping* θ *is a homology isomorphism, that is, there is an isomorphism*

$$\theta_* \colon \operatorname{Tor}^{C_*(G)}(C_*(X), C_*(Y)) \to H_*(X \times_G Y).$$

The proof follows the same outline as the proof of Theorem 7.14 with homology substituted for cohomology. The source of the comparison with the Leray-Serre spectral sequence is the fibration induced by the G-map $X \to *$.

$$X \to X \times_G Y \to Y/G = * \times_G Y.$$

Further details are given in the exercises.

Observe that we have displayed the variance of Tor by writing $C_*(G)$ as a superscript. We follow the conventions described in Chapter 2—all the differentials in sight have degree -1 and so our resolutions are positively graded. In particular, $Q_\bullet \to C_*(X) \to 0$, the resolution, and $\mathrm{Tor}^{C_*(G)}(C_*(X), C_*(Y))$ are bigraded with both integers of the bidegree nonnegative. We can filter $\mathrm{total}(Q_\bullet) \otimes_{C_*(G)} C_*(Y)$ in the same fashion as in the proof of Theorem 7.6 and the same arguments (this time in the first quadrant) yield the following result.

Theorem 7.28. *Suppose G is a connected topological group, X is a right G-space, and Y is the total space of a principal G-bundle where G acts on the left on Y. Then there is a first quadrant spectral sequence with*

$$E^2 \cong \mathrm{Tor}^{H_*(G;k)}(H_*(X;k), H_*(Y;k))$$

and converging strongly to $H_*(X \times_G Y; k)$.

In order to compute Tor in this case, the bar construction, suitably regraded, provides a proper projective resolution,

$$B_n(M, \Gamma, N) = M \otimes_\Gamma \overbrace{\bar{\Gamma} \otimes \cdots \otimes \bar{\Gamma}}^{n \text{ times}} \otimes_\Gamma N,$$

where $\bar{\Gamma}$ denotes the cokernel of the unit of Γ. In the special case of $M = k = N$, Proposition 7.10 still applies and $B_\bullet(k, \Gamma, k)$ is a differential coalgebra with the usual comultiplication. In this case, the spectral sequence of Theorem 7.21 converges to the homology of a space that has a natural coalgebra structure induced by the diagonal mapping. Since this is a homological feature of the homology of a space, it is no surprise that our constructions determine it.

Corollary 7.29. *Let G be a connected topological group. Then the mapping θ induces an isomorphism of coalgebras, $\theta_* \colon \mathrm{Tor}^{C_*(G)}(k, k) \to H_*(BG; k)$. Furthermore, there is a spectral sequence of coalgebras, with*

$$E^2 \cong \mathrm{Tor}^{H_*(G;k)}(k, k)$$

and converging to $H_*(BG; k)$ as a coalgebra.

SKETCH OF PROOF: Regard $B_\bullet(k, C_*(G), k)$ as a coalgebra and filter it as in Theorem 7.14. This filtration respects the coalgebra structure and induces the desired coalgebra structure on $\mathrm{Tor}^{H_*(G;k)}(k, k)$. Thus we have a spectral sequence of coalgebras, converging to $\mathrm{Tor}^{C_*(G)}(k, k)$ as a coalgebra.

To show that $\mathrm{Tor}^{C_*(G)}(k, k)$ is isomorphic to $H_*(BG; k)$ as a coalgebra, it suffices to compare $B_\bullet(k, C_*(G), k)$ with $C_*(BG)$ via the classical description of the comultiplication on a space. The comparison is implicit in the paper of [Milnor56]. It also follows from the simplicial construction of BG given by [Stasheff63] and [Milgram67].   □

The spectral sequence of Corollary 7.29 has received considerable attention in the literature. It has been referred to as the *bar spectral sequence*, the *Milnor spectral sequence*, the *homology Eilenberg-Moore spectral sequence*, the *Moore spectral sequence*, and the *Rothenberg-Steenrod spectral sequence*. In each case, $H_*(BG; k)$ is the target of the spectral sequence. The problem of computing this homology may be approached in two ways. The first is algebraic in nature, due to Eilenberg and Moore, as sketched in this section. The second approach is geometric and begins with Milnor's description of the space BG as the direct limit of quotients of iterated joins of the group G ([Milnor56']). This construction provides a filtration of BG by join coordinates and thus a spectral sequence converging to $H_*(BG; k)$. [Milnor56'] identified the E^1-term of the spectral sequence as

$$E_{n,q}^1 \cong \bigoplus_{i_1 + \cdots + i_n = q} \tilde{H}_{i_1}(G; k) \otimes \cdots \otimes \tilde{H}_{i_n}(G; k).$$

He also gave an explicit expression for the first differential in terms of the iterated diagonal mapping. [Rothenberg-Steenrod65] analyzed Milnor's construction further and computed the E^2-term of the spectral sequence along with the further structure described in Corollary 7.29. In order to show that the algebraic and geometric constructions of these spectral sequences yield the same spectral sequence, a comparison at the level of chains is needed. This comparison was carried out by [Stasheff63] and [Milgram67] who introduced a different construction of BG that compares well with Milnor's along with a filtration that leads to the same spectral sequence and carries enough structure to allow a comparison of chains. Their definition of BG is given in Chapter 6. An equivalent and elegant definition of BG as a *geometric bar construction*, $BG = B(*, G, *)$, is given by [May75]. The space $B(*, G, *)$ is the geometric realization of the simplicial space, $B_\bullet(*, G, *)$, where the space of n-simplices is given by

$$B_n(*, G, *) = G \times G \times \cdots \times G \quad (n \text{ times}),$$

and the face and degeneracy maps are given by

$$\partial_i([g_1, \ldots, g_n]) = \begin{cases} [g_2, \ldots, g_n], & \text{if } i = 0, \\ [g_1, \ldots, g_i \cdot g_{i+1}, \ldots, g_n], & \text{if } 1 \leq i < n, \\ [g_1, \ldots, g_{n-1}], & \text{if } i = n, \end{cases}$$

$$s_i([g_1, \ldots, g_n]) = [g_1, \ldots, g_i, e, g_{i+1}, \ldots, g_n].$$

An explicit chain equivalence $B_\bullet(k, C_*(G), k) \longrightarrow C_*(B(*, G, *))$ can be given. These comparisons imply the identification of the algebraic and geometric spectral sequences as the same.

Recall that the functor $\text{Tor}_-(k, k)$ respects tensor products of algebras. We apply this fact (Lemma 7.11) to give a different proof of Theorem 6.38 of [Borel53]:

Theorem 7.30. *If G is a connected topological group with $H_*(G; k) \cong \Lambda(x_1, \ldots, x_r)$ as an algebra over k, where $\deg x_i$ is odd for all i, then*

$$H^*(BG; k) \cong k[y_1^*, \ldots, y_r^*],$$

as algebras, with $\deg y_i^* = \deg x_i + 1$.

PROOF ([Moore59]): As algebras we have the isomorphisms

$$H_*(G; k) \cong \Lambda(x_1, \ldots, x_r) \cong \Lambda(x_1) \otimes \cdots \otimes \Lambda(x_r).$$

Consider the E^2-term of the spectral sequence of Corollary 7.29,

$$E^2 \cong \text{Tor}^{H_*(G;k)}(k, k) \cong \text{Tor}^{\Lambda(x_1, \ldots, x_r)}(k, k) \cong \bigotimes_{i=1}^r \text{Tor}^{\Lambda(x_i)}(k, k).$$

By the same argument that computes $H^*(\Omega S^{2n+1}; k)$ as a coalgebra, the bar resolution yields

$$\text{Tor}^{\Lambda(x_i)}(k, k) \cong \Gamma(y_i)$$

where $\Gamma(y_i)$ is the divided power algebra on the generator $y_i = [x_i]$ with $\deg y_i = \deg x_i + 1$ (recall the bar resolution is positively graded in homology). Thus, as coalgebras,

$$E^2 \cong \Gamma(y_1, \ldots, y_r)$$

with all y_i are of even total degree. Since every element of $\text{total}(E^2)$ has even degree and differentials decrease total degree by 1, the spectral sequence collapses and $E^0(H_*(BG; k)) \cong \Gamma(y_1, \ldots, y_r)$.

By an argument dual to the one given in Example 1.K for free graded commutative algebras, $\Gamma(y_1, \ldots, y_r)$ plays the role of a free cocommutative coalgebra and so $H_*(BG; k) \cong \Gamma(y_1, \ldots, y_r)$ as a coalgebra. Since all the elements of $H_*(BG; k)$ lie in even dimensions,

$$H^*(BG; k) \cong H_*(BG; k)^{\text{dual}} \cong \Gamma(y_1, \ldots, y_r)^{\text{dual}},$$

as algebras. We leave it to the reader to prove that, as an algebra, the dual of the coalgebra $\Gamma(y_1, \ldots, y_r)$ is the polynomial algebra $k[y_1^*, \ldots, y_r^*]$ with y_j^* dual to y_j. This completes the proof. $\qquad\square$

Exercises

7.1. Prove that every proper projective differential graded module over Γ, a differential graded algebra over a ring R, is a direct summand of an extended module over Γ.

7.2. On $\mathrm{Tor}_\Gamma(M, N)$:

a) Verify that the definition of $\mathrm{Tor}_\Gamma(M, N)$ given in Definition 7.5 does not depend on the choice of proper projective resolution of N.

b) Verify that $\mathrm{Tor}_\Gamma(M, N)$ is the homology of the complex $\mathrm{total}(Q^\bullet) \otimes_\Gamma N$ with differential $D \otimes 1 + (-1)^{\deg} \otimes d_N$ for $Q^\bullet \xrightarrow{\varepsilon} M \to 0$ a proper projective resolution of M in $\mathbf{DGMod}_\Gamma$.

c) Verify that, given a short exact sequence in $\mathbf{DG}_\Gamma\mathbf{Mod}$,

$$0 \to N_1 \to N_2 \to N_3 \to 0,$$

and M in $\mathbf{DGMod}_\Gamma$, we get a long exact sequence of modules over R

$$\to \mathrm{Tor}_\Gamma^n(M, N_1) \to \mathrm{Tor}_\Gamma^n(M, N_2) \to \mathrm{Tor}_\Gamma^n(M, N_3) \xrightarrow{\partial} \mathrm{Tor}_\Gamma^{n-1}(M, N_1) \to$$
$$\cdots \to \mathrm{Tor}_\Gamma^1(M, N_3) \xrightarrow{\partial} M \otimes_\Gamma N_1 \to M \otimes_\Gamma N_2 \to M \otimes_\Gamma N_3 \to 0.$$

d) Verify that, for a proper projective module P over Γ, $\mathrm{Tor}_\Gamma^i(M, P) = \{0\}$ for all M and all $i > 0$.

e) Verify that $\mathrm{Tor}_\Gamma(M, N)$ is a functor in each variable separately and determine what conditions are necessary to consider Tor as a functor in three variables.

7.3. Show that the bar construction, $B^\bullet(\Gamma, N)$, is a resolution when dealing with algebras and modules over any ring R. Show, however, that some further flatness conditions are needed in order to guarantee

$$H(B^\bullet(\Gamma, N), d_{\mathrm{int}}) \cong B^\bullet(H(\Gamma), H(N)).$$

7.4. On the bar construction.

a) Verify that $\delta \delta = 0$ where δ is the external differential for the bar resolution.

b) Verify that $d^{-n} d^{-n} = 0$ where d^{-n} is the internal differential for the bar resolution.

c) Verify that d^{-n} is Γ-linear and that $\delta \circ d^{-n} + d^{-n+1} \circ \delta = 0$.

d) Verify that $d^{-n-1}s + sd^{-n} = 0$ where s is the contracting homotopy on $B^\bullet(\Gamma, N)$.

7.5. In the Eilenberg-Moore spectral sequence to calculate $H^*(F; k)$ for F the fibre of a fibration $F \hookrightarrow E \xrightarrow{\pi} B$, show that

$$E_\infty^{0,*} = \operatorname{im}(i\colon H^*(E; k) \to H^*(F; k)).$$

7.6. Suppose that $\pi\colon E \to B$ is a fibration and $f\colon X \to B$ a continuous mapping. Show that the Eilenberg-Zilber map $EZ\colon C^*(B \times B) \to C^*(B)$ commutes with the $C^*(B \times B)$- and $C^*(B)$-module structures on $C^*(E \times E)$ and $C^*(E)$, respectively, as well as the module structures on $C^*(X \times X)$ and $C^*(X)$, respectively.

7.7. On the homology Eilenberg-Moore spectral sequence:

a) If C is a differential graded coalgebra over a field k with counit ε, then a **comodule** over C is a differential graded module A over k equipped with a comodule mapping $\lambda\colon A \to A \otimes_k C$ so that the following diagrams commute:

Given a fibre square

$$\begin{array}{ccc} E_f & \longrightarrow & E \\ \downarrow & & \downarrow{\scriptstyle \pi} \\ X & \xrightarrow{\ f\ } & B \end{array}$$

verify that $C_*(X; k)$ and $C_*(E; k)$ are comodules over $C_*(B; k)$.

b) Define the **cotensor product**

$$A \square_C B = \ker\left(A \otimes_k B \xrightarrow{\ \lambda_A \otimes 1 - 1 \otimes \lambda_B\ } A \otimes_k C \otimes_k B\right)$$

when A is a right comodule over C and B, a left comodule. Verify $A \square_C C = A$.

c) Give a definition for the notion of an **injective comodule** over C as dual to the idea of a projective module over an algebra. Define the idea of an injective resolution and then define

$$\operatorname{Cotor}^C(A, B) = H(X_* \square_C B; d \otimes 1)$$

when $0 \to A \to X_0 \to X_1 \to X_2 \to \cdots$ is an injective resolution of A by comodules over C.

d) State and prove the homology analogue for Cotor of the first and second Eilenberg-Moore theorems (Theorems 7.6 and 7.14). (The reader is directed to the classic paper of [Eilenberg-Moore66] for complete details.)

7.8. Prove Corollary 7.18.

7.9. Prove Theorem 7.27. The following strategy is helpful: Begin by considering the principal G-bundle, $G \hookrightarrow Y \to * \times_G Y$, that is, prove the theorem first for $X = *$. The general case follows the proof of Theorem 7.14 and applying those ideas to the fibration

$$X \hookrightarrow X \times_G Y \to Y/G = * \times_G Y.$$

7.10. Use the bar construction to compute $H^*(\Omega \mathbb{C}P(n); k)$. Use a Koszul complex to do the same computation.

7.11. Prove that the divided power Hopf algebra $\Gamma(y)$ on the generator y is the free cocommutative coalgebra on y. Prove that the dual of the coalgebra $\Gamma(y)$ is the polynomial algebra $k[y^*]$ where y^* is the element dual to y.

7.12. Suppose there is a fibration, $F \hookrightarrow E \to B$, with E acyclic and $H^*(B; k)$ a polynomial algebra on finitely many generators, all of even degrees. Compute the cohomology $H^*(F; k)$ of F.

7.13. In the divided power algebra $\Gamma(x)$ over $\mathbb{F}_p$ let $(\gamma_1(x), \gamma_p(x), \dots, \gamma_{p^{s-1}}(x))$ denote the ideal generated by $\gamma_{p^t}(x)$ for $0 \le t < s$. Show that

$$\Gamma(x)/(\gamma_1(x), \gamma_p(x), \dots, \gamma_{p^{s-1}}(x)) \cong \Gamma(\gamma_{p^s}(x)).$$

7.14. The homotopy equivalence $B\Omega K(\pi, n) \simeq K(\pi, n)$ shows that we can inductively define the Eilenberg-Mac Lane spaces as $K(\pi, n) = BK(\pi, n-1)$. Set up the homology Eilenberg-Moore spectral sequence over a field k that has E^2-term determined by $H_*(K(\pi, n-1), k)$ and converges to $H_*(K(\pi, n); k)$. Begin with $K(\mathbb{Z}, 1)$ and see how far you can compute $H_*(K(\mathbb{Z}, n); k)$ without needing more information to decide a differential.

8
The Eilenberg-Moore Spectral Sequence II

> "By 'differential algebra' we mean algebra in the differential category; this includes homological algebra which has significance as a point of view beyond the results which constitute the field."
>
> *From* [Stasheff-Halperin70]

In Chapter 7 we constructed the Eilenberg-Moore spectral sequence with which the cohomology of the fibre of a fibration can be computed from the cohomology of the base and total space. In this chapter we consider some substantial applications of this spectral sequence in situations that would be ambiguous at best with the Leray-Serre spectral sequence.

The first third of the chapter concerns the problem of computing the cohomology of homogeneous spaces. If H is a closed subgroup of a Lie group G and $i\colon H \to G$ the inclusion, then we have the fibration

$$G/H \to BH \xrightarrow{Bi} BG,$$

and so the Eilenberg-Moore spectral sequence can be applied to compute $H^*(G/H; k)$. In many cases (Theorem 6.38), $H^*(BG; k)$ is a polynomial algebra, that is, a free, graded commutative algebra. The homological algebra leading to the computation of $E_2 \cong \operatorname{Tor}_{H^*(BG;k)}(k, H^*(BH; k))$ may simplify sufficiently to induce the collapse of the spectral sequence. For the case $k = \mathbb{R}$, this collapse is found in the work of [Cartan50] and it was conjectured to hold in general. The extent to which the conjecture holds is the topic of §8.1.

In the middle third of the chapter, we give a description of the differentials in the Eilenberg-Moore spectral sequence. We begin with the suspension homomorphism (§6.2), which is known to annihilate products. [Uehara-Massey57] and [Massey58] introduced a secondary operation based on the vanishing of cup products, now called the Massey triple product. Generalizations of this secondary operation to higher orders, due to [Kraines66] and [May68], lead to a complete description of the kernel of the suspension homomorphism and the differentials in the Eilenberg-Moore spectral sequence. [Massey68/98] also related the triple product to higher order linking phenomena. In low dimensions Massey products were conjectured by [Stallings65] to be related to the $\bar{\mu}$-invariants of [Milnor57]. These relations were made precise by various authors making Massey products a useful tool in the study of low dimensional topology.

A spectral sequence can be endowed with extra structure that converges to the appropriate structure on the target. For example, the Steenrod algebra structure on the mod p cohomology of the total space is carried along in the associated Leray-Serre spectral sequence (see Theorem 6.15). In the final third of this chapter, we determine a Steenrod algebra structure on the Eilenberg-Moore spectral sequence by giving two geometric constructions of the spectral sequence from which this extra structure follows directly. The first construction is due to [Smith, L70] and follows an investigation, suggested by [Atiyah62] and [Hodgkin68], of the category **Top**/B of continuous functions with codomain B. The Eilenberg-Moore spectral sequence can be viewed as the topological Künneth spectral sequence for this category. The second construction is due to [Rector70] who introduced a cosimplicial space associated to a pair of mappings $\pi \colon E \to B$ and $f \colon X \to B$. Cosimplicial spaces ([Bousfield-Kan72]) as well as the category **Top**/B ([Crabb-James98]) have other applications in homotopy theory.

We close the chapter with a brief description of further applications of the spectral sequence and its algebraic analogues. The question of the strong convergence of the Eilenberg-Moore spectral sequence for spaces that are not simply-connected is taken up in the next chapter.

8.1 Homogeneous spaces

In the applications of spectral sequences to topology, computations may involve deep geometric information (to determine the differentials; see Example 5.H) or may proceed almost miraculously via algebraic features that determine collapse at some calculable stage (see Example 5.F). We find an excellent mixture of these features when we apply the Eilenberg-Moore spectral sequence to a natural class of examples, homogeneous spaces.

To H, a closed subgroup of a Lie group, G, we associate the fibration $G/H \hookrightarrow BH \xrightarrow{Bi} BG$ (as in §6.2). The cohomology of G/H is an invariant of the inclusion map—how H sits geometrically in G. A nontrivial example of the importance of the choice of inclusion is seen by considering two inclusions of U(n) in U$(n+1)$. There is the usual inclusion $i_1 \colon$ U$(n) \to$ U$(n+1)$ given by

$$i_1 \colon A \mapsto \begin{pmatrix} 1 & 0 & \cdots & 0 \\ 0 & & & \\ \vdots & & A & \\ 0 & & & \end{pmatrix}$$

that yields $U(n+1)/U(n) \cong S^{2n+1}$. Another inclusion is given by i_2,

$$i_2 \colon A \mapsto \begin{pmatrix} \det A^{-1} & 0 & \cdots & 0 \\ 0 & & & \\ \vdots & & A & \\ 0 & & & \end{pmatrix}$$

for which we have $U(n+1)/U(n) \cong S^1 \times \mathbb{C}P(n)$ ([Baum68]).

We fix some assumptions that are satisfied in most of the situations of interest. Take k to be a field, H, a closed, simply-connected subgroup of G, a compact, simply-connected Lie group. Suppose further that $H_*(G)$ and $H_*(H)$ have torsion only of orders prime to the characteristic of k. Under these assumptions, by Proposition 6.37, $H^*(G;k)$ and $H^*(H;k)$ are exterior algebras and it follows that $H^*(BG;k) \cong k[x_1, \ldots, x_n]$ and $H^*(BH;k) \cong k[y_1, \ldots, y_m]$, polynomial algebras on generators of even dimension greater than or equal to two. Thus Bi^* is expressible as an algebra homomorphism

$$Bi^* \colon k[x_1, \ldots, x_n] \longrightarrow k[y_1, \ldots, y_m].$$

The Eilenberg-Moore spectral sequence, converging to $H^*(G/H;k)$, has as E_2-term, $E_2 \cong \mathrm{Tor}_{H^*(BG;k)}(k, H^*(BH;k))$ that we may write as

$$E_2 \cong \mathrm{Tor}_{k[x_1, \ldots, x_n]}(k, k[y_1, \ldots, y_m]).$$

From an algebraic viewpoint, this seems to be a tractable computation (see §7.3); perhaps the structure of this E_2-term is tight enough to force the collapse of the spectral sequence and so determine the algebra $H^*(G/H;k)$ up to extension problems. The following collapse theorem will be considered in various forms in this section.

Theorem 8.1 (the collapse theorem). *If k is a field, either of characteristic zero or of characteristic p and $H^*(G)$ has no p-torsion, then the Eilenberg-Moore spectral sequence for $G/H \hookrightarrow BH \xrightarrow{Bi} BG$ collapses at the E_2-term and so, as a graded vector space,*

$$H^*(G/H;k) \cong \mathrm{Tor}_{H^*(BG;k)}(k, H^*(BH;k)).$$

Before we describe the various approaches toward proving such a theorem, we consider the motivating case due to [Cartan50].

Theorem 8.2. *If H is a closed, simply-connected subgroup of G, a compact, simply-connected Lie group, then*

$$H^*(G/H;\mathbb{R}) \cong \mathrm{Tor}_{H^*(BG;\mathbb{R})}(\mathbb{R}, H^*(BH;\mathbb{R})).$$

SKETCH OF PROOF: We exploit the analytic structure of Lie groups and compute their real cohomology using de Rham cohomology (see the books of [Warner83] or [Bott-Tu82] for details). The de Rham theorem tells us that $H_{\text{deR}}^*(M) \cong H^*(M; \mathbb{R})$ as algebras over $\mathbb{R}$ for a manifold M and so no structure is lost in this choice.

Let $\Omega^*(M)$ denote the de Rham algebra of differential forms on the manifold M. Following essentially the same proof as given for Theorem 7.14, it can be shown that

$$H_{\text{deR}}^*(G/H) \cong \text{Tor}_{\Omega^*(BG)}(\mathbb{R}, \Omega^*(BH)).$$

The de Rham algebra makes sense for BG and BH because these spaces can be approximated to any dimension by homogeneous spaces ([Steenrod51]). To prove the theorem, it suffices to show that

$$\text{Tor}_{\Omega^*(BG)}(\mathbb{R}, \Omega^*(BH)) \cong \text{Tor}_{H_{\text{deR}}^*(BG)}(\mathbb{R}, H_{\text{deR}}^*(BH))$$

and so the Eilenberg-Moore spectral sequence collapses.

Since the field $\mathbb{R}$ has characteristic zero, $H^*(G; \mathbb{R})$ is an exterior algebra and $H^*(BG; \mathbb{R})$ is a polynomial algebra. Write $H_{\text{deR}}^*(BG) \cong \mathbb{R}[u_1, \dots, u_n]$. Consider the mapping $h \colon H_{\text{deR}}^*(BG) \to \Omega^*(BG)$ defined by choosing a representative for each u_i (for example, the harmonic representative of Hodge theory). Since $\Omega^*(BG)$ is a graded-commutative, differential graded algebra, we can extend the mapping to an algebra mapping $H_{\text{deR}}^*(BG) \to \Omega^*(BG)$ that induces the identity on homology. We can do the same for BH.

Though the diagram

$$\begin{array}{ccc}
H_{\text{deR}}^*(BG) & \xrightarrow{\;Bi^*\;} & H_{\text{deR}}^*(BH) \\
\Big\downarrow h & & \Big\downarrow h \\
\Omega^*(BG) & \xrightarrow[\;Bi^*\;]{} & \Omega^*(BH)
\end{array}$$

need not commute without further assumptions, the algebra mappings endow $\Omega^*(BH)$ with two $H_{\text{deR}}^*(BG)$-module structures that coincide on homology. By applying Corollary 7.7 a number of times, we establish the desired isomorphism and the theorem. (See the paper of [Baum-Smith67] for further details and a generalization.) $\Box$

Theorem 8.1 refers to any field of coefficients and so it is a generalization of Cartan's theorem. We present three versions of Theorem 8.1. The first version is due to [Baum68] and depends on the explicit algebraic structure of $\text{Tor}_{k[x_1, \dots, x_n]}(k, k[y_1, \dots, y_m])$. The geometric input is a theorem of [Borel53] on the role of the maximal torus of the subgroup H in the computation of $\text{Tor}_{H^*(BG;k)}(k, H^*(BH; k))$. Though Baum's theorem requires a supplementary algebraic condition for collapse, it remains useful and we apply it to compute the mod p cohomology of the real Stiefel manifolds, $H^*(V_k(\mathbb{R}^n); \mathbb{F}_p)$, for p, an odd prime. The range of applicability of Baum's method remains an open question.

In the second version, we present work of [Gugenheim-May74] who exploited the simplification afforded by a theorem of [Baum68] on the maximal torus, and constructed special resolutions as well as explicit mappings for tori, $C^*(BT^n; k) \to H^*(BT^n; k)$, that induce the identity on homology. This implies a collapse theorem in the manner of the proof of Theorem 8.2. The resolutions considered in their work are based on a definition of Tor and Ext for differential algebras that relates more closely to the homology of the modules and algebras involved. These new definitions allow the construction of more computable resolutions.

The third version of Theorem 8.1 is based on the freeness of $H^*(BG; k)$ and examines mappings, $H^*(BG; k) \to C^*(BG; k)$, which can be constructed because we have a free algebra basis, but are obstructed as algebra mappings by the lack of commutativity of the cup product on $C^*(BG; k)$. The system of higher homotopies needed to remove the obstructions is catalogued by the bar resolution and its coalgebra structure. Motivated by a similar phenomena in the study of H-spaces, [Stasheff-Halperin70] began a study of coalgebra mappings of the bar construction as a system of higher homotopies. This line of ideas culminated in the powerful collapse theorem of [Munkholm74]. (The statement of Theorem 8.1 is due to [Wolf77] and follows by similar means.)

With the development of the topics in §8.1, we settle a special case of the collapse of the Eilenberg-Moore spectral sequence. From the examples of homogeneous spaces and the computation of the cohomology of certain two-stage Postnikov systems, one might conjecture the collapse of the Eilenberg-Moore spectral sequence where $E_2 \cong \mathrm{Tor}_A(B, C)$ and A, B and C are free graded commutative algebras. [Hirsch] posed this problem for two-stage Postnikov systems and [Schochet71] gave a counterexample to Hirsch's conjecture. The Eilenberg-Moore spectral sequence is a natural tool for the study of two-stage Postnikov systems (see the papers of [Smith, L67, 71]).

Baum's thesis

The homotopy-theoretic computation of $H^*(G/H; k)$ was begun in the papers of [Samelson41], [Koszul50], [Leray50], and [Cartan50]. These developments were focused on the real cohomology of G/H. Among the structural features revealed in these papers, a key role was played by maximal tori and maximal rank subgroups. Recall that the **rank of a Lie group**, G, may be defined as the number of algebra generators of $H^*(G; \mathbb{Q})$. When G has no torsion of order a power of the characteristic of the field k, this rank is also the number of algebra generators of $H^*(G; k)$ or $H^*(BG; k)$. [Borel53] computed the cohomology of G/H for more general coefficients under the following conditions:

Theorem 8.3. *Suppose H is a closed subgroup of G and* $\operatorname{rank} G = \operatorname{rank} H$ *and that $H^*(G)$ and $H^*(H)$ have no p-torsion where $p = \operatorname{char} k$. Then*

(1) *As an $H^*(BG; k)$-module, induced by the algebra homomorphism Bi^*, $H^*(BH; k)$ is finitely-generated and isomorphic to the extended module $H^*(BG; k) \otimes H^*(G/H; k)$.*

(2) $k \longrightarrow H^*(BG; k) \overset{Bi^*}{\longrightarrow} H^*(BH; k) \overset{j^*}{\longrightarrow} H^*(G/H; k) \longrightarrow k$ *is* **coexact**, *that is, j^* is onto and $\ker j^*$ is generated as an ideal by $Bi^*(H^+(BG; k))$.*

PROOF: The Leray-Serre spectral sequence for the fibration $G/H \hookrightarrow BH \to BG$ implies that $H^*(BH; k)$ is a subquotient of $H^*(BG; k) \otimes H^*(G/H; k)$. Since $k[x_1, \dots, x_n] \cong H^*(BG; k)$ is Noetherian, it follows that $H^*(BH; k)$ is finitely-generated over $H^*(BG; k)$.

Since $\operatorname{rank} G = \operatorname{rank} H$, $H^*(BH; k)$ is a polynomial algebra on the same number of generators as $H^*(BG; k)$ and is also finitely-generated. It follows ([Zariski-Samuel58/60]) that $H^*(BH; k)$ is free as an $H^*(BG; k)$-module (Corollary 3.10 of [Baum68]).

Apply the Eilenberg-Moore spectral sequence to the fibration Bi. Since $H^*(BH; k)$ is free over $H^*(BG; k)$,

$$\operatorname{Tor}_{H^*(BG;k)}(k, H^*(BH; k)) \cong k \otimes_{H^*(BG;k)} H^*(BH; k)$$

and the spectral sequence collapses to the column $E_2^{0,*}$. Thus

$$\begin{aligned} H^*(BH; k) &\cong H^*(BG; k) \otimes_k k \otimes_{H^*(BG;k)} H^*(BH; k) \\ &\cong H^*(BG; k) \otimes_k H^*(G/H; k). \end{aligned}$$

The coexactness of the sequence follows from the fact that $\operatorname{im} j^* = E_\infty^{0,*}$ and $k \otimes_{H^*(BG;k)} H^*(BH; k)$ is the set of indecomposables with respect to the $H^*(BG; k)$-module action, as required. $\qquad\square$

The following aspect of Lie group structure is central to the computations of $H^*(G/H; k)$.

Definition 8.4. *Let G be a compact connected Lie group. A subgroup T is called a* **maximal torus** *of G if T is abelian, compact and connected (from which it follows that T is isomorphic to $S^1 \times \cdots \times S^1$ (n times) for some n) and if U is another such subgroup with $T \subset U \subset G$, then $T = U$. From the classical theory of Lie groups (see* [Borel55] *or* [Mimura-Toda91]*) we know*

(1) *All maximal tori are conjugate to one another.*

(2) *Given a torus subgroup T' of G, there is a maximal torus T with $T' \subset T \subset G$.*

(3) $\operatorname{rank} T = \operatorname{rank} G$.

(4) *If N_T is the normalizer of T in G, then T has finite index in N_T. The finite group, $W(G) = N_T/T$, is called the **Weyl group** of G.*

(5) *The Weyl group acts on $H^*(BT) = \mathbb{Z}[v_1 \dots, v_n]$, $\deg v_j = 2$ for all j. If $i \colon T \hookrightarrow G$ is the inclusion and $H^*(G)$ has no p-torsion for $p = \operatorname{char} k$, then $Bi^* \colon H^*(BG; k) \to H^*(BT; k)$ maps $H^*(BG; k)$ isomorphically onto $k[v_1, \dots, v_n]^{W(G)}$, where $k[v_1, \dots, v_n]^{W(G)}$ denotes the ring of invariants of the Weyl group action on $H^*(BT; k)$.*

[Baum68] reduced the problem of the collapse of the Eilenberg-Moore spectral sequence for computing $H^*(G/H; k)$ to the case of $H^*(G/T_H; k)$ where T_H is a maximal torus in H.

Lemma 8.5. *Let $T_H \subset H$ be a maximal torus of H. Then, for all $i \leq 0$,*

$$\operatorname{Tor}^{i,*}_{H^*(BG;k)}(k, H^*(BH; k)) \neq \{0\}$$

if and only if $\operatorname{Tor}^{i,*}_{H^*(BG;k)}(k, H^*(BT_H; k)) \neq \{0\}$.

PROOF: Since $\operatorname{rank} T_H = \operatorname{rank} H$, by Theorem 8.3, $H^*(BT_H; k)$ is isomorphic to $H^*(BH; k) \otimes H^*(H/T_H; k)$ as an $H^*(BH; k)$-module. The $H^*(BG; k)$-module structure on $H^*(BT_H; k)$ is derived from the inclusion $T_H \subset H \subset G$ and so it factors through $H^*(BH; k)$. Thus, as an $H^*(BG; k)$-module, $H^*(BT_H; k)$ is still isomorphic to $H^*(BH; k) \otimes H^*(H/T_H; k)$ and so, as vector spaces,

$$\operatorname{Tor}^{i,*}_{H^*(BG;k)}(k, H^*(BT_H; k)) \cong \operatorname{Tor}^{i,*}_{H^*(BG;k)}(k, H^*(BH; k)) \otimes H^*(H/T_H; k).$$

Since H/T_H is connected, the result follows. $\square$

Corollary 8.6. *Let $T_H \subset H$ be a maximal torus of H. Then the Eilenberg-Moore spectral sequence, converging to $H^*(G/H; k)$ for the fibration $G/H \hookrightarrow BH \to BG$, collapses at the E_2-term if and only if the corresponding spectral sequence for $G/T_H \hookrightarrow BT_H \to BG$, converging to $H^*(G/T_H; k)$ collapses at the E_2-term.*

PROOF: It suffices to observe that the morphism of spectral sequences induced by $T_H \subset H$ respects the isomorphism of the proof of Lemma 8.5 as well as the splitting as a tensor product. Thus $E_2 = E_\infty$ for $G/T_H \hookrightarrow BT_H \xrightarrow{Bi} BG$ implies $E_2 = E_\infty$ for $G/H \hookrightarrow BH \xrightarrow{Bi} BG$ by naturality. The converse follows from the injection of $\operatorname{Tor}_{H^*(BG;k)}(k, H^*(BH; k))$ into $\operatorname{Tor}_{H^*(BG;k)}(k, H^*(BT_H; k))$ as a factor with the quotient $H^*(H/T_H; k)$ concentrated in $E_2^{0,*}$. $\square$

With Corollary 8.6 we can prove the following technical lemma that leads to Baum's collapse theorem.

Lemma 8.7. $\mathrm{Tor}^{i,*}_{H^*(BG;k)}(k, H^*(BH;k)) = \{0\}$ *if* $i < \mathrm{rank}\, H - \mathrm{rank}\, G$.

PROOF: It suffices to prove this for $\mathrm{Tor}^{i,*}_{H^*(BG;k)}(k, H^*(BT_H;k))$, $T_H \subset H$, a maximal torus. Let $T \subset G$ be a maximal torus of G containing T_H and $\Delta = T/T_H$ with $\pi\colon T \to T/T_H$, the canonical projection. The maps $\Delta \leftarrow T \to G$ induce the maps $B\Delta \leftarrow BT \to BG$, which on cohomology provide $H^*(BT;k)$ a module structure over $H^*(B\Delta;k)$ as well as over $H^*(BG;k)$. As an algebra, $H^*(BT;k)$, is isomorphic to $H^*(B\Delta;k) \otimes H^*(BT_H;k)$ (just count the S^1 factors) and so $H^*(BT;k)$ is a free $H^*(B\Delta;k)$-module. Since $\mathrm{rank}\, G = \mathrm{rank}\, T$, Theorem 8.3 implies that $H^*(BT;k)$ is also a free $H^*(BG;k)$-module.

In the presence of free modules over pairs of related algebras, we have a change-of-rings theorem for Tor. We leave the proof of the following elementary fact to the reader.

Fact: Suppose A, B, and C are algebras over a field k. If M is a right A-, right B-module, N is a left A-, right C-module, and L a left A-, left B-module, and $\mathrm{Tor}^n_A(M,N) = \{0\} = \mathrm{Tor}^n_C(N,L)$ for $n > 0$, then

$$\mathrm{Tor}^*_{B \otimes C}(M \otimes_A N, L) \cong \mathrm{Tor}^*_{A \otimes B}(M, N \otimes_C L).$$

In our case, $A = k$, $B = H^*(B\Delta;k)$, $C = H^*(BG;k)$, $M = N = k$ and $L = H^*(BT;k)$. Because every algebra in sight is graded commutative, we can apply the change of rings to obtain the isomorphisms,

$$\mathrm{Tor}_{H^*(B\Delta;k)}(k, k \otimes_{H^*(BG;k)} H^*(BT;k))$$

$$\uparrow$$

$$\mathrm{Tor}_{H^*(B\Delta;k) \otimes H^*(BG;k)}(k, H^*(BT;k))$$

$$\downarrow$$

$$\mathrm{Tor}_{H^*(BG;k)}(k, k \otimes_{H^*(B\Delta;k)} H^*(BT;k)).$$

Since $k \otimes_{H^*(B\Delta;k)} H^*(BT;k)$ is isomorphic to $H^*(BT_H;k)$ as an $H^*(BG;k)$-module, we finish the proof of the lemma by observing that

$$\mathrm{Tor}^{i,*}_{H^*(B\Delta;k)}(k, k \otimes_{H^*(BG;k)} H^*(BT;k)) = \{0\}$$

when $i < \mathrm{rank}\, H - \mathrm{rank}\, G$. This follows because $H^*(B\Delta;k)$ is a polynomial algebra on $\mathrm{rank}\, G - \mathrm{rank}\, H$ generators. In homological degrees $i < \mathrm{rank}\, H - \mathrm{rank}\, G$, the Koszul resolutions for modules over $H^*(B\Delta;k)$ are trivial (Corollary 7.23). $\qquad\square$

We focus on some homological algebra: What can be said about the structure of $\mathrm{Tor}_{k[x_1,\dots,x_n]}(k, k[y_1,\dots,y_m])$ in the case where $k[y_1,\dots,y_m]$ is a $k[x_1,\dots,x_n]$-module through an algebra homomorphism $Bi^*\colon k[x_1,\dots,x_n] \to k[y_1,\dots,y_m]$?

Definition 8.8. *Suppose A is a connected, commutative, finitely-generated algebra over k that is concentrated in even degrees. A* **presentation of** *A is a coexact sequence*

$$k[w_1,\ldots,w_s] \xrightarrow{g} k[v_1,\ldots,v_t] \xrightarrow{f} A \xrightarrow{\varepsilon} k,$$

for which the induced mapping $\bar{g} = 1 \otimes g$

$$\bar{g}\colon k \otimes_{k[w_1,\ldots,w_s]} k[w_1,\ldots,w_s]^+ \longrightarrow k \otimes_{k[w_1,\ldots,w_s]} \ker f$$

is an isomorphism of graded vector spaces over k.

The sequence being coexact means that $\ker \varepsilon$ is generated as an ideal by $\{f(v_1),\ldots,f(v_t)\}$, and $\ker f$ is generated by $\{g(w_1),\ldots,g(w_s)\}$. In general, the sequence of maps due to the fibration

$$H^*(BG;k) \xrightarrow{Bi^*} H^*(BH;k) \xrightarrow{j^*} H^*(G/H;k),$$

does not comprise a presentation of $H^*(G/H;k)$ nor is it coexact. However, as proved in the next few results, the conditions described in the definition provide some control of the algebra of the E_2-term of the associated spectral sequence.

Given a presentation of an algebra

$$k[w_1,\ldots,w_s] \xrightarrow{g} k[v_1,\ldots,v_t] \xrightarrow{f} A \xrightarrow{\varepsilon} k,$$

the elements $\{g(w_1),\ldots,g(w_s)\}$ in $k[v_1,\ldots,v_t]$ minimally generate the ideal $\ker f$ in the following sense; no proper subset of the $g(w_i)$ generates the ideal $\ker f$. This follows by the vector space isomorphism $\bar{g} = 1 \otimes g$

$$\bar{g}\colon k\{w_1,\ldots,w_s\} \longrightarrow k\otimes_{k[w_1,\ldots,w_s]} \ker f \cong {\ker f}\big/{(k[w_1,\ldots,w_s])^+ \cdot \ker f}.$$

When we have such an isomorphism for an ideal I in $k[v_1,\ldots,v_t]$, we say that the set $\{g(w_1),\ldots,g(w_s)\}$ forms a **nonredundant set of generators** for I.

In the next proposition we connect presentations of an algebra and the homological algebra of polynomial rings. Recall that the graded vector space of indecomposables of a connected algebra B is given by $Q(B) = B^+/B^+ \cdot B^+ \cong k \otimes_B B^+$, where k is a B-module by the augmentation $\varepsilon\colon B \to k$ and $B^+ = \ker \varepsilon$. The vector space of homogeneous elements of degree i in $Q(B)$ is denoted by $Q^i(B)$.

Proposition 8.9. *If A is a connected, commutative, finitely-generated algebra over k that is concentrated in even degrees, and*

$$k[w_1, \dots, w_s] \xrightarrow{g_1} k[v_1, \dots, v_t] \xrightarrow{f_1} A \to k$$

$$and \quad k[x_1, \dots, x_p] \xrightarrow{g_2} k[y_1, \dots, y_q] \xrightarrow{f_2} A \to k$$

are two presentations of A, then

$$\mathrm{Tor}_{k[w_1, \dots, w_s]}(k, k[v_1, \dots, v_t]) \cong \mathrm{Tor}_{k[x_1, \dots, x_p]}(k, k[y_1, \dots, y_q])$$

as bigraded algebras over k. Furthermore, for any i,

$$\dim_k Q^i(k[w_1, \dots, w_s]) - \dim_k Q^i(k[v_1, \dots, v_t])$$
$$= \dim_k Q^i(k[x_1, \dots, x_p]) - \dim_k Q^i(k[y_1, \dots, y_q]).$$

PROOF: We first treat a special case. Suppose that g_1 and g_2 have the same domain and range, that is, $f_1 = f_2 = f$, but $g_1 \neq g_2$.

$$k[w_1, \dots, w_s] \xrightarrow[g_2]{g_1} k[v_1, \dots, v_t] \xrightarrow{f} A \xrightarrow{\varepsilon} k.$$

Following §7.3, we compute $\mathrm{Tor}_{k[w_1, \dots, w_s]}(k, k[v_1, \dots, v_t])$ in each case via the Koszul resolution. Let $L^{*,*} = \Lambda(e_1, \dots, e_s) \otimes k[v_1, \dots, v_t]$ and $\bar{L}^{*,*} = \Lambda(\bar{e}_1, \dots, \bar{e}_s) \otimes k[v_1, \dots, v_t]$ denote the Koszul complexes for the given data. The differentials on each complex are determined by the mappings g_1 and g_2 that induce the $k[w_1, \dots, w_s]$-module structures: $d(1 \otimes v_j) = 0 = \bar{d}(1 \otimes v_j)$,

$$d(e_i \otimes v_j) = 1 \otimes g_1(w_i) \cdot v_j \quad \text{and} \quad \bar{d}(\bar{e}_i \otimes v_j) = 1 \otimes g_2(w_i) \cdot v_j.$$

Since the two pairs of mappings (g_i, f) are presentations of the algebra A, the sets $\{g_1(w_1), \dots, g_1(w_s)\}$ and $\{g_2(w_1), \dots, g_2(w_s)\}$ are nonredundant sets of generators of $\ker f$. Let $\bar{g}_i \colon k\{w_1, \dots, w_s\} \to k \otimes_{k[w_1, \dots, w_s]} \ker f$ denote the reduction of g_i modulo the module actions. Then the sets $\{\bar{g}_i(w_j)\}$, for $i = 1, 2$ and $1 \leq j \leq s$, are vector space bases of $k \otimes_{k[w_1, \dots w_s]} \ker f$. It follows that we can write $g_1(w_j) = \sum_l g_2(w_l) u_{jl}$ for $u_{jl} \in \ker f$.

Using the matrix of elements $u_{jl} \in \ker f$, define a mapping of Koszul complexes given by

$$\theta \colon L^{*,*} \to \bar{L}^{*,*}, \quad \theta(1 \otimes u) = 1 \otimes u, \quad \theta(e_j \otimes V) = \sum_l \bar{e}_l \otimes u_{jl} V,$$

where we extend as a mapping of modules over $k[v_1, \dots, v_t]$. The mapping θ commutes with the differentials:

$$\theta d(e_j \otimes V) = \theta(1 \otimes g_1(w_j) V)$$
$$= 1 \otimes g_1(w_j) V = \sum_l 1 \otimes (g_2(w_l) u_{jl}) V$$
$$= \sum_l 1 \otimes g_2(w_l)(u_{jl} V) = \bar{d} \left(\sum_l \bar{e}_l \otimes u_{jl} V \right)$$
$$= \bar{d} \theta(e_j \otimes V).$$

In $k \otimes_{k[w_1,\dots,w_s]} \ker f$, $\{\bar{g}_i(w_j)\}$ are bases. The change of basis matrix is given by $\bar{g}_1(w_j) = \sum_l \bar{g}_2(w_l)\varepsilon(u_{jl})$ and so $(\varepsilon(u_{jl})$ is an invertible matrix. It follows that $Q\theta(e_j) = \sum_l \bar{e}_l\varepsilon(u_{jl})$ and so $Q\theta$ is onto. Since $L^{*,*}$ is a free graded commutative algebra, $Q\theta$ onto implies θ is onto. As finite dimensional vector spaces over k, $L^{p,q}$ and $\bar{L}^{p,q}$ have the same dimensions and so θ is an isomorphism, and θ induces an isomorphism on homology.

For the general case of two different presentations, build an intermediate complex. First take the composite

$$k[v_1,\dots,v_t] \otimes k[y_1,\dots,y_q] \xrightarrow{f_1 \otimes f_2} A \otimes A \xrightarrow{\varphi} A$$

as f_3 and then choose g_3 for $r = s + p$

$$k[u_1,\dots,u_r] \xrightarrow{g_3} k[v_1,\dots,v_t] \otimes k[y_1,\dots,y_q] \xrightarrow{f_3} A \to k$$

so that the Koszul complex for this presentation has an acyclic factor and a factor with homology $\mathrm{Tor}_{k[w_1,\dots,w_s]}(k, k[v_1,\dots,v_t])$. By a symmetric construction, one can choose a g_4 with the same domain and range and having an acyclic factor and homology given by $\mathrm{Tor}_{k[x_1,\dots,x_p]}(k, k[y_1,\dots,y_q])$. By the first case we have established the theorem.

To prove the assertion about the dimensions of the Q^i, one argues by comparing dimensions of indecomposables for the two Koszul complexes used in the argument for the general case. $\square$

With this proposition, two homological invariants of the algebra A emerge. First, we have the bigraded algebra

$$J^{*,*}(A) = \mathrm{Tor}^{*,*}_{k[w_1,\dots,w_s]}(k, k[v_1,\dots,v_t])$$

constructed from any presentation of A. Though $J(A)$ is not functorial in A, it is determined up to isomorphism and furthermore, $J(A)^{0,*} \cong A$.

Secondly, define the integers

$$\mathrm{df}^i(A) = \dim_k Q^i(k[w_1,\dots,w_s]) - \dim_k Q^i(k[v_1,\dots,v_t])$$

and the **deficiency** of A as $\mathrm{df}(A) = \sum_i \mathrm{df}^i(A)$. [Baum68] identified the class of algebras for which $J(A)^{i,*} = \{0\}$ for $i < 0$ or, equivalently for A finite-dimensional, with $\mathrm{df}(A) = 0$. He refers to such algebras as E-**algebras** and they can be characterized by being generated, as an algebra, by a **regular sequence**, that is, a set of elements $a_1,\dots,a_t$ in A^+ so that, for $i \geq 1$, the class $[a_i]$ in $A/(a_1,\dots,a_{i-1})$ is not a zero divisor. Polynomial algebras are the simplest examples of E-algebras. The class of E-algebras has been studied in algebraic geometry and research on local rings where they are called *complete intersections*. In topology, E-algebras sometimes occur as the cohomology of a space—see the work of [Smith, L82] and [Vigué-Poirrier95].

The new invariants of A, $J^{*,*}(A)$ and $\mathrm{df}(A)$, combine with coexactness to give us a toehold on the structure of $\mathrm{Tor}_{k[w_1,\dots,w_s]}(k, k[v_1,\dots,v_t])$.

Theorem 8.10. *If the sequence of algebra mappings,*

$$k[w_1, \dots, w_s] \xrightarrow{g} k[v_1, \dots, v_t] \xrightarrow{f} A \to k,$$

is coexact (but not necessarily a presentation), then

$$\mathrm{Tor}_{k[w_1,\dots,w_s]}(k, k[v_1, \dots, v_t]) \cong E \otimes J(A),$$

as bigraded algebras, where $E = \Lambda(e_1, \dots, e_r)$, *is an exterior algebra on the elements* e_i *of bidegree* $(-1, s_i)$. *Furthermore, for each* i,

$$\dim_k E^{-1,i} = \dim_k Q^i(k[w_1, \dots, w_s]) - \dim_k Q^i(k[v_1, \dots, v_t]) - \mathrm{df}^i(A).$$

PROOF: Suppose that $a_i = g(w_i)$ and the set $\{a_i\}$ generate ker f, not necessarily in a nonredundant manner. Suppose that $a_s = a_{s-1}u_{s-1} + \cdots + a_1u_1$. Let $(L^{*,*}, d)$ be the Koszul complex $\Lambda(e_1, \dots, e_s) \otimes k[v_1, \dots, v_t]$ with differential given by $d(1 \otimes u) = 0$, $d(e_i \otimes u) = 1 \otimes a_i u$. Let $(\bar{L}^{*,*}, \bar{d})$ be the bigraded algebra with $\bar{L} = \Lambda(\bar{e}_1, \dots, \bar{e}_s) \otimes k[v_1, \dots, v_t]$ and differential given by $\bar{d}(\bar{e}_i \otimes u) = 1 \otimes a_i u$ for $1 \le i < s$, $\bar{d}(\bar{e}_s \otimes 1) = 0 = \bar{d}(1 \otimes u)$. Define a homomorphism $\psi \colon L^{*,*} \to \bar{L}^{*,*}$ by $\psi(1 \otimes u) = 1 \otimes u$, $\psi(e_i \otimes 1) = \bar{e}_i \otimes 1$ for $i < s$ and, finally, $\psi(e_s \otimes 1) = \bar{e}_s \otimes 1 + \bar{e}_{s-1} \otimes u_{s-1} + \cdots + \bar{e}_1 \otimes u_1$. Then ψ commutes with the differentials and it induces an isomorphism on Tor. The theorem follows.   □

Notice that the bigraded algebra $E^{*,*}$ in the theorem measures the failure of the coexact sequence to be a presentation. Thus, if the coexact sequence contains superfluous algebra generators ($\bar{g}$ is not an isomorphism), then E has a generator for each w_i that is superfluous.

In the case of a homogeneous space and the absence of torsion at the characteristic of k, Theorem 8.3 implies that we have the sequence of algebra mappings:

$$H^*(BH; k) \xrightarrow{Bi^*} H^*(BG; k) \xrightarrow{j^*} H^*(G/H; k) \xrightarrow{\varepsilon} k.$$

Here $H^*(BH; k) \cong k[w_1, \dots, w_s]$ and $H^*(BG; k) \cong k[v_1, \dots, v_t]$. The sequence need not be coexact, however. In order to define deficiency in this case, we focus on Bi^*. For any mapping $f \colon A \to B$ of commutative algebras, we define $B /\!/ f$ to be the quotient $B/f(A^+) \cdot B$. With this notation, the sequence

$$H^*(BG; k) \xrightarrow{Bi^*} H^*(BH; k) \to H^*(BH; k) /\!/ Bi^* \xrightarrow{\varepsilon} k$$

is coexact and so Theorem 8.10 applies. We define the *deficiency of H in G* by

$$\mathrm{df}^l(H, G; k) = \mathrm{df}^l(H^*(BH; k) /\!/ Bi^*) \text{ and } \mathrm{df}(H, G; k) = \sum_l \mathrm{df}^l(H, G; k).$$

Theorem 8.11. *For H, a closed, simply-connected subgroup of a compact, simply-connected Lie group, G, if $\mathrm{df}(H, G; k) \leq 2$, then the associated Eilenberg-Moore spectral sequence collapses, that is, as bigraded algebras,*

$$E_0^{*,*}(H^*(G/H; k)) \cong \mathrm{Tor}_{H^*(BG;k)}^{*,*}(k, H^*(BH; k)).$$

PROOF: Theorem 8.10 implies that $\mathrm{Tor}_{H^*(BG;k)}(k, H^*(BH; k))$ is isomorphic as a bigraded algebra to $E \otimes J(H^*(BH; k)//Bi^*)$ where E is an exterior algebra generated by $E^{-1,*}$. Furthermore,

$$\begin{aligned}
\dim_k E^{-1,*} &= \mathrm{rank}\, G - \mathrm{rank}\, H - \mathrm{df}(H, G; k) \\
&\geq \mathrm{rank}\, G - \mathrm{rank}\, H - 2.
\end{aligned}$$

If $J^{l,*}(H^*(BH; k)//Bi^*) \neq \{0\}$ for $l < -2$, then $E \otimes J(H^*(BH; k)//Bi^*)$ will be nontrivial in homological degrees less than $\mathrm{rank}\, H - \mathrm{rank}\, G$ by tensoring with the appropriate numbers of exterior generators. However, this contradicts Lemma 8.7. Therefore, as an algebra, $\mathrm{Tor}_{H^*(BG;k)}(k, H^*(BH; k))$ is generated by elements of bidegree (p, q) with $-2 \leq p \leq 0$. Since the Eilenberg-Moore spectral sequence is a spectral sequence of algebras, the differentials are determined by their values on the generators.

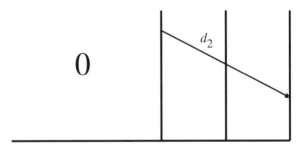

Notice that d_2, of bidegree $(2, -1)$, is zero—the generators all lie in even total degrees, which follows from the form of the Koszul complex. The higher differentials, of bidegree $(r, 1 - r)$, take the generators in $E^{-2,*}$, $E^{-1,*}$ and $E^{0,*}$ to zero. Therefore, $E_2 = E_\infty$ and the theorem follows. $\qquad\square$

As a sample application of Theorem 8.11, we compute $H^*(V_l(\mathbb{R}^n); \mathbb{F}_p)$ for p, an odd prime. Recall (Example 5.H) that $V_l(\mathbb{R}^n)$ denotes the Stiefel manifold of l-frames in $\mathbb{R}^n$. We can represent $V_{n-l}(\mathbb{R}^n)$ as the homogeneous space $\mathrm{SO}(n)/\mathrm{SO}(l)$ where $\mathrm{SO}(l)$ is a subgroup of $\mathrm{SO}(n)$ by the mapping

$$i \colon A \mapsto \begin{pmatrix} I_{n-l} & 0 \\ 0 & A \end{pmatrix}$$

(I_{n-l} is the $(n - l) \times (n - l)$ identity matrix). We take n and l to be even and consider $Bi^* \colon H^*(B\mathrm{SO}(n); \mathbb{F}_p) \to H^*(B\mathrm{SO}(l); \mathbb{F}_p)$. If subscripts indicate dimension, then Bi^* is determined by

$$Bi^* : \mathbb{F}_p[x_4, x_8, \dots, x_{2n-4}, x_n{}'] \longrightarrow \mathbb{F}_p[y_4, y_8, \dots, y_{2l-4}, y_l{}'],$$
$$Bi^*(x_{4i}) = y_{4i}, \text{ for } 1 \le i \le (l/2) - 1,$$
$$Bi^*(x_{2l}) = (y_l{}')^2,$$
$$Bi^*(x_{4i}) = 0, \text{ for } (l/2) + 1 \le i \le (n/2) - 1,$$
$$Bi^*(x_n{}') = 0.$$

It follows directly that $H^*(BSO(l); \mathbb{F}_p)//Bi^* = \mathbb{F}_p[y_l{}']/(y_l{}')^2$ and the presentation

$$\mathbb{F}_p[w_{2l}] \to \mathbb{F}_p[v_l] \to \mathbb{F}_p[y_l{}']/(y_l{}')^2 \to \mathbb{F}_p$$

shows that $\mathrm{df}(SO(l), SO(n); \mathbb{F}_p) = 0$. By Theorems 8.3 and 8.11, we obtain

$$H^*(V_{n-l}(\mathbb{R}^n); \mathbb{F}_p) \cong E^{*,*} \otimes J(H^*(BSO(l); \mathbb{F}_p)//Bi^*)$$
$$\cong \Lambda(e_{2l+3}, \dots e_{2n-5}, e_{n-1}{}') \otimes \mathbb{F}_p[y_l{}']/(y_l{}')^2.$$

The factor $\Lambda(e_{2l+3}, \dots, e_{2n-5}, e_{n-1}{}')$ comes from the kernel of Bi^*. The other factor $J(H^*(BSO(l); \mathbb{F}_p)//Bi^*)$ is $\mathrm{Tor}_{\mathbb{F}_p[x_4, \dots, x_{2l}]}(\mathbb{F}_p, \mathbb{F}_p[y_4, \dots, y_{2l-4}, y_l{}'])$. The Koszul complex for this Tor splits into an acyclic factor and the complex $(\Lambda(e_{2l-1}) \otimes \mathbb{F}_p[y_l{}'], d)$ with $d(e_{2l-1} \otimes 1) = 1 \otimes (y_l{}')^2$. The homology of this small complex contributes $\mathbb{F}_p[y_l{}']/(y_l{}')^2$.

Another approach to differential homological algebra Ⓝ

In the previous section, the Eilenberg-Moore spectral sequence was used to compute $H^*(G/H; k)$ in many cases. Baum's algebraic condition,

$$\mathrm{df}(H, G; k) \le 2,$$

insures that plenty of 'holes' are strewn about the E_2-term of the spectral sequence and so it collapses. The limitations of this approach for proving a more extensive collapse theorem are clear—the condition $\mathrm{df}(H, G; k) \le 2$ is not generic.

In this section and the next we consider more general situations in which the spectral sequence collapses. If H is a closed subgroup of a compact Lie group G, as in Theorem 8.1, then we seek conditions that guarantee

$$\mathrm{Tor}_{H^*(BG; k)}(k, H^*(BH; k)) \cong \mathrm{Tor}_{C^*(BG; k)}(k, C^*(BH; k)).$$

Corollary 7.7 and its application in the proof of Theorem 8.2 provide an approach. When there exist algebra mappings

$$C^*(BG; k) \to H^*(BG; k) \quad \text{and} \quad C^*(BH; k) \to H^*(BH; k)$$
$$\text{or } H^*(BG; k) \to C^*(BG; k) \quad \text{and} \quad H^*(BH; k) \to C^*(BH; k)$$

that induce the identity on homology and commute with respect to Bi^*, we obtain the desired collapse. An obstruction to the existence of such mappings is the fact that $H^*(X; k)$ is a graded commutative algebra, but $C^*(X; k)$ is commutative only up to chain homotopy. The success of the proof of Theorem 8.2 relies on the graded commutativity of the de Rham complex of a manifold that allows us to construct honest algebra mappings.

Our next approach toward a proof of Theorem 8.1 is in three parts; we first redefine the functor Tor, following [Gugenheim-May74] and [Felix-Halperin-Thomas95], in order to exploit resolutions that make clearer the relationship between $\mathrm{Tor}_\Gamma(M, N)$ and $\mathrm{Tor}_{H(\Gamma)}(H(M), H(N))$. Next we consider the failure of commutativity of the singular cochain complex of a space and how it is measured by the cup_1-product. This cup_1-product, together with the free commutative algebra $H^*(BG; k)$, leads to a resolution of the appropriate form. Finally we apply this special resolution to the geometry of homogeneous spaces and obtain a collapse theorem.

Definition 8.12. *Let* (Γ, d_Γ) *be a differential graded algebra over a field k. A Γ-module (X, d) is* **semifree** *if X is an increasing union of submodules*

$$X(1) \subset X(0) \subset X(-1) \subset \cdots \subset X(-n) \subset \cdots \subset X$$

such that $X(1)$ and $X(-n)/X(-n+1)$ are free over Γ on a basis of cycles. A **semifree resolution** *of a Γ-module (M, d) is a semifree module (X, d) together with a homology isomorphism of Γ-modules $\alpha\colon (X, d) \to (M, d)$.*

The key features of semifree resolutions follow from the filtered structure that is part of the definition. We filter (X, d) by $F^{-p}X = X(-p)$. The associated graded module $E_0^{-p,*}(X, F) = X(-p)/X(-p+1) \cong \bar{X}^{-p,*} \otimes \Gamma$ for $p \geq 0$ and for some graded k-vector space $\bar{X}^{-p,*}$. The quotient mapping $X(-p) \to \bar{X}^{-p,*} \otimes \Gamma$ splits to give $X(-p) \cong X(-p+1) \oplus \bar{X}^{-p,*} \otimes \Gamma$. The differential on $E_0^{*,*}$ takes the form $1 \otimes d_\Gamma$ and hence the restriction of d to the basis satisfies $d|\colon \bar{X}^{-p,*} \to X(-p+1)$.

In the differential homological algebra of Chapter 7, the total differential on a resolution has the form $d_{\mathrm{total}} = d_0 + d_1$, that is, the total differential is the sum of an internal differential and a resolution differential. In the case of a semifree resolution, the differential may be further refined; $d = \sum_{r \geq 0} d_r$ where $d_r(X(-p)) \subset X(-p+r)$.

We associate the usual spectral sequences to the filtered module (X, F). By the definitions, $E_1^{1,*}(X) = H^*(M)$ and $E_1^{i,*}(X) \cong \bar{X}^{i,*} \otimes H(\Gamma)$ for $i \leq 0$. If the sequence

$$\cdots \to E_1^{p,*}(X) \to E_1^{p+1,*}(X) \to \cdots \xrightarrow{E_1(d)} E_1^{0,*}(X) \to H^*(M) \to 0$$

is exact, we say that X is a **resolution** of M. Furthermore, if each $E_1^{p,*}(X)$ is a flat $H(\Gamma)$-module, then we define, for (N, d_N), a left differential graded module over Γ,

$$\mathrm{Tor}_\Gamma(M, N) = H(X \otimes_\Gamma N).$$

The induced filtration on $X \otimes_\Gamma N$, given by $F^p(X \otimes_\Gamma N) = F^p X \otimes_\Gamma N$, yields a spectral sequence converging to $\mathrm{Tor}_\Gamma(M, N)$. By the Künneth theorem,

$$E_2 \cong \mathrm{Tor}_{H(\Gamma)}(H(M), H(N)),$$

where Tor is the classical functor for graded modules over $H(\Gamma)$. If we take $Y_\bullet \to M \to 0$ is a proper projective resolution of M over (Γ, d_Γ), we can compare it with a semifree resolution by using the projective property and Corollary 7.7 to obtain an isomorphism between $\mathrm{Tor}_\Gamma(M, N)$ and $\mathrm{Tor}_\Gamma(M, N)$.

We next state a theorem that describes the computational advantage of this new functor Tor. The proof is by a direct construction given in [Gugenheim-May74, p. 12].

Theorem 8.13. *Suppose M is a right differential graded module over Γ and $H(M)$ has a projective resolution over $H(\Gamma)$ of the form*

$$\cdots \to \overline{X}^{p,*} \otimes H(\Gamma) \to \overline{X}^{p+1,*} \otimes H(\Gamma) \to \cdots \to \overline{X}^{0,*} \otimes H(\Gamma) \to H(M) \to 0.$$

Then there is a differential on the filtered Γ-module, $X = \overline{X}^{*,*} \otimes \Gamma$ and a homology isomorphism $\alpha \colon X \to M$ so that X is a resolution of M whose associated E_1-term agrees with the given resolution.

To apply this framework to homogeneous spaces, we study the singular cochain algebra. The cup product on $H^*(X; k)$ is graded commutative, while $C^*(X; k)$ is only commutative up to homotopy. The failure of graded commutativity on $C^*(X; k)$ gives rise to cohomology operations as in the classical construction of the Steenrod operations. We realize the chain homotopy for the cup product by the cup_1-product (§5.3) and generalize this structure to differential graded algebras as follows:

Definition 8.14. *A differential graded algebra (Γ, μ, d) over k is said to have a* **cup$_1$-product** *if there is a k-linear mapping for all p and q*

$$\smile_1 \colon \Gamma^p \otimes_k \Gamma^q \to \Gamma^{p+q-1}$$

satisfying the **Hirsch formulas;** *if we write $\mu(a, b) = a \cdot b$, then*

$$d(a \smile_1 b) = a \cdot b - (-1)^{|a| \, |b|} b \cdot a - d(a) \smile_1 b - (-1)^{|a|} a \smile_1 d(b),$$
$$(a \cdot b) \smile_1 c = (-1)^{|a|} a \cdot (b \smile_1 c) + (-1)^{|b| \, |c|} (a \smile_1 c) \cdot b.$$

For singular cochains with coefficients in $k = \mathbb{F}_2$ the cup$_1$-product determines an operation on the cocycles of X; for each s,

$$Sq_1 \colon Z^s(X; \mathbb{F}_2) \to Z^{2s-1}(X; \mathbb{F}_2),$$

defined by $Sq_1(z) = z \smile_1 z$. This passes to a homomorphism

$$Sq_1 \colon H^s(X; \mathbb{F}_2) \to H^{2s-1}(X; \mathbb{F}_2)$$

and $Sq_1 = Sq^{s-1}$, the operation in the mod 2 Steenrod algebra (see the construction in [Steenrod-Epstein62]).

The first of the Hirsch formulas provides a chain homotopy between the mappings μ and $\mu \circ T \colon \Gamma \otimes \Gamma \to \Gamma$. The second formula can be understood to say that for each $c \in \Gamma$, the mapping $a \mapsto a \smile_1 c$ is a derivation with respect to μ. Note that the homology of a differential graded algebra with cup$_1$-product is graded commutative. Furthermore, if $\nu \colon \Gamma \to H(\Gamma)$ is an algebra homomorphism that induces the identity on homology and commutes with cup$_1$-products, then ν must annihilate cup$_1$-products. We will soon see that the condition that a differential graded algebra mapping $\Gamma \to H(\Gamma)$ annihilate cup$_1$-products is sufficient to induce the isomorphism we are seeking for the collapse theorem.

Suppose Γ is a differential graded algebra with cup$_1$-product such that $H(\Gamma)$ is a polynomial algebra on even degree generators, $H(\Gamma) \cong k[x_1, \ldots, x_n]$. Let $M = k$; by Corollary 7.23, the Koszul resolution, $\mathcal{K}(H(\Gamma)) = \Lambda(\{a_i\}) \otimes H(\Gamma)$, provides a resolution of k of the form given in Theorem 8.13. If we let $\mathcal{K}(\Gamma) = \Lambda(\{a_i\}) \otimes \Gamma$ with the obvious augmentation, $\varepsilon \colon \mathcal{K}(\Gamma) \to k$, then Theorem 8.13 implies that there is a differential on $\mathcal{K}(\Gamma)$ such that $\mathcal{K}(\Gamma)_\varepsilon$ is a resolution of k and $E_1(\mathcal{K}(\Gamma))$ is the Koszul resolution of k over $H(\Gamma)$.

[Gugenheim-May74] give the differential on $\mathcal{K}(\Gamma)$ explicitly. The vector space generators for $\Lambda(a_1, \ldots, a_n)$ are indexed over the set of sequences $I = (i_1 < i_2 < \cdots < i_p)$. We write $a_I = a_{i_1} a_{i_2} \cdots a_{i_p}$. If $H(\Gamma) \cong k[x_1, \ldots, x_n]$ and, for $i = 1, \ldots, n$, $b_i \in \Gamma$ is a representative for $x_i \in H(\Gamma)$, then we associate to each sequence I the element

$$b_I = (\cdots ((b_{i_1} \smile_1 b_{i_2}) \smile_1 b_{i_3}) \smile_1 \cdots) \smile_1 b_{i_p})$$

($b_{(i_1)} = b_{i_1}$ and $b_\emptyset = 0$). The differential on $\mathcal{K}(\Gamma)$ takes the form

$$d(a_I) = -\sum_{J \subset I} \pm(I, J)\, a_{I-J} \otimes b_J,$$

where J is a subsequence of I and $I - J$ is the complementary subsequence. The sign is given by $\pm(I, J) = (-1)^{p-r+\varepsilon(J)}$ where r is the length of J and $\varepsilon(J) = \sum_{t=1}^{r}(J(t) - t)$. With this definition, it is shown that $dd = 0$.

With an explicit semifree resolution, [Gugenheim-May74] prove the following algebraic collapse theorem.

Theorem 8.15. *Suppose Λ and Γ are differential graded algebras over k that have cup_1-products. Suppose $f\colon \Gamma \to \Lambda$ is a differential graded algebra mapping that commutes with the cup_1-products. If Γ is augmented, $H(\Gamma)$ is a polynomial algebra, and there is a mapping $g\colon \Lambda \to H(\Lambda)$ of differential graded algebras with $H(g) = \mathrm{id}$ such that g annihilates cup_1-products, then, as graded vector spaces, $\mathrm{Tor}_\Gamma(k, \Lambda) \cong \mathrm{Tor}_{H(\Gamma)}(k, H(\Lambda))$.*

The proof reduces to examining the induced differential on $\mathcal{K}(\Gamma) \otimes_\Gamma \Lambda$ where the expression for the differential on $\mathcal{K}(\Gamma)$ and the fact that g annihilates cup_1-products imply the isomorphism.

Having developed the algebra to this point, we add a geometric fact that turns the key for the desired collapse theorem for homogeneous spaces. The proof of the next proposition is found in the appendix of [Gugenheim-May74] and it follows by applying the simplicial techniques in [Eilenberg-Mac Lane53].

Proposition 8.16. *Let $T^n = S^1 \times \cdots \times S^1$ (n times). For any commutative ring R with unit, there is a morphism of differential graded algebras over R*

$$\chi\colon C^*(BT^n; R) \to H^*(BT^n; R)$$

that induces the identity on homology and annihilates cup_1-products.

Theorem 8.17. *If H is a closed, simply-connected subgroup of a compact, simply-connected Lie group G, satisfying the conditions given in Theorem 8.1, then, as a graded vector space,*

$$H^*(G/H; k) \cong \mathrm{Tor}_{H^*(BG;k)}(k, H^*(BH; k)).$$

PROOF: Let $T_H \subset H$ be a maximal torus for H. By Corollary 8.6, it suffices to prove that $H^*(G/T_H; k)$ is isomorphic to $\mathrm{Tor}_{H^*(BG;k)}(k, H^*(BT_H; k))$. But $T_H \cong T^n$ for some n and so, by Proposition 8.16, there is a mapping $\chi\colon C^*(BT_H; k) \to H^*(BT_H; k)$ that preserves products and annihilates cup_1-products. By the assumptions on G and H in Theorem 8.1, the rest of the assumptions of Theorem 8.15 are satisfied and so

$$\mathrm{Tor}_{H^*(BG;k)}(k, H^*(BT_H; k)) \cong \mathrm{Tor}_{C^*(BG;k)}(k, C^*(BT_H; k))$$

and the theorem is proved. $\qquad\square$

[Gugenheim-May74] developed their results over a ring and so they apply more generally with the correct flatness assumptions. The dual theory for the functor Ext is also developed. Finally, the assumption of a finite polynomial ring for $H(\Gamma)$ plays no role in the proof and so it is possible to study the Eilenberg-Moore spectral sequence for a two-stage Postnikov system:

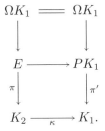

Here K_1 and K_2 are taken to be Eilenberg-Mac Lane spaces and π' the path-loop fibration. When $k = \mathbb{F}_2$ the cohomology of K_i is polynomial (Theorem 6.19) and the homological algebra of [Gugenheim-May74] may be applied.

[Schochet71] studied an important example that turns on the cup$_1$-product structure. Let $\imath$ and $\jmath$ be the fundamental classes in $H^2(K(\mathbb{Z}/2\mathbb{Z}\oplus\mathbb{Z}/2\mathbb{Z}, 2); \mathbb{F}_2)$ and take

$$\kappa = \imath \smile \jmath \colon K(\mathbb{Z}/2\mathbb{Z} \oplus \mathbb{Z}/2\mathbb{Z}, 2) \to K(\mathbb{Z}/2\mathbb{Z}, 4).$$

[Schochet71] showed that there is a class in $E_2^{-2,17}$ of the associated Eilenberg-Moore spectral sequence on which d_2 is nonzero. This gave a counterexample to a conjecture of [Hirsch] that asserted an additive isomorphism $H^*(E; \mathbb{F}_2) \cong \mathrm{Tor}_{H^*(K_1;\mathbb{F}_2)}(\mathbb{F}_2, H^*(K_2; k))$ because $H^*(K_1; \mathbb{F}_2)$ and $H^*(K_2; \mathbb{F}_2)$ are polynomial algebras.

Suppose X is a space with exactly *two* nonvanishing homotopy groups:

$$\pi_r(X) = \begin{cases} G_1, & \text{if } r = n, \\ G_2, & \text{if } r = n + k, \\ \{0\}, & \text{otherwise.} \end{cases}$$

Up to homotopy, X can be realized by the pullback over a mapping of Eilenberg-Mac Lane spaces of a path-loop fibration:

$$\begin{array}{ccc} K(G_2, n + k) & \!\!\!=\!\!\! & K(G_2, n + k) \\ \downarrow & & \downarrow \\ E & \longrightarrow & PK(G_2, n + k + 1) \\ \downarrow & & \downarrow \\ K(G_1, n) & \xrightarrow{\ \kappa\ } & K(G_2, n + k + 1). \end{array}$$

The two-stage Postnikov system, X, depends on the choice of $[\kappa]$;

$$[\kappa] \in [K(G_1, n), K(G_2, n + k + 1)] = H^{n+k+1}(K(G_1, n); G_2).$$

Thus $[\kappa]$ can be interpreted as a primary cohomology operation. In §9.1, we define higher order cohomology operations and find that the space X is the universal example for the secondary operation based on $[\kappa]$. This motivates the study of $H^*(X; R)$. [Kristensen63] studied such spaces using the Leray-Serre spectral sequence and the action of the Steenrod algebra (Theorem 6.15). [Smith, L67'] and [Kraines73] applied the Eilenberg-Moore spectral sequence in their study of two-stage and higher stage Postnikov systems.

Extending the functor Tor Ⓜ

We return again to the main idea of Corollary 7.7—when $f \colon \Lambda \to \Gamma$ is a morphism of differential graded algebras, $g \colon M \to M'$ and $h \colon N \to N'$ are morphisms of modules, and all of these mappings induce isomorphisms on homology, then

$$\mathrm{Tor}_\Lambda(M, N) \cong \mathrm{Tor}_\Gamma(M', N').$$

Thus Tor only 'sees' the chain homotopy type of the algebra and modules. In this section we extend Tor to accept morphisms that are algebra and module mappings up to chain homotopy.

The idea that properties up to homotopy are the more fundamental notion appears crucially in the study of H-spaces. For example, a morphism of chain complexes need not be a morphism of algebras or modules but induces such a morphism on homology. Perhaps the simplest example is the homotopy associativity of the loop multiplication on $\Omega(X, x_0)$. Here the multiplication is associative up to homotopy, has inverses up to homotopy, and a unit up to homotopy. More dramatically, we know that there is a multiplication on S^7 (the unit Cayley numbers), which is not associative. We can ask in reverse whether it can be deformed into an associative multiplication.

[Sugawara57] and [Stasheff63] constructed a homotopy invariant way to study associativity. The notion of A_n-structures ([Stasheff63]) stratifies possible 'degrees' of associativity—an A_3-space is a homotopy associative H-space and an A_∞-space is homotopy equivalent to an associative H-space. [Clark65] extended these notions to study the commutativity of the loop multiplication on ΩX when X is an H-space. The analogous structures for differential graded algebras and modules over them were identified by [Stasheff-Halperin70] who proposed the following approach to the problem of the collapse of the Eilenberg-Moore spectral sequence for homogeneous spaces: Since $H^*(BG; k)$ is a polynomial algebra in the case of Theorem 8.1, it is free as a commutative algebra. One can study a mapping $H^*(BG; k) \to C^*(BG; k)$ that chooses a representative for each generator. If this mapping is systematically homotopy equivalent to an algebra homomorphism, then we have a situation like Cartan's proof of Theorem 8.2 and a chance to prove the collapse result.

To study Tor as a functor 'up to homotopy,' we fix the class of differential objects we plan to consider, and choose the bar construction as the basic object; $\mathrm{Tor}_\Gamma(M, N) = H(\mathrm{B}^\bullet(M, \Gamma, N))$. Let $\bar{\mathrm{B}}(\Gamma) = \mathrm{B}^\bullet(k, \Gamma, k)$ and recall that

$(\bar{B}(\Gamma), d^{\bullet} + \delta)$ is a differential coalgebra where $d^{\bullet}$ is the internal differential (and $\bar{\gamma} = (-1)^{1+\deg \gamma}\gamma$),

$$d^{\bullet}([\gamma_1 \mid \cdots \mid \gamma_n]) = \sum_{i=1}^{n} [\bar{\gamma}_1 \mid \cdots \mid \bar{\gamma}_{i-1} \mid d_{\Gamma}(\gamma_i) \mid \cdots \mid \gamma_n],$$

δ is the external differential,

$$\delta([\gamma_1 \mid \cdots \mid \gamma_n]) = \sum_{i=1}^{n-1} [\bar{\gamma}_1 \mid \cdots \mid \bar{\gamma}_{i-1} \cdot \gamma_i \mid \cdots \mid \gamma_n],$$

and the comultiplication is given by

$$\Delta([\gamma_1 \mid \cdots \mid \gamma_n]) = \sum_{i=0}^{n} [\gamma_1 \mid \cdots \mid \gamma_{i-1}] \otimes [\gamma_i \mid \cdots \mid \gamma_n].$$

If $f \colon \Lambda \to \Gamma$ is a mapping of differential graded algebras, then f induces a mapping of differential coalgebras, $\bar{B}(f) \colon \bar{B}(\Lambda) \to \bar{B}(\Gamma)$. The central observation that makes the extension of Tor possible is that maps, $\bar{B}(\Lambda) \to \bar{B}(\Gamma)$, of differential coalgebras carry the chain homotopy information we need. Furthermore, not every map of differential graded coalgebras, $\bar{B}(\Lambda) \to \bar{B}(\Gamma)$, is $\bar{B}(f)$ for some algebra mapping, $f \colon \Lambda \to \Gamma$. These observations are made precise in the following result of [Stasheff-Halperin70].

Theorem 8.18. *Suppose Λ and Γ are connected differential graded algebras over a field k. Let $\mathbf{DCoalg}(\bar{B}(\Lambda), \bar{B}(\Gamma))$ denote the set of morphisms of differential coalgebras, $\bar{B}(\Lambda) \to \bar{B}(\Gamma)$. Then $\mathbf{DCoalg}(\bar{B}(\Lambda), \bar{B}(\Gamma))$ is in one-to-one correspondence with the set of sequences of k-linear mappings, $(f^1, f^2, \dots)$ where*

$$\overbrace{\qquad}^{n \text{ times}}$$

(1) $f^i \colon \overbrace{\Lambda \otimes \cdots \otimes \Lambda}^{n \text{ times}} \to \Gamma$ *has degree $1 - n$,*

(2) *for all n,*

$$d_{\bar{B}(\Gamma)} f^n(a_1 \otimes \cdots \otimes a_n) - \sum_{i=1}^{n} f^n(\bar{a}_1 \otimes \cdots \otimes \bar{a}_{i-1} \otimes d_{\Lambda}(a_i) \otimes \cdots \otimes a_n)$$

$$= \sum_{i=1}^{n-1} f^{n-1}(\bar{a}_1 \otimes \cdots \otimes \bar{a}_i \cdot a_{i+1} \otimes \cdots \otimes a_n)$$

$$- \sum_{i=1}^{n-1} f^i(\bar{a}_1 \otimes \cdots \otimes \bar{a}_{i-1}) \cdot f^{n-i}(a_{i+1} \otimes \cdots \otimes a_n).$$

PROOF: If $F \colon \bar{B}(\Lambda) \to \bar{B}(\Gamma)$ is a morphism of differential coalgebras, then the result follows by direct calculation. Given such a sequence of mappings as described above, define the mapping $\bar{B}f^{\bullet} \colon \bar{B}(\Lambda) \to \bar{B}(\Gamma)$ by

$$\bar{B}f^{\bullet}([a_1 \mid \cdots \mid a_n])$$

$$= \sum_{k=1}^{n} \sum_{S(n,k)} [f^{i_1}(a_1 \otimes \cdots \otimes a_{i_1}) \mid \cdots \mid f^{i_k}(a_{n-i_k+1} \otimes \cdots \otimes a_n)]$$

where $S(n, k)$ is the set $\{(i_1, \dots, i_k) \mid \sum_1^k i_j = n\}$. Another direct calculation shows that this is a mapping of differential coalgebras. (See the papers of [Clark65] or [Wolf77] for more details.) □

To see how this theorem relates to 'algebras up to homotopy,' observe that, for $n = 2$, the formulas above give

$$d_{\bar{B}(\Gamma)} f^2(a \otimes b) - f^2(d_\Lambda(a) \otimes b + \bar{a} \otimes d_\Lambda(b)) = f^1(a \cdot b) - f^1(a) \cdot f^1(b),$$

that is, f^1 induces an algebra mapping on homology. In order for f^1 to induce an associative multiplication, f^3 is needed to fill in the appropriate chain homotopies. To quote from [Wolf77], " ... f^2 is a chain homotopy measuring how far f^1 deviates from being multiplicative. Thus, in a sense, f^2 atones for the sins of f^1—but adds a few of its own. f^3, in turn, is a chain homotopy of chain homotopies, and atones for the sins of f^1 and f^2—but ... and so on."

Definition 8.19. *A sequence of mappings, $(f^1, f^2, \dots)$, that arises from a differential coalgebra morphism $\bar{B}(\Lambda) \to \bar{B}(\Gamma)$ is called an* **shm** *(strongly homotopy multiplicative)* **map**, *denoted $\Lambda \Longrightarrow \Gamma$. We also say that a mapping of differential graded modules over k, $f \colon \Lambda \to \Gamma$ is an shm map, if there is a sequence as above $(f^1, f^2, \dots)$ with $f^1 = f$.*

The terminology of 'strongly homotopy multiplicative' mappings and the systems of higher homotopies that express the relations implied by associativity were first codified in [Sugawara57] and [Stasheff63].

Extend the category $\mathbf{DGAlg}_k$ of differential graded algebras over k to a new category $\mathbf{DASH}_k$ with the same objects as $\mathbf{DGAlg}_k$ but with the sets of morphisms given by

$$\mathbf{DASH}_k(\Lambda, \Gamma) = \mathbf{DCoalg}(\bar{B}(\Lambda), \bar{B}(\Gamma)).$$

The category $\mathbf{DGAlg}_k$ embeds in $\mathbf{DASH}_k$ by sending an algebra homomorphism $f \colon \Lambda \to \Gamma$ to the sequence $(f, 0, 0, \dots)$. We denote a morphism in $\mathbf{DASH}_k$ by $f \colon \Lambda \Longrightarrow \Gamma$.

To prove Theorem 8.1, we develop the notions of algebras and modules over algebras having sh-structure maps and extend the functor Tor to accept sh-objects and shm maps as variables. This extension was carried out in [Gugenheim-Munkholm74]. [Stasheff-Halperin70] observed that, for the differential graded algebra $C^*(BG; k)$, satisfying the assumptions of Theorem 8.1, there is an shm map, $H^*(BG; k) \Longrightarrow C^*(BG; k)$, inducing the identity mapping on homology. By getting the sh-module structure correct, the desired isomorphism on Tor follows.

The following series of remarks and results, stated without proofs, gives the steps in this program leading to the proof of the powerful collapse theorem of [Munkholm74]. The interested reader can find details in the references cited along the way.

A. If Γ is a differential graded algebra, then a **left sh-module** over Γ is a differential graded module, M, together with a sequence of k-linear mappings

$$g^n: \underbrace{\Gamma \otimes \cdots \otimes \Gamma}_{n \text{ copies}} \otimes M \to M,$$

such that

(1) g^n has degree $1 - n$, and
(2) the adjoint mappings ad $g^n: \Gamma \otimes \cdots \otimes \Gamma \to \mathrm{Hom}(M, M)$ form an shm map.

A similar definition can be given for *right* sh-modules over Γ.

The definition of sh-modules allows us to induce sh-module structures from mappings in $\mathbf{DASH}_k$: If $f: \Lambda \Longrightarrow \Gamma$ is a mapping in $\mathbf{DASH}_k$, then Γ is an sh-module over Λ via the mapping f.

B. If Λ and Γ are differential graded algebras, $f: \Lambda \Longrightarrow \Gamma$, a mapping in $\mathbf{DASH}_k$, M, N are, respectively, right and left differential graded modules over Λ, and M', N' are similarly differential graded modules over Γ, then we write $g: M \Longrightarrow M'$ and $h: N \Longrightarrow N'$ for mappings $g: \mathrm{B}(M, \Lambda, k) \to \mathrm{B}(M', \Gamma, k)$ and $h: \mathrm{B}(k, \Lambda, N) \to \mathrm{B}(k, \Gamma, N')$ that commute with f as comodules over the coalgebras $\bar{\mathrm{B}}(\Lambda)$ and $\bar{\mathrm{B}}(\Gamma)$. Such mappings induce a natural homomorphism

$$\mathrm{Tor}_f(g, h): \mathrm{Tor}_\Lambda(M, N) \to \mathrm{Tor}_\Gamma(M', N'),$$

extending the classical case. The following generalization of Corollary 7.7 is due to [Gugenheim-Munkholm74].

Theorem 8.20. *Suppose* $f: \Lambda \Longrightarrow \Gamma$ *is a mapping in* $\mathbf{DASH}_k$ *and* $g: M \Longrightarrow M'$ *and* $h: N \Longrightarrow N'$ *are sh-module mappings. Then we can write* $f = (f^1, f^2, \dots)$, $g = (g^0, g^1, \dots)$ *and* $h = (h^0, h^1, \dots)$ *with* $f^1: \Lambda \to \Gamma$, $g^0: M \to M'$ *and* $h^0: N \to N'$. *If* $H(f^1)$, $H(g^0)$ *and* $H(h^0)$ *are all isomorphisms, then so is* $\mathrm{Tor}_f(g, h)$.

C. We say that a differential graded algebra Γ is **shc** (strongly homotopy commutative) if the multiplication on Γ, $m: \Gamma \otimes \Gamma \to \Gamma$, is an shm map. Examples of such shc algebras are $C^*(X; k)$, for X a space and multiplication given by the cup-product. When Y is an H-space, then $C_*(\Omega Y; k)$ is an shc algebra ([Sugawara60] and [Clark65]). [Stasheff-Halperin70] observed an important property of shc algebras of a certain type that applies to the problem of computing $H^*(G/H; k)$.

Proposition 8.21. *Suppose* Γ *is an shc algebra and* $H(\Gamma) \cong k[x_1, \dots , x_n]$. *Then there is an shm map,* $\psi: H(\Gamma) \to \Gamma$ *inducing the identity on homology.*

The proof follows by induction on the number of polynomial generators. For $n = 1$, the definition of an shc algebra allows us to construct the mapping by sending the generator to a representative of its class. Generally, the choice of a representative leads to a shm map $k[x_1] \otimes \cdots \otimes k[x_n] \to \Gamma \otimes \cdots \otimes \Gamma$ that can be multiplied down to an shm map $k[x_1, \ldots, x_n] \to \Gamma$ because $m \colon \Gamma \otimes \Gamma \to \Gamma$ is shm.

To apply the proposition to the problem of homogeneous spaces, there are shm maps $H^*(BG; k) \Longrightarrow C^*(BG; k)$ and $H^*(BH; k) \Longrightarrow C^*(BH; k)$ inducing the identities on homology. The desired collapse theorem can be derived from Theorem 8.20 if the sh-module structures can be brought into place. This requires a careful consideration of when diagrams such as

$$
\begin{array}{ccc}
H^*(BG; k) & \longrightarrow & C^*(BG; k) \\
\Big\downarrow{\scriptstyle Bi^*} & & \Big\downarrow{\scriptstyle Bi^*} \\
H^*(BH; k) & \longrightarrow & C^*(BH; k)
\end{array}
$$

commute up to homotopy.

D. Proposition 8.21 does not restrict us to the case of homogeneous spaces but applies to the cochain algebras of spaces with $H^*(X; k)$ a polynomial algebra. Thus the extra structure of Lie theory is unnecessary for the collapse theorem. The general theorem is due to [Munkholm74].

Theorem 8.22. *Suppose we have the pullback diagram with π, a fibration and B, simply-connected:*

$$
\begin{array}{ccc}
E_f & \longrightarrow & E \\
\Big\downarrow & & \Big\downarrow{\scriptstyle \pi} \\
X & \overset{f}{\longrightarrow} & B.
\end{array}
$$

If $H^*(E; k)$, $H^*(B; k)$ and $H^*(X; k)$ are all polynomial algebras over k in at most countably many variables, and if char $k = 2$, we suppose further that Sq_1 vanishes on $H^*(E; k)$ and $H^*(X; k)$, then, as graded k-vector spaces

$$
H^*(E_f; k) \cong \operatorname{Tor}_{H^*(B;K)}(H^*(X; k), H^*(E; k)).
$$

That is, the Eilenberg-Moore spectral sequence, converging to $H^*(E_f; k)$, collapses at the E_2-term.

Theorem 8.1, the collapse result for homogeneous spaces, follows as a corollary. Theorem 8.22 also applies to two-stage Postnikov systems. The restriction on Sq_1, the cup$_1$-product, is best possible by the example of [Schochet71]. [Wolf77] gave another proof of Theorem 8.1, similar to [Munkholm74], but using the geometric properties of homogeneous spaces. [Husemoller-Moore-Stasheff74] developed a general theory of differential homological algebra that applies to the problem of computing $H_*(G/H; k)$ via the homology Eilenberg-Moore spectral sequence and they are able to prove a similar collapse result.

The methods of strongly homotopy multiplicative maps in differential algebra and more generally, the notion of algebraic properties up to homotopy were developed into powerful organizing principles under the rubric of operads during the 1970s ([May72]) and this idea has found its way into mathematical physics and quantum algebra ([Loday96]), [Stasheff97]). The perturbation theory of a differential on a resolution or of a differential on a graded algebra that underlies Theorem 8.15 was introduced by [Gugenheim60], [Liulevicius63], and [Gugenheim-Milgram70], and has enjoyed considerable development (with an especially neat formalism due to [Brown, R67]). For a survey of these advances, see the papers of [Huebschmann-Kadeishvili91] and [Lambe92].

8.2 Differentials in the Eilenberg-Moore spectral sequence

The underlying category for the Eilenberg-Moore spectral sequence has objects differential graded algebras and modules over them where the module structures are induced by algebra mappings and multiplications. The functor Tor, defined by resolutions, encodes nontrivial relations in the classical sense of syzygies. These relations are due to the multiplicative structures involved and the interplay between homology and the product.

When a pair of products vanish, $u \cdot v = 0 = v \cdot w$, a secondary operation, $\langle u, v, w \rangle$ may be defined that was introduced by [Uehara-Massey57]. In this section, we introduce the Massey triple product and its generalizations due to [Massey58], [Kraines66], and [May68]. A motivating example is the loop suspension homomorphism that may be described as follows: The functor $H^n(\ ; k)$ is representable as $[\ , K(k,n)]$ for each $n \geq 0$. This description of cohomology allows us to define a mapping $H^n(X; k) \rightarrow H^{n-1}(\Omega X; k)$ by applying the topological loop functor:

$$\Omega^* : [X, K(k,n)] \rightarrow [\Omega X, \Omega K(k,n)] = [\Omega X, K(k, n-1)], \qquad f \mapsto \Omega(f).$$

[Eilenberg-Mac Lane50] introduced the loop suspension homomorphism in their study of the spaces $K(\Pi, n)$ and [Serre51] developed it in homology (§6.2), relating it to the Leray-Serre spectral sequence. [Whitehead, GW55] proved that Ω^* annihilates products and furthermore, Ω^* is an isomorphism for n less than three times the connectivity of X. [Kraines66] showed that the higher operations based on products, the higher Massey products, are also annihilated by Ω^*. These operations were developed further and shown to determine the differentials in the Eilenberg-Moore spectral sequence. [May68] applied these results to obtain collapse theorems similar to those of §8.1.

To begin we describe the cohomology loop suspension homomorphism for a fibration and, in particular, for the path-loop fibration. We next introduce the classical Massey triple product and describe some of its geometric applications. Then we consider the generalization of the triple product to an arbitrary number of variables. These higher order operations are related to the cohomology loop suspension homomorphism via the differentials in the Eilenberg-Moore spectral sequence. Finally, we consider matric Massey products that generalize the higher order products and a structure theorem that expresses the differentials in the spectral sequence in terms of matric Massey products.

The cohomology loop suspension homomorphism

Dual to the transgression homomorphism (§6.2) is the **suspension** (or *loop suspension*) associated to a fibration, $F \hookrightarrow E \xrightarrow{\pi} B$. It is defined as an additive relation by the homomorphisms

$$H^q(B) \cong H^q(B, *) \xrightarrow{\pi_0^*} H^q(E, f) \xleftarrow{\delta} H^{q-1}(F)$$

as in the diagram:

The homomorphism $\Omega^* : (\pi_0^*)^{-1}(\operatorname{im} \delta) \to H^{q-1}(F)/\ker \delta \cong \operatorname{im} \delta$ is induced by π_0^*. It is a homomorphism from a submodule of $H^q(B)$ to a quotient of $H^{q-1}(F)$.

For the path-loop fibration, $E = PX \simeq *$, so δ is an isomorphism and Ω^* is a homomorphism $H^q(X) \to H^{q-1}(\Omega X)$. [Serre51, I.n° 3] showed for the path-loop fibration, that Ω^* has the same image as the monomorphism $E_q^{0,q-1} \hookrightarrow H^{q-1}(\Omega X)$ and the same kernel as the surjection $H^q(X) \longrightarrow E_q^{q,0}$, where $E_q^{*,*}$ refers to the E_q-term of the Leray-Serre spectral sequence for $\Omega X \hookrightarrow PX \to X$ (see Proposition 6.10). In the Eilenberg-Moore spectral sequence the loop suspension homomorphism will be seen to have a very simple expression in the case of the path-loop fibration. The motivating problem for §8.2 and its subsections is to describe the kernel of Ω^*. Following [Smith, L67], we approach this question by looking at the filtration due to the homological nature of the Eilenberg-Moore spectral sequence.

Fix a fibration $\pi : E \to B$ with connected fibre F and let $f : X \to B$ be a continuous mapping. Let $E_f \to X$ denote the pullback fibration. Recall that $E_f \subset X \times E$. Let

$$\hat{\pi} : X \times E \to X \times B \times E \quad \text{and} \quad \hat{f} : X \times E \to X \times B \times E$$

be given by $\hat{\pi}(x, e) = (x, \pi(e), e)$ and $\hat{f}(x, e) = (x, f(x), e)$. It follows that $\hat{\pi}$ and $\hat{f}$ agree on the subset E_f. These data determine a **difference homomorphism**, $(\hat{f} - \hat{\pi})^* : H^*(X \times B \times E; k) \to H^*(X \times E, E_f; k)$, following

[Steenrod47], that is defined as a lifting at the cochain level in the diagram:

$$0 \longrightarrow C^*(X \times E, E_f; k) \longrightarrow C^*(X \times E; k) \longrightarrow C^*(E_f; k) \longrightarrow 0.$$

$$(\hat{f} - \hat{\pi})^* \qquad\qquad \Big\uparrow \hat{f}^* - \hat{\pi}^*$$

$$C^*(X \times B \times E; k)$$

Fix a field k and write $H^*(X) = H^*(X; k)$. By the Künneth theorem, we have $H^*(X \times B \times E) \cong H^*(X) \otimes H^*(B) \otimes H^*(E)$.

Next consider the boundary homomorphism $\delta \colon H^{q-1}(E_f) \to H^q(X \times E)$ of the long exact sequence for the pair $(X \times E, E_f)$. Together, $(\hat{f} - \hat{\pi})^*$ and δ determine an additive relation Φ similar to the suspension and the transgression:

$$H^*(X) \otimes \tilde{H}^*(B) \otimes H^*(E) \xrightarrow{(\hat{f} - \hat{\pi})^*} H^*(X \times E, E_f) \xleftarrow{\delta} H^*(E_f)$$

where $\Phi \colon ((\hat{f} - \hat{\pi})^*)^{-1}(\operatorname{im} \delta) \to H^*(E_f)/\ker \delta$ is induced by $(f - \hat{\pi})^*$.

For the path-loop fibration, $\mathrm{ev}_1 \colon PB \to B$, we obtain the fibre ΩB as the pullback over a choice of basepoint $\eta \colon * \to B$. The difference homomorphism reduces to ev_1^* in this case because $PB \simeq *$. Thus the following diagram commutes:

$$H^*(B, *) \xrightarrow{\ \mathrm{ev}_1^*\ } H^*(PB, \Omega B)$$
$$\Big\| \qquad\qquad\qquad \Big\|$$
$$k \otimes \tilde{H}^*(B) \otimes k \xrightarrow[(\widehat{\eta - \mathrm{ev}_1})^*]{} H^*(PB, \Omega B)$$

In this case, the additive relation Φ is Ω^*.

For a general fibration, recall that

$$H^*(X) \otimes \tilde{H}^*(B) \otimes H^*(E) = \mathrm{B}^{-1}(H^*(X), H^*(B), H^*(E))$$

is the (-1)-column of the bar construction for $\mathrm{Tor}_{H^*(B)}(H^*(X), H^*(E))$. Let $\phi \colon H^*(X) \otimes \tilde{H}^*(B) \otimes H^*(E) \to H^*(X) \otimes H^*(E)$ be given by $\phi(u \otimes a \otimes u) = \bar{u}a \otimes v + \bar{u} \otimes \bar{a}v$, where $\bar{x} = (-1)^{1 + \deg x} x$. This is the bar differential and so we get an epimorphism

$$T \colon \ker \phi \longrightarrow\!\!\!\!\rightarrow \mathrm{Tor}_{H^*(B)}^{-1,*}(H^*(X), H^*(E)).$$

We next identify $\ker \phi$ with the domain of the additive relation Φ. Consider the diagram:

$$H^*(X \times E, E_f)$$

$$(\hat{f} - \hat{\pi})^* \qquad\qquad \Big\downarrow j^* \qquad\qquad \delta$$

$$H^*(X) \otimes \tilde{H}^*(B) \otimes H^*(E) \xrightarrow{\hat{f}^* - \hat{\pi}^*} H^*(X \times E) \xrightarrow{i^*} H^*(E_f)$$

$$\phi \qquad\qquad\qquad \Big\|$$

$$H^*(X) \otimes H^*(E)$$

The left triangles commute by the definitions of $\hat{f}$ and $\hat{\pi}$ and the Künneth theorem. The right triangle is exact. It follows that:

$$(\hat{f} - \hat{\pi})^*(w) \in \mathrm{im}\, \delta \Longleftrightarrow j^*(\hat{f} - \hat{\pi})^*(w) = 0$$
$$\Longleftrightarrow (\hat{f}^* - \hat{\pi}^*)(w) = 0$$
$$\Longleftrightarrow \phi(w) = 0.$$

Thus we can write $\Phi \colon \ker \phi \to H^{*-1}(E_f)/\ker \delta$.

Proposition 8.23. *Suppose that B is a simply-connected space. Then*

$$F^0 H^*(E_f) = \mathrm{im}\{i^* \colon H^*(X) \otimes H^*(E) \to H^*(E_f)\}.$$

Furthermore, $F^{-1}H^*(E_f)/F^0 H^*(E_f)$ is additively generated by elements of the form $\Phi(w)$ for some $w \in H^*(X) \otimes \tilde{H}^*(B) \otimes H^*(E)$.

PROOF ([Smith, L67]): To prove the assertion about F^0, we consider the mapping $k \to C^*(B; k)$ given by a choice of basepoint. Then there are algebraic spectral sequences with E_2-terms related by the map induced by the unit

$$\mathrm{Tor}_k^{*,*}(H^*(X), H^*(E)) \to \mathrm{Tor}_{H^*(B)}^{*,*}(H^*(X), H^*(E))$$

and converging to $\mathrm{Tor}_k(C^*(X), C^*(E)) \to H^*(E_f)$. However, since k is a field, $\mathrm{Tor}_k(H^*(X), H^*(E)) \cong H^*(X) \otimes_k H^*(E)$, concentrated in the 0-column. Furthermore, the mapping induced by the basepoint is given by the morphism induced by the inclusion $E_f \subset X \times E$, as seen by examining the bar constructions.

In the Eilenberg-Moore spectral sequence converging to $H^*(E_f)$, there is an edge phenomenon similar to the one in Example 5.E for the Leray-Serre spectral sequence. Since d_r has bidegree $(r, 1-r)$, $d_r \equiv 0$ on $E_r^{-1,*}$ for $r \geq 2$. Thus $E_{n+1}^{-1,*} \cong E_n^{-1,*}/\mathrm{im}\, d_n$ and there is a sequence of epimorphisms

$$\ker \phi \xrightarrow{\;T\;} \mathrm{Tor}_{H^*(B)}^{-1,*}(H^*(X), H^*(E)) \longrightarrow E_3^{-1,*} \longrightarrow \cdots \longrightarrow E_\infty^{-1,*}.$$

Thus $E_\infty^{-1,*} = F^{-1}H^*(E_f)/F^0 H^*(E_f)$ is generated by the image of the classes from $\ker \phi$ under T. Notice, however, that Φ factors through T

$$\ker \phi \longrightarrow \mathrm{Tor}_{C^*(B)}^{-1,*}(C^*(X), C^*(E)) \hookrightarrow H^{*-1}(E_f)/\ker \delta.$$

As the diagram identifying the domain of Φ with $\ker \phi$ shows, both T and Φ have the same kernel and the second assertion is proved. $\square$

In fact, we can say more. Recall that we have the short exact sequence

$$0 \to F^0 H^*(E_f) \to F^{-1}H^*(E_f) \to E_\infty^{-1,*} \to 0.$$

Following [Smith, L67], there is an explicit way to represent all of the elements in $F^{-1}H^*(E_f)$ via Φ. If $w \in \ker \phi$, then we can write $w = \sum_i x_i \otimes a_i \otimes y_i$. Let X_i, A_i and Y_i denote choices of cochain representatives for the classes appearing in w. Since $\phi(w) = 0$, there is a class $U \in C^*(X) \otimes C^*(E)$ with $d_\otimes(U) = \sum_i \pm X_i A_i \otimes Y_i + \pm X_i \otimes A_i Y_i$. Form the element

$$Z = \sum_i X_i[A_i]Y_i - U \in \mathrm{B}^\bullet(C^*(X), C^*(B), C^*(E)).$$

The association of w with the class of Z in $\mathrm{Tor}^{-1,*}_{C^*(B)}(C^*(X), C^*(E))$ has an indeterminacy given by $F^0 H^*(E_f)$. We can represent the suspension homomorphism by choosing representatives and varying a class given by $\Phi(w)$ over the indeterminacy. The result is well-defined in the spectral sequence.

Corollary 8.24. *For the path-loop fibration, the image of the loop suspension homomorphism is given by $F^{-1}H^*(\Omega B)$. Furthermore, this homomorphism is given by the rule: $x \in \tilde{H}^n(B)$ goes to $[X] \in \mathrm{Tor}^{-1,n}_{C^*(B)}(k, k)$ for any representative X of the class x.*

Finally, we examine the consequences of having a spectral sequence. Since $\mathrm{Tor}^{-1}_{C^*(B)}(k, k)$ is a quotient of $\mathrm{Tor}^{-1}_{H^*(B)}(k, k)$, we can use the E_2-term to help describe the image of Ω^*. In particular, the bar construction differential $\mathrm{B}^{-2} \to \mathrm{B}^{-1}$ is given by the multiplication on $H^*(B)$ and so $QH^{n+1}(B) \cong \mathrm{Tor}^{-1,*}_{H^*(B)}(k, k)$ by way of the inclusion, $x \mapsto [x]$. We may take the domain of the loop suspension to be $QH^{*+1}(B)$. [Whitehead, GW55] showed that Ω^* annihilates decomposables.

The codomain of Ω^* is $\mathrm{Tor}^{-1}_{C^*(B)}(k, k)$ and is contained in the space of primitives in $H^*(\Omega B)$ with respect to the coproduct on the bar construction. We examine the consequences of connectedness for the loop suspension homomorphism in a result due to [Whitehead, GW55] and [Smith, L67].

Corollary 8.25. *If B is n-connected $(n \geq 1)$, then*

$$\Omega^* : QH^{q+1}(B) \to \mathrm{Prim}\, H^q(\Omega B)$$

is an isomorphism for $q \leq 3n$.

PROOF: By Corollary 7.19, it suffices to show that no element in $E_2^{-1,q}$ for q less than or equal to $3n + 1$ is in the image of a differential, and hence such elements survive to E_∞. Since d_r has bidegree $(r, 1 - r)$, an element in $E_2^{-1,t}$ would be hit by a differential d_r only if there are elements in $E_r^{-(r+1),t+r-1}$. Since $E_2^{-(r+1),s} = \{0\}$ for $s < (n+1)(r+1)$ by the connectedness assumption, $d_r : E_r^{-(r+1),s} \to E_r^{-1,s-r+1}$ is zero for all $r \geq 1$. From $s < (n + 1)(r + 1)$

we get $s - r + 1 < n(r+1) + 2$. Starting at $r = 2$ it follows that $t = s - r + 1 < 3n + 2$. The total degree of such elements is then less than $3n + 1$ and so the result follows.   □

These results make the Eilenberg-Moore spectral sequence the tool of choice to study the loop suspension homomorphism. We take up the problem of describing possible differentials in subsequent sections. We remark that $F^{-1} H^*(E_f; \mathbb{F}_p)$ as described by the partial homomorphism Φ is determined by geometric maps and hence is closed under the action of the Steenrod algebra. [Smith, L67] asked if the higher filtration pieces were also closed under the action of the Steenrod algebra. This problem is taken up in §8.3.

Another interpretation of the filtration submodules in the special case of the Eilenberg-Moore spectral sequence converging to the cohomology of a classifying space was given by [Toomer74] and is related to the Lusternik-Schnirelmann category of a space. Generalizations of this notion are due to [Fadell-Husseini92] and [Strom]—*essential category weight* of a cohomology class, and by [Rudyak99]—*strict category weight*. In both cases the Eilenberg-Moore spectral sequence and Massey products (even matric Massey products) play a role.

Massey's triple product

Let (Γ, d, μ) denote a differential graded algebra over a ring R and denote the multiplication on Γ and $H(\Gamma) = H(\Gamma, d)$ by $\mu(u, v) = u \cdot v$. Suppose $[u]$, $[v]$ and $[w]$ denote classes in $H(\Gamma)$, represented by $u \in \Gamma^p$, $v \in \Gamma^q$ and $w \in \Gamma^r$. If $[u] \cdot [v] = 0 = [v] \cdot [w]$ in $H(\Gamma)$, then we introduce a new cohomology class definable because $[u] \cdot [v] \cdot [w] = 0$ *for two different reasons*. It is constructed as follows: Since $[u] \cdot [v] = 0$, there is an element $s \in \Gamma^{p+q-1}$ with $ds = \bar{u} \cdot v$ (recall that $\bar{u} = (-1)^{1+\deg u} u$). Similarly, there is an element $t \in \Gamma^{q+r-1}$ with $dt = \bar{v} \cdot w$. The element $\bar{s} \cdot w + \bar{u} \cdot t$ determines a cocycle in $\Gamma^{p+q+r-1}$:

$$d(\bar{s} \cdot w + \bar{u} \cdot t) = (-1)^{p+q} ds \cdot w + (-1)^p \bar{u} \cdot dt$$
$$= (-1)^{p+q} \bar{u} \cdot v \cdot w + (-1)^{p+q+1} \bar{u} \cdot v \cdot w = 0.$$

We define the **Massey triple product** of $[u]$, $[v]$ and $[w]$ as the set of all cohomology classes that can be defined in this manner:

$$\langle [u], [v], [w] \rangle = \{ [\bar{s} \cdot w + \bar{u} \cdot t] \in H^{p+q+r-1}(\Gamma), \text{ where } u, v, w \text{ represent}$$
$$[u] \in H^p(\Gamma), [v] \in H^q(\Gamma), \text{ and } [w] \in H^r(\Gamma), \text{ respectively, and}$$
$$s, t \text{ vary over choices that satisfy } ds = \bar{u} \cdot v \text{ and } dt = \bar{v} \cdot w. \}$$

By regarding our choices more carefully, we can identify the indeterminacy.

Proposition 8.26. *The Massey triple product* $\langle[u], [v], [w]\rangle$ *is an element of the quotient group* $H^{p+q+r-1}(\Gamma)/([u] \cdot H^{q+r-1}(\Gamma) + H^{p+q-1}(\Gamma) \cdot [w])$.

PROOF: We need to show that different choices do not affect the coset in $H^{p+q+r-1}(\Gamma)$ given above. For example, if s and s' are chosen with $ds = \bar{u} \cdot v = ds'$ and t is chosen with $dt = \bar{v} \cdot w$, then

$$(\bar{s} \cdot w + \bar{u} \cdot t) - (\bar{s}' \cdot w + \bar{u} \cdot t) = (\bar{s} - \bar{s}') \cdot w,$$

which, on homology, lies in $H^{p+q-1}(\Gamma) \cdot [w]$. Similar arguments complete the proof. □

Some of the formal properties of the triple product (for example, naturality and relations with the product) are special cases of more general results proved later. Therefore, we postpone listing those properties until then and discuss some of the geometric applications of the triple product.

A. If $\alpha \in \pi_p(X)$ and $\beta \in \pi_q(X)$, then the **Whitehead product** of α and β, $[\alpha, \beta] \in \pi_{p+q-1}(X)$ is defined as follows: Let $f\colon (I^p, \partial I^p) \to (X, *)$ and $g\colon (I^q, \partial I^q) \to (X, *)$ represent α and β, respectively. Then $I^p \times I^q$ represents an $(p+q)$-cell and so $\partial(I^p \times I^q) = I^p \times \partial I^q \cup \partial I^p \times I^q \cong S^{p+q-1}$. Define $[\alpha, \beta]$ to be the homotopy class of the mapping $h\colon (S^{p+q-1}, *) \to (X, *)$ given by

$$h(x, y) = \begin{cases} f(x), & \text{if } x \in I^p, y \in \partial I^q, \\ g(y), & \text{if } x \in \partial I^p, y \in I^q. \end{cases}$$

When $p = q = p + q - 1 = 1$, $[\alpha, \beta] = \alpha\beta\alpha^{-1}\beta^{-1}$, the actual commutator, and hence the notation. The elementary properties of the Whitehead product include $[\alpha, \beta] = (-1)^{pq}[\beta, \alpha]$, an indication of the choice of signs for a graded Lie algebra. This operation was introduced by [Whitehead, JHC41] and was conjectured by Weil to satisfy the Jacobi identity for a graded Lie algebra by analogy with the product of [Samelson54] for the homotopy groups of a group-like space. Several approaches to proving the identity appeared about the same time in work of [Whitehead, GW54], [Nakaoka-Toda54], [Hilton55], and [Uehara-Massey57].

[Uehara-Massey57] proved the Jacobi identity by applying the triple product. Consider $X = S^p \vee S^q \vee S^r$ and suppose $\alpha \in \pi_p(X)$, $\beta \in \pi_q(X)$ and $\gamma \in \pi_r(X)$ are generators. Form the adjunction space, $K = X \cup_{[\alpha,[\beta,\gamma]]} e^{p+q+r-1}$. This is the universal example for the triple Whitehead product. In this case, $H^*(K)$ has generators in degrees p, q, r and $p+q+r-1$ that we denote by $[u]$, $[v]$, $[w]$ and $[z]$.

Theorem 8.27. *For* $K = (S^p \vee S^q \vee S^r) \cup_{[\alpha,[\beta,\gamma]]} e^{p+q+r-1}$ *and* $[u]$, $[v]$, $[w]$, $[z]$ *in* $H^*(K)$ *as described,*

$$\langle[u], [v], [w]\rangle = (-1)^p[z], \qquad \langle[v], [w], [u]\rangle = (-1)^{pq+pr+p}[z]$$
$$\langle[w], [u], [v]\rangle = 0.$$

Thus the triple product is an operation that detects the triple Whitehead product, $[\alpha, [\beta, \gamma]]$. In order to prove that the Whitehead product satisfies the Jacobi identity for graded Lie algebras, form the mapping

$$(-1)^{pr}[\alpha, [\beta, \gamma]] + (-1)^{pq}[\beta, [\gamma, \alpha]] + (-1)^{qr}[\gamma, [\alpha, \beta]] = [f]$$

and consider the complex $K_f = (S^p \vee S^q \vee S^r) \cup_f e^{p+q+r-1}$. By the properties of adjunction spaces and Theorem 8.27, [Uehara-Massey57] showed that f is nullhomotopic if the following identity holds for triple products in $H^*(K_f)$;

$$(-1)^{pr}\langle [u], [v], [w] \rangle + (-1)^{qp}\langle [v], [w], [u] \rangle + (-1)^{rq}\langle [w], [u], [v] \rangle = 0.$$

This holds, however, by the basic formal properties of triple products and so the desired identity is established.

B. The cup product is useful geometrically in the classical theory of intersections of subcomplexes of $\mathbb{R}^n$. If S^p and S^q are disjoint spheres in $\mathbb{R}^n$, where $n = p + q + 1$, then, by Alexander duality, $H^*(\mathbb{R}^n - (S^p \cup S^q))$ has infinite cyclic generators in dimensions p, q and $p + q$. The cup product of the generators in dimensions p and q is a multiple of the generator in dimension $p + q$ and this multiple can be shown to be the linking number of the spheres up to sign ([Rolfsen76, p. 132ff.]). If three piecewise unlinked spheres are considered, then a higher order linking number can be introduced by computing the triple product. [Massey68/98] showed that this higher order linking number is nonzero in the case of the 'Borromean rings.'

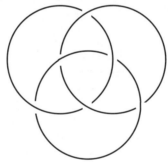

This gives a rigorous proof that, though pairwise unlinked, the three rings cannot be pulled apart. Applications of this higher order link invariant are discussed in the next section along with its generalization to many component links.

C. [Viterbo98] has applied the triple Massey product to the problem of determining the minimal number of critical points of a function on a Hilbert manifold X. Through Morse theory this problem can be related to classical Lusternik-Schnirelmann category and hence to the cup length of a space, that is, the maximal number of cohomology classes with nonzero cup product. Viterbo introduced the notion of tied cohomology classes and tie length to bound the number of critical points. He then used Massey products to give examples of bounds for tie length that are greater than the cup length and hence closer to the desired number of critical points.

Higher order Massey products

When two triple products are defined $\langle [u], [v], [w] \rangle$ and $\langle [v], [w], [x] \rangle$, then it can happen that there is a choice of elements in (Γ, d) for which

$$dY_1 = \bar{t}_0 \cdot v + \bar{u} \cdot t_1, \text{ where } dt_0 = \bar{u} \cdot v \text{ and } dt_1 = \bar{v} \cdot w$$
$$dY_2 = \bar{t}_1 \cdot w + \bar{v} \cdot t_2, \text{ where } dt_2 = \bar{w} \cdot x.$$

In this case, $0 \in \langle [u], [v], [w] \rangle$ and $0 \in \langle [v], [w], [x] \rangle$, and we can form the element

$$\bar{u} \cdot Y_2 + \bar{t}_0 \cdot t_2 + \bar{Y}_1 \cdot x.$$

This expression is a cocycle and as we vary our choices we obtain a subset of $H^{|u|+|v|+|w|+|x|-2}(\Gamma, d)$ that we denote by $\langle [u], [v], [w], [x] \rangle$.

A triple product is formed when certain products among three cohomology classes vanish. A fourfold product is defined when certain triple products among four cohomology classes contain the zero cohomology class. In like manner we could define fivefold Massey products and so on. We next give a uniform definition, for all orders, of higher order Massey products. A definition was first given by [Massey58]. The following definition (with its choice of signs) is due to [Kraines-Schochet72].

Definition 8.28. *Suppose* (Γ, d) *is a differential graded algebra and* $[\gamma_1]$, $[\gamma_2]$, ..., $[\gamma_n]$ *are classes in* $H^*(\Gamma, d)$ *with* $[\gamma_i] \in H^{p_i}(\Gamma, d)$. *A* **defining system**, *associated to* $\langle [\gamma_1], \ldots, [\gamma_n] \rangle$, *is a set of elements* (a_{ij}) *for* $1 \le i \le j \le n$ *and* $(i, j) \ne (1, n)$ *with* $a_{ij} \in \Gamma$ *satisfying*

(1) $a_{ij} \in \Gamma^{p_i + p_{i+1} + \cdots + p_j - j + i}$,
(2) $a_{i,i}$ *is a representative of* $[\gamma_i]$ *in* $H^{p_i}(\Gamma, d)$,
(3) $d(a_{ij}) = \sum_{r=i}^{j-1} \bar{a}_{ir} \cdot a_{r+1,j}$.

To a defining system we associate the cocycle,

$$\sum_{r=1}^{n-1} \bar{a}_{1,r} \cdot a_{r+1,n} \in \Gamma^{p_1 + \cdots + p_n - n + 2}.$$

The n-**fold Massey product**, $\langle [\gamma_1], \ldots, [\gamma_n] \rangle$, *is the set of cohomology classes of cocycles associated to all possible defining systems for* $\langle [\gamma_1], \ldots, [\gamma_n] \rangle$.

First observe how this definition generalizes the Massey triple product. To give a defining system for $\langle [u], [v], [w] \rangle$, we need a "matrix" of values from Γ,

$$\begin{pmatrix} a_{11} & a_{12} & \\ & a_{22} & a_{23} \\ & & a_{33} \end{pmatrix} = \begin{pmatrix} u & s & \\ & v & t \\ & & w \end{pmatrix}.$$

The defining properties of a Massey product can be summarized in a kind of partial matrix multiplication:

$$d \begin{pmatrix} u & s & \\ & v & t \\ & & w \end{pmatrix} = \begin{pmatrix} 0 & \bar{u}v & \\ & 0 & \bar{v}w \\ & & 0 \end{pmatrix},$$

$$\begin{pmatrix} \bar{u} & \bar{s} & \\ & \bar{v} & \bar{t} \\ & & \bar{w} \end{pmatrix} \begin{pmatrix} & v & t \\ & & w \end{pmatrix} = \begin{pmatrix} & \bar{u}v & \bar{u}t + \bar{s}w \\ & & \bar{v}w \end{pmatrix}$$

The cocycle associated with this defining system appears in the upper right hand corner and represents an element in the Massey product. Extending this matrix notation to more classes recovers Definition 8.28.

The n-fold Massey product may be thought of as an nth order cohomology operation because the data provide that the related $(n-1)^{\text{st}}$ order and lower order Massey products must be defined and contain zero in a coherent manner. For example, for $\langle [\gamma_1], [\gamma_2], [\gamma_3], [\gamma_4] \rangle$, the defining system provides the boundaries that show $0 \in \langle [\gamma_1], [\gamma_2], [\gamma_3] \rangle$ and $0 \in \langle [\gamma_2], [\gamma_3], [\gamma_4] \rangle$.

We record some of the properties of the n-fold Massey product. Most of these properties are straightforward consequences of the definition (with the exception of (5)). We encourage the reader to provide proofs of these assertions (one can also refer to the paper of [Kraines66] for details with slightly different signs).

Theorem 8.29. *Let* (Γ, d) *be a differential graded algebra over a ring R and* $[\gamma_1], \ldots, [\gamma_n]$ *be in* $H^*(\Gamma, d)$. *Where the following make sense, they hold:*

(1) *(Linearity) if $\lambda \in R$, then, for all $1 \leq i \leq n$,*

$$\lambda \langle [\gamma_1], \ldots, [\gamma_n] \rangle \subset \langle [\gamma_1], \ldots, \lambda[\gamma_i], \ldots, [\gamma_n] \rangle.$$

(2) *(Naturality) if $f \colon \Gamma \to \Lambda$ is a morphism of differential graded algebras, then*
$$f^* \langle [\gamma_1], \ldots, [\gamma_n] \rangle \subset \langle f^*[\gamma_1], \ldots, f^*[\gamma_n] \rangle.$$

(3) *(Associativity) for $v \in H^*(\Gamma, d)$,*

$$\langle [\gamma_1], \ldots, [\gamma_n] \rangle v \subset \langle [\gamma_1], \ldots, [\gamma_n]v \rangle$$
$$v \langle [\gamma_1], \ldots, [\gamma_n] \rangle \subset (-1)^{\deg v} \langle v[\gamma_1], \ldots, [\gamma_n] \rangle$$

Suppose further that (Γ, d) has a cup_1-product satisfying the Hirsch formulas (Definition 8.12), then

$$\langle [\gamma_1], \ldots, [\gamma_t]v, \ldots, [\gamma_n] \rangle \cap (-1)^{\deg v} \langle [\gamma_1], \ldots, [\gamma_t], v[\gamma_{t+1}], \ldots, [\gamma_n] \rangle \neq \emptyset$$

(4) *(Symmetry) if $p_j = \deg \gamma_j$ and $l = \displaystyle\sum_{1 \leq r \leq s \leq n} p_r p_s + (n-1)(n-2)/2$, then*

$$\langle [\gamma_n], \ldots, [\gamma_1] \rangle = (-1)^l \langle [\gamma_1], \ldots, [\gamma_n] \rangle.$$

(5) *If $k = \mathbb{F}_p$ for p, an odd prime and $[\gamma] \in H^{2m+1}(X; \mathbb{F}_p)$ then*

$$-\beta P^m [\gamma] \in \underbrace{\langle [\gamma], \ldots, [\gamma] \rangle}_{p \text{ times}}.$$

Property (5) plays a role in the analysis of $H^*(\Omega X; \mathbb{F}_p)$ when X is an H-space; see Chapter 10 and the paper of [Kane75]. In the special case of coefficients in $\mathbb{F}_p$ for p, an odd prime, every odd dimensional class has square zero and so one can try to form the triple product. By using a particular choice of defining system based on the cup$_1$-product, [Kraines66] showed that $\langle [\gamma], \ldots, [\gamma] \rangle$ (p many), is defined. Furthermore, since we can use some of the data in the defining system iteratively, we can define a cohomology operation, denoted $\langle [\gamma] \rangle^p$ that lies in $H^{2mp+2}(\Gamma, d)$ when $[\gamma] \in H^{2m+1}(\Gamma, d)$ and satisfies $\langle [\gamma] \rangle^p \subset \langle [\gamma], \ldots, [\gamma] \rangle$ (p many).

[Kraines66] showed that the operation $[\gamma] \mapsto \langle [\gamma] \rangle^p$ is equal to $-\beta P^m [\gamma]$ with zero indeterminacy. The proof is by universal example. This result was thought to be surprising at the time as it relates an unstable operation, the iterated Massey product, to a stable operation $-\beta P^m$. In response to a question posed by [Stasheff68], [Kraines73] applied the Eilenberg-Moore spectral sequence to k stage Postnikov systems to extend property (5) and relate $\langle [\gamma] \rangle^{p^k}$ to $-\beta_k P^{p^{k-1} m} \cdots P^m [\gamma]$ for $[\gamma] \in H^{2m+1}(X; \mathbb{F}_p)$ and X a space for which there is a mapping satisfying certain conditions, $f : X \to E_k$, the universal example for the iterated Massey product operation.

We return to the question of determining $\ker \Omega^*$ with the definition of n-fold Massey products. A connection is made through the following formal lemma, the "staircase argument" for double complexes, which is the argument that led to the discovery of spectral sequences (see Lemme 2 of n° 4 of [Leray45] and the discussion of [Borel98]). In this presentation we follow [Kraines-Schochet72].

Lemma 8.30. *Let $(A_{n,m}, d', d'')$ be a double complex and $a_1, \ldots, a_s$ elements in $A_{*,*}$. Suppose $d'' a_{r+1} - d' a_r = 0$ for $1 \leq r \leq s-1$ and define $a = a_1 - a_2 + \cdots + (-1)^{s-1} a_s$. Then $da = d'a + d''a = d''a_1 + (-1)^{s-1} d' a_s$ and, in the spectral sequence associated to the double complex, if $d'' a_1 = 0$, then a_s survives to E_s and $(-1)^s d_s([a_s]) = [d' a_s]$.*

PROOF: First we show that if a_1 and a_2 exist as assumed, then d_1 is zero on a_1 and d_2 applied to the class determined by a_1 is the same as the class of $d' a_2$. Finally, if a_3 exists as assumed, then $[d_1 a_2] = 0$ and $d_2 [a_1] = 0$.

By the assumptions, it is clear that $a_1, \ldots, a_s$ all have the same total degree so we can picture the double complex with its filtration as in the picture:

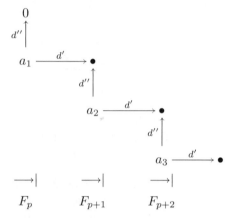

Consider the class $a_1 - a_2$. Note that $[a_1] = [a_1 - a_2]$ in $E_0^p = F^p/F^{p+1}$ and, since $d''a_1 = 0$ and $d'a_1 - d''a_2 = 0$, we have

$$d(a_1 - a_2) = d''a_1 + d'a_1 - d''a_2 - d'a_2 = -d'a_2.$$

In the subquotients $d'a_2$ is a class in the same filtration determined by $d(a_1 - a_2)$ and hence $d_2[a_1] = [-d'a_2]$. Denote $\mathrm{total}_*(A_{*,*})$ by T_*; in the proof of Theorem 2.6, d_2 is described explicitly as

$$F_p T_{p+q} \cap d^{-1}(F_{p+2}T_{p+q+1}) \Big/ {}_{F_{p+1}T_{p+q} \cap d^{-1}(F_{p+1}T_{p+q+1}) + F_p T_{p+q} \cap d(F_{p-1}T_{p+q-1})}$$

$$= E_2^{p,q} \xrightarrow[\text{induced by } d]{d_2} E_2^{p+2,q-1} =$$

$$F_{p+2} T_{p+q+1} \cap d^{-1}(F_{p+4}T_{p+q+2}) \Big/ {}_{F_{p+3}T_{p+q+1} \cap d^{-1}(F_{p+3}T_{p+q+2}) + F_{p+2}T_{p+q+1} \cap d(F_{p+1}T_{p+q})}$$

Observe that $d'a_2$ lies in $F_{p+2}T_{p+q+1}$ and since $d(d'a_2) = 0$, $d'a_2$ is in $E_2^{p+2,q-1}$. Also a_2 lies in $F_{p+1}T_{p+q} \cap d^{-1}(F_{p+1}T_{p+q+1})$ and so $a_1 \sim a_1 - a_2$. If a_3 exists as assumed, then a_3 lies in $F_{p+2}T_{p+q}$ and we get

$$a_1 \sim a_1 - a_2 \sim a_1 - a_2 + a_3$$

and $d(a_1) \sim d(a_1 - a_2) \sim d(a_1 - a_2 + a_3) = d'a_3,$

which is in $F_{p+3}T_{p+q+1} \cap d^{-1}(F_{p+3}T_{p+q+2})$. So, if a_3 exists, $d_2[a_1] = 0$, that is, a_1 persists to E_3. The general case of $s > 3$ is similar. $\square$

We apply this lemma immediately to the algebraic Eilenberg-Moore spectral sequence for computing $\mathrm{Tor}_\Gamma(k, k) = H(\mathrm{B}^\bullet(k, \Gamma, k), d)$ to prove a result of [May68] that was conjectured by Massey. By the results of §7.2, we take the E_1-term of this spectral sequence to be $\mathrm{B}^\bullet(k, H(\Gamma), k)$.

Theorem 8.31. *If* $\langle [u_1], \ldots, [u_s] \rangle$ *is defined in* $H(\Gamma, d)$*, then* $[u_1 \mid \cdots \mid u_s]$
survives to E_{s-1} *and*

$$d_{s-1}([u_1 \mid \cdots \mid u_s]) \in \pm[\langle [u_1], \ldots, [u_s] \rangle].$$

PROOF: Let (b_{ij}) be a defining system for $\langle [u_1] \cdots, [u_s] \rangle$ with $b_{i,i} = u_i$. In $B^\bullet(k, \Gamma, k)$ define

$$a_1 = [b_{11} \mid \cdots \mid b_{ss}]$$

$$a_2 = \sum_r [b_{11} \mid b_{22} \mid \cdots \mid b_{r-1,r-1} \mid b_{r,r+1} \mid b_{r+2,r+2} \mid \cdots \mid b_{ss}]$$

$$\vdots$$

$$a_l = \sum [b_{1i_1} \mid b_{i_1+1,i_2} \mid \cdots \mid b_{i_{l-1}+1,s}]$$

$$\vdots$$

$$a_{s-1} = \sum_r [b_{1r} \mid b_{rk}],$$

where the summation for a_l is over all sequences $1 \leq i_1 \leq i_2 \cdots \leq i_{l-1} \leq s$ such that $|i_1 - 1| \leq l$, $|i_{l-1} - s| \leq l$ and $|i_j - i_{j+1}| \leq l$. We have $d^\bullet a_1 = 0$ (since $d_\Gamma(b_{ii}) = 0$) and, by a routine calculation, $d^\bullet a_{r+1} - \delta a_r = 0$ for each r (where δ is the external and $d^\bullet$ the internal differential on $B^\bullet(k, \Gamma, k)$). By the staircase argument,

$$d_{s-1}[a_1] = \pm[\delta a_{s-1}] = \pm\left[\sum_r \bar{b}_{1r} \cdot b_{rs}\right] \in \pm[\langle u_1, \ldots, u_s \rangle].$$

Since $[a_1] = [u_1 \mid \cdots \mid u_s]$, the theorem follows. $\square$

Corollary 8.32. *If* $\langle [u_1], [u_2], \ldots, [u_s] \rangle$ *is defined in* $H^*(X; k)$*, then*

$$\Omega^*(\langle [u_1], [u_2], \ldots, [u_s] \rangle) = 0.$$

PROOF: In the Eilenberg-Moore spectral sequence, converging to $H^*(\Omega X; k)$, $[\langle [u_1], \ldots, [u_s] \rangle]$ is an element of $E_1^{-1,*}$ and its survival in the spectral sequence determines its image in $H^{*-1}(\Omega X; k)$ under Ω^* by Corollary 8.24. Theorem 8.31, however, shows that $[\langle [u_1], \ldots, [u_s] \rangle]$ is a boundary at E_{s-1} and hence it does not persist to E_∞. Thus Ω^* must annihilate $\langle [u_1], \ldots, [u_s] \rangle$. $\square$

In low dimensions the notion of n-fold Massey products can be used to study links with n components that are unlinked pairwise, three-wise, etc., up to $(n-1)$-wise. The class of *Brunnian links* ([Brunn1892]) includes such links as the Borromean rings, and others that are pairwise unlinked, but linked in some higher order. Higher order Massey products are ideal for defining invariants of such links and considerable work has been done in this direction (see [Cochran90] and the bibliography there).

The classical invariants of knots and links are defined from the fundamental group of the complement in S^3 of the knot or link. [Milnor57] introduced the $\bar{\mu}$-invariants of a link, defined as elements in the quotients of the fundamental group of the complement by its lower central series (Definition §8bis.15). [Stallings65] conjectured a relation between Milnor's $\bar{\mu}$-invariants and Massey products in the cohomology of the complement. This connection was made precise by [Dwyer75"], [Turaev76], and [Porter80], and further developed by [Fenn-Sjerve84]. New link invariants defined by [Cochran90], [Orr91], and [Stein90] are closely related to Massey products and show considerable more subtlety than classical invariants such as the Alexander module.

The techniques of Koszul complexes, Massey products and spectral sequences of Eilenberg-Moore type have also found deep applications in the study of local rings. If R is a local ring with maximal ideal, $\mathfrak{m}$, then the **Poincaré series** of the ring R is given by

$$P_R(t) = \sum_{i=0}^{\infty} (-1)^i \dim_{R/\mathfrak{m}}(\text{Tor}_i^R(R/\mathfrak{m}, R/\mathfrak{m}))t^i.$$

The homological structure of the ring R is closely related to the properties of this series. For example, [Golod62] gave a condition, equivalent to the vanishing of all higher order Massey products in the homology of a Koszul complex, which leads to explicit expressions for $P_R(t)$. [Avramov81] introduced a spectral sequence related to the algebraic Eilenberg-Moore spectral sequence that can be used to define obstructions to multiplicative structures on minimal free resolutions of a certain type. In fact, there is a well-defined "dictionary" between ideas in algebraic topology and local ring theory, described in detail by [Avramov-Halperin86].

Another important topological application of Massey products is to the study of Kähler manifolds. In a fundamental paper [Deligne-Griffiths-Morgan-Sullivan 75] proved, using Sullivan's methods of rational homotopy theory, that the vanishing of Massey products of all orders is a necessary condition for the existence of a Kähler structure. Such a manifold is said to be **formal**, that is, its real homotopy type is a formal consequence of its real cohomology ring. The relation between formality and the vanishing of Massey products has been studied by [Halperin-Stasheff79]. Massey products also play a key role in the classification of rational homotopy types ([Schlessinger-Stasheff85])

To what degree do higher order Massey products determine differentials in the Eilenberg-Moore spectral sequence? To answer this question and to complete the analysis of $\ker \Omega^*$, we need to introduce a more involved generalization of the Massey products.

Matric Massey products Ⓝ

One of the features of the homological invariants we have discussed through-out this book is their independence of choice of projective resolution for com-putation. In particular, we have seen how a carefully constructed resolution can reveal subtle geometric information (for example, in §8.1 and §8.2) or permit straightforward algebraic computation (for example, in §7.3). In unpublished work, Massey constructed a version of the bar construction using matrices of elements in the algebras and modules and differentials defined with the slide products of [Mac Lane55]. [May68, 69] introduced a powerful generalization of the Massey product that was inspired by Massey's bar construction. With this generalization as a tool, May was able to complete the analysis of $\ker \Omega^*$ and to obtain some previously inaccessible collapse theorems for homogeneous spaces. Furthermore, a complete conceptual description of the differentials in the Eilenberg-Moore spectral sequence is possible with the matric Massey products of [May68]. In this section, we define May's matric Massey products and discuss some of their applications.

We begin with (Γ, d), a differential graded algebra over a ring R and let $\mathrm{Mat}(H(\Gamma))$ denote the set of matrices with entries in $H(\Gamma, d) = H(\Gamma)$ and similarly, $\mathrm{Mat}(\Gamma)$, the set of matrices with entries in Γ. If $V = (\gamma_{ij})$ is in $\mathrm{Mat}(\Gamma)$, then we let $dV = (d\gamma_{ij})$ and $\bar{V} = (\bar{\gamma}_{ij}) = ((-1)^{1+\deg \gamma_{ij}}\gamma_{ij})$. For a given matrix V, consider the matrix in $\mathrm{Mat}(\mathbb{Z})$ defined by $D(V) = (\deg \gamma_{ij})$. We say that two matrices, V and W, are **multipliable** if V is a $(p \times q)$-matrix and W is a $(q \times r)$-matrix and the sum $e_{ij} = \deg v_{ik} + \deg w_{kj}$ is independent of k. When V and W are multipliable, we find that VW makes sense and $D(VW) = D(V)D(W) = (e_{ij})$.

We define the matric Massey products inductively. Let $V_1, V_2, \dots, V_n$ be in $\mathrm{Mat}(H(\Gamma))$ and suppose V_i and V_{i+1} are multipliable for $i = 1, \dots, n-1$. We take $A_{i-1,i}$ in $\mathrm{Mat}(\Gamma)$ to be a matrix of representatives for the entries in V_i. When $\bar{V}_i V_{i+1} = (0)$, there are matrices $A_{i-1,i+1} \in \mathrm{Mat}(\Gamma)$ for which $dA_{i-1,i+1} = \bar{A}_{i-1,i}A_{i,i+1}$. As in the case of the Massey triple products,

$$\bar{A}_{i-1,i}A_{i,i+2} + \bar{A}_{i-1,i+1}A_{i+1,i+2}$$

is a matrix of cocycles in Γ. The set of all associated matrices of homology classes defined in this manner give $\langle V_i, V_{i+1}, V_{i+2} \rangle$ a subset of $\mathrm{Mat}(H(\Gamma))$. Inductively, we are seeking A_{ij} in $\mathrm{Mat}(\Gamma)$, for $0 \le i < j \le n$, $(i,j) \ne (0,n)$, so that $[A_{i-1,i}] = V_i$ in $\mathrm{Mat}(H(\Gamma))$ and, for $1 < j - i < n$,

$$dA_{ij} = \sum_{k=i+1}^{j-1} \bar{A}_{ik}A_{kj} \equiv \tilde{A}_{ij}.$$

It is a straightforward calculation to show that $d\tilde{A}_{0n} = d\left(\sum_{k=1}^{n-1} \bar{A}_{0k}A_{kn}\right) = (0)$, and so we say that the homology class $[\tilde{A}_{0n}]$ lies in $\langle V_1, \dots, V_n \rangle$. Such a system of matrices is a **defining system** for the **matric Massey product** $\langle V_1, \ldots, V_n \rangle = \{[\tilde{A}_{0n}] \in \text{Mat}(H(\Gamma)) \mid (A_{ij})$ is a defining system for V_1, $\ldots, V_n\}$.

By reindexing and taking only (1×1)-matrices, we recover the definition of the higher order Massey products. Furthermore, the analogues of the properties listed in Theorem 8.29 hold for matric Massey products. We list some applications.

A. In a first version of his study of the Eilenberg-Moore spectral sequence, [May68'] used a complex (following Massey) built of matrices and with a differential based on slide cycles to serve as the bar construction. By making precise the homology isomorphism between this complex and the bar construction, it followed directly that the differentials in this algebraic Eilenberg-Moore spectral sequence are expressible as matric Massey products. The details, however, are intricate. [Gugenheim-May74] (see §8.1) reframe this kind of differential homological algebra. Their choice of fundamental object is the 'mapping cylinder' associated to a semi-free resolution and a mapping to a module, M. This leads to a resolution

$$\to \overline{X}^{p*} \otimes H(\Gamma) \to \overline{X}^{p-1,*} \otimes H(\Gamma) \to \cdots \to \overline{X}^{0*} \otimes H(\Gamma) \to H(M) \to 0.$$

The modules $\overline{X}^{p*}$ can be taken to be free over R, and so one can introduce ordered bases for each $\overline{X}^{p*}$ and consider each basis as a row vector. In the complex, $\overline{X} \otimes_\Gamma N$, used to compute $\text{Tor}_\Gamma(M, N)$, typical elements can be expressed in terms of a row matrix times a column matrix. Given the particular form of a semifree resolution, we focus attention on certain canonically defined matric Massey products that arise from the expression of the differential on $\bar{X} \otimes_\Gamma N$ in terms of the bases. It follows that the associated differentials in the Eilenberg-Moore spectral sequence are expressible in terms of these canonical matric Massey products.

B. The first immediate corollary of this characterization of the differentials in the Eilenberg-Moore spectral sequence is the collapse theorem of §8.1. For a space with polynomial algebra as cohomology, all matric Massey products vanish, and with them, all potential differentials in the spectral sequence.

C. Another immediate corollary is the determination of the elements in $\ker \Omega^*$ as those classes in $H^*(X; k)$ that are representable as matric Massey products. The proof is similar to the proof of Corollary 8.32. This makes precise the most general notion of when a class is 'decomposable' with regard to the cup product and its associated higher order operations.

D. Finally, these homological methods have applications to the dual problem of computing and describing $\text{Ext}_\Gamma(M, N)$ (see §9.2). Via these descriptions, [May68] obtained results on the stable homotopy of spheres through the study of $\text{Ext}_{A_p}(\mathbb{F}_p, \mathbb{F}_p)$ and the Adams spectral sequence.

8.3 Further structure

The Eilenberg-Moore spectral sequence, as developed in Chapter 7, is principally an algebraic tool that relates $\mathrm{Tor}_\Gamma(M, N)$ to $\mathrm{Tor}_{H(\Gamma)}(H(M), H(N))$ for Γ, M, and N in the appropriate differential categories. The application to topology follows from the second theorem of Eilenberg and Moore (Theorem 7.14) that asserts, for a pullback diagram with π a fibration,

we have an isomorphism $H^*(E_f; k) \cong \mathrm{Tor}_{C^*(B;k)}(C^*(X; k), C^*(E; k))$. The spectral sequence converging to $H^*(E_f; k)$ has been to this point an algebraic artifact of the definition of cohomology.

One of the advantages enjoyed by singular cohomology is its rich structure. Cohomology is an algebra via the cup product and Proposition 7.17 shows the homological product on Tor is compatible with that of $H^*(E_f; k)$ and the spectral sequence. When $k = \mathbb{F}_p$, for p, a prime, $H^*(E_f; \mathbb{F}_p)$ carries an action of the mod p Steenrod algebra $\mathcal{A}_p$. A natural question to ask is whether this extra structure on $H^*(E_f; \mathbb{F}_p)$ is also available in the Eilenberg-Moore spectral sequence.

In the light of this question, the construction of the Eilenberg-Moore spectral sequence appears somewhat ad hoc and the role of the underlying geometric data in the construction becomes muddled. One route to the existence of further structure would be a geometric construction of the spectral sequence where all differentials and subquotients are seen to be consequences of applying an algebraic functor to a system of spaces and so the extra structure follows for free. We present two geometric constructions of the Eilenberg-Moore spectral sequence, both achieving this goal of uncovering some of the deeper structure. We pay particular attention to the Steenrod algebra structure that can be recovered in this manner.

The first construction is due to [Smith, L70] and is motivated by efforts of [Atiyah62], [Landweber66], [Hodgkin68], and [Conner-Smith69] toward proving a Künneth theorem for generalized cohomology theories such as K-theory and bordism theories. The main idea is to develop the properties of the category **Top**/B of spaces over B in which the product in the category is the pullback to which the Eilenberg-Moore spectral sequence applies. Thus the Künneth theorem for spaces, when applied to products in the new category, takes the form of the Eilenberg-Moore spectral sequence. After developing the homotopy theory of the new category **Top**/B, [Smith, L70] constructed a geometric resolution whose associated exact couple gives rise to the desired spectral sequence.

In the second construction, due to [Rector70], we introduce a cosimplicial space associated to a pullback diagram. By defining the appropriate face and degeneracy maps on the bar construction, it is easy to see that it is the condensation of a simplicial module. Rector's geometric cobar construction (here presented as a geometric bar construction) takes this idea and derives the Eilenberg-Moore spectral sequence directly from the geometry of the cosimplicial space. This approach is a special case of constructions of [Bousfield-Kan72] to be discussed in the next chapter.

Both of the constructions sketched below not only provide us with the sought after further structure but are examples of methods that have been successful in homotopy theory. At the end of this section we mention some further work of [Singer75], [Dwyer80], and [Turner98] that captures some of the further subtleties of the action of the Steenrod algebra on a spectral sequence.

The Eilenberg-Moore spectral sequence as a Künneth theorem

For spaces X and Y and a commutative ring R, the Künneth theorem allows us to compare $H^*(X \times Y; R)$ and $H^*(X; R) \otimes_R H^*(Y; R)$. When the flatness assumptions needed for the Künneth theorem fail, then the Künneth spectral sequence (Theorem 2.20) applies to make this comparison. We now recast these results in a way that allows a generalization.

We can write the product $X \times E$ in **Top** as the pullback of the unique mappings $X \to *$ and $E \to *$:

$$
\begin{array}{ccc}
X \times E & \longrightarrow & E \\
\downarrow & & \downarrow \\
X & \longrightarrow & *
\end{array}
$$

If we apply the Eilenberg-Moore spectral sequence to this diagram, then, over a field k,

$$E_2 \cong \mathrm{Tor}_k(H^*(X; k), H^*(E; k)) \cong H^*(X; k) \otimes_k H^*(E; k),$$

and we recover the Künneth theorem. In what follows we will take the Künneth theorem and the Künneth spectral sequence as basic functors on products and develop the geometry of pullbacks to fit this framework.

Definition 8.33. *Let* **Top**$/B$ *denote the category with objects given by continuous functions,* $f: Tf \to B$, *where* Tf *denotes the domain of* f. *A morphism in* **Top**$/B$, $\alpha: f \to g$ *is a mapping* $\alpha: Tf \to Tg$, *so that the following diagram commutes:*

This category has been the focus of considerable development since its introduction ([James71, 89] and [Crabb-James98]). For the Eilenberg-Moore spectral sequence, the key observation is that the categorical product in **Top**/B, $f \times_B g$, is given by the pullback of $f: Tf \to B \leftarrow Tg: g$

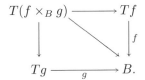

This product generalizes the case of $B = *$ to include fibre squares over B.

In **Top**/B the 'one-point space' (a terminal object) is given by the mapping $\mathrm{id}_B: B \to B$. Our goal is to prove geometrically the following analogue of the Künneth theorem.

Theorem 8.34 (the Künneth theorem for Top/B). *For f, g in* **Top**/B, *there is a spectral sequence with E_2-term given by*

$$E_2 \cong \mathrm{Tor}_{H^*(\mathrm{id}_B;k)}(H^*(f;k), H^*(g;k))$$

and converging to $H^*(f \times g; k)$.

The functor $H^*(f; k)$ associates $H^*(Tf; k)$ to an object f in **Top**/B. The theorem gives the Eilenberg-Moore spectral sequence when f is a fibration. We get the expected collapse to the Künneth theorem when $B = *$. More generally, it is sometimes possible to substitute other generalized cohomology theories for H^*. What makes this approach attractive is an insight of [Atiyah62] that the Künneth theorem for K-theory does not follow as in the case of ordinary singular cohomology but requires a construction of spaces whose K-theory has the correct homological properties. A generalization of this sort of construction is needed to obtain the Eilenberg-Moore spectral sequence.

In order to develop a homotopy theory for **Top**/B with the usual geometric constructions, we require a notion of spaces over B with basepoints.

Definition 8.35. *Let* $(\mathbf{Top}/B)_*$ *denote the category of pointed spaces over B, with objects, (f, s), where $f: Tf \to B$ is in* **Top**/B *and $s: B \to Tf$ is a continuous section, that is, $f \circ s = \mathrm{id}_B$. A morphism in* $(\mathbf{Top}/B)_*$, $\alpha: (f, s) \to (g, t)$, *is a mapping, $\alpha: Tf \to Tg$ so that the following diagrams commute*

Since $B \xrightarrow{s} Tf \xrightarrow{f} B$ is id_B, the one-point space (now an initial and terminal object) in $(\mathbf{Top}/B)_*$ is given by $(\mathrm{id}_B, \mathrm{id}_B)$. Notice that $H^*(f; k)$ is an *augmented* $H^*(B; k)$-module via the mapping f^* and its splitting s^*. We extend $H^*(\ ; k)$ to a functor on $(\mathbf{Top}/B)_*$ by taking the cohomology of the pair,

$$H^*((f, s); k) = H^*(Tf, sB; k).$$

The geometric constructions to follow are based on the fundamental notions of homotopy theory for the category $(\mathbf{Top}/B)_*$.

Proposition 8.36. *In the category* $(\mathbf{Top}/B)_*$, *the following can be defined:*

(1) *homotopies between morphisms;*
(2) *fibrations and cofibrations;*
(3) *mapping cylinders, mapping cones, suspensions, smash products, and*
(4) *the Barratt-Puppe sequence.*

A **homotopy** between two morphisms in $(\mathbf{Top}/B)_*$, say

$$\varphi_0, \varphi_1 \colon (f, s) \longrightarrow (g, t),$$

is a family of mappings, $\psi_r \colon (f, s) \longrightarrow (g, t)$ with $r \in [0, 1]$ such that ψ_r is continuous in r and on Tf, and $\psi_0 = \varphi_0$ and $\psi_1 = \varphi_1$. When f and g are fibrations, this definition coincides with the notion of a fibre homotopy ([Spanier66, p. 99]). This notion of homotopy gives rise to an equivalence relation. Let $[(f, s), (g, t)]$ denote the set of homotopy classes of morphisms $(f, s) \to (g, t)$.

A morphism in $(\mathbf{Top}/B)_*$, $\phi \colon (f, s) \to (g, t)$, is a **cofibration** if it has the homotopy extension property, that is, for any diagram of mappings and a homotopy

$$
\begin{array}{ccc}
(f, s) & \xrightarrow{\phi} & (g, t) \\
{\scriptstyle \psi_r} \downarrow & \swarrow {\scriptstyle \tilde{\psi}_0} & \\
(h, u) & &
\end{array}
$$

that commutes when $r = 0$, there is a homotopy $\tilde{\psi}_r \colon (g, t) \to (h, u)$ such that, for all $r \in [0, 1]$, the following diagram commutes:

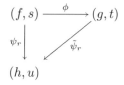

$$
\begin{array}{ccc}
(f, s) & \xrightarrow{\phi} & (g, t) \\
{\scriptstyle \psi_r} \downarrow & \swarrow {\scriptstyle \tilde{\psi}_r} & \\
(h, u) & &
\end{array}
$$

There is an analogous definition for fibrations in $(\mathbf{Top}/B)_*$. The notions of fibration and cofibration enjoy the usual exactness properties with respect to sets of homotopy classes of mappings.

For $\alpha\colon (f,s) \to (g,t)$, let $(M_B(\alpha), s_{M_B(\alpha)})$ denote the **mapping cylinder** defined by

$$T(M_B(\alpha)) = Tg \cup (Tf \times I) \Big/ \begin{array}{l} \alpha(x) \sim (x,0), \text{ if } x \in Tf \\ t(b) \sim (s(b),r), \text{ if } b \in B, r \in I \end{array}$$

where $M_B(\alpha)(x) = g(x)$, if $x \in Tg$, and $M_B(\alpha)(y,r) = f(y)$, if $y \in Tf$ and $r \in I$. The section is defined by $s_{M_B(\alpha)}(b) = t(b) = (s(b),r)$ for $b \in B$. By forming further quotients by subspaces of $(M_B(\alpha), s_{M_B(\alpha)})$, one can construct the **mapping cone**, $(C_B(\alpha), s_{C_B(\alpha)})$, of α.

The **suspension** of an object (f,s), denoted by $(S_B f, S_B s)$, is defined by

$$T(S_B f) = Tf \times I \Big/ \begin{array}{l} (x,0) \sim (x',0), (x,1) \sim (x',1), \text{ if } f(x) = f(x'), \\ (s(b),r) \sim (s(b),r'), \text{ for } b \in B, r, r' \in I, \end{array}$$

with $S_B f(x,r) = f(x)$ and $S_B s(b) = (s(b),r)$. The suspension of a morphism in $(\mathbf{Top}/B)_*$ is also easily defined.

We put these objects together to form the **Barratt-Puppe sequence** of a morphism in $(\mathbf{Top}/B)_*$:

$$(f,s) \xrightarrow{\alpha} (g,t) \to (C_B(\alpha), s_{C_B(\alpha)}) \to (S_B f, S_B s) \xrightarrow{S_B(\alpha)} (S_B g, S_B t) \to,$$

which gives an exact sequence on application of the functor $[-, (h,u)]$.

The product in $\mathbf{Top}/B$ is the pullback. This product can be extended to $(\mathbf{Top}/B)_*$ by giving the section: Given (f,s) and (g,t), then $(f \times_B g, s \times_B t)$ is given by $T(f \times_B g) = \{(x,y) \mid s \in Tf, y \in Tg, f(x) = g(y)\}$, together with the mappings $f \times_B g(x,y) = f(x) = g(y)$ and $s \times_B t(b) = (s(b), t(b))$. The **smash product** $(f,s) \wedge_B (g,t) = (f \wedge_B g, s \wedge_B t)$ is defined by

$$T(f \wedge_B g) = T(f \times_B g) \Big/ (x, t(b)) \sim (s(b), y), (x,y) \in T(f \times_B g),$$

$(f \wedge_B g)(x,y) = f(x) = g(y)$ and $(s \wedge_B t)(b) = s(b) \wedge_B t(b)$.

If we apply a generalized cohomology functor to $(\mathbf{Top}/B)_*$, then the coefficients of the theory are carried by the analogue of the zero-sphere. In $(\mathbf{Top}/B)_*$ this is the object $(S^0 \times B, s_0)$ where $S^0 \times B$ is shorthand for the projection mapping $S^0 \times B \to B$ and $s_0(b) = (1,b)$. [Smith, L70] gave particular care to identifying the conditions on a generalized cohomology functor for which the desired Künneth theorem and spectral sequence results (there are ten conditions). Singular cohomology $H^*(\ ; k)$ with coefficients in a field k determines one such generalized theory on $(\mathbf{Top}/B)_*$.

The constructions of Proposition 8.36 allow us to begin to build a geometric homological algebra, following [Atiyah62].

Lemma 8.37. *Given* (f, s) *in* $(\mathbf{Top}/B)_*$, *there is* (h, u) *in* $(\mathbf{Top}/B)_*$ *and a morphism* $\alpha\colon (f, s) \to (h, u)$ *so that*

(1) $H^*(\alpha)\colon H^*((h, u); k) \to H^*((f, s); k)$ *is an epimorphism,*
(2) $H^*((h, u); k)$ *is a projective* $H^*((S^0 \times B, s_0); k)$*-module.*

PROOF: Let $\alpha\colon Tf \to (Tf/sB) \times B$ be the mapping $\alpha(x) = (\pi(x), f(x))$ where $\pi\colon Tf \to Tf/sB$ is the quotient map. If (h, u) denotes the object in $(\mathbf{Top}/B)_*$ with $h\colon Tf/sB \times B \to B$, the second projection, and $u\colon B \to Tf/sB \times B$ given by $b \mapsto (sB/sB, b)$, then α determines a mapping $\alpha\colon (f, s) \to (h, u)$. Let $\beta\colon (Tf/sB \times B)/uB \to Tf/sB$ be given by $\beta(x, u) = x$. It is well-defined and satisfies $\beta \circ \alpha = \mathrm{id}$. Therefore, $H^*(\alpha)$ is an epimorphism. Furthermore, by the Künneth theorem,

$$H^*((h, u); k) \cong H^*(Tf/sB \times B, uB; k) \cong H^*(Tf, sB; k) \otimes H^*(B; k),$$

and (2) holds.   □

With Lemma 8.37 as a tool, we can iterate the procedure in the proof to obtain a sequence of cofibration sequences:

$$(f, s) \xrightarrow{\ \alpha_0\ } (h_0, u_0) \longrightarrow (f_{-1}, s_{-1})$$

$$(f_{-1}, s_{-1}) \xrightarrow{\ \alpha_{-1}\ } (h_{-1}, u_{-1}) \longrightarrow (f_{-2}, s_{-2})$$

$$\vdots \qquad\qquad \vdots \qquad\qquad \vdots$$

$$(f_{-n}, s_{-n}) \xrightarrow{\ \alpha_{-n}\ } (h_{-n}, u_{-n}) \longrightarrow (f_{-n-1}, s_{-n-1})$$

$$\vdots \qquad\qquad \vdots \qquad\qquad \vdots$$

such that, for all i,

(1) $H^*((h_{-i}, s_{-i}); k)$ is a projective $H^*((S^0 \times B, s_0); k)$-module,
(2) $H^*(\alpha_{-i})$ is an epimorphism.

Such a sequence is called an $H^*(\ ; k)$-**display** of (f, s). To such a display we associate an exact couple via the Barratt-Puppe sequences. First, we note that when (f, s) is a fibration with section over B, such $H^*(\ ; k)$-displays always exist ([Smith, L70, Proposition 3.8]). Furthermore, as required of a resolution in homological algebra, any two displays may be compared with displays that act as free objects in $(\mathbf{Top}/B)_*$ and behave well under $H^*(\ ; k)$ ([Smith, L70, Proposition 3.6]). Therefore, we can standardize our displays. For example, if $k = \mathbb{F}_p$, carefully chosen products of Eilenberg-Mac Lane spaces may be taken.

Consider the Barratt-Puppe sequence for $\alpha_{-i}\colon (f_{-i}, s_{-i}) \to (h_{-i}, u_{-i})$. The cofibration sequences, $(f_{-i}, s_{-i}) \to (h_{-i}, u_{-i}) \to (f_{-i-1}, s_{-i-1})$, induces mappings, $\Delta_{i+1}\colon (f_{-i-1}, s_{-i-1}) \to (S_B f_{-i}, S_B s_{-i})$, and so, for each

n, we get a geometric filtration of $(S_B^n f, S_B^n s)$:

$$(f_{-n}, s_{-n}) \xrightarrow[\Delta_n]{} (S_B f_{-n+1}, S_B s_{-n+1}) \xrightarrow[S_B \Delta_{n-1}]{}$$

$$\cdots \xrightarrow[S_B^{n-1} \Delta_1]{} (S_B^n f, S_B^n s).$$

Notice that the suspensions are imposed on us by the Barratt-Puppe sequence. However, the cohomology is only affected by a shift in degree by suspension— this does not change essentially the module structure over $H^*((S^0 \times B, s_0); k)$ or, in the case $k = \mathbb{F}_p$, over $\mathcal{A}_p$.

Theorem 8.38. *Suppose B is a connected, simply-connected space. Given (f, s) and (g, t), objects in $(\mathbf{Top}/B)_*$, with (f, s) a fibration with section over B, there is a spectral sequence with*

$$E_2 \cong \mathrm{Tor}_{H^*((S^0 \times B, s_0); k)}(H^*((g, t); k), H^*((f, s); k)),$$

and converging to $H^*(g \wedge_B f, t \wedge_B s; k)$, where $g \wedge_B f$ is the smash product of two objects, g and f, in $(\mathbf{Top}/B)_*$.

SKETCH OF PROOF: We first make the observation that the $H^*(; k)$-displays provide partial projective resolutions of $H^*((S_B^n f, S_B^n s); k)$. By running out the Barratt-Puppe sequences for each line of the display, we can apply $H^*(; k)$ and splice together the resulting exact sequences; we get an exact sequence

$$0 \leftarrow H^*((S_B^n f, S_B^n s); k) \leftarrow H^*((S_B^n h_0, S_B^n u_0); k) \leftarrow \cdots$$
$$\leftarrow H^*((S_B^{n-i} h_{-i}, S_B^{n-i} u_{-i}); k) \leftarrow \cdots \leftarrow H^*((h_{-n}, u_{-n}); k).$$

Each $H^*((h_{-i}, u_{-i}); k)$ is a projective module over $H^*((S^0 \times B, s_0); k)$ and the effect of S_B^m is simply a shift in dimension. Therefore the modules, $H^*((S_B^m h_{-i}, S_B^m u_{-i}); k)$, remain projective.

Apply $(g, t) \wedge_B -$ to the display and the other associated spaces. For a fixed positive integer n, we get the space $(g \wedge_B S_B^n f, t \wedge_B S_B^n s)$ filtered by successive images:

$$(g \wedge_B f_{-n}, t \wedge_B s_{-n}) \to (g \wedge_B S_B f_{-n+1}, t \wedge_B S_B s_{-n+1}) \to \cdots \to$$
$$(g \wedge_B S_B^{n-1} f_{-1}, t \wedge_B S_B^{n-1} s_{-1}) \to (g \wedge_B S_B^n f, t \wedge_B S_B^n s).$$

This leads to an exact couple $\mathcal{C}(n)$ given by

$$D^{-p,q} = H^{-p+q-1}((g \wedge_B S_B^{n-p} f_{-p}, t \wedge_B S_B^{n-p} s_{-p}); k)$$

and $\quad E^{-p,q} = H^{-p+q}((g \wedge_B S_B^{n-p} h_{-p}, t \wedge_B S_B^{n-p} u_{-p}); k)$

together with a spectral sequence $\{E_r(n), d_r(n)\}$. We make the observation that, for (f, s) and (g, t), objects in $(\mathbf{Top}/B)_*$,

$$(g, t) \wedge_B S_B(f, s) \simeq S_B((g, t) \wedge_B (f, s)),$$

so $E_r^{p,q}(n) \cong E_r^{p,q+1}(n+1)$, when $p > 1 - n$. Leaving the dependence on n behind, we can define

$$E_r^{p,q} = E_r^{p,q+n}(n), \quad \text{for } p > 1 - n,$$

and obtain a spectral sequence $\{E_r, d_r\}$ associated to (g, t) and (f, s).

In order to identify the E_2-term of this spectral sequence, we note that the usual Künneth theorem implies that there are natural isomorphisms

$$H^*((g, t); k) \otimes_{H^*((S^0 \times B, s_0); k)} H^*((S_B^m h_{-i}, S_B^m u_{-i}); k)$$

$$\downarrow \cong$$

$$H^*((g \wedge_B S_B^m h_{-i}, t \wedge_B S_B^m u_{-i}); k).$$

Furthermore, $E_2(n)$ is the homology of the complex

$$0 \leftarrow H^*((g \wedge_B S_B^n h_0, t \wedge_B S_B^n u_0); k) \leftarrow$$
$$H^*((g \wedge_B S_B^{n-1} h_{-1}, t \wedge_B S_B^{n-1} u_{-1}); k) \leftarrow$$
$$\cdots \leftarrow H^*((g \wedge_B h_{-n}, t \wedge_B u_{-n}); k) \leftarrow 0$$

It follows that

$$E_2^{p,q}(n) \cong \operatorname{Tor}_{H^*((S^0 \times B, s_0); k)}(H^*((g, t); k), H^*(S_B^n(f, s); k).$$

Accounting for the shifts in dimension, we have, for $p > 1 - n$,

$$E_2^{p,q} \cong \operatorname{Tor}_{H^*((S^0 \times B, s_0); k)}^{p,q}(H^*((g, t); k), H^*((f, s); k)).$$

All this was derived from the successive filtrations of $(g, t) \wedge_B S_B^n(f, s)$ with the correction for the dimension shift. A careful argument is needed en route, using the simple connectivity of B, to prove that $H^j((f_{-p}, s_{-p}); k) = \{0\}$ for $j < 2p$. Thus the spectral sequence converges additively to the desired cohomology. □

In the proof of Theorem 8.38, the homomorphisms in the exact couples and the differentials in the derived spectral sequences are all given by applying $H^*(\ ; k)$ to continuous functions in the Barratt-Puppe sequence. When $k = \mathbb{F}_p$, this means that the spectral sequence is a spectral sequence of modules over the Steenrod algebra $\mathcal{A}_p$.

In order to remove the basepoint from the discussion, one can use the functor $(\ )_+ : \mathbf{Top}/B \to (\mathbf{Top}/B)_*$ that is analogous to the topological functor that adds a disjoint basepoint—one associates to f, an object in $\mathbf{Top}/B$ the pointed object (f_+, s) where $T(f_+) = Tf \coprod B$, the disjoint union, $f_+ = f$ on Tf and $f_+ = \operatorname{id}_B$ on B, and $s(b) = b$. With this functor, smash products correspond to regular products in $\mathbf{Top}/B$.

Corollary 8.39. *Suppose p is a prime, B is simply-connected, $H^*(B; \mathbb{F}_p)$ is of finite type and f and g are objects in* **Top**$/B$. *Suppose g is a fibration and $H^*(Tg; \mathbb{F}_p)$ is also of finite type. Then there is a spectral sequence with*

$$E_2 \cong \mathrm{Tor}_{H^*(B; \mathbb{F}_p)}(H^*(Tf; \mathbb{F}_p), H^*(Tg; \mathbb{F}_p))$$

and converging strongly to $H^*(T(f \times g); \mathbb{F}_p)$. Furthermore, for $r \geq 2$ and for each $s \leq 0$, $E_r^{s,*}$ is a module over the mod p Steenrod algebra, $\mathcal{A}_p$, d_r is $\mathcal{A}_p$-linear and $E_\infty^{s,*}$ is isomorphic to $E_0^{s,*}(H^*(T(f \times g); \mathbb{F}_p))$ as $\mathcal{A}_p$-modules.

The corollary provides the Eilenberg-Moore spectral sequence with a vertical Steenrod algebra structure that converges to the $\mathcal{A}_p$ structure on $H^*(E_f; \mathbb{F}_p)$. The result answers the question posed by [Smith, L67] whether the filtration underlying the Eilenberg-Moore spectral sequence is closed under the action of the Steenrod algebra. It is a more delicate issue to prove that this $\mathcal{A}_p$-structure is compatible with the product structure of Corollary 7.18 (for example, that the Cartan formula holds in the spectral sequence). This compatibility does hold, however, and we direct the reader to [Smith, L70, §7] for a proof.

This geometric construction also provides an approach to the Eilenberg-Moore spectral sequence for a class of generalized cohomology theories made precise by [Smith, L70] and [Hodgkin68]. [Yamaguchi86] has computed $h_*(\Omega B)$ in this way for generalized homology theories and deduced positive results for the Morava K-theory of double loop spaces of complex Stiefel manifolds. [Jeanneret-Osse98] have applied the Eilenberg-Moore spectral sequence to compute the generalized cohomology of fibre squares when the base space of the fibration has $h^*(\Omega B)$ an exterior algebra on odd degree generators. This applies to p-compact groups, a generalization of Lie groups suited to homotopy theory methods. The general convergence properties of such spectral sequences is a very delicate issue that remains an open question. Examples due to [Hodgkin75] and [Ghazal89] give indications of the extent to which the Eilenberg-Moore spectral sequence can fail: For example, there are finite complexes X for which $\tilde{K}^*(X) = \{0\}$ while $\tilde{K}^*(\Omega X) \neq \{0\}$.

Cosimplicial techniques

In his geometric construction of the Eilenberg-Moore spectral sequence, [Smith, L70] introduced $H^*(\ ; k)$-displays comprised of sequences of objects in $(\mathbf{Top}/B)_*$, $\{(h_{-i}, u_{-i})\}$, whose cohomology algebras are projective modules over $H^*((S^0 \times B, s_0); k)$ and that lead to projective resolutions. By analogy with the bar construction, it would be convenient to standardize the construction of spaces that lead to the spectral sequence and furthermore, to base the constructions entirely on the initial datum, the pullback diagram:

[Rector70] carried out this analogy exactly by constructing a **geometric cobar construction**. When one applies the homology functor, $H_*(\ ;\mathbb{F}_p)$, to the associated sequence of spaces, one gets an exact couple that gives the homology Eilenberg-Moore spectral sequence with an apparent Steenrod algebra comodule structure. We present this construction for the cohomology spectral sequence.

Definition 8.40. *A* **cosimplicial object** $A^\bullet$ *in a category* **C** *consists of*

(1) *A sequence of objects in* **C**, $A^0, A^1, A^2, \ldots$.
(2) *For each* $n \geq 0$, *a collection of morphisms in* **C**, $d^i \colon A^{n-1} \to A^n$, *for* $0 \leq i \leq n$, *called* **coface maps** *and*
(3) *for each* $n \geq 0$, *a collection of morphisms in* **C**, $s^i \colon A^{n+1} \to A^n$, *for* $0 \leq i \leq n$, *called* **codegeneracy maps**.

The coface and codegeneracy maps satisfy the identities:

$$d^j \circ d^i = d^i \circ d^{j-1}, \ \textit{if } i < j,$$

$$s^j \circ d^i = \begin{cases} d^i \circ s^{j-1}, & \textit{if } i < j, \\ \mathrm{id}, & \textit{if } i = j \textit{ or } i = j+1, \\ d^{i-1} \circ s^j, & \textit{if } i > j+1. \end{cases}$$

$$s^j \circ s^i = s^{i-1} \circ s^j, \ \textit{if } i > j.$$

Examples: A. If $\mathbb{V}_\bullet$ is a simplicial vector space over a field k, then let $\mathbb{V}_\bullet^{\mathrm{dual}}$ be the cosimplicial vector space with $(\mathbb{V}^{\mathrm{dual}})^m = \mathbb{V}_m^{\mathrm{dual}}$. The coface and codegeneracy maps are dual to the face and degeneracy maps of $\mathbb{V}_\bullet$.

B. Denote the category of sets and functions by **Ens**. For Y a set, we can apply levelwise the Hom functor, $\mathbf{Ens}(-, Y)$ to a simplicial set $X_\bullet$ to obtain a cosimplicial set denoted $(X^Y)^\bullet = \mathbf{Ens}(X_\bullet, Y)$.

C. [Eilenberg-Moore65] introduced the idea of a **triple** (R, φ, ψ), which is a functor $R \colon \mathbf{C} \to \mathbf{C}$, together with two natural transformations $\varphi \colon \mathrm{id} \to R$ and $\psi \colon R \circ R \to R$ satisfying

$$(R\varphi)R = (\varphi R)\varphi, \ \psi(R\psi) = \psi(\psi R) \text{ and } \psi(R\varphi) = \mathrm{id} = \psi(\varphi R).$$

If S and T are any pair of adjoint functors

$$T \dashv S \quad S \colon \mathbf{C} \to \mathbf{D}, \quad T \colon \mathbf{D} \to \mathbf{C},$$

then the composite functor, $ST \colon \mathbf{D} \to \mathbf{D}$ is a triple with the structure maps coming from the adjointness relations.

A triple gives rise to a cosimplicial object in **C** for each object in **C**, defined by the sequence of functors, $R^n = R^{n-1} \circ R$ and $R^0 = \mathrm{id}$. Let X be an object in

C. Then $(CoR(R)X)^\bullet$ (the *cosimplicial resolution*) is defined by the sequence of objects $(CoR(R)X)^n = R^{n+1}X$ together with the coface and codegeneracy morphisms defined by the natural transformations, for $0 \le i, j \le n$,

$$d^i = R^i \varphi_{R^{n-i}} : R^n \to R^{n+1}, \qquad s^j = R^j \psi_{R^{n-j-1}} : R^{n+1} \to R^n.$$

The relations satisfied by φ and ψ imply the cosimplicial identities.

Applications of cosimplicial objects derived in this manner to homotopy theory are of considerable importance (see the treatise of [Bousfield-Kan72]).

Suppose we have a pair of mappings, $f : X \to B$ and $\pi : E \to B$. To these data we associate a cosimplicial space, $G^\bullet(X, B, E)$, defined as follows:

$$G^n(X, B, E) = X \times \underbrace{B \times \cdots \times B}_{n \text{ times}} \times E$$

$$d^i(x, b_1, \ldots, b_n, e) = \begin{cases} (x, f(x), b_1, \ldots, b_n, e), & \text{if } i = 0, \\ (x, b_1, \ldots, b_i, b_i, \ldots, b_n, e), & \text{if } 1 \le i \le n, \\ (x, b_1, \ldots, b_n, \pi(e), e), & \text{if } i = n+1. \end{cases}$$

$$s^{i-1}(x, b_1, \ldots, b_{n+1}, e) = (x, b_1, \ldots, b_{i-1}, \widehat{b_i}, b_{i+1}, \ldots, b_{n+1}, e),$$
$$\text{for } 1 \le i \le n+1.$$

By the Künneth theorem, if k is a field, when we apply $H^*(\ ; k)$ to each space in $G^\bullet(X, B, E)$ and let $\delta = \sum_{i=0}^{n} (-1)^i (d^i)^*$, then we get the bar construction, $B^\bullet(H^*(X; k), H^*(B; k), H^*(E; k))$, as desired. We next derive the Eilenberg-Moore spectral sequence from the cosimplicial space $G^\bullet(X, B, E)$.

We first normalize the spaces, $G^i = G^i(X, B, E)$ to get a sequence

$$* \to G_N^0 \xrightarrow{\delta^1} G_N^1 \xrightarrow{\delta^2} G_N^2 \xrightarrow{\delta^3} \cdots .$$

The normalization goes as follows: Let $G_N^0 = G^0(X, B, E)/\emptyset$, which is the disjoint union of $G^0(X, B, E)$ with a new basepoint, $G^0(X, B, E) \sqcup *$. Then define

$$G_N^i = G^i(X, B, E) \Big/ \operatorname{im} d^1 \cup \cdots \cup \operatorname{im} d^i \ .$$

Since $d^0 \circ d^i = d^{i+1} \circ d^0$, the remaining map d^0 induces a map, $\delta^i : G_N^i \to G_N^{i+1}$, and furthermore, $\delta^{i+1} \circ \delta^i = *$. We next construct something like the $H^*(\ ; k)$-displays of [Smith, L70]. Let $X^0 = G_N^0$, and form X^1 as the mapping cone, M_{δ^1}, to get a cofibration sequence, $X^0 \to G_N^1 \to X^1$. Since $\delta^2 \circ \delta^1 = *$, we can form the diagram

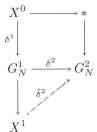

where the lift $\tilde{\delta}^2$ exists by the properties of mapping cones. Inductively, we have

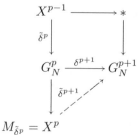

and, from the Barratt-Puppe sequences, we have the sequence of cofibrations

$$G_N^0 \longrightarrow X^1 \longrightarrow SX^0$$
$$G_N^1 \longrightarrow X^2 \longrightarrow SX^1$$
$$\vdots$$
$$G_N^l \longrightarrow X^{l+1} \longrightarrow SX^l$$
$$\vdots$$

Apply $\tilde{H}^*(\ ; k)$ to each line of the sequence to obtain an exact couple:

$$D^{-l,q} = \tilde{H}^q(X^l; k), \qquad E^{-l,q} = \tilde{H}^q(G_N^l; k),$$

with the maps in the couple given by the cohomology of the maps in the cofibrations and the connecting maps from the Barratt-Puppe sequence. The exact couple in turn gives a spectral sequence. Furthermore, if $k = \mathbb{F}_p$, all of the mappings in the couple and hence the differentials in the spectral sequence are $\mathcal{A}_p$-linear.

To see that the spectral sequence converges to $H^*(E_f; k)$, observe that there is a natural inclusion $\varepsilon\colon E_f \hookrightarrow X \times E$, as a pullback. This mapping satisfies $d^1 \circ \varepsilon = d^0 \circ \varepsilon$ and so this induces a geometric augmentation, $\varepsilon\colon E_f/\emptyset \to G_N^0$ with $\delta^1 \circ \varepsilon = *$. Comparing the cofibration sequences for ε and δ^1 we get the diagram:

$$
\begin{array}{ccccccc}
E_f/\emptyset & \xrightarrow{\ \varepsilon\ } & G_N^0 & \longrightarrow & M_\varepsilon & \longrightarrow & S(E_f/\emptyset) \\
{\scriptstyle \varepsilon}\downarrow & & {\scriptstyle \delta^1}\downarrow & & \downarrow & \nearrow & \downarrow{\scriptstyle S\varepsilon} \\
X^0 & \xrightarrow{\ \varepsilon\ } & G_N^1 & \longrightarrow & X^1 & \longrightarrow & SX^0.
\end{array}
$$

The condition $\delta^1 \circ \varepsilon = *$ implies a lift $S(E_f/\emptyset) \to X^1$ that commutes with $S\varepsilon$.

By induction we can continue through the construction of the X^p to obtain mappings $S^l(E_f/\emptyset) \to X^l$ that are compatible with the mappings $X^l \to$

SX^{l-1}. This determines a mapping from the direct limit of the mappings $D^{-l,r} \to D^{-l+1,r-1}$ associated to the exact couple. Using the techniques of Chapter 3, we get a relation between the target of the spectral sequence and $\tilde{H}^*(E_f; k)$.

To summarize in the case $k = \mathbb{F}_p$, we have a spectral sequence, related to $H^*(E_f; \mathbb{F}_p)$ and, for $r \geq 1$, the E_r-terms are modules over $\mathcal{A}_p$ and the differentials are $\mathcal{A}_p$-linear. In order to establish the analogue of Corollary 8.39 for this construction, we need to determine the E_2-term of the spectral sequence.

Lemma 8.41. $\tilde{H}^*(G_N^l; k) \cong H^*(G^l(X, B, E); k)/\langle \mathrm{im}(d^1)^* \oplus \cdots \oplus \mathrm{im}(d^l)^* \rangle.$

PROOF: By induction, we show that

$$\tilde{H}^*\left(G^l(X, B, E)\big/_{\mathrm{im}\, d^{l-j} \cup \cdots \cup \mathrm{im}\, d^l} ; k\right)$$
$$\cong \tilde{H}^*(G^l(X, B, E); k)/\langle \mathrm{im}(d^{l-j})^* \oplus \cdots \oplus \mathrm{im}(d^l)^* \rangle.$$

The cosimplicial identities imply the assertion for $j = 0$. Since $s^n \circ d^i = d^{i-1} \circ s^n$, for $i > n$, s^n induces a mapping

$$G^l(X, B, E)\big/_{\mathrm{im}\, d^{n+1} \cup \cdots \cup \mathrm{im}\, d^l} \xrightarrow{\bar{s}^n} G^l(X, B, E)\big/_{\mathrm{im}\, d^n \cup \cdots \cup \mathrm{im}\, d^{l-1}}$$

such that $\bar{s}^n \circ d^l = \mathrm{id}$. This gives us the cofibration sequence

$$G^{l-1}(X, B, E)\big/_{\mathrm{im}\, d^n \cup \cdots \cup \mathrm{im}\, d^{l-1}} \to G^l(X, B, E)\big/_{\mathrm{im}\, d^{n+1} \cup \cdots \cup \mathrm{im}\, d^l}$$
$$\to G^l(X, B, E)\big/_{\mathrm{im}\, d^n \cup \cdots \cup \mathrm{im}\, d^l}.$$

To complete the proof, apply $\tilde{H}^*(\ ; k)$ and the induction hypothesis. □

Corollary 8.42. *The homology of the complex obtained by applying* $H^*(\ ; k)$ *to*

$$* \to G_N^0 \xrightarrow{\delta^1} G_N^1 \xrightarrow{\delta^2} G_N^2 \xrightarrow{\delta^3} \cdots$$

is isomorphic to the homology of the complex $H^*(G^l(X, B, E); k)$ *with* $\delta = \sum_{i=0}^l (-1)^i (d^i)^*$. *Therefore the* E_2-*term for the spectral sequence associated to the geometric cobar construction is given by*

$$E_2 \cong \mathrm{Tor}_{H^*(B;k)}(H^*(X; k), H^*(E; k)).$$

The rest of the proof follows by comparing this geometrically defined spectral sequence with the algebraic bar construction spectral sequence. In the case that $\pi \colon E \to B$ is a fibration, the second Eilenberg-Moore spectral sequence applies to identify the target of the spectral sequence with $H^*(E_f; k)$ when it converges strongly. Finally, if $k = \mathbb{F}_p$, we have derived the Steenrod operations of Corollary 8.39.

Before leaving the simplicial point of view, we mention other approaches that obtain Steenrod algebra operations in spectral sequences. One way is to construct them directly at the level of the filtered cochains and then keep track of how these operations behave in the spectral sequence. Using the construction of the mod 2 Steenrod algebra operations derived from the diagonal approximation of the Eilenberg-Zilber map (à la [Steenrod53], [Dold62], or [May70]), this program was carried out, first by [Singer73] for first-quadrant spectral sequences and later by [Dwyer80] for second-quadrant spectral sequences. Singer's work yields not only vertical operations

$$Sq^k \colon E_r^{p,q} \to E_r^{p,q+k}$$

but also a diagonal action $Sq^k \colon E_r^{p,q} \to E_r^{p+k-q,2q}$ when $k \geq q$. These new operations fill a deficit of operations that should be available in the target cohomology and they also provide more structure on the spectral sequence that can be exploited in computation. (See the papers of [Singer73] and [Mimura-Mori77] for examples.)

Dwyer's work by contrast does not yield unstable operations and creates another deficit. In order for the unstable axioms for the action of the Steenrod algebra to be satisfied by the target of this second quadrant spectral sequence, [Dwyer80] constructs new operations

$$\delta_i \colon E_2^{-p,q} \to E_2^{-p-i,2q}, \ 2 \leq i \leq p,$$

which interact with the differentials to insure that the unstable axioms hold at E_∞. Analogous operations for odd primes have been constructed by [Mori79] and [Sawka82].

In another direction, the spaces $\Omega^n X$ appear as the fibres of path-loop fibrations, and so their cohomology and homology may be computed by the appropriate Eilenberg-Moore spectral sequences. In his Ph.D. thesis, [Hunter89] computed $H_*(\Omega_0^{n+2} S^{n+1}; \mathbb{F}_2)$ using this method and further structure. For such iterated loop spaces, however, the mod p homology enjoys the action of the Dyer-Lashof algebra ([Araki-Kudo56], [Dyer-Lashof62]). This structure is particularly useful for the class of infinite loop spaces and it would be useful if the Eilenberg-Moore spectral sequence carried this structure. [Ligaard-Madsen75] and [Bahri83] described an action of the Dyer-Lashof algebra on the homology Eilenberg-Moore spectral sequences, and these results have led to successful computations of the homology of certain classifying spaces found in geometric topology. [Turner98] has given a uniform construction of operations on spectral sequences beginning from the point of view of [Dold61] and [May70] and generalizing the results of Singer, Dwyer, and Bahri.

Chapter 7 and 8 do not come near to exhausting the applications of the Eilenberg-Moore spectral sequence. A particularly nice case of a fibre square is the situation where the spaces X, B and E are H-spaces and the mappings

$f\colon X \to B$ and $\pi\colon E \to B$ are H-maps, that is, if m_X and m_B are the multiplications on X and B, respectively, then $m_B \circ (f \times f) \simeq f \circ m_X$ and similarly for π. By the properties of pullbacks, the space E_f pulled back from π over f, is also an H-space. Therefore, the algebraic input for the Eilenberg-Moore spectral sequence consists of Hopf algebras and Hopf algebra morphisms. Also, the target, $H^*(E_f; k)$, of the spectral sequence is a Hopf algebra. In the best situation, the spectral sequence is a spectral sequence of Hopf algebras, converging to $H^*(E_f; k)$ *as a Hopf algebra*. [Moore-Smith68] exploit this ideal situation to show that, under certain assumptions (satisfied by many two-stage Postnikov systems), the only possibly nontrivial differential in the associated Eilenberg-Moore spectral sequence is d_{p-1} where $p = \operatorname{char} k \neq 0$ and p is odd. This fact has been applied by [Kane75] and others in the study of H-spaces.

In the presence of Hopf algebras, Tor computations are much more accessible. [Cohen-Moore-Neisendorfer79] gave examples of the power of these methods in their study of exponents of homotopy groups.

Another source of fibre squares is a Postnikov tower. The $(n-1)$-connective cover of a space X is defined as the homotopy fibres of the natural mappings to the Postnikov tower:

$$X(n, \cdots, \infty) \to X \xrightarrow{f_n} P_n X.$$

In his Ph.D. thesis [Singer68] used the Eilenberg-Moore spectral sequence to compute $H^*(BU(2r, \dots, \infty); \mathbb{F}_p)$ and $H^*(U(2r+1, \dots, \infty); \mathbb{F}_p)$ where p is a prime and U is the infinite unitary group associated to complex K-theory.

Finally, we mention that the Eilenberg-Moore spectral sequence plays an important computational role in the work of [Dror-Farjoun-Smith90] on the cohomology of certain mapping spaces, expression of which is given in terms of the T-functor of [Lannes92].

Exercises

8.1. Prove that the two inclusions $i_1, i_2\colon \mathrm{U}(n) \hookrightarrow \mathrm{U}(n+1)$ described in §8.1 lead to different homogeneous spaces $\mathrm{U}(n+1)/\mathrm{U}(n)$.

8.2. Prove the fact: Suppose A, B, and C are algebras over a field k. If M is a right A-, right B-module, N is a left A-, right C-module, and L a left A-, left B-module, and $\operatorname{Tor}_A^n(M, N) = \{0\} = \operatorname{Tor}_C^n(N, L)$ for $n > 0$, then

$$\operatorname{Tor}^*_{B \otimes C}(M \otimes_A N, L) \cong \operatorname{Tor}^*_{A \otimes B}(M, N \otimes_C L).$$

8.3. Show that, in the case $H^*(G; k)$ and $H^*(H; k)$ are exterior algebras over k, then $\operatorname{Tor}_{H^*(BG;k)}(k, H^*(BH; k))$ is a Poincaré duality algebra.

8.4. Use the Sullivan-de Rham complex for rational cohomology (for definitions and applications, see the book of [Griffiths-Morgan81]) to obtain the rational analogue of Cartan's result, Theorem 8.2;

$$H^*(G/H; \mathbb{Q}) \cong \text{Tor}_{H^*(BG; \mathbb{Q})}(\mathbb{Q}, H^*(BH; \mathbb{Q})).$$

8.5. Show that a polynomial algebra A on n indeterminates that is also a finitely-generated module over another polynomial algebra B on n indeterminates, via an algebra homomorphism $\phi\colon B \to A$, is free as a B-module (Corollary 3.10 of [Baum68]).

8.6. Prove that an E-algebra, A (a graded commutative algebra over a field k, generated by a regular sequence), has zero deficiency, $\text{df}(A) = 0$.

8.7. Prove the assertions in §8.1 about the values of $(Bi)^* = H^*(Bi; \mathbb{F}_p)$ for $i\colon \text{SO}(l) \to \text{SO}(n)$, the usual inclusion and p an odd prime. (Hint: Use the Leray-Serre spectral sequence as needed.)

8.8. Show that the cup_1 product introduced in the last section of Chapter 5 satisfies the Hirsch formulas.

8.9. Complete the proof of Theorem 8.18.

8.10. Compute the cohomology of the complement of the Borromean rings.

8.11. Prove the assertions (1) through (4) of Theorem 8.29 for higher order Massey products.

8.12. Suggested by Claude Schochet: Suppose given a pullback fibration of the path-loop fibration over a space B with respect to a mapping $f\colon B_0 \to B$, as in the diagram

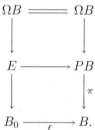

Show that the Eilenberg-Moore spectral sequence converging to $H^*(E; k)$ with $E_2 \cong \text{Tor}_{H^*B; k)}(H^*(B; k), k)$ collapses at E_2 if and only if the Leray-Serre spectral sequence with $E_2 \cong H^*(B_0; H^*(\Omega B; k))$ converging to $H^*(E; k)$ has all differentials arising from transgressions.

8.13. Show that a pair of adjoint functors

$$T \dashv S \quad S\colon \mathbf{C} \to \mathbf{D}, \quad T\colon \mathbf{D} \to \mathbf{C},$$

leads to a triple and hence cosimplicial resolutions of objects in the category $\mathbf{D}$.

8bis
Nontrivial fundamental groups

"One of the advantages of the category of nilpotent spaces over that of simply-connected spaces is that it is closed under certain constructions."

E. Dror-Farjoun

The category of simply-connected spaces is blessed with certain features that make homotopy theory tractable. In the first place, there is the White-head Theorem (Theorem 4.5) that tells us when a mapping of spaces of the homotopy type of CW-complexes is a homotopy equivalence—the necessary condition that the mapping induces an isomorphism of integral homology groups is also sufficient. Secondly, the Postnikov tower of a simply-connected space is a tower of principal fibrations pulled back via the k-invariants of the space (Theorem 8bis.37). This makes cohomological obstruction theory accessible, if not computable ([Brown, E57], [Schön90], [Sergeraert94]). Furthermore, the system of local coefficients that arises in the description of the E_2-term of the Leray-Serre spectral sequence is simple when the base space of a fibration is simply-connected, and the cohomology Eilenberg-Moore spectral sequence converges strongly for a fibration pulled back from such a fibration.

A defect of the category of simply-connected spaces is the fact that certain constructions do not stay in the category. The dishearteningly simple example is the based loop space functor—if (X, x_0) is simply-connected, $\Omega(X, x_0)$ need not be. Furthermore, the graded group-valued functor, the homotopy groups of a space, does not always distinguish distinct homotopy types of spaces that are not simply-connected. A classic example is the pair of spaces $X_1 = \mathbb{R}P^{2m} \times S^{2n}$ and $X_2 = S^{2m} \times \mathbb{R}P^{2n}$; the homotopy groups in each degree k are abstractly isomorphic, $\pi_k(X_1) \cong \pi_k(X_2)$. If we had principal Postnikov towers, we could use the abstract isomorphisms to try to build a weak homotopy equivalence. However, the cohomology rings over $\mathbb{F}_2$ "know" that X_1 is not homotopy equivalent to X_2.

In this chapter we introduce the larger category of nilpotent spaces. These spaces enjoy some of the best homotopy-theoretic properties of simply-connected spaces, like a Whitehead theorem ([Dror71]) and reasonable Postnikov towers.

Furthermore, this category is closed under many constructions such as the formation of function spaces. Group-theoretic functors, like localization and completion, have topological extensions in this category. A subtler result related to nilpotent spaces is due to [Dwyer74] who showed that the Eilenberg-Moore spectral sequence for the fibre of a fibration converges strongly if and only if the action of $\pi_1(B)$ on $H_i(F; A)$ is nilpotent for all $i \geq 0$ (Theorem 8bis.29).

In §8bis.1, we discuss the various actions of a nontrivial fundamental group. The action of a group on a module leads to a right exact functor, the coinvariants of the group action, whose left-derived functors are the homology groups of a group. This theory is developed briefly in §8bis.2 with an eye to its application to nilpotent spaces. In particular, we construct the Lyndon-Hochschild-Serre spectral sequence associated to a group extension, and the Cartan-Leray spectral sequence associated to a group acting freely and properly on a space.

With these tools in place we study the category of nilpotent groups and spaces in §8bis.3. We first prove the generalized Whitehead Theorem of [Dror71]. We then discuss the Postnikov tower of a nilpotent space. This tower characterizes such spaces and provides a tool for making new spaces such as the localization of a space à la [Sullivan71]. Cosimplicial methods offer a functorial route to localization and we give a short survey of the foundational work of [Bousfield-Kan72]. We also prove Dwyer's convergence theorem for the Eilenberg-Moore spectral sequence. The (co)simplicial constructions described here have proved to be fundamental in homotopy theory. We end with a theorem of [Dror73] that, for connected spaces, any homotopy type can be approximated up to homology equivalence by a tower of nilpotent spaces.

§8bis.1 Actions of the fundamental group

We begin with a small digression. Let (G, e, μ) denote a topological group with identity element e and write $g \cdot h$ for $\mu(g, h)$. It is an elementary fact that $\pi_0(G)$ is a group with multiplication induced by μ. There is an action of $\pi_0(G)$ on $\pi_n(G, e)$ defined as follows: If $g \in [g] \in \pi_0(G)$ is a point in a path component of G and $\alpha \colon (S^n, e_1) \to (G, e)$ represents a class $[\alpha]$ in $\pi_n(G, e)$, consider

$$g \cdot \alpha \cdot g^{-1} \colon (S^n, e_1) \to (G, e), \quad \text{defined by } g \cdot \alpha \cdot g^{-1}(x) = g \cdot \alpha(x) \cdot g^{-1}.$$

When we vary the choice of g in $[g]$ or the choice of representative for $[\alpha]$, we get homotopic maps and so this recipe determines a pairing

$$\pi_0(G) \times \pi_n(G, e) \xrightarrow{\nu_n} \pi_n(G, e), \quad ([g], [\alpha]) \mapsto [g \cdot \alpha \cdot g^{-1}].$$

Since the multiplication on $\pi_0(G)$ is determined by $[g] \cdot [h] = [g \cdot h]$, we have $\nu_n([g] \cdot [h], [\alpha]) = \nu_n([g], [h] \cdot [\alpha])$. Furthermore, the addition on $\pi_n(G, e)$ agrees with the operation induced by μ and so we have $\nu_n([g], [\alpha] + [\beta]) = \nu_n([g], [\alpha]) + \nu_n([g], [\beta])$. Thus $\pi_n(G, e)$ is a module over the group $\pi_0(G)$.

When (X, x_0) is a pointed space of the homotopy type of a countable CW-complex, [Milnor56] showed how to replace the based loop space $\Omega(X, x_0)$ with a topological group. In this case we write $\bar{\nu}_n^X = \nu_{n-1} \colon \pi_0(\Omega(X, x_0)) \times \pi_{n-1}(\Omega(X, x_0)) \to \pi_{n-1}(\Omega(X, x_0))$ and we have proved the following theorem.

Theorem 8bis.1. *Given a connected, pointed space (X, x_0), for each $n \geq 1$, $\pi_n(X, x_0)$ is a module over $\pi_1(X, x_0)$ via the pairing*

$$\bar{\nu}_n \colon \pi_1(X, x_0) \times \pi_n(X, x_0) \to \pi_n(X, x_0).$$

Furthermore, when $n = 1$, $\pi_1(X, x_0)$ acts on itself by conjugation, that is, $\bar{\nu}_1([\omega], [\lambda]) = [\omega] \cdot [\lambda] \cdot [\omega]^{-1} = [\omega * \lambda * \omega^{-1}]$ where $*$ denotes composition of paths.

In our discussion of the failure of the homotopy groups to distinguish the pair of spaces $\mathbb{R}P^{2m} \times S^{2n}$ and $S^{2m} \times \mathbb{R}P^{2n}$, we only used the observation that, for all k, $\pi_k(\mathbb{R}P^{2m} \times S^{2n}) \cong \pi_k(\mathbb{R}P^{2n} \times S^{2m})$, *as groups*. They differ as π_1-modules. It is a consequence of the interpretation of the action of the fundamental group as deck transformations that $\pi_1 = \mathbb{Z}/2\mathbb{Z}$ acts nontrivially on the $\mathbb{Z}$ factor in $\pi_{2m}(\mathbb{R}P^{2m} \times S^{2n})$ and $\pi_{2n}(\mathbb{R}P^{2n} \times S^{2m})$ coming from the projective space. Since these factors of $\mathbb{Z}$ as nontrivial $\mathbb{Z}/2\mathbb{Z}$-modules occur in different dimensions, the spaces could not be homotopy equivalent, and the homotopy groups, considered as graded π_1-modules, distinguish the spaces as different.

The fundamental group acts on other groups when we have a fibration $F \hookrightarrow E \xrightarrow{p} B$ with connected fibre F. Consider the long exact sequence of homotopy groups:

$$\cdots \xrightarrow{\partial} \pi_n(F, e) \xrightarrow{i_*} \pi_n(E, e) \xrightarrow{p_*} \pi_n(B, p(e)) \xrightarrow{\partial} \pi_{n-1}(F, e) \xrightarrow{i_*} \cdots.$$

We can induce an action of $\pi = \pi_1(E, e)$ on $\pi_n(B, p(e))$ via the composite

$$\nu_n^{E,B} \colon \pi \times \pi_n(B, p(e)) \xrightarrow{p_* \times 1} \pi_1(B, p(e)) \times \pi_n(B, p(e)) \xrightarrow{\bar{\nu}_n^B} \pi_n(B, p(e)).$$

Thus we can view $\pi_n(B, p(e))$ as a $\pi_1(E, e)$-module and the mapping p_* as a module homomorphism. In fact, more is true.

Proposition 8bis.2. *When $F \hookrightarrow E \xrightarrow{p} B$ is a fibration of connected spaces, the long exact sequence on homotopy is a long exact sequence of $\pi_1(E, e)$-modules and module homomorphisms.*

PROOF: We first construct the action of $\pi = \pi_1(E, e)$ on $\pi_n(F, e)$ here taken to be $\pi_{n-1}(\Omega F, c_e)$. Let $\alpha \colon (S^{n-1}, \vec{e}_1) \to (\Omega F, c_e)$ represent $[\alpha] \in$

$\pi_{n-1}(\Omega F, c_e)$ and $\omega \in [\omega] \in \pi_1(E, e)$. The action of π on $\pi_{n-1}(\Omega E, c_e)$ associates to $[\omega]$ and $[i \circ \alpha]$ the class $[\beta] = \nu^E_{n-1}([\omega], [i \circ \alpha])$, where $\beta(\vec{u}) = \omega * i \circ \alpha(\vec{u}) * \omega^{-1}$. The mapping $\Omega p : \Omega E \to \Omega B$ takes β to

$$p \circ \beta(\ ) = (p \circ \omega) * (p \circ i \circ \alpha(\ )) * (p \circ \omega)^{-1} = (p \circ \omega) * c_{p(e)} * (p \circ \omega)^{-1} \simeq c_{p(e)}.$$

Let $h : I \times I \to B$ denote a pointed homotopy between $p \circ \beta$ and $c_{p(e)}$. Construct the homotopy $H : S^{n-1} \times I \to \Omega B$ given by $H(\vec{u}, t)(r) = h(r, t)$. Thus $H(\vec{u}, 0)(r) = (p \circ \omega) * c_{p(e)} * (p \circ \omega)^{-1}(r)$ and $H(\vec{u}, 1)(r) = p(e)$. Consider the homotopy lifting problem posed by the diagram:

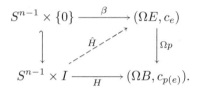

Since Ωp is a fibration, there is a lifting $\hat{H} : S^{n-1} \times I \to (\Omega E, c_e)$ with $\hat{H}(\vec{u}, 0)(r) = \beta(\vec{u})(r)$. Because $p \circ \hat{H} = H$, the mapping $\beta'(\vec{u})(r) = \hat{H}(\vec{u}, 1)(r)$ determines a loop in F. We define the action of π on $\pi_{n-1}(\Omega F, c_e)$ to be the rule that associates $[\beta']$ to $[\omega] \in \pi$ and $[\alpha] \in \pi_{n-1}(\Omega F, c_e)$. By the properties of fibrations this is well-defined and gives a module action. We denote this action by $\nu^{E,F}_n : \pi_1(E, e) \times \pi_{n-1}(\Omega F, c_e) \to \pi_{n-1}(\Omega F, c_e)$.

Notice that $i \circ \beta' \simeq \beta$. Because the class $[\beta]$ represents $\nu^E_{n-1}([\omega], [i \circ \alpha])$, we have that $i_*(\nu^{E,F}_{n-1}([\omega], [\alpha])) = \nu^E_{n-1}([\omega], i_*([\alpha]))$ and the homomorphism $i_* : \pi_{n-1}(\Omega F, c_e) \to \pi_{n-1}(\Omega E, c_e)$ is a homomorphism of π-modules. Furthermore, if $[\omega] = i_*([\omega'])$ for $[\omega'] \in \pi_1(F, e)$, then $i_*(\nu^F_{n-1}([\omega'], [\alpha])) = \nu^E_{n-1}(i_*([\omega']), i_*([\alpha]))$.

Finally, we consider the transgression $\partial : \pi_n(B, p(e)) \to \pi_{n-1}(F, e)$. It is best here to substitute a geometric mapping for ∂. Consider the pullback diagram:

$$
\begin{array}{ccccc}
\Omega B & \xrightarrow{\ j\ } & E_{\mathrm{ev}_1, p} & \longrightarrow & PB \\
& & \downarrow{\scriptstyle q} & & \downarrow{\scriptstyle \mathrm{ev}_1} \\
& & E & \xrightarrow{\ p\ } & B.
\end{array}
$$

Here $\mathrm{ev}_1 : PB \to B$ is the path-loop fibration. Because PB is contractible, $E_{\mathrm{ev}_1, p}$ has the homotopy type of F and $j_* = \partial$. The space $E_{\mathrm{ev}_1, p}$ is the subspace of $E \times PB$ given by $\{(y, \lambda) \mid y \in E, \lambda : I \to B, \text{such that } \lambda(0) = p(e), \lambda(1) = p(y)\}$. The mapping $j : \Omega B \to E_{\mathrm{ev}_1, p}$ is given by $j(\gamma) = (e, \gamma)$. The action of a loop $\omega \in [\omega] \in \pi$ may be expressed at this level as $(e, \gamma) \mapsto (e, (p \circ \omega) * \gamma * (p \circ \omega)^{-1})$. In the definition of the action of π on $\pi_n(\Omega B, c_{p(e)})$, the mapping j_* takes the class $\nu^B_n([\omega], [\alpha])$ to $\bar{\nu}^{F,E}_{n-1}([\omega], j_*([\alpha]))$ after the identification of the fibre with $E_{\mathrm{ev}_1, p}$. Thus $j_* = \partial$ is a π-homomorphism.

After all the relevant identifications are made, we have shown that the long exact sequence

$$\cdots \to \pi_n(F, e) \xrightarrow{i_*} \pi_n(E, e) \xrightarrow{p_*} \pi_n(B, p(e)) \xrightarrow{\partial} \pi_{n-1}(F, e) \to \cdots$$

is an exact sequence of π-modules and π-module homomorphisms.   □

In certain cases the π_1-action is trivial, even for π_1 a nontrivial group. For example, if X is an H-space, we have the following result of [Serre51].

Corollary 8bis.3. *If (X, μ, e) is an H-space, then the action $\bar{\nu}_n^X$ of $\pi_1(X, e)$ on $\pi_n(X, e)$ is trivial for all n.*

PROOF: Let $\omega \in [\omega] \in \pi_1(X, e)$. If $\alpha \colon (S^{n-1}, e_1) \to (\Omega X, c_e)$ represents a class in $\pi_n(X, e)$, then consider the homotopy $H \colon S^{n-1} \times I \to (\Omega X, c_e)$ given by $H(\vec{u}, t)(r) = \mu((c_e * \alpha(\vec{u}) * c_e)(r), h(r, t))$ where $h \colon I \times I \to X$ is a pointed homotopy between c_e and $\omega * c_e * \omega^{-1}$. From the definition $H(\vec{u}, 0)(r) = \mu((c_e * \alpha(\vec{u}) * c_e)(r), e) \simeq \alpha(\vec{u})(r)$ and $H(\vec{u}, 1)(r) = (\omega * \alpha(\vec{u}) * \omega^{-1})(r)$. It follows that $\nu_{n-1}^X([\omega], [\alpha]) = [\alpha]$.   □

This result extends the fact that the fundamental group of an H-space is abelian.

Another rich source of actions of the fundamental group is the notion of bundles of groups (§5.3) over a space. For example, when $F \hookrightarrow E \to B$ is a fibration, then $\pi_1(B, b_0)$ acts on the homology of F_{b_0}, the fibre over b_0. In Chapters 5 and 6 the applications of the Leray-Serre spectral sequence involved simple systems of local coefficients, that is, where $\pi_1(B, b_0)$ acts trivially on $H_i(F_{b_0}; R)$. When the bundle of groups is not simple, then the E^2-term of the spectral sequence need not be a product, even for field coefficients. In the next proposition we give the first case of such a difference. The functor $\Gamma_\pi^2(\ )$ on π-modules defined in the proposition is the first of a family described fully in Definition 8bis.17.

Proposition 8bis.4. *Suppose $\mathcal{G}$ is a bundle of abelian groups over a pointed, path-connected space (X, x_0) and $G_0 = G_{x_0}$. Then $H_0(X; \mathcal{G}) \cong G_0/\Gamma_\pi^2 G_0$, where $\Gamma_\pi^2 G_0$ is the subgroup of G_0 generated by all elements of the form $[\alpha] \cdot g - g$ with $[\alpha] \in \pi = \pi_1(X, x_0)$ and $g \in G_0$.*

PROOF: Consider the mapping $\phi \colon G_0 \to H_0(X; \mathcal{G})$ given by $g \mapsto g \otimes x_0$. Suppose that $\alpha \in [\alpha] \in \pi_1(X, x_0)$. Then we associate to $g \in G_0$ and $\alpha \colon (\Delta^1, \partial \Delta^1) \to (X, x_0)$ the element $g \otimes \alpha$ in $C_1(X; \mathcal{G})$. The boundary of $g \otimes \alpha$ is given by the formula in §5.3:

$$\partial_h(g \otimes \alpha) = h[\alpha^{-1}](g) \otimes \alpha(1) - g \otimes \alpha(0) = (h[\alpha^{-1}](g) - g) \otimes x_0.$$

Since α and g were arbitrary, we see that ϕ takes $\Gamma^2_\pi G_0$ to 0 and so induces a homomorphism $\bar{\phi}\colon G_0/\Gamma^2_\pi G_0 \to H_0(X;\mathcal{G})$.

To show that $\bar{\phi}$ is an isomorphism, we describe its inverse. Suppose $u \in H_0(X;\mathcal{G})$. We can write

$$u = \sum_{i=1}^{n} g_i \otimes x_i + B_0(X;\mathcal{G}).$$

Let $\lambda_i\colon [0,1] \to X$ denote a path in X joining x_0 to x_i. Then we have the isomorphism $h[\lambda_i]\colon G_{x_i} \to G_0$ for each i. We define $\psi\colon H_0(X;\mathcal{G}) \to G_0/\Gamma^2_\pi G_0$ by

$$\psi\left(\sum_{i=1}^{n} g_i \otimes x_i + B_0(X;\mathcal{G})\right) = \sum_{i=1}^{n} h[\lambda_i](g_i).$$

To see that ψ is well-defined, we notice that ψ takes boundaries to zero—let $h[\alpha^{-1}](g) \otimes \alpha(1) - g \otimes \alpha(0)$ denote a generator of $B_0(X;\mathcal{G})$. The homomorphism ψ applied to such an element gives $h[\beta_1](h[\alpha^{-1}](g)) - h[\beta_0](g)$ where β_i is a path in X starting at x_0 and ending at $\alpha(i)$. There is a loop based at x_0 given by $\beta_0 * \alpha * \beta_1^{-1}$ and

$$h[\beta_0 * \alpha * \beta_1^{-1}](h[\beta_1](h[\alpha^{-1}](g)) - h[\beta_0](g))$$
$$= h[\beta_0](g) - h[\beta_0 * \alpha * \beta_1^{-1}](h[\beta_0](g)),$$

which is an element of $\Gamma^2_\pi G_0$. Since $\Gamma^2_\pi G_0$ is closed under the action of $\pi_1(X, x_0)$, ψ takes $B_0(X;\mathcal{G})$ to zero.

We must account for the choices made in the construction of ψ. If λ_i and μ_i are paths joining x_0 to x_i, then we compare $h[\lambda_i](g_i)$ and $h[\mu_i](g_i)$, the images of $g_i \otimes x_i$ with respect to the different paths. The difference between these values can be rewritten

$$h[\lambda_i](g_i) - h[\mu_i](g_i) = h[\mu_i * \lambda_i^{-1}]h[\lambda_i * \mu_i^{-1}](h[\lambda_i](g_i) - h[\mu_i](g_i))$$
$$= h[\mu_i * \lambda_i^{-1}](h[\lambda_i * \mu_i^{-1}](h[\lambda_i](g_i)) - h[\lambda_i](g_i)),$$

which is an element in $\Gamma^2_\pi G_0$. Thus ψ is well-defined and the inverse of $\bar{\phi}$. $\square$

By Theorem 5.1, it follows for an arbitrary fibration with path-connected base and connected fibre, $F \hookrightarrow E \to B$, that the associated Leray-Serre spectral sequence has $E^2_{0,*} \cong H_*(F;k)/\Gamma^2_\pi H_*(F;k)$ as the leftmost column of the E^2-term. Though this seems to put us in murkier waters, we can still see our way to deep results in homotopy theory by studying abstractly the action of groups on abelian groups.

§8bis.2 Homology of groups

If π denotes a group and M a module over a ring R, then we say that π **acts on** M, or M is a π-**module**, if there is a homomorphism $\rho\colon \pi \to \mathrm{Aut}_R(M)$, where $\mathrm{Aut}_R(M)$ is the group of R-linear isomorphisms of M to itself. More generally, π acts on a group G if there is a homomorphism $\rho\colon \pi \to$ $\mathrm{Aut}(G)$. To study modules over a group π we introduce the homology of groups, a homological functor analogous to Tor. The homology groups of a group π with coefficients in a π-module M satisfy the axioms for the left-derived functors ([Cartan-Eilenberg56]) of the functor that associates to a π-module M its **coinvariants**,

$$(\ )_\pi\colon M \mapsto M_\pi = M/\Gamma_\pi^2 M.$$

Here $\Gamma_\pi^2 M$ is the submodule of M generated by elements of the form $am - m$ where $a \in \pi$ and $m \in M$. (More generally, if G is a nonabelian group acted on by π, let $\Gamma_\pi^2 G$ be the normal subgroup generated by elements of the form $(ag)g^{-1}$.) Since $b(am-m) = (bab^{-1})bm-bm$, $\Gamma_\pi^2 M$ is a π-module. Thus a π-equivariant module homomorphism induces a homomorphism on coinvariants. The induced action of π on M_π is trivial. In fact, M_π may be characterized as the *largest* quotient of M on which π acts trivially. By 'largest' quotient we mean that, if M/M' is another quotient of M by a π-submodule M' and π acts trivially on M/M', then $\Gamma_\pi^2 M \subset M'$ and there is an epimorphism $M/\Gamma_\pi^2 M \longrightarrow M/M'$.

To see that the functor $(\ )_\pi$ is right exact, we give another expression for M_π when M is a π-module.

Lemma 8bis.5. *Let $\mathbb{Z}$ denote the ring of integers, taken as a trivial right π-module. If M is a π-module, then $M_\pi \cong \mathbb{Z} \otimes_{\mathbb{Z}\pi} M$, where $\mathbb{Z}\pi$ denotes the integral group ring of π.*

PROOF: Recall that $\mathbb{Z} \otimes_{\mathbb{Z}\pi} M$ is defined as the cokernel in the sequence

$$\mathbb{Z} \otimes \mathbb{Z}\pi \otimes M \xrightarrow{\psi\otimes 1 - 1\otimes\phi} \mathbb{Z} \otimes M \to \mathbb{Z} \otimes_{\mathbb{Z}\pi} M \to 0,$$

where $\phi\colon \mathbb{Z}\otimes\mathbb{Z}\pi \to \mathbb{Z}$ is the trivial right action of π on $\mathbb{Z}$, and $\psi\colon \mathbb{Z}\pi\otimes M \to M$ is the left action of π on M. In this case, $\otimes = \otimes_\mathbb{Z}$ and so the sequence becomes

$$\mathbb{Z}\pi \otimes M \xrightarrow{\eta} M \to \mathbb{Z} \otimes_{\mathbb{Z}\pi} M \to 0,$$

where $\eta(a \otimes m) = m - am$. Thus $\mathbb{Z} \otimes_{\mathbb{Z}\pi} M = M/\operatorname{im}\eta = M/\Gamma_\pi^2 M$. $\square$

The lemma implies that $(\ )_\pi$ is a right exact functor since $\mathbb{Z} \otimes_{\mathbb{Z}\pi} (\ )$ is right exact.

Definition 8bis.6. *The **homology of a group** π **with coefficients in a (left)** π-**module** M is defined by*

$$H_i(\pi, M) = \mathrm{Tor}_i^{\mathbb{Z}\pi}(\mathbb{Z}, M).$$

We write $H_i(\pi)$ for $H_i(\pi, \mathbb{Z})$ when $\mathbb{Z}$ is the trivial left π-module.

To compute group homology we introduce a convenient functorial resolution. Let B_n denote the free abelian group on $(n+1)$-tuples of elements of π. Then B_n is a right π-module with the π-diagonal action

$$(x_0, \dots, x_n)a = (x_0 a, \dots, x_n a).$$

As a $\mathbb{Z}\pi$-module, B_n is free on those $(n+1)$-tuples with last entry 1. Consider the complex

$$\cdots \to B_n \xrightarrow{\partial} B_{n-1} \xrightarrow{\partial} \cdots \xrightarrow{\partial} B_2 \xrightarrow{\partial} B_1 \xrightarrow{\partial} \mathbb{Z}\pi \xrightarrow{\varepsilon} \mathbb{Z} \to 0$$

where $\partial = \sum_i (-1)^i \bar{d}_i$ and $\bar{d}_i(x_0, \dots, x_n) = (x_0, \dots, \widehat{x}_i, \dots, x_n)$; ε is the usual group ring augmentation given by $\varepsilon\left(\sum_i n_i a_i\right) = \sum_i n_i$. This complex is exact since there is the contracting homotopy defined by $s(x_0, \dots, x_n) = (1, x_0, \dots, x_n)$ and $s(1) = 1$. To make the free $\mathbb{Z}\pi$-module structure evident, we introduce the **bar notation** for generators over $\mathbb{Z}\pi$: For $n = 0$, $[\ ] = 1$, and for $n > 0$,

$$[x_1 \mid \cdots \mid x_n] = (x_1 x_2 \cdots x_n, \ x_2 x_3 \cdots x_n, \ \dots, \ x_{n-1} x_n, \ x_n, \ 1) \in B_n.$$

The differential ∂ can be rewritten as $\partial = \sum_i (-1)^i d_i$ where

$$d_i([x_1 \mid \cdots \mid x_n]) = \begin{cases} [x_2 \mid \cdots \mid x_n], & \text{for } i = 0, \\ [x_1 \mid \cdots \mid x_{i-1} \mid x_i x_{i+1} \mid \cdots \mid x_n], & \text{for } 0 < i < n, \\ [x_1 \mid \cdots \mid x_{n-1}]x_n, & \text{for } i = n. \end{cases}$$

This is the familiar **bar construction** on π that gives a functorial free right $\mathbb{Z}\pi$-module resolution of the trivial module $\mathbb{Z}$.

Although it is large, the bar construction can still be used to prove structure results.

Proposition 8bis.7. *If π is a finite group of order $|\pi|$ and M is a π-module, then every element of $H_i(\pi, M)$ for $i > 0$ has order a divisor of $|\pi|$.*

PROOF: Consider the π-module homomorphism , $s \colon B_n \to B_{n+1}$, given on generators by $s([x_1 \mid \cdots \mid x_n]) = \sum y \in \pi[y \mid x_1 \mid \cdots \mid x_n]$. We show that $\partial \circ s + s \circ \partial = |\pi|\, \mathrm{id}$:

$$\partial \circ s([x_1 \mid \cdots \mid x_n]) = \partial\left(\sum_{y \in \pi} [y \mid x_1 \mid \cdots \mid x_n]\right)$$

$$= \sum_{y \in \pi} \sum_i (-1)^i d_i([y \mid x_1 \mid \cdots \mid x_n])$$

$$= |\pi|[x_1 \mid \cdots \mid x_n] - \sum_{y \in \pi} \Big\{ [y x_1 \mid \cdots \mid x_n] - [y \mid x_1 x_2 \mid x_3 \mid \cdots \mid x_n]$$

$$+ \cdots + (-1)^{i+1}[y \mid x_1 \mid \cdots \mid x_i x_{i+1} \mid \cdots \mid x_n]$$

$$+ \cdots + (-1)^{n+1}[y \mid x_1 \mid \cdots \mid x_{n-1}]x_n \Big\}$$

$$= |\pi|[x_1 \mid \cdots \mid x_n] - s \circ \partial([x_1 \mid \cdots \mid x_n]).$$

We have used the fact that $\sum_{y \in \pi} [yx_1 \mid x_2 \mid \cdots \mid x_n] = \sum_{y \in \pi} [y \mid x_2 \mid \cdots \mid x_n]$. If $\sum_j [x_{1j} \mid \cdots \mid x_{nj}] \otimes m_j$ represents a homology class in $H_n(\pi, M)$, then

$$\partial \circ s \left(\sum_j [x_{1j} \mid \cdots \mid x_{nj}] \otimes m_j \right) = |\pi| \sum_j [x_{1j} \mid \cdots \mid x_{nj}] \otimes m_j$$

and so $|\pi|$ times any homology class in $H_n(\pi, M)$ is zero. $\qquad \square$

Particular groups may have smaller resolutions. For $\pi = \mathbb{Z}/m\mathbb{Z}$, the cyclic group of order m, there is a very small resolution: Let t denote a generator of π, thought of as a multiplicative group. Let W_i denote the free $\mathbb{Z}\pi$-module on a single generator w_i. There is an acyclic complex

$$\cdots \to W_{2n} \xrightarrow{N} W_{2n-1} \xrightarrow{T} \cdots \xrightarrow{T} W_2 \xrightarrow{N} W_1 \xrightarrow{T} W_0 \xrightarrow{\varepsilon} \mathbb{Z} \to 0$$

where $T(w_{2n-1}) = tw_{2n-2} - w_{2n-2}$ (trace), and $N(w_{2n}) = w_{2n-1} + tw_{2n-1} + \cdots + t^{m-1} w_{2n-1}$ (norm). This resolution allows us to compute $H_i(\mathbb{Z}/m\mathbb{Z})$ immediately from the complex $W_\bullet \otimes_{\mathbb{Z}\pi} \mathbb{Z}$ which takes the form

$$\cdots \xrightarrow{\times m} \mathbb{Z} \xrightarrow{0} \mathbb{Z} \xrightarrow{\times m} \mathbb{Z} \xrightarrow{0} \mathbb{Z} \xrightarrow{\times m} \mathbb{Z}.$$

Thus $H_{2i+1}(\mathbb{Z}/m\mathbb{Z}) = \{0\}$ and $H_{2i}(\mathbb{Z}/m\mathbb{Z}) = \mathbb{Z}/m\mathbb{Z}$ for all $i \geq 0$.

We close this discussion with a useful lemma.

Lemma 8bis.8. *If M is a free left $\mathbb{Z}\pi$-module, then $H_i(\pi, M) = \{0\}$ for $i > 0$ and $H_0(\pi, M) = M_\pi$. If M is a trivial left $\mathbb{Z}\pi$-module that is free over $\mathbb{Z}$, then $H_i(\pi, M) \cong H_i(\pi) \otimes M$.*

PROOF: The assertion about a free module follows simply from the properties of $\mathrm{Tor}_*^{\mathbb{Z}\pi}(\mathbb{Z}, M)$. For trivial modules we can write $B_\bullet \otimes_{\mathbb{Z}\pi} M = (B_\bullet)_\pi \otimes M$ and the result follows. $\qquad \square$

The Cartan-Leray spectral sequence

Suppose (X, x_0) is a connected, pointed space on which a **group π acts freely and properly**, that is,

(1) for all $x \in X$, the subgroup $G_x = \{g \in G \mid gx = x\}$ is trivial;
(2) every point $x \in X$ has a neighborhood U such that $gU \cap U = \emptyset$ for all $g \in \pi, g \neq 1$.

For example, if (X, x_0) is a connected, locally simply-connected space, then the fundamental group, $\pi = \pi_1(X, x_0)$ acts freely and properly on $\tilde{X}$, the universal covering space of X.

Suppose π acts freely and properly on X. For any abelian group G, it is a classical result that $C_*(X/\pi; G) \cong C_*(X; G)_\pi$, where $C_*(\; ; G)$ denotes the singular chains with coefficients in G and $C_*(X; G)$ is a π-module by viewing each $a \in \pi$ as a mapping $a: X \to X$. Moreover, $C_i(X) = C_i(X; \mathbb{Z})$ is a free $\mathbb{Z}\pi$-module for all i.

Let $F_\bullet \to \mathbb{Z} \to 0$ denote a free, right, $\mathbb{Z}\pi$-module resolution of $\mathbb{Z}$ as a trivial π-module. Consider the double complex $C_{\bullet,*}$ given by

$$C_{p,q} = F_p \otimes_{\mathbb{Z}\pi} C_q(X), \quad \partial_F \otimes 1 + (-1)^q 1 \otimes \partial_X.$$

When we filter $C_{\bullet,*}$ row-wise, we get $E^0_{*,p} = F_\bullet \otimes_{\mathbb{Z}\pi} C_p(X)$, $d^0 = \partial_F \otimes 1$, which is the complex computing $H_*(\pi, C_p(X))$. However, $C_p(X)$ is a free $\mathbb{Z}\pi$-module and so the E^1-term is concentrated in the 0-column where we find $H_0(\pi, C_p(X)) = C_p(X)_\pi = C_p(X/\pi)$. By the appropriate version of Theorem 2.15, $d^1 = 1 \otimes_\pi \partial_X = \partial_{X/\pi}$ and so the spectral sequence collapses at E^2 to $H_*(X/\pi)$.

When we filter by $C_{\bullet,*}$ column-wise, we get $E^0_{p,*} = F_p \otimes_{\mathbb{Z}\pi} C_*(X)$, $d^0 = 1 \otimes \partial_X$. Viewing F_p as an extended module of the form $F_p = A_p \otimes \mathbb{Z}\pi$ with A_p a free abelian group, we get the identification

$$F_p \otimes_{\mathbb{Z}\pi} C_*(X) = A_p \otimes C_*(X)$$

and so $E^1_{p,*} = A_p \otimes H_*(X) = F_p \otimes_{\mathbb{Z}\pi} H_*(X)$. This is the complex computing $H_*(\pi, H_*(X))$ and so we have proved the following result of [Cartan-Leray49].

Theorem 8bis.9 (the Cartan-Leray spectral sequence). *If X is a connected space on which the group π acts freely and properly, then there is a spectral sequence of first quadrant, homological type, with*

$$E^2_{p,q} \cong H_p(\pi, H_q(X))$$

and converging strongly to $H_*(X/\pi)$.

More generally, we can use homology with coefficients in an abelian group G taken as a trivial π-module. For the case of the fundamental group acting on the universal cover, the spectral sequence has $E^2_{p,q} \cong H_p(\pi, H_q(\tilde{X}))$, where $\pi = \pi_1(X, x_0)$ and converges to $H_*(X)$. From this spectral sequence we prove a theorem that relates the fundamental group and its homology groups to the higher homology and homotopy groups of a space. The case of $n = 1$ was first proved by [Hopf42] using other methods. This paper launched the study of the homology and cohomology of groups.

Theorem 8bis.10. *Suppose (X, x_0) is a connected, locally simply-connected space whose universal covering space $\tilde{X}$ is n-connected. Let π denote $\pi_1(X, x_0)$. Then there are isomorphisms $H_i(X) \cong H_i(\pi)$ for $1 \leq i \leq n$, and an exact sequence*

$$H_{n+2}(X) \to H_{n+2}(\pi) \to (H_{n+1}(\tilde{X}))_\pi \to H_{n+1}(X) \to H_{n+1}(\pi) \to 0.$$

PROOF: Since $H_i(\tilde{X}) = \{0\}$ for $1 \leq i \leq n$, we have a big hole in the Cartan-Leray spectral sequence converging to $H_*(X)$. The theorem follows from interpreting the lowest degree information. The isomorphisms $H_i(X) = H_i(\tilde{X}/\pi) \cong H_i(\pi)$ for $1 \leq i \leq n$ follow because there are no differentials involved. The first place there is a possible differential is $d^{n+1} : E^2_{n+2,0} \rightarrow E^2_{0,n+1}$. This leads to the following short exact sequences:

$$H_{n+2}(X) \rightarrow E^\infty_{n+2,0} \rightarrow 0$$

$$0 \rightarrow E^\infty_{n+2,0} \rightarrow H_{n+2}(\pi) \xrightarrow{d^{n+1}} (H_{n+1}(\tilde{X}))_\pi \rightarrow E^\infty_{0,n+1} \rightarrow 0$$

$$0 \rightarrow E^\infty_{0,n+1} \rightarrow H_{n+1}(X) \rightarrow E^\infty_{n+1,0} \rightarrow 0.$$

Splicing these sequences together (as in Example 5.D) and substituting $H_{n+1}(\pi)$ for $E^2_{n+1,0} = E^\infty_{n+1,0}$ we get the desired exact sequence. $\square$

Since $\pi_i(X) \cong \pi_i(\tilde{X})$ for $i \geq 2$, we can substitute $\pi_{n+1}(X)$ for the term in the middle of the exact sequence of Theorem 8bis.10 by the Hurewicz theorem. There is a natural epimorphism of a π-module onto its coinvariants, so we can truncate the short exact sequence to obtain another exact sequence

$$\pi_{n+1}(X) \rightarrow H_{n+1}(X) \rightarrow H_{n+1}(\pi) \rightarrow 0.$$

When $n = 1$, there is no restriction on X except that it have a universal covering space. We conclude from the theorem that $H_1(X) \cong H_1(\pi)$ and so $H_1(\pi) \cong \pi/[\pi, \pi]$ follows from Poincaré's classical isomorphism. We also get the **short exact sequence of Hopf**:

$$\pi_2(X) \rightarrow H_2(X) \rightarrow H_2(\pi) \rightarrow 0.$$

For an **aspherical space**, that is, a space X whose universal cover has trivial higher homotopy groups, the integral homology of X is determined by its fundamental group, $H_i(X) \cong H_i(\pi_1(X, x_0))$ for all i. Examples of aspherical spaces are the Eilenberg-Mac Lane spaces $K(\pi, 1)$. The study of the homology of groups was one of the motivations for [Eilenberg-Mac Lane53] to introduce the spaces $K(\pi, n)$. [Kan-Thurston76] reversed the process of studying groups using spaces by showing that for any pointed, connected space (X, x_0) there is a group GX and a mapping $t_X : K(GX, 1) \rightarrow X$ that induces an integral homology isomorphism.

Using the same filtration that leads to the Cartan-Leray spectral sequence, we can investigate systems of local coefficients further. In particular, for X a connected space and $\mathcal{G}$ a system of groups on X, we restrict our attention to the reduced homology. Consider the group of the reduced chains $\tilde{C}_q(X; \mathcal{G})$ given by sums of expressions $g \otimes u$ where $u : (\Delta^q, (\Delta^q)^{(0)}) \rightarrow (X, x_0)$ and $g \in G_{x_0}$, where the singular simplex u sends all its vertices to the basepoint x_0, and g

lies in the group over the basepoint. By the properties of homology groups with coefficients, the study of $H_q(X;\mathcal{G})$ may be carried out using $\tilde{C}_*(X;\mathcal{G})$. We quote here a result of [Eilenberg47] and refer the reader to [Whitehead, GW78, VI.3] for a proof.

Theorem 8bis.11. *If (X, x_0) is a pointed, path-connected space and $\mathcal{G}$ a bundle of groups over X, then $H_q(X;\mathcal{G})$ is isomorphic to the homology of the complex $G_0 \otimes_\pi C_*(\tilde{X})$ where $G_0 = G_{x_0}$, $\pi = \pi_1(X, x_0)$ and $\tilde{X}$ is the universal covering space of X together with its action of π.*

The proof follows by a direct comparison. We note a useful corollary. Since $\tilde{X}$ is one-connected, the complex $C_*(\tilde{X})$ up to degree two is acyclic and free over π. Thus we could use this complex as part of a free resolution of $\mathbb{Z}$ over π and so identify $H_1(X;\mathcal{G})$ with $H_1(\pi_1(X); G_0)$. This extends Proposition 8bis.4 and the identification of homology groups with local coefficients to include $q = 1$. The homology of the complex $G_0 \otimes_\pi C_*(\tilde{X})$ was termed the **equivariant homology of** X by [Eilenberg47] and it represents one of the basic functors in the study of equivariant homotopy theory.

The Lyndon-Hochschild-Serre spectral sequence

The Leray-Serre spectral sequence expresses the relation between the total space of a fibration and its base and fibre. In the realm of groups, a "fibration" is an extension of the form

$$1 \to H \to \pi \to Q \to 1,$$

where H is normal in π and $Q \cong \pi/H$. The "total space" π is the extension of the "base" Q and "fibre" H. There is a corresponding spectral sequence relating the homology of a group to a normal subgroup and associated quotient.

To describe the spectral sequence we observe that the quotient group Q acts on the homology of H. Let $F_\bullet \xrightarrow{\varepsilon} \mathbb{Z} \to 0$ be a free right $\mathbb{Z}\pi$-module resolution of $\mathbb{Z}$. Then it is also a free right $\mathbb{Z}H$-module resolution. If M is a left π-module, it likewise is a left H-module by restriction. If $g \in \pi$, then define $(g)_*\colon F_\bullet \otimes_{\mathbb{Z}H} M \to F_\bullet \otimes_{\mathbb{Z}H} M$ by $(g)_*(x \otimes m) = xg^{-1} \otimes gm$. It follows formally that $(g)_*$ commutes with the differential on $F_\bullet \otimes_{\mathbb{Z}H} M$ and so $(g)_*$ induces a homomorphism $(g)_*\colon H_*(H, M) \to H_*(H, M)$. From the definition of the tensor product over $\mathbb{Z}H$, $(h)_* = \mathrm{id}$ for all $h \in H$ and so this action of π on $H_*(H, M)$ induces an action of π/H on $H_*(H, M)$. With this bit of structure in place we construct a spectral sequence associated to an extension of groups.

Theorem 8bis.12 (the Lyndon-Hochschild-Serre spectral sequence). *Let* $1 \to$ $H \to \pi \to Q \to 1$ *be a group extension. Suppose M is a module over π. Then there is a first quadrant spectral sequence with*

$$E_{p,q}^2 \cong H_p(Q, H_q(H, M)),$$

and converging strongly to $H_*(\pi, M)$.

PROOF: We begin with some elementary algebraic observations. When we take the coinvariants of a π-module M with respect to the action of H, we make the action of H trivial on M_H. Thus π/H acts on M_H by $gH \cdot (m + \Gamma_H^2 M) = gm + \Gamma_H^2 M$, where $\Gamma_H^2 M$ is the submodule of M generated by elements $hm - m$ for $h \in H$ and $m \in M$.

We next consider the coinvariants of M_H via the $Q = \pi/H$ action and prove the following formula:

$$(M_H)_Q \cong M_\pi.$$

This follows from the relation $\mathbb{Z}Q \otimes_{\mathbb{Z}\pi} M \cong M_H$ by using the isomorphism $\mathbb{Z} \otimes_{\mathbb{Z}\pi} M \cong \mathbb{Z} \otimes_{\mathbb{Z}Q} \mathbb{Z}Q \otimes_{\mathbb{Z}\pi} M$.

Consider the mapping $\phi \colon M_H \to \mathbb{Z}Q \otimes_{\mathbb{Z}\pi} M$ given by $m \mapsto 1 \otimes m$. Since $1 \otimes m = 1 \otimes h^{-1} hm = 1 \cdot h \otimes hm = 1 \otimes hm$, we see that ϕ is defined on M_H. An inverse to ϕ is the mapping $\psi \colon gH \otimes m \mapsto gm + \Gamma_H^2 M$. That ψ is well-defined follows from the fact that M_H is a Q-module.

We put these elementary observations to work and suppose that $F_\bullet \to \mathbb{Z} \to 0$ is a free, right, $\mathbb{Z}\pi$-module resolution of $\mathbb{Z}$. Consider the complex computing $H_*(\pi, M)$, that is, $F_\bullet \otimes_{\mathbb{Z}\pi} M$. Since $F_\bullet \otimes_{\mathbb{Z}} M$ is a $\mathbb{Z}\pi$-module by the diagonal action $(g \cdot (x \otimes m) = xg^{-1} \otimes gm)$, it is a simple exercise to show that

$$(F_\bullet \otimes_{\mathbb{Z}} M)_\pi \cong F_\bullet \otimes_{\mathbb{Z}\pi} M,$$

and so $F_\bullet \otimes_{\mathbb{Z}\pi} M \cong ((F_\bullet \otimes M)_H)_Q$.

We also have that $H(F_\bullet \otimes_{\mathbb{Z}H} M) = H_*(H, M)$. We want to compute the coinvariants of the action of Q on these homology groups. Let $\tilde{F}_\bullet \to \mathbb{Z} \to 0$ be a free right $\mathbb{Z}Q$-module resolution of $\mathbb{Z}$. Form the double complex

$$C_{\bullet *} = \tilde{F}_\bullet \otimes_{\mathbb{Z}Q} (F_* \otimes M)_H,$$

where Q acts on $(F_* \otimes M)_H$ by the diagonal action. When we filter column-wise, we get the complex $(\tilde{F}_\bullet \otimes_{\mathbb{Z}Q} (F_q \otimes_{\mathbb{Z}H} M), d \otimes 1)$. However, treating F_q as an extended $\mathbb{Z}\pi$-module (on the left via the antiautomorphism $g \mapsto g^{-1}$ on $\mathbb{Z}\pi$), we can write $F_q = \mathbb{Z}\pi \otimes A_q$ for some free $\mathbb{Z}$-module A_q. It follows that

$$F_q \otimes_{\mathbb{Z}H} M = \mathbb{Z} \otimes_{\mathbb{Z}H} (F_q \otimes M) = \mathbb{Z} \otimes_{\mathbb{Z}H} (\mathbb{Z}\pi \otimes A_q \otimes M) = \mathbb{Z}Q \otimes A_q \otimes M,$$

because $(\mathbb{Z}\pi)_H = \mathbb{Z}Q$. Thus we have $E^0_{\bullet,q} \cong \tilde{F}_\bullet \otimes_{\mathbb{Z}Q} \mathbb{Z}Q \otimes A_q \otimes M$ and so $E^1_{p,q} \cong H_p(Q, (F_q \otimes_{\mathbb{Z}H} M)) = \{0\}$ if $p > 0$ and if $p = 0$ $E^1_{0,q}$ is isomorphic to $(F_q \otimes_{\mathbb{Z}H} M)_Q = (F_q \otimes_{\mathbb{Z}\pi} M)$. Thus $E^2_{0,*} = H_*(\pi, M)$, and the double complex has total homology given by $H_*(\pi, M)$.

When we filter row-wise, we have $E^0_{p,*} = \tilde{F}_p \otimes_{\mathbb{Z}Q} (F_* \otimes_{\mathbb{Z}H} M)$ and $d^0 = 1 \otimes d$. We argue similarly and write $\tilde{F}_p$ as an extended right $\mathbb{Z}Q$-module, $\tilde{A}_p \otimes \mathbb{Z}Q$, to get $E^0_{p,*} \cong \tilde{A}_p \otimes (F_* \otimes_{\mathbb{Z}H} M)$, $d^0 = 1 \otimes d$ and so $E^1_{p,*} \cong \tilde{A}_p \otimes H_*(H, M) \cong \tilde{F}_p \otimes_{\mathbb{Z}Q} H_*(H, M)$ and $d^1 = d \otimes 1$. Thus $E^2_{p,q} \cong H_p(Q, H_q(H, M))$. $\qquad\square$

The first calculations of some of the relations implied by this spectral sequence appeared in Chicago Ph.D. thesis of [Lyndon48] without the benefit of the structure made apparent in the work of [Leray46]. The cohomology version of the spectral sequence of Theorem 8bis.12 first appeared in the Comptes Rendues note of [Serre50']. [Hochschild-Serre53] introduced the spectral sequence of Theorem 8bis.12 along with a formalism that allowed analogous constructions for Lie algebras. Their work was based on a different filtration on the cochain complex whose homology gives the cohomology of a group. Another point of view that gives rise to this spectral sequence is due to [Grothendieck57]. The composition of functors spectral sequence (see Chapter 11) for the composition of the coinvariants of the H action followed by the coinvariants of the π/H action gives another construction of the Lyndon-Hochschild-Serre spectral sequence. The fact that these alternative constructions give isomorphic spectral sequences from the E_2-term is due to [Beyl81].

A simple example of the use of the spectral sequence is the case of the extension associated to a Sylow p-subgroup when it is normal in a finite group.

Proposition 8bis.13. *Suppose π is a finite group and P is a normal Sylow p-subgroup. Then*

$$H_i(\pi, \mathbb{F}_p) \cong H_i(P, \mathbb{F}_p)_Q,$$

where π acts on $\mathbb{F}_p$ trivially and $Q = \pi/P$.

PROOF: The spectral sequence associated to the extension $1 \to P \to \pi \to Q \to 1$ has E^2-term given by $E^2_{p,q} \cong H_p(Q, H_q(P, \mathbb{F}_p))$. Since the order of Q is relatively prime to p and p times any element in $H_q(P, \mathbb{F}_p)$ is zero, we conclude from Proposition 8bis.7 that $E^2_{p,q} \cong \{0\}$ for $p > 0$. Thus the spectral sequence collapses to a single column given by $E^2_{0,q} \cong H_q(P, \mathbb{F}_p)_Q$ and the result follows. $\qquad\square$

We next consider the "lower left-hand corner" of the spectral sequence. Theorems 8bis.14 and 8bis.16 were first obtained by [Stallings65] and [Stammbach66].

Theorem 8bis.14. *To an extension of groups* $1 \to H \to \pi \to Q \to 1$, *there is an exact sequence*

$$H_2(\pi) \to H_2(Q) \to H/[\pi, H] \to H_1(\pi) \to H_1(Q) \to 0.$$

PROOF: This is simply the five-term exact sequence from splicing together the differential d^2 with the associated graded filtration for $H_1(\pi)$ (Example I.1) as in the diagram:

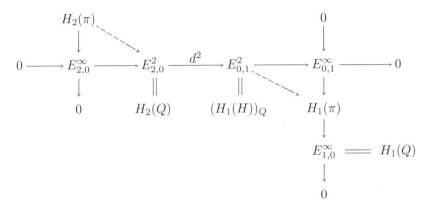

To complete the proof we make the identification $(H_1(H))_Q \cong H/[\pi, H]$. Recall that $H_1(H) = H/[H, H]$ and that Q acts on this group by conjugation. The conjugation action of π on itself has coinvariants $\pi/[\pi, \pi] = H_1(\pi)$ and it is the residue of this action that remains as Q acts on $H_1(H)$. The Q-coinvariants of the conjugation action on $H/[H, H]$ are given by $H/[\pi, H]$ and so the theorem is proved. $\square$

To extend this theorem further we recall the lower central series of a group—a version of this series will play an important role in the study of nilpotent groups and spaces.

Definition 8bis.15. *Let π denote a group. The* **lower central series** *of π is the family of subgroups defined inductively,*

$$\cdots \subset \Gamma^r \pi \subset \Gamma^{r-1}\pi \subset \cdots \subset \Gamma^2 \pi \subset \pi$$
$$\Gamma^2 \pi = [\pi, \pi], \qquad \Gamma^r \pi = [\pi, \Gamma^{r-1}\pi].$$

To the lower central series of π we associate the **completion** *of π:*

$$\widehat{\pi} = \lim_{\leftarrow r} \pi/\Gamma^r \pi.$$

We can extend the lower central series to higher ordinals by letting $\Gamma^\alpha \pi = [\pi, \Gamma^\beta \pi]$ when $\alpha = \beta + 1$ and $\Gamma^\alpha \pi = \bigcap_{\beta < \alpha} \Gamma^\beta \pi$ when α is a limit ordinal.

Theorem 8bis.16. *Suppose $\phi\colon \pi \to \pi'$ is a homomorphism of groups that induces an isomorphism $H_1(\phi)\colon H_1(\pi) \to H_1(\pi')$ and a surjection $H_2(\phi)\colon H_2(\pi) \twoheadrightarrow H_2(\pi')$. Then the induced mapping $\bar{\phi}\colon \pi/\Gamma^r\pi \to \pi'/\Gamma^r\pi'$ is an isomorphism for all positive integers r. Furthermore, ϕ induces an isomorphism of completions $\hat{\phi}\colon \hat{\pi} \to \hat{\pi}'$.*

PROOF: We work by induction. We let $\Gamma^1\pi = \pi$ and $\Gamma^1\pi' = \pi'$, and so the induction begins trivially. Next consider the extension $1 \to \Gamma^r\pi \to \pi \to \pi/\Gamma^r\pi \to 1$ and similarly for π'. By Theorem 8bis.14 we have short exact sequences

$$H_2(\pi) \longrightarrow H_2(\pi/\Gamma^r\pi) \longrightarrow \Gamma^r\pi/[\pi,\Gamma^r\pi] \longrightarrow H_1(\pi) \longrightarrow H_1(\pi/\Gamma^r\pi) \to 0$$

$$\phi_* \downarrow \qquad \bar{\phi}_* \downarrow \qquad\qquad \downarrow \qquad\qquad \phi_* \downarrow \qquad \bar{\phi}_* \downarrow$$

$$H_2(\pi') \to H_2(\pi'/\Gamma^r\pi') \to \Gamma^r\pi'/[\pi',\Gamma^r\pi'] \to H_1(\pi') \to H_1(\pi'/\Gamma^r\pi') \to 0$$

The induction hypothesis together with the Five-lemma implies that ϕ induces an isomorphism $\Gamma^r\pi/[\pi,\Gamma^r\pi] \to \Gamma^r\pi'/[\pi',\Gamma^r\pi']$. Furthermore, by definition, $\Gamma^r\pi/[\pi,\Gamma^r\pi] = \Gamma^r\pi/\Gamma^{r+1}\pi$, and so there is a short exact sequence

$$1 \to \Gamma^r\pi/\Gamma^{r+1}\pi \to \pi/\Gamma^{r+1}\pi \to \pi/\Gamma^r\pi \to 1$$

and similarly for π'. The homomorphism ϕ induces a morphism of short exact sequences and the Five-lemma implies that ϕ induces an isomorphism of $\pi/\Gamma^{r+1}\pi$ with $\pi'/\Gamma^{r+1}\pi'$. $\qquad\qquad\qquad\square$

Applications of the Lyndon-Hochschild-Serre spectral sequence abound. We refer the reader to the book-length treatments of the cohomology of groups by [Brown, K82], [Evens93], [Weibel94], [Thomas86], [Benson91], and [Adem-Milgram94] for more of the algebraic details, applications, and structure of this spectral sequence. [Huebschmann81, 91] has studied the differentials in the cohomology Lyndon-Hochschild-Serre spectral sequence and made successful use of this information to compute the cohomology of some important classes of groups. Some of the exercises at the end of the chapter expose the algebraic possibilities afforded by this tool.

§8bis.3 Nilpotent spaces and groups

When a group fails to be abelian, the lower central series is a composition series whose consecutive quotients are abelian. One direction to generalize the category of abelian groups is to consider groups whose lower central series is bounded in length. A group π is said to be **nilpotent** of **nilpotence class** c if $c = \max\{r \mid \Gamma^r\pi \neq \{1\}\}$ is finite. Thus $\Gamma^{c+k}\pi = [\pi, \Gamma^{c+k-1}\pi] = \{1\}$ for $k \geq 1$.

If we view π as a group on which π acts by conjugation, then $\Gamma^2\pi = [\pi, \pi]$ is the smallest normal subgroup for which the quotient has an induced trivial action. We generalize this idea to the action of a group π on a module M.

Definition 8bis.17. The **lower central series** associated to the action of a group π on a module M is the series of submodules

$$\cdots \subset \Gamma_\pi^r M \subset \Gamma_\pi^{r-1} M \subset \cdots \subset \Gamma_\pi^2 M \subset M$$

where $\Gamma_\pi^2 M$ is the $\mathbb{Z}\pi$-submodule of M generated by elements of the form $am - m$ for $a \in \pi$ and $m \in M$, and $\Gamma_\pi^r M = \Gamma_\pi^2(\Gamma_\pi^{r-1} M)$. We say that π **acts nilpotently on** M of **nilpotence class** c if $c = \max\{r \mid \Gamma_\pi^r M \neq \{0\}\}$ is finite.

To a group or to a group action on a module, we associate various completion functors:

$$\widehat{\pi} = \varprojlim_r \pi/\Gamma^r \pi, \qquad \widehat{M} = \varprojlim_r M/\Gamma_\pi^r M.$$

By the universal property of the inverse limit there are canonical homomorphisms $i\colon \pi \to \widehat{\pi}$ and $i\colon M \to \widehat{M}$. Define

$$\Gamma^\infty \pi = \ker(i\colon \pi \to \widehat{\pi}) \qquad \Gamma_\pi^\infty M = \ker(i\colon M \to \widehat{M})$$

$$\Gamma_\infty' \pi = \operatorname{coker}(i\colon \pi \to \widehat{\pi}) \qquad \Gamma_\infty' M = \operatorname{coker}(i\colon M \to \widehat{M}).$$

By Lemma 3.10, $\Gamma_\pi^\infty M = \bigcap_r \Gamma_\pi^r M$.

Notice that, for a nilpotent group or nilpotent group action, i is an isomorphism and the functors Γ^∞ and Γ_∞' vanish. Before turning to some examples, we introduce one more functor. A submodule N of a π-module M is said to π-**perfect** if $\Gamma_\pi^2 N = N$. If we take the family $\mathcal{P}_M$ of all π-perfect submodules of M, then we can define

$$\Gamma M = \text{ submodule generated by the union of the family } \mathcal{P}_M.$$

We leave it to the reader to show that ΓM is also π-perfect; ΓM is the **maximal π-perfect submodule** of M as it contains all π-perfect submodules. By construction, $\Gamma M \subset \Gamma_\pi^\infty M$.

A nice example of a nonnilpotent group action is given by our example $\pi_n(S^{2m} \times \mathbb{R}P^{2n})$. Here the generator of $\pi_1(S^{2m} \times \mathbb{R}P^{2n}) = \mathbb{Z}/2\mathbb{Z}$ acts on $\mathbb{Z}$ by $1 \mapsto -1$. The lower central series for this action is seen to be

$$\cdots \subset 2^r \mathbb{Z} \subset 2^{r-1} \mathbb{Z} \subset \cdots \subset 2\mathbb{Z} \subset \mathbb{Z}$$

and so the action is nonnilpotent. The completion of $\mathbb{Z}$ is the group of 2-adic integers, $\Gamma_{\mathbb{Z}/2\mathbb{Z}}^\infty \mathbb{Z} = \{0\} = \Gamma\mathbb{Z}$. Notice that $\Gamma_\infty' \mathbb{Z}$ is an uncountable group.

If M is a module over the group π, then we develop another expression for the terms in the lower central series for M. Let $I\pi$ denote the kernel of the augmentation $\varepsilon\colon \mathbb{Z}\pi \to \mathbb{Z}$, given by $\varepsilon(\sum_{g \in \pi} n_g g) = \sum_{g \in \pi} n_g$.

Lemma 8bis.18. $\Gamma_\pi^2 M = I\pi \cdot M$, *that is,* $\Gamma_\pi^2 M$ *is the submodule of M generated by expressions of the form xm with $x \in I\pi$ and $m \in M$.*

PROOF: Since $\Gamma_\pi^2 M$ is the submodule of M generated by $am - m$ with $a \in \pi$ and $m \in M$, we can write

$$am - m = (a - 1)m \in I\pi \cdot M.$$

Thus $\Gamma_\pi^2 M \subset I\pi \cdot M$. To obtain the reverse inclusion, take any element $a = \sum_{g \in \pi} n_g g$ in $\mathbb{Z}\pi$ and consider $\sum_{g \in \pi} n_g g - \sum_{g \in \pi} n_g 1$. This is an element in $I\pi$. For any element m in M we can now write

$$\left(\sum_{g \in \pi} n_g (g - 1) \right) \cdot m = \sum_{g \in \pi} n_g (gm - m) \in \Gamma_\pi^2 M.$$

It remains to show that every element of $I\pi$ has the form $\sum_{g \in \pi} n_g (g - 1)$. To see this, write $a = \sum_{g \in \pi} n_g g = n_1 1 + \sum_{g \in \pi, g \neq 1} n_g g$. When $a \in I\pi$, $\sum_{g \in \pi, g \neq 1} n_g = -n_1$, and so we can write $a = \sum_{g \in \pi, g \neq 1} n_g g - \sum_{g \in \pi, g \neq 1} n_g 1$. Thus $I\pi \cdot M \subset \Gamma_\pi^2 M$. $\square$

Corollary 8bis.19. *If $0 \to K \to M \to Q \to 0$ is a short exact sequence of π-modules and π-module homomorphisms, then M is a nilpotent π-module if and only if K and Q are.*

We next introduce the topological category of interest.

Definition 8bis.20. *A pointed connected CW-complex (X, x_0) is said to be* **nilpotent** *if $\pi_1(X, x_0)$ is a nilpotent group that acts nilpotently on $\pi_n(X, x_0)$ for all $n \geq 2$.*

The category of all nilpotent spaces together with continuous mappings contains all simply-connected spaces, all H-spaces (Corollary 8bis.3), and the Eilenberg-Mac Lane spaces $K(\pi, 1)$ for π nilpotent. As we will soon see, this category is also closed under certain constructions of importance in homotopy theory.

There are spaces that are not nilpotent. Some general constructions are possible: (1) closed surfaces of genus greater than one; (2) the wedge product $X \vee Y$ of any two non-simply connected spaces X and Y; (3) if X has fundamental group π and there is an epimorphism $\pi \to G$ with G finite and G acts nontrivially on the rational homology of the covering space of X corresponding to the kernel of $\pi \to G$, then X is not nilpotent.

We begin our investigation of some of these constructions with a useful analogue of Corollary 8bis.19.

Proposition 8bis.21. *If $F \hookrightarrow E \xrightarrow{p} B$ is a fibration with F connected and E nilpotent, then F is nilpotent.*

PROOF: We can apply Lemma 8bis.18 inductively to show that $\Gamma_\pi^r M = (I\pi)^{r-1} \cdot M$. Suppose that the nilpotence class of $\pi_n(E)$ as a module over $\pi = \pi_1(E)$ is c. Then $\Gamma_\pi^{c+1} \pi_n(E) = \{0\}$, and so $(I\pi)^c \cdot \pi_n(E) = \{0\}$.

Suppose that $\omega \in \pi_1(F)$ and $\alpha \in \pi_n(F)$ for $n \geq 2$. Then $\omega \cdot \alpha = (i_*\omega) \cdot \alpha$ in terms of the $\pi_1(E)$-module structure on $\pi_n(F)$. Suppose that $\omega \in (I\pi)^c$. Since i_* is a $\pi_1(E)$-module homomorphism, $i_*(\omega \cdot \alpha) = (i_*\omega) \cdot (i_*\alpha) = 0$. Thus $\omega \cdot \alpha = \partial\beta$ for some $\beta \in \pi_{n+1}(B)$. Suppose that $\eta \in \pi_1(F)$. Then $i_*\eta \in \pi_1(E)$ and

$$\partial((i_*\eta - 1) \cdot \beta) = (i_*\eta - 1)(\partial\beta) = (i_*\eta - 1)(\omega \cdot \alpha) = (\eta - 1)(\omega \cdot \alpha).$$

However, $\pi_1(E)$ acts on $\pi_{n+1}(B)$ via p_*, and so $(i_*\eta-1)\cdot\beta = (p_*i_*\eta-1)\cdot\beta = -\beta$. Thus $\eta \cdot \omega \cdot \alpha = 0$. This shows that $(I\pi_1(F))^{c+1} \cdot \pi_n(F) = \{0\}$ and so the nilpotence class of $\pi_n(F)$ over $\pi_1(F)$ is less than or equal to $c + 1$.

When $n = 1$ the argument for $\pi_1(F)$ acting on itself by conjugation is similar and left to the reader. $\square$

We next present one of the first results on nilpotent spaces, due to [White-head, GW54] in a paper in which the nilpotence class is related to the Lusternik-Schnirelmann category (for the purposes of proving the Jacobi identity for the Whitehead product!).

Theorem 8bis.22. *Suppose (A, a_0) is a finite, connected, pointed CW-complex and (X, x_0) is a connected, pointed CW-complex. Suppose $f: (A, a_0) \to (X, x_0)$ is a fixed pointed map. In the compact-open topology, the space $\mathrm{map}_f((A, a_0), (X, x_0))$ of pointed maps in the component of f is a nilpotent space. Furthermore, if X is nilpotent, then the space of unpointed maps $\mathrm{map}_f(A, X)$ is nilpotent.*

PROOF: We proceed by induction. We assume that $A^{(0)}$, the 0-skeleton of A, consists of a single point, a_0. Then the theorem is seen to be true for $\mathrm{map}_f((a_0, a_0), (X, x_0))$ and $\mathrm{map}_f(A^{(0)}, X)$.

Consider the cofibration $\bigvee_{i=1}^{k} S^n \to A^{(n)} \to A^{(n+1)}$ that determines the addition of cells to the next skeleton. Apply the functor $\mathrm{map}_f(-, (X, x_0))$ to obtain a fibration (see §4.3):

$$\mathrm{map}_f((A^{(n+1)}, a_0), (X, x_0)) \to \mathrm{map}_{f|}((A^{(n)}, a_0), (X, x_0))$$

$$\to \mathrm{map}_{f|}((\bigvee_{i=1}^{k} S^n, *), (X, x_0)).$$

By induction, we assume that $\mathrm{map}_{f|}((A^{(n)}, a_0), (X, x_0))$ is nilpotent. Proposition 8bis.21 finishes the proof. $\square$

A generalized Whitehead Theorem

When does a mapping $f: X \to Y$ of connected spaces that induces an isomorphism of homology groups induce an isomorphism of homotopy groups here taken as a family of π_1-modules? An answer to this question would generalize the Whitehead theorem (Theorem 4.5) to non-simply connected spaces. The functors introduced to study the action of a group on a module in fact give the key to the answer. The following theorem of [Dror71] helped to establish the category of nilpotent spaces as an appropriate category for homotopy theory.

Theorem 8bis.23 (the generalized Whitehead theorem). *Let $f: X \to Y$ be a map of connected pointed spaces such that $H_*(f): H_*(X) \to H_*(Y)$ is an isomorphism. Then $\pi_*(f)$ is also an isomorphism if, in addition, f satisfies the three conditions:*

(1) $\Gamma^\infty \pi_*(f): \ker(\pi_*(X) \to \widehat{\pi_*(X)}) \to \ker(\pi_*(Y) \to \widehat{\pi_*(Y)})$ *is an epimorphism.*

(2) $\Gamma'_\infty \pi_*(f): \operatorname{coker}(\pi_*(X) \to \widehat{\pi_*(X)}) \to \operatorname{coker}(\pi_*(Y) \to \widehat{\pi_*(Y)})$ *is a monomorphism.*

(3) $\Gamma\pi_*(f): \Gamma\pi_*(X) \to \Gamma\pi_*(Y)$ *is a monomorphism.*

Theorem 8bis.23 fixes the role of the various limit functors. For example, if X and Y are aspherical spaces and $f: X \to Y$ is a map for which $\Gamma\pi_*(f)$ is an isomorphism, then f is a homotopy equivalence. When the functors Γ^∞_π, Γ'_π and Γ all vanish on $\pi_*(X)$, then the completion homomorphism $\pi_*(X) \to \widehat{\pi_*(X)}$ is an isomorphism and the space X is called π-**complete**. When a map between π-complete spaces is a homology isomorphism, it induces an isomorphism of homotopy groups. All nilpotent spaces are π-complete.

Corollary 8bis.24. *If X and Y are connected nilpotent spaces and $f: X \to Y$ is a mapping that induces an isomorphism on integral homology, then f induces an isomorphism of homotopy groups as graded modules over their fundamental groups.*

To prove the theorem we sneak up on it inductively. Let S_n denote the collection of statements:

1_n. $\pi_j(f): \pi_j(X) \to \pi_j(Y)$ is an isomorphism of π_1-modules for $0 \le j \le n-1$.

2_n. $H_n(f): H_n(X) \to H_n(Y)$ is an isomorphism and $H_{n+1}(f)$ is an epimorphism.

3_n. $\Gamma^\infty \pi_n(f): \Gamma^\infty_{\pi_1(X)} \pi_n(X) \to \Gamma^\infty_{\pi_1(Y)} \pi_n(Y)$ is an epimorphism.

4_n. $\Gamma'_\infty \pi_n(f): \Gamma'_\infty \pi_n(X) \to \Gamma'_\infty \pi_n(Y)$ is a monomorphism.

5_n. $\Gamma\pi_n(f): \Gamma\pi_n(X) \to \Gamma\pi_n(Y)$ is a monomorphism.

The condition S_1 holds by the assumptions of Theorem 8bis.23. We claim that S_n implies that $\pi_n(f)$ is an isomorphism and hence that S_{n+1} holds. Proving this claim gives us the theorem. We first prove that, among S_n, the statements 1_n and 2_n imply that $\widehat{\pi_n(f)}: \widehat{\pi_n(X)} \to \widehat{\pi_n(Y)}$ is an isomorphism. This follows from two remarkable lemmas due to [Dror71], the first of which extends Theorem 8bis.16.

Lemma 8bis.25. *Suppose $\phi: M \to M'$ is a homomorphism of modules over a group π. If $H_0(\phi): H_0(\pi, M) \to H_0(\pi, M')$ is an isomorphism and $H_1(\phi): H_1(\pi, M) \to H_1(\pi, M')$ is an epimorphism, then ϕ induces an isomorphism $M/\Gamma_\pi^r M \to M'/\Gamma_\pi^r M'$ for all $r \geq 1$ and hence, an isomorphism $\widehat{\phi}: \widehat{M} \to \widehat{M'}$ of completions.*

PROOF: To the short exact sequence of π-modules

$$0 \to \Gamma_\pi^2 M \to M \to M/\Gamma_\pi^2 M \to 0,$$

there is a long exact sequence of homology groups, ending with

$$H_1(\pi, M) \to H_1(\pi, M/\Gamma_\pi^2 M) \to H_0(\pi, \Gamma_\pi^2 M)$$
$$\to H_0(\pi, M) \to H_0(\pi, M/\Gamma_\pi^2 M).$$

$H_0(\pi, M) = M/\Gamma_\pi^2 M$, and, since $M/\Gamma_\pi^2 M$ has a trivial π-action,

$$H_0(\pi, M/\Gamma_\pi^2 M) \cong H_0(\pi) \otimes M/\Gamma_\pi^2 M \cong \mathbb{Z} \otimes M/\Gamma_\pi^2 M \cong H_0(\pi, M).$$

This shows that the last homomorphism in the exact sequence is the isomorphism of coinvariants $M_\pi \to (M_\pi)_\pi$. The interesting part of the long exact sequence becomes $H_1(\pi, M) \to H_1(\pi, M/\Gamma_\pi^2 M) \to H_0(\pi, \Gamma_\pi^2 M) \to 0$.

Suppose $H_1(\phi): H_1(\pi, M) \to H_1(\pi, M')$ is an epimorphism and that ϕ induces an isomorphism $\phi: M/\Gamma_\pi^2 M \to M'/\Gamma_\pi^2 M'$. Then we have the diagram

$$
\begin{array}{ccccccc}
H_1(\pi, M) & \longrightarrow & H_1(\pi, M/\Gamma_\pi^2 M) & \longrightarrow & \Gamma_\pi^2 M/\Gamma_\pi^2(\Gamma_\pi^2 M) & \longrightarrow & 0 \\
\downarrow & & \cong \downarrow & & \downarrow & & \\
H_1(\pi, M') & \longrightarrow & H_1(\pi, M'/\Gamma_\pi^2 M') & \longrightarrow & \Gamma_\pi^2 M'/\Gamma_\pi^2(\Gamma_\pi^2 M') & \longrightarrow & 0.
\end{array}
$$

The Five-lemma implies an isomorphism $\Gamma_\pi^2 M/\Gamma_\pi^3 M \to \Gamma_\pi^2 M'/\Gamma_\pi^3 M'$. By the same quotient argument as in the proof of Theorem 8bis.16, we see that ϕ induces an isomorphism $M/\Gamma_\pi^3 M \to M'/\Gamma_\pi^3 M'$. The lemma follows by applying the same argument inductively. □

Corollary 8bis.26. *If $\phi\colon M \to M'$ is a π-module homomorphism, M and M' are nilpotent, and ϕ induces an isomorphism $H_0(\phi)$ and an epimorphism $H_1(\phi)$, then M and M' are isomorphic.*

The next lemma provides another step in proving the generalized Whitehead Theorem.

Lemma 8bis.27. *Suppose X is a connected space and $K(\pi_n(X), n) \hookrightarrow P_n X \to P_{n-1} X$ is the nth fibration in the Postnikov tower for X. Then there is an exact sequence, functorial in X, given by*

$$H_{n+2}(P_n X) \to H_{n+2}(P_{n-1}X) \to H_1(\pi_1(X), \pi_n(X)) \to H_{n+1}(P_n X)$$
$$\to H_{n+1}(P_{n-1}X) \to (\pi_n(X))_\pi \to H_n(X) \to H_n(P_{n-1}X) \to 0$$

PROOF: The Leray-Serre spectral sequence for this fibration has E^2-term given by $E^2_{p,q} \cong H_p(P_{n-1}X; \mathcal{H}_q(K(\pi_n(X), n)))$, where the action of $\pi = \pi_1(X)$ on $\pi_n(X)$ determines the local coefficients. Since $H_{n+1}(K(\pi_n(X), n)) = \{0\}$ (a consequence of Lemma 6.2) and $K(\pi_n(X), n)$ is $(n-1)$-connected, we get a lacunary E^2-term in bidegrees $(*, i)$ for $i \leq n+1$—there are only two nonzero stripes in bidegrees $(*, 0)$ and $(*, n)$. As in the derivation of the Gysin sequence (Example 1.D) we get short exact sequences

$$0 \to E^\infty_{n+1,0} \to E^2_{n+1,0} \xrightarrow{d^{n+1}} E^2_{0,n} \to E^\infty_{0,n} \to 0$$
$$0 \to E^\infty_{n+2,0} \to E^2_{n+2,0} \xrightarrow{d^{n+1}} E^2_{1,n} \to E^\infty_{1,n} \to 0$$
$$0 \to E^\infty_{0,n} \to H_n(P_n X) \to E^\infty_{n,0} \to 0$$
$$0 \to E^\infty_{1,n} \to H_{n+1}(P_n X) \to E^\infty_{n+1,0} \to 0$$

Splicing these together we get

$$H_{n+2}(P_n X) \to H_{n+2}(P_{n-1}X) \to H_1(P_{n-1}X, \mathcal{H}_n(K(\pi_n(X), n)))$$
$$\to H_{n+1}(P_n X) \to H_{n+1}(P_{n-1}X) \to H_0(P_{n-1}X; \mathcal{H}_n(K(\pi_n(X), n)))$$
$$\to H_n(P_n X) \to H_n(P_{n-1}X) \to 0$$

However, from Proposition 8bis.4 and Theorem 8bis.10 we know that

$$H_i(P_{n-1}X; \mathcal{H}_n(K(\pi_n, n))) \cong \begin{cases} (\pi_n(X))_\pi, & i = 0, \\ H_1(\pi, \pi_n(X)), & i = 1. \end{cases}$$

By the definition of a Postnikov tower, we have that $H_n(P_n X) = H_n(X)$. Furthermore, $H_n(X) \twoheadrightarrow H_n(P_{n-1}X)$ because $H_n(P_{n-1}X) = E^2_{n,0} = E^\infty_{n,0}$. The lemma follows after we make these substitutions in the exact sequence. $\square$

We now complete the proof of Theorem 8bis.23. Suppose that our map $f: X \to Y$ satisfies the conditions S_n. Then f induces a map of Postnikov towers and by naturality of the short exact sequence of Lemma 8bis.27 we get a morphism of exact sequences

$$H_{n+1}(P_n X) \to H_{n+1}(P_{n-1}X) \dashrightarrow (\pi_n(X))_\pi \to H_n(X) \to H_n(P_{n-1}X) \to 0$$

with vertical maps: $2_n \cdot$ epi, $\cong$, f_*, $2_n \cdot \cong$, $\cong$

$$H_{n+1}(P_n Y) \to H_{n+1}(P_{n-1}Y) \to (\pi_n(Y))_\pi \to H_n(Y) \to H_n(P_{n-1}Y) \to 0.$$

The leftmost horizontal map is seen to be an epimorphism by considering the next stage of the Postnikov tower where we have $H_{n+1}(X) \twoheadrightarrow H_{n+1}(P_n X)$, and similarly for Y. Since $H_{n+1}(f)$ is an epimorphism by 2_n, we get the first vertical epimorphism. By the Five-lemma, $(\pi_n(X))_\pi \to (\pi_n(Y))_\pi$ is an isomorphism. Next consider the other end of the exact sequence:

$$H_{n+2}(P_{n-1}X) \to H_1(\pi_1(X), \pi_n X) \to H_{n+1}(P_n X) \to H_{n+1}(P_{n-1}X) \to \pi_n(X)_\pi$$

with vertical maps: $2_n \cdot \cong$, (blank), $2_n \cdot$ epi, $\cong$, $f_* \cong$

$$H_{n+2}(P_{n-1}Y) \to H_1(\pi_1(Y), \pi_n Y) \to H_{n+1}(P_n Y) \to H_{n+1}(P_{n-1}Y) \to \pi_n(Y)_\pi.$$

The Five-lemma implies that $H_1(\pi_1(X), \pi_n(X)) \to H_1(\pi_1(Y), \pi_n(Y))$ is an epimorphism. By Lemma 8bis.25 we have that $\pi_n(f)$ induces an isomorphism between $\pi_n(X)/\Gamma_\pi^r \pi_n(X)$ and $\pi_n(Y)/\Gamma_\pi^r \pi_n(Y)$ for all r and hence induces an isomorphism $\widehat{\pi_n(f)}$. Finally we use the remaining conditions of S_n.

There are exact sequences of functors given by

$$0 \to \Gamma^\infty \pi_n \to \pi_n \to \widehat{\pi_n} \to \Gamma'_\infty \pi_n \to 0$$

$$0 \to \Gamma \pi_n \to \pi_n \to \pi_n/\Gamma \pi_n \to 0.$$

The Five-lemma and conditions 3_n, 4_n, and 5_n for $\pi_n(f)$ imply that $\pi_n(f)$ is an epimorphism.

To prove that $\pi_n(f)$ is a monomorphism, we use 5_n, that is, $\Gamma \pi_n(f)$ is a monomorphism. We only need to show that $\pi_n X/\Gamma \pi_n(X) \to \pi_n(Y)/\Gamma \pi_n(Y)$ is a monomorphism. The lower central series has the property that $\Gamma_\pi^r M \subset \Gamma_\pi^{r-1} M$ is always strictly decreasing until it becomes stable. This is because $\Gamma_\pi^r M = \Gamma_\pi^2(\Gamma_\pi^{r-1}M)$. We also know that $\Gamma M \subset \Gamma_\pi^r M$ for all r. In fact, this inclusion extends to r, any transfinite ordinal, as follows: If $\alpha = \beta + 1$ are ordinals, then let $\Gamma_\pi^\alpha M = \Gamma_\pi^2(\Gamma_\pi^\beta M)$; if α is a limit ordinal, let $\Gamma_\pi^\alpha M = \bigcap_{\beta < \alpha} \Gamma_\pi^\beta M$. It still follows that $\Gamma M \subset \Gamma_\pi^\alpha M$ for all ordinals α. But the lower central series always decreases so $\Gamma M = \Gamma_\pi^\gamma M$ for some ordinal γ. We have shown already that $\pi_n(X)/\Gamma_\pi^r \pi_n(X) \to \pi_n(Y)/\Gamma_\pi^r \pi_n(Y)$ is an isomorphism for finite r. Introducing the limit ordinals, we get an isomorphism for $r = \omega$ and the argument of Lemma $8^{bis}.25$ works for the higher ordinals. Thus, $\pi_n(f)$ induces an isomorphism $\pi_n(X)/\Gamma \pi_n(X) \to \pi_n(Y)/\Gamma \pi_n(Y)$ and so, by the Five-lemma, $\pi_n(f)$ is a monomorphism. $\square$

A characterization of nilpotent spaces

In Chapter 4 (Theorem 4.35) we constructed the Postnikov tower of a space and stated that, for simply-connected spaces, the fibrations in the tower could be taken to be **principal**, that is, each $p_n \colon P_n X \to P_{n-1} X$ is a pullback of the path-loop fibration over the Eilenberg-Mac Lane space $K(\pi_n(X), n+1)$ via a k-invariant, $k^n \colon P_{n-1} X \to K(\pi_n(X), n+1)$:

$$
\begin{array}{ccc}
K(\pi_n(X), n) & = \!\!=\!\!= & K(\pi_n(X), n) \\
\downarrow & & \downarrow \\
P_n X & \longrightarrow & PK(\pi_n(X), n+1) \\
\scriptstyle p_n \downarrow & & \downarrow \\
P_{n-1} X & \overset{k^n}{\longrightarrow} & K(\pi_n(X), n+1).
\end{array}
$$

We next give a proof of this property of simply-connected spaces and generalize it to nilpotent spaces.

Lemma $8^{bis}.28$. *Let A be a finitely generated abelian group and let E and B be spaces of finite type. A fibration $K(A, n) \hookrightarrow E \overset{p}{\to} B$ is principal if and only if it is simple, that is, the action of $\pi_1(B)$ on $K(A, n)$ is trivial.*

PROOF: Let's assume that $p \colon E \to B$ is principal and it is pulled back over a classifying map $\theta \colon B \to K(A, n+1)$. The relevant part of the long exact sequence of homotopy groups may be written

$$
\begin{array}{ccccccccc}
0 \to \pi_{n+1}(E) & \overset{p_*}{\to} & \pi_{n+1}(B) & \overset{\tau}{\longrightarrow} & \pi_n(K(A, n)) & \longrightarrow & \pi_n(E) & \overset{p_*}{\to} & \pi_n(B) \to 0 \\
= \downarrow & & = \downarrow & & = \downarrow & & = \downarrow & & = \downarrow \\
0 \to \pi_{n+1}(E) & \underset{p_*}{\to} & \pi_{n+1}(B) & \underset{\theta_*}{\to} & \pi_{n+1}(K(A, n+1)) & \to & \pi_n(E) & \to & \pi_n(B) \to 0.
\end{array}
$$

The action of $\pi_1(B)$ on A can be identified in the second row with the action of the fundamental group of the total space of the fibration θ on the base space $K(A, n+1)$. But this factors through the action of the fundamental group of $K(A, n+1)$, which is trivial. Hence, the fibration is simple.

Suppose next that $\pi_1(B)$ acts trivially on $A = H_n(K(A, n))$. Consider the cohomology Leray-Serre spectral sequence for the fibration with coefficients in the abelian group A. Then, $E_2^{0,n} \cong H^n(K(A, n); A)$ contains the fundamental class ι corresponding to the identity map on $K(A, n)$. Since $K(A, n)$ is $(n-1)$-connected, the first differential to arise on ι is the transgression d_{n+1}, and this gives a class $d_{n+1}(\iota) = [\theta] \in H^{n+1}(B; A)$; we can form the pullback over $\theta \colon B \to K(A, n+1)$. This produces a space E_θ together with a mapping $g \colon E \to E_\theta$. Checking the long exact sequence of homotopy groups, g induces an isomorphism on homotopy, and so, in the category of spaces of the homotopy type of CW-complexes of finite type, g is a homotopy equivalence, and p is a principal fibration. $\qquad\square$

It follows immediately from the lemma that a simply-connected space X has a Postnikov tower of principal fibrations. For an arbitrary space X, let $\{P_n X, p_n, f_n\}$ denote its Postnikov tower. We say that $p_n \colon P_n X \to P_{n-1}X$ **admits a principal refinement** if there is a sequence of principal fibrations

$$P_n X = P_{n,c}X \xrightarrow{q_c} P_{n,c-1}X \xrightarrow{q_{c-1}} \cdots \xrightarrow{q_3} P_{n,2}X \xrightarrow{q_2} P_{n,1}X = P_{n-1}X$$

with $p_n = q_2 \circ q_3 \circ \cdots \circ q_c$. With this extension of the notion of a principal fibration, we can now give a characterization of nilpotent spaces.

Theorem 8bis.29. *A space X is nilpotent if and only if every stage of its Postnikov tower admits a principal refinement.*

PROOF: Since each q_j is a principal fibration, we can write its classifying map as $\theta_{n,j} \colon P_{n,j}X \to K(A_{n,j}, n+1)$. We proceed by induction. By the properties of a Postnikov tower, $\pi_n(P_{n,1}X) = \pi_n(P_{n-1}X) = \{0\}$ and so $\pi_1(X)$ acts trivially (hence nilpotently) on $\pi_n(P_{n,1}X)$. Suppose that $\pi_1(X)$ acts nilpotently on $\pi_n(P_{n,j-1}X)$ of nilpotency class $\leq j - 1$. View the k-invariant $\theta_{n,j}$ as a fibration (up to homotopy) and $q_j \colon P_{n,j}X \to P_{n,j-1}X$ as the inclusion of the fibre. By Proposition 8bis.21, $\pi_1(X)$ acts nilpotently on $\pi_n(P_{n,j}X)$ of class $\leq j$. By induction, $\pi_1(X)$ acts nilpotently on $\pi_n(P_{n,c}X) \cong \pi_n(P_n X) \cong \pi_n(X)$.

Suppose that X is a nilpotent space and $\pi = \pi_1(B)$. The lower central series for $\pi_n(X)$ as a π-module has the form

$$\{0\} \subset \Gamma_\pi^c \pi_n(X) \subset \Gamma_\pi^{c-1} \pi_n(X) \subset \cdots \subset \Gamma_\pi^2 \pi_n(X) \subset \pi_n(X).$$

By construction each quotient $\Gamma_\pi^t \pi_n(X)/\Gamma_\pi^{t+1} \pi_n(X)$ is a trivial π-module. Consider the fibration $p_n \colon P_n X \to P_{n-1}X$. The homology Leray-Serre spectral sequence (Lemma 8bis.27) for this fibration gives the exact sequence

$$H_{n+1}(P_n X) \xrightarrow{p_{n*}} H_{n+1}(P_{n-1}X) \xrightarrow{d^{n+1}} \pi_n(X)/\Gamma_\pi^2 \pi_n(X) \to \cdots.$$

Consider the cohomology Leray-Serre spectral sequence with coefficients in $(\pi_n X)_\pi = \pi_n(X)/\Gamma_\pi^2 \pi_n(X)$ for which

$$E_2^{p,q} \cong H^p(P_{n-1}X; \mathcal{H}^q(K(\pi_n(X), n); (\pi_n(X))_\pi)).$$

There is a class $\jmath \in H^n(K(\pi_n(X), n); (\pi_n(X))_\pi)$ that represents the quotient $\pi_n(X) \to (\pi_n(X))_\pi$. This class transgresses to a class $[l_2]$ lying in $H^{n+1}(P_{n-1}X; (\pi_n(X))_\pi)$ which represents $H_{n+1}(P_{n-1}X) \xrightarrow{d^{n+1}} (\pi_n(X))_\pi$ and for which we take a representative $l_2: P_{n-1}X \to K((\pi_n(X))_\pi, n+1)$. Let $q_2: P_{n,2}X \to P_{n-1}X$ be the pullback of the path-loop fibration over l_2 and let $u_2: P_n X \to P_{n,2}X$ be a lifting of p_n through $P_{n,2}X$. Such a lifting exists because $l_2 \circ p_n \simeq *$.

We can modify u_2 to be a fibration and consider a portion of the homotopy exact sequences

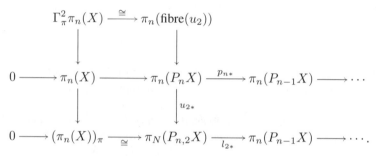

From this diagram we see that the fibre of u_2 is $K(\Gamma_\pi^2 \pi_n(X), n)$. If we repeat this construction with u_2 replacing p_n, then we get a space $P_{n,3}X$ together with a principal fibration $q_3: P_{n,3}X \to P_{n,2}X$. Continuing in this way, if X is nilpotent, we eventually get to $\Gamma_\pi^{c+1} \pi_n(X) = \{0\}$ and the process stops with $u_c = q_c$ and p_n refined by principal fibrations.   □

The sequence of k-invariants that a tower of principal fibrations admits may be applied to many problems in classical homotopy theory. For example, the k-invariants are the data for classical obstruction arguments. Another application was introduced by [Sullivan71] in his work on the Adams conjecture. [Serre53] showed, in his development of classes of abelian groups, that homotopy theory can become simpler when viewed one prime at a time. Making this notion topological rather than algebraic is the goal of localization at a prime. To localize a space X at a prime p, first consider the **ring of integers localized at the prime** p, denoted $\mathbb{Z}_p$, and given by the subring of $\mathbb{Q}$ of fractions a/b with b relatively prime to p. The functor on abelian groups, $A \mapsto A \otimes \mathbb{Z}_p$, is called **localization** at the prime p; it eliminates all torsion prime to p and so leaves only the p-primary data. This functor can be extended to spaces by modifying the refinement of the Postnikov tower by composing the classifying maps $\theta_{n,j}$ with the mapping induced by the localization, $K(A_{n,j}, n) \to K(A_{n,j} \otimes \mathbb{Z}_p, n)$, and then pulling back carefully. The resulting space X_p has homotopy groups $\pi_n(X_p) \cong \pi_n(X) \otimes \mathbb{Z}_p$ and integral homology groups $H_n(X_p) \cong H_n(X) \otimes \mathbb{Z}_p$.

Later in the chapter, we will present an alternate construction of the localization of a space, due to [Bousfield-Kan72] and carried out simplicially.

Convergence of the Eilenberg-Moore spectral sequence Ⓝ

Theorem 8bis.23, the generalized Whitehead theorem, illustrates how the nilpotence condition can control the effect of the fundamental group. The relations between the homotopy groups of a space and their nilpotent completions provide the data for measuring the departure from the simply-connected case of the Whitehead theorem. Another naive situation in which simple connectivity plays a role is the convergence of the Eilenberg-Moore spectral sequence. The goal of this section is to prove the following result of [Dwyer74] that shows how the nilpotence of a certain action of the fundamental group is decisive in generalizing the naive convergence criterion.

Theorem 8bis.30. *Suppose $F \hookrightarrow E \xrightarrow{p} B$ is a fibration with all spaces connected, and A is an abelian group. Then the Eilenberg-Moore spectral sequence for the fibre of p converges strongly to $H_*(F; A)$ if and only if $\pi_1(B)$ acts nilpotently on $H_i(F; A)$ for all $i \geq 0$.*

Following [Rector71] (§8.3) we associate to the pullback data $X \xrightarrow{f} B \xleftarrow{p} E$ the cosimplicial space (the geometric cobar construction) $G^\bullet(X, B, E)$ where $G^n(X, B, E) = X \times B^{\times n} \times E$ for $n \geq 0$ and with coface and codegeneracy maps given by

$$d^i(x, b_1, \ldots, b_n, e) = \begin{cases} (x, f(x), b_1, \ldots, b_n, e) & i = 0, \\ (x, b_1, \ldots, b_i, b_i, \ldots, b_n, e) & 1 \leq i \leq n, \\ (x, b_1, \ldots, b_n, p(e), e) & i = n+1. \end{cases}$$

$$s^j(x, b_1, \ldots, b_n, e) = (x, b_1, \ldots, \widehat{b_{i+1}}, \ldots, b_n, e), \qquad 0 \leq i \leq n-1.$$

In this discussion we take all spaces involved to be simplicial sets. Thus $G^\bullet(X, B, E)$ is a cosimplicial simplicial set. We explore the combinatorial structure of such an object in what follows.

Let A denote an abelian group and X, a simplicial set (§4.2). Then we define the simplicial abelian group $A \otimes X$ by $(A \otimes X)_n = \bigoplus_{x \in X_n} A$, for $n \geq 0$, with face and degeneracy maps induced by the maps on the generators and extended to be A-linear. It follows that $\pi_*(A \otimes X) \cong H_*(X; A)$ and problems concerning homology become open to homotopy methods.

In homological algebra, the basic datum of a resolution of a module M is the augmentation $F_\bullet \xrightarrow{\varepsilon} M \to 0$. We can view a cosimplicial space $Y^\bullet$ as a kind of resolution (for example, when constructed from a triple; [Bousfield-Kan72, I,§5]). We consider all possible augmentations of $Y^\bullet$, that is, maps $\epsilon\colon Z \to Y^0$ satisfying $d^0 \circ \epsilon = d^1 \circ \epsilon$. The **maximal augmentation** associated to $Y^\bullet$ is the subspace $aY^\bullet$ of Y^0 that gives the equalizer (as simplicial sets) of the coface mappings $d^0, d^1\colon Y^0 \to Y^1$. In detail, the space $aY^\bullet$ is given by $aY^\bullet = \{y \in Y^0 \mid d^0(y) = d^1(y)\}$. The maximal augmentation has the following characterization in the category of cosimplicial spaces.

Lemma 8bis.31. *The maximal augmentation $aY^\bullet$ of a cosimplicial space $Y^\bullet$ is the simplicial set* $\mathbf{CoSimp}(*, Y^\bullet)$ *of cosimplicial maps from the constant cosimplicial space* $*$.

We leave the proof of the lemma to the reader. The Hom-set of cosimplicial maps between $X^\bullet$ and $Y^\bullet$, $\mathbf{CoSimp}(X^\bullet, Y^\bullet)$, has the structure of a simplicial set with n-simplices given by the cosimplicial maps $\Delta[n] \times X^\bullet \to Y^\bullet$. Here $\Delta[n]_\bullet$ denotes the standard simplicial n-simplex, whose s-simplices are given by
$$\Delta[n]_s = \{\langle x_0, x_1, \dots, x_s\rangle \mid 0 \le x_0 \le x_1 \le \cdots \le x_s \le n\}.$$
The face and degeneracy maps on $\mathbf{CoSimp}(X^\bullet, Y^\bullet)$ are induced by the standard maps. The inclusions $\varepsilon_i\colon \Delta[n] \to \Delta[n+1]$ are given by $\varepsilon_i(\langle x_0, x_1, \dots, x_s\rangle) = \langle X_0, X_1, \dots, X_s\rangle$, where $X_j = x_j$, if $j < i$, and $X_j = x_j + 1$, if $j \ge i$. The face mapping is given by $d_i\colon \Delta[n] \times X^\bullet \xrightarrow{\varepsilon_i \times 1} \Delta[n+1] \times X^\bullet \to Y^\bullet$. The degeneracy maps are defined by the combinatorial collapse onto the jth face, namely $\eta_j\colon \Delta[n] \to \Delta[n-1]$, given by $\eta_j(\langle x_0, x_1, \dots, x_s\rangle) = \langle X_0, X_1, \dots, X_s\rangle$, where $X_l = x_l$, if $l < j$, and $X_l = x_l - 1$, if $l \ge j$. Thus $s_j\colon \Delta[n] \times X^\bullet \xrightarrow{\eta_j \times 1} \Delta[n-1] \times X^\bullet \to Y^\bullet$.

A desirable property of resolutions is homotopy invariance. For cosimplicial spaces, we want a similar property—if $f\colon Y^\bullet \to Z^\bullet$ is a morphism of cosimplicial spaces that satisfies the condition that $f\colon Y^n \to Z^n$ is a homotopy equivalence of simplicial sets for all n, then f ought to induce a homotopy equivalence of maximal augmentations. However, this is too much to ask for. The fix for this desideratum is to replace the construction of the maximal augmentation with one that is more robust homotopically.

Definition 8bis.32. *Given a cosimplicial space $Y^\bullet$, let* $\mathrm{Tot}(Y^\bullet)$ *denote the simplicial set* $\mathbf{CoSimp}(\Delta^\bullet, Y^\bullet)$ *where $\Delta^\bullet$ denotes the cosimplicial space with $\Delta[n]$ at level n and coface and codegeneracy mappings induced by the canonical face inclusions, ε_i, and projections, η_j, respectively.*

This functor was introduced by [Bousfield-Kan72] and forms the basis for their study of localization and completion. $\mathrm{Tot}(Y^\bullet)$ can be built up canonically from a tower of fibrations. Let $\Delta^{\bullet(s)}$ denote the s-skeleton of the cosimplicial space $\Delta^\bullet$, that is, at level n, one takes the s-skeleton of $\Delta[n]$. Define

$\mathrm{Tot}_s(Y^\bullet) = \mathbf{CoSimp}(\Delta^{\bullet(s)}, Y^\bullet)$. The cofibrations $\Delta^{\bullet(s)} \to \Delta^{\bullet(s+1)}$ induce fibrations $\mathrm{Tot}_{s+1}(Y^\bullet) \to \mathrm{Tot}_s(Y^\bullet)$, whose inverse limit is $\mathrm{Tot}_\infty(Y^\bullet) = \mathrm{Tot}(Y^\bullet)$. Notice that $\mathrm{Tot}_0(Y^\bullet) = Y^0$, and if $\epsilon\colon Z \to Y^\bullet$ is any augmentation, then ϵ induces a mapping $Z \to \mathrm{Tot}_s(Y^\bullet)$, for all $s \leq \infty$.

A tower of fibrations gives rise to an exact couple based on the long exact sequences of homotopy groups. The E^1-term is determined by the homotopy groups of the fibres of $\mathrm{Tot}_s \to \mathrm{Tot}_{s-1}$. A typical fibre takes the form $\Omega^s((NY^\bullet)^s)$ where $(NY^\bullet)^s$ may be written as $Y^s \cap \ker s^0 \cap \cdots \cap \ker s^{s-1}$ when the simplicial sets at each level of $Y^\bullet$ are **fibrant** (that is, $Y^n \to *$ is a fibration for all n). It follows from the grading for the exact couple that this is a second quadrant spectral sequence. There are general conditions for its strong convergence to $\pi_*(\mathrm{Tot}(Y^\bullet))$ (see [Bousfield-Kan72, IX, §5]). We will obtain the Eilenberg-Moore spectral sequence in this manner by taking the homotopy spectral sequence associated to the tower of fibrations $\{\mathrm{Tot}_s(A \otimes G^\bullet(X, B, E))\}$.

In the category of cosimplicial spaces we find the usual notions of homotopy theory such as fibrations, cofibrations, and homotopy equivalences. The case of interest is the following diagram depicting a fibration of cosimplicial spaces along with an augmenting fibration of spaces:

$$
\begin{array}{ccc}
F & \xrightarrow{\ \epsilon\ } & G^\bullet(*, B, E) \\
\downarrow & & \downarrow \\
E & \xrightarrow{\ \epsilon\ } & G^\bullet(B, B, E) \\
{\scriptstyle p}\downarrow & & \downarrow{\scriptstyle q} \\
B & \xrightarrow{\quad = \quad} & B.
\end{array}
$$

Here B denotes the constant cosimplicial space with B at all levels and the identity map for all coface and codegeneracy maps. The maps for $G^\bullet(B, B, E)$ are given by $\mathrm{id}\colon B \to B \leftarrow E\colon p$. The mapping q is given by first projection off the product $B \times B^{\times n} \times E$. Thus, at each level, we have a trivial fibration and so $\pi_1(B)$ acts trivially on each fibre $G^n(*, B, E)$. We next show that the action of $\pi_1(B)$ on $H_i(F; A)$ is compatible via the augmentation with this trivial action.

Proposition 8bis.33. *The augmentation map $\epsilon\colon F \to * \times E = G^0(*, B, E)$ induces a $\pi_1(B)$-equivariant homomorphism $\epsilon_*\colon H_*(F; A) \to H_*(E; A)$, where $\pi_1(B)$ acts trivially on $H_*(E; A)$*

PROOF: We argue with spaces and lifting functions as in §4.3. The simplicial versions of these structures can be found in [May67]. The pullback spaces for the fibrations p and q are given by $\Omega_p = \{(\lambda, e) \mid \lambda \in WB, e \in E, \lambda(0) = p(e)\}$, and $\Omega_q = \{(\lambda, b, e) \mid \lambda \in WB, (b, e) \in B \times E, \lambda(0) = b\}$. The augmentation maps induce a mapping between these pullbacks $\Omega_p \to \Omega_q$, given explicitly by $(\lambda, e) \mapsto (\lambda, p(e), e)$. This gives rise to the diagram

$$
\begin{array}{ccccccc}
\Omega B \times F & \longleftarrow & \Omega_p & \overset{\Lambda}{\longrightarrow} & WE & \overset{\mathrm{ev}_1}{\longrightarrow} & F \\
\downarrow & & \downarrow & & \downarrow & & \downarrow \\
\Omega B \times (* \times E) & \longleftarrow & \Omega_q & \underset{\Lambda'}{\longrightarrow} & W(B \times E) & \underset{\mathrm{ev}_1}{\longrightarrow} & (* \times E).
\end{array}
$$

As described in §4.3, the lifting function for the trivial fibration is given by $\Lambda'(\lambda, b, e) = (\lambda, c_e)$, where c_e denotes the constant path at e. Since the action of $\pi_1(B)$ is induced by these composites, compatibility of the actions is equivalent to the homotopy commutativity of this diagram. Let $H: \Omega B \times F \times I \to (* \times E)$ be given by $H((\omega, y), t) = (*, \Lambda(\omega, y)(t))$. Then H makes the leftmost square commute up to homotopy and so proves the proposition. $\qquad\square$

The fibration of cosimplicial spaces $G^\bullet(*, B, E) \to G^\bullet(B, B, E) \to B$ provides control of the $\pi_1(B)$-action in the tower of fibrations that give rise to the Eilenberg-Moore spectral sequence.

Lemma 8bis.34. *For all $i \geq 1$, $\pi_i(\mathrm{Tot}_s(A \otimes G^\bullet(*, B, E)))$ is a nilpotent $\pi_1(B)$-module.*

PROOF: We prove this by induction over s. When $s = 0$, we have the trivial fibration $E \to B \times E \to B$ that describes the 0-level of the fibration of cosimplicial spaces. Thus $\pi_1(B)$ acts trivially on $\pi_i(\mathrm{Tot}_0(A \otimes G^\bullet(*, B, E)))$.

By induction we consider the fibration

$$
\mathrm{Tot}_n(A \otimes G^\bullet(*, B, E)) \to \mathrm{Tot}_{n-1}(A \otimes G^\bullet(*, B, E)).
$$

[Bousfield-Kan72, X, §6] give an explicit expression for the fibre of this fibration from which we deduce its structure as a $\pi_1(B)$-module. To wit, the fibre of $\mathrm{Tot}_n(Y^\bullet) \to \mathrm{Tot}_{n-1}(Y^\bullet)$, for any cosimplicial space $Y^\bullet$, is given by the function space $\mathrm{Hom}((S^n, *), (NY^n, *))$ where $NY^n = \ker(Y^n \overset{s}{\to} M^{n-1}Y^\bullet)$ and $M^{n-1}Y^\bullet$ is the $(n-1)^{st}$ *matching space* consisting of simplices in $(Y^{n-1})^{\times n}$, written $(x^0, \ldots, x^{n-1})$, that satisfy $s^i x^j = s^{j-1} x^i$ whenever $0 \leq i < j \leq n-1$. The mapping $s: Y^n \to M^{n-1}Y^\bullet$ is given by $y \mapsto (s^0 y, \ldots, s^{n-1} y)$. In the case of a cosimplicial simplicial abelian group, the homotopy groups of the fibre may be written

$$
\pi_i(\mathrm{fibre}(\mathrm{Tot}_n(Y^\bullet) \to \mathrm{Tot}_{n-1}(Y^\bullet))) \cong \pi_{i+n}(NY^n)
$$
$$
= \pi_{i+n}(Y^n) \cap \ker s^0 \cap \cdots \cap \ker s^{n-1}.
$$

When $Y^\bullet = A \otimes G^\bullet(*, B, E)$, the homotopy groups of the fibre have a $\pi_1(B)$-action inherited from the inclusion into $A \otimes G^n(*, B, E)$. However,

$\pi_{i+n}(A \otimes G^n(*, B, E))$ is a trivial $\pi_1(B)$-module, and so then are the homotopy groups of the fibre of $\mathrm{Tot}_n(Y^\bullet) \to \mathrm{Tot}_{n-1}(Y^\bullet)$. By induction, we assume that the groups $\pi_j(\mathrm{Tot}_{n-1}(A \otimes G(*, B, E)))$ are nilpotent $\pi_1(B)$-modules. The long exact sequence of homotopy groups for the fibration $\mathrm{Tot}_n(Y^\bullet) \to \mathrm{Tot}_{n-1}(Y^\bullet)$ and the triviality of the $\pi_1(B)$-action on the homotopy groups of the fibre complete the induction. $\qquad\square$

From the lemma we can deduce half of the proof of Theorem 8bis.30. Suppose that the spectral sequence converges strongly to $H_*(F; A)$. Then there is a filtration of $H_i(F; A)$ for each i with $E^\infty_{p,i-p}$ isomorphic to the associated graded group to this filtration. Strong convergence implies that the direct limit of the sequence

$$\cdots \to \pi_i(A \otimes \mathrm{Tot}_s(A \otimes G^\bullet(*, B, E))) \to \pi_i(\mathrm{Tot}_{s-1}(A \otimes G^\bullet(*, B, E))) \to \cdots$$

vanishes and so there is an injection

$$E^\infty_{p,*} \to R_p = \bigcap_r \mathrm{im}(\pi_*(\mathrm{Tot}_{p+r}(A \otimes G^\bullet) \to \mathrm{Tot}_p(A \otimes G^\bullet))$$
$$\subset \pi_*(\mathrm{Tot}_p(A \otimes G^\bullet(*, B, E))).$$

It follows that each $E^\infty_{p,*}$ is a nilpotent $\pi_1(B)$-module. Strong convergence also implies that the nonzero $\pi_1(B)$-modules $E^\infty_{p,i-p}$ are finite in number. Arguing inductively using Corollary 8bis.19 we have proved that $H_i(F; A)$ is a nilpotent $\pi_1(B)$-module.

To prove the other half of Theorem 8bis.30 we use the towers of fibrations that arise from the application of the functor Tot_s to the cosimplicial fibration $G^\bullet(*, B, E) \to G^\bullet(B, B, E) \to B$. The augmentation from the fibration p may be depicted in the diagram:

$$F \xrightarrow{\epsilon} \mathrm{Tot}_0\, G^\bullet(*, B, E) \longleftarrow \mathrm{Tot}_1\, G^\bullet(*, B, E) \longleftarrow \cdots \longleftarrow \mathrm{Tot}\, G^\bullet(*, B, E)$$

$$E \xrightarrow{\epsilon} \mathrm{Tot}_0\, G^\bullet(B, B, E) \longleftarrow \mathrm{Tot}_1\, G^\bullet(B, B, E) \longleftarrow \cdots \longleftarrow \mathrm{Tot}\, G^\bullet(B, B, E)$$

$$B \xrightarrow{\ =\ } B \xleftarrow{\ =\ } B \xleftarrow{\ =\ } \cdots \longleftarrow_{=} B.$$

Lemma 8bis.35. $\mathrm{Tot}(G^\bullet(B, B, E)) \simeq E$.

PROOF: The projection off the last coordinate $G^n(B, B, E) \to E$ provides an inverse to the augmentation $E \to G^\bullet(B, B, E)$. $\qquad\square$

It follows that we can compare the augmentation fibration with the limit fibration. The nilpotency condition plays a role in the following proposition that is a form of the Zeeman comparison theorem. The proposition was known in the early 1970's—it is stated explicitly by [Hilton-Roitberg76].

Proposition 8bis.36. *Suppose $F \hookrightarrow E \to B$ and $F' \hookrightarrow E' \to B$ are fibrations with B connected, and $f \colon E \to E'$ is a map over B inducing an isomorphism on homology. If $\pi_1(B)$ acts nilpotently on $H_i(F)$ and on $H_i(F')$, for all i, then $f| \colon F \to F'$ induces an isomorphism on homology.*

PROOF: We proceed by induction on the degree i of $H_i(f|)$. In the case $i = 0$, $H_0(f) = H_0(f|)$ because B is connected.

Suppose $H_i(f|)$ is an isomorphism for $0 \le i \le n-1$. This implies that the E^2-terms of the associated Leray-Serre spectral sequences are isomorphic in bidegrees $(*, i)$ for $i \le n-1$. We consider the morphism of spectral sequences in bidegrees $(0, n)$ and $(1, n)$, where we have $E^2_{0,n} \cong H_0(B, \mathcal{H}_n(F)) \to H_0(B, \mathcal{H}_n(F')) \cong E'^2_{0,n}$. By Proposition 8bis.4,

$$H_0(B, \mathcal{H}_n(F)) \cong H_0(\pi, H_n(F)) = (H_n(F))_\pi,$$

where $\pi = \pi_1(B)$. By Theorem 8bis.11, $E^2_{1,n} \cong H_1(\pi, H_n(F))$. On the vertical edge of the spectral sequence the map of spectral sequences gives

$$
\begin{array}{ccccccc}
E^2_{0,n} & \longrightarrow & E^3_{0,n} & \longrightarrow & \cdots & \longrightarrow & E^{n+1}_{0,n} = E^\infty_{0,n} \\
\downarrow & & \downarrow & & & & \cong \downarrow \\
E'^2_{0,n} & \longrightarrow & E'^3_{0,n} & \longrightarrow & \cdots & \longrightarrow & E'^{n+1}_{0,n} = E'^\infty_{0,n}.
\end{array}
$$

Since the E^2-terms are isomorphic in bidegrees $(*, i)$ for $i \le n-1$, the differentials arising to make the successive epimorphisms along the vertical edges are the same in each spectral sequence and so we conclude that $H_0(\pi, H_n(F))$ is isomorphic to $H_0(\pi, H_n(F'))$ via $H_n(f|)$. Similarly, we find that $H_1(\pi, H_n(F))$ maps onto $H_1(\pi, H_n(F'))$ via $H_n(f|)$. Theorem 8bis.16 implies that $H_n(F)$ is isomorphic to $H_n(F')$, and the inductive step follows. $\square$

The second half of the proof of the Theorem 8bis.30 follows because the homotopy spectral sequence for the tower of fibration $\{\mathrm{Tot}_s(A \otimes G^\bullet(*, B, E))\}$ converges to $\pi_*(\mathrm{Tot}(A \otimes G^\bullet(*, B, E)))$. Proposition 8bis.36 implies that $\pi_*(\mathrm{Tot}(A \otimes G^\bullet(*, B, E))) \cong H_*(F; A)$.

Theorem 8bis.30 has been extended to connective generalized homology theories ([Bousfield87]), nonconnected bases B ([Dror-Farjoun–Smith, J90], a useful case when dealing with function spaces) and to pullback fibre squares with data $X \xrightarrow{f} B \xleftarrow{p} Y$ for which the set $\pi_0(X) \times_{\pi_0(B)} \pi_0(Y)$ is finite and, for all $y \in Y$, $\pi_1(B, p(y))$ acts nilpotently on $H_*((Y)_y)$ where $(Y)_y$ denotes the component of Y containing y ([Shipley96]).

The development of convergence criteria for the Eilenberg-Moore spectral sequence is, in fact, a spinoff of the investigation of the general convergence properties of the **Bousfield-Kan spectral sequence**.

Theorem 8bis.37. *Given a fibrant, pointed, cosimplicial space $Y^\bullet$, there is a spectral sequence associated to the tower of fibrations $\{\operatorname{Tot}_n(Y^\bullet) \to \operatorname{Tot}_{n-1}(Y^\bullet)\}$ with*

$$E_1^{s,t}(Y^\bullet) \cong \pi_t(Y^s) \cap \ker s^0 \cap \cdots \cap \ker s^{s-1}, \qquad t \geq s \geq 0$$

and converging under favorable conditions to $\pi_*(\operatorname{Tot}(Y^\bullet))$.

General results indicating favorable conditions were obtained by [Bousfield87], [Shipley96], and [Goodwillie98]. The fundamental example introduced by [Bousfield-Kan72] is the cosimplicial space associated to the completion of a space with respect to a ring R.

The R-**completion** of a pointed space (X, x_0) is obtained by applying the totalization functor, Tot, to the cosimplicial space $R^\bullet X$ obtained from the triple $\{R, \phi, \psi\}$ as follows: If (X, x_0) is a pointed simplicial set, then define the simplicial R-module RX by $(RX)_n = R \otimes X_n / R \otimes x_0$. The natural transformation $\phi_X \colon X \to RX$ is defined by $x \mapsto [1 \otimes x]$, and the natural transformation $\psi_X \colon R^2 X \to RX$ is given by $[r \otimes [s \otimes x]] \mapsto [rs \otimes x]$. The R-completion of X is defined by

$$R_\infty X = \operatorname{Tot}(R^\bullet X) = \operatorname{Hom}(\Delta^\bullet, R^\bullet X),$$

where $R^k X = R(R^{k-1} X)$ and $R^0 X = RX$. The cosimplicial structure is based on the natural transformations ϕ and ψ, with the coface and codegeneracy maps given by

$$d^i \colon R^k X \to R^{k+1} X, \quad d^i = R^i(\phi_{R^{k-i}X}),$$
$$s^j \colon R^k X \to R^{k-1} X, \quad s^j = R^j(\psi_{R^{k-j}X}).$$

It follows from the properties of Tot that $R_\infty X$ is the inverse limit of a tower of fibrations $R_s X \to R_{s-1} X$ where $R_s X = \operatorname{Tot}_s(R^\bullet X)$. This tower of fibrations is augmented by a family of mappings $f_s \colon X \to R_s X$ and it leads to the spectral sequence of [Bousfield-Kan72].

When R is a subring of $\mathbb{Q}$, then, one can prove that, for some set P of primes,

$$R = \mathbb{Z}_P = \{a/b \in \mathbb{Q} \mid p \nmid b, \text{ for all } p \in P\}.$$

The R-completion of a nilpotent space (X, x_0) coincides in this case with its $\mathbb{Z}_P$-localization ([Bousfield-Kan72, V, §4]). Thus (co)simplicial techniques generalize the localization construction via Postnikov towers of [Sullivan71] to general rings. The basic algebraic condition on the ring R that guarantees good completion properties is that R be **solid**, that is, the multiplication on R induces an isomorphism $R \otimes R \to R$.

When $f_{\infty *} \colon \tilde{H}_*(X; R) \to \tilde{H}_*(R_\infty X; R)$ is an isomorphism, then we say that X is R-**good**. For R-good spaces the R-completion, $f_\infty \colon X \to$

$R_\infty X$, enjoys certain universal properties. For example, a mapping $f\colon X \to Y$ induces an isomorphism $H_*(f)\colon H_*(X;R) \to H_*(Y;R)$ if and only if $R_\infty f\colon R_\infty X \to R_\infty Y$ is a homotopy equivalence ([Bousfield-Kan72, I.5.5]). However, there are spaces that are not R-good—for example, an infinite wedge of circles is not $\mathbb{Z}$-good. Nilpotent spaces are $\mathbb{Z}$-good. With this language we can describe the solution to the natural question—what is the target of the Eilenberg-Moore spectral sequence in general? [Dwyer75] found the answer for the Eilenberg-Moore spectral sequence associated to the fibre of a fibration $p\colon E \to B$: The spectral sequence converges to the homology of the **nilpotent completion** of the fibration, that is, to $H_*(\widetilde{F};R)$, where $\widetilde{F}$ is the fibre of the fibration $R_\infty p\colon R_\infty E \to R_\infty B$.

Completion and localization constructions have become fundamental in homotopy theory and a complete exposition of these ideas would take us too far afield. Nice expositions of this circle of ideas may be found in [Sullivan71], [Mimura-Nishida-Toda71], [Hilton75], [Hilton-Mislin-Roitberg75], and [Arkowitz76]. The most complete exposition of these ideas is the work of [Bousfield-Kan72].

A consequence of the cosimplicial construction of the R-completion is a result of [Dror73] that shows the extent to which nilpotent spaces approximate general homotopy types. To state precisely what sort of approximation we mean, we compare a connected space X with the associated tower of fibrations $\{R_s X\}$. By the definition of Tot_s, we have the augmentation mappings $f_s\colon X \to R_s X$ for all $s \geq 0$ and these mappings are compatible with the sequence of fibrations $R_{s+1}X \to R_s X$. Thus the mappings $\{f_s\}$ determine a mapping of towers of spaces $\{X\} \to \{R_s X\}$.

A tower of groups $\{G_s\}$ is a sequence of homomorphisms $G_{s+1} \to G_s$ for $s \geq 0$. A homomorphism of towers of groups, $\xi\colon \{G_s\} \to \{H_s\}$, is a sequence of group homomorphisms $\xi_s\colon G_s \to H_s$, compatible with the tower mappings. The natural maps $f_s\colon X \to R_s X$ determine, for each $i \geq 0$, a homomorphism of towers of groups $f_*\colon \{H_i(X;R)\} \to \{H_i(R_s X;R)\}$.

Definition 8bis.38. *A homomorphism of towers of groups, $\xi\colon \{G_s\} \to \{H_s\}$, is a* **pro-isomorphism** *if, for any group A, ξ induces an isomorphism*

$$\xi^*\colon \varinjlim \mathrm{Hom}_{\mathbf{Grp}}(H_s, A) \to \varinjlim \mathrm{Hom}_{\mathbf{Grp}}(G_s, A).$$

We leave it as an exercise to show that $\xi\colon \{G_s\} \to \{H_s\}$ is a pro-isomorphism if and only if, for each $t \geq 0$, there is a value $t' \geq t$ and a homomorphism $u_t\colon H_{t'} \to G_t$ such that the following diagram commutes:

$$
\begin{array}{ccc}
G_{t'} & \xrightarrow{\ \xi_{t'}\ } & H_{t'} \\[2pt]
{\scriptstyle p^G_{t',t}}\Big\downarrow & {\scriptstyle u_t}\swarrow & \Big\downarrow{\scriptstyle p^H_{t',t}} \\[2pt]
G_t & \xrightarrow[\ \xi_t\ ]{} & H_t.
\end{array}
$$

Here $p_{t',t}^G$ denotes the composition $G_{t'} \to G_{t'-1} \to \cdots G_{t+1} \to G_t$ and likewise for $p_{t',t}^H$. We also leave it to the reader to show that a pro-isomorphism induces an isomorphism of limits:

$$\xi \colon \lim_{\leftarrow s} G_s \cong \lim_{\leftarrow s} H_s \text{ and } \xi \colon \lim_{\leftarrow s}{}^1 G_s \cong \lim_{\leftarrow s}{}^1 H_s.$$

We say that a pointed space (X, x_0) is R-**nilpotent** if X is nilpotent and, for each $n \geq 1$, there is a central series of $\pi_1(X, x_0)$-modules

$$\pi_n(X, x_0) = M_1 \supset M_2 \supset \cdots \supset M_{c-1} \supset M_c \supset M_{c+1} = \{0\},$$

for which each subquotient M_j/M_{j+1} is a trivial $\pi_1(X, x_0)$-module and an R-module. A space is nilpotent when it is $\mathbb{Z}$-nilpotent.

Proposition 8bis.39. *For an arbitrary connected, pointed space (X, x_0), the spaces $R_s X = \mathrm{Tot}_s(R^\bullet X)$ are R-nilpotent for all $s \geq 0$. Furthermore, the natural maps $f_s \colon X \to R_s X$ induce, for all $i \geq 1$, a pro-isomorphism of towers of homology groups $f_* \colon \{H_i(X; R)\} \to \{H_i(R_s X; R)\}$.*

SKETCH OF A PROOF: The space RX is R-nilpotent since it is an H-space and an R-module. According to [Bousfield-Kan72, III.5.5], if $p \colon E \to B$ is a principal fibration with connected fibre F and any two of E, B, and F are R-nilpotent, then so is the third. Their Proposition II.2.5 asserts that $R_s X \to R_{s-1} X$ is a principal fibration with fibre a connected simplicial R-module. Thus the spaces $R_s X$ are R-nilpotent for all $s \geq 0$.

To establish that we have a homology pro-isomorphism, we observe that $H_k(X; R) \cong \pi_k(RX, x_0)$ and so we can compare the tower of homotopy groups $\{\pi_k(RX, x_0)\}$ with $\{\pi_k(RR_s X, x_0)\}$. When comparing the spaces RX and $RR_s X$, we have a triple structure available and hence mappings

$$\phi \colon RX \leftrightarrow RR_s X \colon \psi \qquad \text{with } \psi\phi = \mathrm{id}.$$

[Dror73] interpolates a condition that implies that a pro-isomorphism on homotopy is induced by $\{RX\} \to \{RR_s X\}$, namely, that the map of towers $\{R_n X\} \to \{R_s R_n X\}$ induce a pro-isomorphism. He then uses the convergence of the homotopy spectral sequence associated to the tower $\{R_s X\}$ to obtain the pro-isomorphism $\{H_k(X; R)\} \to \{H_k(R_s X; R)\}$ for $k \geq 1$. $\square$

It follows from the proposition that every connected, pointed homotopy type may represented by a tower of R-nilpotent spaces, up to homology equivalence. This approximation is analogous to the Stone-Weierstrass theorem: Every homotopy type (continuous function) may be represented by a tower of R-nilpotent spaces (a sequence of polynomials) such that $\tilde{H}_*(X; R) \cong \lim_{\leftarrow s} \tilde{H}_*(R_s X; R)$ (limits agree). To study whether a space is R-good, we can focus on the relation between $\lim_{\leftarrow s} \tilde{H}_*(R_s X; R)$ and $\tilde{H}_*(R_\infty X; R)$.

Exercises

8bis.1. Show that $\pi_1(\mathbb{R}P^{2n})$ acts nonnilpotently on $\pi_{2n}(\mathbb{R}P^{2n})$.

8bis.2. Show that the action $\nu_n^{E,F} : \pi_1(E,e) \times \pi_n(F,e) \to \pi_n(F,e)$ is well-defined and that $i_* : \pi_n(F,e) \to \pi_n(E,e)$ is a $\pi_1(E,e)$-module homomorphism.

8bis.3. Suppose that M is a module over a group π. Show that the coinvariants M_π is the largest quotient of M on which π acts trivially. Show directly that the functor $M \mapsto M_\pi$ is right exact.

8bis.4. Let π denote the cyclic group of order m, with generator $t \in \pi$. Show that the complex

$$\cdots \to W_n \to \cdots \to W_2 \xrightarrow{N} W_1 \xrightarrow{T} W_0 \xrightarrow{\varepsilon} \mathbb{Z} \to 0$$

is a resolution of $\mathbb{Z}$ over $\mathbb{Z}\pi$, where W_k the free $\mathbb{Z}\pi$-module on a single generator w_k and boundary homomorphisms $T : W_{2n+1} \to W_{2n}$ given by $T(w_{2n+1}) = tw_{2n} - w_{2n}$ and $N : W_{2n} \to W_{2n-1}$ given by $N(w_{2n}) = w_{2n-1} + tw_{2n-1} + \cdots + t^{m-1}w_{2n-1}$.

8bis.5. Suppose that π is a finitely generated group. Show that $H_i(\pi, M)$ is finitely generated whenever M is finitely generated over $\mathbb{Z}\pi$ and $i \geq 0$.

8bis.6. Prove directly that $H_1(\pi) \cong \pi/[\pi, \pi]$.

8bis.7. Prove Theorem 8bis.11.

8bis.8. Suppose that $1 \to H \to \pi \to Q \to 1$ is an extension of groups. Complete the proof of Theorem 8bis.14 by showing that the Q-coinvariants of the conjugation action on $H/[H,H]$ are given by $H/[\pi, H]$.

8bis.9. Suppose that $1 \to R \to F \to \pi \to 1$ is a presentation of the fundamental group $\pi = \pi_1(X)$ of a space X, where F and R are free groups. Prove the classic result of Hopf that $H_2(X)/h_*(\pi_2(X)) \cong R \cap [F,F]/[F,R]$ where $h_* : \pi_2(X) \to H_2(X)$ denotes the Hurewicz homomorphism.

8bis.10. Suppose that $1 \to H \to \pi \to Q \to 1$ is a central extension, that is, H maps to a subgroup of the center of π. Show that there is an exact sequence:

$$H_2(\pi) \to H_2(Q) \to H \to H_1(\pi) \to H_1(Q) \to 0.$$

8bis.11. Prove Corollary 8bis.19.

8bis.12. Suppose that $F \hookrightarrow E \xrightarrow{p} B$ is a fibration of connected spaces. Suppose that E is nilpotent. Show that $\pi_1(F)$ acts nilpotently on itself by conjugation.

8bis.13. Suppose π acts nilpotently on M and $H_0(\pi, M) = \{0\}$. Conclude that $M = \{0\}$.

$\mathbf{8^{bis}.14.}$ The functor Γ associates to a π-module the submodule ΓM generated by the union of the family of all perfect submodules of M, that is, submodules N with $N = \Gamma_\pi^2 N$. Show that ΓM is also perfect and that it is the maximal π-perfect submodule of M. Show that $\Gamma M \subset \Gamma_\pi^n M$ for all n.

$\mathbf{8^{bis}.15.}$ Show that $\Gamma_\pi^n M = \Gamma_\pi^{n+1} M$ implies that $\Gamma_\pi^n M = \Gamma_\pi^{n+k} M$ for all $k \geq 0$. Thus the lower central series is a sequence of proper inclusions until it stablilizes.

$\mathbf{8^{bis}.16.}$ Show that all nilpotent spaces are π-complete.

$\mathbf{8^{bis}.17.}$ Show that the maximal augmentation of a cosimplicial space $Y^\bullet$ is given by $aY^\bullet = \mathbf{CoSimp}(*, Y^\bullet)$.

$\mathbf{8^{bis}.18.}$ If $R \subset \mathbb{Q}$ is a subring of $\mathbb{Q}$, then show that there is a set of primes P (possibly empty) for which $R = \mathbb{Z}_P$.

$\mathbf{8^{bis}.19.}$ Show that a homomorphism of towers of groups, $\xi \colon \{G_s\} \to \{H_s\}$, is a pro-isomorphism if and only if, for each $t \geq 0$, there is a value $t' \geq t$ and a homomorphism $u_t \colon H_{t'} \to G_t$ such that the following diagram commutes:

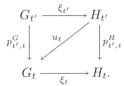

Show further that a pro-isomorphism induces an isomorphism of limits:

$$\xi \colon \lim_{\leftarrow s} G_s \cong \lim_{\leftarrow s} H_s \text{ and } \xi \colon \lim_{\leftarrow s}{}^1 G_s \cong \lim_{\leftarrow s}{}^1 H_s.$$

9

The Adams Spectral Sequence

"In (*various papers*) it is shown that homological algebra
can be applied in stable homotopy-theory."

J. F. Adams

One of the principal unsolved problems in modern mathematics is the determination of the homotopy groups, $\pi_*(X)$, of any nontrivial finite CW-complex X. These groups play a key role in the solution of certain geometric problems and in the classification of CW-complexes up to homotopy. The computation of $\pi_*(X)$, however, remains difficult if not intractable. (For a discussion of the computability of $\pi_*(X) \otimes \mathbb{Q}$, see the paper of [Anick85]; for some interesting progress for $X = S^n$, see the work of [Wu, J]).

A first approximation to $\pi_*(X)$ is provided by the Freudenthal suspension theorem (Theorem 4.10). The limit groups, $\lim_{k \to \infty}[S^{n+k}, S^k X] = \pi_n^S(X)$, are called the **stable homotopy groups** of X and they enjoy some regularity and further structure. Knowledge of these groups may also be sufficient for the solution of geometric problems (§9.4). A classical example is the celebrated theorem of [Adams60] on the nonexistence of elements of Hopf invariant one (Theorem 9.38). As this theorem was part of the motivation for the construction of the Adams spectral sequence, we discuss some of the details.

The question settled by Adams arose with W. R. Hamilton (see [Ebbing-haus90]): For which n, does $\mathbb{R}^n$ have a division algebra structure? That is, for which n is there a bilinear mapping, $\mu \colon \mathbb{R}^n \times \mathbb{R}^n \to \mathbb{R}^n$, so that $\mu(\vec{u}, \vec{v}) = 0$ implies that either $\vec{u} = 0$ or $\vec{v} = 0$. For $n = 1, 2, 4$ or 8, there are the real, complex, quaternionic, and Cayley multiplications, respectively, that were classically known. If one requires further that $\|\mu(\vec{x}, \vec{y})\| = \|\vec{x}\| \|\vec{y}\|$ (a **normed algebra**), then [Radon22] and [Hurwitz23] showed that these classical multiplications are the only examples. [Hopf31, 35] used the classical multiplications to construct mappings, $\eta \colon S^3 \to S^2$, $\nu \colon S^7 \to S^4$, and $\sigma \colon S^{15} \to S^8$, which are not homotopic to the constant map. These are the first examples of nontrivial elements in the homotopy groups of spheres (other than the degree maps in $\pi_n(S^n)$). Hopf's proof is geometric and proceeds, in the modern parlance, by introducing a homomorphism $H \colon \pi_{2n-1}(S^n) \to \mathbb{Z}$, constructed by counting linking numbers of the inverse images of points in S^n. Hopf showed that the linking numbers are a homotopy invariant and then applied the division algebra structure to deduce that the maps derived from the classical multiplications have H-invariant equal to one.

One can also compute linking numbers with the cup-product in cohomology. [Steenrod49] studied the Hopf invariant, $H \colon \pi_{2n-1}(S^n) \to \mathbb{Z}$, using *functional cup products*: Given $\gamma \colon S^{2n-1} \to S^n$, form the mapping cone, $K = S^n \cup_\gamma e^{2n}$. The Hopf invariant can be defined as follows: Let $x_n \in H^n(K)$ and $y_{2n} \in H^{2n}(K)$ be generators for the free abelian group in each dimension. Then $x_n \smile x_n = \pm H(\gamma) y_{2n}$.

When we reduce to mod 2 coefficients, we can make the transition to stable homotopy. If $H(\gamma) = 1$, then $x_n \smile x_n = y_{2n}$. By the unstable axioms for the action of the Steenrod algebra (Theorem 4.45), $Sq^n x_n = y_{2n}$. We can suspend the map $\gamma \colon S^{2n-1} \to S^n$, and form the mapping cone. Then $\Sigma K \simeq S^{n+1} \cup_{\Sigma\gamma} e^{2n+1}$. The suspension isomorphism determines $H^*(\Sigma K; \mathbb{F}_2)$ as a module over the Steenrod algebra. In particular, $Sq^n x_{n+1} = y_{2n+1}$, where $x_{n+1} \in H^{n+1}(\Sigma K; \mathbb{F}_2)$ and $y_{2n+1} \in H^{2n+1}(\Sigma K; \mathbb{F}_2)$ are generators. This implies immediately that $\Sigma\gamma$ is not homotopic to the constant map. By iterating this procedure we see that if $H(\gamma) = 1$, then γ determines a nontrivial element in $\pi_{n-1}^S = \pi_{n-1}^S(S^0) =$ the $(n-1)^{\mathrm{st}}$ *stem of the stable homotopy groups of spheres*. The existence of a division algebra structure on $\mathbb{R}^n$, then, implies a nonzero element $[\gamma]$ exists in π_{n-1}^S with the mapping cone exhibiting a nonzero Sq^n operation in mod 2 cohomology.

This reduction is already useful. According to the Adem relations (Theorem 4.45), Sq^n factors into sums of products of lower degree Steenrod operations, unless $n = 2^k$, for some k. For example, the relation

$$Sq^3 Sq^4 = \sum\nolimits_{j=0}^{1} \binom{4-1-j}{3-2j} Sq^{7-j} Sq^j,$$

implies $Sq^7 = Sq^3 Sq^4$. It follows that Sq^7 cannot act nontrivially on the cohomology of the mapping cone, $H^*(S^n \cup_f e^{n+7}; \mathbb{F}_2)$, because Sq^1 through Sq^6 act trivially for dimensional reasons. We conclude, then, that $\mathbb{R}^7$ cannot carry a division algebra structure. In fact, $\mathbb{R}^n$ is a candidate for a division algebra structure only if $n = 2^k$.

To settle the division algebra problem, we are led to an analysis of Sq^{2^k} and the possible factorizations that might arise through secondary or higher order operations associated to the Steenrod algebra. [Adams60] completed this analysis to prove that the classical examples of $\mathbb{R}$, $\mathbb{R}^2$, $\mathbb{R}^4$, and $\mathbb{R}^8$ provide a complete list of real vector spaces with a division algebra structure. In the course of this work, [Adams60] also introduced his eponymous spectral sequence that has become one of the fundamental tools in the study of stable homotopy theory. The aim of this chapter is the construction and elaboration of this spectral sequence. The first section contains some motivation, a statement of the main theorem, and a discussion of secondary and higher order cohomology operations. Based on the motivating discussion, we take a brief detour into homological algebra to introduce another important tool in the study of categories of modules over an algebra, the Ext functor. During this digression we introduce a change-of-rings spectral sequence.

In §9.3 we construct the spectral sequence and derive its basic properties. In keeping with the spirit of previous chapters, we do not utilize the technology of spectra and the stable homotopy category ([Elmendorf-Kriz-Mandell-May97]). The reasons for this choice are as follows: The approach using spaces is contained in the original papers of [Adams58] and, though cumbersome, it can be understood by the novice. Also, there are now several careful and complete expositions of the spectrum approach, [Adams69], [Switzer75], [Ravenel86] and [Kochman96], on which this author could not improve. The reader may safely skip to §9.4 if he or she accepts the main results and wishes to go quickly to the computations. The references, especially Adams's papers, may also substitute for this material.

In §9.4 we explore some of the geometric consequences of the existence and explicit form of the spectral sequence. We focus on the role played by the Adams spectral sequence in computing cobordism rings (the work of [Thom54], [Milnor60], [Liulevicius62], and [Wall60]). This section is written backwards—we take as basic the spectral sequence and search for applications. This emphasizes technique over the deeper geometric insight of [Thom54] and others. However, it gives a smooth transition into this set of remarkable results and offers a natural motivation for the study of spectra and stable objects.

In §9.5 some of the simpler, low-dimensional calculations are made and the geometric consequences explored. In particular, the first nonzero differential in the spectral sequence at the prime 2 settles one case of the division algebra problem. The low-dimensional stable homotopy groups of spheres at the primes 2 and 3 are also deduced.

In the final section of the chapter, we consider further structure in the spectral sequence. A product structure allows one to define Massey products and these are seen to converge to the secondary composition products of [Toda62]. The structure of the Steenrod algebra as a Hopf algebra imposes homological conditions on the spectral sequence including a large region of the first quadrant where all of the input at E_2 is trivial. There is also a periodicity operator that acts across part of the spectral sequence. The formidable task of determining the E_2-term of the Adams spectral sequence is developed in §9.6. The tool of choice is the May spectral sequence, introduced by [May64] in his doctoral thesis. We describe this spectral sequence in §9.6. We close the chapter with some tables and a discussion of further applications.

9.1 Motivation: What cohomology sees

The computation of $\pi_*(X)$ or $\pi_*^S(X)$ is a special case of the more general problem of determining $[Y, X]$ or $\{Y, X\}_k = \varinjlim[S^{n+k}Y, S^n X]$. Here we assume that our mappings are basepoint preserving, X and Y are connected and of the type of CW-complexes and, finally, that Y is finite-dimensional. A naive "picture" of $[Y, X]$ may be obtained by considering the image of the mod p cohomology functor,

$$H^*(\ ; \mathbb{F}_p) \colon [Y, X] \longrightarrow \mathrm{Hom}(H^*(X; \mathbb{F}_p), H^*(Y; \mathbb{F}_p))$$

([Greenlees88] develops this idea for other cohomology theories). Because the mod p cohomology of a space carries a rich structure, this Hom may be taken to be a set of mappings in various categories. Of course, $H^*(f; \mathbb{F}_p)$ is a homomorphism of graded vector spaces. It is also a graded algebra mapping. This can be used, for example, to distinguish $\mathbb{C}P(2)$ from $S^2 \vee S^4$—the mod p cohomology of these spaces are isomorphic as graded vector spaces but not as graded algebras. Finally, $H^*(f; \mathbb{F}_p)$ is a mapping of unstable modules over the Steenrod algebra—with this structure we distinguish $\Sigma \mathbb{C}P(2)$ and $S^3 \vee S^5$.

In the rest of this section, we fix a prime p and let $H^*(X)$ denote $H^*(X; \mathbb{F}_p)$. Let $\mathcal{A}_p$ denote the mod p Steenrod algebra and M and N graded left $\mathcal{A}_p$-modules.

Let $\mathrm{Hom}^t_{\mathcal{A}_p}(M, N)$ denote the set of $\mathcal{A}_p$-linear mappings of M to N that have degree $-t$, that is, for all q, $f(M^q) \subset N^{q-t}$. The iterated suspension functor on graded modules over $\mathcal{A}_p$ is defined as follows: For $k \in \mathbb{Z}$, $s^k \colon \mathbf{Mod}_{\mathcal{A}_p} \to \mathbf{Mod}_{\mathcal{A}_p}$ is given on objects by $(s^k M)_n \cong M_{n-k}$ and on morphisms $\phi \colon M \to N$ by $(s^k \phi)_n = (-1)^k \phi_{n-k}$. This generalizes the topological suspension isomorphism $s \colon \tilde{H}^l(X) \cong \tilde{H}^{l+1}(SX)$.

We construct a mapping $\{Y, X\}_t \to \mathrm{Hom}^t_{\mathcal{A}_p}(H^*(X), H^*(Y))$ as follows: A mapping, $f \colon S^{n+t}Y \to S^n X$, determines a morphism $H^*(f)$ of modules over the Steenrod algebra. However, as modules over $\mathcal{A}_p$, $\tilde{H}^*(S^n X) \cong s^n \tilde{H}^*(X)$, $\tilde{H}^*(S^{n+t}Y) \cong s^{n+t}\tilde{H}^*(Y)$ and $\tilde{H}^*(f)$ determines a mapping in $\mathrm{Hom}^t_{\mathcal{A}_p}(H^*(X), H^*(Y))$. Furthermore, $Sf \colon S^{n+t+1}Y \to S^{n+1}X$ determines the same mapping as in the diagram

$$
\begin{array}{ccc}
\tilde{H}^*(S^n X) & \overset{s}{\underset{\cong}{\longrightarrow}} & \tilde{H}^{*+1}(S^{n+1}X) \\
{\scriptstyle f^*}\big\downarrow & & \big\downarrow{\scriptstyle sf^*} \\
\tilde{H}^*(S^{n+t}Y) & \overset{s}{\underset{\cong}{\longrightarrow}} & \tilde{H}^{*+1}(S^{n+t+1}Y).
\end{array}
$$

Thus $H^*(\ ) \colon \{Y, X\}_t \to \mathrm{Hom}^t_{\mathcal{A}_p}(H^*(X), H^*(Y))$ is a well-defined mapping (the choice of pointed maps and connected spaces determines the mappings on H^0). The image of this mapping may be taken as an approximation to $\{Y, X\}_t$. It is, however, only a coarse approximation. A worst case is given by $H^*(\ ) \colon \pi^S_t \to \mathrm{Hom}^t_{\mathcal{A}_p}(\mathbb{F}_p, \mathbb{F}_p)$. The only classes in π^S_t that are mapped nontrivially are generated by $\iota \colon S^n \to S^n$, the identity map in π^S_0. The best case, however, suggests a course of action. Consider a free, left $\mathcal{A}_p$-module on one generator of degree n. This can be constructed as an extended module $\mathcal{A}_p \otimes s^n \mathbb{F}_p$ where $s^n \mathbb{F}_p$ is the graded $\mathbb{F}_p$-vector space with one copy of $\mathbb{F}_p$ in degree n and $\{0\}$ in every other degree. We introduce the notation $\cong_t$ for a homomorphism of graded modules that is an isomorphism in degrees less than t. A consequence of the Cartan-Serre theorem on $H^*(K(\mathbb{Z}/p\mathbb{Z}, n); \mathbb{F}_p)$ (Theorem 6.20) is the bounded isomorphism

$$\mathcal{A}_p \otimes s^n \mathbb{F}_p \cong_{2n-1} \tilde{H}^*(K(\mathbb{Z}/p\mathbb{Z}, n); \mathbb{F}_p).$$

This isomorphism leads to an isomorphism in the limit over the system of homomorphisms induced by the loop suspension mapping (Theorem 6.11) $H^{l+1}(K(\mathbb{Z}/p\mathbb{Z}, n + t + 1); \mathbb{F}_p) \xrightarrow{\cong} H^l(K(\mathbb{Z}/p\mathbb{Z}, n + t); \mathbb{F}_p)$, for $n + t \le l \le 2n + 2t - 1$:

$$\mathcal{A}_p \otimes s^n \mathbb{F}_p \cong \lim_{\leftarrow t} s^{-t} \tilde{H}^*(K(\mathbb{Z}/p\mathbb{Z}, n + t); \mathbb{F}_p).$$

For Y a finite dimensional CW-complex and $n < 2 \dim Y$, the fundamental correspondence $[Y, K(\mathbb{Z}/p\mathbb{Z}, n)] \cong H^n(Y; \mathbb{F}_p)$ implies

$$\{Y, K(\mathbb{Z}/p\mathbb{Z}, n)\}_t = \mathrm{Hom}^t_{\mathcal{A}_p}(\mathcal{A}_p \otimes s^n \mathbb{F}_p, H^*(Y; \mathbb{F}_p)).$$

Thus our approximation is on the mark when we consider spaces that carry a **free** $\mathcal{A}_p$-module structure, that is, Eilenberg-Mac Lane spaces. To increase the accuracy of the approximation, we could include the information that measures how far a module M over $\mathcal{A}_p$ differs from being a free module. To do this, we introduce the functors $\mathrm{Ext}^s_{\mathcal{A}_p}(M, -)$, the derived functors of $\mathrm{Hom}^*_{\mathcal{A}_p}(M, -)$, to be discussed in §9.2. The reader should compare this discussion with §7.1 where the derived functors of the tensor product, $M \otimes_\Gamma -$ are considered. The role of these derived functors is seen in the main theorem of this chapter, due to [Adams58].

For an abelian group G, $_{(p)}G$ denotes the quotient

$$_{(p)}G = G/\{\text{elements of finite order prime to } p\}.$$

It is elementary to show that the set $\{\text{elements of finite order prime to } p\}$ forms a subgroup of G. Since $\{Y, X\}_t$ is an abelian group, $_{(p)}\{Y, X\}_t$ makes sense.

Theorem 9.1. *For X and Y spaces of finite type, with Y a finite dimensional CW-complex, there is a spectral sequence, converging to $_{(p)}\{Y, X\}_*$, with E_2-term given by*

$$E_2^{s,t} \cong \mathrm{Ext}^{s,t}_{\mathcal{A}_p}(H^*(X; \mathbb{F}_p), H^*(Y; \mathbb{F}_p)),$$

and differentials d_r of bidegree $(r, r - 1)$.

Before beginning our discussion of Ext and the construction of the spectral sequence, let us consider how one might show that a mapping $f\colon S^{n+t} \to S^n$ is not homotopic to a constant map. Suppose $t > 0$ and we form the mapping cone, $C(f) = S^n \cup_f e^{n+t+1}$. Then there is a cofibration sequence:

$$S^n \xrightarrow{\text{include bottom cell}} C(f) \xrightarrow{\text{smash bottom cell}} S^{n+t+1}.$$

The exact sequence in cohomology,

$$0 \to \tilde{H}^*(S^{n+t+1}) \to \tilde{H}^*(C(f)) \to \tilde{H}^*(S^n) \to 0,$$

determines an extension of $\mathbb{F}_p$ by $\mathbb{F}_p$ over the algebra $\mathcal{A}_p$ and so a class in $\mathrm{ext}_{\mathcal{A}_p}(\mathbb{F}_p, \mathbb{F}_p)$, here of degree $t+1$. This $\mathrm{ext}_{\mathcal{A}_p}(\mathbb{F}_p, \mathbb{F}_p)$ is the classical Ext or Ext^1 group that figures in the Universal Coefficient theorem for cohomology ([Massey91, p. 314]). When one provides an abelian group structure on Ext, the correspondence between a representative of a stable mapping $S^{n+t} \to S^n$ and the extension it determines gives a homomorphism (an e-invariant),

$$e\colon \pi_{n+t}(S^n) \to \mathrm{ext}_{\mathcal{A}_p}(\mathbb{F}_p, \mathbb{F}_p).$$

In the case of the classical Hopf maps, $e(\text{Hopf map}) \neq 0$.

Whenever $e([f]) \neq 0$, $\tilde{H}^*(C(f))$ is a nontrivial module over $\mathcal{A}_p$. In general, the Steenrod operations on $\tilde{H}^*(C(f))$ are trivial on two-cell complexes. It may be the case, however, that a secondary or higher order operation, coming from relations in $\mathcal{A}_p$, is nontrivial on $\tilde{H}^*(C(f))$. This also implies that $f \not\simeq *$. With this in mind, we next discuss higher order cohomology operations.

Higher order cohomology operations

Suppose W is a space and $\theta\colon K(\mathbb{Z}/p\mathbb{Z}, n) \to K(\mathbb{Z}/p\mathbb{Z}, n+t)$ represents an element in the Steenrod algebra. Suppose $x \in H^n(W; \mathbb{F}_p)$ is a cohomology class and that $\theta(x) = 0$. Under these conditions, a secondary operation can be defined. Let E denote the total space of the pullback of the path-loop fibration over $K(\mathbb{Z}/p\mathbb{Z}, n+t)$ with respect to the mapping θ. Let $\alpha\colon E \to K(\mathbb{Z}/p\mathbb{Z}, m)$ represent a class in $H^m(E; \mathbb{F}_p)$. Since $\theta(x) = 0$, x lifts (not necessarily uniquely) to a mapping $\tilde{x}\colon W \to E$. The set of all composites, $\alpha \circ \tilde{x} \in H^m(W; \mathbb{F}_p)$, varied over all liftings $\tilde{x}$, defines the secondary operation, $\Phi_{\alpha,\theta}(x) \subset H^m(W; \mathbb{F}_p)$.

$$
\begin{array}{ccc}
K(\mathbb{Z}/p\mathbb{Z}, n+t-1) \xrightarrow{\ i\ } E & \xrightarrow{\ \alpha\ } & K(\mathbb{Z}/p\mathbb{Z}, m) \\
& \Big\downarrow & \\
W \xrightarrow{\ x\ } K(\mathbb{Z}/p\mathbb{Z}, n) & \xrightarrow{\ \theta\ } & K(\mathbb{Z}/p\mathbb{Z}, n+t).
\end{array}
$$

We make several observations about this construction:

1) If $t < n$, then $[\theta] \in H^{n+t}(K(\mathbb{Z}/p\mathbb{Z}, n); \mathbb{F}_p)$ is primitive and hence a loop map ($\theta = \Omega\theta'$; Corollary 8.25). It follows that E is a loop space, that is, there is a space Z with $E = \Omega Z$. Thus $[W, E]$ is a group and we can identify $\Phi_{\alpha,\theta}(x)$ as a coset of $H^m(W; \mathbb{F}_p)$. In particular, two liftings $\tilde{x}$ and $\hat{x} \colon W \to E$, differ by a mapping of W to $K(\mathbb{Z}/p\mathbb{Z}, n + t - 1)$. (Recall the exactness of $[W, F] \to [W, E] \to [W, B]$ for a fibration $F \hookrightarrow E \to B$.) It follows that $\Phi_{\alpha,\theta}(x)$ determines elements $\{\alpha \circ \tilde{x}\}$ in

$$H^m(W; \mathbb{F}_p) \Big/ \alpha \circ \big(i_* H^{n+t-1}(W; \mathbb{F}_p)\big),$$

where $i_* \colon [W, K(\mathbb{Z}/p\mathbb{Z}, n+t-1)] \to [W, E]$ is (pre-)composition. As always, indeterminacies can be difficult to make explicit. In the best cases, dimensions conspire to make $\Phi_{\alpha,\theta} \colon \ker\theta \to H^m(W; \mathbb{F}_p)$ a well-defined function.

2) This definition can be made for different coefficient groups and more general cohomology operations ([Maunder64]). We will not need this level of generality.

3) If V is a finite dimensional graded vector space over $\mathbb{F}_p$, then we can write $V = \bigoplus_{j=1}^{s} s^{n_j}\mathbb{F}_p$, where the n_j correspond to the dimensions of basis elements for V. Let $K(V) = \prod_{j=1}^{s} K(\mathbb{Z}/p\mathbb{Z}, n_j)$. Then, as graded vector spaces, $\pi_*(K(V)) \cong V$. A class in $[W, K(V)]$ is representable as a vector $(x_1, \ldots, x_s)$ in $\prod_{j=1}^{s} H^{n_j}(W; \mathbb{F}_p)$. Furthermore, the (abelian) addition on $\mathbb{Z}/p\mathbb{Z}$ determines a mapping,

$$+ \colon K(\mathbb{Z}/p\mathbb{Z} \oplus \mathbb{Z}/p\mathbb{Z}, n) \to K(\mathbb{Z}/p\mathbb{Z}, n),$$

which induces the vector addition on such spaces.

We generalize the definition of secondary operations to vectors of classes. Let $A \colon K(V_0) \to K(V_1)$ represent an n-tuple of cohomology classes $(x_1, \ldots, x_n) \in \bigoplus_i H^i(K(V_0); \mathbb{F}_p)$. Then we have the analogous diagram defining a secondary operation associated to A:

$$
\begin{array}{ccccc}
K(s^{-1}V_1) & \xrightarrow{\ i\ } & E & \longrightarrow & K(\mathbb{Z}/p\mathbb{Z}, m) \\
& \nearrow & \downarrow & & \\
W & \longrightarrow & K(V_0) & \xrightarrow{\ A\ } & K(V_1).
\end{array}
$$

For example, consider the Adem relation $Sq^2Sq^2 + Sq^3Sq^1 = 0$. Let A denote the mapping

$$A = \begin{pmatrix} Sq^1 \\ Sq^2 \end{pmatrix} \colon K(V_0) = K(\mathbb{Z}/2\mathbb{Z}, n) \to$$

$$K(\mathbb{Z}/2\mathbb{Z}, n+1) \times K(\mathbb{Z}/2\mathbb{Z}, n+2) = K(V_1).$$

By the naturality of the Leray-Serre spectral sequence we get a mapping of spectral sequences whose source is associated to the path-loop fibration over $K(V_1)$ with target associated to the pullback of the path-loop fibration with respect to the mapping A. The class in the cohomology of the fibre

$$Sq^3 \iota_n + Sq^2 \iota_{n+1} \in H^*(K(s^{-1}V_1); \mathbb{F}_2)$$

goes to zero under the transgression in the target spectral sequence because the Adem relation holds.

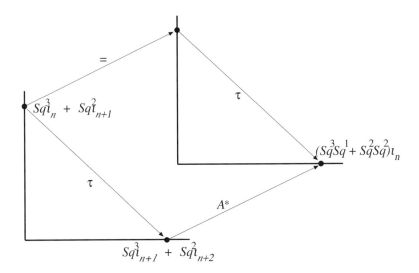

The class $[Sq^3 \iota_n + Sq^2 \iota_{n+1}] \in E_\infty^{n+3,0}$ represents a class $\alpha \in H^{n+3}(E; \mathbb{F}_2)$, where E is the total space of the pullback over A. This gives a secondary operation as in the diagram:

$$W \xrightarrow{\ x\ } K(V_0) \xrightarrow[\ A\ ]{} K(V_1).$$

(with $E \xrightarrow{\ \alpha\ } K(\mathbb{Z}/2\mathbb{Z}, n+3)$ above, and $\tilde{x}$ lifting W to E)

If $x \in H^n(W; \mathbb{F}_2)$ and $Sq^1 x = 0$ and $Sq^2 x = 0$, then lifts $\tilde{x}$ to E of x exist and $\Phi_{\alpha,A}(x)$ is defined. Furthermore, the indeterminacy is the subgroup $Sq^3 H^n(W; \mathbb{F}_2) + Sq^2 H^{n+1}(W; \mathbb{F}_2)$ of $H^{n+3}(W; \mathbb{F}_2)$.

This example was employed by [Adem57] to show the nontriviality of $\eta^2 = \eta \circ \eta \colon S^{n+2} \to S^n$ where η denotes a suspension of the Hopf map $\eta \colon S^3 \to S^2$. On $H^*(C(\eta^2); \mathbb{F}_2)$, the operation $\Phi_{\alpha,A}$ carries the generator in degree n to the generator in degree $n+3$. We say that η^2 is **detected by**

the secondary cohomology operation $\Phi_{\alpha,A}$. More generally, any mapping $f\colon S^{n+r} \to S^n$ for which a secondary cohomology operation acts nontrivially on $H^*(C(f);\mathbb{F}_p)$ is said to be detected by this operation and, in particular, f is not homotopic to the constant map.

4) The Adem relation $Sq^3Sq^1 + Sq^2Sq^2 = 0$ holds universally on the cohomology of any space. Any such quadratic relation between primary operations can be expressed as a composite, $K(V_1) \xrightarrow{\xi} K(V_2) \xrightarrow{\chi} K(V_3)$, with $\chi \circ \xi \simeq *$. For example, in the case of the given Adem relation we have

$$K(s^n\mathbb{F}_2) \xrightarrow{\binom{Sq^1}{Sq^2}} K(s^{n+1}\mathbb{F}_2 \oplus s^{n+2}\mathbb{F}_2) \xrightarrow{+\circ\binom{Sq^3}{Sq^2}} K(s^{n+4}\mathbb{F}_2).$$

Generally, we can construct a diagram of spaces on which to define the associated secondary operation:

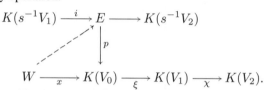

Here E is the pullback of the path-loop fibration over $K(V_1)$ with respect to ξ. The mapping χ, as a cohomology operation, produces classes in $H^*(E;\mathbb{F}_p)$ because classes that transgress from the fibre of the fibration p are annihilated by χ. If $x\colon W \to K(V_0)$ represents a vector of classes in $H^*(W;\mathbb{F}_p)$ with $\xi \circ x \simeq *$, then we obtain the secondary cohomology operation due to the relation $\chi \circ \xi \simeq *$ as a subset of $[W, K(s^{-1}V_2)]$.

5) In order to capture all of the relations between primary operations we turn to a homological description. Let

$$0 \leftarrow H^*(X;\mathbb{F}_p) \xleftarrow{\varepsilon} C_0 \xleftarrow{d_1} C_1 \xleftarrow{d_2} C_2$$

be an exact sequence of $\mathcal{A}_p$-modules, with C_0, C_1 and C_2 free $\mathcal{A}_p$-modules; C_0 can be taken as the free module on a set of $\mathcal{A}_p$-generators of $H^*(X;\mathbb{F}_p)$. We can think of C_1 as the free module on the $\mathcal{A}_p$ relations among the generators of $H^*(X;\mathbb{F}_p)$, and C_2 as the free module on the secondary relations, that is, relations among the relations. If X is $(n-1)$-connected and $C_i \cong \mathcal{A}_p \otimes V_i \cong_{2n-1} H^*(K(V_i);\mathbb{F}_p)$ for V_i, a graded vector space, then we associate a diagram of spaces, where we have written the name of the $\mathcal{A}_p$-module map for a continuous map that induces it. (The identification of algebraic mappings $d_i\colon C_i \to C_{i-1}$ with $d_i\colon K(V_{i-1}) \to K(V_i)$ follows from the representability of mod p cohomology.)

$$K(s^{-1}V_1) \longrightarrow E \longrightarrow K(s^{-1}V_2)$$
$$\downarrow^{p}$$
$$X \xrightarrow{\varepsilon} K(V_0) \xrightarrow{d_1} K(V_1) \xrightarrow{d_2} K(V_2).$$

Applying the mod p cohomology functor to the bottom row gives the sequence of $\mathcal{A}_p$-modules, exact in dimensions less than $2n - 1$,

$$0 \leftarrow H^*(X;\mathbb{F}_p) \xleftarrow{\varepsilon} C_0 \xleftarrow{d_1} C_1 \xleftarrow{d_2} C_2.$$

The secondary operation arising from this diagram lies in $[X, K(s^{-1}V_2)]$. We say that this operation is the **second order operation associated to** $C_0 \xleftarrow{d_1} C_1 \xleftarrow{d_2} C_2$.

6) Paragraph 5) can be generalized to m^{th} order cohomology operations. Continue the exact sequence of free $\mathcal{A}_p$-modules:

$$0 \leftarrow H^*(X;\mathbb{F}_p) \leftarrow C_0 \leftarrow C_1 \leftarrow \cdots \leftarrow C_m.$$

This gives rise to a tower of fibrations:

$$
\begin{array}{ccc}
K(s^{-(m-1)}V_{m-1}) \xrightarrow{\ i\ } E_{m-1} \longrightarrow K(s^{-(m-1)}V_m) \\
\downarrow \\
\vdots \\
\downarrow \\
K(s^{-2}V_2) \xrightarrow{\ i\ } E_2 \longrightarrow K(s^{-2}V_3) \xrightarrow{\ s^{-2}d_4\ } K(s^{-2}V_4) \\
\downarrow \\
K(s^{-1}V_1) \xrightarrow{\ i\ } E_1 \longrightarrow K(s^{-1}V_2) \xrightarrow{\ s^{-1}d_3\ } K(s^{-1}V_3) \\
\downarrow \\
X \xrightarrow{\ \varepsilon\ } K(V_0) \xrightarrow{\ d_1\ } K(V_1) \xrightarrow{\ d_2\ } K(V_2).
\end{array}
$$

In degrees less than $2n - r$, the map in cohomology, induced by the composite $K(s^{-(r-1)}V_{r-1}) \to E_{r-1} \to K(s^{-(r-1)}V_r)$, is $s^{-(r-1)}d_r$. This follows from the Serre exact sequence (Example 5.D), applied inductively to each fibration.

In order to get a lifting of $\varepsilon\colon X \to K(V_0)$ to $\tilde{\varepsilon}\colon X \to E_{m-1}$ and so define an mth order cohomology operation on $[X, K(s^{-(m-1)}V_m)]$, it is necessary that the rth order operations defined by this tower, for $r < m$, all contain the zero class. Only then is a lift to the next stage possible. When $f\colon S^{n+t} \to S^n$ is a mapping and $X = S^n \cup_f e^{n+t+1}$, if the lifts of the class in $H^n(X)$ determine only nonzero classes in $H^{n+t+1}(X)$, then we say that f is **detected by an mth order operation** and f is not homotopic to the constant map.

7) The dependence on the connectivity of X can be removed by considering only **stable m^{th} order operations**. Such an operation Φ determines a commutative

diagram

$$sH^n(X;\mathbb{F}_p) \xrightarrow{\;s\circ\Phi\;} sH^{n+t}(X;\mathbb{F}_p)$$

$$\Big\downarrow{\cong} \qquad\qquad\qquad \Big\downarrow{\cong}$$

$$H^{n+1}(SX;\mathbb{F}_p) \xrightarrow{\quad\Phi\quad} H^{n+t+1}(SX;\mathbb{F}_p)$$

The details of the construction of these operations and their properties can be found in the work of [Maunder63]. The connection between higher order stable cohomology operations and the spectral sequence of Theorem 9.1 lies in the interpretation of the filtration on $_{(p)}\pi_*^S$ and $_{(p)}\{Y, X\}_*$ to which the spectral sequence data converge.

Proposition 9.2. *If an element $u \in {}_{(p)}\pi_*^S$ can be detected by an nth order stable cohomology operation, then, for some $m \le n$, $u \in F^m({}_{(p)}\pi_*^S)$, the mth stage of the filtration of $_{(p)}\pi_*^S$ associated to the Adams spectral sequence.*

A proof of this proposition will emerge with the construction of the spectral sequence. The role of higher order operations in homotopy theory is fundamental and the Adams spectral sequence helps to codify and suggest their further use. The interested reader can consult the papers of [Cohen, R81] and [Lin76] for other possible applications.

9.2 More Homological Algebra; the Functor Ext

Before we construct the spectral sequence, a digression into homological algebra is necessary to secure the algebraic tools. In this section we also construct a spectral sequence associated to an extension of Hopf algebras based on the change-of-rings spectral sequence of [Cartan-Eilenberg56] and elaborated by [Adams58].

In Chapters 3 and 7, we studied the categories of modules and differential modules over rings and over differential algebras. The tool of choice was the functor Tor that measures the deviation from (left) exactness of the functor $M \otimes_R -$; $\text{Tor}_i^R(M, -)$ is the ith left derived functor of $M \otimes_R -$.

One of the fundamental relations in homological algebra is the **Hom-tensor interchange**: When A is a left Λ-module, B is a right Λ-module and B and C are left Γ-modules

$$\text{Hom}_\Lambda(A, \text{Hom}_\Gamma(B, C)) \cong \text{Hom}_\Gamma(B \otimes_\Lambda A, C).$$

This isomorphism plays a key role in the Universal Coefficient theorems in topology. The functors $\text{Hom}_\Gamma(M, -)$ and $\text{Hom}_\Gamma(-, N)$ are half exact. We next study their derived functors.

We begin by identifying the category of interest. Let (Γ, φ) denote a graded algebra, over a field k, with product φ. We generally assume that Γ has a unit,

$\varepsilon\colon k \to \Gamma$, as well as an augmentation, $\eta\colon \Gamma \to k$ (we assume $\eta \circ \varepsilon = \mathrm{id}_k$). The category of graded left Γ-modules is denoted by $_\Gamma\mathbf{Mod}$, and we take the morphisms in this category to be of degree zero. Denote the Γ-linear maps between two left Γ-modules, M and N by $_\Gamma\mathbf{Mod}(M, N)$.

The suspension functor on $_\Gamma\mathbf{Mod}$ is defined as follows: If $M \in {}_\Gamma\mathbf{Mod}$, then sM is the graded vector space, $(sM)_n = M_{n-1}$, with Γ-action given by $\gamma \cdot (sx) = (-1)^{\deg \gamma} s(\gamma \cdot x)$, where $x \in M_n$ and sx is the corresponding element in $(sM)_{n+1}$. Define the iterated suspension by $s^n = s \circ s^{n-1}$ and $s^1 = s$. The graded version of the Hom-functor is given by

$$\mathrm{Hom}_\Gamma^n(M, N) = {}_\Gamma\mathbf{Mod}(M, s^n N).$$

Equivalently, a Γ-module homomorphism in $\mathrm{Hom}_\Gamma^n(M, N)$ can be thought of as a homomorphism $f\colon M \to N$ that *lowers* degree by n.

To study the derived functors of $\mathrm{Hom}_\Gamma^*(M, -)$, we resolve a left Γ-module M by projective Γ-modules. That is, construct a long exact sequence in $_\Gamma\mathbf{Mod}$;

$$0 \leftarrow M \xleftarrow{\varepsilon} P_0 \xleftarrow{d} P_1 \xleftarrow{d} \cdots \xleftarrow{d} P_n \xleftarrow{d} \cdots$$

where each P_i is a projective module over Γ. (The reader should contrast this with the accounts in §2.4 and §7.1 where the presence of a differential is part of the construction.) Let $N \in {}_\Gamma\mathbf{Mod}$ and apply $\mathrm{Hom}_\Gamma^*(-, N)$ to this sequence

$$\mathrm{Hom}_\Gamma^*(P_0, N) \xrightarrow{d} \mathrm{Hom}_\Gamma^*(P_1, N) \xrightarrow{d} \cdots \xrightarrow{d} \mathrm{Hom}_\Gamma^*(P_n, N) \xrightarrow{d} \cdots.$$

The homology of this sequence defines $\mathrm{Ext}_\Gamma^{*,*}(M, N)$. We leave it to the reader to verify the usual properties of this derived functor:

(1) The definition is independent of the choice of projective resolution.

(2) $\mathrm{Ext}_\Gamma^{0,*}(M, N) = \mathrm{Hom}_\Gamma^*(M, N)$.

(3) Given a short exact sequence of left Γ-modules, $0 \to A \to B \to C \to 0$, there are long exact sequences,

$$\to \mathrm{Ext}_\Gamma^{n,*}(C, N) \to \mathrm{Ext}_\Gamma^{n,*}(B, N) \to \mathrm{Ext}_\Gamma^{n,*}(A, N) \xrightarrow{\delta} \mathrm{Ext}_\Gamma^{n+1,*}(C, N) \to$$

and

$$\to \mathrm{Ext}_\Gamma^{n,*}(M, A) \to \mathrm{Ext}_\Gamma^{n,*}(M, B) \to \mathrm{Ext}_\Gamma^{n,*}(M, C) \xrightarrow{\delta} \mathrm{Ext}_\Gamma^{n+1,*}(M, A) \to.$$

(4) Ext is functorial in each of its three variables.

(5) $\mathrm{Ext}_\Gamma^{n,*}(P, M) = \{0\}$ if $n > 0$, P is a projective Γ-module, and M is any left Γ-module. Furthermore, $\mathrm{Ext}_\Gamma^{n,*}(M, J) = \{0\}$, if $n > 0$ and J is an **injective** Γ-module and M is any left Γ-module.

The notion of an **injective module** over Γ is the formal dual (mono replacing epi and arrows reversed) of the notion of a projective module. The key property of an injective module J is the exactness of the functor $\mathrm{Hom}_\Gamma(-, J)$.

We could have defined $\text{Ext}_\Gamma^{*,*}(M, N)$ by forming an injective resolution of the module N and applying the functor $\text{Hom}_\Gamma(M, -)$.

The Ext groups are bigraded. If we write $\text{Ext}^s = \text{Ext}^{s,*}$, the single grading refers to homological degree and forgets the internal degree of the mapping.

In the applications to follow, we focus on the computation of $\pi_*^S(X) = \{S^0, X\}_*$. Theorem 9.1 introduces a spectral sequence, converging to $_{(p)}\pi_*^S(X)$, with the E_2-term given by $\text{Ext}_{\mathcal{A}_p}^{*,*}(H^*(X; \mathbb{F}_p), \mathbb{F}_p)$. Anticipating these computations, we consider the computation of $\text{Ext}_\Gamma^{*,*}(M, k)$ and some convenient resolutions for computation. In the particular case $M = k$, we write

$$\text{Ext}_\Gamma^{*,*}(k, k) = H^{*,*}(\Gamma),$$

and $H^{*,*}(\Gamma)$ is called the **cohomology of the algebra** Γ. The dual situation, given by $\text{Tor}_{*,*}^\Gamma(k, k) = H_{*,*}(\Gamma)$, defines the **homology of the algebra** Γ.

Henceforth, we assume that the algebra Γ is of finite type over k (that is, in each degree n, Γ^n is finite-dimensional over k). For graded vector spaces of finite type, the definition of the **dual** is straightforward: $(\Gamma^{\text{dual}})^n = \text{Hom}_k(\Gamma^n, k)$. Furthermore, if M' and M'' are both of finite type, then so is $M' \otimes_k M''$ and $(M' \otimes_k M'')^{\text{dual}} \cong M'^{\text{dual}} \otimes_k M''^{\text{dual}}$. It follows that an algebra (Γ, φ) and a left Γ-module M, of finite type, with module structure map $\psi\colon \Gamma \otimes M \to M$, yield by duality a coalgebra $(\Gamma^{\text{dual}}, \varphi^*)$ and a comodule M^{dual}, over the coalgebra, Γ^{dual}, with structure map

$$\psi^*\colon M^{\text{dual}} \to \Gamma^{\text{dual}} \otimes_k M^{\text{dual}}.$$

Recall from §7.1 the definition of the bar resolution:

$$0 \leftarrow M \leftarrow \Gamma \otimes M \leftarrow \Gamma \otimes I(\Gamma) \otimes M \leftarrow \Gamma \otimes I(\Gamma) \otimes I(\Gamma) \otimes M \leftarrow \cdots,$$

where $B_n(\Gamma, M) = \Gamma \otimes I(\Gamma)^{\otimes n} \otimes M$ and $I(\Gamma) = \ker(\eta\colon \Gamma \to k)$. The differential and the contracting homotopy are given in Proposition 7.8. To compute $\text{Ext}_\Gamma(M, k)$, we apply $\text{Hom}_\Gamma^*(-, k)$ to the bar resolution. Since k, as a Γ-module, is the $I(\Gamma)$-trivial module concentrated in degree 0, it follows via the Hom-tensor interchange that

$$\text{Hom}_\Gamma^t(N, k) = \text{Hom}_k^t(k \otimes_\Gamma N, k) = ((k \otimes_\Gamma N)^{\text{dual}})^t.$$

The bar resolution becomes the sequence

$$M^{\text{dual}} \xrightarrow{\bar{\psi}^*} I(\Gamma)^{\text{dual}} \otimes M^{\text{dual}} \xrightarrow{d^*} I(\Gamma)^{\text{dual}} \otimes I(\Gamma)^{\text{dual}} \otimes M^{\text{dual}} \xrightarrow{d^*} \cdots$$

where $\bar{\psi}^*$ is the composite

$$M^{\text{dual}} \xrightarrow{\psi^*} \Gamma^{\text{dual}} \otimes M^{\text{dual}} \xrightarrow{i^* \otimes 1} I(\Gamma)^{\text{dual}} \otimes M^{\text{dual}}$$

(i^* is the dual of $I(\Gamma) \hookrightarrow \Gamma$) and d^* is given by

$$d^*([\alpha_1 \mid \cdots \mid \alpha_n]\lambda) = \sum_{i=1}^{n} \sum_{j=1}^{n_i} [\bar\alpha_1 \mid \cdots \mid \bar\alpha_{i-1} \mid \bar\alpha'_{i,j} \mid \alpha''_{i,j} \mid \cdots \mid \alpha_n]\lambda$$
$$+ \sum_{l=1}^{m} [\bar\alpha_1 \mid \cdots \mid \bar\alpha_{n-1} \mid \bar\alpha'_l]\lambda''_l$$

(recall that $\bar\alpha = (-1)^{1+\deg\alpha}\alpha$ for α, an element in a graded module) where $\varphi^*(\alpha_i) = \sum_{j=1}^{n_i} \alpha'_{i,j} \otimes \alpha''_{i,j}$ and $\bar\psi^*(\lambda) = \sum_{l=1}^{m} \alpha_l \otimes \lambda''_l$. Thus

$$\mathrm{Ext}^*_\Gamma(M, k) = H(\bar{\mathrm{B}}(\Gamma)^{\mathrm{dual}} \otimes M^{\mathrm{dual}}, d^*),$$

where $\bar{\mathrm{B}}(\Gamma)^n = I(\Gamma)^{\otimes n}$.

To compute the cohomology of the algebra Γ then, we can use $\bar{\mathrm{B}}(\Gamma)^{\mathrm{dual}}$ which consists of elements $[\alpha_1 \mid \cdots \mid \alpha_n]$ with $\alpha_i \in I(\Gamma)^{\mathrm{dual}}$, and differential

$$d^*([\alpha_1 \mid \cdots \mid \alpha_n]) = \sum_{i=1}^{n} \sum_{j=1}^{n_i} [\bar\alpha_1 \mid \cdots \mid \bar\alpha_{i-1} \mid \bar\alpha'_{i,j} \mid \alpha''_{i,j} \mid \cdots \mid \alpha_n].$$

We point out that such a construction can be made with any augmented coalgebra, (C, Δ, η), where Δ is the coproduct, $J(C) = \mathrm{coker}(\eta\colon k \to C)$ is the cokernel of the augmentation, which is the dual of $I(C^{\mathrm{dual}})$. This functor on coalgebras is sometimes denoted $F_*(C)$ and called the **cobar construction** on (C, Δ, η). It was first introduced by [Adams56] to compute $H_*(\Omega X)$ as a functor of the chains on X, $C_*(X)$, as a coalgebra. We can express the cohomology of Γ in terms of the cobar construction by $H^{*,*}(\Gamma) = H(F_*(\Gamma^{\mathrm{dual}}), d^*)$.

Another application of the Hom-tensor interchange shows the duality between the cohomology and homology of an algebra. Since we are over a field, the Universal Coefficient theorem allows us to interchange the homology operator with Hom_k and obtain the equation, when Γ is of finite type,

$$\mathrm{Ext}_\Gamma(k, k) = H(\mathrm{Hom}_\Gamma(\mathrm{B}(\Gamma), k)) = H(\mathrm{Hom}_k(k \otimes_\Gamma \mathrm{B}(\Gamma), k))$$
$$= \mathrm{Hom}_k(H(k \otimes_\Gamma \mathrm{B}(\Gamma)), k) = \mathrm{Tor}^\Gamma(k, k)^{\mathrm{dual}}.$$

In the next section, the natural coalgebra structure on $\mathrm{Tor}^\Gamma(k, k)$ provides a natural product on $\mathrm{Ext}_\Gamma(k, k)$. Some calculations are eased by working in the dual.

Finally, when speaking of the computation of Ext, we mention another computationally convenient type of resolution.

Definition 9.3. *A homomorphism, $f\colon M \to N$ of left Γ-modules is said to be* **minimal** *if $f(M) \subset I(\Gamma) \cdot N$. A projective resolution of a module M is said to be a* **minimal resolution** *if every mapping in the resolution is minimal.*

Proposition 9.4. Let $0 \xleftarrow{\varepsilon} M \xleftarrow{\varepsilon} P_0 \xleftarrow{d} P_1 \xleftarrow{d} P_1 \xleftarrow{d} P_2 \xleftarrow{d} \cdots$ be a minimal resolution of M by projective Γ-modules. Then $\mathrm{Ext}_\Gamma^s(M, k) \cong \mathrm{Hom}_\Gamma(P_s, k)$.

PROOF: We assume everything in sight is of finite type. We exploit the duality between Ext and Tor for a proof. The duality in this case is between $\mathrm{Ext}_\Gamma(M, k)$ and $\mathrm{Tor}^\Gamma(k, M)$, and the dual statement of the proposition is that $\mathrm{Tor}_s^\Gamma(k, M) \cong k \otimes_\Gamma P_s$. However, for any Γ-module X, $k \otimes_\Gamma X = X/I(\Gamma) \cdot X$, which follows from the definition of $k \otimes_\Gamma X$ as

$$k \otimes X \big/ \{(1 \cdot \gamma) \otimes x - 1 \otimes (\gamma \cdot x)\},$$

where $1 \cdot \gamma = 0$ when γ is in $I(\Gamma)$. Since $d(P_s) \subset I(\Gamma) \cdot P_{s-1}$, $1 \otimes d\colon k \otimes_\Gamma P_s \to k \otimes_\Gamma P_{s-1}$ is the zero homomorphism for every $s \geq 1$ and so $\mathrm{Tor}_s^\Gamma(k, M) = k \otimes_\Gamma P_s$. Passing to the dual, we get a complex with all differentials zero and so $\mathrm{Ext}_\Gamma^s(M, k) \cong \mathrm{Hom}_\Gamma(P_s, k)$.   □

Minimal resolutions come in handy for doing low-dimensional calculations or to begin an induction. As an exercise, the reader should compute $H^{*,*}(\Lambda(x))$ where $\Lambda(x)$ is the exterior algebra on one generator x, here taken to have odd degree. A minimal resolution or the bar construction can be applied to obtain $H^{*,*}(\Lambda(x)) \cong k[y]$ as vector spaces, where y has bidegree $(1, \deg x)$. In the next section, the multiplicative properties of Ext are developed and we find that this isomorphism is true at the algebra level.

Multiplicative structure on Ext

The bigraded $\mathbb{F}_p$-vector space $\mathrm{Ext}_{\mathcal{A}_p}^{*,*}(H^*(X; \mathbb{F}_p), H^*(Y; \mathbb{F}_p))$ enjoys some further structure. There is a product when $X = Y$, and more generally, pairings of Ext groups. We give two constructions of the same operation. We will present the first construction in detail and sketch the second. The first identified by [Yoneda54]. Suppose that Γ is an algebra over a field k.

Theorem 9.5. Let L, M, and N be left Γ-modules. Then there is a bilinear, associative pairing, called the **composition product**, defined for all p, t, q, $t' \geq 0$, $\circ\colon \mathrm{Ext}_\Gamma^{p,t}(L, M) \otimes \mathrm{Ext}_\Gamma^{q,t'}(M, N) \to \mathrm{Ext}_\Gamma^{p+q,t+t'}(L, N)$.

PROOF: Let $0 \leftarrow L \leftarrow P_\bullet$ be a projective resolution of L, and $0 \leftarrow M \leftarrow Q_\bullet$ a projective resolution of M. If $[f]$ lies in $\mathrm{Ext}_\Gamma^{p,t}(L, M)$ and $[g]$ in $\mathrm{Ext}_\Gamma^{q,t'}(M, N)$, then $[f]$ may be represented by $f\colon P_p \to s^t M$ and $[g]$ by $g\colon Q_q \to s^{t'} N$. The following elementary facts about the suspension functor are left to the reader to prove:

(1) If X is projective, then sX is projective. (Hint: $s(\Gamma \otimes V) \cong \Gamma \otimes sV$ for a graded vector space V.)
(2) $_\Gamma\mathbf{Mod}(sW, sX) \cong {_\Gamma}\mathbf{Mod}(W, X)$.
(3) If $0 \leftarrow X \leftarrow W_\bullet$ is a projective resolution of X, then $0 \leftarrow sX \leftarrow sW_\bullet$ is a projective resolution of sX.

We define $[f] \circ [g]$ as follows: Using the defining property of projective modules, lift $f \colon P_p \to s^t M$ up the resolution to $f_q \colon P_{p+q} \to s^t Q_q$ for $q \geq 0$. Suspend g to $s^t g \colon s^t Q_q \to s^{t+t'} N$ and let $[f] \circ [g] = [s^t g \circ f_q]$ in $\operatorname{Ext}_\Gamma^{p+q, t+t'}(L, N)$. The following diagram depicts this construction:

$$
\begin{array}{ccccccccc}
0 \longleftarrow L \longleftarrow P_0 \longleftarrow \cdots \longleftarrow P_p \longleftarrow P_{p+1} \longleftarrow \cdots \longleftarrow P_{p+q} \longleftarrow \\
\quad f \Big/ \quad \Big\downarrow f_0 \qquad \Big\downarrow f_1 \qquad\qquad \Big\downarrow f_q \\
0 \longleftarrow s^t M \longleftarrow s^t Q_0 \longleftarrow s^t Q_1 \longleftarrow \cdots \longleftarrow s^t Q_q \longleftarrow \\
\qquad\qquad\qquad\qquad\qquad\qquad\qquad \Big\downarrow s^t g \\
\qquad\qquad\qquad\qquad\qquad\qquad\qquad s^{t+t'} N
\end{array}
$$

Because we can lift f_q to f_{q+1} and $g \circ d_M = 0$, it follows from the equation $s^t g \circ f_q \circ d_L = s^t g \circ s^t d_M \circ f_{q+1} = 0$ that $s^t g \circ f_q$ is a cycle. To show that all of the choices made in the construction are irrelevant, observe that two choices differ by a chain homotopy and so the difference vanishes on homology. The bilinearity and associativity are elementary to establish.

We remark that speaking of the suspension is the same as speaking of maps that change degree and so, by introducing the appropriate signs, this entire discussion can be carried out without the suspension. We do this later. □

Yoneda's original construction of the composition pairing is useful both conceptually and computationally. The construction depends on the identification of $\operatorname{Ext}_\Gamma^n(L, M)$ with equivalence classes of exact sequences of Γ-modules of the form

$$0 \to M \to E_{n-1} \to E_{n-2} \to \cdots \to E_0 \to L \to 0 \quad (n > 0).$$

Two exact sequences of this form are said be equivalent if there are homomorphisms $\varphi_i \colon E_i \to E_i'$ that commute with the morphisms in the sequences and the identity maps on L and M.

Given classes in $\operatorname{Ext}_\Gamma^p(L, M)$ and $\operatorname{Ext}_\Gamma^q(M, N)$ we can take representative exact sequences,

$$0 \to M \to E_{p-1} \to E_{p-2} \to \cdots \to E_1 \to E_0 \to L \to 0$$

and $$0 \to N \to F_{q-1} \to F_{q-2} \to \cdots \to F_1 \to F_0 \to M \to 0.$$

To represent the product splice these two sequences together at M:

$$
\begin{array}{c}
M \\
\diagup \quad \diagdown \\
0 \to N \to F_{q-1} \to \cdots \to F_0 \longrightarrow E_{p-1} \to \cdots \to E_0 \to L \to 0.
\end{array}
$$

[Yoneda54] showed that this pairing coincides with the composition product. An immediate consequence of this identification is the following result.

Proposition 9.6. *If* $0 \to A \to B \to C \to 0$ *is a short exact sequence of* Γ-*modules representing a class* α *in* $\mathrm{Ext}^1_\Gamma(C, A)$, *then the coboundary maps in the long exact sequences derived from the sequence,*

$$\delta\colon \mathrm{Ext}^s_\Gamma(A, N) \to \mathrm{Ext}^{s+1}_\Gamma(C, N) \quad \text{and} \quad \delta\colon \mathrm{Ext}^s_\Gamma(N, C) \to \mathrm{Ext}^{s+1}_\Gamma(N, A),$$

are given by left and right multiplication by α, *respectively.*

The internal degree of such an n-fold extension is given by the total change in degree from left to right. That the internal degree of a product is the sum of the internal degrees follows immediately.

Corollary 9.7. $\mathrm{Ext}^{*,*}_\Gamma(M, M)$ *is a bigraded algebra over* k *with the composition product as multiplication. Furthermore,* $\mathrm{Ext}^{*,*}_\Gamma(L, M)$ *is a right, and* $\mathrm{Ext}^{*,*}_\Gamma(M, L)$ *a left,* $\mathrm{Ext}^{*,*}_\Gamma(M, M)$-*module.*

With a concrete resolution like the cobar construction on hand, it is reasonable to attempt to represent the composition product on $\mathrm{Ext}^{*,*}_\Gamma(k, k)$ at the level of elements in the cobar resolution. Let $[\alpha_1 \mid \cdots \mid \alpha_p]$ be in $F_p(\Gamma^{\mathrm{dual}})$ and $[\beta_1 \mid \cdots \mid \beta_q]$ be in $F_q(\Gamma^{\mathrm{dual}})$. Notice that the internal degree of $\alpha\colon I(\Gamma) \to k$ is the degree on which α is nonzero, that is, $\deg \alpha = t$ if $\alpha \neq 0$ as a mapping $I(\Gamma)^t \to k$. We define a product on $F_*(\Gamma^{\mathrm{dual}})$ by juxtaposition

$$[\alpha_1 \mid \cdots \mid \alpha_p] \otimes [\beta_1 \mid \cdots \mid \beta_q] \mapsto [\alpha_1 \mid \cdots \mid \alpha_p \mid \beta_1 \mid \cdots \mid \beta_q].$$

This mapping is clearly bilinear and associative. To see that it induces a product on $\mathrm{Ext}^{*,*}_\Gamma(k, k)$, we show that the differential is a derivation. Let $U = [\alpha_1 \mid \cdots \mid \alpha_p]$ and $V = [\alpha_{p+1} \mid \cdots \mid \alpha_{p+q}]$, then

$$d^*(UV) = d^*([\alpha_1 \mid \cdots \mid \alpha_{p+q}])$$

$$= \sum_{i=1}^{p+q} \sum_j [\bar{\alpha}_1 \mid \cdots \bar{\alpha}_{i-1} \mid \bar{\alpha}'_{i,j} \mid \alpha''_{i,j} \mid \cdots \mid \alpha_{p+q}]$$

$$= \sum_{i=1}^{p} \sum_j [\bar{\alpha}_1 \mid \cdots \bar{\alpha}_{i-1} \mid \bar{\alpha}'_{i,j} \mid \alpha''_{i,j} \mid \cdots \mid \alpha_p][\alpha_{p+1} \mid \cdots \mid \alpha_{p+q}]$$

$$+ \sum_{r=1}^{q} \sum_u [\bar{\alpha}_1 \mid \cdots \bar{\alpha}_p][\bar{\alpha}_{p+1} \mid \cdots \mid \bar{\alpha}_{p+r-1} \mid \bar{\alpha}'_{p+r,u} \mid \alpha''_{p+r,u} \mid \cdots \mid \alpha_{p+q}]$$

$$= d^*(U)V + (-1)^{p+t}U d^*(V),$$

where t is the internal degree of U, that is, $t = \sum_{i=1}^{p} \deg \alpha_i$. Thus, this juxtaposition product induces a product on $\mathrm{Ext}^{*,*}_\Gamma(k, k)$ with the correct bidegree.

Theorem 9.8. *The composition product and juxtaposition product coincide on* $\mathrm{Ext}_{\Gamma}^{*,*}(k, k)$.

PROOF: We show how to lift a given mapping through the bar construction. Suppose $[f] \in \mathrm{Ext}_{\Gamma}^{s}(k, k)$ and $[g] \in \mathrm{Ext}_{\Gamma}^{t}(k, k)$. Consider the diagram where $f_i(\gamma[\gamma_1 \mid \cdots \mid \gamma_{s+i}]) = \bar{\gamma}[\bar{\gamma}_1 \mid \cdots \mid \bar{\gamma}_i]f([\gamma_{i+1} \mid \cdots \mid \gamma_{i+s}])$.

$$\cdots \to \Gamma \otimes I(\Gamma)^{s+t} \to \cdots \to \Gamma \otimes I(\Gamma)^{s+1} \to \Gamma \otimes I(\Gamma)^{s} \to \Gamma \otimes I(\Gamma)^{s-1} \to \cdots$$

$$\cdots \longrightarrow \Gamma \otimes I(\Gamma)^{t} \longrightarrow \cdots \longrightarrow \Gamma \otimes I(\Gamma) \longrightarrow \Gamma \longrightarrow k \longrightarrow 0.$$

with vertical maps f_t, f_1, f_0, f and g, k.

This lifting satisfies $d \circ f_i = f_{i-1} \circ d$ and so we have a mapping of part of the one resolution to the other. By the definition of the composition product $[f] \circ [g] = [g \circ f_t]$. The value of $g \circ f_t$ on a typical class $\gamma[\gamma_1 \mid \cdots \mid \gamma_{s+t}]$ is $g(\bar{\gamma}[\bar{\gamma}_1 \mid \cdots \bar{\gamma}_t])f([\gamma_{t+1} \mid \cdots \mid \gamma_{t+s}])$. When we represent $[f]$ and $[g]$ in the cobar complex as tensor products of elements of Γ^{dual} we get exactly the value obtained by applying the juxtaposed dual elements to a typical argument and adjusting signs for the suspensions. Thus the products coincide. □

The simplicity of the product induced by juxtaposition allows one to do computations at the level of the cobar construction. This is especially useful for determining such secondary phenomena as Massey products (see §8.2) and $\smile_1$-products.

Suppose Γ is a Hopf algebra with cocommutative coproduct $\psi \colon \Gamma \to \Gamma \otimes \Gamma$ and counit $\eta \colon \Gamma \to k$. We assume further that Γ is of finite type over k. With these data, there is yet another way to induce a multiplication on $\mathrm{Ext}_{\Gamma}^{*,*}(k, k)$. Suppose we are given a projective resolution of k, $X_{\bullet} \to k \to 0$ with the homomorphism $\varepsilon \colon X_0 = \Gamma \to k \to 0$, the counit of Γ. Let $X_{\bullet} \otimes X_{\bullet}$ be the complex with $(X_{\bullet} \otimes X_{\bullet})_s = \bigoplus_{i+j=s} X_i \otimes X_j$ and differential $d_{\otimes}(x \otimes y) = d(x) \otimes y + (-1)^{p+r} x \otimes d(y)$ $(x \in (X_p)^r)$, then $X_{\bullet} \otimes X_{\bullet}$ can be given a $\Gamma \otimes \Gamma$ action via the twist map and we obtain a projective $\Gamma \otimes \Gamma$ resolution of $k = k \otimes_k k$.

Using properties of the coproduct, counit and projective modules, we can construct a map, $\Delta \colon X_{\bullet} \to X_{\bullet} \otimes X_{\bullet}$, making the following diagram commute:

$$
\begin{array}{ccccccc}
0 & \longleftarrow & k & \xleftarrow{\ \varepsilon\ } & \Gamma & \longleftarrow & X_{\bullet} \\
& & \Big\| {\scriptstyle =} & & \Big\downarrow {\scriptstyle \psi} & & \Big\downarrow {\scriptstyle \Delta} \\
0 & \longleftarrow & k \otimes_k k & \xleftarrow[\varepsilon \otimes \varepsilon]{} & \Gamma \otimes \Gamma & \longleftarrow & X_{\bullet} \otimes X_{\bullet}
\end{array}
$$

Applying the functor $\mathrm{Hom}_\Gamma^*(\ ,k)$ induces a product

$$\mathrm{Hom}_\Gamma(X_\bullet, k) \otimes \mathrm{Hom}_\Gamma(X_\bullet, k) \longrightarrow \mathrm{Hom}_\Gamma(X_\bullet, k)$$

that reduces to a product

$$\mu\colon \mathrm{Ext}_\Gamma^{s,t}(k, k) \otimes \mathrm{Ext}_\Gamma^{s',t'}(k, k) \longrightarrow \mathrm{Ext}_\Gamma^{s+s',t+t'}(k, k).$$

One can construct an explicit Δ on the cobar resolution from which it is easily seen that μ is the same product as the one induced by juxtaposition.

We introduce this construction to prove the following result.

Theorem 9.9. *If the coproduct on Γ is cocommutative, then the multiplication on $\mathrm{Ext}_\Gamma^{*,*}(k, k)$ is graded commutative with signs given by*

$$\alpha \cdot \beta = (-1)^{ss'+tt'}\beta \cdot \alpha$$

for $\alpha \in \mathrm{Ext}_\Gamma^{s,t}(k, k)$ and $\beta \in \mathrm{Ext}_\Gamma^{s',t'}(k, k)$.

PROOF: We extend the diagram in the construction to another row:

$$
\begin{array}{ccccccc}
0 & \longleftarrow & k & \xleftarrow{\ \epsilon\ } & \Gamma & \longleftarrow & X_\bullet \\
& & \Big\downarrow{\scriptstyle =} & & \Big\downarrow{\scriptstyle \psi} & & \Big\downarrow{\scriptstyle \Delta} \\
0 & \longleftarrow & k \otimes_k k & \xleftarrow{\ \epsilon\otimes\epsilon\ } & \Gamma \otimes \Gamma & \longleftarrow & X_\bullet \otimes X_\bullet \\
& & \Big\downarrow{\scriptstyle =} & & \Big\downarrow{\scriptstyle T} & & \Big\downarrow{\scriptstyle T} \\
0 & \longleftarrow & k \otimes_k k & \xleftarrow{\ \epsilon\otimes\epsilon\ } & \Gamma \otimes \Gamma & \longleftarrow & X_\bullet \otimes X_\bullet
\end{array}
$$

Since Γ is cocommutative, $\psi = T\psi$ and Δ is chain homotopic to $T\Delta$. This proves the theorem. $\qquad\square$

On the cobar resolution with its juxtaposition product it is apparent that Δ and $T\Delta$ are not the same mapping. An explicit chain homotopy can be constructed. If we restrict our attention to $k = \mathbb{F}_2$, this chain homotopy allows us to define $\smile_i$-products and hence Steenrod operations on $H^{*,*}(\Gamma)$. For $k = \mathbb{F}_p$, a similar construction over the $(p-1)^{\mathrm{st}}$ iterate of the coproduct, $\psi^p\colon \Gamma \to \Gamma \otimes \cdots \otimes \Gamma$ (p times), allows one to define the mod p Steenrod operations on $H^{*,*}(\Gamma)$. We refer the reader to [Adams58] and [Liulevicius62] for details of these constructions. The elementary properties of these operations are listed next for later applications.

Theorem 9.10. *Let* $(\Gamma, \psi, \Delta, \varepsilon, \eta)$ *be a cocommutative Hopf algebra over* $\mathbb{F}_p$.
(a) $p = 2$. *There are operations* Sq^i *on* $\operatorname{Ext}_\Gamma^{*,*}(\mathbb{F}_2, \mathbb{F}_2)$ *that satisfy the following properties:*

(1) $Sq^i \colon \operatorname{Ext}_\Gamma^{s,t}(\mathbb{F}_2, \mathbb{F}_2) \to \operatorname{Ext}_\Gamma^{s+i,2t}(\mathbb{F}_2, \mathbb{F}_2)$,

(2) $Sq^i(ab) = \sum_{m+n=i} Sq^m(a) Sq^n(b)$,

(3) $Sq^r Sq^s = \sum_{t=0}^{[s/2]} \binom{s-t-1}{r-2t} Sq^{r+s-t} Sq^t$,

(4) $Sq^0(\{[\alpha_1 \mid \cdots \mid \alpha_n]\}) = \{[\alpha_1^2 \mid \cdots \mid \alpha_n^2]\}$,

(5) $Sq^s x = x^2$ *if* $x \in \operatorname{Ext}_\Gamma^s(\mathbb{F}_2, \mathbb{F}_2)$.

(b) p, *an odd prime. There are operations* P^i *and* βP^i *on* $\operatorname{Ext}_\Gamma^{*,*}(\mathbb{F}_p, \mathbb{F}_p)$ *that satisfy the following properties:*

(1) $P^i \colon \operatorname{Ext}_\Gamma^{s,t}(\mathbb{F}_p, \mathbb{F}_p) \to \operatorname{Ext}_\Gamma^{s+(2i-t)(p-1),pt}(\mathbb{F}_p, \mathbb{F}_p)$,
$\quad \beta P^i \colon \operatorname{Ext}_\Gamma^{s,t}(\mathbb{F}_p, \mathbb{F}_p) \to \operatorname{Ext}_\Gamma^{s+(2i-t)(p-1)+1,pt}(\mathbb{F}_p, \mathbb{F}_p)$,

(2) $P^i(ab) = \sum_{i \geq j \geq 0} P^j(a) P^{i-j}(b)$,
$\quad \beta P^i(ab) = \sum_{i \geq j \geq 0} \beta P^j(a) P^{i-j}(b) + (-1)^{\deg a} P^j(a) \beta P^{i-j}(b)$,

(3) $P^r P^s = \sum_{t=0}^{[r/p]} (-1)^{r+t} \binom{(p-1)(s-t)-1}{r-pt} P^{r+s-t} P^t$,

$\quad P^r \beta P^s = \sum_{t=0}^{[r/p]} (-1)^{r+t} \binom{(p-1)(s-t)}{r-pt} (\beta P^{r+s-t}) P^t$
$\qquad + \sum_{t+0}^{[r-1/p]} \binom{(p-1)(s-t)-1}{r-pt-1} P^{r+s-t} (\beta P^t)$,

(4) $P^0(\{[\alpha_1 \mid \cdots \mid \alpha_n]\}) = \{[\alpha_1^p \mid \cdots \mid \alpha_n^p]\}$.

(5) $P^r a = a^p$ *if* $a \in \operatorname{Ext}_\Gamma^{s,t}(\mathbb{F}_p, \mathbb{F}_p)$ *and* $2r = s + t$.

These operations differ from the usual Steenrod operations because Sq^0 and P^0 are not the identity mappings and Sq^1 and β are not Bockstein operators. The explicit expression for Sq^0 on a class in the cobar construction will be useful later.

We consider a simple example over $k = \mathbb{F}_2$: Suppose the Hopf algebra, Γ, is the divided power algebra on a single generator $\Gamma(x)$. Recall that $\Gamma(x)$ is generated as an algebra by generators $\gamma_i(x)$ for $i = 0, 1, \ldots$ with $\gamma_0(x) = 1$, $\gamma_1(x) = x$ and $\deg \gamma_i(x) = i \deg x$. The product is determined by the relations

$$\gamma_i(x) \gamma_j(x) = \binom{i+j}{i} \gamma_{i+j}(x)$$

and the coproduct is given on generators by

$$\psi(\gamma_i(x)) = \sum_{j=0}^{i} \gamma_j(x) \otimes \gamma_{i-j}(x).$$

We choose this Hopf algebra to study because its dual is given by $\Gamma(x)^{\text{dual}} = \mathbb{F}_2[y]$, where y is dual to x. The reader can easily read the Hopf algebra structure on $k[y]$ from the product and coproduct on $\Gamma(x)$ and it agrees with the usual polynomial multiplication and coproduct

$$\psi(y^q) = \sum_{i=0}^{q} \binom{q}{i} y^i \otimes y^{q-i}.$$

To compute $H^{*,*}(\Gamma(x))$, we can apply the cobar construction to the coalgebra $(\mathbb{F}_2[y], \psi)$. The following lemma holds generally.

Lemma 9.11.   Let Γ be a Hopf algebra of finite type over a field k. Then $H^{1,*}(\Gamma) = \text{Ext}_{\Gamma}^{1,*}(k,k) = \text{Prim}^*(\Gamma^{\text{dual}}) \cong Q^*(\Gamma)$.

PROOF: We recall that $\text{Prim}^*(\ )$ is the functor that associates to a Hopf algebra its graded vector subspace (in fact, sub-Lie algebra) of primitive elements (see §4.4). Also the functor $Q^*(\ )$ associates to a Hopf algebra its quotient vector space of indecomposable elements. When Γ is of finite type over k, $\text{Prim}^*(\Gamma^{\text{dual}}) \cong Q^*(\Gamma)$, so it suffices to compute $\text{Prim}^*(\Gamma^{\text{dual}})$. This may be defined as the kernel of the reduced coproduct

$$\varphi^* : I(\Gamma^{\text{dual}}) \longrightarrow I(\Gamma^{\text{dual}}) \otimes I(\Gamma^{\text{dual}}).$$

From the cobar construction, this kernel is exactly $\text{Ext}_{\Gamma}^{1,*}(k,k)$.   □

To obtain $H^{1,*}(\Gamma(x))$ then, we find the primitives in $\mathbb{F}_2[y]$. The arithmetic of binomial coefficients (Lucas's Lemma, see §7.3) determines the primitive classes $l_i = [y^{2^i}] \in \text{Ext}_{\Gamma(x)}^{1,*}(k,k)$, for $i = 0, 1, \ldots$. The products of these classes can be identified in the cobar construction, where they are cycles; $l_i l_j$, for example, corresponds to $[y^{2^i} \mid y^{2^j}]$. Coboundary formulas, such as

$$\delta[y^3] = [y^2 \mid y] + [y \mid y^2] = l_1 l_0 + l_0 l_1,$$

show that the product on the sub algebra determined by the generators $\{l_i\}$ is commutative.

To complete the computation, one must show that the products of the l_i's are not boundaries and furthermore, that no other class can be a cycle. Again the arithmetic of $\binom{q}{i}$ mod 2 can be applied and one deduces that

$$H^{*,*}(\Gamma(x)) \cong \mathbb{F}_2[l_i \mid i = 0, 1, \ldots],$$

as bigraded algebras where each l_i has bidegree $(1, 2^i \deg x)$. A corollary of this computation is the fact that $Sq^0(l_i) = l_{i+1}$ and $Sq^1(l_i) = l_i^2$ in $\text{Ext}_{\Gamma(x)}^{*,*}(\mathbb{F}_2, \mathbb{F}_2)$. The reader should provide any further details needed to feel comfortable with this computation. We will use the results in computing the cohomology of $\mathcal{A}_2$, the Steenrod algebra.

A change-of-rings spectral sequence

We next introduce an analogue of the Lyndon-Hochschild-Serre spectral sequence (Theorem 8bis.12) to compute $H^{*,*}(\Gamma)$, when Γ is an extension of Hopf algebras

$$0 \to \Lambda \to \Gamma \xrightarrow{\pi} \Gamma//\Lambda \to 0.$$

Here $\Gamma//\Lambda = \Gamma/I(\Lambda){\cdot}\Gamma$ and Λ is normal in Γ, that is, $I(\Lambda) \cdot \Gamma = \Gamma \cdot I(\Lambda)$. This spectral sequence can be derived additively by methods of [Cartan-Eilenberg56]; its multiplicative properties were proved by [Adams60].

Theorem 9.12. *Let Λ be a sub-Hopf algebra of a Hopf algebra Γ. Suppose Λ is **central** in Γ, that is, $ab = (-1)^{\deg a \deg b}ba$ for $a \in \Lambda$, $b \in \Gamma$. Then there is a spectral sequence, converging to $H^*(\Gamma)$, with*

$$E_2^{p,q} \cong H^q(\Lambda) \otimes_k H^p(\Gamma//\Lambda),$$

and differentials d_r of bidegree $(r, 1 - r)$. Furthermore, this spectral sequence converges to $H^*(\Gamma)$ as an algebra with the product structure on the E_2-term given in the isomorphism by

$$(x \otimes y) \cdot (x' \otimes y') = (-1)^{pq+tt'}(xx') \otimes (yy')$$

when $y \in H^{p,t}(\Gamma//\Lambda)$ and $x' \in H^{q,t'}(\Lambda)$.

PROOF: Before beginning the proof, we observe that this spectral sequence is actually trigraded—the third grading is given by the internal grading t in $H^{*,t}(\ )$. All of the differentials preserve this grading and so it is carried through to $E_\infty^{*,*}$, where it corresponds to the internal grading on $H^{*,*}(\Gamma)$. This hidden grading plays a key role in computations (§9.6).

The proof exploits duality and so we begin with $\bar{B}(\Gamma)$, the reduced bar construction on Γ. Filter $\bar{B}(\Gamma)$ by

$$F^p\bar{B}(\Gamma) = \{ [\gamma_1 \mid \cdots \mid \gamma_s] \text{ for which at least } s - p \text{ of the } \gamma_i \text{ lie in } I(\Lambda)\}.$$

This is an increasing filtration with $F^0\bar{B}(\Gamma) = \bar{B}(\Lambda)$ and $F^s\bar{B}^s(\Gamma) = \bar{B}^s(\Gamma)$. Since Λ is a sub-Hopf algebra, $d(F^p\bar{B}(\Gamma)) \subset F^p\bar{B}(\Gamma)$. Thus we have a spectral sequence, converging to $\text{Tor}^\Gamma(k, k)$ with E^1-term given by

$$E^1_{p,q} = H_{p+q}(F^p\bar{B}(\Gamma)/F^{p-1}\bar{B}(\Gamma), d^0).$$

To prove the theorem we first establish a chain equivalence between the E^0-term, $(F^p\bar{B}(\Gamma)/F^{p-1}\bar{B}(\Gamma), d^0)$, and $(\bar{B}(\Lambda) \otimes I(\Gamma//\Lambda)^{\otimes p}, d \otimes 1)$.

Consider the mapping, $\nu^p \colon F^p\bar{B}(\Gamma) \to \bar{B}(\Lambda) \otimes I(\Gamma//\Lambda)^{\otimes p}$

$$\nu^p([\gamma_1 \mid \cdots \mid \gamma_s]) = [\gamma_1 \mid \cdots \mid \gamma_{s-p}] \otimes \pi\gamma_{s-p+1} \otimes \cdots \otimes \pi\gamma_s,$$

where $\pi\colon \Gamma \to \Gamma//\Lambda$ is the projection. Since $\pi\lambda = 0$ for $\lambda \in I(\Lambda)$ and $[\gamma_1 \mid \cdots \mid \gamma_s]$ is in $F^p\bar{B}(\Gamma)$ if at least $s-p$ of the γ_i lie in $I(\Lambda)$, $\nu^p([\gamma_1 \mid \cdots \mid \gamma_s]) = 0$ unless exactly γ_1 through γ_{s-p} lie in $I(\Lambda)$. Hence ν^p is well-defined.

We show that the following diagram commutes

$$
\begin{array}{ccc}
F^p\bar{B}(\Gamma) & \xrightarrow{\ \nu^p\ } & \bar{B}(\Lambda) \otimes I(\Gamma//\Lambda)^{\otimes p} \\
\ \downarrow{\scriptstyle d} & & \ \downarrow{\scriptstyle d\otimes 1} \\
F^p\bar{B}(\Gamma) & \xrightarrow[\ \nu^p\ ]{} & \bar{B}(\Lambda) \otimes I(\Gamma//\Lambda)^{\otimes p}.
\end{array}
$$

Since ν^p is zero except on elements of the form $[\lambda_1 \mid \cdots \mid \lambda_{s-p} \mid \gamma_{s-p+1} \mid \cdots \mid \gamma_s]$, we check that $\nu^p \circ d = (d \otimes 1) \circ \nu^p$ on such an expression. We write

$$d([\lambda_1 \mid \cdots \mid \lambda_{s-p} \mid \gamma_{s-p+1} \mid \cdots \mid \gamma_s]) =$$
$$\sum_{i=1}^{s-p-1} [\bar{\lambda}_1 \mid \cdots \mid \bar{\lambda}_i\lambda_{i+1} \mid \cdots \mid \lambda_{s-p} \mid \gamma_{s-p+1} \mid \cdots \mid \gamma_s]$$
$$+ [\bar{\lambda}_1 \mid \cdots \mid \bar{\lambda}_{s-p-1} \mid \bar{\lambda}_{s-p}\gamma_{s-p+1} \mid \cdots \mid \gamma_s]$$
$$+ \sum_{j=1}^{p-1} [\bar{\lambda}_1 \mid \cdots \mid \bar{\lambda}_{s-p} \mid \bar{\gamma}_{s-p+1} \mid \cdots \mid \bar{\gamma}_{s-p+j}\gamma_{s-p+j+1} \mid \cdots \mid \gamma_s].$$

Observe that $\nu^p([\bar{\lambda}_1 \mid \cdots \mid \bar{\lambda}_{s-p-1} \mid \bar{\lambda}_{s-p}\gamma_{s-p+1} \mid \cdots \mid \gamma_s]) = 0$ because $\pi(\bar{\lambda}_{s-p}\gamma_{s-p+1}) = 0$ and that

$$\nu^p([\bar{\lambda}_1 \mid \cdots \mid \bar{\lambda}_{s-p} \mid \bar{\gamma}_{s-p+1} \mid \cdots \mid \bar{\gamma}_{s-p+j}\gamma_{s-p+j+1} \mid \cdots \mid \gamma_s])$$
$$= [\bar{\lambda}_1 \mid \cdots \mid \bar{\lambda}_{s-p-1}] \otimes \pi\bar{\lambda}_{s-p} \otimes \pi\bar{\gamma}_{s-p+1} \otimes \cdots \otimes \pi\gamma_s = 0$$

by the definition of ν^p on $\bar{B}(\Gamma)^{s-1,*}$.

Finally, observe that ν^p takes $F^{p-1}\bar{B}(\Gamma)$ to $\{0\}$ and so we get an induced mapping of complexes

$$\bar{\nu}^p\colon (F^p\bar{B}(\Gamma)/F^{p-1}\bar{B}(\Gamma), d^0) \to (\bar{B}(\Lambda) \otimes I(\Gamma//\Lambda)^{\otimes p}, d \otimes 1).$$

We can rewrite $F^p\bar{B}(\Gamma)/F^{p-1}\bar{B}(\Gamma)$ as

$$k \otimes_\Lambda \Lambda \otimes F^p\bar{B}(\Gamma)/F^{p-1}\bar{B}(\Gamma)$$

and so we plot a circuitous route to showing $\bar{\nu}^p$ is a chain equivalence by showing

$$g^p = 1 \otimes \bar{\nu}^p\colon \Lambda \otimes F^p\bar{B}(\Gamma)/F^{p-1}\bar{B}(\Gamma) \to \Lambda \otimes \bar{B}(\Lambda) \otimes I(\Gamma//\Lambda)^{\otimes p}$$

is a homology equivalence.

We introduce some associated complexes: Let $C(p) = \Lambda \otimes F^p + \Gamma \otimes F^{p-1}$, where we write $F^p = F^p\bar{B}(\Gamma)$. Then the bar construction differential d, defined on $B(\Gamma, \Gamma, k)$, takes $C(p)$ to itself, as does the chain homotopy s. (You'll find the formulas in Proposition 7.8). It follows that $C(p)$ is acyclic. There is a surjection $C(p) \to \Lambda \otimes F^p/F^{p-1}$ that is induced by $1 \otimes pr$ where pr is projection. The kernel of this surjection is $\Gamma \otimes F^{p-1}$. Increase p to $p+1$ and the surjection $C(p+1) \twoheadrightarrow \Lambda \otimes F^{p+1}/F^p$ takes $C(p)$ as a subspace of $C(p+1)$ to 0. This leads to a short exact sequence, which defines another complex $K(p)$ as kernel,

$$0 \to K(p) \to C(p+1)/C(p) \to \Lambda \otimes F^{p+1}/F^p \to 0.$$

The associated long exact sequence on homology implies

$$H_q(K(p)) \cong H_{q+1}(\Lambda \otimes F^{p+1}/F^p),$$

because $C(p+1)/C(p)$ is acyclic. We next analyze the complex $K(p)$ that leads to the desired result via induction.

Up to this point we have been using only part of our hypotheses—that Λ is a normal subalgebra of the algebra Γ. In the case of Hopf algebras we can apply the following remarkable consequence of the Hopf algebra structure due to [Milnor-Moore65] (Exercise 6.12):

Fact. *Suppose Λ is a sub-Hopf algebra of Γ and Λ is normal in Γ. Then Γ has a basis as a Λ-module, consisting of 1 and certain homogeneous elements in $I(\Gamma)$; Γ is free as a right Λ-module on this basis. Furthermore, the basis projects to a vector space basis for $\Gamma//\Lambda$.*

We denote such a basis for Γ by $\{\gamma_i\}$ and its image in $\Gamma//\Lambda$ by $\{\omega_i = \pi(\gamma_i)\}$. It follows that we can write Γ as a right Λ-module by

$$\Gamma \cong (\Gamma//\Lambda) \otimes \Lambda = \Lambda + I(\Gamma//\Lambda) \otimes \Lambda.$$

Since the kernel of $C(p+1) \twoheadrightarrow \Lambda \otimes F^{p+1}/F^p$ is $\Gamma \otimes F^p$, $K(p)$ may be written as the image of the inclusion followed by a quotient:

$$K(p) = \operatorname{im}(\Gamma \otimes F^p \to C(p+1)/C(p)).$$

Replacing our expression for Γ as a right Λ-module, we get

$$
\begin{aligned}
K(p) &= \operatorname{im}(\Gamma \otimes F^p \to C(p+1)/C(p)) \\
&= \operatorname{im}(\Lambda \otimes F^p + I(\Gamma//\Lambda) \otimes \Lambda \otimes F^p \longrightarrow \\
&\qquad \left(\Lambda \otimes F^{p+1} + I(\Gamma//\Lambda) \otimes \Lambda \otimes F^p \big/ \Lambda \otimes F^p + I(\Gamma//\Lambda) \otimes \Lambda \otimes F^{p-1} \right) \big) \\
&\cong I(\Gamma//\Lambda) \otimes \Lambda \otimes F^p/F^{p-1}.
\end{aligned}
$$

We proceed by induction to prove the following assertion:

$$(9.13) \qquad H_q(\Lambda \otimes F^p/F^{p-1}) \cong \begin{cases} I(\Gamma//\Lambda)^{\otimes p}, & \text{if } q = p, \\ 0, & \text{elsewhere.} \end{cases}$$

Consider the composite, denoted by f^p,

$$f^p: \Lambda \otimes F^p/F^{p-1} \xrightarrow{1 \otimes \nu^p} \Lambda \otimes \bar{B}(\Lambda) \otimes I(\Gamma//\Lambda)^{\otimes p} \xrightarrow{\epsilon \otimes 1} I(\Gamma//\Lambda)^{\otimes p}$$

For $p = 0$, f^0 is simply the mapping $\Lambda \otimes \bar{B}(\Lambda) \to k$ given by the augmentation. But this is an equivalence by the properties of the bar construction. Assume (9.13) for p, that is, $H(f^p): H(\Lambda \otimes F^p/F^{p-1}) \to I(\Gamma//\Lambda)^{\otimes p}$ is an isomorphism. By the properties of the complex $K(p)$, we have

$$H(K(p)) \cong H(I(\Gamma//\Lambda) \otimes \Lambda \otimes F^p/F^{p-1}),$$

where the differential on $I(\Gamma//\Lambda)$ is zero. Thus we have

$$H(K(p)) \cong I(\Gamma//\Lambda) \otimes H(\Lambda \otimes F^p/F^{p-1}) \cong I(\Gamma//\Lambda)^{\otimes p+1}.$$

Since $H_{q+1}(\Lambda \otimes F^{p+1}/F^p) \cong H_q(K(p))$, we have shown that the assertion (9.13) holds in case of $p + 1$, and hence for all p by induction.

When we apply $k \otimes_\Lambda -$ to g^p, we obtain the desired chain equivalence between $F^p\bar{B}(\Gamma)/F^{p-1}\bar{B}(\Gamma)$ and $\bar{B}(\Lambda) \otimes I(\Gamma//\Lambda)^{\otimes p}$. The E^1-term of the spectral sequence is given by

$$E^1_{*,p} \cong H_*(\bar{B}(\Lambda) \otimes I(\Gamma//\Lambda)^{\otimes p}, d \otimes 1) \cong \text{Tor}^\Lambda_*(k, k) \otimes I(\Gamma//\Lambda)^{\otimes p}.$$

To compute the E^2-term of this spectral sequence, we introduce a chain mapping $\bar{\mu}^p: \bar{B}(\Lambda) \otimes I(\Gamma//\Lambda)^{\otimes p} \to F^p/F^{p-1}$ that acts as an inverse to $\bar{\nu}^p$ and induces a chain equivalence on $\bar{B}(\Gamma//\Lambda)$. First, use the basis for $\Gamma//\Lambda$, $\{\omega_i\}$, to split the projection π and obtain a map $\sigma: \Gamma//\Lambda \to \Gamma$ of graded vector spaces. Next, introduce a version of shuffle map, $\mu^p: \bar{B}(\Lambda) \otimes I(\Gamma)^{\otimes p} \to F^p\bar{B}(\Gamma)$, given by

$$\mu^p([a_1 \mid \cdots \mid a_r] \otimes [a_{r+1} \mid \cdots \mid a_{r+p}]) = \sum_{(r,p)\text{-shuffles } \sigma} (-1)^{\varepsilon(\sigma)}[a_{\sigma(1)} \mid \cdots \mid a_{\sigma(r+p)}].$$

Recall that an (r, p)-shuffle is an $(r + p)$-permutation that preserves order on 1 through r and on $r+1$ through $r+p$. The sign $\varepsilon(\sigma)$ is the sign of the permutation (see the proof of Lemma 7.11). The key property of the shuffle product in this setting is the formula $d(\mu^p(a \otimes b)) = \mu^p(d(a) \otimes b) \pm \mu^{p-1}(\bar{a} \otimes d(b))$, where $a \in \bar{B}(\Lambda)$ and $b \in \bar{B}(\Gamma)$. This equation depends on the centrality of Λ in Γ.

Let $\bar{\mu}^p$ denote the composite

$$\bar{\mu}^p: \bar{B}(\Lambda) \otimes I(\Gamma//\Lambda)^{\otimes p} \xrightarrow{1 \otimes \sigma^p} \bar{B}(\Lambda) \otimes I(\Gamma)^{\otimes p}$$
$$\xrightarrow{\mu^p} F^p\bar{B}(\Gamma) \to F^p\bar{B}(\Gamma)/F^{p-1}\bar{B}(\Gamma).$$

We leave it to the reader to verify that $\bar{\nu}^p \circ \bar{\mu}^p = \mathrm{id}$. Next consider the diagram

$$
\begin{array}{ccc}
H_q(\bar{B}(\Lambda) \otimes I(\Gamma//\Lambda)^{\otimes p}) & \xrightarrow{\bar{1} \otimes \bar{d}} & H_q(\bar{B}(\Lambda) \otimes I(\Gamma//\Lambda)^{\otimes p-1}) \\
\bar{\mu}^p_* \downarrow & & \uparrow \bar{\nu}^{p-1}_* \\
E^1_{p,q} & \xrightarrow{\quad d^1 \quad} & E^1_{p-1,q}.
\end{array}
$$

For $a \in \bar{B}(\Lambda)$ with $d(a) = 0$ and $b \in I(\Gamma//\Lambda)^{\otimes p}$,

$$
\begin{aligned}
\bar{\nu}^{p-1}_* d^1 \bar{\mu}^p_*(a \otimes b) &= \bar{\nu}^{p-1}_* \bar{d} \mu^p_*(1 \otimes \sigma^p)(a \otimes b) \\
&= \bar{\nu}^{p-1}_* \mu^{p-1}_*(\bar{a} \otimes \bar{d}\sigma^p b) \\
&= (1 \otimes \pi^{p-1})(\bar{a} \otimes \bar{d}\sigma^p b) \\
&= \bar{a} \otimes \bar{d}\pi^p \sigma^p b \\
&= \bar{a} \otimes \bar{d}b.
\end{aligned}
$$

Thus the diagram commutes and $E^2_{p,q} \cong H_q(\Lambda) \otimes H_p(\bar{B}(\Gamma//\Lambda))$.

The last step in proving Theorem 9.12 is the construction of a coproduct structure on the spectral sequence (we are still in the dual spectral sequence for $H_*(\Gamma)$). Recall that $\bar{B}(\Gamma)$ has a natural coproduct given by $\psi \colon \bar{B}(\Gamma) \to \bar{B}(\Gamma) \otimes \bar{B}(\Gamma)$,

$$
\psi([\gamma_1 \mid \cdots \mid \gamma_s]) = \sum_{i=0}^{s} [\gamma_1 \mid \cdots \mid \gamma_i] \otimes [\gamma_{i+1} \mid \cdots \mid \gamma_s]
$$

This coproduct is dual to the juxtaposition product on the cobar construction. We filter $\bar{B}(\Gamma) \otimes \bar{B}(\Gamma)$ with the tensor product filtration,

$$
F^p(\bar{B}(\Gamma) \otimes \bar{B}(\Gamma)) = \sum_{i=0}^{p} F^i \bar{B}(\Gamma) \otimes F^{p-i} \bar{B}(\Gamma).
$$

One can check that

(1) ψ is filtration-preserving and commutes with the differentials;
(2) $\bar{\nu}^p$ and $\bar{\mu}^p$ commute with ψ if we equip $\bar{B}(\Lambda)$ and $\bar{B}(\Gamma//\Lambda)$ with the same coproduct. (This requires that the shuffle product commute with ψ, which it does.)

With these facts the isomorphisms yield that the coproduct on the E^2-term is isomorphic to the tensor product of the coproducts on $H_*(\Lambda)$ and $H_*(\Gamma//\Lambda)$.

Finally, to obtain Theorem 9.12, it suffices to dualize this proof. We have proved the dual of Theorem 9.12 for the homology of Hopf algebras. □

A consequence of the proof of the multiplicative properties of the spectral sequence is the analogue of Corollary 6.9 that the Steenrod operations on $H^*(\Lambda)$ and $H^*(\Gamma//\Lambda)$ commute with the transgression, $d_r \colon E^{0,r-1}_r \to E^{r,0}_r$. These operations, for $k = \mathbb{F}_2$, act classically on the homological degree but double the hidden, internal degree. To prove the analogue of Corollary 6.9, one must work directly with the cobar complex and the explicit chain homotopies that give rise to the Steenrod operations. The interested reader can consult the papers of [Adams60] and [Liulevicius62].

With the necessary algebra all in place, we return to the problem of constructing the spectral sequence of Theorem 9.1.

9.3 The spectral sequence

The goal of the chapter is the computation of the groups, $\{Y, X\}_*$, where X and Y are spaces of finite type and Y is a finite dimensional CW-complex. The tools of choice, in this exposition, are elementary;

(1) the properties of fibrations, in particular, the construction of towers of principal fibrations and the exact sequences that result from these constructions,

(2) the properties of the suspension, including Freudenthal's Theorem (Theorem 4.10) and

(3) exact couples and their subsequent spectral sequences.

We first construct certain towers of fibrations, known as **Adams resolutions** that realize geometrically an algebraic resolution of the mod p cohomology of a space. The reader can compare these constructions with the Postnikov tower of a space (§4.3 and §6.1) and the towers built in §9.1 to describe higher order cohomology operations. By building such a resolution for a space X, then SX, then S^2X, and so forth, the mod p cohomology of the Adams resolutions assembles into better and better approximations of a free $\mathcal{A}_p$-module resolution of $H^*(X; \mathbb{F}_p)$. The application of the functor $[S^mY, -]$ to the system of fibrations results in an exact couple and a spectral sequence. Our analysis of the construction allows us to identify the E_2-term.

We remark that the bigrading on the spectral sequence that results from the exact couple is not standard and, in fact, does not conform to the definitions in Chapter 2. This could be avoided by an unnatural regrading of everything in sight, but this would complicate matters further. The nonstandard grading does not affect the arguments that follow but might seem odd on first exposure.

Next the properties of Adams resolutions are developed to demonstrate the convergence of the spectral sequence. Then a geometric pairing is defined at the E_1-level that provides us with a spectral sequence of algebras. The pairing is so defined as to yield the Yoneda multiplication at the E_2-term.

For the reader who has little interest in the geometric origins of the Adams spectral sequence, but interest in the computations, we suggest you skip on to §9.5 where the homological algebra of §9.2, the Steenrod algebra and the spectral sequence are used to compute some of the stable homotopy groups of spheres.

The construction: Adams resolutions

Let X be a space (of the homotopy type of a CW-complex) that is of finite type over $\mathbb{F}_p$ and suppose X is $(n-1)$-connected. The basic goal of an Adams resolution is the geometric realization of a free $\mathcal{A}_p$-module resolution of $H^*(X; \mathbb{F}_p)$ through dimensions $n \leq t < 2n$ (*the stable range*). The first step is to choose a graded vector space ${}_0V_0$ such that

$$ {}_0V_0 \cong_{2n-1} \tilde{H}^*(X; \mathbb{F}_p)/I(\mathcal{A}_p) \cdot \tilde{H}^*(X; \mathbb{F}_p). $$

By the representability of the mod p cohomology functor we can choose a mapping ${}_0F_0: X \to K({}_0V_0)$ such that $({}_0F_0)^*: H^*(K({}_0V_0); \mathbb{F}_p) \to H^*(X; \mathbb{F}_p)$ is an epimorphism in degrees less than $2n$ realizing the isomorphism when composed with the quotient.

Having chosen the mapping ${}_0F_0$, form the pullback of the path-loop fibration over it:

$$
\begin{array}{ccc}
{}_0X_1 & \longrightarrow & PK({}_0V_0) \\
\downarrow & & \downarrow \\
X & \xrightarrow{\;{}_0F_0\;} & K({}_0V_0).
\end{array}
$$

Consider the long exact sequence on cohomology (Example 5.D), for $t < 2n-1$, where we write $H^*(\ )$ for the cohomology $H^*(\ ; \mathbb{F}_p)$:

$$ \longrightarrow H^{t-1}(X) \longrightarrow H^{t-1}({}_0X_1) \longrightarrow H^{t-1}(\Omega K({}_0V_0)) \longrightarrow H^t(X) \longrightarrow $$

$$ H^t(K({}_0V_0)) $$

with *onto*

The epimorphism, ${}_0F_0: H^t(K({}_0V_0)) \to H^t(X)$, is the transgression associated to the fibration ${}_0X_1 \to X$. Thus, for $t < 2n-1$, $H^t({}_0X_1)$ maps isomorphically onto the kernel of $({}_0F_0)^*$. Observe, also, that ${}_0X_1$ is at least $(n-1)$-connected, and the map $H^t(X) \to H^t({}_0X_1)$ is null for $t < 2n-1$.

Iterate this procedure for ${}_0X_1$; that is, choose ${}_0V_1$ isomorphic in degrees less than $2n-1$ to $\tilde{H}^*({}_0X_1)/I(\mathcal{A}_p) \cdot \tilde{H}^*({}_0X_1)$ and a mapping ${}_0F_1: {}_0X_1 \to K({}_0V_1)$ realizing an epimorphism on mod p cohomology. Pullback the path-loop fibration over $K({}_0V_1)$ with respect to ${}_0F_1$ to get a fibration ${}_0X_2 \to {}_0X_1$.

Continuing, we get a tower of fibrations, $\cdots \to {}_0X_{i+q} \to {}_0X_i \to \cdots \to {}_0X_1 \to X$, with each ${}_0X_i$ at least $(n-1)$-connected, and the mappings $H^t({}_0X_i) \to H^t({}_0X_{i+1})$ null in degrees $t < 2n-1$;

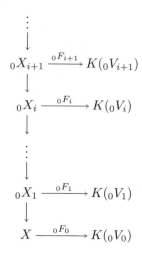

The mappings $\Omega K(_0V_i) \xrightarrow{i} {}_0X_{i+1} \xrightarrow{_0F_{i+1}} K(_0V_{i+1})$ have additional properties. On cohomology we have

$$H^*(K(_0V_{i+1})) \xrightarrow{(_0F_{i+1})^*} H^*(_0X_{i+1}) \xrightarrow{i^*} H^*(\Omega K(_0V_i))$$

where the first map is onto in degrees less than $2n$ and the second map can be taken to be a degree 1 map that is one-one, onto the kernel of $H^{s+1}(K(_0V_i)) \to H^{s+1}(_0X_i)$ in degrees less than $2n - 1$. If we compose and desuspend, the homomorphisms $H^*(K(_0V_{i+1})) \to s^{-1}H^*(K(_0V_i))$ can be assembled into a complex:

$$0 \leftarrow H^*(X) \leftarrow H^*(K(_0V_0)) \leftarrow s^{-1}H^*(K(_0V_1)) \leftarrow s^{-2}H^*(K(_0V_2)) \leftarrow,$$

which, in degrees less than $2n$, *is* a free $\mathcal{A}_p$-module resolution of $H^*(X)$.

The next step in this process is to repeat the previous construction based on X for SX, the suspension of X. Recall that $H^*(SX) \cong sH^*(X)$ as modules over $\mathcal{A}_p$. Since SX is n-connected, the stable range extends to degree $2n + 1$. We can relate the data from the Adams resolution of X to that of SX by choosing the graded vector spaces $_1V_i$ to be $s(_0V_i) \oplus W_i$, where W_i is the additional term needed to obtain an epimorphism up to degree $2n + 1$.

With these details of construction, the long exact sequences of cohomology vector spaces can be examined to show that $_1X_i$ and $S(_0X_i)$ have the same mod p cohomology in degrees less than $2n - i$.

This establishes the inductive step. We continue by building Adams resolutions over S^2X, S^3X, and so on. Over S^mX, we have a tower of principal fibrations:

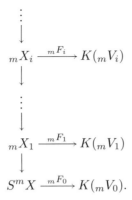

The properties of the construction can be summarized:

(1) each $_mX_{i+1} \to {}_mX_i$ is a fibration with fibre $\Omega K(_mV_i)$,
(2) each $_mX_i$ is at least $(n+m)$-connected,
(3) $H^*(_mX_i; \mathbb{F}_p) \cong H^*(S(_{m-1}X_i); \mathbb{F}_p)$ in degrees less than $n+m-i$,
(4) $H^*(_mX_i; \mathbb{F}_p) \to H^*(_mX_{i+1}; \mathbb{F}_p)$ is null in degrees less than $2(n+m)$.

Finally, desuspending the appropriate number of times leads to the complex, a free $\mathcal{A}_p$-module resolution in degrees less than $2(n+m)$:

$$0 \leftarrow s^{-m}H^*(S^mX) \leftarrow s^{-m}H^*(K(_mV_0)) \leftarrow s^{-m-1}H^*(K(_mV_1)) \leftarrow \cdots .$$

We next prove an important property holds for Adams resolutions that is analogous to the universal property enjoyed by projective resolution.

Lemma 9.14. *Suppose $\{ W_1 \to Y,\ W_{i+1} \to W_i, i \geq 1 \}$ is a tower of principal fibrations and there is an integer $N \geq 0$ so that, if $\Omega K(M_i)$ is the fibre of $W_{i+1} \to W_i$, then M_i is a graded vector space, trivial in degrees greater than N. If X is $(n-1)$-connected, $2n-1 > N$, and $f\colon X \to Y$ is a continuous mapping, then there is a sequence of mappings, $_0X_i \to W_i$, for each i, such that the following diagram commutes,*

PROOF: We construct the mapping from $_0X_1$ to W_1 and leave the inductive step (essentially the same argument) to the reader. Consider the diagram

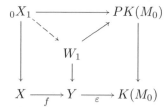

Since $H^t(X) \to H^t({}_0X_1)$ is null in degrees less than $2n - 1$, the composite ${}_0X_1 \to K(M_0)$ is null-homotopic and so there is a lift of the composite to $PK(M_0)$. By the universal property of the pullback, $W_1 \to Y$, we get a mapping ${}_0X_1 \to W_1$, making the diagram commute.   □

Corollary 9.15. *Given two Adams resolutions of a space X, there are maps between them covering the identity.*

We finally derive the spectral sequence. First, form all of the Adams resolutions over X, SX, S^2X, ... From the observation that each tower yields a certain part of an $\mathcal{A}_p$-free resolution of $H^*(X)$, let

$$0 \leftarrow H^*(X; \mathbb{F}_p) \xleftarrow{\varepsilon} P_0 \leftarrow P_1 \leftarrow P_2 \leftarrow \cdots$$

denote the limit of these approximations, which can be realized as

$$P_i = \lim_{m \to \infty} s^{-m-i} H^*(K({}_mV_i)).$$

For a given m, the Adams resolution of $S^m X$ yields the system of fibrations:

$$
\begin{array}{ccccccccc}
& & \vdots & & & & \vdots & & \\
& & \downarrow & & & & \downarrow & & \\
\longrightarrow \Omega^2 K({}_mV_1) & \longrightarrow & \Omega({}_mX_2) & \longrightarrow & \Omega K({}_mV_2) & \longrightarrow & {}_mX_3 & \longrightarrow & K({}_mV_3) \\
& & \downarrow & & & & \downarrow & & \\
\longrightarrow \Omega^2 K({}_mV_0) & \longrightarrow & \Omega({}_mX_1) & \longrightarrow & \Omega K({}_mV_1) & \longrightarrow & {}_mX_2 & \longrightarrow & K({}_mV_2) \\
& & \downarrow & & & & \downarrow & & \\
& & \Omega(S^mX) & \longrightarrow & \Omega K({}_mV_0) & \longrightarrow & {}_mX_1 & \longrightarrow & K({}_mV_1) \\
& & & & & & \downarrow & & \\
& & & & & & S^mX & \longrightarrow & K({}_mV_0).
\end{array}
$$

To this system of spaces and maps, apply the functor $[S^mY, -]$, where Y is a finite dimensional CW-complex. This yields an exact couple:

$$
\begin{array}{ccc}
[S^{m+p}Y, {}_mX_q] & \xrightarrow{\quad i \quad} & [S^{m+p}Y, {}_mX_{q-1}] \\
& & \\
[S^{m+p-1}Y, {}_mX_q] & & \Big\downarrow{j} \\
\xleftarrow{\quad k \quad} & & \\
& [S^{m+p}Y, K({}_mV_{q-1})] &
\end{array}
$$

i is induced by $\Omega^p(_mX_q) \to \Omega^p(_mX_{q-1})$, the p-fold loops on the fibration; j is induced by $\Omega^p(_mX_{q-1}) \to \Omega^pK(_mV_{q-1})$, the p-fold loops on the classifying map of the fibration, $_mF_{q-1}$; and k is induced by $\Omega^pK(_mV_{q-1}) \to \Omega^{p-1}(_mX_q)$, the $(p-1)$-fold loops on the inclusion of the fibre. We have identified $[T, \Omega^r U]$ with $[S^r T, U]$.

To fix the bigrading, let the first degree denote the level in the Adams resolution where the map is found, and the second degree denotes the codegree for the number of suspensions. This yields

$$[S^{m+p}Y, _mX_q] = D^{q,p+q} \qquad [S^{m+p}Y, K(_mV_q)] = E^{q,p+q}.$$

The bidegrees of i, j and k are $(-1,-1)$, $(0,0)$ and $(1,-1)$, respectively. When we display the unrolled exact couple, we get

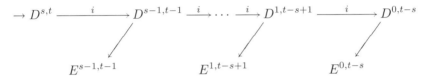

Let X be $(n-1)$-connected, Y of dimension N and suppose that $N + m + p < 2(n+m)$. By the Freudenthal suspension theorem, $[S^{m+p}Y, S^m X] \cong \{Y, X\}_p$. Furthermore, $[S^{m+p+r}Y, _mX_i] \cong [S^{m+p+r+1}Y, _{m+1}X_i]$, if $i < r$. So, we may write the groups, $D^{s,t}$ and $E^{s,t}$ as independent of m when we chose m large enough for a given s and t. In particular, when m is large enough,

$$D^{0,t-s} = [S^{m+t-s}Y, S^m X] = \{Y, X\}_{t-s}.$$

We really have a spectral sequence for each m, but we think of the spectral sequences approaching a limiting value as m grows larger.

We next apply the properties of Adams resolutions to identify the E_2-term that arises from this construction. In the system of fibrations, the mappings j and k arise from the inclusion of the fibre and the classifying map for the fibration $\Omega^t K(_mV_s) \hookrightarrow \Omega^{t-1}(_mX_{s+1}) \to \Omega^{t-1}K(_mV_{s+1})$. If we apply the functor, $[S^m Y, -]$, then the first differential is given by

$$k \circ j = d_1 : E^{s,t+s} = [S^{m+t}Y, K(_mV_s)] \to [S^{m+t-1}Y, K(_mV_{s+1})] = E^{s+1,t+s}.$$

However, for m large enough,

$$\begin{aligned}
[S^{m+t}Y, K(_mV_s)] &= \mathrm{Hom}^0_{\mathcal{A}_p}(H^*(K(_mV_s)), H^*(S^{m+t}Y)) \\
&= \mathrm{Hom}^t_{\mathcal{A}_p}(H^*(K(_mV_s)), H^*(S^m Y)) \\
&= \mathrm{Hom}^t_{\mathcal{A}_p}(s^m P_s, s^m H^*(Y)) \\
&= \mathrm{Hom}^t_{\mathcal{A}_p}(P_s, H^*(Y)).
\end{aligned}$$

Also, the mapping d_1 is simply

$$\mathrm{Hom}^0_{\mathcal{A}_p}(H^*(K(_mV_s)), s^{m+t}H^*(Y))$$

$$\xrightarrow{\mathrm{Hom}(\partial,1)} \mathrm{Hom}^0_{\mathcal{A}_p}(sH^*(K(_mV_{s+1})), s^{m+t}H^*(Y))$$

where ∂ is induced by $\Omega K(_mV_s) \to {_mX_{s+1}} \to K(_mV_{s+1})$, which in turn induces $\partial \colon P_{s+1} \to P_s$. Thus the following diagram commutes (up to the sign introduced by the suspensions):

$$
\begin{array}{ccc}
E^{s,t+s} & \xrightarrow{\quad d^1 \quad} & E^{s+1,t+s} \\
\cong \Big\uparrow & & \Big\uparrow \cong \\
\mathrm{Hom}^t_{\mathcal{A}_p}(P_s, H^*(Y)) & \xrightarrow{\mathrm{Hom}(\partial,1)} & \mathrm{Hom}^t_{\mathcal{A}_p}(P_{s+1}, H^*(Y)).
\end{array}
$$

This proves $E_2^{s,t} \cong \mathrm{Ext}^{s,t}_{\mathcal{A}_p}(H^*(X), H^*(Y))$ in the spectral sequence resulting from the exact couple and the first part of Theorem 9.1.

Convergence

The exact couple that gives rise to the spectral sequence has the "rightmost column" of groups, obtained by applying $[S^mY, -]$ to the system of fibrations, given by $D^{0,t-s} = \{Y, X\}_{t-s}$. Unrolling the exact couple, we have a sequence of maps:

$$\xrightarrow{\ i\ } D^{s,t} \xrightarrow{\ i\ } D^{s-1,t+1} \xrightarrow{\ i\ } \cdots \xrightarrow{\ i\ } D^{0,t-s} = \{Y, X\}_{t-s}.$$

We now apply the methods of Chapter 3 to determine the convergence of the associated spectral sequence.

Lemma 9.16. $d_r \colon E_r^{s,t} \to E_r^{s+r,t+r-1}$.

PROOF: We can factor d_r by

$$d_r \colon E_r^{s,t} \xrightarrow{\ k^{(r)}\ } D_r^{s+1,t} = i^{(r)} D^{s+r,t+r-1} \xrightarrow{\ j^{(r)}\ } E_r^{s+r,t+r-1}. \qquad \square$$

Filter $\{Y, X\}_*$ by $F^s\{Y, X\}_q = \mathrm{im}\, i^s \colon D^{s,q+s} \to D^{0,q} = \{Y, X\}_q$.

Lemma 9.17. *The filtration of $\{Y, X\}_*$ arising from a system of Adams resolutions does not depend on the choice of Adams resolutions.*

PROOF: Apply Corollary 9.15 to cover the identity map $S^m X \to S^m X$ between Adams resolutions. This system of mappings shows that each filtration is contained in the other. □

We can relate this filtration to the spectral sequence by using Corollary 2.10, which yields a short exact sequence, for each r,

$$0 \to {}^{D^{s,*}}\big/{\ker(i^r \colon D^{s,*} \to D^{s-r,*}) + iD^{s+1,*}} \xrightarrow{\bar{\jmath}} E_{r+1}^{s,*}$$

$$\xrightarrow{\bar{k}} \operatorname{im}(i^r \colon D^{s+r+1,*} \to D^{s+1,*}) \cap \ker(i \colon D^{s+1,*} \to D^{s,*}) \to 0.$$

Let r go to infinity and observe that the left hand term of the short exact sequence stabilizes when $r = s$, since $i^s \colon D^{s,*} \to D^{0,*}$.

Lemma 9.18. *There are monomorphisms*

$$0 \to {}^{F^s\{Y, X\}_q}\big/{F^{s+1}\{Y, X\}_q} \to E_\infty^{s,q+s}.$$

PROOF: It suffices to show that

$${}^{F^s\{Y, X\}_q}\big/{F^{s+1}\{Y, X\}_q} \cong {}^{D^{s,*}}\big/{\ker(i^s \colon D^{s,*} \to D^{0,*}) + iD^{s+1,*}}.$$

Notice $F^s\{Y, X\}_q = i^s D^{s,q+s}$ and $F^{s+1}\{Y, X\}_q = i(i^s D^{s+1,q+s+1})$. There are short exact sequences

$$
\begin{array}{ccccccccc}
0 & \longrightarrow & \ker i^s + iD^{s+1,*} & \longrightarrow & D^{s,*} & \longrightarrow & {}^{D^{s,*}}\big/{\ker i^s + iD^{s+1,*}} & \longrightarrow & 0 \\
 & & \downarrow{\scriptstyle i^s} & & \downarrow{\scriptstyle i^s} & & \downarrow{\scriptstyle \bar{\imath}^s} & & \\
0 & \longrightarrow & i^{s+1}D^{s+1,*} & \longrightarrow & i^s D^{s,*} & \longrightarrow & F^s/F^{s+1} & \longrightarrow & 0
\end{array}
$$

These maps are onto by the Five-lemma. We show that $\bar{\imath}^s$ is also a monomorphism. Let $[a], [b]$ lie in $D^{s,*}/\ker i^s + iD^{s+1,*}$. If $\bar{\imath}^s[a] = \bar{\imath}^s[b]$, then $\bar{\imath}^s[a-b] = [\bar{\imath}^s(a-b)] = 0$. This implies either $i^s(a-b) = 0$, and so $a \equiv b \ (\ker i^s)$, or $i^s(a-b)$ lies in $i^{s+1}D^{s+1,*}$, which implies $(a-b)$ lies in $iD^{s+1,*}$. In both of these cases $[a] = [b]$. □

From Lemma 9.18, we have the exact sequence;

$$0 \to {}^{F^s\{Y, X\}_*}\big/{F^{s+1}\{Y, X\}_*} \to E_\infty^{s,*}$$

$$\to \bigcap_r \operatorname{im}(i^r \colon D^{s+r+1,*} \to D^{s+1,*}) \cap \ker(i \colon D^{s+1,*} \to D^{s,*}) \to 0.$$

To complete our proof of Theorem 9.1, we must show that the right term of this short exact sequence is trivial. We first develop some properties of the filtration.

Lemma 9.19. *If α is in $\{Y, X\}_q$ and α is divisible by p^n, then α is in $F^n\{Y, X\}_q$.*

PROOF: The term 'divisible by p^n' makes sense in an abelian group, that is, there is a β in $\{Y, X\}_q$ so that $\alpha = p^n\beta$. We proceed by induction on n for all spaces. The result for $n = 0$ is trivial.

The case of $n = 1$ follows by observing that if $f\colon S^{m+q}Y \to S^m X$ is such that $pf \simeq *$, then the composite $S^{m+q}Y \to S^m X \to K({}_mV_0)$ is null-homotopic since it represents classes in mod p cohomology. Therefore f lifts to ${}_mX_1$ and is in $F^1\{Y, X\}_q$.

Assume the result for $n-1$. Since $\alpha = p^n\beta = p^{n-1}(p\beta)$, we have that $p\beta$ lies in $F^1\{Y, X\}_q$, that is, $p\beta = iu$ where $i\colon {}_mX_1 \to S^m X$ and $u\colon S^{m+q}Y \to {}_mX_1$. However, $\alpha = p^{n-1}(p\beta) = p^{n-1}(iu) = i(p^{n-1}u)$, and so $p^{n-1}u$ lies in $F^{n-1}\{S^m Y, {}_mX_1\}_q$. Thus there is a map, $w\colon S^{m+q}Y \to {}_mX_n$ such that $i^{n-1}w = p^{n-1}u$. Thus $\alpha = i(i^{n-1}w) = i^n w$ and α lies in $F^n\{Y, X\}_q$. $\square$

Lemma 9.20. *If α is in $\{Y, X\}_q$ and α is not divisible by p^n, then α is not in $F^s\{Y, X\}_q$, for some s.*

PROOF: We introduce an auxiliary space constructed as follows: In $\{X, X\}_0$ consider the element given by p^n times the identity. Let $h\colon S^m X \to S^m X$ represent this map. Pullback the path-loop fibration over $S(h)$ to obtain the space U:

$$
\begin{array}{ccc}
\Omega S^{m+1}X & = \!\!= & \Omega S^{m+1}X \\
\downarrow & & \downarrow \\
U & \longrightarrow & PS^{m+1}X \\
\downarrow & & \downarrow \\
S^{m+1}X & \xrightarrow{S(h)} & S^{m+1}X
\end{array}
$$

By construction, if X is n-connected, then U is $(n+m+1)$-connected. If W is a CW-complex of dimension less than $2(n+m+1)$, then by the Freudenthal suspension theorem (Theorem 4.10) $[W, \Omega S^{m+1}X] \cong [W, S^m X]$. We also have the exact sequence

$$[W, S^m X] \xrightarrow{p^n} [W, S^m X] \to [W, U] \to [W, S^{m+1}X] \xrightarrow{p^n} [W, S^{m+1}X],$$

which follows from the long exact sequence for the fibration. This traps $[W, U]$ between the coker p^n and ker p^n in the short exact sequence:

$$0 \to \operatorname{coker} p^n \to [W, U] \to \ker p^n \to 0.$$

Since p^n annihilates both ends of this extension, $[W, U]$ is a $\mathbb{Z}/p^{2n}\mathbb{Z}$-module.

Let j be the first nonzero dimension with $\pi_j(U) \neq \{0\}$ ($j \geq n+m+1$). If $j < 2(n+m+1)$, then $\pi_j(U)$ is a $\mathbb{Z}/p^{2n}\mathbb{Z}$-module by the argument above. Consider the quotient map, $\pi_j(U) \to \pi_j(U)/p\pi_j(U) = \pi_j(U) \otimes \mathbb{Z}/p\mathbb{Z}$. Since $\pi_j(U) \otimes \mathbb{Z}/p\mathbb{Z}$ is a vector space over $\mathbb{F}_p$, we can realize this quotient map by $U \to K(\pi_j(U) \otimes \mathbb{Z}/p\mathbb{Z}, j)$. Let U_1 be the total space of the principal fibration induced by this map:

$$
\begin{array}{ccc}
U_1 & \longrightarrow & PK(\pi_j(U) \otimes \mathbb{Z}/p\mathbb{Z}, j) \\
\downarrow & & \downarrow \\
U & \longrightarrow & K(\pi_j(U) \otimes \mathbb{Z}/p\mathbb{Z}, j).
\end{array}
$$

The short exact sequence, $0 \to \pi_j(U_1) \to \pi_j(U) \to \pi_j(U) \otimes \mathbb{Z}/p\mathbb{Z} \to 0$, implies that $\pi_j(U_1)$ is $p\pi_j(U)$. But $\pi_j(U)$ is a $\mathbb{Z}/p^{2n}\mathbb{Z}$-module, so if we repeat this procedure enough times, we get U_{i_1} with $\pi_j(U_{i_1}) = \{0\}$. Starting on $\pi_{j+1}(U_{i_1})$, we can iterate the procedure until we eventually get to U' with $\pi_k(U') = \{0\}$ for $0 \leq k < 2(n+m+1)$. The tower of fibrations $\{U_i \to U_{i-1}\}$, satisfies the conditions of Lemma 9.14 and so we have a mapping over the inclusion $S^m X \hookrightarrow \Omega S^{m+1} X \hookrightarrow U$:

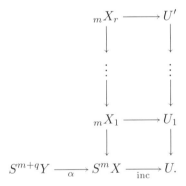

Recall α is not divisible by p^n and suppose $\alpha \in F^s\{Y, X\}_q$ for all $s \leq r$. Then $\alpha = i^r v$ for $v \colon S^{m+q} Y \to {}_m X_r$. This factors through $[S^{m+q}Y, U'] = \{0\}$ for $S^{m+q}Y$ in the stable range and so α is in $\ker(\mathrm{inc}) = p^n[S^{m+q}Y, S^m X]$. But then $\alpha = p^n \beta$, contradicting our assumption. Therefore, $\alpha \notin F^s\{Y, X\}_q$ for some $s \leq r$. $\qquad \square$

Corollary 9.21. $F^\infty\{Y, X\}_q = \bigcap_n F^n\{Y, X\}_q = \{$*elements of finite order prime to p in $\{Y, X\}_q\}$.*

Notice that the assumptions that X is of finite type and Y of finite dimension play a role in the corollary. By Proposition 5.17 and induction over skeleta, we know that $\{Y, X\}_q$ is finitely generated and so there are no elements in the group of infinite divisibility by p.

Corollary 9.21 points out, however, that our filtration of $\{Y, X\}_q$ is not Hausdorff (§3.1) and so the spectral sequence cannot be expected to have $\{Y, X\}_*$ as its target—after all, all our constructions were done mod p. We determine the actual target as follows: For an abelian group G let $_{(p)}G$ be the *p*-**component** of G, that is, the quotient of G by the subgroup of elements of finite order prime to p. Then $_{(p)}\{Y, X\}_* = \{Y, X\}_*/F^\infty\{Y, X\}_*$. We induce a filtration on the *p*-component, $_{(p)}\{Y, X\}_*$ by

$$\cdots \subset F^s\{Y, X\}_*\big/F^\infty\{Y, X\}_* \subset \cdots \subset F^2\{Y, X\}_*\big/F^\infty\{Y, X\}_*$$
$$\subset F^1\{Y, X\}_*\big/F^\infty\{Y, X\}_* \subset {}_{(p)}\{Y, X\}_*.$$

This filtration is exhaustive and convergent (Hausdorff), that is,

$$\bigcup_s F^s/F^\infty = {}_{(p)}\{Y, X\}_*, \qquad \bigcap_n F^s/F^\infty = \{0\}.$$

Furthermore we still have monomorphisms

$$0 \to F^s/F^\infty\Big/F^{s+1}/F^\infty \to E_\infty^{s,*}.$$

To complete our discussion of convergence, we prove the following lemma.

Lemma 9.22. $E_\infty^{s,*} \cong F^s{}_{(p)}\{Y, X\}_*\Big/F^{s+1}{}_{(p)}\{Y, X\}_*.$

PROOF: It suffices to show that

$$\left[\bigcap_r (\operatorname{im} i^r : D^{s+r+1,*} \to D^{s+1,*})\right] \cap \left[\ker(i : D^{s+1,*} \to D^{s,*})\right] = \{0\}.$$

Consider the relevant piece of one of the towers;

$$\Omega K({}_mV_s)$$
$$\downarrow$$
$$\longrightarrow {}_mX_{s+n+1} \longrightarrow \cdots \longrightarrow {}_mX_{s+1} \longrightarrow {}_mX_s$$

$$i^n$$

For a finite complex Z we have the exact sequence

$$\to [Z, \Omega K({}_mV_s)] \to [Z, {}_mX_{s+1}] \xrightarrow{i} [Z, {}_mX_s] \to [Z, K({}_mV_s)] \to$$

for which $\ker i = \operatorname{im}[Z, \Omega K({}_mV_s)]$ and p annihilates ${}_mV_s$. Therefore, $\ker i$ is *p*-torsion.

Apply $[Z, -]$ again to the tower, and we can filter $[Z, {}_mX_{s+1}]$ by the images of the i^n. But $\bigcap_n \operatorname{im} i^n = F^\infty[Z, {}_mX_{s+1}]$ and the argument of Lemma 9.20 carries over to show that $\bigcap_n \operatorname{im} i^n$ contains only elements of finite order prime to p. Therefore, $\bigcap_n \operatorname{im} i^n \cap \ker i = \{0\}$ since an element in a finitely generated abelian group cannot be *p*-torsion and have finite order prime to p unless it is zero. $\square$

Corollary 9.23. *There is a spectral sequence, converging to* $_{(p)}\pi_*^S$, *the p-components of the stable homotopy groups of spheres, with E_2-term given by*
$$E_2^{s,t} \cong H^{s,t}(\mathcal{A}_p) = \mathrm{Ext}_{\mathcal{A}_p}^{s,t}(\mathbb{F}_p, \mathbb{F}_p).$$

In §9.5 we explore the consequences of Corollary 9.23 and compute some of the groups $_{(p)}\pi_*^S$ for $p = 2$ and $p = 3$.

Multiplicative structure on the spectral sequence

The multiplicative structure on a spectral sequence is often pivotal in the computations. We next introduce the composition pairing, $\circ\colon \{Y, Z\}_s \otimes \{X, Y\}_t \to \{X, Z\}_{s+t}$, that is reflected in a pairing of the spectral sequences converging to these groups. We sketch how the pairing at each level arises. Complete proofs of the existence and properties of the pairing are given in detail by [Douady58] or [Moss68] where different tools are to hand. We identify this pairing at the E_2-term as the Yoneda composition pairing. When $X = Y = Z$, the pairing becomes a product and so we have a product structure on the relevant spectral sequence.

Definition 9.24. *Suppose $\alpha \in \{X, Y\}_s$ and $\beta \in \{Y, Z\}_t$ and suppose that $f\colon S^{m+s}X \to S^m Y$ and $g\colon S^{n+t}Y \to S^n Z$ represent α and β, respectively. Define the* **composition product** *of α and β, $\beta \circ \alpha$ to be the class in $\{X, Z\}_{s+t}$ given by*
$$S^{n+m+s+t}X \xrightarrow{S^{n+t}f} S^{m+n+t}Y \xrightarrow{S^m g} S^{n+m}Z.$$

Proposition 9.25. *The composition product is bilinear, associative and functorial. The composition product induces the structure of a ring on $\{X, X\}_*$ and furthermore, $\{Y, X\}_*$ is a left $\{X, X\}_*$-module and a right $\{Y, Y\}_*$-module. In fact, $\{Y, X\}_*$ has the structure of a $\{X, X\}_*$-$\{Y, Y\}_*$-bimodule.*

PROOF: These properties follow directly from the analogous properties of the unstable composition product, $\circ\colon [V, W] \times [U, V] \to [U, W]$. In particular, the pairings
$$[V, W] \times [SU, V] \xrightarrow{\circ} [SU, W]$$
$$[SV, W] \times [U, V] \to [SV, W] \times [SU, SV] \xrightarrow{\circ} [SU, W]$$
are additive in the second and first factors, respectively. (For the reader who is unfamiliar with these properties, we suggest Chapter 3 of the classic book of [Whitehead, GW78].) ☐

In the particular case of $X = S^0$, the ring structure on $\{S^0, S^0\}_* = \pi_*^S$ has better properties—π_*^S is graded commutative. This follows by comparing the composition product with the smash product:

$$S^{p+k} \wedge S^{q+l} \xrightarrow{(-1)^{(q+l)k} E^{q+l}\alpha} S^p \wedge S^{q+l}$$

with $E^{p+k}\beta$ on the left vertical, $\alpha \wedge \beta$ on the diagonal, $E^p\beta$ on the right vertical, and

$$S^{p+k} \wedge S^q \xrightarrow{(-1)^{qk} E^q\alpha} S^p \wedge S^q.$$

We have $(-1)^{qk} E^{p+k}\beta \circ E^q\alpha = (-1)^{(q+l)k} E^{q+l}\alpha \circ E^q\beta$. This result was first proved by [Barratt-Hilton53]. The relationship between this smash product and the E_2-term, Ext, is through the external tensor product on Ext as defined by [Cartan-Eilenberg56]. We refer the reader to the blue book of [Adams74] for a thorough treatment of products.

Let X be a space and construct a system of Adams resolutions for X. We construct 'pairings' on $D^{*,*}$ and $E^{*,*}$ of the resultant exact couple, that agree with the composition pairing on $\{X, X\}_*$. Suppose $[f]$ is a class in

$$D^{s,t} = [S^{m+t-s+t'-s'} X, {}_{m+t'-s'} X_s]$$

and $[g]$ is a class in $D^{s',t'} = [S^{m+t'-s'} X, {}_m X_{s'}]$. By Lemma 9.14 we can lift g through the Adams resolution:

$$S^{m+t-s+t'-s'} X \xrightarrow{f} {}_{m+t'-s'} X_s \xrightarrow{g_s} {}_m X_{s'+s}$$

$${}_{m+t'-s'} X_{s-1} \xrightarrow{g_{s-1}} {}_m X_{s'+s-1}$$

$${}_{m+t'-s'} X_1 \xrightarrow{g_1} {}_m X_{s'+1}$$

$$S^{m+t'-s'} X \xrightarrow{g} {}_m X_{s'}.$$

We define $[g] \circ [f] = \{[g_s \circ f]$ for all choices of $g_s\} \subset D^{s+s',t+t'}$. Notice that if $s = s' = 0$, then this is the composition product on $\{X, X\}_*$.

Similarly we define such a 'pairing' on $E^{*,*}$; let

$$[f] \in E^{s,t} = [S^{m+t-s+t'-s'} X, K({}_{m+t'-s'} V_s)]$$
$$\text{and } [g] \in E^{s',t'} = [S^{m+t'-s'} X, K({}_m V_{s'})].$$

Because the $K(V)$'s are generalized Eilenberg-Mac Lane spaces and because $S^{m+t'-s'} X \to K({}_{m+t'-s'} V_0)$ induces an epimorphism on mod p cohomology, we can define a mapping $k_0 \colon K(_{m+t'-s'}V_0) \to K(_mV_{s'})$ so that the following diagram commutes:

$$
\begin{array}{ccc}
_{m+t'-s'}X_1 & \xrightarrow{\ \ g_1\ \ } & _mX_{s'+1} \\
\downarrow & & \downarrow \\
S^{m+t'-s'}X & \xrightarrow{\ \ g\ \ } & _mX_{s'} \\
\downarrow & & \downarrow \\
K(_{m+t'-s'}V_0) & \xrightarrow{\ \ k_0\ \ } & K(_mV_{s'}).
\end{array}
$$

The mapping g_1 exists since the composite

$$_{m+t'-s'}X_1 \to S^{m+t'-s'}X \to K(_{m+t'-s'}V_0) \xrightarrow{\ k_0\ } K(_mV_{s'})$$

is null homotopic.

Inductively, if g_{i+1} and k_i exist, then we can find g_{i+2} and k_{i+1} by choosing k_{i+1} so that the following diagram commutes

(9.26)
$$
\begin{array}{ccc}
\Omega K(_{m+t'-s'}V_i) & \xrightarrow{\ \ \Omega k_i\ \ } & \Omega K(_mV_{s'+i}) \\
\downarrow & & \downarrow \\
_{m+t'-s'}X_{i+1} & \xrightarrow{\ \ g_{i+1}\ \ } & _mX_{s'+i+1} \\
\downarrow & & \downarrow \\
K(_{m+t'-s'}V_{i+1}) & \xrightarrow{\ \ k_{i+1}\ \ } & K(_mV_{s'+i+1})
\end{array}
$$

and choosing an appropriate lifting of g_{i+1}. We can define the composition of $[g]$ and $[f]$ as $[g] \circ [f] = \{[k_s \circ f] \mid \text{ all choices of } k_s\} \subset E^{s+s',t+t'}$. We make some observations.

I. The choices made in all of the constructions differ by elements in the groups, $[S^q X, F]$, of homotopy classes of mappings to the fibres of the fibrations in the towers. If $[f]$ and $[g]$ are in $E^{s,t}$ and $E^{s',t'}$ and they are cycles under d_1, then their product $[g] \circ [f]$ is also made up of cycles in $E^{s+s',t+t'}$ and the differences vanish as an element in $E_2^{s+s',t+t'}$. Thus the defined 'pairing' is an actual pairing on $E_2^{*,*}$.

II. When we apply mod p cohomology to the diagrams (9.26), and compute the effect of the maps k_i^*, then the construction is seen to be a geometric realization of the Yoneda composition product and, in the isomorphism of $E_2^{*,*}$ with $\mathrm{Ext}_{\mathcal{A}_p}^{*,*}(H^*(X), H^*(Y))$, the products go over isomorphically.

III. The composition product is filtration-preserving on $\{X, X\}_*$. To see this, examine the commutative diagram:

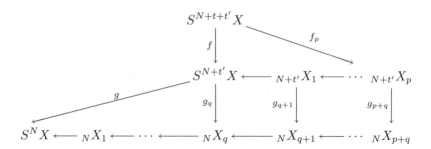

This shows that $\circ\colon F^p\{X, X\}_{t'} \otimes F^q\{X, X\}_t \to F^{p+q}\{X, X\}_{t+t'}$.

IV. With some change in notation, we could have defined the 'pairings' for $\circ\colon \{Y, Z\}_* \otimes \{X, Y\}_* \to \{X, Z\}_*$ just as easily.

We now state the full theorem on products in the Adams spectral sequence, due to [Adams58] and [Moss68].

Theorem 9.27. *There exist associative, bilinear pairings, functorial in the spaces X, Y, Z, all finite dimensional CW-complexes, and, for $r \geq 2$,*

$$E_r^{s,t}(Y, Z) \otimes E_r^{s',t'}(X, Y) \to E_r^{s+s',t+t'}(X, Z)$$

such that

(1) *for $r = 2$, the pairing agrees with the Yoneda composition pairing*

$$\mathrm{Ext}_{\mathcal{A}_p}^{s,t}(H^*(Z), H^*(Y)) \otimes \mathrm{Ext}_{\mathcal{A}_p}^{s',t'}(H^*(Y), H^*(X))$$
$$\to \mathrm{Ext}_{\mathcal{A}_p}^{s+s',t+t'}(H^*(Z), H^*(X)).$$

(2) *The differentials, d_r are derivations with respect to these pairings, that is, in $E_r^{*,*}(X, Z)$, $d_r(uv) = (d_ru)v + (-1)^{t-s}u(d_rv)$, if bideg $u = (s, t)$.*

(3) *The pairings commute with the isomorphisms $E_{r+1} \cong H(E_r, d_r)$.*

(4) *The pairings converge to the composition pairing*

$$\circ\colon \{Y, Z\}_* \otimes \{X, Y\}_* \to \{X, Z\}_*,$$

that is, this pairing is filtration-preserving and the induced pairing on the associated bigraded modules is isomorphic to the pairing on the E_∞-terms of the spectral sequences.

Of course, these pairings not only allow us to compute more easily but they also allow us to define Massey products in the spectral sequence. We take up this notion in §9.5.

The simplest computation reveals the role of the composition pairing. For any prime, p, the beginning of a minimal resolution of $\mathbb{F}_p$ over $\mathcal{A}_p$ may be presented as in the diagram:

$$
\begin{array}{ccccc}
& & & \mid & \\
& & & P^1 & \mid \\
& & & \mid & \\
& & & & \beta\imath_1 \\
& & & \beta \longleftarrow & \imath_1 \\
0 \longleftarrow & \mathbb{Z}/p\mathbb{Z} \longleftarrow & \mathbb{Z}/p\mathbb{Z} & & \\
0 \longleftarrow & \mathbb{Z}/p\mathbb{Z} \longleftarrow & \mathcal{A}_p \longleftarrow & P_1 &
\end{array}
$$

We identify an element a_0 in $\mathrm{Ext}^{1,1}_{\mathcal{A}_p}(\mathbb{F}_p, \mathbb{F}_p)$ generated by the dual of the Bockstein, and occurring in this resolution as the element $\imath_1$. Because the Bockstein has degree 1, we show that it detects the mapping $p\colon S^n \to S^n$, the degree p map. Consider the complex $S^n \cup_p e^{n+1}$, the mod p Moore space, and the sequence

$$S^n \hookrightarrow S^n \cup_p e^{n+1} \xrightarrow{\text{pinch}} S^{n+1}.$$

On cohomology, we have the extension

$$0 \to \tilde{H}^*(S^{n+1}) \to \tilde{H}^*(S^n \cup_p e^{n+1}) \to \tilde{H}^*(S^n) \to 0.$$

Since $\tilde{H}^*(S^n \cup_p e^{n+1}) = \mathbb{F}_p\{x_n, \beta x_n\}$ is a nontrivial module over $\mathcal{A}_p$, this sequence identifies the only possible extension that can represent a_0 in $\mathrm{Ext}^{1,1}_{\mathcal{A}_p}(\mathbb{F}_p, \mathbb{F}_p)$.

Suppose $l \in \mathrm{Ext}^{s,t}_{\mathcal{A}_p}(\mathbb{F}_p, \mathbb{F}_p)$ is a nonzero permanent cycle in the Adams spectral sequence, that is, l lives to E_∞. Also suppose $a_0 l$ is nonzero. If l represents λ in $_{(p)}\pi^S_{t-s}$, then, by Theorem 9.27, $a_0 l$ represents the composition of l with the element that detects the degree p mapping. We conclude that $p\lambda \neq 0$ in $_{(p)}\pi^S_{t-s}$.

Much more is known about the product structure on π^S_*. In a classic paper [Nishida73] studied the global properties of the composition product. Using the extended power construction on a space and work of [Kahn-Priddy78], he proved a conjecture of Barratt that any element in π^S_i for $i > 0$ is nilpotent.

9.4 Other geometric applications

The workhorse pulling the Adams spectral sequence along is the notion of an Adams resolutions together with convergence of the associated spectral sequence. The construction is applied to a sequence of spaces X, SX, $S^2 X$, ... The resolution for $S^m X$ is obtained through a range of dimensions dependent on the connectivity of X. Taking these ideas as basic, we can enlarge the compass of application of the Adams spectral sequence to sequences of spaces with similar properties for which we can carry out the construction of an Adams resolution and identify the E_2-term of the associated spectral sequence.

Definition 9.28. *A sequence of spaces $X = \{X_1, X_2, \ldots, X_n, \ldots\}$ constitutes a **spectrum** if, for all n, there is a mapping $f_n: SX_n \to X_{n+1}$. A spectrum $X = \{X_n\}$ is called a **stable object** ([Adams64]) if, for each n, X_n is $(n-1)$-connected and the mapping $f_n: SX_n \to X_{n+1}$ is a $(2n-1)$-equivalence.*

Spectra were introduced by [Lima59] and [Whitehead, GW62] to study Spanier-Whitehead duality and generalized homology theories. For our purposes, a full discussion of spectra and stable homotopy theory is not needed. The student of homotopy theory needs some exposure to these ideas. The books of [Adams74], [Switzer75], and [Ravenel86, 92] are excellent introductions.

If $X = \{X_n\}$ is a spectrum, then the cohomology and homotopy of X are defined as the limits:

$$H^q(X; k) = \varprojlim_r H^{q+r}(X_r; k), \quad \text{and} \quad \pi_q(X) = \varinjlim_r \pi_{q+r}(X_r).$$

When the spectrum is a stable object, these limits are achieved at some finite stage (dependent on q). Furthermore, $\pi_q(X)$ is the qth stable homotopy group of X_N, for some $N = N(q)$ and $H^q(X; k) = s^{-r} H^{q+r}(X_r; k)$ for some $r \gg q$. The construction of Adams resolutions may be applied without change in the case of a stable object.

Proposition 9.29. *Let X be a stable object. Then there is a spectral sequence with $E_2^{s,t} \cong \mathrm{Ext}_{\mathcal{A}_p}^{s,t}(H^*(X; \mathbb{F}_p), \mathbb{F}_p)$ converging strongly to $_{(p)}\pi_{t-s}(X)$ under mild conditions.*

Though this appears to be a machine in search of a problem, in fact, these remarks apply broadly to the computation of cobordism groups as found implicitly in the work of [Thom54] and explicitly in the work of [Milnor60], [Liulevicius62], and others. Recall that two compact differentiable n-dimensional manifolds M_1, M_2 are (unoriented) **cobordant** if there is a compact differentiable $(n+1)$-dimensional manifold W with $\partial W = M_1 \amalg M_2$ (disjoint union). Being cobordant is an equivalence relation and the set of equivalence classes of n-manifolds is denoted by $\mathfrak{N}_n$. Disjoint union provides $\mathfrak{N}_n$ with the structure of an abelian group. The cartesian product provides the direct sum of the cobordism groups $\mathfrak{N}_*$ with a ring structure.

Lemma 9.30. $\mathfrak{N}_*$ *is a vector space over* $\mathbb{F}_2$.

PROOF: Given an n-dimensional manifold M, consider the $(n+1)$-dimensional manifold with boundary $M \times I$. Since the boundary of $M \times I$ is $M \amalg M$, we have that twice the cobordism class of M is zero in $\mathfrak{N}_n$. □

Cobordism was introduced in 1895 by [Poincaré1895]. The homotopy-theoretic study of cobordism was begun by [Pontrjagin55] who showed that the study of cobordism classes of framed manifolds is related to the study of π_*^S. By a theorem of [Whitney36] an n-dimensional manifold can be embedded in $\mathbb{R}^{n+k}$ for k large enough, and so a manifold is equipped with a normal bundle to the embedding. A manifold $M^n \subset \mathbb{R}^{n+k}$ is said to be **framed** if the normal bundle to the embedding is trivial. If we restrict our attention to framed manifolds only, then cobordism remains an equivalence relation and we denote the framed cobordism ring by Ω_*^{fr}.

A sufficiently small tubular neighborhood of M in $\mathbb{R}^{n+k}$ is homeomorphic to $M \times \mathbb{R}^k$. Projection off the second factor gives a mapping to $\mathbb{R}^k$. Taking the one-point compactifications of $\mathbb{R}^{n+k}$ and $\mathbb{R}^k$, we construct a mapping $f_M \colon S^{n+k} \to S^k$ by sending the complement of the tubular neighborhood of M in S^{n+k} to ∞ and going by the composite of the homeomorphism and second projection on the tubular neighborhood. This construction is well-defined up to homotopy and determines a class in $\pi_{n+k}(S^k)$. When we embed the manifold into $\mathbb{R}^{n+k+1}$ via the canonical inclusion $M \subset \mathbb{R}^{n+k} \subset \mathbb{R}^{n+k+1}$, the construction yields Σf_M. Continuing in this way, we determine a class in π_n^S. [Pontrjagin55] proved that all the choices made in this construction remain within the homotopy class and so the mapping $\Omega_*^{\mathrm{fr}} \to \pi_*^S$ is well-defined. Furthermore, it is easy to see that it is a homomorphism.

An inverse mapping may be constructed by using some facts of differential topology. If $g \colon S^{n+k} \to S^k$ represents a class in π_n^S, then we can choose g to be smooth. Let $p \in S^k$ be a regular value of g, that is, the differential $dg_x \colon TS_x^{n+k} \to TS_p^k$ is of maximal rank for all $x \in g^{-1}(\{p\})$. Regular points exist in abundance by the theorem of [Sard42]. The Implicit Function theorem implies that $M(g) = g^{-1}(p)$ is an n-dimensional manifold whose normal bundle is trivial by comparing it with TS_p^k, a single vector space. To show that $\Omega_*^{\mathrm{fr}} \cong \pi_*^S$ it remains to show that the construction from framed manifold to homotopy class provides the same element in π_n^S, that is, $f_{M(g)} \simeq g$. For complete details, see the classic book of [Stong68].

This construction was significantly generalized by [Thom54] in his thesis. The normal bundle ν_M over an n-dimensional manifold M embedded in $\mathbb{R}^{n+k}$ is classified by a homotopy class of a mapping $F \colon M \to BO(k)$. From the embedding, we can talk of a unit disk subbundle $D_{\leq 1}(\nu_M)$ of the normal bundle, as well as its boundary, the unit sphere bundle $S(\nu_M)$ in ν_M. The key to the generalization is the **Thom space** associated to the normal bundle,

$$\mathrm{Th}(\nu_M) = D_{\leq 1}(\nu_M)/S(\nu_M).$$

When M is compact, $\mathrm{Th}(\nu_M)$ is homeomorphic to the one-point compactification of ν_M and so this construction may be carried out for any vector bundle over a compact space. When a vector bundle η is given by a Whitney sum, $\eta = \xi_1 \oplus \xi_2$, then $\mathrm{Th}(\eta) \cong \mathrm{Th}(\xi_1) \wedge \mathrm{Th}(\xi_2)$.

By taking a limit over Grassman manifolds, there is a universal Thom space, $MO(k) = \mathrm{Th}(\gamma_k)$, associated to the universal dimension k vector bundle γ_k over $BO(k)$. The Thom space construction is functorial and so the classifying map provides a mapping $\mathrm{Th}(F)\colon \mathrm{Th}(\nu_M) \to MO(k)$. The canonical inclusion $O(k) \subset O(k+1)$ corresponds to the addition of a trivial bundle to γ_k and this provides a mapping $f_k\colon \Sigma MO(k) = \mathrm{Th}(\gamma_k \oplus \mathbb{R}) \to MO(k+1)$. Thus the sequence

$$\mathbf{MO} = \{MO(1), MO(2), MO(3), \dots\}$$

constitutes a spectrum called the **Thom spectrum**. ([Rudyak98] has written an excellent book on the properties of such spectra.)

The passage from cobordism groups $\mathfrak{N}_n$ to the homotopy groups of the spectrum **MO** is made by taking a sufficiently small tubular neighborhood of M in $\mathbb{R}^{n+k}$ that we denote by N. This space is homeomorphic to the open unit disk bundle $D_{<1}(\nu_M)$ with boundary $S(\nu_M)$. Suppose $N \subset \mathbb{R}^{n+k} \subset \mathbb{R}^{n+k} \cup \infty = S^{n+k}$; we define a mapping $f_M\colon S^{n+k} \to \mathrm{Th}(\nu_M)$ as follows: Send N to $D_{<1}(\nu_M)$ via the homeomorphism $N \cong D_{<1}(\nu_M)$. Send the complement of N in S^{n+k} to the basepoint of $\mathrm{Th}(\nu_M) = D_{\leq 1}(\nu_M)/S(\nu_M)$. Composition with $\mathrm{Th}(F)$ determines a mapping $t_M\colon S^{n+k} \to MO(k)$. When we embed M into $\mathbb{R}^{n+k+1}$ by the canonical inclusion $M \subset \mathbb{R}^{n+k} \subset \mathbb{R}^{n+k+1}$, the inclusion adds a trivial bundle to the normal bundle and the construction results in the suspension Σt_M. Thus we can pass from an embedded n-manifold M to a homotopy class in $\pi_n(\mathbf{MO})$.

[Thom54] proved that this procedure is well-defined and defines a homomorphism, $\Theta\colon \mathfrak{N}_n \to \pi_n(\mathbf{MO})$. The differential topology developed by [Thom54] leads to a description of the inverse homomorphism: Consider the zero section of the universal bundle as an inclusion $BO(k) \hookrightarrow MO(k)$. If $f\colon S^{n+k} \to MO(k)$ represents a class in $\pi_n(\mathbf{MO})$ as a smooth mapping, then the inverse image of the zero section generically gives an n-dimensional manifold in S^{n+k} and varying the representative remains in the cobordism class. The Whitney sum operation, as the mapping

$$\mathrm{Wh}\colon BO(r) \times BO(s) \to BO(r+s),$$

provides a mapping of Thom spaces $MO(r) \wedge MO(s) \to MO(r+s)$ that gives rise to a product on $\pi_*(\mathbf{MO})$. Since the Whitney sum of normal bundles represents the normal bundle to the product embedding, this product on homotopy groups corresponds to the product on cobordism groups.

Theorem 9.31 ([Thom54]). *As rings, $\mathfrak{N}_* \cong \pi_*(\mathbf{MO})$.*

It follows from Lemma 9.30 that $_{(2)}\pi_*(\mathbf{MO}) = \pi_*(\mathbf{MO})$ and so we can apply the mod 2 Adams spectral sequence and hope to compute directly the cobordism ring $\mathfrak{N}_*$. We turn next to the input to the spectral sequence.

Homology and cohomology of Thom spaces

In order to study the mod 2 cohomology of the spectrum $\mathbf{MO}$ as an algebra over the Steenrod algebra, we work on the individual Thom spaces in the spectrum. One of the main results of [Thom54] is the following computational toehold.

Theorem 9.32 (the Thom isomorphism theorem). *If $\xi \to B$ is an oriented k-dimensional vector bundle over a space B of the homotopy type of a finite complex, then, for any ring R, $H^{n+k}(\mathrm{Th}(\xi); R)$ is isomorphic to $H^n(B; R)$ for $n \geq 0$. Furthermore, there is a class $U_k \in H^k(\mathrm{Th}(\xi); R)$, corresponding in the isomorphism to $1 \in H^0(B; R)$, such that, for all n, $H^{n+k}(\mathrm{Th}(\xi); R) \cong H^n(B; R) \smile U_k$.*

SKETCH OF A PROOF: Recall that $\mathrm{Th}(\xi) = D_{\leq 1}(\xi)/S(\xi)$. When B has the homotopy type of a finite complex, we can write

$$\tilde{H}^*(\mathrm{Th}(\xi); R) = H^*(D_{\leq 1}(\xi), S(\xi); R).$$

We next apply the Leray-Serre spectral sequence for pairs (Exercise 5.6) to the fibration $(e^k, S^{k-1}) \hookrightarrow (D_{\leq 1}(\xi), S(\xi)) \to B$. The E_2-term is concentrated in the kth row where we find $E_2^{n,k} \cong H^n(B; H^k(e^k, S^{k-1}; R)) \cong H^n(B; R)$. The orientation allows us to make this isomorphism globally. The theorem follows from convergence of the Leray-Serre spectral sequence and the cup-product structure on the spectral sequence. (See [Milnor-Stasheff74] for a more geometric proof.) □

We apply this result to the universal $\mathbb{R}^k$-bundle, γ_k over $BO(k)$. Following the discussion in §6.3 we know that $H^*(BO(k); \mathbb{F}_2)$ is isomorphic to $\mathbb{F}_2[w_1, \dots, w_k]$ where the w_i are the universal Stiefel-Whitney classes and $\deg w_i = i$. The w_i may be defined from the symmetric functions on classes $y_1, \dots, y_k$ of dimension one in $H^*((BO(1))^{\times k}; \mathbb{F}_2)$ where $O(1) \times \cdots \times O(1) \hookrightarrow O(k)$ is the inclusion of the diagonal matrices with entries ± 1. ($O(1) \cong \mathbb{Z}/2\mathbb{Z}$.)

The Thom isomorphism theorem gives

$$H^*(M O(k); \mathbb{F}_2) \cong H^*(BO(k); \mathbb{F}_2) \smile U_k \cong \mathbb{F}_2[w_1, \dots, w_k] \smile U_k.$$

The zero section provides a map $BO(k) \hookrightarrow MO(k)$ that is compatible with the structure maps induced by the inclusions, $i \colon O(k) \hookrightarrow O(k+1)$. It follow that there is a commutative diagram:

$$
\begin{array}{ccc}
s^{-(k+1)} H^*(MO(k+1); \mathbb{F}_2) & \longrightarrow & s^{-k} H^*(MO(k); \mathbb{F}_2) \\
{\scriptstyle (t^*)^{-1}} \downarrow & & \downarrow {\scriptstyle (t^*)^{-1}} \\
H^*(BO(k+1); \mathbb{F}_2) & \xrightarrow[Bi^*]{} & H^*(BO(k); \mathbb{F}_2)
\end{array}
$$

where the vertical maps are the inverses of the Thom isomorphism and the top horizontal map is induced by the spectrum map $\Sigma MO(k) \rightarrow MO(k+1)$. It follows that we can identify $H^*(\mathbf{MO}; \mathbb{F}_2)$ with $\lim_{\leftarrow r} H^*(BO(k); \mathbb{F}_2) = H^*(BO; \mathbb{F}_2)$ as a vector space. There is a coproduct structure on $H^*(BO; \mathbb{F}_2)$ that is induced by the Whitney product Wh: $BO(r) \times BO(s) \rightarrow BO(r+s)$ and $H^*(BO; \mathbb{F}_2)$ is a Hopf algebra with this coproduct. As an algebra

$$
H^*(BO; \mathbb{F}_2) \cong \mathbb{F}_2[w_1, w_2, \dots, w_k, \dots].
$$

The coproduct formula is given by the Whitney sum formula (Lemma 6.42). This yields a commutative product on the dual of $H^*(BO; \mathbb{F}_2)$ and the Hopf algebra in this case is self-dual, that is, as algebras,

$$
H_*(BO; \mathbb{F}_2) \cong \mathbb{F}_2[a_1, a_2, \dots, a_k, \dots].
$$

We turn to homology in order to avoid the noncommutative product on the Steenrod algebra. Mod 2 homology is endowed with the structure of a comodule over $\mathcal{A}_2^{\mathrm{dual}}$, that is, there are homomorphisms

$$
\psi_X \colon H_*(X; \mathbb{F}_2) \rightarrow \mathcal{A}_2^{\mathrm{dual}} \otimes H_*(X; \mathbb{F}_2)
$$

satisfying the dual axioms for the Steenrod algebra action. Thus the following diagram commutes for all spaces X:

$$
\begin{array}{ccc}
H_*(X; \mathbb{F}_2) & \xrightarrow{\ \psi_X\ } & \mathcal{A}_2^{\mathrm{dual}} \otimes H_*(X; \mathbb{F}_2) \\
{\scriptstyle \psi_X} \downarrow & & \downarrow {\scriptstyle 1 \otimes \psi_X} \\
\mathcal{A}_2^{\mathrm{dual}} \otimes H_*(X; \mathbb{F}_2) & \xrightarrow[\psi \otimes 1]{} & \mathcal{A}_2^{\mathrm{dual}} \otimes \mathcal{A}_2^{\mathrm{dual}} \otimes H_*(X; \mathbb{F}_2)
\end{array}
$$

where $\psi \colon \mathcal{A}_2^{\mathrm{dual}} \rightarrow \mathcal{A}_2^{\mathrm{dual}} \otimes \mathcal{A}_2^{\mathrm{dual}}$ is the coproduct on the dual of the Steenrod algebra.

To study $H_*(\mathbf{MO}; \mathbb{F}_2)$, we notice that the Whitney sum induces a pairing $MO(r) \wedge MO(s) \rightarrow MO(r+s)$ that commutes with the zero sections and the Whitney sum map on the classifying spaces. With some care in identifying generators ([Stong68, Chapter VI]), this induces a product on $H_*(\mathbf{MO}; \mathbb{F}_2)$ and gives an algebra isomorphism

$$
H_*(\mathbf{MO}; \mathbb{F}_2) \cong \mathbb{F}_2[a_1, a_2, \dots, a_k, \dots].
$$

Recall the theorem of [Milnor58] on the structure of $\mathcal{A}_2^{\text{dual}}$ (Theorem 4.47): At the prime 2 we have

$$\mathcal{A}_2^{\text{dual}} \cong \mathbb{F}_2[\xi_1, \xi_2, \dots, \xi_k, \dots],$$

where $\deg \xi_i = 2^i - 1$. Consider the naive splitting:

$$H_*(\mathbf{MO}; \mathbb{F}_2) \cong \mathbb{F}_2[a_i \mid i = 2^j - 1, j = 0, 1, \dots] \otimes \mathbb{F}_2[a_k \mid k \neq 2^j - 1, k \geq 2].$$

This decomposition suggests that $H_*(\mathbf{MO}; \mathbb{F}_2)$ may be isomorphic to an extended comodule over $\mathcal{A}_2^{\text{dual}}$, that is, $H_*(\mathbf{MO}; \mathbb{F}_2) \cong \mathcal{A}_2^{\text{dual}} \otimes \mathbb{F}_2[a_k \mid k \neq 2^j - 1, k \geq 2]$. An interpretation of the results of [Thom54] by [Liulevicius62] in the setting of Hopf algebras leads to a proof of this splitting.

Theorem 9.33. $H_*(\mathbf{MO}; \mathbb{F}_2) \cong \mathcal{A}_2^{\text{dual}} \otimes \mathbb{F}_2[a_k \mid k \neq 2^j - 1, k \geq 2]$ *as extended comodules over* $\mathcal{A}_2^{\text{dual}}$.

SKETCH OF PROOF: Recall that $H_*(BO(1); \mathbb{F}_2) \cong \Gamma(x_1)$, the divided power algebra on a generator of dimension one. The main ingredients in the proof are: (1) The fact that $Bi_* \colon H_*(BO(1); \mathbb{F}_2) \to H_*(BO; \mathbb{F}_2)$ is given by $\gamma_k(x_1) \mapsto a_k$. This follows from the dual representation of the universal Stiefel-Whitney classes as symmetric polynomials in the one-dimensional classes in $H^*(BO(1)^{\times k}; \mathbb{F}_2)$. (2) The determination of the $\mathcal{A}_2^{\text{dual}}$-comodule structure on $H_*(BO(1); \mathbb{F}_2)$ and hence, by virtue of the commutative diagram,

$$
\begin{array}{ccc}
H_*(BO(1); \mathbb{F}_2) & \xrightarrow{\psi_{BO(1)}} & \mathcal{A}_2^{\text{dual}} \otimes H_*(BO(1); \mathbb{F}_2) \\
{\scriptstyle Bi_*}\downarrow & & \downarrow{\scriptstyle 1 \otimes Bi_*} \\
H_*(BO; \mathbb{F}_2) & \xrightarrow{\psi_{BO}} & \mathcal{A}_2^{\text{dual}} \otimes H_*(BO; \mathbb{F}_2),
\end{array}
$$

the $\mathcal{A}_2^{\text{dual}}$-comodule structure on $H_*(BO; \mathbb{F}_2)$ can be determined. Since the algebra structure on $H_*(BO; \mathbb{F}_2)$ is compatible with the $\mathcal{A}_2^{\text{dual}}$-comodule structure, it suffices to check on generators. The comodule structure may be written $\psi_{BO}(a_n) = \sum_{i=0}^n \boldsymbol{o}_i^n \otimes a_i$ where $\boldsymbol{o}_i^n \in (\mathcal{A}_2^{\text{dual}})_{n-i}$. (4) There is another commutative diagram that allows us to determine the $\mathcal{A}_2^{\text{dual}}$-comodule structure on $H_*(\mathbf{MO}; \mathbb{F}_2)$:

$$
\begin{array}{ccc}
H_*(BO; \mathbb{F}_2) & \xrightarrow{\psi_{BO}} & \mathcal{A}_2^{\text{dual}} \otimes H_*(BO; \mathbb{F}_2) \\
{\scriptstyle t_*}\downarrow & & \downarrow{\scriptstyle 1 \otimes t_*} \\
H_*(\mathbf{MO}; \mathbb{F}_2) & \xrightarrow{\psi_{MO}} & \mathcal{A}_2^{\text{dual}} \otimes H_*(\mathbf{MO}; \mathbb{F}_2).
\end{array}
$$

Since the classes coming from $MO(1)$ are identifiable with $\gamma_k(x_1) \smile a_1$, we get a map $H_*(BO(1); \mathbb{F}_2) \rightarrow H_*(\mathbf{MO}; \mathbb{F}_2)$ given by $\gamma_k(x_1) \mapsto a_{k-1}$. This gives the crucial step as we can compute

$$\psi_{MO}(a_k) = \begin{cases} \xi_r \otimes 1 + \sum_{l>0} \xi_{I_l} \otimes a_l & \text{if } k = 2^r - 1, \\ 1 \otimes a_k + \text{decomposables} & \text{if } k \neq 2_1^r. \end{cases}$$

(5) Finally, we can check that $\psi_{MO}(a_{2^r-1}) = \sum_{s=0}^{r} \xi_{r-s}^{2^s} \otimes a_{2^s-1}$. This follows from the representation of the dual classes w_k as symmetric polynomials and the pairing of the Steenrod algebra and its dual (Proposition 4 of [Liulevicius62]). A complete exposition of all these details may be found in [Schochet71'] or from the cohomological point of view in [Stong68]. $\square$

It follows from the theorem that $H^*(\mathbf{MO}; \mathbb{F}_2)$ is a free module over $\mathcal{A}_2$ and so $\pi_*(\mathbf{MO})$ is computable immediately from $\text{Ext}_{\mathcal{A}_2}(H^*(\mathbf{MO}; \mathbb{F}_2), \mathbb{F}_2)$. But this is the dual to the generating module for the free module.

Corollary 9.34 ([Thom54]). $\pi_*(\mathbf{MO}) \cong \mathbb{F}_2[a_k \mid k \neq 2^r - 1, k \geq 2]$ as algebras.

Thom proved that there is a weak homotopy equivalence between $\mathbf{MO}$ and a product of Eilenberg-Mac Lane spectra $\Sigma^{|s(\omega)|} K\mathbb{Z}/2\mathbb{Z}$ with the $|s(\omega)|$ given by the degrees of homogeneous polynomials $s(\omega)$ in $\mathbb{F}_2[a_k \mid k \neq 2^r - 1, k \geq 2]$. The role of characteristic classes in distinguishing nontrivial classes is crucial—one of the main theorems of [Thom54] is the sufficiency of the mod 2 characteristic numbers in classifying cobordism classes of unoriented manifolds.

Thom's construction of the Thom spaces admits considerable generalization. In particular, we can define the **oriented cobordism ring** Ω_*^{SO} by admitting only oriented manifolds M_1 and M_2 and requiring that a cobordism W be orientable with boundary $M_1 \amalg (-M_2)$, where $-M_2$ is the manifold M_2 with the opposite orientation. The normal bundles of such manifolds have a lifting of their classifying map $f_\nu : M \rightarrow BO(n)$ to $\bar{f}_\nu : M \rightarrow BSO(n)$. The universal n-dimensional vector bundle over $BSO(n)$ has Thom space $MSO(n)$ and the same argument for unoriented cobordism can be made to prove that the spaces $MSO(n)$ form a spectrum $\mathbf{MSO}$ and $\pi_*(\mathbf{MSO}) \cong \Omega_*^{SO}$.

If one focuses on the lifting of the normal bundle to some classifying space $BG \rightarrow BO(n)$, then a very general notion of cobordism is possible ([Stong68]). For almost complex manifolds, there is a lifting of the normal bundle to $BU(n) \rightarrow BO(2n)$ and Chern classes and numbers distinguish cobordism classes. In a classic paper that introduced the application of the Adams spectral sequence to cobordism, [Milnor60] proved that Ω_*^{SO} has torsion only at the prime 2, and that Ω_*^U is torsion-free. Furthermore, Milnor computed $\pi_*(\mathbf{MU})$ by determining the structure of $H^*(\mathbf{MU}; \mathbb{F}_p)$ for all primes p, and then using the Adams spectral sequence. In fact, $H^*(\mathbf{MU}; \mathbb{F}_p)$ is free over a quotient Hopf algebra of $\mathcal{A}_p$ and a change-of-rings argument (Theorem 9.12) allows the straightforward calculation of $_{(p)}\pi_*(\mathbf{MU})$. The homological simplicity of complex cobordism led [Novikov67] in his study of a generalization of the Adams spectral sequence founded on $\mathbf{MU}$. See the book of [Ravenel86] for more details.

Similar arguments were first carried out by [Ray72] and [Kochman80] to obtain partial results for symplectic cobordism ([Kochman96] gave a different approach to calculating $\pi_*(\mathbf{MSp})$).

9.5 Computations

Our point of departure is Corollary 9.23—there is a spectral sequence with E_2-term isomorphic to the cohomology of the Steenrod algebra $\mathcal{A}_p$, and converging to $_{(p)}\pi_*^S$. Thus the problem of computing $_{(p)}\pi_*^S$ breaks into the problems of computing $H^{*,*}(\mathcal{A}_p)$, and then the differentials in the Adams spectral sequence. Finally, there is the problem of determining the extensions.

We first construct a small part of a minimal resolution for $\mathcal{A}_3$, the mod 3 Steenrod algebra. The computation begins easily enough and you even get some of $_{(3)}\pi_*^S$, but it quickly gets complicated. We then consider the case of $p = 2$ more systematically. Following [Adams60], we are able to describe $H^{s,*}(\mathcal{A}_2)$ in some detail for $s \le 3$. Next we put these computations to work and find the first nontrivial differential in the spectral sequence. A corollary is the first case of the Hopf invariant one problem. We continue the hands-on computations with a discussion of Massey products and their relation to Toda brackets.

Low-dimensional calculations

We begin by constructing the beginning of a minimal resolution,

$$0 \leftarrow \mathbb{F}_3 \xleftarrow{\varepsilon} \mathcal{A}_3 \xleftarrow{d_0} P_1 \xleftarrow{d_1} P_2,$$

up to internal degree 9. For the most part, the discussion will be descriptive; the reader should construct a chart of everything that is happening.

By Lemma 9.11 and Theorem 4.45, P_1 has $\mathcal{A}_3$-module generators a_0 of degree 1 and h_i of degree $4 \cdot 3^i = 2 \cdot 3^i \cdot (3 - 1)$ for $i = 0, 1, 2 \dots$. These generators correspond to β and P^{3^i}; the homomorphism $d_0 \colon P_1 \to \mathcal{A}_3$ is given by $d_0(a_0) = \beta$ and $d_0(h_i) = P^{3^i}$. In the kernel of d_0 there is already βa_0 since $\beta^2 = 0$. We put a generator $\bar{a}_0$ in P_2 with $d_1(\bar{a}_0) = \beta a_0$. The next phenomenon to arise in the kernel of d_0 that is not accounted for by $d_0(a_0)$ occurs in degree 9. The diligent reader making a chart will find $P^2\beta$, βP^2, $P^1\beta P^1$ and $\beta P^1 P^1$ in $\mathcal{A}_3$ in degree 9. The Adem relations imply that only two of these expressions are independent. In degree 9 in P_1 there are three independent generators, $P^2 a_0$, $\beta P^1 h_0$ and $P^1 \beta h_0$. The Adem relation

$$P^1 \beta P^1 = \beta P^2 + P^2 \beta$$

implies that the element $P^2 a_0 - P^1 \beta h_0 - \beta P^1 h_0$ is in the kernel of d_0. Let g_1 be in P_2 in degree 9 with $d_1(g_1) = P^2 a_0 - P^1 \beta h_0 - \beta P^1 h_0$.

The chart should be getting a bit complicated by now. However, two simple patterns emerge:

(1) Except for the element $\bar{\bar{a}}_0$ in P_3 with $d_2(\bar{\bar{a}}_0) = \beta \bar{a}_0$, the first generator to appear in P_3 is in degree greater than 11. This implies that we have computed $H^{s,t}(\mathcal{A}_3)$ for $s \leq 2$, $t \leq 11$;

(2) The recurring Bockstein that arises at each stage behaves systematically and so, if we remove the chain of generators due to $\beta^2 = 0$, the connectivity of this minimal resolution implies we have computed $H^{s,*}(\mathcal{A}_3)$ for $t \leq 11$ and all s.

Because the resolution is minimal, $\mathrm{Ext}_{\mathcal{A}_3}^{s,t}(\mathbb{F}_3, \mathbb{F}_3) = (\mathbb{F}_3 \otimes_{\mathcal{A}_3} P_s)^{\mathrm{dual}}$, and so we have computed

$$\mathrm{Ext}_{\mathcal{A}_3}^{s,t}(\mathbb{F}_3, \mathbb{F}_3) = \begin{cases} \mathbb{F}_3, & (s,t) = (0,0), (1,1), (2,2), \\ & \qquad\quad (3,3), (1,4), (2,9), \\ \{0\}, & \text{elsewhere for } s \leq 3 \text{ and } t \leq 11. \end{cases}$$

Since $E_\infty^{s,t}$ is related to $_{(3)} \pi_{t-s}^S$, we display $\mathrm{Ext}^{s,t}$ with $t - s$ running horizontally and s vertically. The differentials d_r then lower $t - s$ degree by 1 and raise s degree by r (that is, d_r goes left one space and up r spaces in the $(t-s, s)$-plane). In the spectral sequence, we can display these data in the diagram:

With this chart, the connectivity of the resolution, and the evident lack of differentials, we have computed the following stable homotopy groups.

Proposition 9.35.

$$_{(3)}\pi_n^S = \begin{cases} \mathbb{Z}, & \text{if } n = 0, \\ \mathbb{Z}/3\mathbb{Z}, & \text{if } n = 3, 7, \\ \{0\}, & \text{if } n = 1, 2, 4, 5, 6, 8, 9. \end{cases}$$

The reader should compare this method with the method of killing homotopy groups in Chapter 6 (Corollary 6.27).

If the reader has been creating a chart to keep track of the minimal resolution, it should be clear that a systematic method of computation is desirable.

We change to the prime 2 and study $H^{*,*}(\mathcal{A}_2)$ more systematically with the tool of choice—the change-of-rings spectral sequence (Theorem 9.12).

Theorem 9.36. *Let $\mathcal{G}$ denote the bigraded algebra*

$$\mathcal{G} = \mathbb{F}_2[x_0, x_1, x_2, \dots,] \left/ \begin{array}{l} x_i x_{i+1} = 0 \\ x_i^2 x_{i+2} + x_{i+3}^3 = 0 \\ x_i x_{i+2}^2 = 0 \end{array} \right.$$

with $\text{bideg } x_i = (1, 2^i)$. *There are elements h_i in $H^{1,*}(\mathcal{A}_2)$ for $i = 0, 1, 2, \dots$ such that* $\text{bideg } h_i = (1, 2^i)$ *and a mapping $\alpha \colon \mathcal{G} \to H^{*,*}(\mathcal{A}_2)$, determined on generators by $x_i \mapsto h_i$; α is a well-defined mapping of bigraded algebras and α restricts to isomorphisms, $\mathcal{G}^{1,*} \to H^{1,*}(\mathcal{A}_2)$ and $\mathcal{G}^{2,*} \to H^{2,*}(\mathcal{A}_2)$, and to a monomorphism $\mathcal{G}^{3,*} \to H^{3,*}(\mathcal{A}_2)$. All relations among products of the generators h_i are consequences of this mapping for $H^{s,*}(\mathcal{A}_2)$, $s \leq 3$.*

PROOF ([Adams60]): To compute $H^{*,*}(\mathcal{A}_2)$, consider the supporting cochain complex for this algebra, $F_*(\mathcal{A}_2^{\text{dual}})$ given by the cobar construction on $\mathcal{A}_2^{\text{dual}}$.

Recall the coproduct for $\mathcal{A}_2^{\text{dual}} = \mathbb{F}_2[\xi_1, \xi_2, \dots]$ on the generators is given by the formula of [Milnor58]:

$$\varphi^*(\xi_i) = \sum_{j+k=i} \xi_j^{2^k} \otimes \xi_k \qquad (\xi_0 = 1).$$

We proceed by a series of remarks:

I. $H^{1,*}(\mathcal{A}_2) = \mathbb{F}_2\{h_i \mid i = 0, 1, 2, \dots,$ and $\deg h_i = 2^i\}$. This follows from Lemma 9.11. To determine $\text{Prim}(\mathcal{A}_2^{\text{dual}})$ observe that ξ_1 is primitive and so, because we are working mod 2, $\xi_1^{2^i}$ is also primitive for $i > 0$. The formula for φ^* shows that these are the only primitives. Let h_i denote the class $[\xi_1^{2^i}]$ in the cobar construction.

II. If $h_1 h_0 = 0$, $h_0^2 h_2 + h_1^3 = 0$ and $h_0 h_2^2 = 0$, then, for all $i \geq 0$, $h_{i+1} h_i = 0$, $h_i^2 h_{i+2} + h_{i+1}^3 = 0$ and $h_i h_{i+2}^2 = 0$.

Because $\mathcal{A}_2^{\text{dual}}$ is commutative, we can apply Theorem 9.10: Sq^0 exists in the Steenrod algebra that acts on $H^{*,*}(\mathcal{A}_2)$ and the Cartan formula shows that Sq^0 is multiplicative. Since $Sq^0 h_i = Sq^0([\xi_1^{2^i}]) = [\xi_1^{2^{i+1}}] = h_{i+1}$, repeated applications of Sq^0 to the initial identities obtains the identities for all $i \geq 0$.

III. $h_1 h_0 = 0$, $h_0^2 h_2 + h_1^3 = 0$, and $h_0 h_2^2 = 0$.

Consider the formulas in the cobar construction:

$$d^*([\xi_2]) = [\xi_1^2 \mid \xi_1],$$

$$d^*([\xi_2 \mid \xi_2] + [\xi_1^2 \mid \xi_1\xi_2] + [\xi_1^2\xi_2 \mid \xi_1]) = [\xi_1^4 \mid \xi_1 \mid \xi_1] + [\xi_1^2 \mid \xi_1^2 \mid \xi_1^2],$$

$$d^*([\xi_3 \mid \xi_1^2] + [\xi_2^2 \mid \xi_2] + [\xi_1^4 \mid \xi_1^2\xi_2] + [\xi_2^2 \mid \xi_1^3]) = [\xi_1^4 \mid \xi_1^4 \mid \xi_1].$$

These formulas imply that $\alpha \colon \mathcal{G} \to H^{*,*}(\mathcal{A}_2)$ is well-defined in degrees $s \leq 3$. It remains to prove that α on $\mathcal{G}^{1,*}$ and on $\mathcal{G}^{2,*}$ gives isomorphisms and on $\mathcal{G}^{3,*}$ a monomorphism.

Because we can describe $\mathcal{A}_2^{\text{dual}}$ so explicitly, we can find sub-Hopf algebras of $\mathcal{A}_2^{\text{dual}}$ of particularly simple form. Let $B'_n = \mathbb{F}_2[\xi_1, \dots, \xi_n]$; then B'_n is a sub-Hopf algebra of $\mathcal{A}_2^{\text{dual}}$. Furthermore, we have the short exact sequence for each n,

$$0 \to B'_{n-1} \to B'_n \to A'_n \to 0$$

where $A'_n = B'_n // B'_{n-1} \cong \mathbb{F}_2[\xi_n]$, the Hopf algebra with ξ_n primitive. By dualizing, we obtain the extensions of Hopf algebras

$$0 \to A_n \to B_n \to B_{n-1} \to 0.$$

Notice that $(B'_n)^r \cong (\mathcal{A}_2^{\text{dual}})^r$ for $r < 2^n - 1$. Thus $H^{*,t}(B_n) \cong H^{*,t}(\mathcal{A}_2)$ for $t < 2^n - 1$. We will use these extensions with the change-of-rings spectral sequence to compute $H^{*,*}(\mathcal{A}_2)$ in the desired range. To apply Theorem 9.12, we need a further remark.

IV. A_n is central in B_n.

Because we are not giving explicit descriptions of A_n and B_n, we consider the dual situation and ask: Does the following diagram commute?

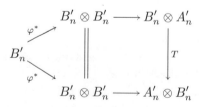

Because φ^* is multiplicative, it suffices to check the commutativity on the algebra generators, $\xi_1, \xi_2, \dots, \xi_n$. The explicit formula for the coproduct implies that the diagram commutes.

By Theorem 9.12, for each n, there is a spectral sequence, converging to $H^*(B_n)$, with $E_2^{p,q}$ given by $H^q(A_n) \otimes H^p(B_{n-1})$. Because A_n is $(2^n - 1)$-connected, $\mathcal{A}_2 = \varprojlim B_n$, and so we get better and better approximations to $H^*(\mathcal{A}_2)$ with each $H^*(B_n)$.

V. $H^*(A_n) \cong \mathbb{F}_2[h_{n,i} \mid i = 0, 1, 2, \ldots]$, where $h_{n,i}$ denotes the class $[\xi_n^{2^i}]$ in the cobar construction on A'_n. This follows because $A'_n \cong \mathbb{F}_2[\xi_n]$.

With these data, we begin an induction. For $n = 1$,

$$H^*(B_1) \cong H^*(A_1) \cong \mathbb{F}_2[h_i \mid i = 0, 1, 2, \ldots]$$

where the h_i corresponds to $[\xi_1^{2^i}]$ of degree 2^i.

For $n = 2$, $E_2^{p,q} \cong H^q(A_2) \otimes H^p(B_1)$. First examine $d_2 \colon E_2^{0,1} \to E_2^{2,0}$; $d_2([\xi_2]) = [\xi_1^2 \mid \xi_1]$ or $d_2(h_{2,0}) = h_1 h_0$. When we apply Sq^0, we obtain $d_2(h_{2,i}) = h_{i+1} h_i$. Thus d_2 is monic on $E_2^{0,1}$ and, since $H^1(B_2)$ comes from $E_\infty^{0,1}$ and $E_\infty^{1,0} = E_2^{1,0} = H^1(B_1)$, we have shown $H^1(B_2) \cong H^1(B_1)$ with the isomorphism coming from the projection $B_2 \to B_1$. Notice further that we have introduced the identities $h_{i+1} h_i = 0$ into $H^*(B_2)$.

Consider next $d_2 \colon E_2^{1,1} \to E_2^{3,0}$; $E_2^{1,1}$ is isomorphic to $H^1(A_2) \otimes H^1(B_1)$ $\cong \mathbb{F}_2\{h_{2,j} \otimes h_k\}$. Since d_2 is a derivation, $d_2(h_{2,j} \otimes h_k) = h_{j+1} h_j h_k$. This differential is a monomorphism except when the dimensions conspire to give a kernel. In particular,

$$d_2(h_{2,i} \otimes h_{i+2} + h_{2,i+1} \otimes h_i) = h_{i+1} h_i h_{i+2} + h_{i+2} h_{i+1} h_i = 0.$$

Because no other differential is defined on $E_2^{1,1}$, these classes live to $E_\infty^{1,1}$ and determine elements in $H^2(B_2)$ that we denote by

$$g_{2,i} = \{h_{2,i} \otimes h_{i+2} + h_{2,i+1} \otimes h_i\}.$$

These classes lie in t degrees $3 \cdot 2^i + 2^{i+2} = 7 \cdot 2^i$, and so they are linearly independent.

To finish our description of $H^2(B_2)$, we must determine $E_\infty^{0,2}$. The differential $d_2 \colon E_2^{0,2} \to E_2^{2,1}$ is given by $d_2(h_{2,i} h_{2,j}) = h_{2,j} \otimes h_{i+1} h_i + h_{2,i} \otimes h_{j+1} h_j$, and so is nonzero except when $i = j$. Thus only the classes $(h_{2,i})^2$ survive to $E_3^{0,2}$.

To determine $d_3 \colon E_3^{0,2} \to E_3^{3,0}$, we consider $(h_{2,0})^2$. In the cobar construction, this is the name for $[\xi_2 \mid \xi_2]$; by III, this element is congruent modulo the filtration to $[\xi_2 \mid \xi_2] + [\xi_1^2 \mid \xi_1 \xi_2] + [\xi_1^2 \xi_2 \mid \xi_1]$ which is carried by d^* to $[\xi_1^4 \mid \xi_1 \mid \xi_1] + [\xi_1^2 \mid \xi_1^2 \mid \xi_1^2]$. Thus $d_3((h_{2,0})^2) = h_0^2 h_2 + h_1^3$. Repeated application of Sq^0 gives

$$d_3((h_{2,i})^2) = h_i^2 h_{i+2} + h_{i+1}^3.$$

Therefore, $E_\infty^{0,2} = \{0\}$.

Because the spectral sequence converges to $H^*(B_2)$, we have the short exact sequence

$$0 \to E_\infty^{2,0} \to H^2(B_2) \to E_\infty^{1,1} \to 0.$$

We can describe $E_\infty^{2,0}$ as $\mathbb{F}_2\{h_i h_j \mid j \neq i+1\}$ and $E_\infty^{1,1}$ as $\mathbb{F}_2\{g_{2,i} \mid i = 0, 1, \ldots\}$. Recall the t degrees of all of these elements; $h_i h_j \in E_\infty^{2,0,2^i+2^j}$ and $g_{2,j} \in E_\infty^{1,1,7\cdot 2^i}$. We deduce that $H^2(B_2)$ is the direct sum of $E_\infty^{2,0}$ and $E_\infty^{1,1}$.

Finally, we determine $H^3(B_2)$. In $H^3(B_1)$ we have introduced the relations $h_{i+1}h_i h_k = 0$ and $h_i^2 h_{i+2} + h_{i+1}^3 = 0$. We compute $d_2 \colon E_2^{1,2} \to E_2^{3,1}$. Again, d_2 is a derivation, so on elements $h_{2,i} h_{2,j} \otimes h_k$ we have

$$d_2(h_{2,i} h_{2,j} \otimes h_k) = h_{2,j} \otimes h_{i+1} h_i h_k + h_{2,i} \otimes h_{j+1} h_j h_k.$$

Unless $i = j$, d_2 is nonzero, and so $E_3^{1,2} = \mathbb{F}_2\{h_{2,i} h_{2,i} \otimes h_k\}$. A subtlety enters here: $d_3(h_{2,i} h_{2,i} \otimes h_k) = h_i^2 h_{i+2} h_k + h_{i+1}^3 h_k$ follows from the derivation property of d_3. It is possible for a relation to produce elements in the kernel of d_3 if that relation is induced by d_2. Consider

$$d_3(h_{2,i} h_{2,i} \otimes h_{i+3} + h_{2,i+1} h_{2,i+1} \otimes h_{i+1})$$
$$= h_i^2 h_{i+2} h_{i+3} + h_{i+1}^3 h_{i+3} + h_{i+1}^2 h_{i+3} h_{i+1} + h_{i+2}^3 h_{i+3}.$$

The first and last terms are 0 since $h_{l+1} h_l = 0$ and the middle vanishes as a pair. Since no other differential affects $E_3^{1,2}$, we have determined a flock of classes in $E_\infty^{1,2}$, denoted by

$$f_{2,i} = \{h_{2,i} h_{2,i} \otimes h_{i+3} + h_{2,i+1} h_{2,i+1} \otimes h_{i+1}\}$$
$$= \{[\xi_2^{2^i} \mid \xi_2^{2^i} \mid \xi_1^{2^{i+3}}] + [\xi_2^{2^{i+1}} \mid \xi_2^{2^{i+1}} \mid \xi_1^{2^{i+1}}]\}.$$

Once again, the t degrees of the $f_{2,i}$ show that they are linearly independent.

Next consider $d_2 \colon E_2^{0,3} \cong H^3(A_2) \to H^2(A_2) \otimes H^2(B_1) \cong E_2^{2,2}$. The formula

$$d_2(h_{2,i} h_{2,j} h_{2,k}) = h_{2,j} h_{2,k} \otimes h_{i+1} h_i + h_{2,i} h_{2,k} \otimes h_{j+1} h_j + h_{2,i} h_{2,j} \otimes h_{k+1} h_k$$

is enough to show that d_2 is a monomorphism; therefore, $E_\infty^{0,3} = \{0\}$. To finish off $H^3(B_2)$, we consider $d_2 \colon E_2^{2,1} \to E_2^{4,0}$. A class in $E_2^{2,1}$ is a sum of classes of the form $h_{2,i} \otimes h_j h_k$, and $d_2(h_{2,i} \otimes h_j h_k) = h_{i+1} h_i h_j h_k$. Many relations can be obtained by manipulating subscripts; however, most of these are generated by the image of $d_2 \colon E_2^{0,2} \to E_2^{2,1}$ and so are known. The exceptions are of the form

$$h_{2,i} \otimes h_{i+2} h_k + h_{2,i+1} \otimes h_i h_k,$$

which is seen to be $g_{2,i} h_k$. These classes are missed by d_2 and so give permanent cycles. However, they are not linearly independent in the rest of $H^3(B_2)$.

VI. $g_{2,i} h_{i+1} = h_{i+2}^2 h_i$ in $H^3(B_2)$.

To prove this, observe that in the cobar complex,

$$d^*([\xi_2^2 \mid \xi_2] + [\xi_1^4 \mid \xi_1^2 \xi_2] + [\xi_2^2 \mid \xi_1^3]) = [\xi_1^4 \mid \xi_1^4 \mid \xi_1] + [\xi_1^4 \mid \xi_2 \mid \xi_1^4] + [\xi_2^2 \mid \xi_1 \mid \xi_1^2],$$

which represents $h_2^2 + g_{2,0}h_1$. Since $Sq^0 g_{2,i} = g_{2,i+1}$, this formula proves VI.

If the reader has kept track of the new elements in $H^3(B_2)$, then, because the internal degrees of these elements all differ, we have shown that $H^3(B_2)$ contains $\mathbb{F}_2\{h_i h_j h_k \mid i \neq j+1, j \neq k+1\}$ modulo $h_i^2 h_{i+2} = h_{i+1}^3$, as well as $\mathbb{F}_2\{f_{2,i} \mid i = 0, 1, \dots\}$ and $\mathbb{F}_2\{g_{2,i}h_j \mid g_{2,i}h_{i+1} = h_{i+2}^3 h_i\}$. We next proceed in our induction to $H^*(B_3)$.

For $n = 3$, $E_2^{p,q} \cong H^q(A_3) \otimes H^p(B_2)$. First examine $d_2 \colon E_2^{0,1} \to E_2^{2,0}$.

$$d_2(h_{3,0}) = \{d^*([\xi_3])\} = \{[\xi_2^2 \mid \xi_1] + [\xi_1^4 \mid \xi_2]\} = \{h_{2,1} \otimes h_0 + h_{2,0} \otimes h_2\} = g_{2,0}.$$

When we apply Sq^0, we obtain $d_2(h_{3,i}) = g_{2,i}$. Thus d_2 is monic on $E_2^{0,1}$ and therefore, $E_\infty^{0,1} = \{0\}$. This shows $H^1(B_3) \cong H^1(B_2) \cong H^1(B_1)$ with the isomorphisms induced by the projections $B_3 \to B_2 \to B_1$.

Consider next $d_2 \colon E_2^{1,1} \to E_2^{3,0}$. By the derivation property of d_2, we have $d_2(h_{3,i} \otimes h_j) = g_{2,i}h_j$. If $j = i + 1$, this still makes sense and we get $d_2(h_{3,i} \otimes h_{i+1}) = g_{2,i}h_{i+1} = h_{i+2}^2 h_i$, which introduces a new relation among the 3-fold products of elements in $H^1(B_3)$. Looking at sums in $E_2^{1,1}$ we consider classes $h_{3,i} \otimes h_{i+3} + h_{3,i+1} \otimes h_i$. Apply the differential to get

$$
\begin{aligned}
&d_2(h_{3,i} \otimes h_{i+3} + h_{3,i+1} \otimes h_i) \\
&= g_{2,i}h_{i+3} + g_{2,i+1}h_i \\
&= \{h_{2,i} \otimes h_{i+2} + h_{2,i+1} \otimes h_i\}h_{i+3} + \{h_{2,i+1} \otimes h_{i+3} + h_{2,i+2} \otimes h_{i+1}\}h_i \\
&= 0, \quad \text{since } h_{l+1}h_l = 0.
\end{aligned}
$$

This determines classes $g_{3,i} = \{h_{3,i} \otimes h_{i+3} + h_{3,i+1} \otimes h_i\}$ in $E_\infty^{1,1}$ and so in $H^2(B_3)$. The classes $g_{3,i}$ lie in t degrees $2^i(2^3 - 1) + 2^{i+3} = 15 \cdot 2^i$ and so are linearly independent.

To finish the description of $H^2(B_3)$, consider $d_2 \colon E_2^{0,2} \to E_2^{2,1}$:

$$d_2(h_{3,i}h_{3,j}) = h_{3,j} \otimes g_{2,i} + h_{3,i} \otimes g_{2,j},$$

which is nonzero unless $i = j$. Next consider $d_3 \colon E_3^{0,2} \to E_3^{3,0}$ on the remaining classes $(h_{3,i})^2$. In the cobar construction for B_3, $[\xi_3 \mid \xi_3]$ is identified modulo the filtration with $[\xi_3 \mid \xi_3] + [\xi_2\xi_3 \mid \xi_1] + [\xi_1^4\xi_3 \mid \xi_2] + [\xi_2^2 \mid \xi_1\xi_3] + [\xi_1^4 \mid \xi_2\xi_3]$. The differential on this sum is $[\xi_1^8 \mid \xi_2 \mid \xi_2] + [\xi_2^2 \mid \xi_2^2 \mid \xi_1^2]$ which represents $\{h_{2,0}^2 \otimes h_3 + h_{2,1}^2 \otimes h_1\} = f_{2,0}$. By applying Sq^0, we have shown $d_3((h_{3,i})^2) = f_{2,i}$.

Thus $H^2(B_3)$ is the direct sum,

$$\mathbb{F}_2\{h_i h_j \mid j \neq i + 1\} \oplus \mathbb{F}_2\{g_{3,i} \mid i = 0, 1, \dots\}.$$

Next, let us consider $H^3(B_3)$. Notice that in $H^3(B_3)$ the relations

$$h_{i+1}h_i h_k = 0, \quad h_i h_{k+1}h_k = 0, \quad h_i^2 h_{i+2} + h_{i+1}^3 = 0 \text{ and } h_{i+2}^2 h_i = 0$$

hold for 3-fold products of generators of $H^1(B_3)$. In order to show that these are the only relations, we continue the induction and we determine more of $H^3(B_3)$.

Consider $d_2 \colon E_2^{1,2} \to E_2^{3,1}$. The formula,

$$d_2(h_{3,i}h_{3,j} \otimes h_k) = h_{3,j} \otimes g_{2,i}h_k \otimes g_{2,j}h_k,$$

shows d_2 is nonzero unless $i = j$, so we have $E_3^{1,2} = \mathbb{F}_2\{h_{3,i}h_{3,i} \otimes h_k\}$. Furthermore, $d_3((h_{3,i})^2 \otimes h_k) = f_{2,i}h_k$. This produces a kernel, however, when we consider the representative expressions for the $f_{2,i}$:

$$\begin{aligned}
d_3(h_{3,i}&h_{3,i} \otimes h_{i+4} + h_{3,i+1}h_{3,i+1} \otimes h_{i+1}) \\
&= f_{2,i}h_{i+4} + f_{2,i+1}h_{i+1} \\
&= \{h_{2,i}h_{2,i} \otimes h_{i+3} + h_{2,i+1}h_{2,i+1} \otimes h_{i+1}\}h_{i+4} \\
&\quad + \{h_{2,i+1}h_{2,i+1} \otimes h_{1+4} + h_{2,i+2}h_{2,i+2} \otimes h_{i+2}\}h_{i+1} = 0.
\end{aligned}$$

We denote the classes $\{h_{3,i}h_{3,i} \otimes h_{i+4} + h_{3,i+1}h_{3,i+1} \otimes h_{i+1}\}$ by $f_{3,i}$ in $E_\infty^{1,2}$. The t degrees of the $f_{3,i}$ are $15 \cdot 2^{i+1}$ and so they are linearly independent.

Next, we leave it to the reader to check that d_2 and d_3 map $E_2^{0,3}$ and $E_3^{0,3}$ in such a way as to leave $E_\infty^{0,3} = \{0\}$. In $E_2^{2,1}$, we find the classes $g_{3,i}h_j$ left over after clearing the image of $d_2 \colon E_2^{0,2} \to E_2^{2,1}$ and the classes mapped nontrivially by $d_2 \colon E_2^{2,1} \to E_2^{4,0}$. If we check t degrees, however, it is possible for more than one $g_{3,i}h_j$ to inhabit the same degree. The classes that require comparison are $g_{3,i}h_{i-1}$ and $g_{3,i-1}h_{i+3}$. Writing them out, we see that

$$g_{3,i}h_{i-1} = \{h_{3,i} \otimes h_{i+3}h_{i-1} + h_{3,i+1} \otimes h_i h_{i-1}\} = \{h_{3,i} \otimes h_{i+3}h_{i-1}\},$$

$$g_{3,i-1}h_{i+3} = \{h_{3,i-1} \otimes h_{i+2}h_{i+3} + h_{3,i} \otimes h_{i-1}h_{i+3}\} = \{h_{3,i} \otimes h_{i-1}h_{i+3}\},$$

and so we must introduce the relation $g_{3,i}h_{i-1} = g_{3,i-1}h_{i+3}$. Some amusing number theory can be employed to show that the other classes occur in differing t degrees and so we find classes $g_{3,i}h_j$, for $i \neq j + 1$, all linearly independent. Thus $H^3(B_3)$ is seen to contain $\mathbb{F}_2\{h_ih_jh_k \mid i \neq j + 1, j \neq k + 1\}$ modulo the relations $h_ih_{i+2} = h_{i+1}^3$ and $h_{i+2}^2h_i = 0$, $\mathbb{F}_2\{g_{3,i} \mid i = 0, 1, \dots\}$ and $\mathbb{F}_2\{g_{3,i}h_j \mid i \neq j + 1\}$.

We can now summarize the inductive step in a series of formulas that extend the pattern above:

(1) There is a spectral sequence, converging to $H^*(B_n)$ with

$$E_2^{p,q} \cong H^q(A_n) \otimes H^p(B_{n-1}).$$

(2) $H^1(B_{n-1}) \cong \mathbb{F}_2\{h_i \mid i = 0, 1, \dots\}$
$H^2(B_{n-1}) \cong \mathbb{F}_2\{h_ih_j \mid i \neq j + 1\} \oplus \mathbb{F}_2\{g_{n-1,i} \mid i = 0, 1, \dots\}$
$H^3(B_{n-1})$ contains $\mathbb{F}_2\{h_ih_jh_k\}$ modulo the relations $h_{i+1}h_ih_k = 0$,

$h_i^2 h_{i+2} + h_{i+1}^3 = 0$ and $h_{i+2}^2 h_i = 0$ plus the direct sum $\mathbb{F}_2\{f_{n-1,i}$
$i = 0, 1, \dots\} \oplus \mathbb{F}_2\{g_{n,i} h_j \mid i \neq j + 1\}$.

(3) $d_2(h_{n,i}) = g_{n-1,i}$, $d_3((h_{n,i})^2) = f_{n-1,i}$.

(4) There are new classes in $E_\infty^{1,1}$ and so in $H^2(B_n)$ given by

$$g_{n,i} = \{h_{n,i} \otimes h_{i+n} + h_{n,i+1} \otimes h_i\}.$$

These classes are a result of the relation $g_{n-1,i+1} h_i = g_{n-1,i} h_{n+i}$. The
relation $g_{n,i+1} h_i = g_{n,i} h_{i+n+1}$ is seen to hold for the new $g_{n,i}$.

(5) There are new classes in $E_\infty^{1,2}$ and so in $H^3(B_n)$ given by

$$f_{n,i} = \{h_{n,i} h_{n,i} \otimes h_{i+n} + h_{n,i+1} h_{n,i+1} \otimes h_{i+1}\}.$$

These classes are a result of the relation $h_{k+1} h_k = 0$, and the expressions
for $f_{n-1,i}$.

The reader can check that the t degrees of the new classes grows to infinity
as n goes to infinity. What is left in the limit, $H^*(\mathcal{A}_2)$, is the set of classes
$\{h_i \mid i = 0, 1, \dots\}$ and the two and three fold products, subject to the relations
we have derived. This proves Theorem 9.36. □

The chart on the next page summarizes the data for $H^*(\mathcal{A}_2)$ given in
Theorem 9.36 and for a small range of $t - s$. We follow the convention of
writing $H^{s,t}(\mathcal{A}_2)$ on the lattice point $(t - s, s)$. We also use the convention
(due to [Tangora66]) of joining two classes together by a vertical line if one is
the product of the other with h_0, and by a line of slope 1 if one is the product
of the other with h_1.

The diligent reader can write out the first few stages of a minimal resolution
to see that nothing occurs above filtration degree 3 for $0 < t - s < 5$, and the
tower of h_0^r continues to infinity (this also follows from the discussion at the
end of §9.3 and the fact that $_{(2)}\pi_0^S = \mathbb{Z}$).

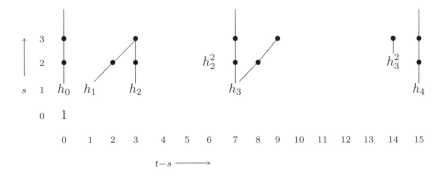

Corollary 9.37.

$$
_{(2)}\pi_n^S \cong
\begin{cases}
\mathbb{Z}, & \text{if } n = 0, \\
\mathbb{Z}/2\mathbb{Z}, & \text{if } n = 1, 2, \\
\mathbb{Z}/8\mathbb{Z}, & \text{if } n = 3, \\
\{0\}, & \text{if } n = 4, 5.
\end{cases}
$$

PROOF: From the diagram we must first dispense with the possibility that $d_r(h_1) = h_0^r$ for some r. We use the fact that d_r is a derivation and compute

$$
0 = d_r(h_1 h_0) = h_0 d_r(h_1) = h_0^{r+1} \neq 0,
$$

a contradiction. Therefore, $d_r(h_1) = 0$ for all r and h_1 is a permanent cycle.

As for the $\mathbb{Z}/8\mathbb{Z} \cong {}_{(2)}\pi_3^S$, the discussion at the end of §9.3 shows that the relations $h_2 \neq 0$, $h_0 h_2 \neq 0$ and $h_0^2 h_2 \neq 0$ describe an element in ${}_{(2)}\pi_3^S$ with 4 times that element nonzero. This forces the composition series for ${}_{(2)}\pi_3^S$ to be

$$
0 \subset \mathbb{Z}/2\mathbb{Z} \subset \mathbb{Z}/4\mathbb{Z} \subset \mathbb{Z}/8\mathbb{Z} = {}_{(2)}\pi_3^S. \qquad \square
$$

In order to do further calculation of ${}_{(2)}\pi_*^S$, more of $H^{*,*}(\mathcal{A}_2)$ is necessary than computed in Theorem 9.36. In fact, not all of $H^{3,*}(\mathcal{A}_2)$ is given by 3-fold products, as we see later. A more powerful technique for computing $H^{*,*}(\mathcal{A}_2)$ is given in §9.6. In the meantime, we obtain some geometric consequences of our computations.

The first nontrivial differentials

The problem of the existence of elements in π_*^S with Hopf invariant one has been reduced to the study of the Steenrod operation Sq^{2^i} and whether it acts nontrivially on a two-cell complex. Because Sq^{2^i} is dual to $\xi_1^{2^i}$ in $\mathcal{A}_2^{\text{dual}}$, the Adams spectral sequence further reduces the question to the survival of h_i in $E_2^{1,2^i}$ to $E_\infty^{1,2^i}$.

If h_i survives to $E_\infty^{1,2^i}$, it detects a class in ${}_{(2)}\pi_{2^i-1}^S$. The stable homotopy ring π_*^S is graded commutative and so the square of the class detected by h_i has order 2 because $[h_i] \circ [h_i] = (-1)^{2^i-1}[h_i] \circ [h_i]$ and so $2[h_i] \circ [h_i] = 0$. This implies that the class $h_0 h_i^2$ that represents $2[h_i] \circ [h_i]$, which is nonzero if $i \geq 3$, cannot survive to E_∞ by the identification of products with h_0 in Ext with the doubling map on π_*^S. The kind reader will forgive the following algebraic *deus ex machina*:

Fact. $\text{Ext}_{\mathcal{A}_2}^{s,s+13}(\mathbb{F}_2, \mathbb{F}_2) = \{0\}$ for all s.

With our present techniques, the proof of this is a good day's work constructing a minimal resolution. More elegant and streamlined techniques will be presented later. Notice that this implies ${}_{(2)}\pi_{13}^S = 0$.

Proposition 9.38 *([Toda55], [Adams58]). There is no element of Hopf invariant one in $\pi_{31}(S^{16}) = \pi_{15}^S$ and so there is no division algebra structure on $\mathbb{R}^{16}$.*

PROOF: We consider $h_0 h_3^2$ in $E_2^{3,17}$. Because $E_2^{s,s+13} = \{0\}$ for all s, no differential on $h_0 h_3^2$ has a nonzero image and $h_0 h_3^2$ is an infinite cycle. Since it cannot survive to E_∞, it must be a boundary. There is only one possible nonzero differential, that is, $d_2(h_4) = h_0 h_3^2$. Hence h_4 does not survive to E_∞. □

We remark on the extraordinary blend of topology and algebra in this proof. The graded commutativity of π_*^S and an Ext computation together imply a deep result.

The next question to consider is whether this technique propagates through the spectral sequence to settle completely the Hopf invariant one problem? A quick glance forward to the charts in §9.6 shows that we cannot naively proceed even to h_5 (in particular, $H^{s,s+29}(\mathcal{A}_2) \neq \{0\}$ for some s). Something more is needed. Our goal for the rest of this section is to outline how to prove:

Theorem 9.39. *$d_2(h_i) = h_0 h_{i-1}^2$, for $i \geq 4$.*

In a celebrated paper, [Adams60] gave the first proof of this theorem based on a generalization of the factorization of Steenrod operations of [Adem52]. The equation $d_2(h_4) = h_0 h_3^2$ is equivalent to a nontrivial factorization of Sq^{16} into products of primary and secondary cohomology operations. Therefore, Sq^{16} cannot act nontrivially on $S^{16} \cup_\alpha e^{32}$ for any α in $\pi_{31}(S^{16})$. The factorization is based on secondary operations, $\Phi_{i,j}$ that arise from the Adem relation $Sq^{2^i} Sq^{2^j} = \sum_{0 \leq s < j} b_s Sq^{2^s}$ $(i \leq j, i \neq j - 1)$. [Adams60] showed that a decomposition, $Sq^{2^n} = \sum_{i,j} a_{i,j} \Phi_{i,j}$ with $a_{i,j} \in \mathcal{A}_2$ and modulo some indeterminacy, holds for all $n \geq 4$. This settles the Hopf invariant one problem.

We note that [Bott-Milnor58] and [Kervaire58] had also settled the division algebras question shortly after Adams by using K-theory techniques. [Maunder63, 64] developed the notion of higher order cohomology operations and related them generally to differentials in the Adams spectral sequence. The connections between these operations, Theorem 9.2 and the operations in §9.1 were clarified by [Maunder63, 64].

Another way to show that $d_2(h_i) = h_0 h_{i-1}^2$ is to consider the Steenrod algebra that acts on $\text{Ext}_{\mathcal{A}_2}^{*,*}(\mathbb{F}_2, \mathbb{F}_2)$ and the fact that this action intertwines with the differentials in the spectral sequence. A priori, there seems to be little connection as this action is a formal feature of the cohomology of a Hopf algebra. The missing geometric link was forged by [Kahn70] using the geometric construction called the **quadratic construction** on a pointed space, (X, x_0): First form the space $S^n \diamond (X \times X)$, where $U \diamond V = U \times V / * \times V$, and then take the quotient $\Gamma^n(X) = S^n \diamond (X \times X)/(\theta, x, x') \sim (-\theta, x', x)$.

The inclusion of $S^n \hookrightarrow S^{n+1}$ as equator determines a natural transformation $\Gamma^n(X) \to \Gamma^{n+1}(X)$ and the quadratic construction is the direct limit of the natural transformations:

$$\Gamma^\infty(X) = \lim_{n \to \infty} \Gamma^n(X).$$

[Kahn70] showed that this construction carries the chain homotopy that gives rise to the Steenrod operations on $H^{*,*}(\mathcal{A}_2)$ and so these operations can be related to the differentials.

The theorem that applies to the question at hand is given in the formula of [Milgram72]:

$$d_2(Sq^i(\alpha)) = h_0 Sq^{i+1}(\alpha),$$

which holds if $\alpha \in \mathrm{Ext}^{s,t}_{\mathcal{A}_2}(\mathbb{F}_2, \mathbb{F}_2)$ and $i \equiv t \pmod 2$. Since h_i is in $\mathrm{Ext}^{1,2^i}_{\mathcal{A}_2}(\mathbb{F}_2, \mathbb{F}_2)$ and $2^i \equiv 0 \pmod 2$, $d_2(Sq^0 h_i) = h_0 Sq^1(h_i)$, that is, $d_2(h_{i+1}) = h_0 h_i^2$, which holds for $i \geq 1$ by Theorem 9.10. This technique has been employed with great success to determine many of the known differentials in the Adams spectral sequence ([Kahn70], [Milgram72], and [Bruner84]).

In §9.6 we return to the question of differentials in the Adams spectral sequence and discuss some other methods to determine them.

Massey products

Before we leave the computations that can be done by hand, we fill in more of our chart by considering the analogue of Massey products for a bigraded differential algebra. Ordinary Massey products and their higher order analogues are discussed in §8.2.

$\mathrm{Ext}^{*,*}_\Gamma(k,k) = H^{*,*}(\Gamma)$ is computed from a differential bigraded algebra, $(B^{*,*}, d) = (F_*(\mathcal{A}_2^{\mathrm{dual}}), d^*)$, the cobar construction, with its differential of bidegree $(1,0)$. In such a bigraded algebra, suppose $[u]$ is a class in $H^{s,t}(B^{*,*}, d)$, $[v]$ in $H^{s',t'}(B^{*,*}, d)$ and $[w]$ in $H^{s'',t''}(B^{*,*}, d)$, and furthermore, $[u][v] = 0 = [v][w]$. Then we can define the **Massey triple product**

$$\langle [u], [v], [w] \rangle \subset H^{s+s'+s''-1, t+t'+t''}(B^{*,*}, d)$$

by taking elements a in $B^{s+s'-1, t+t'}$ and b in $B^{s'+s''-1, t'+t''}$ such that $da = uv$ and $db = vw$, where $u \in [u]$, $v \in [v]$, and $w \in [w]$. As in §8.2,

$$\langle [u], [v], [w] \rangle = \{[aw \pm ub] \mid \text{all possible choices of } a, b, u, v, w\},$$

where we denote the homology class of an element t by $[t]$. The indeterminacy for $\langle [u], [v], [w] \rangle$ is given by $[u] H^{s'+s''-1, t'+t''} + H^{s+s'-1, t+t'}[w]$. The higher order analogues of the Massey product can be defined as in §8.2 with the extra index kept in tow.

Theorem 9.36 presents many trivial products that can give rise to triple products in $H^{*,*}(\mathcal{A}_2)$. We next record some nontrivial Massey products. In particular, we compute in detail that $c_0 = \langle h_2^2, h_0, h_1 \rangle$ and $h_1 c_0$ are nontrivial. Other computations are left as exercises or given in the references. The identification of Massey products in Ext also follows by special spectral sequence arguments first given by [Ivanovskiĭ64] and by [May64].

We begin with an exercise for the reader. These relations were identified by [Adams60] and follow from the formulas in the proof of Theorem 9.36:

$$\langle h_i, h_{i+1}, h_i \rangle = h_{i+1}^2, \quad \langle h_{i+1}, h_i, h_{i+1} \rangle = h_{i+2} h_i, \quad \langle h_{i+2}, h_{i+1}, h_i \rangle = g_{2,i}.$$

These dispose of the most obvious choices for Massey products.

The next relations to try are $h_2^2 h_0 = 0$ and $h_0 h_1 = 0$. In the cobar construction, these products vanish because the following formulas hold:

$$[\xi_1^4 \mid \xi_1^4 \mid \xi_1] = d^*([\xi_3 \mid \xi_1^2] + [\xi_2^2 \mid \xi_2] + [\xi_1^4 \mid \xi_1^2 \xi_2] + [\xi_2^2 \mid \xi_1^3])$$
$$[\xi_1 \mid \xi_1^2] = d^*([\xi_2 + \xi_1^3]).$$

Thus, in $H^{3,11}(\mathcal{A}_2)$, $\langle h_2^2, h_0, h_1 \rangle$ contains the class

$$\{[\xi_3 \mid \xi_1^2 \mid \xi_1^2] + [\xi_2^2 \mid \xi_2 \mid \xi_1^2] + [\xi_1^4 \mid \xi_1^2 \xi_2 \mid \xi_1^2] + [\xi_2^2 \mid \xi_1^3 \mid \xi_1^2] + [\xi_1^4 \mid \xi_1^4 \mid \xi_2 + \xi_1^3]\}.$$

In fact, this class is the unique representative for $\langle h_2^2, h_0, h_1 \rangle$ since the indeterminacy of $\langle h_2^2, h_0, h_1 \rangle$ is $h_2^2 \operatorname{Ext}^{1,3} + \operatorname{Ext}^{2,9} h_1 = \{0\}$.

To show $\langle h_2^2, h_0, h_1 \rangle$ is nonzero, it suffices to show that it is nonzero in $H^3(B_4)$ since $11 < 2^4 - 1$. We first identify it in $H^3(B_3)$. Observe that the class can be identified with $[\xi_3 \mid \xi_1^2 \mid \xi_1^2]$ modulo the filtration in the cobar construction for B_3 and so it names the class $h_{3,0} \otimes h_1 h_1$ in $E_2^{2,1} = H^1(A_3) \otimes H^2(B_2)$ in the spectral sequence converging to $H^*(B_3)$.

$$d_2(h_{3,0} \otimes h_1 h_1) = g_{2,0} h_1 h_1 = h_2^2 h_0 h_1 = 0$$

and so $h_{3,0} \otimes h_1 h_1$ gives a class in $E_3^{2,1}$ that persists to $E_\infty^{2,1}$. Checking the degrees of the other nonzero classes in $H^3(B_3)$, we find $\{h_{3,0} \otimes h_1 h_1\}$, of t degree 11, is not accounted for by classes $h_i h_j h_k$, $f_{3,i}$ or $g_{3,i} h_j$.

To see that $\{h_{3,0} \otimes h_1 h_1\}$ determines a nonzero class in $H^3(B_4)$, we consider the next spectral sequence and

$$d_2 \colon H^1(A_4) \otimes H^1(B_3) = E_2^{1,1} \rightarrow E_2^{3,0} = H^3(B_3).$$

Since $E_2^{1,1}$ is 15-connected in this case, $\langle h_2^2, h_0, h_1 \rangle$ lives to $H^3(B_4)$ and hence to a nonzero class in $H^{3,11}(\mathcal{A}_2)$.

A similar dimension counting argument can be given for the class

$$\langle h_2^2, h_0, h_1 \rangle h_1 = \{h_{3,0} \otimes h_1 h_1 h_1\}.$$

Thus we have identified two new elements in $\text{Ext}^{3,11}$ and $\text{Ext}^{4,13}$, denoted by c_0 and $h_1 c_0$ by [May64].

The cobar construction is a very large complex to use for computing Massey products. [Ivanovskiĭ64] and [May64] worked in more manageable complexes from which to compute $H^{*,*}(\mathcal{A}_2)$. We summarize some of their computations in low degrees. We picture only the Massey product elements.

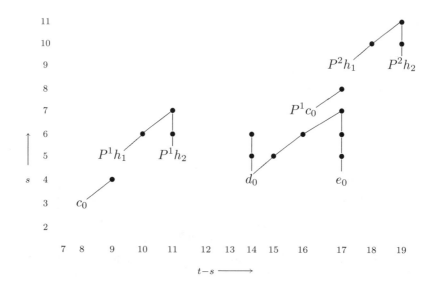

Theorem 9.40. *For $t - s \leq 19$, the following Massey products are nonzero and their products with h_0 and h_1 are given in the chart. (The operators P^1 and P^2 are periodicity operators that will be defined in §9.6.)*

(1) $c_0 = \langle h_2^2, h_0, h_1 \rangle$ in $\text{Ext}^{3,11}$,
(2) $P^1 h_1 = \langle h_1, h_0^4, h_3 \rangle$ in $\text{Ext}^{5,14}$,
(3) $P^1 h_2 = \langle h_2, h_0^4, h_3 \rangle$ in $\text{Ext}^{5,16}$,
(4) $d_0 = \langle h_0, h_2^2, h_2^2, h_0 \rangle$ in $\text{Ext}^{4,18}$,
(5) $e_0 = \langle h_2, c_0, h_2, h_1 \rangle$ in $\text{Ext}^{4,21}$,
(6) $f_0 = \langle h_0^2, h_3^2, h_2 \rangle$ modulo $h_0^2 h_2 h_4$ in $\text{Ext}^{4,22}$,
(7) $c_1 = \langle h_2, h_1, h_3^2 \rangle$ in $\text{Ext}^{3,22}$,
(8) $P^1 c_0 = \langle c_0, h_0^4, h_3 \rangle$ in $\text{Ext}^{7,23}$,
(9) $P^2 h_1 = \langle h_1, h_0^8, h_4 \rangle = \langle \langle h_3, h_0^4, h_1 \rangle, h_0^4, h_3 \rangle$ in $\text{Ext}^{9,26}$,
(10) $P^2 h_2 = \langle h_2, h_0^8, h_4 \rangle = \langle \langle h_3, h_0^4, h_2 \rangle, h_0^4, h_3 \rangle$ in $\text{Ext}^{9,26}$.

For a more extensive and complete description of Massey products in the cohomology of the Steenrod algebra, the reader can consult the paper of [Tangora94].

If the reader adds this chart to the results of Theorem 9.36, then the description of $\mathrm{Ext}^{s,t}$ for $t - s \leq 19$ is almost complete. The missing data are given by a theorem of [Novikov59] that adds the relations

$$h_1^3 h_4 = h_0^2 h_2 h_4 \neq 0 \text{ and } h_0^k h_m \neq 0 \text{ for } k < 2^{m-1}.$$

[Novikov59] showed that $h_0^{2^m} h_{m+1} = 0$ for $m \geq 2$. [Maunder65] showed the nontriviality of the products $h_0^l h_{m+1}$ for $l < 2^m$ by comparing the Adams spectral sequence converging to $_{(2)}\pi_*^S$ with the Adams spectral sequence converging to $_{(2)}\pi_*^S(BU(2q, \ldots, \infty))$, the stable homotopy groups of the $2q$-coconnected cover of BU. With these facts, the description in these limited degrees is complete. We discuss some of the differentials in §9.6.

Massey products are a formal consequence of the structure of a differential graded algebra. In cohomology they capture higher order linking phenomena. We can ask if there is a topological interpretation of the Massey products in the Adams spectral sequence. The product structure on stable homotopy groups of spheres is identified with the composition product that shows up as the Yoneda product on the E_2-term. We need the notion of secondary products for the composition product, introduced by [Toda59].

Definition 9.41. *Let $\gamma \in [X, Y]$, $\beta \in [Y, Z]$ and $\alpha \in [Z, W]$. Suppose $\beta \circ \gamma$ and $\alpha \circ \beta$ are null-homotopic in $[X, W]$. Let c, b, a be mappings representing γ, β and α, respectively. There are extensions of $b \circ c$ and $a \circ b$ to $CX \to Z$ and $CY \to W$, which we denote by B and C, respectively. Write $SX = C^+ X \cup C^- X$ and consider the mapping $SX \to W$ given on $C^+ X$ as $a \circ B$ and on $C^- X$ as $C \circ c$. The set of all such mappings is denoted $\langle \alpha, \beta, \gamma \rangle \subset [SX, W]$, called the* **Toda bracket** *of α, β and γ. It has indeterminacy given by the subset $\alpha_{\#}[SW, Y] + (S\gamma)^{\#}[SX, Z]$.*

The definition, like the definition for Massey products, can be generalized to n-fold Toda brackets and matric Toda brackets. Furthermore, if we apply the definition to representatives of mappings in π_*^S, by careful suspension, we can define Toda brackets of stable maps for which the Toda brackets represent cosets in π_*^S. [Toda62] gave extensive computations of $\pi_{n+k}(S^n)$ for $k \leq 19$, using this secondary bracket product to determine and name many elements. [Cohen, J68] proved that *all of π_*^S* can be represented by higher order Toda brackets applied to integer multiples of the classes $\iota \in \pi_0^S$, $\eta \in \pi_1^S$, $\nu \in \pi_3^S$, $\sigma \in \pi_7^S$ and $\alpha_1 \in \pi_{2p-3}^S$ for each odd prime p.

Massey products of all orders may be defined in $\mathrm{Ext} \cong F_2$. Their relationship to differentials can lead to a connection with Toda brackets. [Moss70] provided a description in some cases. Broadly stated, he proved that Massey products of permanent cycles, under certain conditions, converge to Toda brackets in π_*^S (in fact, the main result of [Moss70] applies more generally to the composition product $\circ \colon \{Y, W\}_* \times \{X, Y\}_* \to \{X, W\}_*$).

Theorem 9.42 ([Moss70]). *Let* $(E_r^{s,t}, d_r)$ *denote the* E_r-*term of the Adams spectral sequence converging to* $_{(p)}\pi_*^S$.

(1) *Suppose* $a \in E_r^{s,t}$, $b \in E_r^{s't'}$, *and* $c \in E_r^{s'',t''}$ *satisfy* $ab = 0$ *and* $bc = 0$. *Then*

$$d_r(\langle a, b, c \rangle) \subset \left\langle d_r a, \; a, \; \begin{matrix} b & 0 \\ (-1)^i d_r b & b \end{matrix}, \; \begin{matrix} c \\ (-1)^{i+i'} d_r c \end{matrix} \right\rangle,$$

the matric Massey product, where $i = t - s$, $i' = t' - s'$.

(2) *If* $ad_r b = 0$ *and* $bd_r c = 0$, *then*

$$d_r \langle a, b, c \rangle \subset -\langle d_r a, b, c \rangle - (-1)^i \langle a, d_r b, c \rangle - (-1)^{i+i'} \langle a, b, d_r c \rangle.$$

(3) *If* a, b *and* c *are permanent cycles representing* α *in* π_i^S, β *in* $\pi_{i'}^S$ *and* γ *in* $\pi_{i''}^S$ *with* $\alpha \circ \beta = 0$ *and* $\beta \circ \gamma = 0$, *then, under certain technical assumptions on the filtrations of* a, b *and* c, *the Massey product* $\langle a, b, c \rangle$ *contains a permanent cycle that is realized by an element of the Toda bracket* $\langle \alpha, \beta, \gamma \rangle$.

The simplest example of this is given by the element c_0 in $H^{3,11}(\mathcal{A}_2)$. Looking ahead to the more complete table, c_0 is a permanent cycle and h_2^2, h_0 and h_1 satisfy the unspoken filtration conditions of the theorem. Thus $\langle h_2^2, h_0, h_1 \rangle$ represents a Toda bracket that is given in Toda's notation as $\langle \nu^2, 2\iota, \eta \rangle$.

9.6 Further structure

The element-by-element arguments of §9.5 led to many useful results; in this section we take a different point of view and discuss the spectral sequence for $_{(p)}\pi_*^S$ in more global terms. We begin with two deep theorems of [Adams66]. The first determines conditions on s and t for which $\mathrm{Ext}_{\mathcal{A}_2}^{s,t}(\mathbb{F}_2, \mathbb{F}_2) = \{0\}$. The second reveals portions of $\mathrm{Ext}_{\mathcal{A}_2}^{s,t}(\mathbb{F}_2, \mathbb{F}_2)$ that are isomorphic via periodicity operators; this periodic phenomenon determines infinitely many nontrivial values of the E_2-term of the spectral sequence.

We then return to the computation of $H^{*,*}(\mathcal{A}_p)$ to exploit the fact that the Steenrod algebra is a graded Hopf algebra. The method of computation is called the May spectral sequence and was introduced by [May64]. Tables for $H^{*,*}(\mathcal{A}_2)$ and $H^{*,*}(\mathcal{A}_3)$ in a range are given. We then discuss some of the techniques to determine differentials in the Adams spectral sequence. We close the chapter with some remarks on further developments involving the Adams spectral sequence.

The vanishing line

The main theorem of this section gives a description of the cohomology of the Steenrod algebra in the large. The proof is due to [Adams61, 66] for $p = 2$, and to [Liulevicius63] for p, an odd prime.

Theorem 9.43. *Let $U(n)$ denote the function on natural numbers given by* $U(4s) = 12s - 1$, $U(4s + 1) = 12s + 2$, $U(4s + 2) = 12s + 4$, $U(4s + 3) = 12s + 6$. *Then* $\mathrm{Ext}_{\mathcal{A}_2}^{s,t}(\mathbb{F}_2, \mathbb{F}_2) = \{0\}$, *for* $0 < s < t < U(s)$.
For p, an odd prime, $\mathrm{Ext}_{\mathcal{A}_p}^{s,t}(\mathbb{F}_p, \mathbb{F}_p) = \{0\}$, *for* $0 < s < t < (2p - 1)s - 2$.

The vanishing condition in terms of $n = t - s$, the stem dimension, is given by $0 < t - s < U(s) - s$. By definition, $U(s) - s < 2s$ and so the condition $n < U(s) - s$ is satisfied when $s > n/2$. In other words, above the line of slope $1/2$ in the $(t - s, s)$ plane, $\mathrm{Ext}_{\mathcal{A}_2}^{s,t}(\mathbb{F}_2, \mathbb{F}_2) = \{0\}$. This line is called the **vanishing line**.

Recall that the **exponent** of a group G is the least natural number, m, such that all m^{th} powers of elements in G are zero. The vanishing line of Theorem 9.43 puts an upper bound on the length of a composition series for the stable homotopy groups of spheres.

Corollary 9.44. *For $n \geq 1$, the exponent of $_{(2)}\pi_n^S$ is less than $2^{f(n)}$ where $f(n)$ is the minimum of $\{s \mid n < U(s) - s - 1\}$. The exponent of $_{(p)}\pi_n^S$ is less than $p^{g(n)}$ where $g(n)$ is the minimum of $\{s \mid n < (2p - 1)s - s - 3\}$.*

We give a proof of Theorem 9.43 in the case $p = 2$. We proceed by a series of lemmas to prove a slightly weaker result that is strengthened later by the periodicity isomorphisms.

Consider another numerical function, $T(n)$, given by

$$T(4s) = 12s, \quad T(4s+1) = 12s+2, \quad T(4s+2) = 12s+4, \quad T(4s+3) = 12s+7.$$

Let $A(r)$ denote the subalgebra of $\mathcal{A}_2$ generated by $\{Sq^1, Sq^2, \ldots, Sq^{2^r}\}$.

Lemma 9.45. $\mathrm{Ext}_{\mathcal{A}_2}^{s,t}(A(0), \mathbb{F}_2) = \{0\}$ *when $s \leq 4$ and $0 < s < t < T(s)$.*

PROOF: Observe first that the suspension isomorphism on graded modules over a graded algebra satisfies the relation for $t \geq r$: $\mathrm{Hom}_\Gamma^t(s^r M, N) = {}_\Gamma\mathbf{Mod}(s^r M, s^t N) \cong {}_\Gamma\mathbf{Mod}(M, s^{t-r} N) = \mathrm{Hom}_\Gamma^{t-r}(M, N)$. It follows that

$$\mathrm{Ext}_\Gamma^{s,t}(s^r M, N) \cong \mathrm{Ext}_\Gamma^{s,t-r}(M, N).$$

Following [Adams66], we consider the extension over $\mathcal{A}_2$:

$$0 \to s\mathbb{F}_2 \to A(0) \to \mathbb{F}_2 \to 0.$$

This extension determines a class in $\mathrm{Ext}_{\mathcal{A}_2}^{1,1}(s\mathbb{F}_2, \mathbb{F}_2) \cong \mathrm{Ext}_{\mathcal{A}_2}^{1,0}(\mathbb{F}_2, \mathbb{F}_2)$, which is given by h_0 and corresponds to Sq^1. The short exact sequence leads to a long exact sequence of Ext groups, here abbreviated as $\mathrm{Ext}^{s,t}_{\mathcal{A}_2}(s\mathbb{F}_2, \mathbb{F}_2) = \mathrm{Ext}^{s,t}(s\mathbb{F}_2)$, $\mathrm{Ext}^{s,t}_{\mathcal{A}_2}(\mathbb{F}_2, \mathbb{F}_2) = H^{s,t}$, and $\mathrm{Ext}^{s,t}_{\mathcal{A}_2}(A(0), \mathbb{F}_2) = \mathrm{Ext}^{s,t}(A(0))$:

$$\to \mathrm{Ext}^{s-1,t}(s\mathbb{F}_2) \overset{\partial}{\to} H^{s,t} \to \mathrm{Ext}^{s,t}(A(0)) \to \mathrm{Ext}^{s,t}(s\mathbb{F}_2) \overset{\partial}{\to} H^{s+1,t} \to .$$

The reader can consult the charts in §9.5 to prove the lemma in the bidegrees stated (there are 11 cases). Whenever the boundary homomorphism ∂ is nonzero, it is given by multiplication by h_0, and the result follows. $\qquad\square$

We next restrict our attention to a particular class of $\mathcal{A}_2$-modules. An $\mathcal{A}_2$-module L is an $A(r)$-module for any r because $A(r)$ is a subalgebra of $\mathcal{A}_2$. Therefore we can speak of L being a free $A(r)$-module. The relation $Sq^1 Sq^1 = 0$ provides a neat criterion for a module to be free over $A(0)$: Take $Sq^1 \colon L \to L$ as a differential and compute the homology $H(L, Sq^1)$; an $\mathcal{A}_2$-module L is free over $A(0)$ if and only if $H(L, Sq^1) = \{0\}$. This follows because $Sq^1 x = 0$ if and only if $x = Sq^1 y$. (For generalizations of this idea see the work of [Adams-Margolis71], [Margolis83], and [Palmieri92].)

Lemma 9.46. *Suppose L is an $\mathcal{A}_2$-module, free over $A(0)$, and L is $(n-1)$-connected (that is, $L^t = \{0\}$ for $t < n$). Then $\mathrm{Ext}^{s,t}_{\mathcal{A}_2}(L, \mathbb{F}_2) = \{0\}$ for $s \leq 4$ and $0 < s < t < n + T(s)$.*

PROOF: Let $\{b_1, b_2, \dots, b_j, \dots\} \subset L$ be an $A(0)$ basis for L and let $L(m)$ be the submodule over $A(0)$ generated by the b_i of degree $\geq m$. Notice that $L(n) = L$, that $L(m)$ is an $\mathcal{A}_2$-submodule of L and that $L(m)/L(m+1)$ is a free $A(0)$-module on basis elements of degree m. Therefore, we can write

$$L(m)/L(m+1) \cong \bigoplus s^m A(0).$$

We proceed by induction on m. Consider the short exact sequence

$$0 \to L(m)/L(m+1) \to L/L(m+1) \to L/L(m) \to 0.$$

Lemma 9.45 applies, with a dimension shift, to $L(m)/L(m+1)$ and so we have

$$\mathrm{Ext}^{s,t}_{\mathcal{A}_2}(L(m)/L(m+1), \mathbb{F}_2) = \{0\}$$

for $s \leq 4$, $0 < s < t < T(s) + m$. If $m = n$, then $L/L(m) = \{0\}$ and the lemma holds trivially. If the lemma holds for values up to m, then the long exact sequence,

$$\to \mathrm{Ext}^{s,t}_{\mathcal{A}_2}(L/L(m), \mathbb{F}_2) \to \mathrm{Ext}^{s,t}_{\mathcal{A}_2}(L/L(m+1), \mathbb{F}_2)$$
$$\to \mathrm{Ext}^{s,t}_{\mathcal{A}_2}(L(m)/L(m+1), \mathbb{F}_2) \to ,$$

provides the inductive step. Finally, for a given s and t one can find m large enough that $\mathrm{Ext}^{s,t}_{\mathcal{A}_2}(L, \mathbb{F}_2) \cong \mathrm{Ext}^{s,t}_{\mathcal{A}_2}(L/L(m), \mathbb{F}_2)$. This proves the lemma. $\square$

Lemma 9.47. *Suppose L is an $\mathcal{A}_2$-module, free over $A(0)$, and $(n-1)$-connected. Then, for $0 < s < t < n + T(s)$, $\mathrm{Ext}_{\mathcal{A}_2}^{s,t}(L, \mathbb{F}_2) = \{0\}$.*

PROOF: Suppose we are given a short exact sequence of $\mathcal{A}_2$-modules

$$0 \longrightarrow L_1 \longrightarrow L_2 \longrightarrow L_3 \longrightarrow 0,$$

and suppose two of the three modules is $A(0)$-free. When we apply the functor $H(\ , Sq^1)$ to the short exact sequence we get a long exact sequence on homology. It follows immediately that the third module is $A(0)$-free since its Sq^1-homology must vanish. We use this observation in what follows.

Lemma 9.46 gives us the lemma for $s \leq 4$, that is, for $k = 0$ and $s = 4k+i$, for $i = 1, \dots, 4$. Suppose the lemma holds for all modules and for values of s less than $4k + 5$. Let

$$0 \longleftarrow L \overset{\varepsilon}{\longleftarrow} C_0 \overset{d_0}{\longleftarrow} C_1 \overset{d_1}{\longleftarrow} C_2 \overset{d_2}{\longleftarrow} C_3 \overset{d_3}{\longleftarrow} C_4$$

be an $\mathcal{A}_2$-free resolution of L. Since the C_i are $\mathcal{A}_2$-free, they are $A(0)$-free. It follows that the modules $\ker d_i$ are $A(0)$-free, for $i = 0$ to 4. Let $M = \ker d_2 = \mathrm{im}\, d_3$. For a minimal resolution, Lemma 9.46 implies that M is $(n+11)$-connected. Therefore

$$\mathrm{Ext}_{\mathcal{A}_2}^{s,t}(M, \mathbb{F}_2) = \{0\},$$

for $s \leq 4k + 4, 0 < s < t < n + 12 + T(s)$. However, because $M = \mathrm{im}\, d_3$,

$$\mathrm{Ext}_{\mathcal{A}_2}^{s,t}(M, \mathbb{F}_2) = \mathrm{Ext}_{\mathcal{A}_2}^{s+4,t}(L, \mathbb{F}_2)$$

and so the lemma holds for s less than $4(k+1) + 5$. $\qquad\square$

Corollary 9.48. $\mathrm{Ext}_{\mathcal{A}_2}^{s,t}(\mathbb{F}_2, \mathbb{F}_2) = \{0\}$ *for $0 < s < t < V(s)$ where $V(s)$ is the function given by $V(4s) = 12s-3$, $V(4s+1) = 12s+2$, $V(4s+2) = 12s+4$, $V(4s+3) = 12s+6$.*

PROOF: Consider the short exact sequence

$$0 \longrightarrow I(\mathcal{A}_2)/I(\mathcal{A}_2) \cdot Sq^1 \longrightarrow \mathcal{A}_2/I(\mathcal{A}_2) \cdot Sq^1 \longrightarrow \mathbb{F}_2 \longrightarrow 0.$$

The minimal resolution for $\mathcal{A}_2/I(\mathcal{A}_2) \cdot Sq^1$ is particularly simple and implies that, for $s > 0, t \neq s$,

$$\mathrm{Ext}_{\mathcal{A}_2}^{s,t}(\mathcal{A}_2/I(\mathcal{A}_2) \cdot Sq^1, \mathbb{F}_2) = \{0\}.$$

It follows from the long exact sequence associated to the extension that

$$\mathrm{Ext}_{\mathcal{A}_2}^{s,t}(\mathbb{F}_2, \mathbb{F}_2) \cong \mathrm{Ext}_{\mathcal{A}_2}^{s-1,t}(I(\mathcal{A}_2)/I(\mathcal{A}_2) \cdot Sq^1, \mathbb{F}_2)$$

for $t > s > 0$. The module $I(\mathcal{A}_2)/I(\mathcal{A}_2) \cdot Sq^1$ is 1-connected and free over $A(0)$—this follows from the formula $Sq^1 Sq^I = Sq^{I'}$ when $I = (i_1, \ldots, i_r) \neq (0)$, and i_1 is even, while $Sq^1 Sq^I = 0$ when i_1 is odd. The corollary follows from Lemma 9.47 and the appropriate expression for $V(n)$.   □

Notice that $V(n) = U(n)$ for $n \not\equiv 0 \pmod 4$; to prove Theorem 9.43 we only need to settle one further case. We do this next. The proof for odd primes is similar and the appropriate homological algebra was developed by [Liulevicius62].

Periodicity

The use of subalgebras of $\mathcal{A}_2$ in proving structure theorems about Ext was very fruitful in §9.5. We continue to study the family of subalgebras $A(n) = \langle Sq^1, Sq^2, \ldots, Sq^{2^n} \rangle$, the subalgebra of $\mathcal{A}_2$ generated by the indecomposables Sq^1 to Sq^{2^n}. The first such subalgebra, $A(0)$ is isomorphic to an exterior algebra on a single generator of degree one. We can picture $A(1)$ as in the diagram, where 1 is the bottom element, and each circle is a basis element. The straight line connections are given by multiplying (on the left) by Sq^1, and all curved line connections are given by multiplying by Sq^2.

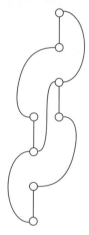

We construct a minimal resolution of $A(0)$ as an $A(1)$-module, from the following short exact sequence:

$$0 \leftarrow A(0) \leftarrow A(1) \xleftarrow{d_1} A(1)a \oplus A(1)b \xleftarrow{d_2}$$
$$A(1)u \oplus A(1)v \xleftarrow{d_3} A(1)t \leftarrow s^{12} A(0) \leftarrow 0,$$

where $d_1(a) = Sq^2$, $d_1(b) = Sq^2 Sq^1$, $d_3(u) = Sq^2 a + Sq^1 b$, $d_3(v) = Sq^2 b$, and $d_4(t) = Sq^2 Sq^1 u + Sq^2 v$ (The reader is encouraged to make a chart of this resolution using pictures of $A(1)$). We can extend this sequence to a complete minimal resolution of $A(0)$ over $A(1)$ that is periodic of order 4 and degree 12. This implies the following result.

Lemma 9.49. $\mathrm{Ext}_{A(1)}^{s,t}(A(0),\mathbb{F}_2) \cong \mathrm{Ext}_{A(1)}^{s+4,t+12}(A(0),\mathbb{F}_2)$, for $s > 0$.

The quibble at $s = 0$ has to do with the units in $A(0)$ and $A(1)$. An immediate corollary of the lemma is the isomorphism for $A(1)$-modules, free over $A(0)$:

$$\mathrm{Ext}_{A(1)}^{s,t}(L,\mathbb{F}_2) \cong \mathrm{Ext}_{A(1)}^{s+4,t+12}(L,\mathbb{F}_2).$$

The proof follows the same lines as the proof of Lemma 9.46.

Ext groups have an interpretation in terms of equivalence classes of finite exact sequences; the piece of a minimal resolution we constructed gives a class in $\mathrm{Ext}_{A(1)}^{4,12}(A(0),A(0))$ and the isomorphism in Lemma 9.49 can be obtained by splicing (Yoneda multiplication) with this class. There is also the composition pairing that determines a right action:

$$\mathrm{Ext}_{A(1)}^{s,t}(L,\mathbb{F}_2) \otimes H^{s',t'}(A(1)) \to \mathrm{Ext}_{A(1)}^{s+s',t+t'}(L,\mathbb{F}_2).$$

Our first goal is to describe the isomorphism between $\mathrm{Ext}_{A(1)}^{s,t}(L,\mathbb{F}_2)$ and $\mathrm{Ext}_{A(1)}^{s+4,t+12}(L,\mathbb{F}_2)$ as multiplication on the right with a class in $H^{4,12}(A(1))$. From this viewpoint we will see how to generalize.

Lemma 9.50. $\mathrm{Ext}_{A(1)}^{4,12}(\mathbb{F}_2,\mathbb{F}_2) \cong \mathrm{Ext}_{A(1)}^{4,12}(A(0),\mathbb{F}_2) \cong \mathbb{F}_2$. *Furthermore, right multiplication by the unique class* $\{u\} \in \mathrm{Ext}_{A(1)}^{4,12}(\mathbb{F}_2,\mathbb{F}_2)$ *gives the isomorphism* $\mathrm{Ext}_{A(1)}^{s,t}(A(0),\mathbb{F}_2) \cong \mathrm{Ext}_{A(1)}^{s+4,t+12}(A(0),\mathbb{F}_2)$.

PROOF: Consider the short exact sequence of $A(1)$-modules:

$$0 \to s\mathbb{F}_2 \to A(0) \to \mathbb{F}_2 \to 0.$$

This induces a long exact sequence of Ext groups:

$$\to H^{3,11}(A(1)) \xrightarrow{\partial} H^{4,12}(A(1)) \to \mathrm{Ext}_{A(1)}^{4,12}(A(0),\mathbb{F}_2) \to H^{4,11}(A(1)) \to .$$

The six-term exact sequence that starts a minimal resolution of $A(0)$ over $A(1)$ determines a class in $\mathrm{Ext}_{A(1)}^{4,12}(A(0),\mathbb{F}_2)$. By constructing a minimal resolution of $\mathbb{F}_2$ over $A(1)$, the reader will find that $H^{3,11}(A(1)) \cong \{0\} \cong H^{4,11}(A(1))$, and that the isomorphism at bidegree $(4,12)$ is induced by the quotient $\eta: A(0) \to \mathbb{F}_2$.

If $0 \leftarrow \mathbb{F}_2 \leftarrow Q_{\bullet}$ is a minimal resolution of $\mathbb{F}_2$ over $A(1)$, then there is a morphism of exact sequences:

Let $u\colon Q_4 \to s^{12}\mathbb{F}_2$ denote the map representing the unique class $\{u\} \in H^{4,12}(A(1))$. To complete the proof of the lemma, we show that right multiplication by the cohomology class of u determines the isomorphism given by left multiplication by the class of the six term short exact sequence given by the minimal resolution of $A(0)$ over $A(1)$.

A minimal resolution of $A(0)$ can be constructed from the six-term exact sequence,

$$0 \leftarrow A(0) \leftarrow A(1) \leftarrow P_1 \leftarrow P_2 \leftarrow P_3 \leftarrow s^{12}A(1) \leftarrow s^{12}P_1 \leftarrow \cdots .$$

Given a class in $\mathrm{Ext}^{s,t}_{A(1)}(A(0), \mathbb{F}_2)$, it may be represented by a cohomology class of a homomorphism $f\colon P_s \to s^t\mathbb{F}_2$. Lift this mapping through the minimal resolution $0 \leftarrow \mathbb{F}_2 \leftarrow Q_\bullet$, suspended t times.

The commutativity of the diagram

$$
\begin{array}{ccc}
P_{s+4} & \xrightarrow{\ v_s\ } & s^{12}P_s \\
{\scriptstyle f_4}\downarrow & & \downarrow{\scriptstyle s^{12}f} \\
s^t Q_4 & \xrightarrow[\ s^t u\ ]{} & s^{t+12}\mathbb{F}_2,
\end{array}
$$

where v_s is the isomorphism that carries the periodicity of the resolution of $A(0)$ over $A(1)$, follows by the properties of the resolutions and liftings. The different composites imply that multiplication on the right by $\{u\} \in \mathrm{Ext}^{4,12}_{A(1)}(\mathbb{F}_2, \mathbb{F}_2)$ is isomorphic to left multiplication by the element determined by the periodicity of the resolution. $\qquad\square$

We next generalize the isomorphism of Lemma 9.49 to the other subalgebras $A(n)$ of $\mathcal{A}_2$. Our point of departure is the observation that $H^{4,12}(\mathcal{A}_2) = \{0\} = H^{5,12}(\mathcal{A}_2)$, while $H^{4,12}(A(1)) \cong \mathbb{F}_2$. More generally, we observe that, for $r \geq 3$, we have $h_{r+1}h_0^{2^r} = 0$ in $H^{*,*}(\mathcal{A}_2)$: Consider the composite of Steenrod operations acting on $H^{*,*}(\mathcal{A}_2)$: $Sq^{2^r} Sq^{2^{r-1}} \cdots Sq^2 Sq^1(h_1 h_0)$. By Theorems 9.10 and 9.36, one finds that, for $r \geq 2$,

$$Sq^{2^r} \cdots Sq^2 Sq^1(h_1 h_0) = h_{r+1}h_0^{2^r}.$$

Since $h_3 h_0^4 = 0$, it follows that $h_{r+1}h_0^{2^r} = 0$ for $r \geq 3$.

These relations suggest the following construction: The dual of the inclusion $i\colon A(r) \hookrightarrow \mathcal{A}_2$, annihilates $\xi_1^{2^{r+1}}$, the class dual to $Sq^{2^{r+1}}$. Let c_r be a class in $F_*(A_2^{\mathrm{dual}})$ such that

$$d^*(c_r) = [\underbrace{\xi_1 \mid \cdots \mid \xi_1}_{2^r \text{ times}} \mid \xi_1^{2^{r+1}}],$$

which represents $h_0^{2^r} h_{r+1}$ in $F_*(A(r)^{\mathrm{dual}})$. The class $i^* c_r$ in $F_*(A(r)^{\mathrm{dual}})$ is a cycle. We let $\omega_r = \{i^* c_r\}$ denote the cohomology class of $i^* c_r$ in $H^{2^r, 3\cdot 2^r}(A(r))$. [Adams66] described explicit representatives for the classes ω_r using the $\smile_1$-product on the cobar construction. We need the following facts from this explicit construction:

Fact 1. ω_2 in $H^{4,12}(A(2))$ maps nontrivially to the class $\{u\}$ in $H^{4,12}(A(1))$ that induces the isomorphism in Lemma 9.49.

To see this, consider the short exact sequence of finite Hopf algebras,

$$0 \to A(1) \to A(2) \to A(2)//A(1) \to 0 :$$

$A(2)//A(1)$ is an exterior algebra that is 3-connected, and so the associated long exact sequence provides an isomorphism $H^{4,12}(A(2)) \to H^{4,12}(A(1))$. [Adams66] gives a nonzero class for ω_2 and so it must go over to the unique generator for $H^{4,12}(A(1))$.

Fact 2. Under the inclusion $A(r) \to A(r+1)$, the class ω_{r+1} maps to $(\omega_r)^2$ in cohomology, $H^*(A(r+1)) \to H^*(A(r))$.

This also follows from the explicit representative given by [Adams66]. The cobar classes c_r may be used to define Massey products; let $\ker(h_0^{2^r})$ denote the subset of $\mathrm{Ext}_{\mathcal{A}_2}^{s,t}(L, \mathbb{F}_2)$ of elements whose product (on the right) with $h_0^{2^r}$ vanishes. In the cobar construction, $L^{\mathrm{dual}} \otimes F_*(\mathcal{A}_2^{\mathrm{dual}})$, let a represent a class $\alpha = \{a\} \in \mathrm{Ext}_{\mathcal{A}_2}^{s,t}(L, \mathbb{F}_2)$ satisfying $\alpha h_0^{2^r} = 0$; let y be such that $d^* y = a[\xi_1 \mid \cdots \mid \xi_1]$. Define the homomorphism

$$P^{r-1} \colon \ker(h_0^{2^r}) \to \frac{\mathrm{Ext}_{\mathcal{A}_2}^{s+2^r, t+3\cdot 2^r}(L, \mathbb{F}_2)}{\mathrm{Ext}_{\mathcal{A}_2}^{s+2^r-1, t+2^r}(L, \mathbb{F}_2) h_{r+1}}$$

by $P^{r-1}(\alpha) = \{ac_r + y[\xi_1^{2^{r+1}}]\}$. Notice that P^{r-1} is well-defined because the choices are absorbed into the quotient. Since c_r is a specific choice of element with $d^*(c_r) = [\xi_1 \mid \xi_1 \mid \cdots \mid \xi_1^{2^{r+1}}]$, this class $P^{r-1}(\alpha)$ can be further projected to $\mathrm{Ext}_{\mathcal{A}_2}^{s+2^r, t+3\cdot 2^r}(L, \mathbb{F}_2)$ modulo indeterminacy where the indeterminacy is given by

$$\mathrm{Ext}_{\mathcal{A}_2}^{s, t+2^{r+1}}(L, \mathbb{F}_2) h_0^{2^r} + \mathrm{Ext}_{\mathcal{A}_2}^{s+2^r-1, t+2^r}(L, \mathbb{F}_2) h_{r+1}.$$

Thus $P^{r-1}(\alpha)$ represents the Massey product $\langle \alpha, h_0^{2^r}, h_{r+1} \rangle$.

Consider the mapping

$$- \circ \omega_r \colon \mathrm{Ext}_{A(r)}^{s,t}(L, \mathbb{F}_2) \to \mathrm{Ext}_{A(r)}^{s+2^r, t+3\cdot 2^r}(L, \mathbb{F}_2),$$

given by the composition product on the right with ω_r.

Lemma 9.51. The following diagram commutes:

$$
\begin{array}{ccc}
\ker(h_0^{2^r}) & \xrightarrow{\ P^{r-1}\ } & \dfrac{\mathrm{Ext}_{\mathcal{A}_2}^{s+2^r, t+3\cdot 2^r}(L, \mathbb{F}_2)}{\mathrm{Ext}_{\mathcal{A}_2}^{s+2^r-1, t+2^r}(L, \mathbb{F}_2) h_{r+1}} \\[2em]
\Big\downarrow{\scriptstyle i^*} & & \Big\downarrow{\scriptstyle i^*} \\[1em]
\mathrm{Ext}_{A(r)}^{s,t}(L, \mathbb{F}_2) & \xrightarrow{\ -\circ\omega_r\ } & \mathrm{Ext}_{A(r)}^{s+2^r, t+3\cdot 2^r}(L, \mathbb{F}_2)
\end{array}
$$

PROOF: $i^*(ac_r + y[\xi_1^{2^{r+1}}]) = (i^*a)(i^*c_r) = (i^*a) \circ \omega_r.$ $\square$

This shows that the mapping $- \circ \omega_r$ is computable in terms of Massey products. This identification will be sharpened as we proceed.

Theorem 9.52. *If L is an $(n-1)$-connected module over $A(r)$ that is free over $A(0)$, then $- \circ \omega_r \colon \mathrm{Ext}_{A(r)}^{s,t}(L, \mathbb{F}_2) \to \mathrm{Ext}_{A(r)}^{s+2^r, t+3 \cdot 2^r}(L, \mathbb{F}_2)$ is an isomorphism for $s \geq 0$ and $t < n + 4s$.*

PROOF: For $s = 0$, the result follows because the Ext groups vanish in these degrees. We proceed by induction on s and $t - n$. Consider the short exact sequence

$$0 \to K \to A(r) \otimes_{A(1)} L \to L \to 0.$$

We make some useful observations:

Fact 3 (A change-of-rings theorem). *If H is a sub-Hopf algebra of a cocommutative Hopf algebra Γ over a field k, then Γ is free as an algebra over H, and furthermore, for all Γ-modules L,*

$$\mathrm{Ext}_H^{s,t}(L, k) \cong \mathrm{Ext}_\Gamma^{s,t}(\Gamma \otimes_H L, k).$$

Fact 4. *If L is an $A(r)$-module that is free over $A(0)$, then, for $r \leq \rho \leq \infty$, $A(\rho) \otimes_{A(r)} L$ is free over $A(0)$.*

Fact 3 follows from a theorem of [Milnor-Moore65]. Fact 4 is proved by [Adams66, p. 368] from an explicit choice of representatives for the dual comodules.

Since L is free over $A(0)$ and, by Fact 4, $A(r) \otimes_{A(1)} L$ is free over $A(0)$, we have that K is free over $A(0)$. Notice also that K is $(n+3)$-connected. Consider the commutative diagram, where we have written $H^{s,t}(M)$ for $\mathrm{Ext}_{A(r)}^{s,t}(M, \mathbb{F}_2)$:

$$
\begin{array}{ccccc}
H^{s-1,t}(A(r) \otimes_{A(1)} L) & \longrightarrow & H^{s-1,t}(K) & \longrightarrow & H^{s,t}(L) \longrightarrow \\
\downarrow{\scriptstyle \omega_r} & & \downarrow{\scriptstyle \omega_r} & & \downarrow{\scriptstyle \omega_r} \\
H^{s+2^r-1,t+3 \cdot 2^r}(A(r) \otimes_{A(1)} L) & \to & H^{s+2^r-1,t+3 \cdot 2^r}(K) & \to & H^{s+2^r,t+3 \cdot 2^r}(L) \to
\end{array}
$$

$$
\begin{array}{ccc}
H^{s,t}(A(r) \otimes_{A(1)} L) & \longrightarrow & H^{s,t}(K) \longrightarrow \\
\downarrow{\scriptstyle \omega_r} & & \downarrow{\scriptstyle \omega_r} \\
H^{s+2^r,t+3 \cdot 2^r}(A(r) \otimes_{A(1)} L) & \to & H^{s+2^r,t+3 \cdot 2^r}(K) \to
\end{array}
$$

By induction, we assume the results hold up to $s - 1$ and for all $(N - 1)$-connected modules and $t - N < 4(s - 1)$. The first and fourth vertical arrows are isomorphisms by applying the change-of-rings to get $H^{s,t}(A(r) \otimes_{A(1)} L) \cong$

$\mathrm{Ext}_{A(1)}^{s,t}(L,\mathbb{F}_2)$. This isomorphism takes ω_r to $(\omega_2)^{2^{r-2}}$, which is an isomorphism for $t-(n+4) < 4(s-1)$ by Lemma 9.49. Since K is $(n+3)$-connected, the second and last ω_r are isomorphisms for $t-(n+4) < 4(s-1)$, that is, $t-n < 4s$. The Five-lemma implies that the third ω_r is an isomorphism and the theorem is proved. $\qquad\square$

In order to extend this result about subalgebras of $\mathcal{A}_2$ to the entire Hopf algebra, we need the following approximation result.

Theorem 9.53. *Suppose $r \le \rho \le \infty$ and $i\colon A(r) \to A(\rho)$ denotes the inclusion of Hopf algebras. Then, if L is an $\mathcal{A}_2$-module that is free over $A(0)$ and $(n-1)$-connected, then the induced homomorphism*

$$i^*\colon \mathrm{Ext}_{A(\rho)}^{s,t}(L,\mathbb{F}_2) \to \mathrm{Ext}_{A(r)}^{s,t}(L,\mathbb{F}_2)$$

is an isomorphism for $0 < s < t < n + 2^{r-1} + T(s-1)$.

PROOF: We consider once more the short exact sequence of $A(\rho)$-modules, each free over $A(0)$:

$$0 \to K \to A(\rho) \otimes_{A(r)} L \to L \to 0.$$

By definition, K is $(n+2^{r+1}-1)$-connected. By Lemma 9.47, $\mathrm{Ext}_{A(\rho)}^{s,t}(K,\mathbb{F}_2)$ vanishes when $t < n + 2^{r+1} + T(s-1)$ and theorem follows from the change-of-rings isomorphism. $\qquad\square$

We combine theorems 9.52 and 9.53 with the results on vanishing to prove the main result of this section.

Theorem 9.54. *Suppose L is an $\mathcal{A}_2$-module that is free over $A(0)$ and $(n-1)$-connected. For $r \ge 2$ and $s > 0$, the mapping P^{r-1} induces an isomorphism*

$$P^{r-1}\colon \mathrm{Ext}_{\mathcal{A}_2}^{s,t}(L,\mathbb{F}_2) \to \mathrm{Ext}_{\mathcal{A}_2}^{s+2^r,t+3\cdot 2^r}(L,\mathbb{F}_2)$$

whenever $t < n + \min(4s, 2^{r+1} + T(s-1))$.

PROOF: If $t < n + 2^{r+1} + T(s-1)$, then

$$\begin{aligned}
t + 2^r &< n + 2^{r+1} + 2^r + T(s-1)\\
&< n + 12 \cdot 2^r + T(s-1)\\
&= n + T(s + 2^r - 1).
\end{aligned}$$

By Lemma 9.47, $\text{Ext}_{\mathcal{A}_2}^{s+2^r,t+2^r}(L,\mathbb{F}_2) = \{0\}$ and so $\text{Ext}_{\mathcal{A}_2}^{s,t}(L,\mathbb{F}_2)h_0^{2^r} = \{0\}$. Similarly $\text{Ext}_{\mathcal{A}_2}^{s-2^r-1,t+2^r}(L,\mathbb{F}_2)h_{r+1} = \{0\}$. Thus we have the following diagram from Lemma 9.51:

$$
\begin{array}{ccc}
\text{Ext}_{\mathcal{A}_2}^{s,t}(L,\mathbb{F}_2) & \xrightarrow{P^{r-1}} & \text{Ext}_{\mathcal{A}_2}^{s+2^r,t+3\cdot 2^r}(L,\mathbb{F}_2) \\
\downarrow{\scriptstyle i^*} & & \downarrow{\scriptstyle i^*} \\
\text{Ext}_{\mathcal{A}_2}^{s,t}(L,\mathbb{F}_2) & \xrightarrow{-\circ w_r} & \text{Ext}_{\mathcal{A}_2}^{s+2^r,t+3\cdot 2^r}(L,\mathbb{F}_2).
\end{array}
$$

By Theorem 9.53, i^* is an isomorphism in both cases. Since $t < n + 4s$, it is also the case that $-\circ w_r$ is an isomorphism. Thus P^{r-1} is an isomorphism. $\square$

Corollary 9.55. *For $r \geq 2$, the mapping P^{r-1} induces an isomorphism*

$$
P^{r-1}\colon\ \text{Ext}_{\mathcal{A}_2}^{s,t}(\mathbb{F}_2,\mathbb{F}_2) \longrightarrow \text{Ext}_{\mathcal{A}_2}^{s+2^r,t+3\cdot 2^r}(\mathbb{F}_2,\mathbb{F}_2)
$$

for $1 < s < t < \min(4s-2, 2+2^{r+1}+T(s-2))$.

The proof of this follows as in the proof of Corollary 9.48. Notice that the isomorphism,

$$
\text{Ext}_{\mathcal{A}_2}^{s,t}(\mathbb{F}_2,\mathbb{F}_2) \cong \text{Ext}_{\mathcal{A}_2}^{s-1,t}(I(\mathcal{A}_2)/I(\mathcal{A}_2)\cdot Sq^1,\mathbb{F}_2)
$$

is induced by multiplication on the left by the equivalence class of the extension

$$
0 \rightarrow I(\mathcal{A}_2)/I(\mathcal{A}_2)\cdot Sq^1 \rightarrow \mathcal{A}_2/I(\mathcal{A}_2)\cdot Sq^1 \rightarrow \mathbb{F}_2 \rightarrow 0
$$

lying in $\text{Ext}_{\mathcal{A}_2}^{1,0}(\mathbb{F}_2, I(\mathcal{A}_2)/I(\mathcal{A}_2)\cdot Sq^1)$.

We have showed already that $H^{4,12}(\mathcal{A}_2) = \{0\}$. Applying the operator $-\circ w_2$ repeatedly, we get $H^{4k,12k}(\mathcal{A}_2) = \{0\}$ for all k. This completes the proof of Theorem 9.43. It also shows that $H^{s,t+2^{r+1}}(\mathcal{A}_2)h_0^{2^r} = \{0\}$ for $s > 0$. This group appears in the indeterminacy of the Massey product of P^{r-1}. By carefully tracking through the isomorphisms one concludes that, for $1 < s < t < \min(4s-2, 2+2^{r+1}+T(s-2))$ and $\alpha \in H^{s,t}(\mathcal{A}_2)$,

$$
P^{r-1}(\alpha) = \langle \alpha, h_0^{2^r}, h_{r+1}\rangle \text{ modulo } \{0\}.
$$

We leave it as an exercise in the definition of Massey products to show $P^{r-1}\circ P^{r-1} = P^r$.

The reader should look ahead to the tables to see how the periodicity interacts with the vanishing line to determine that the vanishing is best possible. We note further that the results of [Moss70] show how the periodicity operator interacts with the differentials in the spectral sequence.

The May spectral sequence

In spite of our understanding of some of the global features of $H^{*,*}(\mathcal{A}_p)$ and our ability to compute in low dimensions, we still need an effective technique to compute large parts of the E_2-term of the Adams spectral sequence. The Princeton thesis of [May64] provided such a method. The point of departure for this work is the observation of [Milnor-Moore65] that Hopf algebras are endowed with certain natural filtrations. In this space we cannot present all of the details of May's work. We can give the thread of the argument. For the relevant definitions and further details, we refer the reader to the papers of [Milnor-Moore65] and [May64, 66].

Let $(\Gamma, \phi, \psi, \varepsilon, \eta)$ denote a graded Hopf algebra over a field k with product φ, coproduct ψ, augmentation ε and counit η. Following [Milnor-Moore65], Γ is filtered: Let $I(\Gamma)$ denote the augmentation ideal, then we filter Γ by letting $F_n\Gamma = \Gamma$, if $n \geq 0$, and $F_{-n}\Gamma = \mathrm{im}(\varphi^n : I(\Gamma)^{\otimes n} \to I(\Gamma))$, where φ^n is the iterated product. Denote the associated bigraded object by

$$(E^0\Gamma)_{q,r} = (F_q\Gamma/F_{q-1}\Gamma)_{q+r}.$$

Fact: $E^0\Gamma$ is a primitively generated Hopf algebra, that is, the natural mapping $\mathrm{Prim}(E^0\Gamma) \to Q(E^0\Gamma)$ is an isomorphism.

We assume henceforth that $k = \mathbb{F}_p$ for p, a prime, the case of interest for $\mathcal{A}_p$. Let $\mathrm{Prim}(\Gamma)$ denote the space of primitives in the Hopf algebra Γ.

Fact: $\mathrm{Prim}(\Gamma)$ is a restricted Lie algebra.

A **restricted Lie algebra** is a graded Lie algebra over $\mathbb{F}_p$, say L, together with a map, $\beta \colon L_n \to L_{pn}$, defined for n, even, if p is an odd prime, and for all n, if $p = 2$, such that, for some graded algebra A, there is a monomorphism of Lie algebras, $f \colon L \to A$, such that the diagram for each n

$$
\begin{array}{ccc}
L_n & \xrightarrow{\ \beta\ } & L_{pn} \\
{\scriptstyle f_n}\downarrow & & \downarrow{\scriptstyle f_{pn}} \\
A_n & \xrightarrow[\ \xi\ ]{} & A_{pn}
\end{array}
$$

commutes, where ξ is the Frobenius map $\xi(x) = x^p$. The Lie bracket product on $\mathrm{Prim}(\Gamma)$ is the canonical graded commutator.

To each Lie algebra L, one can associate an algebra $U(L)$ called the **universal enveloping algebra**. If L is restricted, an algebra $V(L)$ can be defined that has a compatible Frobenius map. Both $U(L)$ and $V(L)$ are Hopf algebras.

Fact: If Γ is a primitively generated Hopf algebra over $\mathbb{F}_p$, then Γ is isomorphic to $V(\mathrm{Prim}(\Gamma))$ as a Hopf algebra.

A corollary of this fact is that $E^0\Gamma$ is isomorphic to $V(\mathrm{Prim}(E^0\Gamma))$ for any Hopf algebra Γ over $\mathbb{F}_p$. With these definitions and facts, we can now state the main theorem of [May66].

Theorem 9.56. *Given a filtered, augmented algebra* Γ *over a field* k, *there is a spectral sequence, converging strongly to* $\mathrm{Ext}^{*,*}_{\Gamma}(k,k)$, *with differentials* d_r *of bidegree* $(r, 1 - r)$, *and* $E^{*,*}_2 \cong \mathrm{Ext}^{*,*}_{E^0\Gamma}(k,k)$.

For a graded Hopf algebra Γ over $\mathbb{F}_p$, $E^0\Gamma \cong V(\mathrm{Prim}(E^0\Gamma))$ and so we can turn to the theory of restricted Lie algebras for tools to determine the E_2-term of this spectral sequence. In particular, the cohomology of a Lie algebra is defined to be the cohomology of its universal enveloping algebra. In this context, [May66] introduced small resolutions, resembling Cartan's constructions and Koszul resolutions, to compute $H^{*,*}(V(\mathrm{Prim}(E^0\Gamma)))$.

By determining $E^0 A_p$ and applying these methods, [May66] proved the following computational result.

Theorem 9.57. *For* $p = 2$, $\mathrm{Ext}^{*,*}_{E^0 A_2}(\mathbb{F}_2, \mathbb{F}_2) \cong H^*(\mathcal{R}, d)$, *where* $\mathcal{R}$ *is the bigraded polynomial algebra* $\mathbb{F}_2[R_{i,j} \mid i \geq 0, j \geq 1]$ *on generators* $R_{i,j}$ *of bidegree* $(1, 2^i(2^j - 1))$ *and the differential* d *is given on generators by*

$$d(R_{i,j}) = \sum_{k=1}^{j-1} R_{i,k} R_{i+k,j-k} \, .$$

The product on $\mathrm{Ext}^{*,*}_{E^0 A_2}(\mathbb{F}_2, \mathbb{F}_2)$ *is induced by the polynomial product.*

For p, *an odd prime,* $\mathrm{Ext}^{*,*}_{E^0 A_p}(\mathbb{F}_p, \mathbb{F}_p) \cong H^*(\mathcal{S}, d)$, *where* $\mathcal{S}$ *is the bigraded commutative algebra*

$$\bigotimes_{i \geq 0, j \geq 1} \Lambda(R_{i,j}) \otimes \mathbb{F}_p[S_i] \otimes \mathbb{F}_p[\tilde{R}_{i,j}]$$

on generators $R_{i,j}$, $\tilde{R}_{i,j}$ *and* S_k, *of bidegree* $(1, 2p^i(p^j - 1))$, $(2, 2p^{i+1}(p^j - 1))$ *and* $(1, 2p^k - 1)$, *respectively, and the differential is given on generators by*

$$d(\tilde{R}_{i,j}) = 0, \quad d(R_{i,j}) = \sum_{k=1}^{j-1} R_{i,k} R_{i+k,j-k}, \quad d(S_i) = \sum_{k=0}^{i-1} R_{i-k,k} S_k.$$

The product on $\mathrm{Ext}^{*,*}_{E^0 A_p}(\mathbb{F}_p, \mathbb{F}_p)$ *is induced by the product on this algebra.*

Though the E_2-terms of these spectral sequences appear simple, there are many differentials and the product structure on $E^0 \mathrm{Ext}^*_{A_p}(\mathbb{F}_p, \mathbb{F}_p)$ is not the one induced by the spectral sequence. These obstacles can be overcome and [May64] computed $H^{s,t}(A_p)$ for $t - s \leq 2(p - 1)(2p^2 + p + 2) - 4$, when p is an odd prime, and for $t - s \leq 42$, when $p = 2$. [Tangora70] extended these techniques to compute $H^{s,t}(A_2)$ for $t - s \leq 70$. Other filtrations of the initial datum, A_p, are possible leading to other versions of the May spectral sequence with computational aspects better suited to a given problem. For a thorough discussion of these ideas, see Appendix A.1 of the book of [Ravenel86].

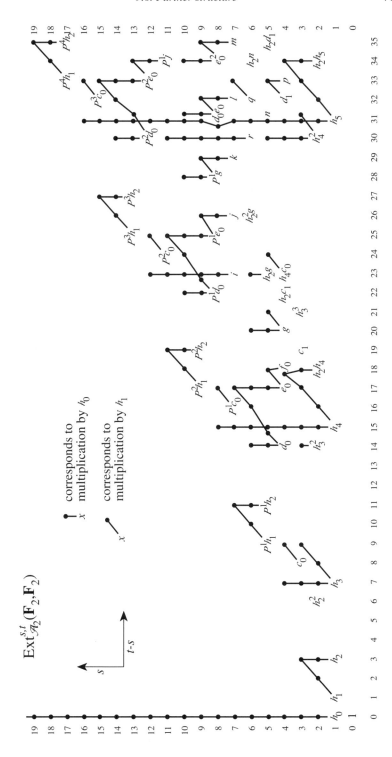

$$\text{Ext}^{s,t}_{\mathcal{A}_2}(\mathbf{F}_2, \mathbf{F}_2)$$

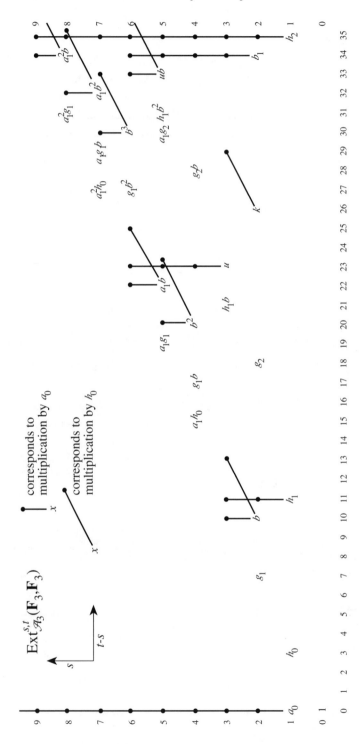

Machine calculations of $H^{*,*}(\mathcal{A}_p)$ have been made since 1964 ([Liule-vicius66]). [Bruner93] used minimal resolutions and considerable computing expertise to push the tables of known Ext groups for $H^{s,t}(\mathcal{A}_2)$ to $t - s \leq 88$, with $t \leq 116$ and $s < 38$.

We have given the tables for $H^{*,*}(\mathcal{A}_2)$ and $H^{*,*}(\mathcal{A}_3)$ for $t - s \leq 35$. The reader is referred to [May64] and [Tangora70] for the origins and naming of the elements in the May spectral sequence and the relations between them. Other tables of Ext have been prepared by [Shick93], [Bruner93], by [Nassau] (an internet page that features h_2 connections), and by [Hatcher] (another internet page that features different axes for which h_2 connections are horizontal lines). These charts are the raw data from which we will compute some of π_*^S.

Extensions and differentials

Having computed a portion of the E_2-term of the Adams spectral sequence we next determine the differentials in this range. As you have come to expect, this can be a difficult task. Furthermore, once we have determined even part of the E_∞-term, we only have a composition series for each $_{(p)}\pi_{t-s}^S$. There can be extension problems. In this section, we discuss techniques that help determine differentials. Having done this, we settle some extension problems in order to give the reader an idea of how one can approach them.

The most successful methods for constructing differentials are those that arise from geometric properties. The first example of this is the graded com-mutativity of π_*^S (Theorem 9.38). This forced classes $h_0 h_i^2$, for $i \geq 3$, to be in the image of a differential. The difference between the ring structure on $_{(2)}\pi_*^S$ and the ring structure on $E_0^{*,*}(_{(2)}\pi_*^S) \cong E_\infty^{*,*}$ induced through the spectral sequence from E_2 must be accounted for by differentials.

Another geometric idea is the nontriviality of secondary and higher order cohomology operations.[Maunder64] showed how higher order operations can be related to differentials, the primary examples being the decompositions of Sq^{2^n} by [Adams60] and the decomposition of P^{p^n} by [Liulevicius62]. These decompositions correspond to $d_2(h_{i+1}) = h_0 h_i^2$, for $p = 2$ and $i > 3$, and $d_2(h_i) = a_0 b_{i-1}$, for p an odd prime.

Theorem 9.58. *Let X and Y be CW-complexes of finite type over $\mathbb{F}_p$ with Y finite dimensional and let*

$$0 \leftarrow H^*(X; \mathbb{F}_p) \leftarrow C_0 \leftarrow C_1 \leftarrow C_2 \leftarrow \cdots$$

be an $\mathcal{A}_p$-free resolution of $H^*(X; \mathbb{F}_p)$. Then there is a family of higher order cohomology operations, $\{\Phi^{r,s}\}$, associated to $C_0 \leftarrow C_1 \leftarrow C_2 \leftarrow \cdots$ such that, in the Adams spectral sequence converging to $_{(p)}\{Y, X\}_*$, the differential $d_r \colon E_r^{s,t} \to E_r^{s+r,t+r-1}$ is given by $\Phi^{s+r,s}$ acting on $H^*(Y; \mathbb{F}_p)$.

That is, if $u \in E_r^{s,t}$, then it has a representative

$$\bar{u} \in E_1^{s,t} = \mathrm{Hom}_{\mathcal{A}_p}^{s,t}(C_*, H^*(Y; \mathbb{F}_p))$$

on which $d_1, d_2, \ldots, d_{r-1}$ vanish, on which $\Phi^{s+r,s}$ is defined, and $\Phi^{s+r,s}(\bar{u})$ is a coset of $E_1^{s+r,t+r-1}$ identified with $d_r(u)$.

Another source of differentials are the known stable groups themselves. If the E_2-term of the Adams spectral sequence lies before us and a known group $_{(2)}\pi_n^S$ does not agree with the initial data, a differential must be nontrivial to correct the discrepancy. The computations of [Toda62] provide a geometric 'priming' for the Adams spectral sequence with explicit groups $_{(p)}\pi_n^S$ for $0 \leq n \leq 19$. For convenience, we list these data for the prime 2 (we write $(\mathbb{Z}/n\mathbb{Z})^{\oplus k}$ for $\mathbb{Z}/n\mathbb{Z} \oplus \mathbb{Z}/n\mathbb{Z} \oplus \cdots \mathbb{Z}/n\mathbb{Z}$, k times):

TODA'S TABLES

n	$_{(2)}\pi_n^S$	generators	comments
0	$\mathbb{Z}$	ι	
1	$\mathbb{Z}/2\mathbb{Z}$	η	$\{h_1\}$
2	$\mathbb{Z}/2\mathbb{Z}$	η^2	$\{h_1^2\}$
3	$\mathbb{Z}/8\mathbb{Z}$	ν	$\{h_2\}, 4\nu = \eta^3$
4	$\{0\}$		
5	$\{0\}$		
6	$\mathbb{Z}/2\mathbb{Z}$	ν^2	$\{h_2^2\}$
7	$\mathbb{Z}/16\mathbb{Z}$	σ	$\{h_3\}$
8	$(\mathbb{Z}/2\mathbb{Z})^{\oplus 2}$	$\bar{\nu}, \varepsilon$	$\eta\sigma = \bar{\nu} + \varepsilon$
9	$(\mathbb{Z}/2\mathbb{Z})^{\oplus 3}$	$\nu^3, \mu, \eta \circ \varepsilon$	$\{h_2^3 = h_1^2 h_3\}, \eta\bar{\nu} = \nu^3$
10	$\mathbb{Z}/2\mathbb{Z}$	$\eta \circ \mu$	
11	$\mathbb{Z}/8\mathbb{Z}$	ζ	$\{P^1 h_2\}$
12	$\{0\}$		
13	$\{0\}$		
14	$(\mathbb{Z}/2\mathbb{Z})^{\oplus 2}$	σ^2, κ	$\{h_3^2\}$
15	$\mathbb{Z}/32\mathbb{Z} \oplus \mathbb{Z}/2\mathbb{Z}$	$\rho, \eta \circ \kappa$	
16	$(\mathbb{Z}/2\mathbb{Z})^{\oplus 2}$	$\eta^*, \eta \circ \rho$	$\eta\rho = \sigma\mu$
17	$(\mathbb{Z}/2\mathbb{Z})^{\oplus 4}$	$\eta \circ \eta^*, \nu \circ \kappa, \eta^2 \circ \rho, \bar{\mu}$	$\eta^2\rho = \varepsilon\mu$
18	$\mathbb{Z}/8\mathbb{Z} \oplus \mathbb{Z}/2\mathbb{Z}$	$\nu^*, \eta \circ \bar{\mu}$	$\eta^2\eta^* = 4\nu^*, \eta\bar{\mu} = \mu^2$
19	$\mathbb{Z}/8\mathbb{Z} \oplus \mathbb{Z}/2\mathbb{Z}$	$\zeta_2, \bar{\sigma}$	$P^2 h_2, c_1$

Theorem 9.59 ([May64]). *There is only one pattern of differentials consistent with Toda's data on $E_2^{s,t}$ for $t - s \le 19$. This pattern is given by*

$$d_2(h_4) = h_0 h_3^2$$
$$d_3(h_0 h_4) = h_0 d_0, \; d_3(h_0^2 h_4) = h_0^2 d_0$$
$$d_2(e_0) = h_1^2 d_0$$
$$d_2(f_0) = h_0 h_2 d_0, \; d_2(h_0 f_0) = h_0^2 h_2 d_0.$$

The reader will find proving this theorem quite straightforward and instructive. Notice that a relation in the E_2-term has been made part of the statement of the theorem, that is, $h_0 e_0 = h_2 d_0$. This can be shown using identities with Massey products.

These differentials immediately imply later ones by virtue of the product structure and the relations between differentials and the periodicity operators. This allows us to compute stable stems.

Corollary 9.60. *The following differentials are implied by the previous ones. For $i \ge 0$,*

$$d_r(P^i d_0) = 0, \; \text{for all } r, \qquad d_r(P^i g^n) = 0, \; \text{for all } r,$$
$$d_2(P^i e_0) = P^i h_1^2 d_0, \quad d_2(P^i j) = P^{i+1} h_2 d_0, \quad d_2(P^i k) = P^{i+1} h_0 g.$$

PROOF: We show the case for $P^i j$. First of all, $d_2 P^i = P^i d_2$ can be shown to follow from a homotopy computation or Theorem 9.42. Among the relations that hold in Ext (see [Tangora70]), we find $h_0 j = h_2 i = P^1 f_0$. Also, from the structure of the May spectral sequence, $P^1 xy = xP^1 y = yP^1 x$, where it applies. Thus $h_0^2 j = P^1 h_0 f_0 = P^1 h_1 e_0 = e_0 P^1 h_1$ and so $d_2(h_0^2 j) = d_2(e_0 P^1 h_1) = h_1^2 d_0 P^1 h_1 = P^1 h_1^3 d_0 = P^1 h_0^2 h_2 d_0 = h_0^2 P^1 h_2 d_0$. It follows that $d_2(j) = P^1 h_2 d_0$.

The other relations that enter this proof include $h_2 e_0 = h_0 g$, $P^{i+1} h_1 h_3 = P^i h_1^2 d_0$, $P^1 h_4 = h_2 g$, $P^1 g = d_0^2$, $d_0 g = e_0^2$. We add that $i = P^1 h_0^2 h_4$. $\qquad\square$

This corollary allows one to compute $_{(2)}\pi_{t-s}^S$ for $20 \le t - s \le 28$.

MAY'S TABLES

n	$_{(2)}\pi_n^S$	generators
20	$\mathbb{Z}/8\mathbb{Z}$	$\{g\}$
21	$(\mathbb{Z}/2\mathbb{Z})^{\oplus 2}$	$\{h_3^3\}, \{h_1 g\}$
22	$(\mathbb{Z}/2\mathbb{Z})^{\oplus 2}$	$\{h_2 c_1\}, \{P^1 d_0\}$
23	$(\mathbb{Z}/2\mathbb{Z})^{\oplus 2} \oplus \mathbb{Z}/8\mathbb{Z} \oplus \mathbb{Z}/16\mathbb{Z}$	$\{h_4 c_0\}, \{P^1 h_1 d_0\}, \{h_2 g\}, \{P^2 h_3\}$
24	$(\mathbb{Z}/2\mathbb{Z})^{\oplus 2}$	$\{h_1 h_4 c_0\}, \{P^2 c_0\}$
25	$(\mathbb{Z}/2\mathbb{Z})^{\oplus 2}$	$\{P^2 h_1 c_0\}, \{P^3 h_1\}$

26	$(\mathbb{Z}/2\mathbb{Z})^{\oplus 2}$	$\{h_2^2 g\}, \{P^3 h_1^2\}$
27	$\mathbb{Z}/8\mathbb{Z}$	$\{P^3 h_2\}$
28	$\mathbb{Z}/2\mathbb{Z}$	$\{P^1 g\}$

In order to extend these computations further, we employ the naturality of the Adams spectral sequence. [Maunder65] considered the mapping $f\colon S^{2q} \to F$ where F is the homotopy fibre of the mapping $p\colon BU(2q,\dots,\infty) \to K(\mathbb{Z}, 2q + 2^m)$, $BU(2q,\dots,\infty)$ is the $2q$-coconnected space associated to the classifying space of the infinite unitary group BU (that is, the homotopy groups $\pi_i(BU(2q,\dots,\infty)) = \{0\}$ for $i \geq 2q$) and p represents the Chern character $ch_{q,2^m-1}$. The mapping $f\colon S^{2q} \to F$ is induced by the generator of $\pi_{2q}(BU)$, which is given by the qth iterate of the Bott map. The mapping f induces a homomorphism of spectral sequences:

$$\operatorname{Ext}_{\mathcal{A}_2}^{s,t}(H^*(S^{2q};\mathbb{F}_2),\mathbb{F}_2) \to \operatorname{Ext}_{\mathcal{A}_2}^{s,t}(H^*(F;\mathbb{F}_2),\mathbb{F}_2),$$

and by naturality we have, for all $r \geq 2$, $d_r(f^*(x)) = f^*(d_r(x))$. [Maunder65] computed $H^*(F;\mathbb{F}_2)$ as a module over $\mathcal{A}_2$ from which he computed the relevant parts of $\operatorname{Ext}_{\mathcal{A}_2}^{*,*}(H^*(F;\mathbb{F}_2),\mathbb{F}_2)$. The main result of this paper is that the classes $h_0^n h_m$, for $n < 2^m$, in $H^{*,*}(\mathcal{A}_2)$ are never in the image of any differential in the Adams spectral sequence converging to $_{(2)}\pi_*^S$.

The papers of [Mahowald67], [Mahowald-Tangora67], and [Barratt-Mahowald-Tangora70] use stable cofibration sequences of small complexes, $S^0 \to X \to X'$ to determine differentials. Such cofibrations induce a short exact sequence on cohomology,

$$0 \to H^*(X';\mathbb{F}_2) \to H^*(X;\mathbb{F}_2) \to \mathbb{F}_2 \to 0$$

and so long exact sequences of Ext groups.

$$\xrightarrow{\delta} \operatorname{Ext}_{\mathcal{A}_2}^{s,t}(\mathbb{F}_2,\mathbb{F}_2) \to \operatorname{Ext}_{\mathcal{A}_2}^{s,t}(H^*(X;\mathbb{F}_2),\mathbb{F}_2) \xrightarrow{p^*} \operatorname{Ext}_{\mathcal{A}_2}^{s,t}(H^*(X';\mathbb{F}_2),\mathbb{F}_2) \longrightarrow$$

If the complexes are chosen carefully, the coboundary operator in this sequence has a nice form and computation of the stable homotopy of X and X' in low dimensions is possible.

Examples of such sequences are

$$S^0 \xrightarrow{i} S^0 \cup_\eta e_2 \xrightarrow{p} S^2 \quad \text{or} \quad S^0 \xrightarrow{i} S^0 \cup_\nu e^4 \cup_\sigma e^8 \xrightarrow{p} S^4 \vee S^8.$$

The following proposition (Lemma 3.4.1 of [Mahowald-Tangora67]) gives a general method for applying such sequences.

Proposition 9.61. *Consider a sequence $S^0 \xrightarrow{i} X \xrightarrow{p} X'$ such that $p_*i_* = 0$ on stable homotopy. Suppose $\alpha \in \mathrm{Ext}$ for S^0 is such that $i^*\alpha$ in Ext for X survives and, for any $\bar{\alpha} \in \{i^*\alpha\}$, the class represented by $i^*\alpha$, $p_*\bar{\alpha}$ is essential, then α is not a permanent cycle.*

PROOF: Let $f\colon S^q \to S^0$ represent $\{\alpha\}$. Then $[if]$ is in $\{i^*\alpha\}$ and so $[pif]$ is essential. But $p_*i_* = 0$, therefore α cannot be a surviving cycle. Suppose α is in the image of some differential, $\alpha = d_r\beta$. By naturality, $i^*\alpha = d_ri^*\beta$, which is impossible since $i^*\alpha$ is a surviving cycle. Therefore, some differential must originate on α. ☐

This lemma can be applied to the element e_0 with the sequence given by $S^0 \to S_0 \cup_\eta e^2 \to S^2$. [Mahowald-Tangora67] showed that i^*e_0 survives and $p_*\{i^*e_0\} = \eta\kappa \neq 0$. Thus a differential arises on e_0.

The last source of differentials to be considered here is the interaction of differentials with the Steenrod algebra action on Ext (Theorem 9.10). Developed first by [Kahn70], it has been extended by [Milgram72], [Maakinen73] and [Bruner84] to a powerful tool in this enterprise. The reader is encouraged to read the contributions of Bruner in [Bruner-May-McClure-Steinberger86] for an overview of this method.

Let us now consider the E_∞-term of the spectral sequence.

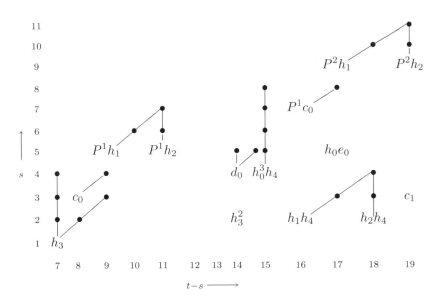

We suppose that Toda's tables are to be deduced from this table with as little input as possible. The interpretation of multiplication by h_0 settles the 7, 10, 11, 15, 16, 18 and 19 stems. Consider the 8-stem. The element h_1h_3 satisfies

$h_0 h_1 h_3 = 0$ and furthermore, $h_0 h_1 h_3$ is in filtration 3 where c_0 lies. Thus we cannot have c_0 as an element representing $2\{h_1 h_3\}$ because in this filtration it would be $h_0 h_1 h_3$. Also no further classes lie in the 8-stem in higher filtrations. Thus $_{(2)}\pi_8^S \cong \mathbb{Z}/2\mathbb{Z} \oplus \mathbb{Z}/2\mathbb{Z}$. An easier argument follows from the relations $2\{h_1\} = 0$ and $\{h_1\}\{h_3\} = \{h_1 h_3\}$. However, such relations are only true up to filtration and indeed, [Toda62] has shown $\eta\nu = \bar{\nu} + \varepsilon$, where $\bar{\nu} = \{h_1 h_3\}$ and $\varepsilon = \{c_0\}$. One needs to be careful.

The same argument works for $\{h_1^2 h_3\}$ and $\{h_1 P^1 c_0\}$. By filtration 4 we can see that $2\{h_1 P^1 c_0\} = 2\{h_1\}\{P^1 c_0\} = \{h_0 h_1\}\{P^1 c_0\} = 0$. Thus $_{(2)}\pi_9^S \cong (\mathbb{Z}/2\mathbb{Z})^{\oplus 3}$. Similarly, $_{(2)}\pi_{17}^S \cong (\mathbb{Z}/2\mathbb{Z})^{\oplus 4}$.

We finally turn to the determination of the composition product structure on $_{(2)}\pi_*^S$ from the spectral sequence. It is here that some geometric input is needed. A 'hidden extension' can be found when we consider the ideal $\eta \circ \pi_*^S$. Let ρ be the generator in the 15-stem of the factor $\mathbb{Z}/32\mathbb{Z}$. From the data in E_∞, $\eta \circ \rho$ appears to be zero. However, we know the following (deep) facts: ρ generates the image of the J-homomorphism $\pi_{15}(SO) \to \pi_{15}^S$ and $\eta \operatorname{im} J$ is nonzero (see [Switzer75, p. 488]). Because $\eta \circ \rho$ must appear in a higher filtration than $h_1 h_0^3 h_4$, it happens that $\eta \circ \rho = \{P^1 c_0\}$, the only other choice.

[Mahowald-Tangora67] and [Tangora70'] consider more difficult extension problems. The interplay between geometric and algebraic data is complicated and extended by numerous identities in Ext and in π_*^S. Knowledge of $_{(2)}\pi_n^S$ can be derived, as far as anyone has tried using these methods, for $n \leq 45$ (see [Bruner84]).

Other approaches to computing stable homotopy groups of spheres have been developed that have features similar to the Adams spectral sequence. One attractive way is via a kind of reverse Adams spectral sequence that was introduced by [Cohen, J70] and applied with great success (and computer aid) by [Kochman90]. Using these methods, information on stable stems out to dimension 64 were obtained (and corrected in [Kochman-Mahowald95]).

Epilogue

Where do we go from here with the Adams spectral sequence?

Work on the classical Adams spectral sequence continues. The stem-by-stem calculations have given way to the determination of regular phenomena such as infinite families of elements in π_*^S (for example, [Mahowald81], [Cohen, R81], and [Lin, WH85]), global structures in Ext ([Singer81]), and the identification of geometric phenomena like the EHP-sequence or the image of the J-homomorphism at the level of the spectral sequence ([Mahowald82]). Recent surveys of this work can be found in the book of [Kochman96] and the paper of [Miller-Ravenel93].

The notion of spectrum discussed in §9.4 was introduced in order to study Spanier-Whitehead duality and generalized (co)homology theories. All generalized theories are represented by spectra ([Brown, E62]) and among the most important have been the bordism theories represented by Thom spectra ([Rudyak98]). [Novikov67] carried out a program initiated by [Adams64] to construct and compute the stable homotopy of spheres using complex cobordism theory. The development of this line of ideas now forms a major part of homotopy theory. The excellent books of [Switzer75] and [Ravenel86] offer a basic introduction. The viewpoint of complex cobordism has led to many new algebraic tools (for example, formal group laws [Quillen69"]) and deep global results ([Devinatz-Hopkins-Smith88]). The book of [Ravenel92] gives a sketch of this work and its place in the emerging global picture of homotopy theory.

As a tool the Adams spectral sequence has been applied with considerable success to various geometric problems. Beyond the solution to the Hopf invariant one problem we discuss two further spectacular examples.

The **Kervaire invariant** of an almost framed manifold was introduced by [Kervaire60]. It is based on the Arf invariant of a quadratic form. In a classic paper, [Browder69] identified the Kervaire invariant with the value of a cohomology operation defined for Poincaré duality spaces M^{2q} with extra structure. That extra structure is a lifting of the classifying map of the normal bundle to a fibration with vanishing Wu class v_{q+1}. This gives a cobordism theory based on the vanishing of the Wu class (for a discussion of the Wu class see [Milnor-Stasheff74]). By analyzing orientations with respect to this cobordism theory one can show that there is a structure on $S^q \times S^q$ with $v_{q+1} = 0$ and Arf invariant one. To complete the analysis one has to know if this structure comes from a framed manifold. The identification of framed cobordism with π_*^S and the Adams spectral sequence allow one to ask this question at the level of the E_2-term of the spectral sequence. [Browder69] proved that the only dimensions in which a Kervaire invariant one manifold may exist are of the form $2^i - 2$ and that there is a manifold of Kervaire invariant one if and only if the class h_{i-1}^2 in $\mathrm{Ext}_{\mathcal{A}_2}^{2,2^i}(\mathbb{F}_2, \mathbb{F}_2)$ represents a nontrivial element in $\pi_{2^i-2}^S$. At this time, calculations of [Barratt-Jones-Mahowald84] and [Kochman90] have shown that there are manifolds of Kervaire invariant one of dimensions 2, 6, 14, 30, and 62. It is still open whether there are Kervaire invariant one manifolds in dimensions $2^i - 2$ for $i \geq 7$.

A differential geometric question one can ask of a manifold is whether it admits a Riemannian metric of positive scalar curvature. Using methods of surgery this question can be reduced to a cobordism problem for which the property of being a Spin manifold or not breaks the problem into two parts. The nonSpin case for simply-connected manifolds was studied by [Gromov-Lawson80]. There all obstructions to admitting a positive scalar metric vanish and examples in each cobordism class are given. There are obstructions in the Spin case studied first by [Lichnerowicz63] and extended by [Hitchin74]. [Stolz92] showed that the vanishing of Hitchin's obstruction was sufficient for the existence of a positive scalar metric. The argument requires the identification of the image in the Spin cobordism ring of the Spin bordism groups of a classifying space of a certain group. Since the Spin bordism groups are the stable homotopy groups of a particular spectrum, then one can compute these groups explicitly via the Adams spectral sequence.

Exercises

9.1. If $f\colon S^n \to Y$ is a mapping and $M_f = Y \cup_f e^{n+1}$ is the mapping cone, show that

$$\Sigma(M_f) = \Sigma(Y \cup_f e^{n+1}) \simeq \Sigma Y \cup_{\Sigma f} e^{n+2} = M_{\Sigma f}.$$

9.2. Use the Cartan-Serre theorem (Theorem 6.20) to prove

$$\mathcal{A}_p \otimes s^n \mathbb{F}_p \cong_{2n-1} \tilde{H}^*(K(\mathbb{Z}/p\mathbb{Z}, n); \mathbb{F}_p).$$

9.3. For an abelian group G, show that the set $\{$elements of finite order prime to $p\}$ forms a subgroup of G. This shows that $_{(p)}G$ is well-defined.

9.4. Carry out the construction in §9.1 to construct a secondary operation associated to the Adem relation $Sq^2 Sq^2 + Sq^3 Sq^1 = 0$. Use this to prove Adem's theorem that $\eta \circ \eta \not\simeq *$.

9.5. Prove that the functors defined on graded left Γ-modules satisfy: $\mathrm{Hom}_\Gamma(M, -)$ is left exact, and $\mathrm{Hom}_\Gamma(-, N)$ is right exact when M and N are fixed Γ-modules.

9.6. Let $\Gamma(x)$ denote the divided power Hopf algebra over $\mathbb{F}_2$ on a single generator x. Prove that $\Gamma(x)^{\mathrm{dual}}$, as a Hopf algebra, is isomorphic to $\mathbb{F}_2[y]$ where y is the dual of the generator $\gamma_1(x)$. Finish the proof, begun in §9.2, that

$$H^{*,*}(\Gamma(x)) \cong \mathbb{F}_2[l_i \mid i = 0, 1, \dots],$$

where the bidegree of l_i is $(1, 2^i \deg x)$.

9.7. On $\mathrm{Ext}_\Gamma(M, N)$.

(1) Verify that the definition of $\mathrm{Ext}_\Gamma(M, N)$ given in §9.2 does not depend on the choice of projective resolution of M.
(2) Verify that $\mathrm{Ext}_\Gamma^{0,*}(M, N) = \mathrm{Hom}_\Gamma^{0,*}(M, N)$.
(3) Verify that, if given a short exact sequence in $_\Gamma\mathbf{Mod}$,

$$0 \to A \to B \to C \to 0,$$

then there is a long exact sequence

$$\to \mathrm{Ext}_\Gamma^{n,*}(M, A) \to \mathrm{Ext}_\Gamma^{n,*}(M, B) \to$$

$$\mathrm{Ext}_\Gamma^{n,*}(M, C) \xrightarrow{\delta} \mathrm{Ext}_\Gamma^{n+1,*}(M, A) \to .$$

(4) Verify that, for a projective module over Γ, P,

$$\mathrm{Ext}_\Gamma^{n,*}(P, M) = \{0\},$$

for $n > 0$ and M any left Γ-module.
(5) Verify that $\mathrm{Ext}_\Gamma(M, N)$ is a functor in each variable separately.

9.8. Show that the composition product on Ext is bilinear and associative. Show further that Yoneda's product induced by splicing agrees with the composition product defined via resolutions.

9.9. In categories of modules, the dual of the notion of a projective module is that of an **injective module**. Give the definition by formally inverting the definition of a projective module (remember epi becomes mono). Prove that a module J is injective if and only if to each monomorphism, $i\colon A \to B$, the mapping $\mathrm{Hom}(B, J) \to \mathrm{Hom}(A, J)$ is an isomorphism. If given a module, N, then an **injective resolution** of N is an exact sequence

$$0 \to N \to J^0 \to J^1 \to J^2 \to \cdots$$

with each J^i injective. Show that any two injective resolutions of the module N can be compared by a lift of the identity mapping between the resolutions. Show that one can define $\mathrm{Ext}_\Gamma(M, N)$ by constructing an injective resolution of N and applying the functor $\mathrm{Hom}(M, -)$ to form a complex and then taking the homology.

9.10. Prove the assertion that $H^{*,*}(\Lambda(x)) \cong k[y]$ as algebras, where $\Lambda(x)$ denotes an exterior algebra over k on a single generator x and y has bidegree $(1, \deg x)$.

9.11. Prove the following facts about the suspension functor on graded Γ-modules and projective modules:

(1) If X is projective, then sX is projective. (Hint: $s(\Gamma \otimes V) \cong \Gamma \otimes sV$ for a graded vector space V.)
(2) $_\Gamma\mathbf{Mod}(sW, sX) \cong {}_\Gamma\mathbf{Mod}(W, X)$.
(3) If $0 \leftarrow X \leftarrow W_\bullet$ is a projective resolution of X, then $0 \leftarrow sX \leftarrow sW_\bullet$ is a projective resolution of sX.

9.12. Suppose that Γ is a cocommutative Hopf algebra. Show that the cobar construction $F_*(\Gamma^{\mathrm{dual}})$ supports a $\smile_1$-product defined by

$$[\alpha_1 \mid \alpha_2 \mid \cdots \mid \alpha_p] \smile_1 [\beta_1 \mid \beta_2 \mid \cdots \mid \beta_q] =$$
$$\sum\nolimits_{1 \le r \le p} [\alpha_1 | \alpha_2 | \cdots | \alpha_{r-1} | \alpha_r^{(1)} \beta_1 | \alpha_r^{(2)} \beta_2 | \cdots | \alpha_r^{(q)} \beta_q | \alpha_{r+1} | \cdots | \alpha_p],$$

where the elements $\alpha_r^{(j)}$ are determined by the iterated coproduct

$$\psi^{q-1}(\alpha_r) = \sum \alpha_r^{(1)} \otimes \alpha_r^{(2)} \otimes \cdots \otimes \alpha_r^{(q)}.$$

Show that this $\smile_1$-product satisfies a Hirsch formula:

$$d^*(x \smile_1 y) = d^*(x) \smile_1 y + x \smile_1 d^*(y) + xy + yx.$$

9.13. Prove the theorem of [Milnor-Moore65]: Given Λ a normal sub-Hopf algebra of a Hopf algebra Γ, then Γ has a basis as a Λ-module consisting of 1 and certain homogeneous elements in $I(\Gamma)$ and Γ is free as a right Λ-module on this basis. Furthermore, this basis projects to a vector space basis for $\Gamma//\Lambda$.

9.14. Prove the *Fact* from §9.5 that, for all s,

$$\mathrm{Ext}_{\mathcal{A}_2}^{s,s+13}(\mathbb{F}_2, \mathbb{F}_2) = \{0\}.$$

9.15. Prove the following relations in the cohomology of $\mathcal{A}_2$:

$$\langle h_i, h_{i+1}, h_i \rangle = h_{i+1}^2, \quad \langle h_{i+1}, h_i, h_{i+1} \rangle = h_{i+2}h_i \quad c_0 h_1 \neq 0.$$

9.16. For the periodicity operator of §9.6, prove that

$$P^r \circ P^r = P^{r+1}.$$

9.17. Consider the following Toda bracket construction suggested by the Adem relation $Sq^3 Sq^1 + Sq^2 Sq^2 = 0$:

$$X \xrightarrow{x} K(\mathbb{Z}/2\mathbb{Z}, n) \xrightarrow{Sq^1, Sq^2} K(\mathbb{Z}/2\mathbb{Z}, n+1) \times K(\mathbb{Z}/2\mathbb{Z}, n+2)$$
$$\xrightarrow{+\circ(Sq^3, Sq^2)} K(\mathbb{Z}/2\mathbb{Z}, n+4).$$

Such a Toda bracket is defined when $Sq^1 x = 0$ and $Sq^2 x = 0$. Show that the elements in the Toda bracket comprise a secondary cohomology operation based on the Adem relation. Thus Toda brackets may be used to express such operations.

9.18. Compute $\mathrm{Ext}_{\mathcal{A}_2}^{s,t}(\mathcal{A}_2/I(\mathcal{A}_2 \cdot Sq^1), \mathbb{F}_2)$.

9.19. Suppose that L is a graded Lie algebra over a field k. Let $(U(L), i_L \colon L \to U(L))$ denote the *universal enveloping algebra* of L, defined by the universal property that if $f \colon L \to A$ is any morphism of Lie algebras where A is an algebra endowed with the bracket product $[a, b] = ab - (-1)^{\deg a \deg b} ba$, then there is a unique morphism of algebras $\tilde{f} \colon U(L) \to A$ such that $\tilde{f} \circ i_L = f$. Show that $U(L)$ may be defined as the quotient of the tensor algebra on L by the ideal generated by elements of the form $x \otimes y - (-1)^{\deg x \deg y} y \otimes x - [x, y]$ for $x, y \in L$. The product of two graded Lie algebras is given by $(L \times L')_n = L_n \times L'_n$. Using these facts show that $U(L)$ is a graded Hopf algebra with the coproduct induced by the diagonal mapping.

9.20. Prove Theorem 9.59.

10

The Bockstein Spectral Sequence

"Unlike the previous proofs which made strong use of
the infinitesimal structure of Lie groups, the proof given
here depends only on the homological structure and can
be applied to H-spaces ... "

W. Browder

In the early days of combinatorial topology, a topological space of finite type (a polyhedron) had its integral homology determined by sequences of integers—the Betti numbers and torsion coefficients. That this numerical data ought to be interpreted algebraically is attributed to Emmy Noether (see [Alexandroff-Hopf35]).

The torsion coefficients are determined by the the Universal Coefficient theorem; there is a short exact sequence

$$0 \to H_n(X) \otimes \mathbb{Z}/r\mathbb{Z} \xrightarrow{\rho} H_n(X; \mathbb{Z}/r\mathbb{Z}) \to \mathrm{Tor}_{\mathbb{Z}}(H_{n-1}(X), \mathbb{Z}/r\mathbb{Z}) \to 0.$$

To unravel the integral homology from the mod r homology there is also the **Bockstein homomorphism**: Consider the short exact sequence of coefficient rings where red_r is reduction mod r:

$$0 \to \mathbb{Z} \xrightarrow{-\times r} \mathbb{Z} \xrightarrow{\mathrm{red}_r} \mathbb{Z}/r\mathbb{Z} \to 0.$$

The singular chain complex of a space X is a complex, $C_*(X)$, of free abelian groups. Hence we obtain another short exact sequence of chain complexes

$$0 \to C_*(X) \xrightarrow{-\times r} C_*(X) \xrightarrow{\mathrm{red}_r} C_*(X) \otimes \mathbb{Z}/r\mathbb{Z} \to 0,$$

and this gives a long exact sequence of homology groups,

$$\cdots \to H_n(X) \xrightarrow{-\times r} H_n(X) \xrightarrow{\mathrm{red}_{r*}} H_n(X; \mathbb{Z}/r\mathbb{Z}) \xrightarrow{\partial} H_{n-1}(X) \to \cdots .$$

When an element $u \in H_{n-1}(X)$ satisfies $ru = 0$, then, by exactness, there is an element $\bar{u} \in H_{n+1}(X; \mathbb{Z}/r\mathbb{Z})$ with $\partial(\bar{u}) = u$. To unpack what is happening here, we write $\bar{u} = \{c \otimes 1\} \in H_n(X; \mathbb{Z}/r\mathbb{Z})$. Since $\partial(c \otimes 1) = 0$ and $\partial(c) \neq 0$, we see that $\partial(c) = rv$ and the boundary homomorphism takes $\bar{u}$ to $\{v\} \in H_{n-1}(X)$. The Bockstein homomorphism is defined by

$$\beta \colon H_n(X; \mathbb{Z}/r\mathbb{Z}) \to H_{n-1}(X; \mathbb{Z}/r\mathbb{Z}), \quad \bar{u} = \{c \otimes 1\} \mapsto \{v \otimes 1\} = \{\tfrac{1}{r}\partial c \otimes 1\}.$$

This mapping was introduced by [Bockstein43]. The Bockstein spectral sequence is derived from the long exact sequence when we treat it as an exact couple (§10.1).

One of the motivating problems for the development of the Bockstein spectral sequence comes from the study of Lie groups. Recall that a space X is **torsion-free** when all its torsion coefficients vanish, that is, when $H_i(X)$ is a free abelian group for each i. A remarkable result due to [Bott54, 56] identifies a particular class of torsion-free spaces.

Theorem 10.1. *If (G, e, μ) denotes a connected, simply-connected, compact Lie group, then ΩG is torsion-free.*

Bott's proof of this theorem is a tour-de-force in the use of the analytic structure of a Lie group. The transition to topological consequences is via Morse theory. The essential ingredient is the study of the **diagram D associated to G**—a system of subspaces of the tangent space to a maximal torus $T \subset G$ which may be described in terms of "root forms" on G. The fundamental chambers in D carry indices that determine the Poincaré series of the based loop space ΩG. In fact, the Poincaré series has nonzero entries only in even degrees. From this condition for all coefficient fields, it follows that ΩG is torsion-free.

By way of contrast, we recall a celebrated result of [Hopf41]. H-spaces and Hopf algebras made their first appearance in this landmark paper where results about the algebraic topology of Lie groups were shown to depend only on the more fundamental notion of an H-space structure.

Suppose (X, x_0, μ) is an H-space. The commutativity of the diagram

$$
\begin{array}{ccccc}
X \times X & \xrightarrow{\Delta \times \Delta} & X \times X \times X \times X & \xrightarrow{1 \times T \times 1} & X \times X \times X \times X \\
\mu \downarrow & & & & \downarrow \mu \times \mu \\
X & & \xrightarrow{\hspace{4cm} \Delta \hspace{4cm}} & & X \times X.
\end{array}
$$

implies that the coproduct on homology,

$$\Delta_* \colon H_*(X; k) \to H_*(X; k) \otimes H_*(X; k),$$

is an algebra map with respect to the product μ_*. Thus $(H_*(X; k), \mu_*, \Delta_*)$ satisfies the defining property of a Hopf algebra. This algebraic observation implies the following structure result.

Theorem 10.2 ([Hopf41]). *If (X, x_0, μ) is an H-space of the homotopy type of a finite CW-complex and k is a field of characteristic zero, then $H^*(X; k)$ is an exterior algebra on generators of odd degree.*

PROOF: Consider the graded vector space of indecomposable elements in $H^*(X; k)$:

$$Q(H^*(X; k)) = H^+(X; k)/H^+(X; k) \smile H^+(X; k).$$

Let $Q(H^*(X; k)) = k\{x_1, x_2, \dots, x_q\}$ with the generators ordered by degree, $\deg x_1 \leq \deg x_2 \leq \cdots \leq \deg x_q$. Let $x = x_j$ denote first even-dimensional generator, of degree $2m$, and A_x denote the sub-Hopf algebra generated by the odd-dimensional classes x_1 through x_{j-1}.

Recall that if $C \subset B$ is a normal sub-Hopf algebra of C, that is, $I(C) \cdot B = B \cdot I(C)$, then $C//B = B/I(C) \cdot B$ is the quotient Hopf algebra and $I(C)$ and $I(B)$ denote the kernels of the augmentation.

Consider the short exact sequence of Hopf algebras:

$$0 \to A_x \to H^*(X; k) \to H^*(X; k)//A_x \to 0.$$

Since $H^*(X; k)$ is commutative, A_x is normal in $H^*(X; k)$. The class x goes to a primitive class $\bar{x}$ in $H^*(X; k)//A_x$, that is, $\mu^*(\bar{x}) = \bar{x} \otimes 1 + 1 \otimes \bar{x}$. Since $H^*(X; k)//A_x$ is also a Hopf algebra, we have that μ^* is a homomorphism of algebras and so $\mu^*((\bar{x})^n) = (\mu^*(\bar{x}))^n = (1 \otimes \bar{x} + \bar{x} \otimes 1)^n$. It follows, as in the proof of the binomial theorem, that, for all $n > 0$,

$$\mu^*((\bar{x})^n) = \sum_{i=0}^{n} \binom{n}{i} (\bar{x})^i \otimes (\bar{x})^{n-i} \qquad \text{where } (\bar{x})^0 = 1.$$

Since X has the homotopy type of a finite CW-complex, for some N, $H^s(X; k) = \{0\}$ for $s \geq N$. It follows that $(\bar{x})^i = 0$ whenever $2mi \geq N$. However, for the first such t,

$$\mu^*((\bar{x})^t) = \sum_{i=0}^{t} \binom{t}{i} (\bar{x})^i \otimes (\bar{x})^{t-i} \neq 0$$

because $\binom{t}{i} \neq 0$ in k and $(\bar{x})^i \otimes (\bar{x})^{t-i} \neq 0$ when $i \geq 1$. Thus, the appearance of $\bar{x} \neq 0$, a primitive of even degree in $H^*(X; k)//A_x$, implies that $(\bar{x})^t \neq 0$ for all $t \geq 1$, and $H^*(X; k)//A_x$ is of infinite dimension over k. Since $H^*(X; k)//A_x$ is a quotient of $H^*(X; k)$, this contradicts the finiteness assumption on X. It follows that $H^*(X; k)$ has only odd degree algebra generators. The theorem follows from Theorem 6.36—a graded commutative Hopf algebra on odd generators is an exterior algebra on those generators. □

The interplay between the homotopy-theoretic properties of H-spaces and the analytic properties of Lie groups has deepened our understanding of such spaces considerably. At first it was believed that H-spaces with nice enough properties need be Lie groups ([Curtis, M71] reviewed this program), but the powerful methods of localization at a prime soon revealed a much richer field of examples including the so-called "Hilton-Roitberg criminal" ([Hilton-Roitberg69]), a manifold and H-space of non-Lie type. The generalization of properties of Lie groups to H-spaces of the homotopy type of a finite complex fueled considerable efforts that include the development of the Bockstein spectral sequence ([Browder61]), the introduction of A_n-structures ([Stasheff63]), new applications of localization ([Zabrodsky70], [Hilton-Mislin-Roitberg75]), and the solution of the torsion conjecture ([Lin82], [Kane86]), which states that ΩX is torsion-free for X a finite, simply-connected H-space. [Dwyer-Wilkerson94] have applied the methods of homotopy fixed point sets developed by [Miller84] and [Lannes92] to recover the algebraic topology of Lie groups from a completely homotopy-theoretical viewpoint ([Dwyer98]).

In this chapter we develop the properties of the Bockstein spectral sequence, especially for applications to H-spaces. We introduce the remarkable notion of ∞-implications due to [Browder61] and apply it to derive certain finiteness results. We then consider some unexpected applications of the Bockstein spectral sequence to differential geometry and to the Adams spectral sequence. The short exact sequence of coefficients that characterizes the Bockstein spectral sequence can also be generalized to other homology theories and to homotopy groups with coefficients (introduced by [Peterson56]). This leads to other Bockstein spectral sequences—for mod r homotopy groups, and for Morava K-theory—whose properties have played a key role in some of the major developments in homotopy theory. These ideas are discussed in §10.2.

10.1 The Bockstein spectral sequence

Although it has a modest form, the Bockstein spectral sequence has led to some remarkable insights, particularly in the study of H-spaces. We recall the construction of the Bockstein spectral sequence here (§2.2). Fix a prime p and carry out the construction of the long exact sequence associated to the exact sequence of coefficients, $0 \to \mathbb{Z} \to \mathbb{Z} \to \mathbb{F}_p \to 0$. Following a suggestion of John Moore, [Browder61] interpreted the long exact sequence as an exact couple:

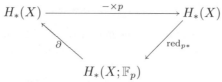

We denote the E^1-term by $B_n^1 \cong H_n(X; \mathbb{F}_p)$. The first differential is given by $d^1 = \partial \circ \mathrm{red}_{p*} = \beta$, the Bockstein homomorphism. The spectral sequence is singly-graded and the results of Chapter 2 apply to give the following theorem.

Theorem 10.3. *Let X be a connected space of finite type. Then there is a singly-graded spectral sequence $\{B^r_*, d^r\}$, natural with respect to spaces and continuous mappings, with $B^1_n \cong H_n(X; \mathbb{F}_p)$, $d^1 = \beta$, the Bockstein homomorphism, and converging strongly to $(H_*(X)/\text{torsion}) \otimes \mathbb{F}_p$.*

PROOF: Suppose G is a finitely generated abelian group. Then we can write

$$G \cong \bigoplus_i \mathbb{Z} \oplus \bigoplus_j \mathbb{Z}/p^{e_j}\mathbb{Z} \oplus \bigoplus_t \mathbb{Z}/q_t^{r_t}\mathbb{Z},$$

where the q_t are primes not equal to p. The *times p* homomorphism is an isomorphism on $\bigoplus_t \mathbb{Z}/q_t^{r_t}\mathbb{Z}$ and a monomorphism on $\bigoplus_i \mathbb{Z}$. Recall the *p-component of G* is the quotient group

$$_{(p)}G = G/\{\text{elements of torsion order prime to } p\} \cong \bigoplus_i \mathbb{Z} \oplus \bigoplus_j \mathbb{Z}/p^{e_j}\mathbb{Z}.$$

An nonzero element u in G is *p-divisible* if $u = pv$ for some v in G. The elements in $\bigoplus_t \mathbb{Z}/q_t^{r_t}\mathbb{Z}$ are *infinitely p-divisible* since $- \times p$ is an isomorphism on this summand. No elements in the rest of G can be infinitely p-divisible without violating the condition that G is finitely generated. With these observations we prove the convergence assertion of the theorem.

By Corollary 2.10 we have the short exact sequence

$$0 \to H_n(X)\big/(pH_n(X) + \ker p^r) \to B^{r+1}_n \to p^r H_{n-1}(X) \cap \ker p \to 0.$$

Notice that $B^{r+1}_n = \{0\}$ implies $H_n(X) = pH_n(X) + \ker p^r$. If $u \in H_n(X)$ generates a copy of $\mathbb{Z}$, the $u \notin \ker p^r$. But if $u \in pH_n(X)$, then u is p-divisible. Writing $u = pv_1$, it follows that v_1 is also p-divisible. Continuing in this manner, we conclude that u is infinitely p-divisible, a contradiction to finite generation. It follows that $_{(p)}H_n(X) = \ker p^r$ and so $_{(p)}H_n(X)$ has exponent less than or equal to p^r.

Let r go to infinity to obtain the short exact sequence

$$0 \to H_n(X)\big/(pH_n(X) + p\text{-torsion}) \to B^\infty_n \to \nabla^{\infty,p}_{n-1} \to 0,$$

where $\nabla^{\infty,p}_{n-1}$ is the subgroup of $H_{n-1}(X)$ of infinitely p-divisible elements that vanish when multiplied by p. Because $H_{n-1}(X)$ is finitely generated, $\nabla^{\infty,p}_{n-1}$ is trivial and so

$$B^\infty_n \cong H_n(X)\big/(pH_n(X) + p\text{-torsion}) \cong (H_n(X)/\text{torsion}) \otimes \mathbb{F}_p \qquad \square$$

Some immediate consequences of the existence and convergence of the Bockstein spectral sequence are the following inequalities. Suppose that X is a space of finite type. Then, in each dimension i, we have

$$\dim_{\mathbb{F}_p} H_i(X; \mathbb{F}_p) \geq \text{free rank } H_i(X)$$
$$= \dim_{\mathbb{Q}} H_i(X; \mathbb{Q})$$
$$= \dim_{\mathbb{F}_p}((H_i(X)/\text{torsion}) \otimes \mathbb{F}_p).$$

This follows from the Universal Coefficient theorem and the fact that $H_i(X)$ is finitely generated. Thus the Bockstein spectral sequence for X collapses at B^r if and only if $\dim_{\mathbb{F}_p} B_i^r(X) = \dim_{\mathbb{Q}} H_i(X; \mathbb{Q})$ for all i.

There is an alternate description of the differential that identifies the Bockstein homomorphism directly. Consider the short exact sequence of coefficients

$$0 \to \mathbb{Z}/p\mathbb{Z} \to \mathbb{Z}/p^2\mathbb{Z} \to \mathbb{Z}/p\mathbb{Z} \to 0$$

where we have written $\mathbb{Z}/p\mathbb{Z} \cong p\mathbb{Z}/p^2\mathbb{Z}$ as the kernel. The associated long exact sequence on homology for a space X is given by

$$\cdots \to H_n(X; \mathbb{Z}/p\mathbb{Z}) \xrightarrow{-\times p} H_n(X; \mathbb{Z}/p^2\mathbb{Z})$$
$$\to H_n(X; \mathbb{Z}/p\mathbb{Z}) \xrightarrow{\beta} H_{n-1}(X; \mathbb{Z}/p\mathbb{Z}) \to \cdots$$

and has $d^1 = \beta$, the connecting homomorphism. This can be seen by comparing the short exact sequences of coefficients

$$
\begin{array}{ccccccccc}
0 & \longrightarrow & \mathbb{Z} & \xrightarrow{-\times p} & \mathbb{Z} & \longrightarrow & \mathbb{Z}/p\mathbb{Z} & \longrightarrow & 0 \\
& & \downarrow{\scriptstyle \text{red}_p} & & \downarrow{\scriptstyle \text{red}_{p^2}} & & \| & & \\
0 & \longrightarrow & \mathbb{Z}/p\mathbb{Z} & \xrightarrow{-\times p} & \mathbb{Z}/p^2\mathbb{Z} & \longrightarrow & \mathbb{Z}/p\mathbb{Z} & \longrightarrow & 0.
\end{array}
$$

The associated homomorphism of long exact sequences carries β to $\text{red}_p \circ \partial$.

When we consider the short exact sequence of coefficients

$$0 \to \mathbb{Z}/p^r\mathbb{Z} \to \mathbb{Z}/p^{2r}\mathbb{Z} \to \mathbb{Z}/p^r\mathbb{Z} \to 0,$$

we obtain the rth **order Bockstein operator** as connecting homomorphism. Taking all of the short exact sequences of coefficients for all $r \geq 1$, the following more refined picture of the Bockstein spectral sequence emerges.

Proposition 10.4. B_n^r can be identified with the subgroup of $H_n(X; \mathbb{Z}/p^r\mathbb{Z})$ given by the image of $H_n(X; \mathbb{Z}/p^r\mathbb{Z}) \xrightarrow{-\times p^{r-1}} H_n(X; \mathbb{Z}/p^r\mathbb{Z})$ and $d^r : B_n^r \to B_{n-1}^r$ can be identified with the connecting homomorphism, the rth order Bockstein homomorphism.

PROOF: Write $G_n^r = \text{im}(-\times p^{r-1} : H_n(X; \mathbb{Z}/p^r\mathbb{Z}) \to H_n(X; \mathbb{Z}/p^r\mathbb{Z}))$ and consider the sequence of homomorphisms

$$p^{r-1}H_n(X) \xrightarrow{-\times p} p^{r-1}H_n(X) \xrightarrow{\alpha} G_n^r \xrightarrow{\zeta} p^{r-1}H_{n-1}(X) \to p^{r-1}H_{n-1}(X)$$

defined by $\alpha\left(\left\{\sum_i p^{r-1}u_i\right\}\right) = \left\{\sum_i u_i \otimes (p^{r-1}+p^r\mathbb{Z})\right\} \in G_n^r$. This homomorphism is well-defined and has $\text{im}(-\times p)$ as its kernel. If a homology class $\left\{\sum_i v_i \otimes (p^{r-1}+p^r\mathbb{Z})\right\} \in H_n(X; \mathbb{Z}/p^r\mathbb{Z})$ is in G_n^r, then define

$$\zeta\left(\left\{\sum_i v_i \otimes (p^{r-1}+p^r\mathbb{Z})\right\}\right) = \left\{\frac{1}{p}\sum_i \partial(v_i)\right\},$$

where ∂ is the chain boundary operator. Since $\partial\left(\sum v_i \otimes (p^{r-1}+p^r\mathbb{Z})\right) = 0$, it follows that $\sum \partial v_i = p\left(p^{r-1}\sum_j x_j\right)$ and so dividing by p determines a class in $p^{r-1}H_{n-1}(X)$. It is easy to see that $\ker \zeta = \text{im }\alpha$ and we have exactness at G_n^r. We compare this sequence with the rth derived couple

$$\to p^{r-1}H_n(X) \xrightarrow{p} p^{r-1}H_n(X) \to G_n^r \to p^{r-1}H_{n-1}(X) \xrightarrow{p} p^{r-1}H_{n-1}(X) \to$$

$$\to p^{r-1}H_n(X) \underset{\overline{p}}{\to} p^{r-1}H_n(X) \to B_n^r \to p^{r-1}H_{n-1}(X) \underset{\overline{p}}{\to} p^{r-1}H_{n-1}(X) \to.$$

The Five-lemma implies $B_n^r \cong G_n^r$.

To identify the differential d^r with the higher Bockstein

$$\beta_r : H_n(X; \mathbb{Z}/p^r\mathbb{Z}) \to H_{n-1}(X; \mathbb{Z}/p^r\mathbb{Z})$$

it suffices to compare the connecting homomorphism that defines β_r with the definition of the homomorphism ζ. □

This representation of the terms in the Bockstein spectral sequence can be completed by embedding the data for all $r \geq 1$ into a Cartan-Eilenberg system, a general technique to construct a spectral sequence (also known as a **spectral system** in [Neisendorfer80] or a **coherent system of coalgebra/algebras/Lie algebras** in [Anick93]). The definition and relation between a Cartan-Eilenberg

system and its associated spectral sequence are explored in Exercises 2.2 and
2.3. For a prime p and a pair (s, t) with $-\infty < s \le t < \infty$ we define

$$H(s, t) = H_*(X; \mathbb{Z}/p^{t-s}\mathbb{Z}).$$

If $s \le s'$ and $t \le t'$, let $H(s, t) \to H(s', t')$ be the homomorphism induced
by the map of coefficients, $\mathbb{Z}/p^{t-s}\mathbb{Z} \to \mathbb{Z}/p^{t'-s'}\mathbb{Z}$, that is determined by
$1 \mapsto p^{t'-t} \colon H_*(X; \mathbb{Z}/p^{t-s}\mathbb{Z}) \to H_*(X; \mathbb{Z}/p^{t'-s'}\mathbb{Z})$. If $r \le s \le t$, then
let $\partial \colon H(r, s) \to H(s, t)$ be the connecting homomorphism associated to the
coefficient sequence

$$0 \to \mathbb{Z}/p^{t-s}\mathbb{Z} \to \mathbb{Z}/p^{t-r}\mathbb{Z} \to \mathbb{Z}/p^{s-r}\mathbb{Z} \to 0,$$

a homomorphism $H_*(X; \mathbb{Z}/p^{s-r}\mathbb{Z}) \to H_{*-1}(X; \mathbb{Z}/p^{t-s}\mathbb{Z})$. In this context
the limit terms of the Cartan-Eilenberg system are given by $H(q) = H(q, q) = H_q(X)$ and $H(q, \infty) = H_q(X; \lim_{r \to \infty} \mathbb{Z}/p^r\mathbb{Z})$. The exact couple determined
by the long exact sequence

$$\cdots \to H(q-1) \to H(q) \to H(q-1, q) \xrightarrow{\partial} H(q-1) \to H(q) \to \cdots$$

gives the Bockstein spectral sequence.

   With this added structure the (co)multiplicative properties of the spectral
sequence may be studied. We refer the reader to the work of [Neisendorfer80]
and [Anick93] for more details.

   Though we developed the Bockstein spectral sequence for homology, it is
just as easy to make the same constructions and observations for cohomology.
The Bockstein homomorphism for cohomology has degree 1,

$$\beta \colon H^n(X; \mathbb{F}_p) \to H^{n+1}(X; \mathbb{F}_p),$$

and is identified with the stable cohomology operation β in the Steenrod algebra
$\mathcal{A}_p$, when p is odd, and Sq^1 in $\mathcal{A}_2$, when $p = 2$. This leads to a spectral sequence
of algebras since β is a derivation with respect to the cup product.

When X is an H-space

   The naturality of the Bockstein spectral sequence applies to the diagonal
mapping to give a morphism of spectral sequences $B_*^r(X) \to B_*^r(X \times X)$.
When (X, x_0, μ) is an H-space, the multiplication mapping induces $B_*^r(\mu) \colon$
$B_*^r(X \times X) \to B_*^r(X)$. Our goal in this section is to identify $B_*^r(X \times X)$
with $B_*^r(X) \otimes B_*^r(X)$ and so obtain a spectral sequence of coalgebras for the
homology Bockstein spectral sequence. Dually, we obtain a spectral sequence
of algebras for the cohomology Bockstein spectral sequence; and for H-spaces,
a spectral sequence of Hopf algebras.

Following [Browder61], we introduce small models of chain complexes whose structure makes explicit the key features of the Bockstein spectral sequence. Suppose n and s are nonnegative integers. Define the chain complex $(A(n, s), d)$, free over $\mathbb{Z}$, where

$$A(n, s)_m = \begin{cases} \{0\}, & m \neq n, n+1, \\ \mathbb{Z} \cong \langle u \rangle, & m = n, \\ \mathbb{Z} \cong \langle v \rangle, & m = n+1 \, (= \{0\} \text{ if } s = 0). \end{cases}$$

The differential is given on generators by $d(v) = su$, and so $H_n(A(n, s), d) \cong \mathbb{Z}/s\mathbb{Z}$ and $H_r(A(n, s), d) = \{0\}$ for $r \neq n$. This chain complex can be realized cellularly by the **mod s Moore space** $P^{n+1}(s) = S^n \cup_s e^{n+1}$ where s here denotes the degree s map on S^n. The reduced integral homology of $P^{n+1}(s)$ is $H_*(A(n, s), d)$.

The *times p* map, denoted $- \times p$, on $A(n, s)$ fits into the short exact sequence

$$0 \to (A(n, s), d) \xrightarrow{- \times p} (A(n, s), d) \xrightarrow{\mathrm{red}_p} (A(n, s) \otimes \mathbb{F}_p, \bar{d}) \to 0,$$

where red_p denotes reduction mod p. The long exact sequence in homology is the Bockstein exact couple. We consider the Bockstein spectral sequence associated to this exact couple.

Proposition 10.5. *If $\gcd(s, p) = 1$, then $H_*(A(n, s) \otimes \mathbb{F}_p, \bar{d}) = \{0\}$. If $s = 0$, then $B^1 \cong B^\infty \cong \mathbb{Z}/p\mathbb{Z}$ in degree n. If $s = ap^k$ with $k > 0$ and $\gcd(a, p) = 1$, then $B^1 \cong B^2 \cong \cdots \cong B^k$ and $B^{k+1} \cong B^\infty = \{0\}$.*

PROOF: The first assertion follows from the Universal Coefficient theorem and the fact that $\mathbb{Z}/s\mathbb{Z} \otimes \mathbb{F}_p = \{0\}$. When $s = 0$, $A(n, 0) \otimes \mathbb{F}_p$ is simply $\mathbb{F}_p$ concentrated in degree n and the spectral sequence collapses.

By the fundamental theorem for finitely generated abelian groups, we can split $\mathbb{Z}/ap^k\mathbb{Z}$ as $\mathbb{Z}/a\mathbb{Z} \oplus \mathbb{Z}/p^k\mathbb{Z}$. Since the contribution by $\mathbb{Z}/a\mathbb{Z}$ vanishes, we only need to consider the case $s = p^k$ with $k > 0$. Since $A(n, p^k) \otimes \mathbb{F}_p \cong A(n, p^k)/pA(n, p^k)$, we have that $\bar{d} = 0$ and so

$$H_r(A(n, p^k) \otimes \mathbb{F}_p, \bar{d}) \cong \begin{cases} \mathbb{F}_p, & \text{when } r = n \text{ or } n+1, \\ \{0\}, & \text{otherwise.} \end{cases}$$

We write $(u)_p$ and $(v)_p$ for the mod p reductions of u and v. The mapping $\partial \colon H_{n+1}(A(n, p^k) \otimes \mathbb{F}_p, \bar{d}) \to H_n(A(n, p^k), d)$ in the exact couple is given by $\partial((v)_p) = p^{k-1}u$ for reasons of exactness. We can peel away powers of p from $p^{k-1}u$ until it becomes the generator of $p^{k-1}(\mathbb{Z}/p^k\mathbb{Z}) \cong \mathbb{Z}/p\mathbb{Z}$, and so $d^1 = d^2 = \cdots = d^{k-1} = 0$. At B^k we have

$$\begin{array}{ccc} B^k_{n+1} \cong \langle (v)_p \rangle & \xrightarrow{\partial} p^{k-1}(\mathbb{Z}/p^k\mathbb{Z}) \to & B^k_n \cong \langle (u)_p \rangle \\ (v)_p \mapsto & p^{k-1}u & \mapsto (u)_p. \end{array}$$

Thus $B^{k+1} \cong B^\infty = \{0\}$.   □

In fact, more can be deduced from the small complexes.

Lemma 10.6. *If* $s = ap^k$ *with* $k > 0$ *and* $\gcd(a, p) = 1$, *then there is an isomorphism of exact couples* $(q, \bar{q})$:

$$
\begin{array}{ccccccc}
H(A(n, ap^k), d) & \xrightarrow{-\times p} & H(A(n, ap^k), d) & \xrightarrow{\mathrm{red}_{p*}} & H(A(n, ap^k) \otimes \mathbb{F}_p, \bar{d}) & \xrightarrow{\partial} & \\
\downarrow{\scriptstyle q} & & \downarrow{\scriptstyle q} & & \downarrow{\scriptstyle \bar{q}} & & \\
pH(A(n, ap^{k+1}), d) & \xrightarrow{-\times p} & pH(A(n, ap^{k+1}), d) & \xrightarrow{\mathrm{red}_{p*}} & B^2(A(n, ap^{k+1})) & \xrightarrow{\partial'} &
\end{array}
$$

PROOF: Write $A = A(n, ap^{k+1})$ with generators u and v and $A' = A(n, ap^k)$ with generators U and V. Consider the mapping $q \colon A' \to A$ and its reduction $\bar{q} \colon A' \otimes \mathbb{F}_p \to A \otimes \mathbb{F}_p$ given by

$$
\begin{aligned}
q(U) &= pu, & \bar{q}((U)_p) &= (u)_p, \\
q(V) &= v, & \bar{q}((V)_p) &= (v)_p.
\end{aligned}
$$

By the linearity of the differentials, q is a chain map. By the definition of q, $q_* H(A(n, ap^k)) = pH(A(n, ap^{k+1}))$. If $k > 0$, then $\bar{q}_*$ is an isomorphism at $B^1(A) \cong B^2(A)$.

It is left to show that the mapping pair $(q_*, \bar{q}_*)$ is a morphism of exact couples. Since q is a chain map, it commutes with $- \times p$. The class $\{U\}$ generates $H_n(A')$. The mapping j on $H(A')$ is given by $\{U\} \mapsto (U)_p$, the reduction mod p of $\{U\}$. Therefore, $\bar{q}_* \circ j(\{U\}) = (u)_p$. By the definition of a derived couple and the fact that $j(\{u\}) = (u)_p$, we have $j' \circ q_*(\{U\}) = j'(p\{u\}) = j(\{u\})$. Thus $j' \circ q_* = \bar{q}_* \circ j$.

For dimensional reasons, $\partial((U)_p) = 0 = \partial'((u)_p)$. For $k > 0$, $(V)_p \neq 0$ and, by exactness, $\partial((V)_p) = \{ap^{k-1}U\}$ and $\partial'((v)_p) = \{ap^k u\}$. Since $q_*(\{U\}) = \{pu\}$, we have that $q_* \circ \partial = \partial' \circ q_*$ and so $(q_*, \bar{q}_*)$ is a morphism of exact couples.   □

With this lemma, we prove a structure result.

Proposition 10.7. *Consider the Bockstein spectral sequence for* $C_1 \otimes C_2$ *where* $C_1 = (A(n, ap^k), d)$ *and* $C_2 = (A(m, bp^l), d)$, $k \geq l > 0$ *and* $\gcd(a, p) = 1 = \gcd(b, p)$. *Then* $B^2(C_1 \otimes C_2)$ *may be taken to be* $B^2(C_1) \otimes B^2(C_2)$.

PROOF: By Lemma 10.6 we can take $B^2(C_1) = H(A(n, ap^{k-1}) \otimes \mathbb{F}_p, \bar{d})$ and $B^2(C_2) = H(A(m, bp^{l-1}) \otimes \mathbb{F}_p, \bar{d})$. We write $B^2(C_i) = C_i'$; denote the generators of C_i by u_i, v_i, and the generators of C_i' by u_i', v_i' for $i = 1, 2$.

Assume that $k \geq l$ and let

$$\gamma = \mathrm{lcm}(a, b) = ag = bh,$$
$$\delta = \gcd(ap^{k-l}, b) = Nap^{k-l} + Mb,$$
$$x = g(v_1 \otimes u_2) - (-1)^{\deg u_1} hp^{k-l}(u_1 \otimes v_2),$$
$$y = N(v_1 \otimes u_2) + (-1)^{\deg u_1} M(u_1 \otimes v_2).$$

It follows that $\{x, y\}$ is a basis for $(C_1 \otimes C_2)_{n+m+1}$. Putting primes on x, y, u_i and v_i, we get a basis $\{x', y'\}$ for $(C_1' \otimes C_2')_{n+m+1}$. By the definitions, $dx = 0 = dx'$, $dy = \delta p^l(u_1 \otimes u_2)$, and $dy' = \delta p^{l-1}(u_1' \otimes u_2')$. Define the morphism of exact couples by letting $q \colon C_1' \otimes C_2' \to C_1 \otimes C_2$ be given by

$$q(u_1' \otimes u_2') = p(u_1 \otimes u_2), \qquad q(x') = px,$$
$$q(y') = y, \qquad q(v_1' \otimes v_2') = v_1 \otimes v_2.$$

Then q is a chain map and $q_* H(C_1' \otimes C_2') = pH(C_1 \otimes C_2)$. On the reductions mod p, define the map $\bar{q}_i \colon C_i' \otimes \mathbb{F}_p \to C_i \otimes \mathbb{F}_p$ by $\bar{q}_i((u_i')_p) = (u_i)_p$ and $\bar{q}_i((v_i')_p) = (v_i)_p$. Let $\bar{q} = \bar{q}_1 \otimes \bar{q}_2$. Then

$$\bar{q}_* \colon H(C_1' \otimes C_2' \otimes \mathbb{F}_p) \xrightarrow{\cong} H(C_1 \otimes C_2 \otimes \mathbb{F}_p) \cong B^2(C_1 \otimes C_2).$$

The morphism $(q_*, \bar{q}_*)$ is a morphism of exact couples and, as in the proof of Lemma 10.6, an isomorphism. $\qquad\square$

We put the small models to work after we state two results of [Browder61] that follow from the properties of free and torsion-free chain complexes. We leave the proofs to the reader.

Proposition 10.8. *Let (A, d) be a chain complex, free over $\mathbb{Z}$; let (A', d') be a torsion-free chain complex, and p, a prime. If $(\phi, \bar{\phi})$ is a morphism of the associated Bockstein exact couples,*

$$
\begin{array}{ccccccc}
\longrightarrow H_n(A) & \xrightarrow{-\times p} & H_n(A) & \xrightarrow{\mathrm{red}_{p*}} & H_n(A \otimes \mathbb{F}_p) & \xrightarrow{\partial} & H_{n-1}(A) \longrightarrow \\
\phi \downarrow & & \phi \downarrow & & \bar{\phi} \downarrow & & \phi \downarrow \\
\longrightarrow H_n(A') & \xrightarrow[-\times p]{} & H_n(A') & \xrightarrow[\mathrm{red}_{p*}]{} & H_n(A' \otimes \mathbb{F}_p) & \xrightarrow[\partial']{} & H_{n-1}(A') \longrightarrow.
\end{array}
$$

Then there is a chain map $f \colon (A, d) \to (A', d')$ such that $H(f) = \phi$ and $H(f \otimes \mathbb{F}_p) = \bar{\phi}$.

Lemma 10.9. *Let (A, d) and (A', d') be torsion-free chain complexes. Then, for all r, $B^r(A \oplus A', d + d') \cong B^r(A, d) \oplus B^r(A', d')$.*

Assume that (A', d') is a torsion-free chain complex whose homology groups are finitely generated in each dimension. Using Proposition 10.8 and Lemma 10.9 we can replace (A', d') with another complex (A, d) which is free and of the form $\bigoplus_i (A_i, d_i)$ with each (A_i, d_i) of the form $(A(n_i, a_i p^{k_i}), d)$. By Lemma 10.9, $B^r(A', d') \cong \bigoplus_i B^r(A_i, d_i)$.

Suppose X is a space of finite type. The homology Bockstein spectral sequence for X is the Bockstein spectral sequence for $(C_*(X), \partial)$ and this spectral sequence is functorial in X. The diagonal mapping on X gives a morphism of spectral sequences

$$B^r(\Delta) \colon B^r(X) \to B^r(X \times X).$$

Replacing the chains on X with a direct sum of small models, we can apply Proposition 10.7 to the Alexander-Whitney map to prove the following result.

Theorem 10.10. *For (X, x_0) a pointed space of finite type, the homology Bockstein spectral sequence is a spectral sequence of coalgebras.*

When X is an H-space of finite type, the same argument applied to the multiplication, along with the compatibility of the multiplication with the diagonal, gives the following key result.

Theorem 10.11. *For X, an H-space of finite type, the homology Bockstein spectral sequence for X is a spectral sequence of Hopf algebras.*

The cohomology Bockstein spectral sequence admits a dual analysis using the small complexes $\operatorname{Hom}(A(n, ap^k), \mathbb{Z})$. In fact, $\operatorname{Hom}(A(n, ap^k), \mathbb{Z})$ is simply $A(n, s)$ with the differential upside down. Its single nontrivial homology group is $H^{n+1}(A(n, ap^k), d^{\text{dual}})$. Using these complexes and carrying out the same kinds of arguments as for the homology Bockstein spectral sequence we obtain the theorem:

Theorem 10.12. *For (X, x_0) a pointed space of finite type, the cohomology Bockstein spectral sequence is a spectral sequence of algebras. Suppose (A_*, d) is a chain complex with homology of finite type. Let $\{B_r^* = B_r^*(\operatorname{Hom}(A_*, \mathbb{Z})), d_r\}$ denote the cohomology spectral sequence for the dual of (A_*, d). Then $B_r^* \cong \operatorname{Hom}(B_*^r(A_*), \mathbb{F}_p)$ and d_r is the differential adjoint to d^r. If X is an H-space of finite type, then the cohomology Bockstein spectral sequence for X, $B_r^*(X) = B_r^*(C^*(X), \delta)$, is a spectral sequence of Hopf algebras dual to the homology Hopf algebras $B_*^r(X)$.*

Having established these structural results, we turn to some examples. The universal examples for cohomology are the Eilenberg-Mac Lane spaces for which we have complete descriptions of the mod p cohomology according to the theorems of Cartan and Serre (Theorem 6.16). We reinterpret these known data to give a complete description of the Bockstein spectral sequence in a range of dimensions.

We note that the limit of the Bockstein spectral sequences for $K(\mathbb{Z}/p^k\mathbb{Z}, n)$ has $B^\infty \cong \{0\}$. To see this, suppose $\tilde{H}_*(K(\mathbb{Z}/p^k\mathbb{Z}, n))$ contained a torsion-free summand. Then $\tilde{H}_*(K(\mathbb{Z}/p^k\mathbb{Z}, n); \mathbb{Q})$ would have a nonzero lowest degree generator. By the Hurewicz-Serre theorem over $\mathbb{Q}$ (Theorem 6.25), this would imply a torsion-free summand in the homotopy of $K(\mathbb{Z}/p^k\mathbb{Z}, n)$ which does not happen. Hence $B^\infty \cong \{0\}$.

Suppose p is an odd prime. The cohomology of $K(\mathbb{Z}/p\mathbb{Z}, n)$ with coefficients in the field $\mathbb{F}_p$ is a free graded commutative algebra (exterior on odd-dimensional classes, tensor polynomial on even-dimensional classes) generated by classes $St^I \iota_n$ where $I = (\varepsilon_0, s_1, \varepsilon_1, \dots, s_m, \varepsilon_m)$ is an admissible sequence ($\varepsilon_i = 0$ or 1, $s_i \geq ps_{i+1} + \varepsilon_i$, for $m > i \geq 1$; Definition 6.17) of excess less than or equal to n. Notice that the excess, $e(I) = 2ps_1 + 2\varepsilon_0 - |I|$, is such that, if $I = (1, s_1, \varepsilon_1, \dots, s_m, \varepsilon_m)$ and $e(I) \leq n$, then $e(I') \leq n$ for $I' = (0, s_1, \varepsilon_1, \dots, s_m, \varepsilon_m)$. Thus, the generators pair off. Since this pairing is given by $\beta St^{I'} \iota_n = St^I \iota_n$ and $d_1 = \beta$, we are looking at two sorts of differential graded algebras:

$$\Lambda(St^{I'} \iota_n) \otimes \mathbb{F}_p[St^I \iota_n], \quad d_1(St^{I'} \iota_n) = St^I \iota_n, \quad \deg St^{I'} \iota_n \text{ odd},$$

$$\mathbb{F}_p[St^{I'} \iota_n] \otimes \Lambda(St^I \iota_n), \quad d_1(St^{I'} \iota_n) = St^I \iota_n, \quad \deg St^{I'} \iota_n \text{ even}.$$

When $St^{I'} \iota_n$ has odd degree, the complex $\Lambda(St^{I'} \iota_n) \otimes \mathbb{F}_p[St^I \iota_n]$ has the same form as the Koszul complex for $\Lambda(x_{\text{odd}})$ and so its homology is trivial. When $St^{I'} \iota_n$ has even degree, the complex has homology $H(\mathbb{F}_p[St^{I'} \iota_n] \otimes \Lambda(St^I \iota_n), d_1)$ $\cong \Lambda(\{(St^{I'} \iota_n)^{p-1} \otimes St^I \iota_n\}) \otimes \mathbb{F}_p[\{(St^{I'} \iota_n)^p\}]$, where $\{U\}$ denotes the homology class of U with respect to the the differential d_1. This follows because d_1 is a derivation and so $d_1((St^{I'} \iota_n)^p) = p(St^{I'} \iota_n)^{p-1} = 0$. Notice how the class $\{(St^{I'} \iota_n)^{p-1} \otimes St^I \iota_n\}$ encodes the transpotence element that figures in Cartan's constructions and Kudo's transgression theorem (§6.2).

Suppose $n = 2m$. Recall that $P^m \iota_{2m} = (\iota_{2m})^p$. In dimensions less than $2mp = \deg \iota_{2m}^p$, we find classes coming from the paired algebras:

$$(\mathbb{F}_p[\iota_{2m}] \otimes \Lambda(\beta\iota_{2m})) \otimes (\mathbb{F}_p[P^1\iota_{2m}] \otimes \Lambda(\beta P^1\iota_{2m})) \otimes \cdots$$
$$\otimes (\mathbb{F}_p[P^{m-1}\iota_{2m}] \otimes \Lambda(\beta P^{m-1}\iota_{2m})).$$

Computing the homology of this product as a differential graded algebra with differential β, we are left with the first nonzero classes, $\{\iota_{2m}^{p-1} \otimes \beta\iota_{2m}\} \in$

B_2^{2mp-1} and $\{\iota_{2m}^p\} \in B_2^{2mp}$. The next indecomposable class in B_2^* corresponds to $\{(P^1\iota_{2m})^p\} \in B_2^{p(2m+2(p-1))}$. Thus, for $q < p(2m + 2(p-1))$,

$$B_2^q(K(\mathbb{Z}/p\mathbb{Z}, 2m)) \cong (\Lambda(\{\beta\iota_{2m} \smile (\iota_{2m})^{p-1}\}) \otimes \mathbb{F}_p[\{\iota_{2m}^p\}])^q.$$

The case of $K(\mathbb{Z}/p^k\mathbb{Z}, n)$ for $k > 1$ yields to a similar analysis of admissible sequences except in the lowest degrees. Here the contributing classes are ι_n and $\beta_k\iota_n$, the Bockstein of kth order associated to the short exact sequence of coefficients $0 \rightarrow \mathbb{Z}/p\mathbb{Z} \rightarrow \mathbb{Z}/p^k\mathbb{Z} \rightarrow \mathbb{Z}/p^{k-1}\mathbb{Z} \rightarrow 0$. In dimensions $q < p(n + 2(p-1))$ we have that $B_l^q(K(\mathbb{Z}/p^k\mathbb{Z}, n)) \cong$

$$\begin{cases} B_1^q, & \text{if } l \leq k, \\ \{0\}, & \text{if } l > k \text{ and } n \text{ is odd}, \\ (\Lambda(\{\beta_k\iota_n \smile (\iota_n)^{p-1}\}) \otimes \mathbb{F}_p[\{\iota_n^p\}])^q, & \text{if } l = k+1 \text{ and } n \text{ is even}. \end{cases}$$

We complete the analysis for the lower dimensions of the Bockstein spectral sequence when n is even. The input is part of the computation of [Cartan54] of the integral cohomology of the Eilenberg-Mac Lane spaces.

Proposition 10.13. *If p is any prime and $k \geq 1$, then $H_{2mp}(K(\mathbb{Z}/p^k\mathbb{Z}, 2m))$ contains a subgroup isomorphic to $\mathbb{Z}/p^{k+1}\mathbb{Z}$ as summand. Furthermore, there are no summands isomorphic to $\mathbb{Z}$ or $\mathbb{Z}/p^{k+j}\mathbb{Z}$ with $j > 1$.*

Corollary 10.14. *Suppose that p is an odd prime. Let ι_{2m} denote the fundamental class in $B_1^{2m}(K(\mathbb{Z}/p^k\mathbb{Z}, 2m))$. Then, for some $c \in \mathbb{F}_p$,*

$$d_{k+1}(\{\iota_{2m}^p\}) = c\{\beta_k\iota_{2m} \smile (\iota_{2m})^{p-1}\} \neq 0.$$

The proof of Proposition 10.13 is a direct computation using the method of constructions ([Cartan54]). This method applies integrally and so one can compute the desired homology group by hand and discover the p-torsion height.

The corollary follows from the convergence of the Bockstein spectral sequence. Since there are no other classes in the degree involved, the formula for $d_{k+1}(\{\iota_{2m}^p\})$ follows without choice. [Browder62, Theorem 5.11] gave a more general chain level computation that obtains the formula directly.

For the prime 2, a new phenomenon occurs in the Bockstein spectral sequence for $K(\mathbb{Z}/2\mathbb{Z}, n)$. [Serre53] showed that $H^*(K(\mathbb{Z}/2\mathbb{Z}, n); \mathbb{F}_2)$ is a polynomial algebra on generators $St^I \iota_n$ where I is an admissible sequence (mod 2) of excess less than or equal to n (Theorem 6.20). However, when $x = St^I \iota_n$ has odd degree $2m + 1$, then

$$x^2 = Sq^{2m+1}x = Sq^1 Sq^{2m}x = Sq^1 St^{(2m,I)}\iota_n,$$

that is, the squares of certain classes are the image under the Bockstein of other generators. The pairing of classes that occurs in the case of odd primes does not occur here and new cycles are produced. We write $Sq^1\iota_{2m} = \eta_{2m+1}$. Because $Sq^1 = \beta$ is a derivation, we have

$$\eta_{2m+1}^2 = Sq^{2m+1}\iota_{2m} = Sq^1 Sq^{2m}\eta_{2m+1} = Sq^1(\iota_{2m}\eta_{2m+1}).$$

Thus $Sq^{2m}\eta_{2m+1} + \iota_{2m} \smile \eta_{2m+1}$ is a cycle under d_1. By the same analysis of the low degrees of $H^*(K(\mathbb{Z}/2\mathbb{Z}, 2m); \mathbb{F}_2)$ and Cartan's integral computation we have the following result.

Corollary 10.15. *Suppose that* $p = 2$. *Let* $\iota_{2m} \in B_1^{2m}(K(\mathbb{Z}/2\mathbb{Z}, 2m))$ *and* $\eta_{2m+1} = Sq^1\iota_{2m}$. *Then*

$$d_2(\{\iota_{2m}^2\}) = \{Sq^{2m}\eta_{2m+1} + \iota_{2m} \smile \eta_{2m+1}\}.$$

We leave the remaining case of $K(\mathbb{Z}/2^k\mathbb{Z}, n)$ for $k > 1$ to the reader. In this case, Corollary 10.14 for odd primes goes over analogously.

We next explore some of the consequences of these calculations.

Infinite implications and their consequences

The proof of Theorem 10.2 for fields of characteristic zero shows that the presence of a primitive element x of even degree implies the condition $x^n \neq 0$ for all n. For fields of characteristic $p > 0$, it can happen that a primitive element x of even degree can satisfy $x^{p^r} = 0$ for some r, and so the finiteness of the H-space need not be violated. For example, the exceptional Lie group F_4 has mod 3 cohomology given by

$$H^*(F_4; \mathbb{F}_3) \cong \mathbb{F}_3[x_8]/(x_8^3) \otimes \Lambda(x_3, x_7, x_{11}, x_{15}),$$

where x_8 is clearly primitive ([Borel54]). The rational cohomology is given by $H^*(F_4; \mathbb{Q}) \cong \Lambda(X_3, X_{11}, X_{15}, X_{23})$. The Bockstein spectral sequence mod 3 requires $\beta(x_7) = x_8$; subsequently the class X_{23} is represented by the product $[x_7 \smile x_8^2]$.

The E_∞-term of the Bockstein spectral sequence of a finite H-space is fixed by Hopf's theorem. The appearance of even-dimensional primitive elements in $H^*(X; \mathbb{F}_p)$ forces some nontrivial differentials in the Bockstein spectral sequence in order to realize this target. The consequences of such differentials are organized by the phenomenon of implications due to [Browder61].

Definition 10.16. *Let* A_* *denote a Hopf algebra of finite type over the finite field* $\mathbb{F}_p$ *and denote its dual by* A^*. *An element* $x \in A_m$ *is said to have* r-**implications** *if there are elements* $x_i \in A_{mp^i}$, *for* $i = 0, 1, 2, \ldots, r$, *with* $x_0 = x$, $x_i \neq 0$ *for all* i, *and either* $x_{i+1} = x_i^p$ *or there exists an element* $\bar{x}_i \in A^{mp^i}$ *such that* $\bar{x}_i(x_i) \neq 0$ *and* $\bar{x}_i^p(x_{i+1}) \neq 0$. *An element has* ∞-**implications** *if it has* r-*implications for all* r.

Lemma 10.17. *If A_* is a Hopf algebra over $\mathbb{F}_p$ that contains an element which has ∞-implications, then A_* is infinite dimensional as a vector space over $\mathbb{F}_p$.*

The Hopf algebras that we want to study are the terms of the Bockstein spectral sequence for an H-space which are, in fact, *differential* Hopf algebras. Before stating Browder's theorem on ∞-implications we collect a few basic lemmas about Hopf algebras and differential Hopf algebras.

Lemma 10.18. *Suppose (A_*, μ, Δ) is a Hopf algebra and $x \in A_{2m}$ is a primitive element. Then*

$$\Delta(x^n) = \sum_{i=0}^{n} \binom{n}{i} x^{n-i} \otimes x^i.$$

This follows like the binomial theorem for the algebra $A_* \otimes A_*$ using the fact that the comultiplication Δ is an algebra map. (We do not need to assume associativity of μ if we define x^n inductively by $x^0 = 1$ and $x^n = x^{n-1} \cdot x$, and pay careful attention to parentheses.)

Lemma 10.19. *Suppose A_* is a Hopf algebra over a field k and A^* is its dual. If $x \in A_{2m}$ is a primitive element and $\bar{x} \in A^*$, then $\bar{x}^n(x^n) = n!(\bar{x}(x))^n$.*

PROOF: We compute

$$\bar{x}^n(x^n) = \Delta^*(\bar{x}^{n-1} \otimes \bar{x})(x^n) = (\bar{x}^{n-1} \otimes \bar{x})(\Delta(x^n))$$

$$= (\bar{x}^{n-1} \otimes \bar{x})\left(\sum_i \binom{n}{i} x^{n-i} \otimes x^i\right)$$

$$= (\bar{x}^{n-1} \otimes \bar{x})(nx^{n-1} \otimes x).$$

Thus $\bar{x}^n(x^n) = n(\bar{x}^{n-1}(x^{n-1}) \cdot \bar{x}(x))$ and so, by induction, we get $\bar{x}^n(x^n) = n!(\bar{x}(x))^n$. □

Lemma 10.20. *Suppose that (A_*, μ, Δ, d) is a connected, differential graded Hopf algebra over the field $\mathbb{F}_p$, $x \in A_{2m}$ is primitive, $x = d(y)$ for some $y \in A_{2m+1}$, and $\bar{x} \in A^{2m}$ satisfies $\bar{x}(x) \neq 0$. Set $\bar{y} = d^*(\bar{x})$ where d^* is the dual differential on A^*. Then $(\bar{x}^{p-1} \cdot \bar{y})(x^{p-1} \cdot y) \neq 0$.*

PROOF: First notice that $\bar{y}(y) = (d^*(\bar{x}))(y) = \bar{x}(d(y)) = \bar{x}(x) \neq 0$ and so $\bar{y} \neq 0$. We next compute

$$(\bar{x}^{p-1} \cdot \bar{y})(x^{p-1} \cdot y) = \Delta^*(\bar{x}^{p-1} \otimes \bar{y})(x^{p-1} \cdot y) = (\bar{x}^{p-1} \otimes \bar{y})(\Delta(x^{p-1})\Delta(y))$$

By Lemma 10.18, we can write

$$\Delta(x^{p-1})\Delta(y) =$$

$$= \left(\sum_{i=0}^{p-1} \binom{p-1}{i} x^{p-1-i} \otimes x^i \right) \left(y \otimes 1 + 1 \otimes y + \sum_j y'_j \otimes y''_j \right)$$

$$= \left(x^{p-1} \otimes y + \sum_{\dim y''_j = 1} (p-1)x^{p-2}y'_j \otimes xy''_j + \text{stuff} \right),$$

where the "stuff" is a sum of tensor products of classes $u \otimes v$ where $\deg u \neq (p-1)\deg \bar{x}$ or $\deg v \neq \deg \bar{y}$. Since $(\bar{x}^{p-1} \otimes \bar{y})(x^{p-1} \otimes y) = \bar{x}^{p-1}(x^{p-1}) \cdot \bar{y}(y) \neq 0$, it suffices to show that $\bar{y}(xy''_j) = 0$ for $y''_j \in A_1$. Consider

$$\bar{y}(xy''_j) = (d^*(\bar{x}))(xy''_j) = \bar{x}(d(x)y''_j + xd(y''_j)).$$

Since $x = d(y)$, $d(x) = 0$. Thus $d(xy''_j) = xd(y''_j)$. If $d(y''_j) \neq 0$, then there is an element $w_j \in A^0$ with $d^*(w_j) \neq 0$. Since A_* is taken to be connected, $w_j = \alpha_j \cdot 1$ for some $\alpha_j \neq 0 \in \mathbb{F}_p$. But $d^*(1) = d^*(1 \cdot 1) = d^*(1) \cdot 1 + 1 \cdot d^*(1) = 2d^*(1)$ and so $d^*(1) = 0$. Thus $d^*(w_j) = d^*(\alpha_j \cdot 1) = \alpha_j d^*(1) = 0$. This implies that $\bar{y}(xy''_j) = 0$ for all j. $\qquad\square$

Lemma 10.21. *If A_* is a differential graded Hopf algebra and $x \in H(A_*)$ satisfies $x^p \neq 0$, then for any $y \in A_*$ with $\{y\} = x$, we have $y^p \neq 0$. If x has r-implications in $H(A_*)$ for some $r \leq \infty$, then y has r-implications in A_*.*

PROOF: Since $x^p = \{y\}^p = \{y^p\} \neq 0$, then $y^p \neq 0$. We can apply this argument at each power of p. Thus, if x has ∞-implications in $H(A_*)$, then y has ∞-implications in A_*. $\qquad\square$

Lemma 10.22. *Suppose that A_* is a differential graded Hopf algebra over $\mathbb{F}_p$. Suppose further that $x \in A_{2m}$ is primitive, that $x^p = 0$, and there is an element y with $d(y) = x$. If $\{x^{p-1}y\} \neq 0$ in $H(A_*)$, then it is primitive.*

PROOF: By definition, $H(\Delta)(\{x^{p-1}y\}) = \{\Delta(x^{p-1}y)\}$. By assumption we have $d(\Delta(y)) = \Delta(d(y)) = \Delta(x) = 1 \otimes x + x \otimes 1$. This implies that $d(\Delta(y) - y \otimes 1 - 1 \otimes y) = 0$. Furthermore, $\Delta(x^{p-1}) = \sum_{i=0}^{p-1} \binom{p-1}{i} x^{p-1-i} \otimes x^i$.

From elementary number theory we know that $\binom{p-1}{i} \equiv (-1)^i \mod p$, and so we can write

$$\Delta(x^{p-1}y) = \Delta(x^{p-1})\Delta(y)$$

$$= \left(\sum\nolimits_{i=0}^{p-1}(-1)^i x^{p-1-i} \otimes x^i\right)(y \otimes 1 + 1 \otimes y + (\Delta(y) - 1 \otimes y - y \otimes 1))$$

$$= x^{p-1}y \otimes 1 + 1 \otimes x^{p-1}y + \sum\nolimits_{i=0}^{p-2}(-1)^i x^{p-1-i} \otimes x^i y$$

$$\quad + \sum\nolimits_{i=1}^{p-1}(-1)^i x^{p-1-i}y \otimes x^i + \Delta(x^{p-1})(\Delta(y) - 1 \otimes y - y \otimes 1)$$

$$= x^{p-1}y \otimes 1 + 1 \otimes x^{p-1}y + d\left(\sum\nolimits_{i=1}^{p-1}(-1)^{i+1}(x^{p-1-i}y \otimes x^{i-1}y)\right)$$

$$\quad + d(\Delta(x^{p-2}y)(\Delta(y) - 1 \otimes y - y \otimes 1)).$$

It follows that $\{\Delta(x^{p-1}y)\} = \{x^{p-1}y\} \otimes 1 + 1 \otimes \{x^{p-1}y\}.$ $\square$

The last lemma we need before we state and prove the main theorem of [Browder61] is a technical fact about the mod 2 Steenrod algebra and H-spaces. While the previous lemmas followed for purely algebraic reasons, this lemma requires that we are working with the mod 2 cohomology of an H-space.

Lemma 10.23. *If (X, x_0, μ) is an H-space, $x \in H_{2m}(X; \mathbb{F}_2)$ is a primitive element, $y \in H_{2m+1}(X; \mathbb{F}_2)$, and $\bar{z} \in H^{2m+1}(X; \mathbb{F}_2)$, then $(Sq^{2m}\bar{z})(xy) = 0$.*

PROOF: In terms of the induced operations we can write

$$(Sq^{2m}\bar{z})(xy) = (\mu^*(Sq^{2m}\bar{z}))(x \otimes y) = (Sq^{2m}(\mu^*\bar{z}))(x \otimes y).$$

We write $\mu^*(\bar{z}) = \sum_i \bar{z}'_i \otimes \bar{z}''_i$ and the Cartan formula gives

$$Sq^{2m}(\bar{z}'_i \otimes \bar{z}''_i) = \sum\nolimits_{q+r=2m} Sq^q(\bar{z}'_i) \otimes Sq^r(\bar{z}''_i).$$

By the unstable axiom for the action of the Steenrod algebra, if $q > \dim \bar{z}'_i$, then $Sq^q(\bar{z}'_i) = 0$, and similarly if $r > \dim \bar{z}''_i$. Let $c = \deg \bar{z}'_i$, $d = \deg \bar{z}''_i$. Then $c + d = 2m + 1$ and it follows by examining the solutions to $q + r = 2m$ that

$$Sq^{2m}(\bar{z}'_i \otimes \bar{z}''_i) = Sq^c \bar{z}'_i \otimes Sq^{d-1}\bar{z}''_i + Sq^{c-1}\bar{z}'_i \otimes Sq^d \bar{z}''_i.$$

Since $Sq^c \bar{z}'_i = (\bar{z}'_i)^2$ and x is primitive, $(\bar{z}'_i)^2(x) = 0$. It follows that $(Sq^c \bar{z}'_i \otimes Sq^{d-1}\bar{z}''_i)(x \otimes y) = 0$. Similarly, $Sq^d \bar{z}''_i = (\bar{z}''_i)^2$, a class of even degree. Since y has odd degree, $(\bar{z}''_i)^2(y) = 0$ and so the lemma follows from $(Sq^{c-1}\bar{z}'_i \otimes Sq^d \bar{z}''_i)(x \otimes y) = 0$. $\square$

Theorem 10.24. *Suppose* (X, x_0, μ) *is a connected, path-connected H-space of finite type and* $\{B^r_*(X)\}$ *is its homology Bockstein spectral sequence. If* $x \in B^r_{2m}$ *is a nonzero primitive element and, for some* $y \neq 0$, $x = d^r(y)$, *then* x *has* ∞-*implications.*

PROOF: We may assume that $x^p = 0$, for otherwise we can take $x_1 = x^p$, also a primitive, with $d^r(x^{p-1}y) = x^p$. Thus x_1 satisfies the hypotheses of the theorem, and if this process never stops, we have obtained the sequence of ∞-implications of x. Assuming $x^p = 0$, we will produce $x_1 \in B^r_{2mp}$ such that $\bar{x}^p(x_1) \neq 0$ for any $\bar{x} \in B^{2m}_r(X)$ for which $\bar{x}(x) \neq 0$. The x_1 produced will be neither primitive nor a boundary, but its homology class $\{x_1\} \in B^{r+1}_{2mp}$ will be both primitive and a boundary. By Lemma 10.21 it suffices to check that there is the 1-implication x_1 at the next stage of the Bockstein spectral sequence and then take a representative in B^r.

In the cohomology Bockstein spectral sequence suppose that $\bar{x} \in B^{2m}_r$ satisfies $\bar{x}(x) \neq 0$. Set $\bar{y} = d_r(\bar{x})$. Then

$$\bar{y}(y) = (d_r(\bar{x}))(y) = \bar{x}(d^r(y)) = \bar{x}(x) \neq 0.$$

It follows that $\bar{y} \neq 0$ and, by Lemma 10.20, that $(\bar{x}^{p-1}\bar{y})(x^{p-1}y) \neq 0$. Furthermore, if $p \neq 2$, $d_r(\bar{x}^{p-1}\bar{y}) = (p-1)\bar{x}^{p-2}\bar{y}^2 = 0$. If $p = 2$ and $r > 1$,

$$\bar{y}^2 = \{Sq^{2m+1}z\} = \{Sq^1 Sq^{2m}z\} = \{d_1(Sq^{2m}z)\} = 0$$

in B_2 where $z \in H^{2m+1}(X; \mathbb{F}_2)$ is such that $\{z\} = \bar{y}$. That is, squares of odd degree classes vanish in B_2. If $p = 2$ and $r = 1$, then

$$d_1(Sq^{2m}\bar{y} + \bar{x}\bar{y}) = \bar{y}^2 + \bar{y}^2 = 0$$

and, by Lemmas 10.20 and 10.23, $(Sq^{2m}\bar{y} + \bar{x}\bar{y})(xy) \neq 0$.

We check that the class $\{\bar{x}^{p-1}\bar{y}\}$ (or $\{Sq^{2m}\bar{y}+\bar{x}\bar{y}\}$ when $r = 1$ and $p = 2$) is nontrivial in B_{r+1}. Suppose that $\bar{x}^{p-1}\bar{y} = d_r(\bar{z})$. Then

$$0 \neq (\bar{x}^{p-1}\bar{y})(x^{p-1}y) = d_r(\bar{z})(x^{p-1}y) = \bar{z}(d^r(x^{p-1}y))$$
$$= \bar{z}(x^p) = \bar{z}(0) = 0,$$

a contradiction. Thus $\bar{x}^{p-1}\bar{y} \neq d_r(\bar{z})$. Similarly, $(Sq^{2m}\bar{y} + \bar{x}\bar{y}) \neq d_1(\bar{z})$.

To complete the proof we show that the class $\{\bar{x}^{p-1}\bar{y}\} \in B_{r+1}$ satisfies $d_{r+1}(\{\bar{x}^p\}) = c\{\bar{x}^{p-1}\bar{y}\} \neq 0$ when $p \neq 2$ or $p = 2$ and $r > 1$. In the case $p = 2$ and $r = 1$, we show that the class $\{Sq^{2m}\bar{y} + \bar{x}\bar{y}\} \in B_2$ satisfies $d_2(\{\bar{x}^2\}) = \{Sq^{2m}\bar{y} + \bar{x}\bar{y}\}$. Recall $d_r(\bar{x}) = \bar{y}$. Then there is a class $\bar{u} \in H^{2m}(X; \mathbb{Z}/p^r\mathbb{Z})$ such that $\{\mathrm{red}^*_p \bar{u}\} = \bar{x} \in B_r$ where $\mathrm{red}_p \colon \mathbb{Z}/p^r\mathbb{Z} \to \mathbb{Z}/p\mathbb{Z}$ is reduction mod p. Let $f \colon X \to K(\mathbb{Z}/p^r\mathbb{Z}, 2m)$ represent $\bar{u}$, that is, $f^*(\iota_{2m}) =$

$\bar{u}$ where $\iota_{2m} \in H^{2m}(K(\mathbb{Z}/p^r\mathbb{Z}, 2m); \mathbb{Z}/p^r\mathbb{Z})$ is the fundamental class. Let $\bar{\iota} = \mathrm{red}_p^*(\iota_{2m})$. It follows that

$$f^*(\bar{\iota}) = f^*(\mathrm{red}_p^* \iota_{2m}) = \mathrm{red}_p^*(f^*(\iota_{2m})) = \mathrm{red}_p^*(\bar{u}) = \bar{x}.$$

Let $f_r^*\colon B_r(K(\mathbb{Z}/p^r\mathbb{Z}, 2m)) \to B_r(X)$ denote the homomorphism induced by f on the cohomology Bockstein spectral sequences. If $\bar{\eta} = d_r(\bar{\iota})$, then we have

$$f_r^*(\bar{\eta}) = f_r^*(d_r(\bar{\iota})) = d_r(f_r^*(\bar{\iota})) = d_r(\bar{x}) = \bar{y}$$

and so $f_r^*(\bar{\iota}^{p-1}\bar{\eta}) = \bar{x}^{p-1}\bar{y}$. Since $\{\bar{x}^{p-1}\bar{y}\} \neq 0$ in B_{r+1}, $f_{r+1}^*(\{\bar{\iota}^{p-1}\bar{\eta}\}) = \{\bar{x}^{p-1}\bar{y}\}$. By naturality and the calculation of the cohomology Bockstein spectral sequence for $K(\mathbb{Z}/p^r\mathbb{Z}, 2m)$, $f_{r+1}^*(d_{r+1}(\{\bar{\iota}^p\})) \neq 0$ and $f_{r+1}^*(\{\bar{\iota}^p\}) = \{\bar{x}^p\} \neq 0$. Thus

$$d_{r+1}(\{\bar{x}^p\}) = d_{r+1}(f_{r+1}^*(\{\bar{\iota}^p\})) = f_{r+1}^*(c\{\bar{\iota}^{p-1}\bar{\eta}\}) = c\{\bar{x}^{p-1}\bar{y}\}.$$

The analogous argument mod 2 gives $d_2(\{\bar{x}^2\}) = \{Sq^{2m}\bar{y} + \bar{x}\bar{y}\}$.

In order to continue the argument, we show that there is an element $v \in B_{2mp}^{r+1}$ that is primitive with $\{\bar{x}^p\}(v) \neq 0$ and $v = d^{r+1}(w)$ for some w. Consider the element $w = \{x^{p-1}y\}$. We compute:

$$\{\bar{x}^p\}(d^{r+1}(\{x^{p-1}y\})) = c\{\bar{x}^{p-1}\bar{y}\}(\{x^{p-1}y\}) \neq 0.$$

By Lemma 10.22, w is primitive. Also, $v = d^{r+1}(w)$ is primitive. In the sequence of elements making up the ∞-implications of x we take x_1 to be a choice of representative of v in B^r. Then, $\bar{x}^p(x_1) = \{\bar{x}^p\}(v) \neq 0$, and, since $\bar{x}(x) \neq 0$, x_1 is the next element in the sequence making up the ∞-implications for x. To obtain x_2, either take x_1^p if nonzero, or repeat the argument using the primitive $v \in B_{2mp}^{r+1}$ with $v = d^{r+1}(w)$. $\square$

Notice that if $x^p = 0$, then the choice of $\bar{x}$ with $\bar{x}(x) \neq 0$ was arbitrary in the construction. It follows from $\bar{x}^p(x_1) \neq 0$ that, if x is a primitive in B_r^{2m} with $0 \neq d^r(y) = x$ and $x^p = 0$, then $\bar{x}^p \neq 0$ for all $\bar{x} \in B_r^{2m}$ with $\bar{x}(x) \neq 0$.

We turn to applications of Theorem 10.24. A space X is said to be a **mod p finite H-space** if it is a connected, path-connected H-space of finite type for which the mod p homology ring is finite-dimensional over $\mathbb{F}_p$. By Theorem 10.2, for a mod p finite H-space X, $B_\infty(X)$ is an exterior algebra on finitely many odd-dimensional generators.

A shorthand statement of Theorem 10.24 is the expression for X, a mod p finite H-space,

$$\mathrm{Im}\, d^r \cap \mathrm{Prim}(H_{\mathrm{even}}(X; \mathbb{F}_p)) = \{0\}.$$

A dual formulation of Theorem 10.24 depends on a fundamental theorem due to [Milnor-Moore65]:

Theorem 10.25. *If* (A, μ, Δ) *is an associative, commutative, connected Hopf algebra over the field* $\mathbb{F}_p$, *then there is an exact sequence*

$$0 \to \mathrm{Prim}(\xi A) \to \mathrm{Prim}(A) \to Q(A)$$

where $\xi \colon A \to A$ *is the* **Frobenius homomorphism** $\xi(a) = a^p$.

SKETCH OF A PROOF: The reader can check that the theorem holds for A a monogenic Hopf algebra. For a finitely generated Hopf algebra A and A', a normal sub-Hopf algebra, there are short exact sequences:

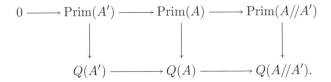

We leave it to the reader to show that, if $A' = \xi(A)$, then the mapping $\mathrm{Prim}(A//A') \to Q(A//A')$ is injective. The theorem follows from the diagram of short exact sequences. $\qquad \square$

Suppose X is a mod p finite H-space, $\bar{x} \in B_r^{2m}$ is a primitive element and $d_r(\bar{x}) = \bar{y} \neq 0$. Since $H^*(X; \mathbb{F}_p)$ is an associative, commutative connected Hopf algebra, Theorem 10.25 implies that $\bar{y}$, a primitive of odd degree, is not a pth power ($\bar{y}^2 = 0$) and hence $\bar{y}$ is indecomposable. Thus there is an element y in B_{2m+1} with $\bar{y}(y) \neq 0$ and y primitive. Then $\bar{y}(y) = d_r(\bar{x})(y) = \bar{x}(d^r(y)) = \bar{x}(x) \neq 0$, and so $x \in B_{2m}$ is a primitive in the image of d^r. Since $H^*(X; \mathbb{F}_p)$ is a finite vector space, there cannot be ∞-implications, and so the assumption that there is an $\bar{x} \in B_r^{2m}$ with $d_r(\bar{x}) \neq 0$ must fail. Thus, the dual version of Theorem 10.24 for mod p finite H-spaces may be written

$$\mathrm{Im}\, d_r \cap \mathrm{Prim}(B_r^{2m+1}) = \{0\}, \text{ for all } m.$$

From the structure of an exact couple, an element in the image of the descending homomorphism is always a cycle (Proposition 2.9). In the case of the Bockstein spectral sequence, the descending homomorphism is reduction mod p. Thus, for a mod p finite H-space, the image of $\mathrm{red}_{p*} \colon H_*(X) \to H_*(X; \mathbb{F}_p)$ cannot contain an even-dimensional primitive element. If $x \in \mathrm{Im}\, \mathrm{red}_{p*} \cap \mathrm{Prim}(H_{\mathrm{even}}(X; \mathbb{F}_p))$, then $d^r(x) = 0$ for all r and since x cannot persist to B^{∞}, then $x = d^s(y)$ for some s and y. But then x has ∞-implications and $H_*(X; \mathbb{F}_p)$ has infinite dimension over $\mathbb{F}_p$.

A consequence of this discussion is the theorem of [Browder61] generalizing the classical result of [Cartan, E36] that $\pi_2(G) = \{0\}$ for simply-connected Lie groups.

Theorem 10.26. *If X is a mod p finite H-space, then the least $m > 1$ for which $\pi_m(X) \otimes \mathbb{F}_p \neq \{0\}$ is odd.*

PROOF: Consider the mod p Hurewicz homomorphism $h \otimes \mathbb{F}_p \colon \pi_m(X) \otimes \mathbb{F}_p \to H_m(X) \otimes \mathbb{F}_p \to H_m(X; \mathbb{F}_p)$. This factors through red_{p*} and takes its image in the primitive elements. It follows that this mapping is trivial when m is even.

When X is simply-connected, the Hurewicz-Serre theorem for mod p coefficients (Theorem 6.25) implies that the first nonvanishing homology group $H_m(X; \mathbb{F}_p)$ is isomorphic via $h \otimes \mathbb{F}_p$ to the first nonvanishing homotopy $\mathbb{F}_p$-module $\pi_m(X) \otimes \mathbb{F}_p$. Since this must happen in an odd dimension, the theorem holds.

When X is not simply-connected we can argue using the universal cover $\tilde{X}$. [Browder59] showed that the universal cover of a mod p finite H-space is again a mod p finite H-space. Since $\pi_m(X) \cong \pi_m(\tilde{X})$ for $m > 1$, we reduce to the simply-connected case. $\qquad\square$

In developments that grew out of the study of torsion in H-spaces, [Jeanneret92] and [Lin93] have shown that the first nonvanishing homotopy group of a mod 2 finite H-space, whose mod 2 homology ring is associative, must be in degree 1, 3, or 7.

An H-space with the homotopy type of a finite CW-complex is called a **finite H-space**. The compact Lie groups offer a large class of examples of finite H-spaces. A guiding principle in the study of such spaces is that the topological properties of compact Lie groups have their origin at the homotopical level of structure. That is to say, what is true homotopically of a compact Lie group G ought to be true because G is a finite H-space. Hopf's theorem (10.2) and Browder's theorem (10.26) lend considerable support to this principle. That the class of finite H-spaces is larger than the examples of compact Lie groups is a result of the development of localization and the mixing of homotopy types due to [Zabrodsky72]. [Hilton-Roitberg71] used mixing to exhibit examples of finite H-spaces not of the homotopy type of any compact Lie group.

A major theme in the development of finite H-spaces is the application of the guiding principle to Bott's theorem (10.1)—if X is a simply-connected finite H-space, then $H_*(\Omega X)$ has no torsion.

Under the assumption that X is a simply-connected finite H-space and $H_*(\Omega X)$ has no torsion [Browder63] showed that $H_*(\Omega X) = H_{\mathrm{even}}(\Omega X)$, strengthening Bott's theorem considerably. This paper introduces a family of spectral sequences based on the natural filtrations on a Hopf algebra that interpolate between the terms in the Bockstein spectral sequence and enjoy a particularly nice algebraic expression.

[Kane77] applied work of [Browder63] and [Zabrodsky71] to obtain a necessary and sufficient condition that $H_*(\Omega X)$ have no p-torsion when X is a simply-connected finite H-space. The condition is given in terms of the action of the Steenrod algebra on the cohomology of the finite H-space:

$$Q(H^{\mathrm{even}}(X; \mathbb{F}_p)) = \sum\nolimits_{m \geq 1} \beta P^m Q(H^{2m+1}(X; \mathbb{F}_p)).$$

Notice, in the case that $p = 2$, this condition holds only when $H^*(X; \mathbb{F}_2)$ has no even-dimensional indecomposables. When $p = 2$, $P^m = Sq^{2m}$ and $\beta Sq^{2m} = Sq^1 Sq^{2m} = Sq^{2m+1}$, which is the squaring map on H^{2m+1}. [Lin76, 78] established that $Q(H^{\mathrm{even}}) = \sum_{m \geq 1} \beta P^m Q(H^{\mathrm{odd}})$ holds for odd primes by extending work of [Zabrodsky71] on secondary operations.

Building on work of [Thomas63] on the action of the Steenrod algebra on the cohomology of an H-space, [Lin82] established the absence of 2-torsion in $H_*(\Omega X)$ when X is a mod 2 finite H-space and $H_*(X; \mathbb{F}_2)$ is an associative Hopf algebra. [Kane86] studied the presence of 2-torsion in $H_*(\Omega X)$ by using a version of the Bockstein spectral sequence for the extraordinary cohomology theory $k(n)^*$ introduced by [Morava85]. Putting together all of these developments, the goal of generalizing Bott's theorem was realized.

Theorem 10.27. *If X is a simply-connected finite H-space, then $H_*(\Omega X)$ has no torsion.*

The proof of Theorem 10.27 generated a number of powerful methods in algebraic topology. Accounts of these developments and much more can be found in [Kane88] and [Lin95].

Other applications of the Bockstein spectral sequence Ⓝ

Away from the study of H-spaces, the results of [Browder61] may be applied to obtain some general results about $H^*(\Omega X; \mathbb{F}_p)$. In particular, using ∞-implications, [McCleary87] proved a generalization of the results of [Serre51] (Proposition 5.16) and [Sullivan73] on the nontriviality of $H_*(\Omega X; k)$ for k a field.

Theorem 10.28. *Suppose M is a simply-connected compact finite-dimensional manifold and $\dim_k Q(\tilde{H}^*(M; k)) > 1$, then the set $\{\dim_k H^i(\Omega M; k) \mid i = 1, 2, \dots\}$ is unbounded.*

The assumption that $\dim_{\mathbb{F}_p} Q(\tilde{H}^*(X; \mathbb{F}_p)) > 1$ together with the results over $\mathbb{Q}$ of [Sullivan73] force the existence of ∞-implications on two elements. The intertwining of the ∞-implications of these elements in a Hopf algebra gives a subspace of $H^*(\Omega M; \mathbb{F}_p)$ that is isomorphic as a vector space to a polynomial algebra on two generators. The vector space $\mathbb{F}_p[x, y]$ has subspaces $\mathbb{F}_p\{x^{lm}, x^{(l-1)m}y^n, \dots, x^{(l-i)m}y^{in}, \dots, x^m y^{(l-1)n}, y^n\}$ where $m \deg x = n \deg y = \mathrm{lcm}(\deg x, \deg y)$. This subspace has dimension $l + 1$ and so grows unbounded with l.

This theorem, like Proposition 5.16, implies geometric results about the geodesics on the manifold M. Under the assumptions of the theorem, the number of geodesics on M joining two nonconjugate points of length less than λ grows at least quadratically in λ.

Another place where p-torsion makes a key appearance is in the Adams spectral sequence. Following the discussion in §9.3, the *times* p map is detected in the Adams spectral sequence by multiplication by a class $a_0 \in \mathrm{Ext}_{A_p}^{1,1}(\mathbb{F}_p, \mathbb{F}_p)$. For an A_p-module M, [May-Milgram81] say that an element $x \in \mathrm{Ext}_{A_p}(M, \mathbb{F}_p)$ **generates a spike** if $x \neq a_0 x'$ and $a_0^i x \neq 0$ for all i. There is a single spike in $\mathrm{Ext}_{A_2}(\mathbb{F}_2, \mathbb{F}_2)$ as the charts (pp. 443-444) in Chapter 9 show—the picture explains the terminology.

[Adams69] wrote of the Adams spectral sequence, "Whenever a chance has arisen to show that a differential d_r is non-zero, the experts have fallen on it with shouts of joy—'Here is an interesting phenomenon! Here is a chance to do some nice, clean research!'—and they have solved the problem in short order." The Bockstein spectral sequence interacts with the Adams spectral sequence to produce differentials that form a coherent pattern. The function $T(s)$ used in the statement of the following theorem refers to Lemma 9.45: When p is odd, then $T(s) = (2p-1)s - 1$; when $p = 2$, then $T(s)$ is defined by $T(4s) = 12s$, $T(4s+1) = 12s + 2$, $T(4s+2) = 12s + 4$, and $T(4s+3) = 12s + 7$.

Theorem 10.29. *Suppose X is an $(n-1)$-connected space of finite type. For $r \geq 1$, suppose that C_r is a basis for $B_*^r(X)$, the homology Bockstein spectral sequence. Assume that C_r is chosen so that $C_r = D_r \cup \beta_r D_r \cup C_{r+1}$ where D_r, $\beta_r D_r$, and C_{r+1} are disjoint, linearly independent subsets of $B_*^r(X)$ such that $\beta_r D_r = \{\beta_r w \mid w \in D_r\}$ and C_{r+1} is a set of cycles with respect to β_r that projects onto the chosen basis for $B_*^{r+1}(X)$. Then*

(1) *The set of spikes in $E_r(X)$, $2 \leq r \leq \infty$, is in one-to-one correspondence with C_r. If $c \in C_r$ has degree q and $\gamma \in E_r^{s,t}(X)$ generates the corresponding spike, then $T(s) - s + n \leq q = t - s$.*

(2) *If $d \in D_r$ and $\delta \in E_r^{s,t}(X)$ and $\epsilon \in E_r^{u,v}(X)$, with $v - u = t - s - 1$, generate spikes corresponding to d and $\beta_r d$, then*

$$d_r(a_0^i \delta) = a_0^{i+r+s-u} \epsilon$$

provided $n + T(i+s) \geq t$.

PROOF: Since X is taken to be of finite type, $H_*(X)$ is a direct sum of torsion prime to p, summands of the form $\mathbb{Z}/p^k\mathbb{Z}$, and summands $\mathbb{Z}$ whose generators reduce mod p to the elements of C_∞. We may use this decomposition to construct mappings

$$\phi_i \colon X \to K(H_i(X), i)$$

that induce isomorphisms on integral homology in degree i. Let Y denote the space $\bigvee_i K(H_i(X), i)$ and $\phi = \bigvee_i \phi_i \colon X \to Y$ denote the wedge product of all of these mappings. On homology with coefficients in $\mathbb{F}_p$, ϕ_* induces a monomorphism from $H_i(X)$ for all i. This gives rise to a short exact sequence

$$0 \to H_*(X; \mathbb{F}_p) \to H_*(Y; \mathbb{F}_p) \to M_* \to 0,$$

where M_* is seen to be $\bigoplus_{q \geq i+2} H_q(K(H_i(X), i); \mathbb{F}_p)$.

Ignoring the contribution to torsion at primes not equal to p, we know from theorems of Cartan and Serre that the dual of M_* is $A(0)$-free (§9.6), that is, the Bockstein homomorphism on M_*^{dual}, as a differential, is exact. We next examine the long exact sequence of Ext groups associated to the short exact sequence:

$$\to \operatorname{Ext}_{\mathcal{A}_p}^{s-1,t}(M_*^{\text{dual}}, \mathbb{F}_p) \to E_2^{s,t}(X) \to E_2^{s,t}(Y) \to \operatorname{Ext}_{\mathcal{A}_p}^{s,t}(M_*^{\text{dual}}, \mathbb{F}_p) \to .$$

Lemma 9.47 implies that $\operatorname{Ext}_{\mathcal{A}_p}^{s,t}(M_*^{\text{dual}}, \mathbb{F}_p) = \{0\}$ when $0 < s < t \leq n + T(s)$. It follows that $E_2^{s,t}(X) \to E_2^{s,t}(Y)$ is onto in this range and an isomorphism when $s \geq 2$ and $0 < s < t \leq n + T(s-1)$. By the naturality of the Adams spectral sequence, that it suffices to examine the case of Eilenberg-Mac Lane spaces to prove the theorem. We leave it to the reader to show that a factor of $K(\mathbb{Z}/p\mathbb{Z}, i)$ introduces a single copy of $\mathbb{F}_p$ that persists to E_∞; a factor of $K(\mathbb{Z}/p^k\mathbb{Z}, i)$ introduces a pair of spikes at E_2 on generators z and y with $d_k(a_0^i z) = a^{i+k} y$, leaving a basis of $\{a_0^i y \mid 0 \leq i \leq k\}$ at E_∞; finally, a factor of $K(\mathbb{Z}, i)$ introduces a permanent spike at E_2. □

This argument requires that spikes have the right Adams filtration to work. Spikes in $E_2(X)$ could be generated by elements lying in lower filtration degree than in the range of the isomorphism. Such generators might have nontrivial differentials earlier than predicted by the theorem. Such differentials could occur on the bottoms of spikes whose top parts survive to $E_\infty(X)$.

Plugging this argument into a dual setting via Spanier-Whitehead duality, [Meyer98] has used the resulting differentials to compute certain cohomotopy groups and these groups force Euler classes associated to geometric bundles to vanish. These data imply an estimate of certain numerical invariants of lens spaces. Let

$$v_{p,2}(m) = \min\{n \mid \text{ there is a } \mathbb{Z}/p\mathbb{Z}\text{-equivariant } f \colon L^{2m-1}(p) \to S^{2n-1}\}.$$

Here the action of $\mathbb{Z}/p\mathbb{Z}$ on $L^{2m-1}(p)$ is induced by the multiplication by a primitive root of unity of order p^2 on $\mathbb{C}^m$ and on S^{2n-1} by the standard action. The estimates of [Meyer, D98] generalize work of [Stolz89] at the prime 2.

10.2 Other Bockstein spectral sequences

Consider the cofibration sequence

$$S^{n-1} \xrightarrow{r} S^{n-1} \xrightarrow{\beta} P^n(r) \xrightarrow{\eta} S^n \xrightarrow{r} S^n$$

where $P^n(r) = S^{n-1} \cup_r e^n$ is a mod r Moore space and r denotes the degree r map on S^{n-1}. Following [Peterson56], these spaces may be used to define the **mod r homotopy groups,**

$$\pi_n(X; \mathbb{Z}/r\mathbb{Z}) = [P^n(r), X].$$

The properties of cofibration sequences lead to an exact couple

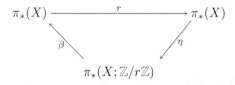

and hence a Bockstein spectral sequence, denoted by $_\pi B_*^r(X)$, with $_\pi B_*^1(X) \cong \pi_*(X; \mathbb{Z}/r\mathbb{Z})$. When $r = p$, a prime, the spectral sequence converges to $(\pi_*(X)/\text{torsion}) \otimes \mathbb{F}_p$ for X of finite type. (Some care has to be taken when $p = 2$ because $\pi_3(X; \mathbb{Z}/2\mathbb{Z})$ need not be abelian.) This spectral sequence was studied by [Araki-Toda65] for applications to generalized cohomology theories, by [Browder78] for applications to algebraic K-theory, and by [Neisendorfer72] for its relations to unstable homotopy theory.

Among the properties of the Moore spaces is the following result of [Neisendorfer72]. The proof requires careful bookkeeping in low dimensions (for details see the memoir of [Neisendorfer80]).

Proposition 10.30. *If m, $n \geq 2$ and r, s are positive integers for which $d = \gcd(r, s)$ is odd, then there is a homotopy equivalence:*

$$\alpha_{m,n} \colon P^{m+n}(d) \vee P^{m+n-1}(d) \to P^m(r) \wedge P^n(s).$$

When $r = s = p$, an odd prime, this homotopy equivalence may be used to define pairings on mod p homotopy groups. In particular, given $f \colon P^m(r) \to X$ and $g \colon P^n(s) \to Y$, we can use the canonical injection, $x \mapsto (x, *)$, $P^{m+n}(d) \to P^{m+n}(d) \vee P^{m+n-1}(d)$ to obtain a mapping $P^{m+n}(d) \to X \wedge Y$ as the composite

$$P^{m+n}(d) \to P^{m+n}(d) \vee P^{m+n-1}(d) \xrightarrow{\alpha_{m,n}} P^m(r) \wedge P^n(s) \xrightarrow{f \wedge g} X \wedge Y.$$

A mapping $\sigma\colon X \wedge Y \to Z$ induces a pairing $\pi_m(X;\mathbb{Z}/r\mathbb{Z}) \otimes \pi_n(Y;\mathbb{Z}/s\mathbb{Z}) \to$
$\pi_{m+n}(Z;\mathbb{Z}/d\mathbb{Z})$ and this pairing for $X = Y = Z = B\mathrm{Gl}(\Lambda)^+$ was developed
by [Browder78] to study the algebraic K-theory with coefficients of a ring Λ
via the homotopy Bockstein spectral sequence.

When (G, μ, e) is a **grouplike space**, that is, G is a homotopy associative
H-space with a homotopy inverse (for example, a based loop space ΩX), then
the commutator mapping $[\ ,\ ]\colon G \times G \to G$, given by $(x,y) \mapsto (xy)(x^{-1}y^{-1})$,
determines a mapping $[\ ,\ ]\colon G \wedge G \to G$, since, up to homotopy, the commutator
mapping restricted to $G \vee G$ is homotopic to the constant mapping to e. This
mapping may be applied to the homotopy groups of G with coefficients to define
a pairing for $d = \gcd(r,s)$:

$$[\ ,\ ]\colon \pi_m(G;\mathbb{Z}/r\mathbb{Z}) \otimes \pi_n(G;\mathbb{Z}/s\mathbb{Z}) \to \pi_{m+n}(G;\mathbb{Z}/d\mathbb{Z}).$$

The pairing is given by the composite

$$P^{m+n}(d) \to P^{m+n}(d) \vee P^{m+n-1}(d) \to P^m(r) \wedge P^n(s) \xrightarrow{f \wedge g} G \wedge G \xrightarrow{[\ ,\ ]} G.$$

The pairing induced on homotopy groups by the commutator mapping is the
Samelson product. The properties of the generalized Samelson product for ho-
motopy groups with coefficients are extensively developed by [Neisendorfer80].
In particular, we have the following result.

Proposition 10.31. *If* $r = s = d$, $\gcd(r,6) = 1$, *and* G *is a 2-connected,*
grouplike space, then $\pi_*(G;\mathbb{Z}/r\mathbb{Z})$ *is a graded Lie algebra.*

When G is grouplike, $H_*(G;\mathbb{Z}/r\mathbb{Z})$ is an associative algebra and hence
enjoys a Lie algebra structure given by $[z, w] = zw - (-1)^{|z| \cdot |w|} wz$. The
Hurewicz map, $h_*\colon \pi_*(X;\mathbb{Z}/r\mathbb{Z}) \to H_*(X;\mathbb{Z}/r\mathbb{Z})$, is induced by $h_*([f]) =$
$f_*(y)$, where $y \in H_m(P^m(r);\mathbb{Z}/r\mathbb{Z})$ is the canonical generator. This mapping for $r = p$, an odd prime, induces a mapping $_\pi B^1_*(X) \to B^1_*(X)$,
where $\{B^s_*(X), d^s\}$ denotes the mod p homology Bockstein spectral sequence.
[Neisendorfer72] showed that both the homotopy and homology Bockstein
spectral sequences are spectral sequences of Lie algebras for $p > 3$, and that the
Hurewicz homomorphism induces a Lie algebra homomorphism on B^s-terms
for all s.

It is possible to develop the properties of differential Lie algebras by analogy with the development of differential Hopf algebras for the Bockstein spectral
sequence. This development makes up the first few sections of [Cohen-Moore-
Neisendorfer79], especially applied to the case of free Lie algebras. These
results may be used to study the spaces $\Omega P^n(p^r)$ and $\Omega F^n(p^r)$, where $F^n(p^r)$
is the homotopy fibre of the pinch map $P^n(p^r) \to S^n$, defined by collapsing the
bottom cell. The main results of [Cohen-Moore-Neisendorfer79] are homotopy

equivalences between the space $\Omega P^n(p^r)$ (suitably localized) and products of countable wedges of known spaces whose structure may be read off the behavior of Bockstein spectral sequence. A similar result holds for $\Omega F^n(p^r)$. The comparison of the homotopy and homology Bockstein spectral sequences via the Hurewicz homomorphism allows one to obtain representative mappings that go into the construction of the homotopy equivalences. Finally, the decompositions are used to establish the inductive argument that goes from the theorem of [Selick78], that p annihilates the p-component $_{(p)}\pi_k(S^3)$ for $k \neq 3$ and $p > 3$, to prove the following result.

Theorem 10.32. *If $p > 3$, and $n > 0$, then p^{n+1} annihilates $_{(p)}\pi_k(S^{2n+1})$, for all $k > 2n + 1$.*

The final generalization of the Bockstein spectral sequence that we present is best framed in the language of spectra and generalized cohomology theories. If X is a spectrum and $f \colon X \to X$ is a selfmap of degree k, then we can form the cofibre of f in the category of spectra and obtain an exact couple:

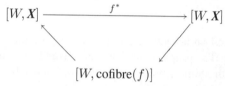

The mapping f may be thought of as a cohomology operation and $[W, X] = X^*(W)$ as the value of the associated generalized cohomology theory on W. If $X = H\mathbb{Z}$, the Eilenberg-Mac Lane spectrum for integer coefficients, and f represents the *times p* map, then the cofibre represents the Eilenberg-Mac Lane spectrum $H\mathbb{F}_p$ and we obtain the usual Bockstein spectral sequence.

Let $k(n)^*(\ )$ denote the generalized cohomology functor known as **connective Morava K-theory** (see the work of [Würgler77] for the definition and properties). This theory has certain remarkable properties:

(1) $k(n)^*(\text{point}) \cong \mathbb{F}_p[v_n]$ where v_n has degree $-2p^n + 2$.
(2) $k(n)^*(W)$ has a direct sum decomposition into summands $\mathbb{F}_p[v_n]$ and $\mathbb{F}_p[v_n]/(v_n^s)$.

Property (2) is analogous to the result for a finitely generated abelian group modulo torsion away from a prime p where the summands are $\mathbb{Z}$ and $\mathbb{Z}/p^s\mathbb{Z}$. We choose the mapping of the representing spectrum for Morava K-theory that induces the *times v_n* map. The cofibre is represented by $H\mathbb{F}_p$ and the exact couple for a finite H-space X may be presented as

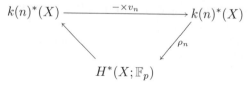

where ρ_n is mod v_n reduction. The Bockstein spectral sequence in this case has $B_1^* = H^*(X; \mathbb{F}_p)$ and the first differential d^1 is identifiable with Q_n, the Milnor primitive in $\mathcal{A}_p^{\text{dual}}$ ([Milnor58], [Kane86]). The limit term, B_∞, is given by $(k^*(n)(X)/v_n\text{-torsion}) \otimes_{\mathbb{F}_p[v_n]} \mathbb{F}_p$. The v_n-torsion subgroup of $k(n)^*(X)$ consists of elements annihilated by some power of v_n. [Johnson, D73] identified this spectral sequence with an Atiyah-Hirzebruch spectral sequence (Theorem 11.16). It follows from this observation that the spectral sequence supports a commutative and associative multiplication. [Kane86] developed many properties of this spectral sequence for the prime 2 including a notion of infinite implications that played a key role in a proof of Theorem 10.27. [Kane86] conjectured that, for a mod 2 finite H-space (X, μ, e), the Bockstein spectral sequence for Morava K-theory should satisfy the following two properties:

(1) The even degree algebra generators of $H^*(X; \mathbb{F}_2)$ can be chosen to be permanent cycles in B_r.
(2) In degrees greater than or equal to 2^{n+1}, the even degree generators can be chosen to be boundaries in B_r.

If these conjectures were to hold, a simple proof of the absence of 2-torsion in $H_*(\Omega X)$ for a mod 2 finite H-space (X, μ, e) would be possible (as outlined by [Kane86]).

Exercises

10.1. Show that the condition, $H^{\text{odd}}(\Omega G; k) = \{0\}$ for all fields k, implies that $H^*(\Omega G)$ is torsion-free.

10.2. Prove that a commutative, associative Hopf algebra over a field of characteristic zero that is generated by odd-dimensional generators is an exterior algebra.

10.3. From the structure of $H^*(\mathbb{R}P^n; \mathbb{F}_2)$ as a module over the Steenrod algebra, determine completely the mod 2 Bockstein spectral sequence for $\mathbb{R}P^n$.

10.4. The mod 2 cohomology of the exceptional Lie group G_2 is given by

$$H^*(G_2; \mathbb{F}_2) \cong \mathbb{F}_2[x_3, x_5]/\langle x_3^4, x_5^2 \rangle.$$

The rational cohomology of G_2 is given by $H^*(G_2; \mathbb{Q}) \cong \Lambda(X_3, X_{11})$. From these data determine the mod 2 Bockstein spectral sequence for G_2.

10.5. Prove Proposition 10.8 and Lemma 10.9.

10.6. Prove the analogue of Corollary 10.14 for $K(\mathbb{Z}/2^k\mathbb{Z}, n)$.

10.7. Suppose X is an H-space and $\pi: \bar{X} \to X$ a covering space of X. Then $\bar{X}$ is an H-space and π a multiplicative mapping. Use the Cartan-Leray spectral sequence (Theorem $8^{\text{bis}}.9$) which is a spectral sequence of Hopf algebras in this case to prove that if X is a mod p finite H-space, then $\bar{X}$ is a mod p finite H-space ([Browder59]).

10.8. Show that if A' is a normal sub-Hopf algebra of the Hopf algebra A, then there is a diagram of short exact sequences:

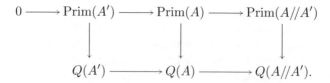

Use this fact to give a complete proof of Theorem 10.25.

10.9. Show that the universal examples of $K(\mathbb{Z}/p^k\mathbb{Z}, n)$, for $k > 0$, and $K(\mathbb{Z}, n)$ lead to the spikes and differentials in the Adams spectral sequence as predicted by Theorem 10.29.

10.10. Suppose that M is compact, closed manifold (or more generally a Poincaré duality space). If M has dimension $4m + 1$, then prove the following result due to [Browder62']: either (1) $H_{2m}(M) \cong F \oplus T \oplus T$, where F is a free abelian group and T is a torsion group, or (2) $H_{2m}(M) \cong F \oplus T \oplus T \oplus \mathbb{Z}/2\mathbb{Z}$ and in this case, $Sq^{2m} : H^{2m+1}(M; \mathbb{F}_2) \to H^{4m+1}(M; \mathbb{F}_2)$ is nonzero.

Part III

Sins of Omission

11
More Spectral Sequences in Topology

> "Topologists commonly refer to this apparatus as 'machinery'."
>
> *J. F. Adams*

The examples developed in Chapters 5 through 10 by no means exhaust the significant appearances of spectral sequences in mathematics. A recent search on the keyword `spectral sequence` in the database MATHSCINET delivered more than 2800 reviews in which the words are mentioned. In this chapter and the next, we present a kind of catalogue, by no means complete or self-contained, meant to offer the reader a glimpse of the scope of the applications of spectral sequences. (Similar catalogues are found in the books of [Griffiths-Harris78], [Benson91], [Weibel94], and the fundamental paper of [Boardman99].) I hope that the reader will find a useful example in this collection or at least the sense in which spectral sequences can be applied in his or her field of interest. The algebraic foundations supplied in Chapters 1, 2, and 3 are sufficient to understand the constructions found in the cited references.

In this chapter we concentrate on diverse applications of spectral sequences in algebraic and differential topology. The examples are organized loosely under the rubrics of spectral sequences associated to a mapping or space of mappings (§11.1), spectral sequences derived for the computation of generalized homology and cohomology theories (§11.2), other Adams spectral sequences (§11.3), spectral sequences that play a role in equivariant homotopy theory (§11.4), and finally, miscellaneous examples (§11.5).

11.1 Spectral sequences for mappings and spaces of mappings

The Leray-Serre spectral sequence is associated to a fibration, $\pi \colon E \to B$. Its success owes much to the right definition of fibration, due to [Serre51]. In this section we discuss some spectral sequences also associated to particular types of mappings or to spaces of mappings.

We first consider the dual of a fibration and present two spectral sequences related to cofibrations. The first is due to [Quillen69] and appears in his foundational paper on rational homotopy theory. The key piece of structure in the following theorem is the fact that the rational homotopy groups $\pi_*(\Omega X) \otimes \mathbb{Q}$ form a graded Lie algebra with the product induced by the Samelson product. The following is a kind of dual to the rational Leray-Serre spectral sequence.

Theorem 11.1. *Suppose $A \to X \to X/A$ is a cofibration sequence. Then there is a spectral sequence of graded Lie algebras with*

$$E^2_{p,q} \cong (\pi_p(\Omega A) \otimes \mathbb{Q}) \vee (\pi_q(\Omega(X/A) \otimes \mathbb{Q}),$$

and converging to $\pi_*(\Omega X) \otimes \mathbb{Q}$. Here $\ell \vee \ell'$ is the direct sum of the graded Lie algebras ℓ and ℓ'.

[Neumann99] has given a parallel derivation of an analogous spectral sequence for loops on a cofibre sequence by filtering the cobar construction.

The next spectral sequence is roughly dual to the homology Eilenberg-Moore spectral sequence of §7.4. The derivation is due to [Barratt62] and the spectral sequence generalizes the suspension phenomena that occur in the EHP sequence of [Whitehead, GW53].

Theorem 11.2 (the Barratt spectral sequence). *Given a cofibration sequence $A \to X \to X/A$, there is a spectral sequence with*

$$E^{p,q}_1 \cong \begin{cases} \pi_q(X), & \text{if } p = 0, \\ \pi_{q+2}(X/A), & \text{if } p = 1, \\ \tilde{\pi}_{2p+q}(X/A \vee \underbrace{SA \vee \cdots \vee SA}_{p-1 \text{ times}}), & \text{if } p > 1, \end{cases}$$

where $\tilde{\pi}_*(X/A \vee SA \vee \cdots \vee SA) \subset \pi_*(X/A \vee SA \vee \cdots \vee SA)$ is the subgroup of cross terms. The spectral sequence converges to $\pi_*(A)$.

[Barratt62] studied d_1 and showed that the E_2-term of this spectral sequence can be expressed in terms of the cohomology of an **analyzer** as defined by [Lazard55]. Generalizations of and computations using this spectral sequence are found in work of [Goerss93].

Just as the Künneth spectral sequence generalizes the Künneth theorem, other classical constructions in homotopy theory admit a generalization by a spectral sequence. A tool in deriving these generalizations is a result due to [Quillen66].

Theorem 11.3. *Let $A_{\bullet\bullet}$ denote a bisimplicial group. There is a natural first quadrant spectral sequence of homological type with $E^2_{p,q} \cong \pi^h_p \pi^v_q(A_{\bullet\bullet})$ and converging to $\pi_{p+q}(\Delta A_{\bullet\bullet})$, where $\Delta A_{\bullet\bullet}$ is the diagonal simplicial group with $(\Delta A_{\bullet\bullet})_n = A_{nn}$.*

The vertical homotopy groups $\pi_q^v(A_{\bullet\bullet})$ denote the homotopy groups of the simplicial groups $A_{n\bullet}$, and the resulting groups form another simplicial group whose homotopy groups are the horizontal groups $\pi_p^h\pi_q^v(A_{\bullet\bullet})$.

When the bisimplicial group $A_{\bullet\bullet} = (GX)_\bullet * (GY)_\bullet$, for spaces X and Y, G the loop group functor of [Kan58], and $*$ the free product of groups, then [Hirschhorn87] has analyzed the resulting spectral sequence:

Theorem 11.4. *For spaces X and Y, there is a natural first quadrant spectral sequence of homological type, converging to $\pi_*(X \vee Y)$. When Y is $(n-1)$-connected, $E_{p,0}^2 \cong \pi_{p+1}(X)$ and $E_{p,q}^2 \cong H_p(\Omega X; \pi_{q+1}(Y))$ for $1 \leq q \leq 2n - 3$. If X is $(k-1)$-connected, then $E_{0,q}^2 \cong \pi_{q+1}(Y)$, and $E_{p,q}^2 = \{0\}$ for $1 \leq p \leq k - 2$.*

More can be said in this case by adding the subtleties that are organized by the notion of a Π-algebra, introduced by Kan and developed by [Stover88]. The homotopy groups of a space X, as a graded set, enjoy the action by the primary homotopy operations, namely,

(1) Composition: $\alpha \in \pi_r(X) \mapsto \alpha \circ \zeta \in \pi_k(X)$ where $\zeta \in \pi_k(S^r)$ and $k > r > 1$.

(2) Whitehead products: $[\alpha, \beta] \in \pi_{p+q-1}(X)$ for any $\alpha \in \pi_p(X)$, $\beta \in \pi_q(X)$; elements of the form $[\lambda]\alpha - \alpha \in \Gamma_\pi^2(\pi_r(X))$ where $[\lambda] \in \pi = \pi_1(X)$ and $\alpha \in \pi_r(X)$ (see Chapter 8bis); and commutators $[\alpha, \beta] = \alpha\beta\alpha^{-1}\beta^{-1} \in \pi_1(X)$, for $\alpha, \beta \in \pi_1(X)$.

The free objects in the category of Π-algebras correspond to wedge products of spheres. A simplicial resolution of a space X, $V_\bullet X$ may be constructed whose homotopy groups constitute a free Π-algebra resolution of the homotopy of a space. Forming the wedge product $V_\bullet X \vee V_\bullet Y$, [Stover90] proved the following generalization of the van Kampen theorem.

Theorem 11.5. *There is a natural first quadrant spectral sequence of homological type, converging strongly to $\pi_*(X \vee Y)$, with $E_{p,*}^2 \cong D_p(\pi_*(X), \pi_*(Y))$, where D_0 denotes the coproduct of $\pi_*(X)$ and $\pi_*(Y)$ in the category of Π-algebras, and D_p is the pth derived functor of the coproduct functor.*

The van Kampen theorem follows from the lower left corner of the spectral sequence where $\pi_1(X \vee Y) \cong E_{0,1}^\infty \cong E_{0,1}^2 \cong \pi_1(X) * \pi_1(Y)$, as expected.

If we view the resolution of [Stover90] as a bisimplicial set, then, for a commutative ring with unit R, the functor $X \mapsto RX$ of [Bousfield-Kan72] may be applied to $V_\bullet X$ to obtain a bisimplicial R-module $RV_\bullet X$. The spectral sequence of [Quillen66] leads to the Hurewicz spectral sequence introduced by [Blanc90]:

Theorem 11.6. *Given a pointed, connected space X and a ring R, there is a spectral sequence that converges strongly to $\tilde{H}_*(X; R)$, with $E^2_{n,k} \cong L_n(Q_k(-) \otimes R)(\pi_*(X))$. Here L_n denotes the derived functors of the functor $Q_k(-) \otimes R$ on Π-algebras that associates to the Π-algebra $\pi_*(X)$ the quotient $\pi_*(X)/P(\pi_*(X))$ for $P(\pi_*(X))$ the subgroup generated by the image of the primary homotopy operations.*

The edge homomorphism $E^2_{0,*} \longrightarrow\!\!\!\!\!\rightarrow E^\infty_{0,*} \subset \tilde{H}_*(X; R)$ is given by the R-Hurewicz homomorphism $Q(\pi_*(X)) \otimes R \to \tilde{H}_*(X; R)$. By developing the homological algebra of the category of Π-algebras, [Blanc90] showed that the E^2-term has a vanishing line of slope 1/2. [Blanc94] enriched this spectral sequence by considering operations on resolutions in abelian categories. He computed the case of $X = K(\mathbb{Z}/2\mathbb{Z}, n)$ and was able to relate the differential d^2 to Toda brackets. An interesting corollary of the new operations is a nonrealization result: There is no space X with $\pi_*(X) \cong \pi_*(S^r) \otimes \mathbb{Z}/2\mathbb{Z}$ as Π-algebras for $r \geq 6$.

The next examples of spectral sequences apply to the problem of computing the homotopy groups of spaces of mappings. Given spaces X and Y, endow the set of continuous functions from X to Y, denoted $\mathrm{map}(X, Y)$, with the compact-open topology. Suppose $f\colon X \to Y$ is a choice of mapping as a basepoint in $\mathrm{map}(X, Y)$. The following theorem is due to [Federer56].

Theorem 11.7 (the Federer spectral sequence). *Suppose X is a finite dimensional CW-complex and Y is a space on which the fundamental group acts trivially on the higher homotopy groups. Then there is a spectral sequence with*

$$E_2^{p,q} \cong H^p(X; \pi_{p+q}(Y)),$$

converging to $\pi_*(\mathrm{map}(X, Y), f)$.

Federer applied the spectral sequence to the cases $Y = K(\pi, m)$ and $Y = S^m$. [Smith, S98] investigated the Federer spectral sequence in rational homotopy theory where models for spaces can be taken to be algebraic objects, such as the Quillen model given by a free Lie algebra. In this context [Smith, S98] made some explicit computations with surprising corollaries about the inequality of homotopy types of components of a mapping space in the general case.

When X and Y have the homotopy type of CW-complexes, there is a natural mapping in the category of Π-algebras

$$b\colon \pi_*(\mathrm{map}_f(X, Y)) \to \hom^\Pi_{\pi_*(f)}(\pi_*(X), \pi_*(Y)),$$

which is an isomorphism when X has the homotopy type of a wedge of spheres. In the general case, there is a spectral sequence that relates $\pi_*(\mathrm{map}_f(X, Y))$ and $\hom^\Pi_{\pi_*(f)}(\pi_*(X), \pi_*(Y))$ due to [Dwyer-Kan-Smith, J-Stover94].

Theorem 11.8. *Let X and Y have the homotopy type of CW-complexes. There is a second quadrant spectral sequence with $E_2^{p,q} \cong \hom_{\pi_*(f)}^{\Pi\,(p)}(\pi_*(X), \pi_*(Y))_q$, for $q \geq p \geq 1$, converging conditionally to $\pi_*(\mathrm{map}_f(X,Y))$. The edge homomorphism for this spectral sequence is the natural homomorphism b.*

Here $\hom_{\pi_*(f)}^{\Pi\,(0)}(\pi_*(X), \pi_*(Y)) = \hom_{\pi_*(f)}^{\Pi}(\pi_*(X), \pi_*(Y))$, the hom-set functor in the category of Π-algebras, and the functor $\hom^{\Pi\,(p)}$ is the pth derived functor of $\hom^{\Pi}(-, \pi_*(Y))$. When Y has only finitely many non-trivial homotopy groups, or $\pi_*(X)$ has finite cohomological dimension as a Π-algebra, then the authors show that the spectral sequence converges strongly to $\pi_*(\mathrm{map}_f(X,Y))$.

Another source of examples of spectral sequences is the problem of computing the homology or cohomology of mapping spaces. The particular case of pointed maps of spheres, $\Omega^n X = \mathrm{map}((S^n, \vec{e}_1), (X, x_0))$, was solved by [Adams56] for the functor $X \mapsto \Omega X$ with the introduction of the cobar construction and its associated spectral sequence. [Baues98] has developed the structure of the cobar construction further so that it may be iterated (compare the work of [Drachman67] and [Smith, Ju94]). The homology Eilenberg-Moore spectral sequence (Chapter 7) provides a generalization of the cobar construction. Using cosimplicial methods, [Anderson72] constructed a spectral sequence that may be used to compute $H_*(\mathrm{map}(X,Y))$:

Theorem 11.9. *Given a Kan complex Y and a finite CW-complex X for which the connectivity of Y is greater than or equal to the dimension of X, there is a spectral sequence, converging to $H_*(\mathrm{map}(X,Y))$, with $E_{p,q}^2 \cong H^q(X; H_p(Y))$.*

A complete proof of this theorem and some considerable generalizations are given by [Bendersky-Gitler91], who show how configuration spaces appear in the computation of the E_1-terms of the associated spectral sequences and relate these results to the computation of the Gelfand-Fuks cohomology of manifolds. [Bousfield87] greatly generalized the construction of [Anderson72] by deriving a dual version of the Bousfield-Kan spectral sequence (Theorem $8^{\mathrm{bis}}.37$) for homology.

Unstable Adams spectral sequences

The existence of the Adams spectral sequence to compute the stable homotopy groups of a space leads one to wonder if similar machinery can be constructed to compute $[X, Y]$ or $\pi_*(X)$, the unstable sets of mappings. In this section we present several variants of the unstable Adams spectral sequence that converge to these unstable homotopy groups.

In the next theorem, the information that determines all of the homotopy of a simply-connected space, its Postnikov system, is used to obtain the homology of the space. Introduced by [Kahn, DW66], the spectral sequence provides a kind of dual to the Adams spectral sequence.

Theorem 11.10. *Let X be a space of the homotopy type of a 1-connected, countable CW-complex. There is a spectral sequence, converging to $H_*(X)$, with E^1-term partially given by $E^1_{p,q} \cong H_{p+q}(K(\pi_p(X), p))$ for $0 \le q \le p$.*

Though practically incalculable, this spectral sequence enjoys many geometric features that make it a useful tool. For example, the first differential, d_1, can be interpreted in terms of the k-invariants of the space. Also the edge homomorphism is the Hurewicz homomorphism. [Kahn, DW66] developed this spectral sequence in order to study composition products in π_*^S and it was used in this context by [Cohen, J68] to prove his celebrated theorem on the decomposition of the stable homotopy groups of spheres in terms of Toda brackets of Hopf maps.

The first spectral sequence to generalize the Adams spectral sequence to unstable computations is due to [Massey-Peterson67]. The construction is based upon their study of the cohomology of spaces satisfying a certain algebraic condition. The action of the mod 2 Steenrod algebra, $\mathcal{A}_2$, on $H^*(X; \mathbb{F}_2)$ satisfies the **unstable axioms**; (U1) $Sq^n x = x^2$, if $\deg x = n$, and (U2) $Sq^n x = 0$ if $n > \deg x$. Suppose M is a module over $\mathcal{A}_2$ such that (U2) holds for M. We define an algebra $U(M)$, satisfying the unstable axioms, by letting $U(M)$ be the quotient of the tensor algebra on M modulo the relations of graded commutativity and $Sq^n x = x^2$ for $\deg x = n$. If X is a space and $H^*(X; \mathbb{F}_2) = U(M)$ for some unstable $\mathcal{A}_2$-module M, then we say that X is **very nice** (following [Bousfield-Curtis70]).

Theorem 11.11. *Suppose Y is a simply-connected, very nice space with $H^*(X; \mathbb{F}_2) = U(M)$. If K is a finite complex, then there is spectral sequence, converging to $_{(2)}[S^m K, Y]$ for $m > 1$, with*

$$E_2^{p,q} \cong \mathrm{Unext}^{p,q}(M, H^*(K; \mathbb{F}_2)),$$

the extension functor derived from $\mathrm{Hom}(M, -)$ in the category of unstable modules over the Steenrod algebra.

For $K = *$ and $Y = S^{2n+1}$, this spectral sequence can be applied to compute $_{(2)}\pi_*(S^{2n+1})$. However, the calculation of the unstable Ext groups remains difficult, if not intractable. This spectral sequence was developed for odd primes by [Barcus68] and further properties, like a vanishing line, have been proved by [Bousfield70].

For more general spaces, we turn to simplicial methods to compute $\pi_*(X)$. If X is a simplicial set, then there is a simplicial free group, GX, that is a model for the loop space on the realization of X, $\Omega|X|$. It follows that

$\pi_i(X) \cong \pi_{i-1}(GX)$. For any simplicial group, $W_\bullet$, there is a filtration of $W_\bullet$, given by the lower central series in $W_\bullet$:

$$W_\bullet = \Gamma_1 W_\bullet \supset \Gamma_2 W_\bullet \supset \Gamma_3 W_\bullet \supset \cdots$$

where $\Gamma_2 W_\bullet = [W_\bullet, W_\bullet]$ and $\Gamma_n W_\bullet = [\Gamma_{n-1} W_\bullet, W_\bullet]$.

[Curtis65] introduced a spectral sequence based on this filtration of GX and converging to $\pi_*(X)$; [Rector66] generalized this with the mod p lower central series and he obtained a spectral sequence converging to $_{(p)}\pi_*(X)$. Finally, [Bousfield et al.66] (a group of six authors; A.K. Bousfield, E.B. Curtis, D.M. Kan, D.G. Quillen, D.L. Rector, and J.W. Schlesinger, then at MIT) analyzed Rector's spectral sequence to prove the following result.

Theorem 11.12 (the Λ-algebra). *Let (Λ, d) denote the associative differential graded algebra with unit given by*

(1) *Λ is generated by $\{\, \lambda_i \mid i = 0, 1, \dots \,\}$ with $\deg \lambda_i = i$.*

(2) *Products are subject to the relations that follow from*

$$\lambda_i \lambda_{2i+1+n} = \sum_{j \geq 0} \binom{n-j-1}{j} \lambda_{i+n-j} \lambda_{2i+1+j} \qquad i \geq 0, n \geq 0.$$

(3) *The differential is given by*

$$d(\lambda_n) = \sum_{j \geq 1} \binom{n-j}{j} \lambda_{n-j} \lambda_{j-1} \qquad n \geq 0.$$

*Then there is a spectral sequence with $(E_1, d_1) \cong (\Lambda, d)$ converging to $_{(2)}\pi_*^S$. If $I = (i_1, \dots, i_r)$, then I is said to be **admissible** if $2i_s \geq i_{s+1}$ for $1 \leq s < r$. Let $\lambda_I = \lambda_{i_1} \cdots \lambda_{i_r}$; we say that λ_I is an admissible monomial if I is admissible. Let $\Lambda(n)$ be generated by admissible monomials with $i_1 < n$. There is a spectral sequence with $(E_1(n), d_1(n)) \cong (\Lambda(n), d|_{\Lambda(n)})$ converging to $_{(2)}\pi_*(S^n)$.*

The odd primary version of the Λ-algebra was also given by [Bousfield et al.66] (and corrected by [Bousfield-Kan72]). Furthermore, by writing the adjoint of the Steenrod algebra action as

$$H_n(X; \mathbb{F}_2) = \mathrm{Hom}(H^n(X; \mathbb{F}_2), \mathbb{F}_2)$$
$$\xrightarrow[(Sq^i)^{\mathrm{dual}}]{} \mathrm{Hom}(H^{n-i}(X; \mathbb{F}_2), \mathbb{F}_2) = H_{n-i}(X; \mathbb{F}_2),$$

there is a differential on $H_*(X; \mathbb{F}_2) \otimes \Lambda$ that gives the E_1-term of a spectral sequence converging to $_{(2)}\pi_*^S(X)$ (see [Bousfield et al.66]). Extensive calculations of the unstable homotopy groups of spheres using the Λ-algebra were done by [Whitehead, GW70] and [Tangora85].)

A special case of the Bousfield-Kan spectral sequence (Theorem $8^{\text{bis}}.37$) for a ring R gives a general spectral sequence converging to the homotopy groups of $R_\infty X$, the R-completion of X under good conditions. When $R = \mathbb{F}_p$, $\mathbb{F}_{p\infty} X$ is the mod p completion of X when X is mod p good, and $\pi_*(\mathbb{F}_{p\infty} X) = \pi_*(X) \otimes \widehat{\mathbb{Z}}_{(p)}$, where $\widehat{\mathbb{Z}}_{(p)}$ denotes the p-adic integers.

Theorem 11.13. *For an $\mathbb{F}_p$-good space X, there is a spectral sequence with*

$$E_2^{*,*} \cong \text{Unext}^{*,*}(\mathbb{F}_p, H_*(X; \mathbb{F}_p)).$$

and converging to $\pi_*(X) \otimes \widehat{\mathbb{Z}}_{(p)}$.

The E_2-term is expressed in terms of the 'derived' functors of the functor Hom in the category of unstable coalgebras over $\mathcal{A}_p$. Since this category is not abelian, we must take the derived functors of Hom in the extended sense of [André67]. The spectral sequence was derived and developed by [Bousfield-Kan72']. In the case of the homotopy groups of a mapping space, [Goerss90] has made considerable progress in identifying the E_2-term of this spectral sequence using André-Quillen cohomology.

The (co)simplicial techniques of [Bousfield-Kan72] can be generalized to derive an unstable Adams spectral sequence associated to the spectrum BP ([Bendersky-Curtis-Miller78]).

Finally, we mention a spectral sequence that relates the unstable homotopy groups of spheres and the stable groups. [James56] identified a fibration of spaces localized at the prime 2:

$$S^n \to \Omega S^{n+1} \to \Omega S^{2n+1},$$

whose long exact sequence of homotopy groups, the **EHP sequence**, is given by

$$\cdots {}_{(2)}\pi_k(S^n) \xrightarrow{E} {}_{(2)}\pi_{k+1}(S^{n+1}) \xrightarrow{H} {}_{(2)}\pi_{k+1}(S^{2n+1}) \xrightarrow{P} {}_{(2)}\pi_{k-1}(S^n) \cdots.$$

[Toda62] extended the EHP sequence to odd primes by introducing p primary fibrations

$$S^{2n-1} \to \Omega \tilde{S}^{2n} \to \Omega S^{2np-1}, \qquad \tilde{S}^{2n} \to \Omega S^{2n+1} \to \Omega S^{2np+1},$$

where $\tilde{S}^{2n}$ is a modified version of the $2n$-sphere that has $p-1$ cells, one in each dimension divisible by $2n$ up to $2n(p-1)$.

The exact couple associated to the resulting long exact sequences of homotopy groups gives the EHP spectral sequence:

Theorem 11.14. *There are spectral sequences for each prime p, converging to $_{(p)}\pi_k^S$, indexed so that $d_r \colon E_r^{k,n} \to E_r^{k-1,n-r}$, and with E^1-terms given by*

$$E_{k,n}^1 \cong {}_{(2)}\pi_{n+k}(S^{2n-1}) \ for \ p = 2,$$

$$E_{k,2m+1}^1 \cong {}_{(p)}\pi_{2m+1+k}(S^{2mp+1}), E_{k,2m}^1 \cong {}_{(p)}\pi_{2m+k}(S^{2mp-1}), \ for \ p, \ odd.$$

[Toda62] was able to compute $\pi_{n+k}(S^n)$ through a range of n and k using the EHP sequence inductively, together with the secondary composition operations (the Toda bracket) that he introduced. The EHP spectral sequence ties together all of the EHP sequences and codifies the 'birth' and 'death' of elements in the homotopy groups of spheres—an element in π_k^S is represented in the E^1-term by the Hopf invariant of the a maximal desuspension of the element; each differential represents a Whitehead product. The EHP sequence may be approached at an algebraic level through the Λ-algebra (see the work of [Whitehead, GW70], [Singer75], [Lin, WH92], and [Mahowald-Thompson95]). A thorough discussion of the EHP spectral sequence may be found in [Ravenel86, §1.5].

11.2 Spectral sequences and spectra

The focus of this book has been on the computation of the classical homotopy invariants of a space—ordinary homology, cohomology, and homotopy groups. However, there are many other homotopy invariants associated to a space—in particular, there are the generalized homology and cohomology functors. These functors satisfy all but one of the Eilenberg-Steenrod axioms for a homology or cohomology theory.

[Brown, E62] proved that the generalized cohomology functors, $X \mapsto E^*(X)$, were **representable**, that is, for each n, there is a space, W_n, such that $E^n(X) = [X, W_n]$. This generalizes the fact that the ordinary cohomology groups are represented by the Eilenberg-Mac Lane spaces. The system of spaces, $\{W_i\}$, satisfies certain relations that had been identified by [Lima58] and [Whitehead, GW62]. In particular, they constitute a spectrum (Definition 9.28).

Definition 11.15. *Given a spectrum $\boldsymbol{E} = \{E_n\}$, with structure mappings $\{\varepsilon_n \colon SE_n \xrightarrow{\simeq} E_{n+1}\}$, the **generalized cohomology theory** associated to $\boldsymbol{E}$ of a space X, is denoted by $E^*(X)$ and defined by $E^n(X) = [X, E_n]$. The **generalized homology theory** associated to $\boldsymbol{E}$ is denoted $E_*(X)$ and defined by $E_k(X) = \lim_{n,\varepsilon_n} \pi_{n+k}(E_k \wedge X)$. The **coefficients** of the generalized theories determined by $\boldsymbol{E}$ are given by the graded group $E^k(*) \cong E_k(*) = \lim_{n,\varepsilon_n} \pi_{n+k}(E_n)$. The analogue of the Steenrod algebra for the cohomology theory $E^*(-)$ is the algebra, E^*E, $(E^*E)_r = \lim_{n,\varepsilon_n}[E^{n+r}, E^n]$.*

Generalized homology and cohomology theories satisfy most of the axioms of Eilenberg-Steenrod for homology and cohomology; the exception is the coefficient axiom. A further axiom, the **wedge axiom** may be introduced: If X is a (possibly infinite) wedge of spaces, $X = \bigvee_\alpha Y_\alpha$, then $E^*(X) \cong \prod_\alpha E^*(Y_\alpha)$ and $E_*(X) \cong \bigoplus_\alpha E_*(Y_\alpha)$. A generalized theory is said to be **connective** if there is an integer N such that $E_k = \{0\}$ for all $k < N$.

How do we compute $E^*(X)$ for a given space X? The most general answer to this question is a spectral sequence relating the classical invariants of the space X and the coefficients of the theory $E^*(-)$ to $E^*(X)$. The first published version is due to [Atiyah-Hirzebruch69], though the spectral sequence was known to exist by G.W. Whitehead and by E.L. Lima.

11.16 (the Atiyah-Hirzebruch spectral sequence). *Suppose E is a spectrum and X is a space of the homotopy type of a CW-complex. Then there are half-plane spectral sequences with*

$$E_2^{p,q} \cong H^p(X; E_q(*)), \quad E^2_{p,q} \cong H_p(X; E_q(*)),$$

converging conditionally to $E^*(X)$ and strongly to $E_*(X)$, respectively.

The construction is based on the cell decomposition and is similar to the proof of Theorem 4.13. [Davis-Lück98] have generalized the Atiyah-Hirzebruch spectral sequence to the framework of spectra over a category, which allows one to use it in many contexts including equivariant homotopy theory, and for algebraic K-theory.

The classic book of [Adams74] is a good starting place for the study of spectra. Other good references include [Switzer75], [Margolis83], [Ravenel92], and [Kochman96].

In the special case of $X = BG$, G a finite group, and $E^* = KU^*$, complex K-theory, the computation of $KU^*(BG)$ is aided by the interpretation of its input ([Atiyah61]):

Theorem 11.17. *For G a finite group, there is a spectral sequence with $E_2^* \cong H^*(G)$ and converging strongly to $KU^*(BG)$.*

The input of the spectral sequence is the cohomology of the group G with coefficients in the trivial G-module $\mathbb{Z}$, an algebraic invariant of the group. A filtration of the complex representation ring of the group G leads to the same associated graded ring for $KU^*(BG)$ related by interpreting representations as vector bundles. The result shows that the complex K-theory of BG is given by the completion of the representation ring of G with respect to this filtration.

The Atiyah-Hirzebruch spectral sequence also plays a key role in computations of the homotopy groups of spheres. In this case the spectrum E is the sphere spectrum and the space X is replaced by a spectrum X. The E^2-term is given by $H^*(X; \pi_*^S)$ and the spectral sequence converges to $\pi_*(X)$. [Cohen, J68] used this when $X = K\mathbb{Z}$, the integral Eilenberg-Mac Lane spectrum. Then

$\pi_*^S(K\mathbb{Z}) \cong H_*(S^0)$ determines a sparse target for the spectral sequence and the homology of the Eilenberg-Mac Lane spaces is well-known, so computations of π_*^S, the coefficient ring for the sphere spectrum may be made. This becomes cumbersome quickly, however. [Kochman90] applied this technique for $X = BP$, the (mod 2) Brown-Peterson spectrum for which both $H_*(BP)$ and $\pi_*(BP)$ are well-known and algebraically tractable. Furthermore, it is known that the Hurewicz homomorphism $h\colon \pi_*(BP) \to H_*(BP)$ is a monomorphism and so $E_{n,0}^\infty \cong h(\pi_n(BP))$ is also known. [Kochman90] pushed the calculation of $_{(2)}\pi_n^S$ to $n \leq 66$ by automating the computation. [Ray72] used this method with $X = \mathbf{MSU}$ and $\mathbf{MSp}$ instead of the sphere spectrum. Since $MSp_*(\mathbf{MSU})$ and $H_*(\mathbf{MSU})$ are known, [Ray72] was able to compute MSp_k for $k \leq 19$.

Finally, we mention work of [Arlettaz92] analyzing the differentials in the Atiyah-Hirzebruch spectral sequence. He proved that there are integers R_r such that $R_r d_{s,t}^r = 0$ for all $r \geq 2$, s and t for any connected space X. The key ingredient of the proof is the structure of the integral homology of Eilenberg-Mac Lane spectra.

By exploiting the analogue of the Steenrod algebra for a generalized cohomology theory, [Novikov67] generalized the Adams spectral sequence to other cohomology theories. We will discuss this advance separately.

A spectrum equipped with a multiplication, $\mu\colon \mathbf{E} \wedge \mathbf{E} \to \mathbf{E}$, (here μ is a map of spectra where the smash product is appropriately defined) is called a **ring spectrum**. If $\mathbf{F}$ is another spectrum and there is a mapping of spectra $\psi\colon \mathbf{E} \wedge \mathbf{F} \to \mathbf{F}$ with good properties, then we say that $\mathbf{F}$ is an $\mathbf{E}$**-module spectrum** ([Elmendorf-Kriz-Mandell-May95]). The following theorem generalizes the Universal Coefficient theorem.

Theorem 11.18 (the Universal Coefficient spectral sequence). *Suppose $\mathbf{E}$ is a ring spectrum, $\mathbf{F}$ is an $\mathbf{E}$-module spectrum, and X is a space. Under certain conditions, there are spectral sequences with*

$$E^2 \cong \mathrm{Tor}^{E^*(*)}(E^*(X), F^*(*)), \quad E_2 \cong \mathrm{Ext}_{E^*(*)}(E^*(X), F^*(*)),$$

converging to $F^*(X)$ and to $F_*(X)$, respectively.

For appropriate conditions, the reader can consult the book of [Adams69] or the paper of [Boardman99] where there is a derivation and applications of this spectral sequence. The unstated technical conditions are satisfied by many of the geometric spectra (the sphere spectrum, mod p Eilenberg-Mac Lane spectrum, the Thom spectra $\mathbf{MO}$, $\mathbf{MU}$, $\mathbf{MSp}$, and the K-theory spectra BU and BO) and this leads to many interesting applications.

Another approach to the computation of $E_*(X)$ is via the Adams spectral sequence. We can carefully define the spectrum $\mathbf{E} \wedge X$ whose homotopy groups are analyzed in the same manner as the stable homotopy groups of a space. This approach figures in the classical computation of $\mathbf{MU}_*$ of [Milnor60] that has served as a paradigm for many computations of generalized homology (for example, [Davis78], [Davis et al.86], [McClure-Staffeldt93]).

For a generalized homology theory $E_*(-)$ and a fibration $\pi: Y \to B$ with fibre F, there is a version of the Leray-Serre spectral sequence that at once generalizes the classical spectral sequence for singular theory and the Atiyah-Hirzebruch spectral sequence. (One can consult the book of [Switzer75] for a derivation.)

Theorem 11.19. *Given a generalized homology theory $E_*(-)$ that satisfies the wedge axiom for CW-complexes and a fibration $F \hookrightarrow Y \overset{\pi}{\to} B$ that is orientable with respect to the theory for which B is connected, there is a spectral sequence, natural with respect to maps of fibrations, converging to $E_*(Y)$, and with $E_{p,q}^2 \cong H_p(B; E_q(F))$.*

There is also a version of the Eilenberg-Moore spectral sequence for generalized theories that was set up by [Hodgkin75] and [Smith, L70] (see §8.3). For this spectral sequence to have an identifiable E_2-term and to converge, however, many conditions must be placed on the generalized theory. In their study of the K-theory of p-compact homogeneous spaces, [Jeanerret-Osse99] gave a tidy statement of a useful case of this tool:

Theorem 11.20. *Suppose $E^*(-)$ is a generalized, multiplicative, cohomology theory such that $E^*(*)$ is a graded field. Suppose B is connected and*

$$
\begin{array}{ccc}
X \times_B Y & \longrightarrow & X \\
\downarrow & & \downarrow{\scriptstyle p} \\
Y & \underset{f}{\longrightarrow} & B
\end{array}
$$

is a pullback diagram. Then there is a spectral sequence of algebras, compatible with the stable operations associated to $E^*(-)$, with

$$E_2^{i,*} \cong \widehat{\mathrm{Tor}}_{E^*(B)}^{-i}(E^*(X), E^*(Y))$$

where $\widehat{\mathrm{Tor}}^{-i}$ denotes the ith derived functor of the completed tensor product. When $p: X \to B$ is a fibration and $E^*(\Omega B)$ is isomorphic to an exterior algebra on odd degree generators, the spectral sequence converges strongly to $E^*(X \times_B Y)$.

The main examples considered by [Jeanneret-Osse99] are p-compact groups and $E^*(-) = H^*(-; k)$, $KU^*(-; \mathbb{Z}/p\mathbb{Z})$, or $K(n)^*(-)$ for which these hypotheses are appropriate. [Tanabe95] has also applied a version of the Eilenberg-Moore spectral sequence for generalized theories to compute the Morava K-theories of Chevalley groups. [Seymour78] also studied the convergence question for generalized theories under more general circumstances.

We close this section with a spectral sequence that computes an invariant of a spectrum $E = \{E_n\}$, its mod p **stable homology**,

$$H_*^s(E; \mathbb{F}_p) \equiv \lim_{\longrightarrow} H_{*+n}(E_n; \mathbb{F}_p).$$

Suppose $E = \{E_n\}$ is an Omega-spectrum, that is, $E_n \simeq \Omega E_{n+1}$, and that E is (-1)-connected, that is E_n is $(n-1)$-connected for all $n \geq 0$. In this case, the space E_0 is an **infinite loop space**; $E_0 \simeq \Omega E_1 \simeq \Omega^2 E_2 \simeq \cdots$. The mod p homology of an infinite loop space is endowed with the action of the Dyer-Lashof algebra, $\Re$ ([Araki-Kudo56], [Dyer-Lashof62]).

Let $Q(-)$ denote the functor that assigns the space of indecomposables to an algebra.

Theorem 11.21 (the Miller spectral sequence). *Given an Omega-spectrum* $\{E_n\}$ *that is* (-1)*-connected, there is a spectral sequence with*

$$E_{s,*}^2 = L_s(\mathbb{F}_p \otimes_\Re Q)(H_*(E_0; \mathbb{F}_p)),$$

the left derived functors of $\mathbb{F}_p \otimes_\Re Q(-)$, *and converging to* $H_*^s(E; \mathbb{F}_p)$.

[Miller78] analyzed the left derived functors in the theorem and expressed them in terms of an unstable Tor functor. The spectral sequence has been applied by [Kraines-Lada82] and [Kuhn82].

11.3 Other Adams spectral sequences

The Adams spectral sequence begins with the algebraic information encoding the action of the Steenrod algebra on the cohomology of the spaces involved. The output is geometric—the groups of stable mappings between the spaces. The construction presented in Chapter 9 is based on the properties of mod p cohomology and focuses on the Eilenberg-Mac Lane spaces for their homological properties on cohomology. [Adams66] introduced a variant of the Adams spectral sequence based on K-theory and posed the question of the existence of an Adams spectral sequence for any generalized cohomology theory. [Novikov67] introduced the appropriate generalization and applied it to the spectrum **MU** representing complex cobordism.

Theorem 11.22 (the Adams-Novikov spectral sequence). *Suppose* E *is a spectrum and* $E_*(E)$ *is flat as a right module over* $E_*(*)$. *Suppose further that* E *is a direct limit of finite spectra that satisfy good duality properties, then there is a spectral sequence with*

$$E_2^{s,t} \cong \mathrm{Ext}_{E_*(E)}^{s,t}(E_*(*), E_*(*)),$$

and converging to π_*^S. *In particular, this theorem holds for the sphere spectrum, the Thom spectra* **MO**, **MU**, **MSp** *and the connective K-theory spectra,* bu *and* bo.

For the full generality of the Adams spectral sequence (which converges to a subgroup of $\{Y, X\}_*$), one needs to introduce the localization of the stable group $\{Y, X\}_*$ with respect to the generalized theory E_*. The details of this localization are due to [Bousfield75] and a complete derivation of the spectral sequence is presented in [Ravenel86].

When we consider **MU** one prime at a time, then we are led to consider the localization **MU**$_{(p)}$ in Novikov's variant of the Adams spectral sequence. [Quillen69"] showed that the mod p part of the **MU** spectrum splits into a wedge of suspensions of another mod p spectrum, constructed by [Brown-Peterson66], now denoted by BP; in particular, there is a retraction **MU**$_{(p)} \to BP$ and so Theorem 11.22 may be localized mod p:

Theorem 11.23. *There is a spectral sequence with*

$$E_2^{s,t} \cong \mathrm{Ext}_{BP_*(BP)}^{s,t}(BP_*(*), BP_*(*)),$$

converging to $_{(p)}\pi_*^S$.

The algebraic properties of this spectrum and the algebra of operations associated to it are considerably more manageable than the analogous case of the Steenrod algebra. In particular, $BP_*(*) \cong \mathbb{Z}_{(p)}[V_1, \dots, V_t, \dots]$, a polynomial algebra on generators $V_i \in BP_{2p^i-2}$, and $BP_*(BP) \cong BP_*(*)[t_1, \dots, t_n, \dots]$ where $t_i \in BP_{2p^i-2}(BP)$. The subsequent further structure that has been developed for the E_2-term of the associated Adams-Novikov spectral sequence has led to great deal of progress in the understanding of the stable groups $_{(p)}\pi_*^S$ (see, for example, the papers of [Thomas-Zahler74], [Miller-Ravenel-Wilson77], and [Devinatz-Hopkins-Smith88]). For a good introduction to this point of view, see the books of [Ravenel86] and [Kochman96].

Another consequence of the study of formal group laws is the possibility of constructing new cohomology theories with particular rings of coefficients as $E_*(*)$. The principal theorem in such constructions is the Landweber exact functor theorem ([Landweber76]). Of particular interest is the case of **elliptic homology**, $\mathrm{Ell}_*(X) = \mathrm{Ell}_*(*) \otimes_{MU_*} MU_*(X)$ where $\mathrm{Ell}_*(*) \cong \mathbb{Z}[1/6][\delta, \varepsilon, \Delta^{-1}]$ where $\Delta = (1/1728)(\delta^3 - \varepsilon^2)$, with $\delta \in \mathrm{Ell}_8$, $\varepsilon \in \mathrm{Ell}_{12}$ and $\Delta \in \mathrm{Ell}_{24}$. This ring is isomorphic to the ring of modular forms of level 1 and there is a genus $MU_*(*) \to \mathrm{Ell}_*(*)$ giving the module structure. The ring of cooperations has been worked out by [Clarke-Johnson92] and so the input for the Adams-Novikov spectral sequence is known. [Hopkins95], [Laures99], and [Baker99] have used methods from number theory to identify parts of the

E_2-term of the spectral sequence converging to the stable stems. This gives information in both directions—number theory to topology, and stable homotopy to number theory.

11.4 Spectral sequences in equivariant homotopy theory

Suppose that G is a topological group of the homotopy type of a CW-complex and X is a space on which G acts. The **equivariant cohomology** of X is defined using the **Borel construction** ([Borel60]):

$$H_G^*(X; R) \cong H^*(EG \times_G X; R),$$

that is, the ordinary cohomology with coefficients in R of the space $EG \times_G X$ that can be thought of as first making X into a free G-space by forming the product with the G-free contractible space EG and then taking the quotient. There is a fibration $EG \times_G X \to EG_G \times * = BG$, induced by the G-mapping $X \to *$, and so we can apply the Leray-Serre spectral sequence:

Theorem 11.24. *There is a first quadrant spectral sequence converging to* $H_G^*(X; R)$ *with* $E_2^{p,q} \cong H^p(BG; H^q(X; R))$.

When G is a discrete group, the E_2-term is the cohomology of the group G with coefficients in the G-module $H^*(X; R)$. The coefficients of equivariant cohomology are given by $H_G^*(*; R) = H^*(BG; R)$. The spectral sequence has an induced action of this ring on its terms, making it more tractable. Applications of this spectral sequence abound in equivariant homotopy theory.

Another invariant of a G-space X is the **Bredon homology** ([Bredon67]) associated to a functor $\mathcal{H} \colon G\text{-}\mathbf{Mod} \to \mathbf{Ab}$, from G-modules to abelian groups, which preserves arbitrary direct sums. When we apply $\mathcal{H}$ to the G-module of n-chains on X, we obtain a chain complex $C_n^G(X; \mathcal{H}) = \mathcal{H}(C_n(X))$. The Bredon homology of X with coefficients in $\mathcal{H}$ is defined as the homology groups $\mathbb{H}_n^G(X; \mathcal{H}) = H_n(C_*^G(X; \mathcal{H}), \mathcal{H}(\partial))$. One can identify the category of G-modules with the orbit category, $\mathcal{O}(G)$, consisting of subgroups of G together with inclusions. A similar definition can be given for a coefficient functor taking values in modules over a given ring.

Suppose G is a finite group and $f \colon X \to Y$ is a G-fibration, that is, f is G-equivariant and has the homotopy lifting property for all G-space. Then there a version of the Leray-Serre spectral sequence, derived by [Moerdijk-Svensson93] using the cohomology of categories, and by [Honkasalo98] using the locally constant cohomology of [Spanier92]. Let $\mathbb{H}_G^*(X, M)$ denote the **Bredon cohomology** of the G-space X with coefficients in a G-module M.

Theorem 11.25. *Given a finite group G and a G-fibration $f: X \to Y$, then there is a natural, first quadrant, spectral sequence converging to $\mathbb{H}_G^*(X, M)$ with $E_2^{p,q} \cong \mathbb{H}_G^p(Y, \mathbb{H}_G^q(f, M))$, where M may be taken to be a G-coefficient system determined by a functor $M: \mathcal{O}(G)^{\mathrm{op}} \to \mathbf{Ab}$ and $\mathbb{H}_G^*(f, M)$ is another G-coefficient system that is induced by the fibration.*

Other spectral sequences useful in ordinary homotopy theory have equivariant versions as well. [Intermont97, 99] has extended the notion of Π-algebras to the equivariant case and derived versions of the spectral sequence of [Stover90] for computing $\pi_{W+n}^G(X \vee Y)$ and $\pi_{W+n}^G(X \wedge Y)$ where G is a finite group and W is a finite dimensional representation of G.

There is a version of the Eilenberg-Moore spectral sequence as it resembles the Universal Coefficient and Künneth spectral sequences for Borel homology and cohomology developed by [Greenlees92]. Equivariant versions of the Federer spectral sequence have been derived by [Møller90] and by [Fieux-Solotar98] converging to $\pi_*(\mathrm{map}_f^G(X, Y))$ under certain conditions. Finally, there are Adams spectral sequences for which the target is the appropriate completion of $\{X, Y\}^G$, the group of homotopy classes of G-equivariant stable mappings. This method has been developed extensively by [Greenlees88', 92']. A nice overview of these ideas and their relation to classical homotopy theory is found in [Greenlees88].

Homotopy limits and colimits spectral sequences

One of the most general topological situations in which a spectral sequence arises is when a homotopy limit or colimit is constructed. Following [Bousfield-Kan72] and [Dwyer98], we associate to a small category $\mathbf{D}$ a simplicial set, $\mathrm{nerve}(\mathbf{D})_\bullet$, given by

$$\mathrm{nerve}(\mathbf{D})_n = \mathrm{Hom}_{\mathbf{Cat}}(\mathbf{n}, \mathbf{D})$$
$$= \{\sigma(0) \xrightarrow{\alpha_1} \sigma(1) \to \cdots \xrightarrow{\alpha_n} \sigma(n) \mid \sigma(i) \in \mathrm{Obj}(\mathbf{D}), \ \alpha_i \in \mathrm{Mor}(\mathbf{D})\},$$

where $\mathbf{Cat}$ is the category of small categories with functors and $\mathbf{n}$ is the category $(0 \to 1 \to \cdots \to n)$ with a single morphism between objects $i \to j$ whenever $i \leq j$. Thus, $\mathrm{nerve}(\mathbf{D})_n$ consists of the length n strings of composable morphisms in $\mathbf{D}$. The face maps are given by omission or composition in $\mathbf{D}$: $d_0(\sigma(0) \xrightarrow{\alpha_1} \cdots \xrightarrow{\alpha_n} \sigma(n)) = \sigma(1) \xrightarrow{\alpha_2} \cdots \xrightarrow{\alpha_n} \sigma(n)$, and if $i > 0$,
$d_i(\sigma(0) \xrightarrow{\alpha_1} \cdots \xrightarrow{\alpha_n} \sigma(n)) = \sigma(0) \to \cdots \to \sigma(i-1) \xrightarrow{\alpha_{i+1} \circ \alpha_i} \sigma(i+1) \to \cdots \to \sigma(n)$. Degeneracies are given by inserting the identity morphism on the objects in the sequence.

If $F: \mathbf{D} \to \mathbf{Simp}$ is a functor, then the **simplicial replacement of F** is the bisimplicial set $(\coprod F)_\bullet$ given by $(\coprod F)_n = \coprod_{\sigma \in \mathrm{nerve}(\mathbf{D})_n} F(\sigma(0))$ (the disjoint union) with d_i determined by $F(\sigma(0)) \to F((d_i\sigma)(0))$, which is the identity if $i > 0$ and $F(\alpha_1): \sigma(0) \to \sigma(1)$ if $i = 0$. Since $F: \mathbf{D} \to \mathbf{Simp}$, $(\coprod F)_\bullet$ is a bisimplicial set.

Definition 11.26. *Given a small category* **D** *and a functor* $F: \mathbf{D} \to \mathbf{Simp}$, *the* **homotopy colimit of** F *is the diagonal simplicial set of the simplicial replacement of* F, *that is,* $\operatorname{hocolim} F = \Delta((\coprod F)_\bullet)$ *with* $(\operatorname{hocolim} F)_n = (\coprod F)_{n,n}$.

Given an abelian group A as coefficients, the spectral sequence of a bisimplicial group (Theorem 11.3) gives the **Bousfield-Kan homology spectral sequence**.

Theorem 11.27. *There is a spectral sequence, converging to* $H_*(\operatorname{hocolim} F; A)$, *with* $E^2_{p,q} \cong \operatorname{colim}_p H_q(F; A)$, *where* $\operatorname{colim} \mathbf{Ab}^{\mathbf{D}} \to \mathbf{Ab}$ *is the colimit functor,* colim_i *the* i^{th} *left derived functor of* colim, *and* $H_q(F; A)$ *is the composite functor* $H_q(-; A) \circ F$.

Dually,[Bousfield-Kan72] defined the homotopy limit of a functor $F: \mathbf{D} \to \mathbf{Simp}$. We first form the **cosimplicial replacement** of F, $(\prod F)^\bullet$, which consists of the product $(\prod F)^n = \prod_{u \in \operatorname{nerve}(\mathbf{D}^{\mathrm{op}})_n} F(u(0))$ and the coface and codegeneracy mappings given by $d^0 = F(\alpha_1): F(\sigma(1)) \to F(\sigma(0))$, $d^j = \operatorname{id}$, for $j > 0$; $s^i = \operatorname{id}$, for $0 \leq i \leq n$. The **homotopy limit of** F is given by $\operatorname{holim} F = \operatorname{Tot}((\prod F)^\bullet)$. There is a natural mapping $\varprojlim_{d \in \mathbf{D}} F(d) \to \operatorname{holim} F$, which may not be a homotopy equivalence. This is called *the homotopy limit problem* ([Thomason83]). The Bousfield-Kan spectral sequence associated to the tower of fibrations built from Tot (Theorem $8^{\mathrm{bis}}.37$) implies the following result.

Theorem 11.28. *Suppose that* $F: \mathbf{D} \to \mathbf{Simp}$ *is such that* $F(d)$ *is fibrant for all* $d \in \operatorname{Obj} \mathbf{D}$. *There is a spectral sequence, with* $E^{p,q}_2 \cong \varprojlim^p \pi_q(F)$, *for* $0 \leq p \leq q$, *where* $\varprojlim \mathbf{Ab}^{\mathbf{D}^{\mathrm{op}}} \to \mathbf{Ab}$ *is the inverse limit functor,* $\varprojlim^i$ *the* i^{th} *derived functor of* $\varprojlim$, *and* $\pi_q(F)$ *is the composite functor* $d \mapsto \pi_q(F(d))$. *The spectral sequence converges to groups related to* $\pi_*(\operatorname{holim} F)$.

[Thomason83] showed how the homotopy limit problem included certain deep problems in homotopy theory. In particular, we can view a group G as a category, **G**, with objects the elements of G and a unique morphism $g \to h$ for all $g, h \in G$. The nerve of this category has the homotopy type of EG. A space (simplicial set) on which G acts determines a functor $X: \mathbf{G} \to \mathbf{Simp}$ for which $\operatorname{hocolim} X = EG \times_G X$ and $\operatorname{holim} X = \operatorname{map}_G(EG, X)$, the space of equivariant mappings $EG \to X$.

The Bousfield-Kan homology spectral sequence in this case can be identified with the homology version of the spectral sequence of Theorem 11.24, which in turn may be identified as the Leray-Serre spectral sequence associated to the fibration $X \hookrightarrow \operatorname{hocolim} X \to \operatorname{hocolim} * = BG$.

The homotopy limit problem for this setting is the comparison of $\varprojlim X = X$ and $\mathrm{map}_G(EG, X)$. If X has a trivial action of the group G, then the comparison leads to a comparison of fixed point sets

$$X = X^G = \mathrm{map}_G(*, X)^G \to \mathrm{map}_G(EG, X)^G = \mathrm{map}(BG, X).$$

[Sullivan70] conjectured that if G is finite and X is a finite CW-complex, then the based mapping space $\mathrm{map}_*(BG, X) = \mathrm{map}((BG, *), (X, x_0))$ would be weakly contractible. Through the use of group theory, the relevant cases to check are $G = \mathbb{Z}/p\mathbb{Z}$ for p, a prime. [Miller84] proved the Sullivan conjecture with a remarkable argument: The target of the Bousfield-Kan spectral sequence for this problem is $\pi_*(\mathrm{map}(B\mathbb{Z}/p\mathbb{Z}, X))$ for which the E^2-term has been identified as $\mathrm{Ext}^p_{\mathrm{CA}}(\tilde{H}_*(\Sigma^q B\mathbb{Z}/p\mathbb{Z}; \mathbb{F}_p), \tilde{H}_*(X; \mathbb{F}_p))$, where the Ext is taken over the category of unstable coalgebras over the mod p Steenrod algebra $\mathcal{A}_p$. The analysis of $H_*(B\mathbb{Z}/p\mathbb{Z}; \mathbb{F}_p)$ as an object in the category $\mathbf{U}$ of unstable comodules over $\mathcal{A}_p$ reveals that $\mathrm{Ext}_{\mathbf{U}}(H_*(B\mathbb{Z}/p\mathbb{Z}; \mathbb{F}_p), H_*(X; \mathbb{F}_p))$ vanishes when $H_*(X; \mathbb{F}_p)$ is bounded above. A version of the EHP spectral sequence for the algebraic functors $\Sigma \colon \mathbf{U} \to \mathbf{U}$ and its adjoint Ω extend the vanishing of $\mathrm{Ext}_{\mathbf{U}}$ to $\mathrm{Ext}_{\mathbf{U}}(H_*(\Sigma B\mathbb{Z}/p\mathbb{Z}; \mathbb{F}_p), H^*(X; \mathbb{F}_p))$. A Grothendieck spectral sequence for composite functors (Theorem 12.9) allows the passage from the category $\mathbf{U}$ to the category $\mathbf{CA}$. Thus the vanishing of the initial term of the spectral sequence converging to the initial term of the Grothendieck spectral sequence that converges to the initial term of the Bousfield-Kan spectral sequence gives the proof of the Sullivan conjecture.

The homotopy limit problem for spectra plays a role in the descent spectral sequence of [Thomason82] (see Chapter 12) and in the analysis of the Segal conjecture ([Carlsson87]).

11.5 Miscellanea

We add to our catalogue a few entries that are not in the mainstream of homotopy theory. The first focuses on manifolds and Poincaré duality; the second has led to considerable progress in the classification problem for knots, and the last appears in the study of singularities of mappings.

The first example is due to [Zeeman62] from his Cambridge thesis. The spectral sequence is derived from a double complex that is defined for a homology theory based on pairs of simplices instead of single cells. If given two simplicial complexes, K and L, then a **facing relation** on $K \times L$ is a set $\mathcal{F}$ of cells in $K \times L$, such that, whenever $\sigma \times \tau \in \mathcal{F}$ and $\sigma' \times \tau' \prec \sigma \times \tau$, then $\sigma' \times \tau' \in \mathcal{F}$. Let $\mathcal{H}^*(\mathcal{F})$ denote the system of local coefficients on K induced by $\mathcal{F}$.

Given a facing relation, let $L_{\mathcal{F}} = \{\sigma \mid \sigma \times \tau \in \mathcal{F}\}$ and we say that $\mathcal{F}$ is **left acyclic** if all $\sigma \in L_{\mathcal{F}}$ are acyclic.

Theorem 11.29 (the dihomology spectral sequence). *A left acyclic facing relation $\mathcal{F}$ on K and L gives rise to a spectral sequence with*

$$E^2_{p,q} \cong H_p(K; \mathcal{H}_q(\mathcal{F}))$$

and converging to $H_*(L)$.

This spectral sequence can be used to relate various homology theories (Čech to Vietoris, simplicial to singular, etc.) and to relate the spectral sequence of Leray to that of Serre (Čech to singular). Also, if

$$\mathcal{D} = \{(\sigma^p, \tau_q) \text{ for } \sigma^p \in K \text{ and } \tau_q \prec \sigma^p\},$$

then the resulting dihomology spectral sequence collapses to the isomorphism of Poincaré duality if K is a closed, orientable, combinatorial n-manifold. Thus the spectral sequence measures the failure of Poincaré duality for an arbitrary complex. Generalizations of this spectral sequence were derived by [Cain74] and [Sklyarenko92].

[Arnol'd70] introduced a spectral sequence to study the space of entire complex functions. [Vassiliev92] has applied the motivating idea of [Arnol'd70] to many different settings including the complexity of algorithms, the cohomology of braid groups, classical Lie groups, spaces of generalized Morse functions, loop spaces, and most dramatically, spaces of knots and links. The key object of study is a function space $\mathcal{F}$, such as the space of monic real polynomials of fixed degree d. This particular space contains a subspace Σ, consisting of polynomials with a multiple root. The subspace Σ is called a **discriminant** and since $\mathcal{F}$ is finite-dimensional, then the space $\mathcal{F}\backslash\Sigma$ consists of real polynomials without multiple roots. The Spanier-Whitehead dual of $\mathcal{F}\backslash\Sigma$ is the one-point compactification $\hat{\Sigma}$. Thus, the cohomology of $\mathcal{F}\backslash\Sigma$ is calculable from the homology of $\hat{\Sigma}$. This space admits a filtration that is well-behaved when we resolve $\hat{\Sigma}$ geometrically by inserting simplices whenever higher multiplicities of roots occur. The filtration leads to a spectral sequence of (Borel-Moore) homology groups. By an index shift we get a spectral sequence converging to the cohomology of the complementary space $\mathcal{F}\backslash\Sigma$.

In the various settings considered by [Vassiliev92], the E_1-term of the associated spectral sequence may be given in terms suited to the problem. For knots, the space $\mathcal{F}$ consists of all smooth maps of S^1 into $\mathbb{R}^3$ and the discriminant consists of maps that have singularities or self-intersections. The complement of this discriminant has path components that correspond to knot types and so its topology is important for the classification problem for knots. Thus the invariants of the space of knots appear as the groups $E^{-i,i}_\infty$ of the spectral sequence. [Vassiliev92] identified a combinatorial procedure for the determination of $E^{-i,i}_1$, giving invariants of a knot diagram. The analysis of the rest of the spectral sequence leads to the **Vassiliev invariants** of knots.

There has been considerable development of these invariants, relating them to classical and more recent knot invariants ([Birman-Lin, X.-S.93]), and giving a combinatorial description, independent of the spectral sequence origins ([Bar-Natan95]).

Another spectral sequence inspired by the work of [Arnol'd70] and used to study singularities was introduced by [Goryunov-Mond93]. The ingredients are a continuous, proper and finite mapping $f\colon X \to Y$ to which we associate the kth **multiple point space**

$$D^k(f) = \mathrm{cls}\{(x_1,\dots,x_k) \mid f(x_1) = \cdots = f(x_k), x_i \neq x_j \text{ for } i \neq j\}.$$

These spaces are equipped with natural mappings $\varepsilon_{i,k}\colon D^k(f) \to D^{k-1}(f)$ defined by $\varepsilon_{i,k}(x_1,\dots,x_k) = (x_1,\dots,\widehat{x_i},\dots,x_k)$, and with an action of the symmetric group Σ_k given by permuting the entries. To any cellular Σ_k-space Z for which the Σ_k-action is cellular, we associate the **alternating chain complex and homology**:

$$C_n^{\text{alt}}(Z) = \{c \in \mathrm{Cell}_n(Z) \mid \sigma c = \mathrm{sign}(\sigma)c \text{ for all } \sigma \in \Sigma_k\},$$

where $\mathrm{sign}\colon \Sigma_k \to \{\pm 1\}$ is the canonical sign representation. Since the action is cellular, the differential on $\mathrm{Cell}_n(Z)$ determines a differential on $C_n^{\text{alt}}(Z)$ and so we can define $H_*^{\text{alt}}(Z)$.

We also associate the kth **image multiple point space** $M^k(f) = \epsilon(D^k(f))$, where $\epsilon\colon D^k(f) \to Y$ is given by $\epsilon(x_1,\dots,x_k) = f(x_1)$.

The following spectral sequence appeared in this form in the paper of [Goryunov95]. The expository paper of [Houston99] is a very nice introduction to its applications.

Theorem 11.30 (the image computing spectral sequence). *Given a continuous, finite, and proper mapping* $f\colon X \to Y$ *for which the* k^{th} *multiple point spaces* $D^k(f)$ *have the* Σ_k-*homotopy type of a* Σ_k-*cellular complex for all* $k > 1$, *and for which each* k^{th} *image multiple point space* $M_k(f)$ *has the homotopy type of a cell complex for* $k > 1$, *there is a spectral sequence, converging to* $H_{p+q+1}(f(X))$, *with*

$$E_{p,q}^1 \cong H_q^{\text{alt}}(D^{p+1}(f)), \quad d^1 = (\varepsilon_{1,p+1})_*\colon H_q^{\text{alt}}(D^{p+1}(f)) \to H_q^{\text{alt}}(D^p(f)).$$

When the spectral sequence collapses at E^1 (for example, when f is a corank-1 map-germ $\mathbb{C}^n \to \mathbb{C}^{n+1}$ with finite $\mathcal{A}$-dimension; [Goryunov95]), the rational homology of the image is the sum of alternating homologies of the multiple point spaces, which is useful in the study of mixed Hodge structures on the image ([Goryunov-Mond93]). [Houston97] applied the spectral sequence to study the singularities of finite analytic mappings and to obtain relations between the fundamental groups of the domain and image of such mappings.

12

Spectral Sequences in Algebra,
Geometry and Analysis

"During the last decade the methods of algebraic topology have invaded extensively the domain of pure algebra,
and initiated a number of internal revolutions."

From [Cartan-Eilenberg56]

Spectral sequences arise from filtered differential modules, from double complexes, and from exact couples (Chapter 2). These basic structures may be found in almost any situation where homological methods are used—many examples of spectral sequences have become essential tools in fields outside of topology.

In this chapter, we continue the catalogue begun in Chapter 11. The examples here fall into three broad classes: those of homological origin (§12.1); those based on algebraic or differential geometric structures (§12.2), and those whose origin is chiefly topological but whose interpretation is algebraic (§12.3). We close the chapter with a short discussion of the notion of derived categories (§12.4), a formalism that lurks behind the 'unreasonable effectiveness' of spectral sequences.

The reader is expected to be acquainted with the categories of discourse for the examples presented in this chapter——definitions can be found in the cited references. Furthermore, this catalogue is quite far from complete (though some might argue that inclusion of Grothendieck's composite functor spectral sequence excludes very few examples). The hope remains that the reader will find a useful example in this collection or at least the sense in which spectral sequences can be applied in his or her field of interest. A search of the review literature in mathematics will provide a bounty of details to the curious reader.

12.1 Spectral sequences for rings and modules

Suppose R and S are commutative rings with unit. Denote the category of left (right) modules over R by $_R\mathbf{Mod}$ ($\mathbf{Mod}_R$) and similarly for the ring S. If we are given a homomorphism of rings, $\varphi \colon R \to S$, then modules over S obtain the structure of modules over R, where $r \cdot m = \varphi(r) \cdot m$. Under these conditions, we would like to relate the homological invariants of modules over the given rings in terms of the homomorphism φ. The following theorem describes these relations in terms of spectral sequences due to [Cartan-Eilenberg56].

Theorem 12.1 (the change-of-rings spectral sequences). *Suppose $\varphi \colon R \to S$ is a homomorphism of commutative rings with unit, and $M \in \mathbf{Mod}_R$, $N \in {}_S\mathbf{Mod}$, $M' \in \mathbf{Mod}_S$ and $N' \in {}_R\mathbf{Mod}$.*

(1) *There is a spectral sequence with $E^2_{p,q} \cong \operatorname{Tor}^S_p(\operatorname{Tor}^R_q(M, S), N)$, and converging to $\operatorname{Tor}^R_*(M, N)$.*

(2) *There is a spectral sequence with $E^2_{p,q} \cong \operatorname{Tor}^S_p(M', \operatorname{Tor}^R_q(S, N'))$, and converging to $\operatorname{Tor}^R_*(M', N')$.*

(3) *There is a spectral sequence with $E_2^{p,q} \cong \operatorname{Ext}^q_S(\operatorname{Tor}^R_p(S, N'), N)$, and converging to $\operatorname{Ext}^*_R(N', N)$.*

(4) *There is a spectral sequence with $E_2^{p,q} \cong \operatorname{Ext}^p_S(N, \operatorname{Ext}^q_R(S, N'))$, and converging to $\operatorname{Ext}^*_R(N, N')$.*

These spectral sequences can be derived in the manner of Theorem 2.20 (the Künneth spectral sequence) by judicious choices of double complexes. They are also special cases of the Grothendieck spectral sequence for derived functors on abelian categories (Theorem 12.10).

The change-of-rings spectral sequence applies in the special case of an extension of algebras over a field, $0 \to B \to A \to A/\!/B \to 0$, where $A/\!/B = A/I(A) \cdot B$.

Theorem 12.2. *Suppose A is an augmented algebra over a field, k. Suppose B is a normal subalgebra of A and A is projective over B. If $M \in \mathbf{Mod}_A$ and $N \in {}_{A/\!/B}\mathbf{Mod}$, then there is a spectral sequence with*

$$E^2_{p,q} \cong \operatorname{Tor}^{A/\!/B}_p(\operatorname{Tor}^B_q(M, k), N)$$

and converging to $\operatorname{Tor}^A_*(M, N)$. Also, for $M' \in {}_A\mathbf{Mod}$, there is a spectral sequence with

$$E_2^{p,q} \cong \operatorname{Ext}^p_{A/\!/B}(N, \operatorname{Ext}^q_B(k, M'))$$

and converging to $\operatorname{Ext}^*_A(N, M')$.

The reader can compare this theorem with Theorem 9.12 for central extensions of Hopf algebras.

The next three examples represent special cases of extra structures on rings or algebras that lead to spectral sequences. The first bears a strong relation to the Eilenberg-Moore spectral sequence. The second example treats some other homological invariants of a ring, namely the Hochschild homology and the cyclic homology of a ring. The third example is quite general and focuses on the consequences of a filtration on an algebra and a module over it; a special case appears as Theorem 9.56.

In the study of local rings, various homological invariants have played a key role. In particular, the Poincaré series for a local ring, $(R, \mathfrak{m})$, is defined by

$$P_R(t) = \sum_{i=0}^{\infty} (\dim_{R/\mathfrak{m}} \operatorname{Tor}_i^R(R/\mathfrak{m}, R/\mathfrak{m}))t^i.$$

Several spectral sequences have been useful in the study of this series (reviewed by [Avramov-Halperin86]). The following example was derived by [Avramov81] to study the problem of whether a minimal free resolution can be given the structure of an algebra.

Theorem 12.3. *Given a diagram of commutative ring homomorphisms*

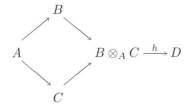

and a module M over B, then $\operatorname{Tor}^A(M, C)$ is a $\operatorname{Tor}^A(B, C)$-module, and D is a $\operatorname{Tor}^A(B, C)$-module via the homomorphism h. Furthermore, there is a spectral sequence with $E_{p,q}^2 \cong \operatorname{Tor}_{p,q}^{\operatorname{Tor}^A(B,C)}(\operatorname{Tor}^A(M, C), D)$, and converging to $\operatorname{Tor}_*^B(M, D)$.

The spectral sequence leads to an obstruction theory for the existence of multiplicative structures on resolutions. [Avramov81] also explicated the relationship of this spectral sequence to the Eilenberg-Moore spectral sequence.

An invariant of associative algebras over a fixed ring R was introduced by [Hochschild45] to study the classification of extensions of algebras. One expression for the Hochschild homology of an algebra, taken to be projective as a module over R, is given by

$$HH_*(A) \cong \operatorname{Tor}_*^{A^{\operatorname{op}} \otimes A}(A, A),$$

where $A^{\operatorname{op}} \otimes A$ acts on A by $(\alpha \otimes \beta)(a) = \beta a \alpha$. [Hochschild45] introduced a functorial complex, resembling the bar construction, to compute $HH_*(A)$. When A is a regular affine k-algebra over a perfect field k, [Hochschild-Konstant-Rosenberg62] proved that $HH_*(A)$ is isomorphic to the algebraic de Rham complex $\Omega_{A/k}^* = \Lambda(J/J^2)$, where $J = \ker \mu \colon A \otimes A \to A$ is the kernel of the multiplication mapping on A.

[Connes85] introduced a variant of the Hochschild homology in order to extend the Chern character to the algebraic K-theory of a C^*-algebra. For a field k of arbitrary characteristic, the *cyclic homology* $HC_*(A)$ of an algebra A over k is defined from a double complex in which each column is the Hochschild complex, and in each row we find the homology of a cyclic group. This leads to two spectral sequences ([Connes85]).

Theorem 12.4. *Given an algebra A over a field k, there are two spectral sequences converging to $HC_*(A)$. In the first spectral sequence, $E^1_{p,q} \cong HH_{q-p}(A)$ and $d^1 = B$, the Connes boundary map. In the second, $E^1_{p,q} \cong H_p(\mathbb{Z}/(q+1)\mathbb{Z}, A^{\otimes q+1})$, the homology of the group $\mathbb{Z}/(q+1)\mathbb{Z}$ with coefficients in the $\mathbb{Z}/(q+1)\mathbb{Z}$-module $A^{\otimes q+1}$, where the generator of $\mathbb{Z}/(q+1)\mathbb{Z}$ acts on $A^{\otimes q+1}$ by the cyclic permutation $a_0 \otimes \cdots \otimes a_q \mapsto (-1)^q a_q \otimes a_0 \otimes \cdots \otimes a_{q-1}$.*

Cyclic homology figures in the computation of the algebraic K-theory of rings, in noncommutative differential geometry, and in mathematical physics. For a comprehensive and comprehensible survey of these ideas, see the excellent book of [Loday98].

Suppose we begin with a filtered augmented algebra (A, μ, F, ε) over a field k, and a filtered A-module, M, satisfying either of the following conditions:

$$I : \begin{cases} F_p A = A, \ F_p M = M, & \text{if } p \geq 0, \\ F_{-1}A = I(A) = \ker \varepsilon & \text{and} \quad \bigcap_p F_p A = \{0\} = \bigcap_p F_p M. \end{cases}$$

$$II : \begin{cases} F_p A = \{0\} = F_p M, & \text{if } p < 0, \\ F_0 A = k & \text{and} \quad \bigcup_p F_p A = A, \ \bigcup_p F_p M = M. \end{cases}$$

Theorem 12.5 (the May spectral sequence). *For a filtered k-algebra, A, and filtered A-module, M, satisfying I or II, there is a spectral sequence with $E^2 \cong \mathrm{Tor}^{E^0 A}(k, E^0 M)$, and converging to $\mathrm{Tor}^A(k, M)$. Dually, there is a spectral sequence with $E_2 \cong \mathrm{Ext}_{E^0 A}(k, (E^0 M)^{\mathrm{dual}})$, and converging to $\mathrm{Ext}^*_A(k, M^{\mathrm{dual}})$ as an $\mathrm{Ext}^*_A(k, k)$-module.*

When A and M are graded, the spectral sequences are trigraded, where the first two gradings sum to the homological degree and the last two sum to the internal degree. Applications of this spectral sequence to the case where A is a Hopf algebra were pioneered by [May64, 66]. A natural generalization of this spectral sequence is presented in Exercise 3.4.

The May spectral sequence may be applied in computations of group cohomology. [Bajer94] established a collapse result for the May spectral sequence converging to $\mathrm{Ext}_{k[G]}(k, k)$ when $A = k[G]$, the group algebra for a finite p-group G and a field k of characteristic $p > 0$.

Other algebraic structures

The homological algebra of other algebraic structures, such as Hopf algebras, Lie algebras, and Leibniz algebras, was developed to obtain invariants that would aid in the classification problem of such structures and, more generally, lead to a deeper understanding of the structures themselves. The notion of an extension of Hopf algebras, Lie algebras, etc., plays the role of a fibration in topology, linking the members of the extension together. The case of a group extension, $1 \to K \to G \to Q \to 1$, is paradigmatic—the homological invariants of the constituent groups are linked together in the behavior of the Lyndon-Hochschild-Serre spectral sequence (Theorem $8^{\text{bis}}.12$).

The analogue of the Lyndon-Hochschild-Serre spectral sequence for Lie algebras was introduced by [Hochschild-Serre53'].

Theorem 12.6 (the Hochschild-Serre spectral sequence). *Let $\mathfrak{h}$ be a Lie ideal in the Lie algebra $\mathfrak{g}$ and M, a $\mathfrak{g}$-module. Then there is a spectral sequence, converging to $H^{p+q}(\mathfrak{g}, M)$, with $E_2^{p,q} \cong H^p(\mathfrak{g}/\mathfrak{h}, H_q(\mathfrak{h}, M))$.*

This theorem generalized the results of [Koszul50] who worked with a relative version of cohomology for pairs of Lie algebras and over fields of characteristic zero, where the geometric theory of Lie groups provided motivation. Koszul pioneered the homological algebra of Lie algebras as a tool independent of the topology and geometry involved.

When a Lie group is present, it is possible to view it as a manifold, a group, and its associated Lie algebra. [van Est58] introduced a spectral sequence in which all of these structures play a role. Let G denote a Lie group and H a compact subgroup of G. Suppose $\pi\colon G \to V$ is a representation of G into V a real vector space. Let $\mathfrak{g}$ and $\mathfrak{h}$ denote the Lie algebras of G and H respectively. With these assumptions there are three cohomology algebras that can be defined; $H^*_{\text{deR}}(G/H)$, the de Rham cohomology of the homogeneous space G/H, $H^*_{\text{alg}}(G, V, \pi)$, the smooth group cohomology of G with respect to the representation π in V, and $H^*_{\text{Lie}}(\mathfrak{g}, \mathfrak{h}; V)$, the relative Lie algebra cohomology with coefficients in V.

The spectral sequence is based on a double complex with elements of bidegree (r, s) given by V-valued functions of which the first r variables are in $\mathfrak{g}$ and the last s are in G. These functions are alternating multilinear on the first r variables and smooth on the last s.

Theorem 12.7 (the van Est spectral sequence). *With G, H, V and π as above, there is a spectral sequence with*

$$E_2^{*,*} \cong H^*_{\text{deR}}(G/H) \otimes H^*_{\text{alg}}(G, V, \pi)$$

and converging to $H^*_{\text{Lie}}(\mathfrak{g}, \mathfrak{h}; V)$.

This spectral sequence has proved useful in the study of the cohomology of Lie algebras. [Tillman93] has related the edge homomorphism in the van Est spectral sequence to the boundary map between Hochschild homology of a Banach algebra A and the cyclic homology of A.

A more general class of algebras that extends the notion of a Lie algebra was identified by [Loday93]: A **Leibniz algebra** $\mathfrak{g}$ is a k-module together with a bilinear mapping $[\ ,\ ] \colon \mathfrak{g} \times \mathfrak{g} \to \mathfrak{g}$ satisfying the Leibniz relation

$$[x, [y, z]] = [[x, y], z] - [[x, z], y].$$

The definition leaves out the antisymmetric relation expected of Lie algebras and so gives a noncommutative version of a Lie algebra. A **Leibniz module** (or representation) is a k-module M together with a bilinear mapping $M \times \mathfrak{g} \to M$, written $(m, g) \mapsto [m, g]$, satisfying $[m, [x, y]] = [[m, x], y] - [[m, y], x]$.

[Loday-Pirashvili93] defined the Leibniz cohomology, $HL^*(\mathfrak{g}, M)$, of a Leibniz algebra $\mathfrak{g}$ with coefficients in a Leibniz module M as the homology of the complex

$$\mathbb{R} \xrightarrow{0} C^1(\mathfrak{g}, M) \xrightarrow{d} C^2(\mathfrak{g}, M) \xrightarrow{d} \cdots \xrightarrow{d} C^k(\mathfrak{g}, M) \xrightarrow{d} \cdots$$

where $C^k(\mathfrak{g}, M) = \mathrm{Hom}_{\mathbb{R}}(\mathfrak{g}^{\otimes k}, M)$ and the differential d is defined

$$d(\alpha)(g_1 \otimes \cdots \otimes g_{k+1}) =$$

$$\sum_{1 \le i < j \le k} (-1)^{j+1} \alpha(g_1 \otimes \cdots \otimes g_{i-1} \otimes [g_i, g_j] \otimes g_{i+1} \otimes \cdots \otimes \widehat{g_j} \otimes \cdots \otimes g_{k+1})$$

$$+ [g_1, \alpha(g_2 \otimes \cdots \otimes g_{k+1})] + \sum_{i=2}^{k+1} (-1)^{i+1} [g_i, \alpha(g_1 \otimes \cdots \otimes \widehat{g_i} \otimes \cdots \otimes g_{k+1})].$$

When $\mathfrak{g}$ is a Lie algebra, the canonical mapping $\mathfrak{g}^{\otimes k} \to \mathfrak{g}^{\wedge k}$ induces a homomorphism $H^*_{\mathrm{Lie}}(\mathfrak{g}, M) \to HL^*(\mathfrak{g}, M)$. [Pirashvili94] and [Lodder98] have introduced a spectral sequence that computes the relative theory defined as

$$H^*_{\mathrm{rel}}(\mathfrak{g}) = H(s^2 C^*(\mathfrak{g})/\Omega^*(\mathfrak{g}), \bar{d}),$$

where s is the operator that shifts degree, and $\Omega^*(\mathfrak{g})$ is the complex that defines Lie algebra cohomology. The spectral sequence measures the difference between the Lie algebra cohomology and the Leibniz cohomology when they are applied to a Lie algebra, that is, it reveals the importance of the anticommutative condition on a Lie algebra. The E_2-term of the spectral sequence is made up of the Leibniz cohomology of the Lie algebra $\mathfrak{g}$ and another term defined as follows: There is a left $\mathfrak{g}$-module structure on $\mathfrak{g}^{\mathrm{dual}}$ given by $g\gamma(h) = \gamma([h, g])$. We can define morphisms $i_1 \colon \Omega^{n+1}(\mathfrak{g}) \to \Omega^n(\mathfrak{g}; \mathfrak{g}^{\mathrm{dual}})$ and $i_2 \colon \Omega^n(\mathfrak{g}, \mathfrak{g}^{\mathrm{dual}}) \to C^{n+1}(\mathfrak{g}, \mathbb{R})$ by the formulas

$$i_1(\alpha)(g_1, g_2, \ldots, g_n)(g_0) = (-1)^n \alpha(g_0, g_1, \ldots, g_n)$$
$$i_2(\beta)(g_0, g_1, \ldots, g_n) = (-1)^n \beta(g_1, \ldots, g_n)(g_0).$$

The complex $CR^*(\mathfrak{g})$ is defined by the short exact sequence

$$0 \to \Omega^*(\mathfrak{g}) \to \Omega^{*-1}(\mathfrak{g}, \mathfrak{g}^{\mathrm{dual}}) \to sCR^*(\mathfrak{g}) \to 0.$$

The homology of $CR^*(\mathfrak{g})$ is denoted by $HR^*(\mathfrak{g})$.

Theorem 12.8. *Let $\mathfrak{g}$ be a Lie algebra and M a $\mathfrak{g}$-module. Then there is a spectral sequence with $E^2_{p,q} \cong HL^p(\mathfrak{g}, M) \otimes HR^q(\mathfrak{g})$, and converging to $H^{p+q}_{\mathrm{rel}}(\mathfrak{g}, M)$.*

The spectral sequence was derived for Leibniz homology by [Pirashvili94] and for cohomology by [Lodder98], who has extended the Leibniz cohomology groups to diffeomorphism invariants of a manifold, and related them to the Gelfand-Fuks cohomology of smooth vector fields. He has also identified the Godbillon-Vey invariant of foliations as a Leibniz cohomology class.

The category of connected Hopf algebras over a ring R shares a great deal with the category of groups. One of the uses of group cohomology is the classification of extensions of groups. Suppose $1 \to K \to G \to Q \to 1$ is an extension. Then Q acts on K by conjugation, giving K a Q-module structure. There is also a twisting function (a factor set) $\tau: Q \times Q \to K$. The Q-module structure and τ together determine the extension G up to a coboundary condition. This identifies $H^2(Q, K)$ as the group that classifies extensions with the given Q-module structure. [Gugenheim62] and [Singer72] carried out a similar development of the structure of an extension of connected Hopf algebras. The notion of a Q-module structure with a twisting function is replaced with the notion of an *abelian matched pair* of Hopf algebras, (A, B). The definition may be found in [Singer72, Definition 3.1]. This leads to cohomology groups $H^n(B, A)$, definable as the derived functors of an appropriate hom functor, or via a cotriple. [Henderson97] has studied the problem of computing the groups $H^n(B, A)$ for which he has introduced a spectral sequence.

Theorem 12.9. *If (A, B) is an abelian matched pair of graded connected Hopf algebras over a ring R, then there is a spectral sequence for each integer $r > 0$ with $_rE_2^{s,t} \cong \mathrm{Ext}_B^{s,r}(R, \mathrm{Cotor}_A^{t,r}(R, R))$, and converging to $E_1^{r,s+t-r}$, the E_1-term of a spectral sequence that converges to $H^*(B, A)$.*

The method of construction is to use cosimplicial objects and interpret the various filtration quotients. [Henderson97] computed the spectral sequences in the case of a truncated monogenic tensor algebra, from which he determined the nature of certain extensions over $\mathbb{F}_p$ that occur in the study of finite H-spaces ([Lin78]).

Abelian categories

The structure of a spectral sequence requires a homological algebra that includes filtrations, subquotients, and additivity of morphisms. The minimal requirements of a category in which spectral sequences may be constructed and studied were identified by [Grothendieck57], who introduced the notion of an **abelian category**. General results about spectral sequences in abelian categories soon followed in papers of [Dold62], [Eilenberg-Moore62], and [Eckmann-Hilton66]. For a thorough introduction to foundations and the homological algebra of abelian categories, see the books of [Tamme94] and [Gelfand-Manin96].

Among the most important results of [Grothendieck57] is a general spectral sequence whose instances include many classical results. We begin with abelian categories, $\mathbf{AbelCat}_1$, $\mathbf{AbelCat}_2$, and $\mathbf{AbelCat}_3$, along with functors

$$F\colon \mathbf{AbelCat}_1 \longrightarrow \mathbf{AbelCat}_2 \quad \text{and} \quad G\colon \mathbf{AbelCat}_2 \longrightarrow \mathbf{AbelCat}_3.$$

We relate the derived functors F and G to the derived functors of $(G \circ F)$.

Theorem 12.10 (the composite functor spectral sequence). *Suppose the functors F and G are covariant, G is left exact and F takes injective objects in $\mathbf{AbelCat}_1$ to G-acyclic objects in $\mathbf{AbelCat}_2$ (G-acyclic objects have the property that the derived functors of G vanish on them). Then there is a spectral sequence with*

$$E_2^{p,q} \cong (R^p G)(R^q F(A)),$$

and converging to $R^*(G \circ F)(A)$ for A in $\mathbf{AbelCat}_1$.

The homological invariants we have considered are instances of the derived functors of such functors as $M \otimes_\Gamma -$ or $\mathrm{Hom}_\Gamma(-, C)$. The interested reader can review the spectral sequences of §12.1, §2.4, §7.1, and §9.2 and try to derive these spectral sequences as instances of the composite functor spectral sequence.

When the categories involved are not abelian, it is still possible to set up a Grothendieck spectral sequence. Using simplicial methods and homotopy theory, [Blanc-Stover92] have generalized the composite functor spectral sequence to categories of universal algebras (such as groups, rings, Lie algebras, etc.) and more general functors.

An example of the Grothendieck spectral sequence is a generalization of the change-of-rings spectral sequence. Suppose we have a family of functors, indexed over $\mathbb{Z}$,

$$\{T_n\}\colon \mathbf{AbelCat}_1 \longrightarrow \mathbf{AbelCat}_2$$

that act like the derived functors of T_0. That is, they are additive, left or right exact, and to a short exact sequence in $\mathbf{AbelCat}_1$, $0 \to A \to B \to C \to 0$, there is a long exact sequence in $\mathbf{AbelCat}_2$

$$\cdots \to T_n A \to T_n B \to T_n C \to T_{n \pm 1} A \to \cdots$$

(where ± 1 depends on the variance of the T's).

Theorem 12.11 (the Universal Coefficient spectral sequence). *Suppose A is an object in* **AbelCat**$_1$ *and* $M \in {}_A\mathbf{Mod}$. *Suppose*

(1) *projdim $M < \infty$ or $T_{-N} = 0$ for sufficiently large N,*
(2) *A is Noetherian and M is finitely generated or, for $q \le n$, T_q commutes with arbitrary direct sums.*

Then there is a spectral sequence with $E^2_{p,q} \cong \mathrm{Tor}^A_p(T_q A, M)$, and converging to $T_* M$. Dually, for $T^* = T_{-*}$, contravariant, there is a spectral sequence with $E_2^{p,q} \cong \mathrm{Ext}^p_A(M, T^q A)$, and converging to $T^* M$.

This theorem is proved in this generality in the paper of [Dold62]. For the abelian categories of chain and cochain complexes, the familiar Universal Coefficient theorem can be recovered. For $T_n = \mathrm{Ext}_A^{-n}(B, -)$, this gives another spectral sequence relating the homological invariants of rings and modules.

Another application may be made to compute the **hypercohomology** of a complex, (A^*, d), of objects in **AbelCat**$_1$ with respect to a left exact functor $F \colon \mathbf{AbelCat}_1 \to \mathbf{AbelCat}_2$. Suppose **AbelCat**$_1$ has enough injectives. Then the **hyperderived** functors of F can be defined: Suppose $(I^\bullet, \partial)$ is a complex of injective objects in **AbelCat**$_1$ with $H(I^\bullet, \partial) \cong H(A^*, d)$. The hypercohomology of A^* is defined by $\mathbb{H}F^*(A^*) = H(FI^\bullet, F\partial)$. This definition can be shown to be independent of the choice of injective object.

Theorem 12.12 (the hypercohomology spectral sequence). *If $\{R^j F\}$ denotes the sequence of right derived functors of F, then there is a spectral sequence, with $E_2^{p,q} \cong (R^p F)(H^q(A^*, d))$, and converging to $\mathbb{H}F^*(A^*)$.*

This theorem can be proved from the composite functor spectral sequence or from the construction of a double complex of injective objects whose total complex has homology $H(A^*, d)$ (in the manner of the proof of Lemma 2.19). A direct proof for rings and modules appears in the classic books of [Cartan-Eilenberg56] and [Mac Lane63].

12.2 Spectral sequences in Geometry

The basic objects that are studied in algebraic geometry, varieties and schemes, carry many different structures. Similarly, the basic objects in differential geometry, manifolds, are rich with structure. There is an underlying topological space (sometimes with the nonHausdorff Zariski topology), possible analytic structure, and, for varieties, the underlying structure of polynomial rings; the interaction between these structures and with the homological invariants of such objects leads to many useful spectral sequences.

The first example is historically the first spectral sequence. [Leray46], in a series of *Comptes Rendues* notes, introduced the notions of a sheaf over a topological space, cohomology with coefficients in a sheaf, and the Leray spectral sequence associated to a mapping (§6.4). Suppose X and Y are spaces with $f\colon X \to Y$, a continuous mapping. Suppose $\Phi = \{U_\alpha \mid \alpha \in J\}$ is an open cover of Y. Such a covering gives rise to a left exact functor from the category of sheaves on X to the category of sheaves on Y, which is constructed from the presheaf of sections. For a sheaf S on X,

$$\Gamma_\Phi(S) = \text{ sheaf derived from the presheaf } U_\alpha \mapsto \{\Gamma(f^{-1}(U_\alpha), S) \mid U_\alpha \in \Phi\}.$$

The theorem of [Leray46] relates the sheaves S on X and $\Gamma_\Phi(S)$ on Y. The category of sheaves (of abelian groups) is an abelian category and so there is a notion of homological algebra for sheaves.

Theorem 12.13 (the Leray spectral sequence). *Let $R^*\Gamma_\Phi$ denote the right derived functors of Γ_Φ. If $\check{H}^*$ denotes the sheaf cohomology of a space in a given sheaf, then there is a spectral sequence with $E_2^{p,q} \cong \check{H}^p(Y, R^q\Gamma_\Phi(S))$, and converging to $\check{H}^*(X, S)$.*

An application of the Leray spectral sequence is the case of a complex variety. Algebraically, the variety has the Zariski topology. Analytically, it carries a topological manifold structure. The sheaves of germs of functions on the variety (analytic and algebraic) and the cohomology of the variety in these sheaves are related by the spectral sequence in the theorem and the continuous function $X_{\mathbb{C}} \to X_{\text{Zar}}$. This example also reveals a role played by spectral sequences in algebraic geometry—patching local data into global data. Another example of a spectral sequence focusing on patching is the **local-to-global** spectral sequence.

Theorem 12.14. *Suppose X is a topological space and S, a sheaf of rings on X. Suppose that $\mathcal{M}$ and $\mathcal{N}$ are sheaves of left S-modules. Then there is a spectral sequence with $E_2^{p,q} \cong \check{H}^p(X, \mathcal{E}xt_S^q(\mathcal{M}, \mathcal{N}))$, and converging to $\text{Ext}_S^*(\mathcal{M}, \mathcal{N})$, which denotes the derived functors of*

$$\mathcal{H}om_S(\mathcal{M}, \mathcal{N}) = \prod_{x \in X} \text{Hom}_{S(x)}(\mathcal{M}(x), \mathcal{N}(x)).$$

We refer the reader to a classic text on sheaves and spectral sequences by [Godement58], for a discussion of local-to-global spectral sequences and a proof of this theorem. See the book of [Griffiths-Harris78] for applications of the Leray spectral sequence, especially in complex algebraic geometry. A generalization of the Leray spectral sequence has been derived by [Paranjape96] in the context of abelian categories and filtered complexes.

Spectral sequences and the de Rham complex

Other sources of spectral sequences in algebraic and differential geometry are filtrations of the de Rham complex that derive from some structural feature of the particular situation. As a first example, we mention a spectral sequence introduced by [Grothendieck61].

If X is a nonsingular algebraic variety of dimension n over a field k of arbitrary characteristic, then define

$$\Omega^1_{X/k} = \text{the sheaf of differential 1-forms on } X \text{ over } k.$$

We form the exterior algebra, $\Omega^i_{X/k} = \Lambda^i(\Omega^1_{X/k})$. Then there is a natural derivation $d \colon \mathcal{O}_X \to \Omega^1_{X/k}$ and so an exterior derivative giving a complex

$$\mathcal{O}_X \xrightarrow{d} \Omega^1_{X/k} \xrightarrow{d} \cdots \xrightarrow{d} \Omega^{n-1}_{X/k} \xrightarrow{d} \Omega^n_{X/k},$$

called the **algebraic de Rham complex** on X. The hypercohomology of this complex is called the **algebraic de Rham cohomology** of X, and denoted by $H^*_{\text{deR}}(X)$.

Theorem 12.15 (the Hodge-de Rham spectral sequence). *There is a spectral sequence with $E_1^{p,q} \cong \check{H}^p(X, \Omega^q_{X/k})$, the cohomology of X in the sheaf $\Omega^q_{X/k}$, and converging to $H^*_{\text{deR}}(X)$.*

[Grothendieck61] related the algebraic de Rham cohomology of a finite dimensional variety X over $\mathbb{C}$ to its singular cohomology by using the spectral sequence to prove

$$H^*_{\text{deR}}(X, \mathbb{C}) \cong H^*_{\text{sing}}(X^{\text{an}}; \mathbb{C})$$

where X^{an}, is the analytic space associated to X. The filtration on $H^*_{\text{deR}}(X)$ derived from this spectral sequence is related to the system of weights due to the Hodge structure on a compact complex variety. This relation has been studied thoroughly by [Deligne71].

When X is a scheme, smooth and proper over a perfect field k of characteristic $p > 0$, then there are other invariants that reflect the p-adic structure of the scheme. In particular, there is the crystalline cohomology of X, $H^*_{\text{cris}}(X/W)$, where W denotes the ring of Witt vectors, defined by [Grothendieck68] and [Bertholet76]. [Bloch78] studied the relations between the various cohomological invariants of a variety over a perfect field of characteristic $p > 2$. He introduced a spectral sequence to compute $H^*_{\text{cris}}(X/W)$ using a complex $C_X^\bullet$ of typical curves on K-theory. The hypercohomology spectral sequence in this case is called the **slope spectral sequence** and has $E_1^{p,q} \cong H^q(X, C_X^p)$ and converges to $H^*_{\text{cris}}(X/W)$. By analogy with the Hodge-de Rham spectral sequence, Deligne ([Illusie79]) introduced the *de Rham-Witt complex* for a scheme X, $W\Omega_X^\bullet$, which agrees with $C_X^\bullet$ when the latter is defined.

Theorem 12.16 (the slope spectral sequence). *Let X be a scheme, smooth and proper over a perfect field k of characteristic $p > 0$, and $W\Omega_X^\bullet$ the associated de Rham-Witt complex. Then there is a spectral sequence with $E_1^{p,q} \cong H^q(X, W\Omega_X^p)$, and converging to $H_{cris}^*(X/W)$.*

[Ekedahl86] has written a booklength account of the structure theory of the slope spectral sequence. The applications of this spectral sequence are numerous in algebraic geometry ([Illusie79], [Ekedahl86]).

Suppose M is a finite dimensional complex manifold. The cotangent bundle T^*M of M admits a decomposition, $T^*M = T^{*\prime}M \oplus T^{*\prime\prime}M$, into holomorphic forms (sums $\sum f_i\, dz_i$ with f_i a holomorphic function on M) and antiholomorphic forms (sums $\sum F_i\, d\bar{z}_i$ with $\bar{F}_i$ holomorphic). This decomposition induces a bigrading on the de Rham complex of $\mathbb{C}$-valued differential forms on M, $\Omega^*(M, \mathbb{C})$, where $\Omega^{p,q}(M, \mathbb{C}) = \Lambda^p(T^{*\prime}M) \otimes \Lambda^q(T^{*\prime\prime}M)$. An n-form with $n = p + q$ in $\Omega^{p,q}(M, \mathbb{C})$ is called a (p, q)-**form**. The exterior differential on $\Omega^*(M, \mathbb{C})$ takes a form $\omega \in \Omega^{p,q}(M, \mathbb{C})$ to the direct sum $\Omega^{p+1,q}(M, \mathbb{C}) \oplus \Omega^{p,q+1}(M, \mathbb{C})$. Composing with the projections we get the expression $d = \partial + \bar{\partial}$ with ∂ of bidegree $(1, 0)$ and $\bar{\partial}$ of bidegree $(0, 1)$. Furthermore, $\partial \circ \partial = 0 = \bar{\partial} \circ \bar{\partial}$.

It follows that the data $(\Omega^{p,q}(M, \mathbb{C}), \partial, \bar{\partial})$ determine a double complex whose total complex is the de Rham complex. The vertical differential, $\bar{\partial}$, leads to the **Dolbeault cohomology of** M,

$$H_{\bar{\partial}}^{p,q}(M) = \Omega^{p,q}(M, \mathbb{C}) \cap \ker \bar{\partial}/\bar{\partial}(\Omega^{p,q-1}(M, \mathbb{C})).$$

Theorem 12.17 (the Frölicher spectral sequence). *Given a complex manifold M, there is a spectral sequence, converging strongly to $H_{deR}^*(M, \mathbb{C})$, with E_2-term given by $E_2^{p,q} \cong H_{\bar{\partial}}^{p,q}(M)$.*

[Frölicher55] introduced the spectral sequence to relate the geometric invariants of the Dolbeault complex to the topological invariants of the de Rham cohomology. He observed that a complex manifold with a positive definite Kähler metric has $E_1 \cong E_\infty$, and so the spectral sequence gives a necessary condition for the existence of a Kähler structure. [Cordero-Fernández-Gray91, 93] have given examples of complex manifolds for which the spectral sequence does not collapse at E_1.

When X is a smooth projective complex variety, there is also a Hodge filtration on the cohomology of X. Because it has an underlying Kähler manifold, the Frölicher spectral sequence collapses for X. In the more general case of a quasi-projective complex variety V (that is, $V = X - Y$ for X and Y complex varieties), [Deligne68] has proved that there is a different filtration on $H^*(X; \mathbb{Q})$, called a **weight filtration**. Such a filtration is increasing

$$0 = W_{-1} \subset W_0 \subset W_1 \subset \cdots \subset W_{2m} = H^m(X; \mathbb{Q})$$

for which the complexified associated graded group $E_l^0 \otimes \mathbb{C} = (W_l/W_{l-1}) \otimes \mathbb{C}$ has a decomposition of Hodge type $E_l^0 = \bigoplus_{p+q=l} H^{p,q}$. Such a structure is called a **mixed Hodge structure**. When X is a quasiprojective algebraic variety, [Deligne71] proved that there is a weight filtration on $H^m(X; \mathbb{Q})$ and a decreasing Hodge filtration on $H^m(X; \mathbb{C})$ such that the filtration induced by the Hodge filtration on the complexified associated graded module from the weight filtration was a mixed Hodge structure. The existence of two filtrations of this sort can lead to the collapse of the associated spectral sequences. For an introduction to mixed Hodge structures, see the 'naive guide' of [Durfee83]. The appearance and uses of spectral sequences from mixed Hodge structures is developed in the book of [El Zein91].

If (M, g) is an n-dimensional Riemannian manifold and $\pi: E \to M$ is a flat vector bundle, then a smooth distribution of k-planes $A \subset TM$ together with its orthogonal complement B leads to a decomposition of the metric $g = g_A \oplus g_B$. If we vary the metric by $g_\delta = g_A + \delta^{-2} g_B$ for $0 < \delta \leq 1$, then we obtain a family of Laplacians for (M, g_δ) and a corresponding exterior derivative on $\Omega^*(M; E)$. There is a filtration on the L^2-completion of $\Omega^p(M; E)$ given by $\omega \in F^k$ when there is a j with $d_\delta(\omega + \delta\omega + \cdots + \delta^j\omega) \in \delta^k \Omega^{p+1}[\delta]$. A spectral sequence results that has been shown to be isomorphic to the Leray spectral sequence associated to the splitting $A \oplus B = TM$ ([Mazzeo-Melrose90], [Forman95]). This spectral sequence is related to the behavior of the spectrum of the Laplacians involved and is called the **adiabatic spectral sequence**. For a general discussion of these ideas, see the paper of [Forman94].

The bigrading of the de Rham complex in Hodge theory has a striking relative in the calculus of variations. Here one wants to study differential equations as sections of jet bundles associated to a smooth vector bundle $\pi: E \to M$. Let $J^\infty(E) \to M$ denote the infinite order jet bundle associated to π. Let I denote the *contact ideal*, the differential ideal of the de Rham complex $\Omega^*(J^\infty(E))$ of forms that pull back to zero under any extension to infinite jets of a section $s: M \to E$. A bigrading results on $\Omega^*(J^\infty(E))$ by counting the number of forms from I needed to express a given form. The exterior derivative can be decomposed into horizontal and vertical components giving a double complex, known as the **variational bicomplex**. The associated spectral sequence was identified by [Vinogradov78] and a clear presentation can be found in the monograph of [Krasil'shchik-Verbotevsky98].

Theorem 12.18 (the $\mathcal{C}$-spectral sequence). *Let $\pi: E \to M$ be a smooth vector bundle over an n-dimensional manifold M. Then the spectral sequence associated to the variational bicomplex converges to $H^*_{\mathrm{deR}}(J^\infty(E))$ and has $E_1^{0,q}$ isomorphic to the horizontal cohomology associated to π; $E_1^{p,n}$ isomorphic to the module $L_p^{\mathrm{alt}}(\pi)$ for $p > 0$, where L_p^{alt} is the homology of the complex of alternating differential operators associated to π, and $E_1^{r,s} \cong \{0\}$, otherwise.*

The differential $d_1 \colon E_1^{0,n} \to E_1^{1,n}$ *can be identified with the operator that associates to a Lagrangian its Euler-Lagrange equation.*

In the case of a specific distribution on a subspace of $J^\infty(E)$, the horizontal cohomology is based on the associated Cartan submodule determined by the equation (hence the $\mathcal{C}$-spectral sequence) and the E_1-term is more complicated to describe. Applications of the variational bicomplex are presented by [Anderson-Thompson92]; applications of the $\mathcal{C}$-spectral sequence by [Krasil'shchik98].

Finally, to close this section we mention work of [Dixon91] on the computation of BRS cohomology for gauge systems ([Henneaux-Teitelboim92]). The BRS operator determines a differential on the Fock space of integrated local polynomial functions of a Yang-Mills field and a Fadeev-Popov ghost field. The resulting cohomology determines invariants of a gauge system, such as the ghost numbers, the Lorentz character, and discrete symmetries. [Dixon91] filtered the space on which the BRS operator acts and deduced the associated spectral sequence. The induced grading from the E_∞-term of the spectral sequence decomposes the desired complicated cohomology in simpler pieces that are computable.

12.3 Spectral sequences in algebraic K-theory

Algebraic K-theory assigns a sequence of invariants to a ring R and these invariants may be constructed as the homotopy groups of a certain space (or a certain spectrum). The tools for the study of algebraic K-theory are as varied as the appearances of rings throughout mathematics and so there are many structures at play, interwoven and interacting in algebraic K-theory.

To a ring R we associate the scheme $\operatorname{Spec} R$ with the Zariski topology. [Brown, K-Gersten73] and [Quillen73] derived a spectral sequence, defined for cohomology groups related to simplicial sheaves, and applicable to algebraic K-theory: Suppose X is a Noetherian space (that is, the open sets in X satisfy the ascending chain condition) and suppose that the irreducible closed subsets of X also satisfy the ACC (for example, if $X = \operatorname{Spec} R$ for R a regular, commutative ring). A **simplicial sheaf** on X is a sheaf with values in **SimpEns**, the category of simplicial sets. If K is a simplicial sheaf on X, then we say that K is **flasque** if the mapping $t \colon K \to *$ satisfies the property that, for U, V open in X,

$$\Gamma(V, K) \xrightarrow{\; (\Gamma(V,t),\mathrm{restr}) \;} \Gamma(V, *) \otimes_{\Gamma(U,*)} \Gamma(U, K)$$

is a simplicial fibration.

Replace the functors $\Gamma(U, -)$, for U open in X, by a functor $R\Gamma(U, -)$ defined on the *homotopy* category of sheaves over X,

$$R\Gamma(U, -)\colon \mathbf{HoSimpSheaves}_X \to \mathbf{HoSimpEns}$$

such that, when K is flasque, there is natural isomorphism of $R\Gamma(U, K)$ with $\Gamma(U, K)$. Define the **generalized sheaf cohomology groups**, $H^*(X, K)$ by

$$H^q(X, K) = \pi_{-q}(R\Gamma(X, K)).$$

Theorem 12.19 (the Brown-Gersten-Quillen spectral sequence). *Suppose* X *is a Noetherian space of finite Krull dimension and* K *is a simplicial sheaf with basepoint that satisfies* $\pi_0(K) = \{0\}$ *and* $\pi_1(K)$ *and* $H^{-1}(X, K)$ *are abelian. Suppose* $H^p(X, \pi_n(K)) = \{0\}$ *for* $p \geq n$. *Then there is a fourth quadrant cohomological spectral sequence with* $E_2^{p,q} \cong H^p(X, \pi_{-q}(K))$, *and converging to* $H^*(X, K)$.

If $X = \operatorname{Spec} R$ for R regular and such that every coherent sheaf on X is a quotient of a locally free sheaf, then one builds a simplicial sheaf K on X from the classifying construction for a category with $\pi_{-q}(K) = \mathcal{K}_{-q}$, the abelian sheaf of local K-groups of R and satisfying, $H^*(X, K) = K_*(R)$. In this case,

$$E_2^{p,q} \cong H^p(\operatorname{Spec} R, \mathcal{K}_{-q})$$

and the spectral sequence converges to $K_*(R)$. A construction and discussion of the applications of the Brown-Gersten-Quillen spectral sequence may be found in the book of [Srinivas96]. [Gillet81] has given an alternate derivation of the Brown-Gersten-Quillen spectral sequence as the solution to a homotopy limit problem (following [Thomason83]).

Another example of an invariant of a scheme (a ringed space) is its étale cohomology ([Milne80], [Tamme94]). In this example, we relate the étale cohomology of a scheme with coefficients in various cyclic groups, to the localized algebraic K-theory of the scheme.

In order to introduce $\mathbb{Z}/m\mathbb{Z}$ coefficients on homotopy groups of a spectrum, one smashes the spectrum with the appropriate Moore spectrum for $\mathbb{Z}/m\mathbb{Z}$ ([Browder78]). The algebraic K-theory of a scheme with coefficients in $\mathbb{Z}/m\mathbb{Z}$ can be defined analogously by taking the spectrum associated to the scheme and smashing it with the Moore spectrum; its homotopy groups are denoted by $(K/m)_*(X)$.

Suppose $b \in (K/m)_*(X)$ and we consider the direct limit of the system

$$(K/m)_*(X) \xrightarrow{\; b\times -\;} (K/m)_*(X) \xrightarrow{\; b\times -\;} \cdots$$

given by left multiplication by b. This direct limit is called the **localization** of $(K/m)_*(X)$ with respect to b (or by inverting the element b), and it is denoted by $(K/m)_*(X)[b^{-1}]$.

In connection with the Lichtenbaum-Quillen conjecture, [Thomason85] introduced a descent spectral sequence associated to schemes X satisfying certain technical conditions:

Theorem 12.20 (the descent spectral sequence). *Suppose* l *is a fixed prime and* ν, *a natural number. Let* X *be a separated, Noetherian, regular scheme of finite Krull dimension, with sufficiently nice residue fields of characteristic different* *from l. Suppose β is the Bott element in $(K/l^\nu)_2(X)$, that is, the element such that the Bockstein of β is an appropriate power of an lth root of unity in $K_1(X)$. Then there is a spectral sequence with differentials of bidegree $(r, r-1)$,*

$$E_2^{p,q} \cong \begin{cases} H_{\text{et}}^p(X; \mathbb{Z}_l(i)), & \text{if } q = 2i, \\ \{0\}, & \text{if } q, \text{ odd}, \end{cases}$$

where $\mathbb{Z}_l(i) = \mathbb{Z}_l(1)^{\otimes i}$ is the ith Tate twist of the l-adic integers. The spectral sequence converges to $(K/l^\nu)_*(X)[\beta^{-1}]$.

The Lichtenbaum-Quillen conjecture relates the order of the K-groups to values of zeta functions for certain arithmetic number fields. [Thomason85] proved it for this localized version of algebraic K-theory. The result also applies to the case of X, a variety over an algebraically closed field k of characteristic $\neq l$ and so allows computation of these algebraic K-groups for such varieties. [Mitchell97] presented a proof of Thomason's theorem in terms of hypercohomology spectra in which he exposes many of the details and conceptual underpinnings of this result, as well as the applications. [Thomason82, 83] described a context where this theorem is a case of a homotopy limit problem, here for diagrams of spectra (see §11.4).

The next spectral sequence has played a key role in recent developments of Voevodsky in his proof of the the Lichtenbaum-Quillen conjecture at 2 for fields of characteristic zero ([Friedlander97]). An important tool in algebraic K-theory is the *motivic cohomology* of a field. Motivic cohomology is a functor on schemes that plays the role of singular cohomology for spaces. For a topological space X, the Atiyah-Hirzebruch spectral sequence has $E_2^{p,q} \cong H^p(X; K_{\text{top}}^q)$ and converges to $K_{\text{top}}^{p+q}(X)$. Beilinson conjectured that there should be a spectral sequence of Atiyah-Hirzebruch type from the motivic cohomology of a scheme with coefficients in the algebraic K-theory of a point $(\text{Spec}(k))$ to the algebraic K-theory of the scheme. Furthermore, tensored with the rational numbers, this spectral sequence would collapse determining the algebraic K-theory groups mod torsion. [Bloch86] has proposed that motivic cohomology of $\text{Spec}(k)$, $H_{\mathcal{M}}^r(\text{Spec}(k), \mathbb{Z}(s))$, may be identified with the higher Chow groups $CH^s(\text{Spec } k, 2s - r)$ and with this definition, there is a spectral sequence derived by [Bloch-Lichtenbaum94].

Theorem 12.21 (the Bloch-Lichtenbaum spectral sequence). *Let k denote a field. There is a fourth quadrant spectral sequence with*

$$E_2^{p,q} \cong H_{\mathcal{M}}^{p-q}(\text{Spec}(k), \mathbb{Z}(-q)),$$

and converging to $K_{-p-q}(\text{Spec } k)$.

Voevodsky used his proof of the Milnor conjecture ([Voevodsky96], [Kahn, B97], [Morel98]), together with this spectral sequence to obtain his proof of the Lichtenbaum-Quillen conjecture.

Finally, we close this section with another analogue of the Atiyah-Hirze-bruch spectral sequence, this time for a different K-theory and a different filtration. If A is a C^*-algebra, then there is a K-theory of A, defined and developed by [Brown-Douglas-Fillmore77], [Pimsner-Popa-Voiculescu79], and [Kasparov79]. [Schochet81] introduced a spectral sequence that applies when A is a filtered C^*-algebra, that is, there is a sequence of closed ideals,

$$A_0 \subset A_1 \subset \cdots \subset A_n \subset A_{n+1} \subset \cdots \subset A = \mathrm{cls}(\bigcup_n A_n).$$

Theorem 12.22. *Suppose given a filtered C^*-algebra, $(A, \{A_n\})$. Then there is a spectral sequence with $E^1_{p,q} \cong K_{p+q}(A_p/A_{p-1})$, and converging to $K_*(A)$. The spectral sequence is natural with respect to filtration-preserving maps of C^*-algebras.*

This result and other results of Schochet bring the technical tools of algebraic topology to bear on the study of C^*-algebras.

12.4 Derived categories

The functors of homological algebra such as Tor and Ext are defined as the homology of chain complexes that are built in a noncanonical manner. In order to obtain homological invariants, the chain complexes must be carefully chosen. For example, a projective resolution of a right A-module,

$$\cdots \to P^{-i+1} \to P^{-i} \to \cdots \to P^{-1} \to P^0 \to M \to 0,$$

gives $\mathrm{Tor}^A_*(M, N)$ by computing $H(P^\bullet \otimes_A N)$ for a left A-module N. Other choices of projective resolution can be compared with this particular choice to give isomorphic Tor groups, that is, groups that depend on A, M, and N only. In the case of modules, flat modules have the property of exactness on tensoring over A and so the axiomatic properties of Tor can be achieved by a choice of a flat resolution. However, it may be difficult to compare two flat resolutions.

Grothendieck and [Verdier63/97] defined the notion of the derived category of an abelian category $\mathbf{A}$ in an effort to establish a framework in which to extend the duality results of [Serre54]. Let $\mathbb{C}(\mathbf{A})$ denote the category of chain complexes of objects and degree zero maps of complexes in $\mathbf{A}$. Let $\mathbb{C}^+(\mathbf{A})$ ($\mathbb{C}^-(\mathbf{A})$) denote the subcategory of chain complexes that are bounded below (above). A morphism of complexes $P^\bullet \to Q^\bullet$ is a **quasi-isomorphism** if it induces an isomorphism $H(P^*) \to H(Q^*)$ of graded objects. The **derived category** of $\mathbf{A}$, $\mathbb{D}(\mathbf{A})$, is obtained by formally inverting the class of quasi-isomorphisms in $\mathbb{C}(\mathbf{A})$. This formal inversion can be made concrete by using a calculus of fractions developed by [Verdier63/97] and [Gabriel-Zisman67].

If $F\colon \mathbf{A} \to \mathbf{B}$ is an additive functor between abelian categories, then we can ask if there is an extension of F to a functor $\mathbb{D}^+(\mathbf{A}) \to \mathbb{D}^+(\mathbf{B})$. A minimal requirement is that F, extended levelwise to $\mathbb{C}^+(\mathbf{A}) \to \mathbb{C}^+(\mathbf{B})$, preserve quasi-isomorphisms. This is true if F is left exact. The right derived functors of F determine the extension of F to $RF \colon \mathbb{D}^+(\mathbf{A}) \to \mathbb{D}^+(\mathbf{B})$. This extension is proved to exist by analyzing the mapping cylinder construction in abelian categories, a construction formalized in the notion of a *triangulated category*.

When the abelian category $\mathbf{A}$ has enough injectives, then the value of $H^s(RF(K^\bullet))$ is called the sth **hyperderived functor** of F with respect to the complex $K^\bullet$. The computation of $H^*(RF(K^\bullet))$ may be carried out by replacing $K^\bullet$ with a double complex of injective objects, from which there is a spectral sequence with $E_2^{p,q} \cong (R^pF)(H^q(K^\bullet))$, converging weakly to $H^*(RF(K^\bullet))$.

The point of derived categories, however, is to argue with the objects up to equivalence and the derived functors as functors on a particular category. An example of this principle in action is the following basic result.

Theorem 12.23. *Given three abelian categories $\mathbf{A}$, $\mathbf{B}$, and $\mathbf{C}$, and additive left exact functors $F \colon \mathbf{A} \to \mathbf{B}$ and $G \colon \mathbf{B} \to \mathbf{C}$ such that F takes injective objects in $\mathbf{A}$ to G-acyclic objects in $\mathbf{B}$, then the extensions of F, G and $G \circ F$ to the derived categories are naturally isomorphic, that is, $R(G \circ F) \cong R(G) \circ R(F)$.*

(Proofs of this theorem can be found in the book of [Weibel94, 10.8.2], or [Gelfand-Manin96, III.7.1] or in the survey paper of [Keller96].) When the spectral sequence is applied to compute the hyperderived functors of the product, we recover the Grothendieck spectral sequence (Theorem 12.9). The underlying equivalence is more revealing than the spectral sequence and the derived category provides the framework to make such insights.

The language of derived categories is based on the basic structures of stable homotopy theory. [May94] has given a dictionary between algebra and topology that illuminates the analogies. The homological algebra of rings and modules can be carried back to stable homotopy through the foundational work of [Elmendorf-Kriz-Mandell-May97].

Derived categories have spread throughout mathematics wherever homological algebra has developed. As derived categories provide organization, spectral sequences will provide computations.

Bibliography

Adams, J.F., On the cobar construction, Proc. Nat. Acad. Sci. U.S.A. **42**(1956), 409–412. (379, 491)

Adams, J.F., On the structure and applications of the Steenrod algebra, Comm. Math. Helv. **32**(1958), 180–214. (368, 370, 376, 384, 406, 425)

Adams, J.F., On the non-existence of elements of Hopf invariant one, Ann. of Math. **72**(1960), 20–104. (191, 207, 366, 367, 387, 392, 415, 417, 425, 427, 445)

Adams, J.F., A finiteness theorem in homological algebra, Proc. Camb. Philos. Soc. **57**(1961), 31–36. (430)

Adams, J.F., A spectral sequence defined using K-theory, Colloque de topologie algèbrique Bruxelles, 1964, 146–166. (408, 451)

Adams, J.F., A periodicity theorem in homological algebra, Proc. Camb. Philos. Soc. **62**(1966), 365–377. (430, 431, 436ff., 499)

Adams, J.F., Stable Homotopy Theory. Lectures delivered at the University of California at Berkeley, 1961. Notes by A. T. Vasquez. Third edition. Springer Lecture Notes in Mathematics, **3**(1969), 78 pp. (368, 430, 478, 497)

Adams, J.F., Stable Homotopy and Generalised Homology. Chicago Lectures in Mathematics, University of Chicago Press, Chicago, Ill.-London, 1974, x+373 pp. (76, 368, 404, 408, 496)

Adams, J.F., Margolis, H.R., Modules over the Steenrod algebra, Topology **10**(1971), 271–282. (432)

Adem, J., The iteration of the Steenrod squares in algebraic topology, Proc. Nat. Acad. Sci. U.S.A. **38**(1952), 720–726. (128, 425)

Adem, J., The relations on Steenrod powers of cohomology classes, in Algebraic Geometry and Topology, Essays in Honor of S. Lefschetz, Princeton Univ. Press 1957, 191–238. (373)

Adem, A., Milgram, R.J., Cohomology of Finite Groups. Grundlehren der Mathematischen Wissenschaften, **309** Springer-Verlag, Berlin, 1994. viii+327 pp. (344)

Ahlfors, L.V., Complex analysis. An introduction to the theory of analytic functions of one complex variable, McGraw-Hill Book Company, Inc., New York-Toronto-London, 1966 xiii+317 pp. (202)

Alexander, J.W., On the chains of a complex and their duals; On the ring of a compact metric space, Proc. Nat. Acad. Sci. U.S.A. **21**(1935), 509–511 and 511–512. (133)

Alexandroff, P., Hopf, H., Topologie, Springer-Verlag, Berlin, 1935. (455)

Anderson, D.W., A generalization of the Eilenberg-Moore spectral sequence, Bull. Amer. Math. Soc. **78**(1972), 784–788. (491)

Anderson, I., Thompson, G., The inverse problem of the calculus of variations for ordinary differential equations, Mem. Amer. Math. Soc. **98**(1992), no. 473, vi+110 pp. (520)

André, M., Méthode simpliciale en algèbre homologique et algèbre commutative, Springer Lecture Notes in Mathematics, **32**(1967), iii+122 pp. (109, 494)

Anick, D.J., The computation of rational homotopy groups is $\#\mathcal{P}$-hard. Computers in geometry and topology (Chicago, IL, 1986), 1–56, Lecture Notes in Pure and Appl. Math., 114, Dekker, New York, 1989. (366)

Anick, D.J., Differential Algebras in Topology, A K Peters, Wellesley, MA, 1993. (461, 462)

Araki, S., Steenrod reduced powers in the spectral sequence associated to a fibering I, II, Mem. Fac. Sci. Kyusyu Univ. Series (A) Math. **11**(1957); 15–64, 81–97. (194)

Araki, S., Kudo, T., Topology of H_n-spaces and H-squaring operations, Mem. Fac. Sci. Kyusyu Univ. Ser. A. **10**(1956), 85–120. (326, 499)

Araki, S., Toda, H., Multiplicative structures in mod q cohomology theories. I; II, Osaka J. Math. **2**(1965), 71–115; **3**(1966), 81–120. (480)

Arkowitz, M., Localization and H-spaces. Lecture Notes No. 44, Aarhus Universitet, Aarhus, 1976, iii+143 pp. (362)

Arlettaz, D., The order of the differentials in the Atiyah-Hirzebruch spectral sequence, K-Theory **6**(1992), 347–361. (497)

Arnol'd, V.I., Certain topological invariants of algebraic functions (Russian), Trudy Moskov. Mat. Obšč. **21** 1970 27–46 (English transl. in Trans. Moscow Math. Soc. **21**(1970), 30–52); II. Funkcional. Anal. i Priložen. **4**(1970), 1–9 (English transl. in Functional Anal. Appl. **4**(1970). (505ff.)

Artin, M., Mazur, B., Etale Homotopy, Springer Lecture Notes in Mathematics **100**(1969). (82)

Assmus, E., On the homology of local rings, Ill. J. Math. 3(1959). 187–199. (14, 246)

Atiyah, M.F., Characters and cohomology of finite groups, Publ. Math. IHES, Paris **9**(1961), 247–289. (496)

Atiyah, M.F., Vector bundles and the Künneth formula, Topology **1**(1962), 245–248. (274, 313, 315, 317)

Atiyah, M.F., Geometry of Yang-Mills Fields, Lezioni Fermiane, Scuola Normale Superiore, Pisa, 1979. (220)

Atiyah, M.F., Hirzebruch, F., Vector bundles and homogeneous spaces, Proc. Symp. Pure Math. **3**(1969), 7–38. (222, 496)

Avramov, L., Obstructions to the existence of multiplicative structures on minimal free resolutions, Amer. J. Math. **103**(1981), 1–31. (310, 509)

Avramov, L., Local rings over which all modules have rational Poincare series, J. Pure Appl. Algebra **91**(1994), no. 1-3, 29–48. (14)

Avramov, L., Halperin, S., Through the looking glass: A Dictionary between rational homotopy theory and local algebra, in Algebra, Algebraic Topology and their Interactions, Springer Lecture Notes in Mathematics **1183**(1986), 3–27. (310, 509)

Bahri, A.P., Operations in the second quadrant Eilenberg-Moore spectral sequence, J. Pure Appl. Algebra **27**(1983), 207–222. (326)

Bajer, A.M., The May spectral sequence for a finite p-group, J. Algebra **167**(1994), 448–459. (510)

Baker, A., Hecke operations and the Adams E_2-term based on elliptic cohomology, Canad. Math. Bull. 42 (1999), 129–138. (500)

Bar-Natan, D., On the Vassiliev knot invariants, Topology 34 (1995), 423–472. (506)

Barcus, W.D., On a theorem of Massey and Peterson, Quart. J. Math. Oxford, **19**(1968), 33–41. (492)

Barnes, D.W., Spectral sequence constructors in algebra and topology. Mem. Amer. Math. Soc. **53**(1985), no. 317, viii+174 pp. (181, 229)

Barratt, M.G., Track groups, I, II, Proc. London Math. Soc. **5**(1955), 71–106, 285–329. (97)

Barratt, M.G., The spectral sequence of an inclusion, Proc. Coll. Alg. Topology Aarhus (1962), 22–27. (488)

Barratt, M.G., Hilton, P.J., On join operations in homotopy groups, Proc. London Philos. Soc. (3)**3**(1953), 430–445. (404)

Barratt, M.G., Jones, J.D.S., Mahowald, M.E., Relations amongst Toda brackets and the Kervaire invariant in dimension 62, J. London Math. Soc. **30** (1984), 533–550. (451)

Barratt, M.G., Mahowald, M.E., Tangora, M.C., Some differentials in the Adams spectral sequence II, Topology **9**(1970), 309–316. (448)

Baues, H.J., Obstruction Theory, Springer Lecture Notes in Mathematics **628** (1977). (96)

Baues, H.-J., The cobar construction as a Hopf algebra, Invent. Math. **132**(1998), 467–489. (491)

Baum, P.F., On the cohomology of homogeneous spaces, Topology **7**(1968), 15–38. (275ff., 328)

Baum, P.F., Smith, L., The real cohomology of differentiable fibre bundles, Comm. Math. Helv. **42**(1967), 171–179. (276)

Bendersky, M., Curtis, E.B., Miller, H.R., The unstable Adams spectral sequence for generalized homology, Topology **17**(1978), 229–248. (494)

Bendersky, M., Gitler, S., The cohomology of certain function spaces, Trans. Amer. Math. Soc. **326**(1991), 423–440. (491)

Benson, D., Representations and Cohomology II, Cambridge University Press, Cambridge, UK, 1991. (344, 487)

Berthelot, P., Cohomologie cristalline des schémas de caractéristique $p > 0$, Springer Lecture Notes in Mathematics,**407**(1974). 604 pp. (517)

Beyl, R., The spectral sequence of a group extension, Bull. Sc. Math. **105**(1981), 417–434. (342)

Birman, J.S., Lin, X.-S., Knot polynomials and Vassiliev's invariants, Invent. Math. **111**(1993), 225–270. (506)

Blanc, D.A., A Hurewicz spectral sequence for homology, Trans. Amer. Math. Soc. **318**(1990), 335–354. (489, 490)

Blanc, D.A., Operations on resolutions and the reverse Adams spectral sequence, Trans. Amer. Math. Soc. **342**(1994), 197–213. (490)

Blanc, D., Stover, C., A generalized Grothendieck spectral sequence, Adams Memorial Symposium on Algebraic Topology, 1 (Manchester, 1990), London Math. Soc. Lecture Note Ser., **175**(1992), 145–161. (514)

Bloch, S., Algebraic K-theory and crystalline cohomology, Inst. Hautes Études Sci. Publ. Math., **47**(1977), 187–268 (1978). (517)

Bloch, S., Algebraic cycles and higher K-theory, Adv. Math. **61**(1986), 267–304. (522)

Bloch, S., Lichtenbaum, S., A spectral sequence for motivic cohomology, preprint, University of Illinois, Champaign-Urbana, K-theory archive, 1995. (522)

Boardman, J.M., Conditionally convergent spectral sequences, in Homotopy Invariant Algebraic Structures: A Conference in Honor of J. Michael Boardman, Edited by: Jean-Pierre Meyer, Jack Morava, and W. Stephen Wilson, AMS, Contemporary Mathematics **239**(1999) 49–84. (40, 76, 79, 80, 487, 497)

Bockstein, M., A complete system of fields of coefficients for the ∇-homological dimension, Doklady Acad. Sci. URSS **38**(1943), 187–189. (456)

Borel, A., Impossibilité de fibrer une sphère par un produit de sphères, C.R. Acad. Sci. Paris **231**(1950), 943–945. (230)

Borel, A., Cohomologie des espaces localement compacts d'après J. Leray, Springer Lecture Notes in Mathematics **2**(1964). Exposés faits au Séminaire de Topologie algébrique de l'Ecole Polytechnique Fédérale au printemps 1951. Troisième edition, 1964. (134, 139)

Borel, A., Sur la cohomologie des espaces fibrés principaux et des espaces homogenes de groupes de Lie compacts, Ann. of Math. **57**(1953), 115–207. (20, 23, 85, 86, 131, 134, 136, 150, 154, 181, 197, 198, 208, 213, 222, 269, 276)

Borel, A., La cohomologie mod 2 de certains espaces homogenes, Comm. Math. Helv. **27**(1953'), 165–197. (219)

Borel, A., Topics in the Homology Theory of Fibre Bundles. Lectures given at the University of Chicago, 1954. Notes by Edward Halpern. Springer Lecture Notes in Mathematics, **36**(1967), 95 pp. (149, 469)

Borel, A., Topology of Lie groups and characteristic classes, Bull. Amer. Math. Soc. **61**(1955), 397–432. (149, 208, 278)

Borel, A., Seminar on Transformation Groups, with contributions by G. Bredon, E.E. Floyd, D. Montgomery, R. Palais; Annals of Mathematics Studies, No. 46 Princeton University Press, Princeton, N.J., 1960. (501)

Borel, A., Jean Leray and algebraic topology, in Leray, Jean: Selected Papers. Œuvres Scientifiques. Vol. I. Topologie et théorème du point fixe. Edited by Paul Malliavin. Springer-Verlag, Berlin; Société Mathématique de France, Paris, 1998. x+507 pp. (198, 222, 307)

Borel, A., Serre, J.-P., Impossibilité de fibrer un espace euclidien par des fibres compactes, C.R. Acad. Sci. Paris **230**(1950), 2258–2260. (141-2)

Bott, R., On torsion in Lie groups, Proc. Nat. Acad. Sci. U. S. A. **40**(1954), 586–588. (456)

Bott, R., An application of the Morse theory to the topology of Lie-groups, Bull. Soc. Math. France **84**(1956), 251–281. (456)

Bott, R., On some recent interactions between mathematics and physics, Canad. Math. Bull. **28**(1985), 129–164. (220)

Bott, R., Milnor, J.W., On the parallelizability of the sphere, Bull. Amer. Math. Soc. **64**(1958), 87–89. (425)

Bott, R., Tu, L.W., Differential Forms in Algebraic Topology. Graduate Texts in Mathematics, 82. Springer-Verlag, New York-Berlin, 1982. xiv+331 pp. (139, 208, 222, 278)

Bousfield, A.K., A vanishing theorem for the unstable Adams spectral sequence, Topology **9**(1970), 337–344. (492)

Bousfield, A.K., The localization of spaces with respect to homology, Topology **14**(1975), 133–150. (500)

Bousfield, A.K., On the homology spectral sequence of a cosimplicial space, Amer. J. Math. **109**(1987), 361–394. (82, 360, 361, 491)

Bousfield, A.K., Curtis, E.B., Kan, D.M., Quillen, D.G., Rector, D.L., Schlesinger, J.W., The mod *p* lower central series and the Adams spectral sequence, Topology **5**(1966), 331–342. (493)

Bousfield, A.K., Curtis, E.B., A spectral sequence for the homotopy of nice spaces, Trans. Amer. Math. Soc. **151**(1970), 457–479. (492)

Bousfield, A.K. and Kan, D., Homotopy Limits, Completions and Localizations, Springer Lecture Notes in Mathematics **304**(1972). (82, 109, 274, 330, 355, 356, 358, 361, 362, 363, 357, 489, 493, 494, 502, 503)

Bousfield, A.K., Kan, D.M., The homotopy spectral sequence of a space with coefficients in a ring, Topology **11**(1972'), 79–106. (494)

Bredon, G., Equivariant Cohomology Theories, Springer Lecture Notes in Mathematics **34** (1967). (501)

Browder, W., The cohomology of covering spaces of H-spaces, Bull. Amer. Math. Soc. **65**(1959), 140–141. (476, 483)

Browder, W., Torsion in H-spaces, Ann. of Math. (2) **74**(1961), 24–51. (458, 463, 465, 469, 472, 475, 477)

Browder, W., Homotopy commutative H-spaces, Ann. of Math. (2) **75**(1962), 283–311. (468)

Browder, W., Remark on the Poincaré duality theorem, Proc. Amer. Math. Soc. **13**(1962'), 927–930. (484)

Browder, W., On differential Hopf algebras, Trans. Amer. Math. Soc. **107**(1963), 153–176. (476)

Browder, W., The Kervaire invariant of framed manifolds and its generalization, Ann. of Math. (2) **90**(1969), 157–186. (451)

Browder, W., Algebraic K-theory with coefficients $\tilde{Z}/p$., in *Geometric applications of homotopy theory (Proc. Conf., Evanston, Ill., 1977), I*, Springer Lecture Notes in Math., **657**(1978), 40–84. (480, 481, 521)

Brown, E.H., Finite computability of Postnikov complexes, Ann. of Math. **65**(1957), 1–20. (180, 330)

Brown, E.H., Twisted tensor products I, Ann. of Math. **69**(1959), 223–246. (110, 181, 223, 224)

Brown, E.H., Cohomology theories, Ann. of Math. **75**(1962), 467–484. (450, 495)

Brown, E.H., The Serre spectral sequence theorem for continuous and ordinary cohomology, Top. and its Appl. **56**(1994), 235–248. (136, 139, 163)

Brown, E.H., Peterson, F.P., A spectrum whose Z_p cohomology is the algebra of reduced p^{th} powers, Topology 5 1966 149–154. (500)

Brown, K., Cohomology of Groups, Springer-Verlag, New York, 1982. (344)

Brown, K., Gersten, S., Algebraic K-theory as generalized sheaf theory, Springer Lecture Notes in Math. **341**(1973), 226–292 (520)

Brown, L.G., Douglas, R.G., Fillmore, P.A., Extensions of C^*-algebras and K-homology, Ann. of Math. (2) **105**(1977), 265–324. (523)

Brown, R., Two examples in homotopy theory, Proc. Cambridge Philos. Soc., **62**(1966) 575–576. (110, 297)

Bruner, R.R., A new differential in the Adams spectral sequence, Topology **23**(1984), 271–276. (426, 449, 450)

Bruner, R.R., Ext in the nineties, Algebraic topology (Oaxtepec, 1991), Contemp. Math. **146**(1993), 71–90. (445)

Bruner, R.R., May, J P., McClure, J.E., Steinberger, M., H_∞ ring spectra and their applications. Springer Lecture Notes in Mathematics **1176**(1986). viii+388 pp. (449)

Brunn, H., Über Verkettung, S.-B. Math. Phys. Kl. Bayer. Akad. Wiss. **22**(1892), 77–99. (309)

Bullett, S.R. and MacDonald, I.G., On the Adem relations, Topology **21**(1982), 329–332. (130)

Cain, R.N., A spectral sequence for the intersection of subspace pairs. Proc. Amer. Math. Soc. **43**(1974), 229–236. (505)

Carlsson, G., Segal's Burnside ring conjecture and the homotopy limit problem. Homotopy theory (Durham, 1985), London Math. Soc. Lecture Note Ser., **117**(1987), 6–34. (504)

Cartan, E., La topologie des espaces représentatifs des groupes de Lie, Acualités Scientifiques et Industrielles, no. 358, Paris, Hermann, 1936. (475)

Cartan, H., Sur la cohomologie des espaces où opère un groupe: I. Notions algébriques préliminaires; II. étude d'un anneau différentiel où opère un groupe, C.R. Acad. Sci. Paris **226**(1948), 148–150, 303–305. (34, 45, 134, 160)

Cartan, H., Une théorie axiomatique des carrés de Steenrod, C.R. Acad. Sci. de Paris **230**(1950), 425–427. (128, 273, 275, 277)

Cartan, H., Notions d'algèbre différentielle; applications aux groupes de Lie et aux variétés où opère un groupe de Lie; La transgression dans une groupe de Lie et dans un espace fibré principal, Colloque de Topologie, Bruxelles (1950), CBRM, Liège, 1951, 15–27, 51–71. (221, 222)

Cartan, H., Algèbres d'Eilenberg-Mac Lane, Seminaire Cartan, ENS, 1954-55, exposés 2 to 11. (128, 180, 194, 197, 224, 232, 468)

Cartan, H., Eilenberg, S., Homological Algebra, Princeton Univ. Press, 1956. (25, 58, 63, 69, 335, 376, 387, 404, 507, 508, 515)

Cartan, H., Leray, J., Relations entre anneaux d'homologie et groupes de Poincaré, Topologie algébrique, pp. 83–85. Colloques Internationaux du Centre National de la Recherche Scientifique, no. 12. Centre de la Recherche Scientifique, Paris, 1949. (338)

Cartan, H., Serre, J.-P., Espaces fibrés et groupes d'homotopie, I. Constructions générales; II. Applications, C.R. Acad. Sci. Paris **234**(1952), 288–290, 393–395. (203)

Čech, E., Théorie générale de l'homologie dans un espace quelconque, Fund. Math. **19**(1932), 149–183. (133)

Chen, K.T., Iterated path integrals, Bull. Amer. Math. Soc. **83**(1977), 831–879. (225)

Chern, S.-S., On the multiplication in the characteristic ring of a sphere bundle, Ann. of Math. **49**(1948), 362–372. (185, 207, 220)

Chern, S.S., From triangles to manifolds, Amer. Math. Monthly **86**(1979), 339–349.

Clark, A., Homotopy commutativity and the Moore spectral sequence. Pacific J. Math. **15**(1965), 65–74. (262, 292, 294, 295)

Clarke, F., Johnson, K., Cooperations in elliptic homology, Adams Memorial Symposium on Algebraic Topology, 2 (Manchester, 1990), 131–143, London Math. Soc. Lecture Note Ser., 176, Cambridge Univ. Press, Cambridge, 1992. (500)

Cochran, T.D., Derivatives of links: Milnor's concordance invariants and Massey's products, Mem. Amer. Math. Soc. **84**(1990), no. 427, x+73 pp. (310)

Cohen, F.R., Moore, J.C., Neisendorfer, J.A., The double suspension and exponents of the homotopy groups of spheres, Ann. of Math. (2) **110**(1979), 549–565. (327, 481)

Cohen, J.M., The decomposition of stable homotopy, Ann. of Math. **87**(1968), 305–320. (429, 492, 496)

Cohen, J.M., Stable Homotopy. Springer Lecture Notes in Mathematics, **165**(1970), v+194 pp. (450)

Cohen, R.L., Odd primary infinite families in stable homotopy theory, Mem. Amer. Math. Soc. **30**(1981), no. 242, viii+92 pp. (376, 450)

Conner, P. E., Smith, L., On the complex bordism of finite complexes, I.H.E.S. Publ. Math. **37**(1969), 117–221. (313)

Connes, A., Non-commutative differential geometry, Publ. I.H.E.S. **62**(1985), 41–144. (510)

Cordero, L.A., Fernández, M., Gray, A., The Frölicher spectral sequence for compact nilmanifolds, Ill. J. Math. **35**(1991), 56–67. (518)

Cordero, L.A., Fernández, M., Gray, A., The failure of complex and symplectic manifolds to be Kählerian, Differential geometry: geometry in mathematical physics and related topics (Los Angeles, CA, 1990), Proc. Sympos. Pure Math., **54** Part 2 (1993), 107–123. (518)

Crabb, M., James, I.M., Fibrewise Homotopy Theory, Springer Monographs in Mathematics, Springer-Verlag London, Ltd., London, 1998. viii+341 pp. (274, 315)

Curtis, E.B., Some relations between homotopy and homology, Ann. of Math. **82**(1965), 386–413. (493)

Curtis, E.B., Simplicial homotopy theory, Adv. in Math. **6**(1971), 107–209. (109, 121)

Curtis, M., Finite dimensional H-spaces, Bull. Amer. Math. Soc. **77**(1971), 1–12, and 1120. (458)

Damay, A.-S., Introduction aux suites spectrales: théorie générale et la suite spectrale de Serre, Projet de diplôme, École Polytechnique Fédérale de Lausanne, 1996. (229)

Davis, D.M., The BP-coaction for projective spaces, Can. Jour. Math. **30**(1978) 45–53. (498)

Davis, D.M., Johnson, D.C., Klippenstein, J., Mahowald, M.E., Wegmann, S.A., The spectrum $(P \wedge BP<2>)_{-\infty}$, Trans. Amer. Math. Soc. **296**(1986) 95–110. (498)

Davis, J.F., Lück, W., Spaces over a category and assembly maps in isomorphism conjectures in algebraic K- and L- theory, K-theory **15**(1998), 201–252. (496)

Deligne, P., Théorème de Lefschetz et critères de dégénérescence de suites spectrales. Inst. Hautes Études Sci. Publ. Math. No. **35**(1968), 259–278. (518)

Deligne, P., Théorie de Hodge. II, Inst. Hautes Études Sci. Publ. Math. **40**(1971), 5–57. (517, 519)

Deligne, P., Griffiths, P., Morgan, J., Sullivan, D., Real homotopy theory of Kähler manifolds, Invent. Math. **29**(1975), 245–274. (310)

Devinatz, E.S., Hopkins, M.J., Smith, J.H., Nilpotence and stable homotopy theory. I, Ann. of Math. **128**(1988), 207–241. (451, 500)

Dieudonné, J., Grothendieck, A., Éléments de géométrie algébrique. II. Étude globale élémentaire de quelques classes de morphismes, Inst. Hautes Études Sci. Publ. Math, **8**(1961), 222 pp. (79)

Dirac, P.A.M., The Principles of Quantum Mechanics, Oxford University Press, Oxford, England, 1935. (220)

Dixon, J. A., Calculation of BRS cohomology with spectral sequences, Comm. Math. Phys. **139**(1991), 495–526. (520)

Dold, A., Universelle Koeffizienten, Math. Z. **80**(1962), 63–88. (326, 514, 515)

Dold, A., Lashof, R., Principal quasifibrations and fibre homotopy equivalences of bundles, Ill. J. Math. **3**(1959), 285–305. (211)

Douady, A., La suite spectrale d'Adams, Séminaire Henri Cartan (1958-59), exposé 18, 19. (403)

Dowker, C.H., Topology of metric complexes, Amer. J. Math. 74, (1952), 555–577. (94)

Drachman, B., A diagonal map for the cobar construction, Bol. Soc. Mat. Mexicana (2) **12**(1967), 81–91. (491)

Dress, A., Zur Spectralsequenz von Faserungen, Inv. Math. **3**(1967), 172–178. (163, 181, 221, 225, 228)

Dror, E., A generalization of the Whitehead Theorem, Springer Lecture Notes in Mathematics **249**(1971), 13–22. (330, 331, 348)

Dror, E., Pro-nilpotent representation of homology types, Proc. Amer. Math. Soc. **38**(1973), 657–660. (330, 362, 363)

Dror Farjoun, E., Smith, J., A geometric interpretation of Lannes' functor T, International Conference on Homotopy Theory (Marseille-Luminy, 1988), Astérisque **191**(1990), 87–95. (327, 360)

Dupont, J.L., Curvature and Characteristic Classes, Lecture Notes in Mathematics, Vol. 640. Springer-Verlag, Berlin-New York, 1978. viii+175 pp. (208, 220)

Dupont, J.L., Algebra of polytopes and homology of flag complexes, Osaka J. Math. **19**(1982), 599–641. (23)

Durfee, A.H., A naive guide to mixed Hodge theory, Singularities, Part 1 (Arcata, Calif., 1981), Proc. Sympos. Pure Math., **40**(1983), 313–320. (519)

Dwyer, W.G., Strong convergence of the Eilenberg-Moore spectral sequence, Topology **13**(1974), 255–265. (331, 355)

Dwyer, W.G., Exotic convergence of the Eilenberg-Moore spectral sequence, Illinois J. Math. **19**(1975), 607–617. (362)

Dwyer, W.G., Vanishing homology over nilpotent groups, Proc. Amer. Math. Soc. 49 (1975), 8–12. (82)

Dwyer, W.G., Homology, Massey products and maps between groups, J. Pure Appl. Algebra **6**(1975), 177–190. (310)

Dwyer, W.G., Higher divided squares in second-quadrant spectral sequences, Trans. Amer. Math. Soc. **260**(1980), 437–447. (326)

Dwyer, W.G., Classifying spaces and homology decompositions, preprint (Notre Dame) 1998. (458, 502)

Dwyer, W.G., Kan, D.M., Smith, J.H., Stover, C.R., A Π-algebra spectral sequence for function spaces, Proc. Amer. Math. Soc. **120**(1994), 615–621. (490)

Dwyer, W.G., Wilkerson, C.W., Homotopy fixed-point methods for Lie groups and finite loop spaces. Ann. of Math. 139 (1994), no. 2, 395–442. (149, 458)

Dyer, E., Cohomology Theories. Mathematics Lecture Note Series W.A. Benjamin, Inc., New York-Amsterdam, 1969 xiii+183 pp. (222)

Dyer, E., Lashof, R.K., Homology of iterated loop spaces, Amer. J. Math. **84**(1962), 35–88. (326, 499)

Ebbinghaus, H.-D., with Hermes, H.; Hirzebruch, F.; Koecher, M.; Mainzer, K.; Neukirch, J.; Prestel, A.; Remmert, R., *Numbers*. With an introduction by K. Lamotke. Translated from the second German edition by H. L. S. Orde. Translation edited and with a preface by J. H. Ewing. Graduate Texts in Mathematics, 123. Readings in Mathematics. Springer-Verlag, New York, 1990. xviii+395 pp. (366)

Eckmann, B., Zur Homotopietheorie gefaserter Räume, Comm. Math. Helv. **14**(1942), 141–192. (134)

Eckmann, B. and Hilton, P.J., Groupes d'homotopies et dualité, C. R. Acad. Sci. Paris **246**(1958), (Groups absolus) 2444–2447, (Suites exactes) 2555–2558, (Coefficients) 2991–2993. Transgression homotopique et cohomologique, C. R. Acad. Sci. Paris **247**(1958), 620–623. (118)

Eckmann, B., Hilton, P. J., Exact couples in an abelian category, J. Algebra, **3**(1966), 38–87. (28, 41, 88, 514)

Ehresmann, Ch., Feldbau, J., Sur les propriétés d'homotopie des espaces fibrés, C.R. Acad. Sci. Paris **212**(1941), 945–948. (133)

Eilenberg, S., Singular homology theory, Ann. of Math. **45**(1944) 407–447. (104)

Eilenberg, S., Homology of spaces with operators. I. Trans. Amer. Math. Soc. **61**(1947) 378–417; errata, **62**(1947), 548. (120, 340)

Eilenberg, S., La suite spectrale. I: Construction générale; II: Espaces fibrés, Exposés 8 et 9, Séminaire Cartan, E.N.S. 1950/51, Cohomologie des groupes, suite spectrale, faisceaux. (139, 190, 229)

Eilenberg, S., Mac Lane, S., Relations between homology and homotopy groups of spaces. I. Ann. of Math. **51**(1945), 480–509. (134)

Eilenberg, S., Mac Lane, S., Relations between homology and homotopy groups of spaces. II. Ann. of Math. **51**(1950), 514–533. (297)

Eilenberg, S., Mac Lane, S., On the groups of $H(\Pi, n)$. I. Ann. of Math. (2) **58**(1953), 55–106. (242, 247, 290, 339)

Eilenberg, S., Moore, J.C., Limits and spectral sequences, Topology **1**(1962), 1–24. (28, 68, 69, 81, 88, 514)

Eilenberg, S., Moore, J.C., Limits and spectral sequences II, unpublished manuscript. (88)

Eilenberg, S., Moore, J.C., Adjoint functors and triples, Ill. J. Math. **9**(1965), 381–395. (322)

Eilenberg, S., Moore, J.C., Homology and fibrations. I. Coalgebras, cotensor product and its derived functors. Comment. Math. Helv. **40**(1966), 199–236. (229, 232, 233, 252, 271)

Eilenberg, S., Steenrod, N.E., Axiomatic approach to homology theory, Proc. Nat. Acad. Sci. U.S.A. **31**(1945), 177–180. (133)

Ekedahl, T., Diagonal complexes and F-gauge structures, With a French summary. Travaux en Cours. [Works in Progress] Hermann, Paris, 1986. xii+122 pp. (518)

Elmendorf, A.D., Kriz, I., Mandell, M.A., May, J.P., Modern foundations for stable homotopy theory, in Handbook of Algebraic Topology, edited by I.M. James, Elsevier Science, Amsterdam, 1995, 213–253. (497)

Elmendorf, A.D., Kriz, I., Mandell, M.A., May, J.P., Rings, Modules, and Algebras in Stable Homotopy Theory. With an appendix by M. Cole. Mathematical Surveys and Monographs, **47**. A.M.S., Providence, RI, (1997) xii+249 pp. (368, 524)

El Zein, F., Introduction à la théorie de Hodge mixte, Actualités Mathématiques. Hermann, Paris, 1991. 238 pp. (519)

Evens, L., The Cohomology of Groups. Oxford Mathematical Monographs. Oxford University Press, New York, 1991. xii+159 pp. (344)

Fadell, E., Review of [Brown, E59], Math. Rev. **21**(1960) MR 21 4423. (110)

Fadell, E., Hurewicz, W., On the structure of higher differentials in spectral sequences, Ann. of Math., **68**(1958), 314–347. (110, 181, 222, 223, 225)

Fadell, E., Husseini, S., Category weight and Steenrod operations, Boletin del la Soc. Mat. Mex. **37**(1992), 151–161. (302)

Federer, H., A study of function spaces by spectral sequences, Trans. Amer. Math. Soc. **82**(1956), 340–361. (490)

Feldbau, J., Sur la classification des espaces fibrés, C.R. Acad. Sci. Paris **208**(1939), 1621–1623. (112)

Felix, Y., Halperin, S., Thomas, J.-C., Differential graded algebras in topology, Handbook of Algebraic Topology, edited by I.M. James, 829–865, North-Holland, Amsterdam, 1995. (287)

Fenn, R., Sjerve, D., Basic commutators and minimal Massey products. Canad. J. Math. **36**(1984), 1119–1146. (310)

Fieux, E., Solotar, A., Une suite spectrale pour les groupes d'homotopie des espaces d'applications équivariantes, Bull. Belg. Math. Soc. Simon Stevin **5**(1998), 565–582. (502)

Forman, R., Hodge theory and spectral sequences, Topology **33**(1994), 591–611. (519)

Forman, R., Spectral sequences and adiabatic limits, Comm. Math. Phys. **168**(1995), 57–116. (519)

Freudenthal, H., Über die Klassen der Sphärenabbildungen, Comp. Math. **5**(1937), 299–314. (98, 205)

Friedlander, E.M., Motivic complexes of Suslin and Voevodsky, Exp. No. 833, Séminaire Bourbaki, 1996/97. Astérisque **245**(1997), 355–378. (522)

Fritsch, R., Piccinini, R.A., Cellular Structures in Topology. Cambridge Studies in Advanced Mathematics, 19. Cambridge University Press, Cambridge, 1990. xii+326 pp. (94, 102)

Frölicher, A., Relations between the cohomology groups of Dolbeault and topological invariants, Proc. nat. Acad. Sci. **41**(1955), 641–644. (518)

Gabriel, P., Zisman, M., Calculus of Fractions and Homotopy Theory. Springer-Verlag, Berlin, Heidelberg, New York, 1967. (523)

Gelfand, S.I., Manin, Yu.I., Methods of Homological Algebra. Springer-Verlag Berlin, Heidelberg, New York, 1996. (514, 524)

Ghazal, T., A new example in K-theory of loopspaces, Proc. Amer. Math. Soc. **107**(1989), 855–856. (321)

Gillet, H., Comparison of K-theory spectral sequences, with applications, Algebraic K-theory, Evanston 1980 (Proc. Conf., Northwestern Univ., Evanston, Ill., 1980), Springer Lecture Notes in Math., **854**(1981), 141–167. (521)

Gitler, S., Spaces fibered by H-spaces, Bol. Soc. Mat. Mexicana (2) **7**(1962), 71–84.(262)

Godement, R., Topologie algébrique et théorie des faisceaux, Publ. de l'Inst. Math. de Strasbourg, XII; Hermann, Paris, 1958. (165, 516)

Goerss, P.G., André-Quillen cohomology and the Bousfield-Kan spectral sequence, International Conference on Homotopy Theory (Marseille-Luminy, 1988). Astérisque **191**(1990), 109–209. (494)

Goerss, P.G., Barratt's desuspension spectral sequence and the Lie ring analyzer, Quart. J. Math. Oxford (2)**44**(1993), 43–85. (488)

Goerss, P.G., Jardine, J.F., Simplicial Homotopy Theory. Progress in Mathematics, 174, Birkhäuser Verlag, Basel, 1999. xvi+510 pp. (109)

Golod, E.S., On the homology of some local rings, Soviet Math. Doklady **3**(1962), 745–748. (14, 310)

Goodwillie, T.G., A remark on the homology of cosimplicial spaces, J. Pure Appl. Alg. **127**(1998), 167–175. (361)

Goryunov, V.V., Semi-simplicial resolutions and homology of images and discriminants of mappings, Proc. London Math. Soc. (3) **70**(1995), 363–385. (506)

Goryunov, V.V., Mond, D.M.Q., Vanishing cohomology of singularities of mappings, Compositio Math. **89**(1993), 45–80. (506)

Gottlieb, D.H., Fiber bundles with cross sections and non-collapsing spectral sequences, Ill. J. Math. **21**(1977), 176–177. (149)

Greenlees, J.P.C., How blind is your favourite cohomology theory?, Expos. Math. **6**(1988), 193–208. (17, 369, 502)

Greenlees, J.P.C., Stable maps into free G-spaces, Trans. Amer. Math. Soc. **310**(1988'), 199–215. (502)

Greenlees, J.P.C., Generalized Eilenberg-Moore spectral sequences for elementary abelian groups and tori, Math. Proc. Cambridge Philos. Soc. **112**(1992), 77–89. (502)

Greenlees, J.P.C., Homotopy equivariance, strict equivariance and induction theory., Proc. Edinburgh Math. Soc. (2) **35**(1992'), 473–492. (502)

Greenlees, J.P.C. and May, J.P., Equivariant stable homotopy theory, Handbook of algebraic topology, 277–323, North-Holland, Amsterdam, 1995. (17)

Griffiths, P., Harris, J., Principles of Algebraic Geometry, Reprint of the 1978 original, Wiley Classics Library, John Wiley & sons, New York, 1994, xiv + 813 pp. (487, 516)

Griffiths, P.A., Morgan, J.W., Rational Homotopy Theory and Differential Forms, PM16, Birkhäuser, Basel, 1981. (328)

Grivel, P.-P., Formes différentielles et suites spectrales, Ann. Inst. Fourier (Grenoble) **2**(1979), 17–37. (226)

Gromov, M., Lawson, H.B., The classification of simply connected manifolds of positive scalar curvature, Ann. of Math. **111**(1980), 423–434. (451)

Grothendieck, A., Sur quelques points d'algèbre homologique, Tôhoku Math. J. (2)**9**(1957), 119–221. (342, 514)

Grothendieck, A., Éléments de Géométrie Algébrique. III. Étude cohomologique des faisceaux cohérents. I. Inst. Hautes Études Sci. Publ. Math. **11**(1961), 167 pp. (517)

Grothendieck, A., Crystals and the de Rham cohomology of schemes, Dix Exposés sur la Cohomologie des Schémas, North-Holland, Amsterdam; Masson, Paris, 1968, 306–358. (517)

Gugenheim, V.K.A.M., On a theorem of E. H. Brown, Illinois J. Math. **4**(1960), 292–311. (224, 297)

Gugenheim, V.K.A.M., On extensions of algebras, co-algebras and Hopf algebras. I, Amer. J. Math. **84**(1962), 349–382. (513)

Gugenheim, V.K.A.M., On a perturbation theory for the homology of the loop-space, J. Pure Appl. Algebra **25**(1982), 197–205. (225)

Gugenheim, V.K.A.M., Lambe, L.A., and Stasheff, J.D., Perturbation theory in differential homological algebra. II. Illinois J. Math. **35**(1991), 357–373. (224)

Gugenheim, V.K.A.M., May, J.P., On the theory and applications of differential torsion products, Memoirs Amer. Math. Soc. **142**, 1974. (277, 287ff., 288, 312)

Gugenheim, V.K.A.M., Milgram, R.J., On successive approximations in homological algebra. Trans. Amer. Math. Soc. **150**(1970), 157–182. (297)

Gugenheim, V.K.A.M., Moore, J.C., Acyclic models and fibre spaces, Trans. Amer. Math. Soc. **85**(1957), 265–306. (222)

Gugenheim, V.K.A.M., Munkholm, H.-J., On the extended functoriality of Tor and Cotor, J. P. App. Alg., **4**(1974), 9–29. (294, 295)

Gysin, W., Zur Homologietheorie der Abbildungen und Faserungen der Mannigfaltigkeiten, Comm. Math. Helv. **14**(1941), 61–121. (144)

Halperin, S., Stasheff, J.D., Obstructions to homotopy equivalences, Adv. in Math. **32**(1979), 233–279. (310)

Hardy, G.H., Wright, E.M., An Introduction to the Theory of Numbers. Fifth edition. The Clarendon Press, Oxford University Press, New York, 1979. xvi+426 pp. First edition, 1938. (263)

Hatcher, A., Ext tables, http://math.cornell.edu/ hatcher/stemfigs/stems.html. (445)

Henderson, G.D., Spectral sequences for the classification of extensions of Hopf algebras. J. Algebra **193**(1997), 12–40. (513)

Henneaux, M., Teitelboim, C., Quantization of Gauge Systems. Princeton University Press, Princeton, NJ, 1992. (520)

Hilton, P.J., On the homotopy groups of the union of spheres, J. London Math. Soc. **30**(1955), 154–172. (303)

Hilton, P.J., Localization in topology, Amer. Math. Monthly, **82**(1975), 113–131. (362)

Hilton, P.J., Mislin, G., Roitberg, J., Localization of Nilpotent Groups and Spaces. North-Holland Mathematics Studies, No. 15 .North-Holland Amsterdam-Oxford, New York, 1975. x+156 pp. (362, 458)

Hilton, P.J., Roitberg, J., On principal S^3-bundles over spheres, Ann. of Math. (2) **90**(1969), 91–107. (458, 476)

Hilton, P.J., Roitberg, J., On the Zeeman comparison theorem for the homology of quasi-nilpotent fibrations, Quart. J. Math. Oxford Ser. **(2) 27**(1976), 433–444. (359)

Hilton, P.J., Wylie, S., Homology theory: An Introduction to Algebraic Topology. Cambridge University Press, New York, NY 1960. (139)

Hirsch, G., Cohomologie d'un espace de Postnikov (cas non stable), unpublished preprint. (277, 291)

Hirsch, G., Sur les groupes d'homologie des espaces fibrés, Bull. Soc. Math. Belg. **2**(1948-49), 24–33. (185)

Hirsch, G., Sur les groupes d'homologie des espaces fibrés, Bull. Soc. Math. Belg. **6**(1950), 79–96. (181, 223, 225)

Hirschhorn, P.S., A spectral sequence for the homotopy groups of a wedge, Amer. J. of Math. **109**(1987), 783–786. (489)

Hirzebruch, F., Topological Methods in Algebraic Geometry. Springer-Verlag, Berlin, New York, Heidelberg, 1966. (217, 220)

Hitchin, N., Harmonic spinors, Adv. Math. **14**(1974), 1–55. (451)

Hochschild, G., On the cohomology groups of an associative algebra, Annals of Math. **46**(1945), 58–67. (509)

Hochschild, G., Konstant, B., Rosenberg A., Differential forms on regular affine algebras, Trans. Amer. Math. Soc. **102**(1962), 383–408. (509)

Hochschild, G., Serre, J.-P., Cohomology of group extensions, Trans. Amer. Math. Soc. **74** (1953), 110–134. (342)

Hochschild, G., Serre, J.-P., Cohomology of Lie algebras, Annals of Math. **57**(1953'), 591–603. (511)

Hodgkin, L., Notes toward a geometric Eilenberg-Moore spectral sequence, (mimeograph) E. T. H. Zurich 1969. (274, 313, 321)

Hodgkin, L., The equivariant Künneth theorem in K-theorem, Topics in K-theory. Two independent contributions, Springer Lecture Notes in Math.,**496**(1975), 1–101. (321, 498)

Honkasalo, H., The equivariant Serre spectral sequence as an application of a spectral sequence of Spanier, Topology and it Appl. **90**(1998), 11–19. (501)

Hopf, H., Über die Abbildungen der dreidimensionalen Sphäre auf die Kugelfläche, Math. Ann. **104**(1931), 637–665. (366)

Hopf, H., Über die Abbildungen von Sphären auf Sphären von niedriger Dimension, Fund. Math. **35**(1935), 427–440. (366)

Hopf, H., Über die Topologie der Gruppenmannigfaltigkeiten und ihre Verallgemeinerungen, Annals of Math. **42**(1941), 22–52. (134, 213, 456, 457)

Hopf, H., Fundamentalgruppe und zweite Bettische Gruppe, Comm. Math. Helv. **17**(1942), 257–309. (134, 338)

Hopf, H., Rinow, W., Über den Begriff der vollständigen differentialgeometrischen Flächen, Comm. Math. Helv. **3**(1931), 209–225. (158)

Hopkins, M.J., Topological modular forms, the Witten genus, and the theorem of the cube, Proceedings of the International Congress of Mathematicians, (Zürich, 1994, 554–565, Birkhäuser, Basel, 1995. (500)

Houston, K., Local topology of images of finite complex analytic maps, Topology **36**(1997), 1077–1121. (506)

Houston, K., An introduction to the image computing spectral sequence. Singularity theory (Liverpool, 1996), 305–324, London Math. Soc. Lecture Note Ser., 263, Cambridge Univ. Press, Cambridge, 1999. (506)

Houzel, C., Les débuts de la théorie des faisceaux, in Sheaves on manifolds by Masaki Kashiwara, Pierre Schapira, Berlin ; New York : Springer-Verlag, 1990. (134)

Hu, S.T., Homotopy Theory,.

Huebschmann, J., Cohomology of metacyclic groups, Trans. Amer. Math. Soc. **328**(1991), 1–72. (344)

Huebschmann, J., Automorphisms of group extensions and differentials in the Lyndon-Hochschild-Serre spectral sequence, J. Algebra **72**(1981), 296–334. (344)

Huebschmann, J., Kadeishvili, T., Small models for chain algebra, Math. Z. **207**(1991), 245–280. (225, 297)

Hunter, T.J., On $H_*(\Omega^{n+2}S^{n+1};F_2)$, Trans. Amer. Math. Soc. **314**(1989), 405–420. (326)

Hurewicz, W., Beiträge zur Topologie der Deformationen, I–IV, Proc. Akad. Wetensch. Amsterdam **38**(1935) 112–119, 521–528; **39**(1936) 117–126, 215–224. (133-4)

Hurewicz, W., On the concept of fiber space, Proc. Nat. Acad. Sci. U.S.A. **41**(1955), 956–961. (110)

Hurewicz, W., Steenrod, N.E., Homotopy relations in fibre spaces, Proc. Amer. Math. Sci. **27**(1941), 60–64.

Hurwitz, A., Über die Komposition der quadratischen Formen, Math. Ann. **88**(1923),1–25. (366)

Husemoller, D., Fibre Bundles. Third edition. Graduate Texts in Mathematics, 20. Springer-Verlag, New York, 1994. xx+353 pp. (154, 208, 210, 211)

Husemoller, D., Moore, J.C., Stasheff, J.D., Differential homological algebra and homogeneous spaces, J. P. Appl. Alg. **5**(1974), 113–185. (224, 296)

Illusie, L., Complexe de de Rham-Witt et cohomologie cristalline, Ann. Sci. École Norm. Sup. (4) **12**(1979), 501–661. (517, 518)

Intermont, M., An equivariant van Kampen spectral sequence, Topology Appl. **79**(1997), 31–48. (502)

Intermont, M., An equivariant smash spectral sequence and an unstable box product, Trans. Amer. Math. Soc. **351**(1999), 2763–2775. (502)

Ivanovskiĭ, I.N., Cohomology of the Steenrod algebra, Dokl. Akad. Nauk USSR **157**(1964), 1284–1287 (Russian). (427, 428)

James, I.M., The suspension triad of a sphere, Ann. of Math., **63**(1956), 407–429. (494)

James, I.M., The transgression and Hopf invariant of a fibration, Proc. London Math. Soc. **11**(1961), 588–600. (185)

James, I.M., Ex-homotopy theory. I, Illinois J. Math. **15**(1971), 324–337. (315, 328)

James, I.M., Fibrewise Topology, Cambridge Tracts in Mathematics, 91. Cambridge University Press, Cambridge-New York, 1989. x+198 pp. (315, 328)

Jeanneret, A., Algebras over the Steenrod algebra and finite H-spaces, Adams memorial symposium on algebraic topology, vol. 2, Proc. Symp., Manchester/UK 1990, Lond. Math. Soc. Lect. Note Ser. **176**(1992), 187-202. (476)

Jeanneret, A., Osse, A., The Eilenberg-Moore spectral sequence in K-theory, Topology **38**(1999), 1049–1073. (321, 498)

Johnson, D.C., A Stong-Hattori spectral sequence, Trans. Amer. Math. Soc. **179**(1973), 211–225. (483)

Jozefiak, T., Tate resolutions for commutative graded algebras over a local ring, Fund. Math. **74**(1972), 209–231. (261)

Kahn, B., La conjecture de Milnor d'après V. Voevodsky, Exp. No. 834, Séminaire Bourbaki, 1996/97. Astérisque **245**(1997), 379–418. (522)

Kahn, D.S., Cup-i products and the Adams spectral sequence, Topology **9**(1970), 1–9. (425, 426, 449)

Kahn, D.S., Priddy, S.B., On the transfer in the homology of symmetric groups; The transfer and stable homotopy theory, Math. Proc. Cambridge Philos. Soc. **83**(1978), 91–101; 103–111. (407)

Kahn, D.W., The spectral sequence of a Postnikov system, Comm. Math. Helv. **40**(1966), 169–198. (492)

Kan, D.M., A combinatorial definition of homotopy groups, Ann. of Math. **67**(()1958), 288–312. (103, 109, 489)

Kan, D.M., Adjoint functors, Trans. Amer. Math. Soc. **87**(1958), 294–329. (108)

Kan, D.M., Thurston, W.P., Every connected space has the homology of a $K(\pi,1)$, Topology **15**(1976), 253–258. (339)

Kane, R., On loop spaces without p torsion, Pac. J. Math. **60**(1975), 189–201. (265, 307, 327)

Kane, R., On loop spaces without p torsion. II, Pacific J. Math. **71**(1977), 71–88. (476)

Kane, R.M., Implications in Morava K-theory, Mem. Amer. Math. Soc. **59**(1986), no. 340, iv+110 pp. (458, 477, 483)

Kane, R.M., The Homology of Hopf Spaces. North-Holland, Amsterdam, 1988. (477)

Karoubi, M., Formes différentielles non commutatives et opérations de Steenrod, Topology **34** (1995), 699–715. (128)

Kasparov, G.G., The K-functors in the theory of extensions of C^*-algebras, Funktsional. Anal. i Prilozhen. **13**(1979), 73–74. (523)

Keller, B., Derived categories and their uses, Handbook of Algebra, vol. 1, edited by M. Hazewinkel, 671–701. (524)

Kervaire, M., Non-parallelizability of the n-sphere for $n>7$, Proc. Nat. Acad. Sci. U.S.A. **44**(1958), 280–283. (425)

Kervaire, M., A manifold which does not admit any differentiable structure, Comment. Math. Helv. **34**(1960), 257–270. (451)

Kochman, S.O., The symplectic cobordism ring. I, Mem. Amer. Math. Soc. **24**(1980), no. 228, ix+206 pp. (415)

Kochman, S.O., Stable homotopy groups of spheres. A computer-assisted approach, L. N. M. **1423**, Springer-Verlag, Berlin, 1990. viii+330 pp. (450, 451, 497)

Kochman, S.O., Bordism, Stable Homotopy and Adams Spectral Sequences. Fields Institute Monographs **7**, AMS, Providence, RI, 1996. (120, 368, 415, 450, 496, 500)

Kochman, S.O., Mahowald, M.E., On the computation of stable stems, The Čech centennial (Boston, MA, 1993), Contemp. Math., **181**(1995), 299–316. (450)

Kolmogoroff, A.N., Über die Dualität im Aufbau der kombinatorischen Topologie, Math. Sbornik **43**(1936), 97–102. (133)

Koszul, J.-L., Sur le opérateurs de derivation dans un anneau, C. R. Acad. Sci. Paris **225** (1947), 217–219. (34, 45, 134, 222)

Koszul, J.-L., Homologie et cohomologie des algèbres de Lie, Bull. Soc. Math. France **78**(1950), 65–127. (185, 221, 222, 277, 511)

Kraines, D., Massey higher products. Trans. Amer. Math. Soc. **124**(1966), 431–449. (273, 297, 306, 307)

Kraines, D., The $\mathcal{A}(p)$ cohomology of some K stage Postnikov systems. Comment. Math. Helv. **48**(1973), 56–71. Corrigendum: **48**(1973), 194. (292, 307)

Kraines, D., Lada, T., Applications of the Miller spectral sequence, Current trends in algebraic topology, Part 1 (London, Ont., 1981), pp. 479–497, CMS Conf. Proc., 2, Amer. Math. Soc., Providence, R.I., 1982. (499)

Kraines, D., Schochet, C., Differentials in the Eilenberg-Moore spectral sequence, J. P. App. Alg. **2**(1972), 131–148. (305, 307)

Krasil'shchik, J., Cohomology background in geometry of PDE, Cont. Math. **219**(1998), 121–139. (520)

Krasil'shchik, J., Verbovetsky, A., Homological Methods in Equations of Mathematical Physics. Open Education & Sciences, P.O. Box 84, 746 20 Opava, Czech Republic, 1998. (519)

Kristensen, L., On the cohomology of two-stage Postnikov systems, Acta Math. **107**(1962), 73–123. (195)

Kristensen, L., On the cohomology of spaces with two non-vanishing homotopy groups. Math. Scand. **12**(1963), 83–105. (292)

Kudo, T., A transgression theorem, Mem. Fac. Sci. Kyusyu Univ. (A)**9**(1956), 79–81. (192)

Kuhn, N.J., A Kahn-Priddy sequence and a conjecture of G. W. Whitehead, Math. Proc. Cambridge Philos. Soc. **92**(1982), 467–483. Corrigenda: **95**(1984), 189–190. (499)

Kuo, T.-C., Spectral operations for filtered simplicial sets, Topology **4**(1965), 101–107. (195)

Lambe, L.A., Homological perturbation theory, Hochschild homology, and formal groups, In Deformation theory and quantum groups with applications to mathematical physics (Amherst, MA, 1990), Contemp. Math. **134**(1992), 183–218. (225, 297)

Lamotke, K., Semisimpliziale Algebraische Topologie, Die Grundlehren der mathematischen Wissenschaften, Band 147 Springer-Verlag, Berlin-New York 1968 viii+285 pp. (109)

Landweber, P.S., Künneth formulas for bordism theories, Trans. Amer. Math. Soc. **121**(1966), 242–256. (313)

Landweber, P.S., Homological properties of comodules over $MU_*(MU)$ and $BP_*(BP)$, Amer. J. Math. **98**(1976), 591–610. (500)

Lannes, J., Sur les espaces fonctionnels dont la source est le classifiant d'un p-groupes abélien élementaire, Publ. I.H.E.S. **75**(1992), 135–244. (327, 458)

Laures, G., The topological q-expansion principle, Topology 38 (1999), 387–425. (500)

Lazard, M., Lois de groupes et analyseurs, Ann. Sci. Ecole Norm. Sup. (3) **72**(1955), 299–400. (488)

Lefschetz, S., On singular chains and cycles, Bull. A.M.S. **39**(1933), 124–129. (104)

Leray, J., Sur la forme des espaces topologiques et sur les points fixes des représentations; Sur la position d'un ensemble fermé de points d'un espace topologique; Sur les équations et les transformations, J. Math. Pures Appl. (9)**24**(1945), 95–167, 169–199, and 201–248. (134, 307)

Leray, J., L'anneau d'une representation; Propértés de l'anneau d'homologie d'une representation; Sur l'anneau d'homologie de l'espace homogène, quotient d'un grooupe clos par un sousgroupe abélien, connexe, maximum, C. R. Acad. Sci. Paris **222**(1946), 1366–1368; 1419–1422; **223**(1946), 412–415. (34, 45, 134, 221, 222, 342, 515, 516)

Leray, J., L'anneau spectral et l'anneau filtré d'homologie d'un espace localement compact et d'une application continue: L'homologie d'un espace fibré dont la fibre est connexe, J. Math. Pures Appl. **29**(1950). 1–80, 81–139; 169–213. (139, 142, 181, 221, 222, 277)

Lichnerowicz, A., Spineurs harmoniques, C.R. Acad. Sci. Paris **257**(1963), 7–9. (451)

Ligaard, H., Madsen, I.,Homology operations in the Eilenberg-Moore spectral seuqence, Math. Z. **143**(1975), 45–54. (326)

Lima, E.L., The Spanier-Whitehead duality in new homotopy categories, Summa Brasil. Math. **4**(1959), 91–148. (408, 495)

Lin, J.P., Torsion in H-spaces. I, Ann. of Math. **103**(1)976, 457–487. (376, 477)

Lin, J.P., Torsion in H-spaces. II, Ann. Math. (2) **107**(1978), 41–88. (477, 513)

Lin, J.P., Two torsion and the loop space conjecture, Ann. of Math. (2) **115**(1982), 35–91. (458, 477)

Lin, J.P., The first homotopy group of a finite H-space, J. Pure Appl. Algebra **90**(1993), 1–22. (476)

Lin, J.P., H-spaces with finiteness conditions, in the *Handbook of Algebraic Topology*, edited by I. M. James, Elsevier Science, Amsterdam, 1995, 1095–1141. (477)

Lin, W.H., Some elements in the stable homotopy of spheres, Proc. Amer. Math. Soc. **95**(1985), 295–298. (450)

Lin, W.H., EHP spectral sequence in the lambda algebra, in Papers in honor of José Adem (Spanish). Bol. Soc. Mat. Mexicana (2) **37**(1992), 339–353. (495)

Liulevicius, A., The factorization of cyclic reduced powers by secondary cohomology operations, Memoirs AMS **42**(1962). (368, 384, 392, 408, 413, 434, 445)

Liulevicius, A., A theorem in homological algebra and stable homotopy of projective spaces, Trans. Amer. Math. Soc. **109**(1963), 540–552. (297, 430)

Liulevicius, A., Coalgebras, resolutions, and the computer, Math. Algorithms **1**(1966), 4–11. (445)

Liulevicius, A., Characteristic classes and cobordism. Part I. Matematisk Institut, Aarhus Universitet, Aarhus 1967 vi+152 pp. (225, 227)

Loday, J.-L., Une version non commutative des algèbres de Lie: les algèbres de Leibniz, L'Enseignement Math. **39**(1993), 269–293. (512)

Loday, J.-L., La renaissance des opérades, Séminaire Bourbaki, Vol. 1994/95. Astérisque **237**(1996), Exp. No. 792, 47–74. (297)

Loday, J.-L., Cyclic Homology. Appendix E by María O. Ronco. Grundlehren der Mathematischen Wissenschaften 301, first edition, 1992. Second edition. Chapter 13 by the author in collaboration with Teimuraz Pirashvili. Springer-Verlag, Berlin, 1998. (510)

Loday, J.-L., Pirashvili, T., Universal enveloping algebras of Leibniz algebras and (co)-homology, Math. Annalen **296**(1993), 139–158. (512)

Lodder, J.M., Leibniz cohomology for differentiable manifolds, Ann. Inst. Fourier, Grenoble **48**(1998), 73–95. (512, 513)

Lubkin, S., Cohomology of Completion. North-Holland, Amsterdam, 1980. (28)

Lundell, A.T., Weingram, S., The Topology of CW-complexes. Van Nostrand Reinhold Co., New York, 1969. (102)

Lyndon, R.C., The cohomology theory of group extensions, Duke Math. J. **15**(1948), 271–292. (342)

Maakinen, J., Boundary formulae for reduced powers in the Adams spectral sequence, Ann. Acad. Sci. Fennicae **(A-1)562**(1973). (449)

MacLane, S., Slide and torsion products for modules, Univ. e Politec. Torino. Rend. Sem. Mat. **15**(1955–56), 281–309. (311)

MacLane, S., Homology. First ed. Springer-Verlag, Heidelberg, 1963. (50, 122, 515)

Mac Lane, S., The Milgram bar construction as a tensor product of functors. 1970 The Steenrod Algebra and its Applications (Proc. Conf. to Celebrate N. E. Steenrod's Sixtieth Birthday, Battelle Memorial Inst., Columbus, Ohio, 1970) pp. 135–152 Lecture Notes in Mathematics, Vol. 168. (243, 247)

Madsen, I., Milgram, R.J., The Classifying Spaces for Surgery and Cobordism of Manifolds. Princeton Univ., Princeton, NJ, 1979. (212)

Mahowald, M.E., The metastable homotopy of S^n, Mem. AMS**72**(1967). (448)

Mahowald, M.E., The primary v_2-periodic family, Math. Z. **177**(1981), 381–393. (450)

Mahowald, M.E., The image of J in the EHP sequence, Ann. of Math. (2) **116**(1982), 65–112. (450)

Mahowald, M.E., Tangora, M.C., Some differentials in the Adams spectral sequence, Topology **6**(1967), 349–369. (448, 449, 450, 484)

Mahowald, M.E., Thompson, R.D., The EHP sequence and periodic homotopy, in Handbook of Algebraic Topology, edited by I.M. James, Elsevier Science B.V., Amsterdam, 1995, 397–423. (495)

Margolis, H.R., Spectra and the Steenrod Algebra. Modules over the Steenrod Algebra and the Stable Homotopy Category. North-Holland Mathematical Library, 29. North-Holland Publishing Co., Amsterdam-New York, 1983. xix+489 pp. (432, 496)

Massey, W.S., Exact couples in algebraic topology, I, II, III, Ann. of Math. **56**(1950), 363–396. (37)

Massey, W.S., Products in exact couples, Ann. of Math. **59**(1954), 558–569. (60)

Massey, W.S., Some problems in algebraic topology and the theory of fibre bundles, Annals of Math. **62**(1955), 327–359. (133, 194)

Massey, W.S., Some higher order cohomology operations, 1958 Symposium internacional de topología algebraica, pp. 145–154 Universidad Nacional Autónoma de México and UNESCO,

Mexico City. (273, 297, 305)

Massey, W.S., Obstructions to the existence of almost complex structures, Bull. Amer. Math. Soc. **67**(1961), 559–564. (221)

Massey, W.S., Algebraic Topology: An Introduction. Harcourt, Brace & World, Inc., New York 1967 xix+261 pp. (159)

Massey, W.S., Higher order linking numbers, J. Knot Theory Ramifications **7**(1998), 393–414. (273, 304)

Massey, W.S., A Basic Course in Algebraic Topology. Graduate Texts in Mathematics 127, Springer-Verlag, New York, 1991. (100, 102, 371)

Massey, W.S., Peterson, F.P., The mod 2 cohomology of certain fibre spaces, Mem. Amer. Math. Soc. **74**(1967). (492)

Maunder, C.R.F., Cohomology operations of the Nth kind, Proc. London Math. Soc., (2)**13**(1963) 125–154. (376, 425)

Maunder, C.R.F., Chern characters and higher-order cohomology operations, Camb. Proc. Philos. Soc., **60**(1964) 751–764. (376, 425, 445)

Maunder, C.R.F., On the differentials in the Adams spectral sequence for the stable homotopy groups of spheres. I.; II, Proc. Cambridge Philos. Soc. **61**(1965), 53–60; 855–868. (429, 448)

May, J.P., The cohomology of restricted Lie algebras and of Hopf algebras; applications to the Steenrod algebra, Ph. D. thesis, Princeton Univ., 1964. (87, 368, 427, 428, 430, 441, 441, 442, 445, 447, 510)

May, J.P., The cohomology of restricted Lie algebras and of Hopf algebras, J. Alg. **3**(1966), 123–146. (441, 442, 510)

May, J.P., Simplicial Objects in Algebraic Topology. Van Nostrand Reinhold Co., New York, 1967. (106, 107, 109, 224, 243, 357)

May, J.P., The algebraic Eilenberg-Moore spectral sequence, preprint, 1968. (312)

May, J.P., The cohomology of principal bundles, homogeneous spaces, and two-stage Postnikov systems, Bull. Amer. Math. Soc. **74**(1968), 334–339. (311)

May, J.P., Matric Massey products, J. Algebra **12**(1969), 533–568. (273, 297, 311, 312)

May, J.P., A general approach to Steenrod operations, in The Steenrod algebra and its Applications, Springer Lecture Notes in Mathematics **168**(1970), 153–231. (128, 193, 326)

May, J.P., The geometry of iterated loop spaces. Springer Lectures Notes in Mathematics **271**(1972) viii+175 pp. (297)

May, J.P., Classifying spaces and fibrations, Memoirs Amer. Math. Soc. **155**(1975). (212, 268)

May, J.P., Derived categories in algebra and topology, Proceedings of the Eleventh International Conference of Topology (Trieste, 1993). Rend. Istit. Mat. Univ. Trieste **25**(1993), 363–377 (1994). (524)

May, J.P., A Concise Course in Algebraic Topology. The University of Chicago Press, Chicago, IL, 1999. (102)

May, J.P., Milgram, R.J., The Bockstein and the Adams spectral sequences., Proc. Amer. Math. Soc. **83**(1981), 128–130. (478)

Mazzeo, R.R., Melrose, R.B., The adiabatic limit, Hodge cohomology and Leray's spectral sequence for a fibration, J. Diff. Geom. **31**(1990), 185–213. (519)

McCleary, J., Cartan's cohomology theories and spectral sequences, Current trends in algebraic topology, Part 1 (London, Ont., 1981), pp. 499–506, CMS Conf. Proc., 2, Amer. Math. Soc.,

Providence, R.I., 1982. (227)

McCleary, J., On the mod p Betti numbers of loop spaces, Invent. Math. **87**(1987), 643–654. (477)

McCleary, J., Homotopy theory and closed geodesics, Homotopy theory and related topics (Kinosaki, 1988), 86–94, Lecture Notes in Math., 1418, Springer, Berlin, 1990. (178, 224)

McCleary, J., A topologist's account of Yang-Mills theory, Expos. Math. **10**(1992), 311–352. (220)

McCleary, J., A history of spectral sequences: Origins to 1953, in History of Topology, ed. James, I.M., North-Holland, Amsterdam, 1999, 631–663. (45, 134)

McClure, J.E., Staffeldt, R.E., On the topological Hochschild homology of bu. I, Amer. J. Math. **115**(1993), 1–45. (498)

McGibbon, C.A., Neisendorfer, J.A., On the homotopy groups of a finite-dimensional space, Comm. Math. Helv. **59**(1984), 253–257. (205)

Meyer, D.M., Z/p-equivariant maps between lens spaces and spheres, Math. Ann. **312**(1998), 197–214. (479)

Milgram, R.J., The bar construction and abelian H-spaces, Ill. J. Math. **11**(1967), 242–250. (211, 212, 268)

Milgram, R.J., Group representations and the Adams spectral sequence, Pac. J. Math. **41**(1972), 157–182. (426, 449)

Miller, H.R., A spectral sequence for the homology of an infinite delooping, Pac. J. Math. **79**(1978), 139–155. (499)

Miller, H.R., The Sullivan conjecture on maps from classifying spaces, Ann. of Math. (2) **120**(1984), 39–87. Correction: Ann. of Math. (2) **121**(1985), 605–609. (205, 458, 504)

Miller, H.R., Ravenel, D.C., Mark Mahowald's work on the homotopy groups of spheres. Algebraic topology (Oaxtepec, 1991), Contemp. Math. **146**(1993), 1–30. (450)

Miller, H.R., Ravenel, D.C., Wilson, W. S., Periodic phenomena in the Adams-Novikov spectral sequence, Ann. Math. (2) **106**(1977), 469–516. (500)

Milne, J.S., Etale Cohomology. Princeton Mathematical Series. 33. Princeton, New Jersey: Princeton University Press. XIII, 323 p.(1980). (521)

Milnor, J.W., Construction of universal bundles I, II, Ann. of Math. (2)**63**(1956), 272–284, 430–436. (211, 268, 331)

Milnor, J.W., The geometric realization of a semi-simplicial complex, Ann. of Math. **65**(1957), 357–362. (107, 274, 310)

Milnor, J.W., The Steenrod algebra and its dual, Ann. of Math. **67**(1958), 150–171. (131, 413, 417, 483)

Milnor, J.W., On spaces having the homotopy type of a CW–complex, Trans. Amer. Math. Soc. **90**(()1959), 272–280. (94)

Milnor, J.W., On the cobordism ring Ω^* and a complex analogue. I, Amer. J. Math. **82**(1960), 505–521. (368, 408, 414, 497)

Milnor, J.W., On axiomatic homology theory, Pac. J. Math. **12**(1962), 337–341. (69)

Milnor, J.W., Morse Theory (Notes by M. Spivak and R. Wells). Princeton Univ. Press, Princeton, 1963. (95)

Milnor, J.W., Moore, J.C., On the structure of Hopf algebras, Ann. of Math. **81**(1965), 211–264. (123, 131, 389, 438, 441, 454, 474)

Milnor, J.W., Stasheff, J.D., Characteristic Classes. Annals of Mathematics Studies, No. 76. Princeton University Press, Princeton, N. J.; University of Tokyo Press, Tokyo, 1974. vii+331 pp. (149, 151, 154 208, 218, 220, 411, 451)

Mimura, M., Mori, M., The squaring operations in the Eilenberg-Moore spectral sequence and the classifying space of an associaive H-space. I, Publ. Res. Math. Inst. Sci. **13**(1977/78), 755–776. (326)

Mimura, M., Nishida, G., Toda, H., Mod p decompositions of compact Lie groups, Publ. RIMS, Kyoto Univ., **13**(1971), 627–680. (362)

Mimura, M., Toda, H., Topology of Lie groups. I, II. Translated from the 1978 Japanese edition by the authors. Translations of Mathematical Monographs, 91. American Mathematical Society, Providence, RI, 1991. iv+451 pp. (149, 278)

Mitchell, S.A., Hypercohomology spectra and Thomason's descent theorem, Algebraic K-theory (Toronto, ON, 1996), Fields Inst. Commun., **16**(1997), 221–277. (522)

Moerdijk, I., Svensson, J.-A., The equivariant Serre spectral wequence, Proc. AMS **118**(1993), 263–278. (501)

Møller, J.M., On equivariant function spaces, Pacific J. Math. **142**(1990), 103–119. (502)

Moore, J.C., Some applications of homology theory to homotopy problems, Ann. of Math. **58**(1953), 325–350. (66)

Moore, J.C., On homotopy groups of spaces with a single non-vanishing homology group, Ann. of Math. **59**(1954), 549–557.

Moore, J.C., Algebraic homotopy theory, mimeographed notes, Princeton, 1956. (108, 224)

Moore, J.C., Semi-simplicial complexes and Postnikov systems, Symposium internacional de topologia algebraica, Univ. Nac. Aut. de Mexico and UNESCO, Mexico City, 1958, 232–247. (122)

Moore, J.C., Algébre homologique et homologie des espace classificants, Séminaire Cartan, 1959/60, exposé 7. (232, 233, 235, 241, 266, 269)

Moore, J.C., Cartan's constructions, the homology of $\mathcal{K}(\pi,n)$'s, and some later developments, Colloque "Analyse et Topologie" en l'Honneur de Henri Cartan (Orsay, 1974), pp. 173–212. Asterisque, No. 32-33, Soc. Math. France, Paris, 1976. (194)

Moore, J.C., Smith, L., Hopf algebras and multiplicative fibrations I, II, Amer. J. Math. **90**(1968), 752–780, 1113–1150. (327)

Morava, J., Noetherian completions of categories of cobordism comodules, Ann. Math. **121** (1985), 1–39. (477)

Morel, F., Voevodsky's proof of Milnor's conjecture, Bull. Amer. Math. Soc. (N.S.) **35**(1998), 123–143. (522)

Mori, M., The Steenrod operations in the Eilenberg-Moore spectral sequence, Hirosh. Math. J. **9**(1979), 17–34. (326)

Morse, M., The Calculus of Variations in the Large. AMS Colloquium Series 18, Providence, RI, 1934. (158)

Mosher, R., Tangora, M.C., Cohomology Operations and Applications in Homotopy Theory. Harper and Row, 1968. (128)

Moss, R.M.F., The composition pairing of Adams spectral sequences, Proc. London Philos. Soc. (3)**18**(1968), 179–192. (403, 406)

Moss, R.M.F., Secondary compositions and the Adams spectral sequence, Math. Zeit. **115**(1970), 283–310. (429, 430, 440)

Munkholm, H.J., The Eilenberg-Moore spectral sequence and strongly homotopy multiplicative maps, J. P. App. Alg. **5**(1974), 1–50. (277, 294, 296)

Nakaoka, M., Toda, H., On Jacobi identity for Whitehead products. J. Inst. Polytech. Osaka City Univ. Ser. A. **5**(1954), 1–13. (303)

Nassau, C. An internet page: http://www.math.uni-frankfurt.de/ nassau/Ext2. (445)

Neisendorfer, J.A., Homotopy theory modulo an odd prime, Princeton University thesis, 1972. (480, 481)

Neisendorfer, J.A., Primary homotopy theory, Mem. AMS **25**, no. 232, (1980). (461, 462, 480, 481)

Neumann, F., Quillen spectral sequence in rational homotopy theory, in Algebraic K-theory and its applications, Proceedings on the Workshop and Symposium, ICTP, Trieste (ed. H. Bass, A.O. Kuku, C. Pedrini), World Scientific Publishing Co. 1999. (488)

Nishida, G., The nilpotency of elements of the stable homotopy groups of spheres, J. Math. Soc. Japan **25**(1973), 707–732. (407)

Novikov, S.P., Cohomology of the Steenrod algebra, Dokl. Akad. Nauk SSSR **128**(1959), 893–895. (429)

Novikov, S.P., The methods of algebraic topology from the viewpoint of cobordism theory, Izvestia Akad. Nauk SSSR Ser. Mat. **31**(1967), 855–951 (Russian). Translation, Math USSR-Izv. (1967), 827–913. (415, 451, 497, 499)

Orr, K.E., Link concordance invariants and Massey products, Topology **30**(1991), 699–710. (310)

Palmieri, J.H., Self-maps of modules over the Steenrod algebra, J. Pure Appl. Algebra **79**(1992), 281–291. (432)

Paranjape, K.H., Some spectral sequences for filtered complexes and applications, J. Algebra **186**(1996), 793–806. (516)

Peterson, F.P., Generalized cohomotopy groups, Amer. J. Math. **78**(1956), 259–281. (458, 480)

Pimsner, M., Popa, S., Voiculescu, D., Homogeneous C^*-extensions of $C(X) \otimes K(H)$. I, J. Operator Theory **1**(1979), 55–108. (523)

Pirashvili, T., On Leibniz homology, Ann. Inst. Fourier, Grenoble **44**(1994), 401-411. (512, 513)

Poincaré, H., Analysis situs, J. Ecole Poly. **1**(1895), 1–121. (409)

Poincaré, J.H., Mémoire sur les courbes définies par une équation différentielle, Jour. de Math. **7**(3) (1881), 375–442. (207)

Pontryagin, L., Characteristic classes on differentiable manifolds, Mat. Sbornik N.S. **21(63)**(1947), 233–284. (207)

Pontryagin, L.S., Smooth manifolds and their applications in homotopy theory, Trudy Mat. Inst. im. Steklov. **45**(1955), Izdat. Akad. Nauk SSSR, Moscow, 139 pp. (409)

Porter, R., Milnor's $\bar{\mu}$-invariants and Massey products, Trans. Amer. Math. Soc. **257**(1980), 39–71. (310)

Postnikov, M.M., Determination of the homology groups of a space by means of the homotopy invariants, Dokl. Akad. Nauk SSSR **76**(1951), 359–362. (120)

Postnikov, M.M., On Cartan's theorem, Russ. Math. Surveys **21**(1966), 25–36. (197)

Prieto, C., The relative spectral sequence of Leray-Serre for fibration pairs, Monografías del Instituto de Matemáticas [Monographs of the Institute of Mathematics], 8. Universidad Nacional Autónoma de México, Mexico City, 1979. i+102 pp. (222)

Puppe, D., Homotopiemenge und ihre induzierten Abbildungen I, Math. Zeit. **69**(1958), 299-344. (97)

Quillen, D.G., Spectral sequences of a double semi-simplicial group, Topology, **5**(1966), 155–157. (488, 489)

Quillen, D.G., Homotopical Algebra, Springer Lecture Notes in Mathematics, **43**(1967). (109)

Quillen, D.G., Rational homotopy theory, Ann. of Math. **90**(1969), 205–295. (487)

Quillen, D.G., An application of simplicial profinite groups, Comm. Math. Helv. **44**(1969'), 45–60. (82)

Quillen, D.G., On the formal group laws of unoriented and complex cobordism theory, Bull. Amer. Math. Soc. **75**(1969"), 1293–1298. (451, 500)

Quillen, D.G., Higher algebraic K-theory. I, in Algebraic K-theory, I: Higher K-theories (Proc. Conf., Battelle Memorial Inst., Seattle, Wash., 1972), Springer Lecture Notes in Math. **341** (1973), 85–147. (520)

Radon, J., Lineare Scharen orthogonaler Matrizen, Abh. Math. Sem. Univ. Hamburg **1**(1922), 1–14. (366)

Ravenel, D.C., Complex Cobordism and Stable Homotopy Groups of Spheres. Pure and Applied Mathematics, 121. Academic Press, Inc., Orlando, Fla., 1986. xx+413 pp. (368, 408, 415, 442, 451, 495, 500)

Ravenel, D.C., Nilpotence and Periodicity in Stable Homotopy Theory. Appendix C by Jeff Smith, Annals of Mathematics Studies, **128**. Princeton University Press, Princeton, NJ, 1992. xiv+209 pp. (408, 451, 496)

Ray, N., The symplectic bordism ring, Proc. Cambridge Philos. Soc. **71**(1972), 271–282. (415, 497)

Rector, D.L., An unstable Adams spectral sequence, Topology **5**(1966), 343–346. (493)

Rector, D.L., Steenrod operations in the Eilenberg-Moore spectral sequence, Comment. Math. Helv. **45**(1970), 540–552. (274, 314, 322, 355)

Rolfsen, D., Knots and Links. Mathematics Lecture Series, No. 7. Publish or Perish, Inc., Berkeley, Calif., 1976. ix+439 pp. (304)

Rothenberg, M., Steenrod, N.S., The cohomology of classifying spaces of H-spaces, Bull. Amer. Math. Soc. **71**(1965), 872–875. (268)

Rotman, J.J., An Introduction to Homological Algebra. Academic Press, New York, 1979.

Rudyak, Y.B., On Thom Spectra, Orientability, and Cobordism. With a foreword by Haynes Miller. Springer Monographs in Mathematics. Springer-Verlag, Berlin, 1998. xii+587 pp. (302, 410)

Rudyak, Y.B., On category weight and its applications, Topology **38**(1999), 37–55.

Samelson, H., Beiträge zur Topologie der Gruppenmannigfaltigkeiten, Annals of Math. **42**(1941), 1091–1137. (134, 148, 277)

Samelson, H., Groups and spaces of loops, Comment. Math. Helv. **28**(1954), 278–287. (303)

Sard, A., The measure of the critical values of differentiable maps, Bull. Amer. Math. Soc. **48**(1942), 883–890. (409)

Sawka, J., Odd primary Steenrod operations in first-quadrant spectral sequences, Trans. Amer. Math. Soc. **273**(1982), 737–752. (326)

Schlessinger, M., Stasheff, J.D., The Lie algebra structure of tangent cohomology and deformation theory, J. Pure Appl. Algebra **38**(1985), 313–322. (310)

Schochet, C., A two-stage Postnikov system where $E_2 \neq E_\infty$ in the Eilenberg-Moore spectral sequence, Trans. Amer. Math. Soc. **157**(1971), 113–118. (277, 291, 296)

Schochet, C., Cobordism from an algebraic point of view. Lecture Notes Series, No. 29. Matematisk Institut, Aarhus Universitet, Aarhus, 1971. iv+190 pp. (414)

Schochet, C., Topological methods for C^*-algebras. I. Spectral sequences. Pacific J. Math. **96**(1981), 193–211. (523)

Schön, R., Effective algebraic topology, Mem. Amer. Math. Soc. **92**(1991), no. 451, vi+63 pp. (330)

Selick, P.S., Odd primary torsion in $\pi_k(S^3)$, Topology **17**(1978), 407–412. (482)

Sergeraert, F., The computability problem in algebraic topology, Adv. Math., **104**(1994), 1–29. (330)

Serre, J.-P., Cohomologie des extensions de groupes, C.R. Acad. Sci. Paris **231**(1950), 643–646. (342)

Serre, J.-P., Homologie singulière des espaces fibrés. I. La suite spectrale; II. Les espaces de lacets; III. Applications homotopiques, C.R. Acad. Sci. Paris **231**(1950), 1408–1410; **232**(1951), 31–33 and 142–144. (134)

Serre, J.-P., Homologie singulière des espaces fibrés, Ann. of Math.**54**(1951), 425–505. (45, 134ff., 190, 205, 206, 222, 297, 298, 331, 477, 487)

Serre, J.-P., Cohomologie modulo 2 des complexes d'Eilenberg-MacLane , Comm. Math. Helv. **27**(1953) , 198–232. (18, 128, 180, 197, 202, 205, 354, 468)

Serre, J.-P., Cohomologie et géoétrie algébrique, Congés International d'Amsterdam **3**(1954), 515–520. (523)

Seymour, R.M., On the convergence of the Eilenberg-Moore spectral sequence, Proc. London Math. Soc. (3)**36**(1978), 141–162. (498)

Shick, P.L., Adams spectral sequence chart, in Algebraic Topology: Oaxtepec 1991, ed. by M.C. Tangora, Cont. Math. **146**(1993), 479–481. (445)

Shih, W., Homologie des espaces fibrés, Publ. I. H. E. S. **13**(1962), 88 pp. (223)

Shipley, B.E., Pro-isomorphisms of homology towers, Math. Z. **220**(1995), 257–271. (82, 360, 361)

Shipley, B.E., Convergence of the homology spectral sequence of a cosimplicial space, Amer. J. Math. **118**(1996), 179–207.

Shnider, S., Sternberg, S., Quantum groups. From coalgebras to Drinfel'd algebras. A guided tour. Graduate Texts in Mathematical Physics, II. International Press, Cambridge, MA, 1993. xxii+496 pp. (126)

Singer, W.M., Connective fiberings over BU and U, Topology **7**(1968), 271–303. (327)

Singer, W.M., Extension theory for connected Hopf algebras, J. Algebra **21**(1972), 1–16. (513)

Singer, W., Steenrod squares in spectral sequences I, II, Trans. AMS **175**(1973), 327–336, 337–353. (195, 221, 326)

Singer, W.M., The algebraic EHP sequence, Trans. Amer. Math. Soc. **201**(1975), 367–382. (495)

Singer, W.M., A new chain complex for the homology of the Steenrod algebra, Math. Proc. Cambridge Philos. Soc. **90**(1981), 279–292. (450)

Sklyarenko, E.G., Zeeman's filtration in homology, Mat. Sb. **183**(1992), 103–116. (505)

Smith, J.R., Iterating the cobar construction, Mem. Amer. Math. Soc. **109**(1994), no. 524, viii+141 pp. (491)

Smith, L., Homological algebra and the Eilenberg-Moore spectral sequence, Trans. Amer. Math. Soc. **129**(1967), 58–93. (235, 243, 277, 298, 300, 301, 321)

Smith, L., The cohomology of stable two stage Postnikov systems, Illinois J. Math. **11**(1967), 310–329. (292)

Smith, L., On the Künneth theorem. I. The Eilenberg-Moore spectral sequence, Math. Z. **116**(1970), 94–140. (274, 313, 317, 318, 321, 323, 328, 498)

Smith, L., On the Eilenberg-Moore spectral sequence, Algebraic topology (1970), Proc. Sympos. Pure Math., **22**(1971), 231–246. (277)

Smith, L., On the characteristic zero cohomology of the free loop space, Amer. J. Math. **103**(1981), 887–910. (178)

Smith, L., A note on the realization of graded complete intersection algebras by the cohomology of a space, Quart. J. Math. Oxford Ser. (2) **33**(1982), 379–384. (283)

Smith, S.B., A based Federer spectral sequence and the rational homotopy of function spaces, Manuscripta Math. **93**(1997), 59–66. (490)

Spanier, E., Algebraic Topology. McGraw-Hill Book Co., New York-Toronto, Ont.-London, 1966 xiv+528 pp. Corrected reprint. Springer-Verlag, New York-Berlin, 1981. xvi+528 pp. (222, 316)

Spanier, E.H., Locally constant cohomology, Trans. Amer. Math. Soc. **329**(1992), 607–624. (501)

Srinivas, V., Algebraic K-theory. Second edition, Birkhäuser, Boston, 1996. (96, 521)

Stallings, J., Homology and central series of groups, J. Algebra **2**(1965), 170–181. (274, 310, 342)

Stammbach, U., Anwendungen der Homologietheorie der Gruppen auf Zentralreihen and auf Invarianten von Präsentierungen, Math. Z. **94**(1966), 157–177. (342)

Stasheff, J.D., Homotopy associativity of H-spaces I, II, Trans. Amer. Math. Soc. **108**(1963), 275–312. (212, 268. 292, 294, 458)

Stasheff, J.D., A classification theorem for fibre spaces, Topology **2**(1963), 239–246. (211, 212)

Stasheff, J.D., *H*-spaces from a homotopy point of view. 1968 Conference on the Topology of Manifolds (Michigan State Univ., E. Lansing, Mich., 1967) pp. 135–146 Prindle, Weber & Schmidt, Boston, Mass. (307)

Stasheff, J.D., Deformation theory and the little constructions of Cartan and Moore, in Algebraic Topology and Algebraic K-theory, edited by W. Browder, Princeton University Press, Princeton, NJ, 1987, 322–331. (194)

Stasheff, J.D., The pre-history of operads, in Operads: Proceedings of Renaissance Conferences (Hartford, CT/Luminy, 1995), 9–14, Contemp. Math. **202**(1997), 9–14. (297)

Stasheff, J.D., Halperin, S., Differential algebra in its own rite, Proceedings of the Advanced Study Institute on Algebraic Topology (Aarhus Univ., Aarhus 1970), Vol. III, pp. 567–577. Various Publ. Ser., No. 13, Mat. Inst., Aarhus Univ., Aarhus, 1970. (273, 277, 292ff.)

Steenrod, N.E., Homology with local coefficients, Ann. of Math. **44**(1943), 610–627. (163)

Steenrod, N.E., Classification of sphere bundles, Ann. of Math. **45**(1944), 294–311. (135)

Steenrod, N.E., Products of cocycles and extensions of mappings, Ann. of Math. (2) **48**(1947), 290–320. (128, 299)

Steenrod, N.E., Cohomology invariants of mappings, Ann. of Math. (2) **50**(1949). 954–988. (367)

Steenrod, N.E., Topology of Fibre Bundles. Princeton Univ. Press, Princeton, NJ, 1951. (113, 149, 207, 211, 278)

Steenrod, N.E., Reduced powers of cohomology classes, Ann. of Math. (2)**56**(1952), 47–67. (128, 134)

Steenrod, N.E., Cohomology operations derived from the symmetric group, Comment. Math. Helv. **31**(1957), 195–218. (193)

Steenrod, N.E., Homology groups of symmetric groups and reduced power operations. Proc. Nat. Acad. Sci. U. S. A. **39**(1953), 213–217. (326)

Steenrod, N.E., Epstein, D.B.A., Cohomology Operations. Princeton Univ. Press, 1962. (128, 130, 289,)

Steenrod, N.E., Cooke, G.E., Finney, R.L., Homology of Cell Complexes. Based on lectures by Norman E. Steenrod Princeton University Press, Princeton, N.J.; University of Tokyo Press, Tokyo 1967 xv+256 pp. (102)

Stein, D., Massey products in the cohomology of groups with applications to link theory, Trans. Amer. Math. Soc. **318**(1990), 301–325. (310)

Stiefel, E., Richtungsfelder und Fernparallelismus in Mannigfaltigkeiten, Comm. Math. Helv. **8**(1936), 3–51. (207)

Stolz, S., The level of real projective spaces, Comm. Math. Helv. **64**(1989), 661–674. (479)

Stolz, S., Simply connected manifolds of positive scalar curvature, Ann. of Math. **136**(1992), 511–540. (451)

Stong, R.E., Notes on Cobordism Theory. Mathematical notes Princeton University Press, Princeton, N.J.; University of Tokyo Press, Tokyo 1968 v+354+lvi pp. (409, 412, 413, 414)

Stover, C.R., A van Kampen spectral sequence for higher homotopy groups, MIT Ph.D. thesis, 1988. (489)

Stover, C.R., A van Kampen spectral sequence for higher homotopy groups, Topology **29**(1990), 9–26. (489, 502)

Strom, J.A., Essential category weight and phantom maps, to appear in the Proceedings of the BCAT98. (302)

Sugawara, M., A condition that a space is group-like, Math. J. Okayama Univ. **7**(1957), 123–149. (292, 294, 295)

Sullivan, D., Geometric Topology. Part I. Localization, Periodicity, and Galois Symmetry. Revised version. Massachusetts Institute of Technology, Cambridge, Mass., 1971. 432 pp. (330, 354, 361, 362, 504)

Sullivan, D., Differential forms and the topology of manifolds, Manifolds—Tokyo 1973 (Proc. Internat. Conf., Tokyo, 1973), 37–49, Univ. Tokyo Press, Tokyo, 1975. (477)

Switzer, R.M., Algebraic Topology—Homotopy and Homology. Die Grundlehren der mathematischen Wissenschaften, Band 212. Springer-Verlag, New York-Heidelberg, 1975. xii+526 pp. (222, 368, 408, 450, 451, 496, 498)

Tamme, G., Introduction to Étale Cohomology. Translated from the German by Manfred Kolster. Universitext. Springer-Verlag, Berlin, 1994. x+186 pp. (514, 521)

Tanabe, M., On Morava K-theories of Chevalley groups, Amer. J. Math. **117**(1995), 263–278. (498)

Tangora, M.C., On the cohomology of the Steenrod algebra, Ph.D. dissertation, Northwestern University, 1966. (423)

Tangora, M.C., On the cohomology of the Steenrod algebra, Math. Zeit. **116**(1970), 18–64. (442, 445, 447, 428)

Tangora, M.C., Some extension questions in the Adams spectral sequence, Proceedings of the Advanced Study Institute on the Algebraic Topology (Aarhus Univ., Aarhus, 1970), Vol.III, pp. 578–587. Various Publ. Ser. **13** Math. Inst., Aarhus Univ., Aarhus, 1970. (450)

Tangora, M.C., Computing the homology of the lambda algebra, Mem. Amer. Math. Soc. **58**(1985), no. 337, v+163 pp. (493)

Tangora, M.C., Some Massey products in Ext, Topology and representation theory (Evanston, IL, 1992), Contemp. Math. **158**(1994), 269–280.

Tate, J., Homology of Noetherian rings and local rings, Ill. J. Math. **1**(1957), 14–27. (261)

Thom, R., Espaces fibrés en sphères et carrés de Steenrod, Ann. ENS **69**(1952), 109–181. (217)

Thom, R., Quelques propriétés globales des variétés différentiables, Comment. Math. Helv. **28** (1954), 17–86. (368, 408, 409, 410, 413)

Thomas, C.B., Characteristic Classes and the Cohomology of Finite Groups. Cambridge University Press, Cambridge, UK, 1986. (344)

Thomas, P.E., Steenrod squares and H-spaces, Ann. Math. **77**(1963), 306–317. (477)

Thomas, P.E., Zahler, R.S., Nontriviality of the stable homotopy element γ_1, J. Pure Appl. Algebra **4**(1974), 189–203. (500)

Thomason, R.W., First quadrant spectral sequences in algebraic K-theory via homotopy colimits, Comm. Algebra **10**(1982), 1589–1668. (504, 522)

Thomason, R.W., The homotopy limit problem, Proceedings of the Northwestern Homotopy Theory Conference (Evanston, Ill., 1982), Contemp. Math. **19**(1983), 407–419. (503, 521, 522)

Thomason, R.W., Algebraic K-theory and étale cohomology, Ann. Sci. École Norm. Sup. (4) **18**(1985), 437–552. Erratum: (4) **22**(1989), 675–677. (521, 522)

Tillmann, U., Relation of the van Est spectral sequence to K-theory and cyclic homology, Illinois J. Math. **37**(1993), 589–608. (512)

Toda, H., Le produit de Whitehead et l'invariant de Hopf, C. R. Acad. Sci. Paris **241**(1955), 849–850. (425)

Toda, H., p-primary components of homotopy groups IV, Compositions and toric constructions, Mem. Coll. Sci. Kyoto Ser. A Math. **32**(1959), 103–119. (429)

Toda, H., Composition Methods in Homotopy Groups of Spheres. Annals of Mathematics Studies, No. 49 Princeton University Press, Princeton, N.J. 1962 v+193 pp. (368, 429, 446, 450, 494, 495)

Toomer, G.H., Lusternik-Schnirelmann category and the Moore spectral sequence, Math. Z. **138** (1974), 123–143. (302)

Turaev, V.G., The Milnor invariants and Massey products, (Russian) Studies in topology, II. Zap. Naučn. Sem. Leningrad. Otdel. Mat. Inst. Steklov. (LOMI) **66**(1976), 189–203, 209–210. (310)

Turner, J.M., Operations and spectral sequences. I, Trans. Amer. Math. Soc. **350**(1998), 3815–3835. (221, 326)

Uehara, H., Massey, W.S., The Jacobi identity for Whitehead products, in Algebraic geometry and topology. A symposium in honor of S. Lefschetz, pp. 361–377. Princeton University Press, Princeton, N. J., 1957. (273, 297, 303, 304)

Umeda, Y., A remark on a theorem of J.-P. Serre, Proc. Japan Acad. **35**(1959), 563–566. (201, 205)

van Est, W.T., A generalization of the Cartan-Leray spectral sequence. I, II., Indag. Math. **20**(1958), 399–413. (511)

Vassiliev, V.A., Complements of Discriminants of Smooth Maps: Topology and Applications. Translated from the Russian by B. Goldfarb. Translations of Mathematical Monographs, 98. American Mathematical Society, Providence, RI, 1992. vi+208 pp. (505)

Vázquez, R., Nota sobre los cuadrados de Steenrod en la sucesion espectral de un espacio fibrado, Bol. Soc. Mat. Mexicana **2**(1957), 1–8. (194)

Verdier, J.-L., Des catégories dérivées des catégories abéliennes. With a preface by Luc Illusie. Edited and with a note by Georges Maltsiniotis. Astérisque No. 239, (1996), xii+253 pp. (1997). (523)

Vigué-Poirrier, M., Homologie et K-théorie des algèbres commutatives: caractérisation des intersections complètes, J. Algebra **173**(1995), 679–695. (283)

Vinogradov, A.M., A spectral sequence associated with a nonlinear differential equation and algebro-geometric foundations of Lagrangian field theory with constraints, Soviet Math. Dokl. **19**(1978), 144–148. (519)

Viterbo, C., Some remarks on Massey products, tied cohomology classes, and the Lusternik-Shnirelman category, Duke Math. J. **86**(1997), 547–564. (304)

Voevodsky, V., The Milnor conjecture, preprint, 1996, Algebraic K-theory preprint server, http://www.math.uiuc.edu/K-theory/. (522)

Wall, C.T.C., Determination of the cobordism ring, Ann. of Math. (2) **72**(1960), 292–311. (368)

Wang, H.C., The homology groups of the fiber bundles over a sphere, Duke Math. J. **16**(1949), 33–38. (145)

Warner, F.W., Foundations of Differentiable Manifolds and Lie Groups. Scott, Foresman and Co., Glenview, Ill.-London, 1971. viii+270 pp. Corrected reprint of the 1971 edition. Graduate Texts in Mathematics, 94. Springer-Verlag, New York-Berlin, 1983. (165, 276)

Weibel, C.A., An Introduction to Homological Algebra. Cambridge University Press, New York, NY, 1994. (28, 44, 344, 487, 524)

Whitehead, G.W., The $(n+2)^{nd}$ homotopy group of the n-sphere. Ann. of Math. **52**(1950), 245–247. (180, 205)

Whitehead, G.W., Fiber spaces and the Eilenberg homology groups, Proc. Nat. Acad. Sci. U.S.A., **38**(1952), 426–430. (203)

Whitehead, G.W., On the Freudenthal theorems, Ann. of Math. **57**(1953), 209–228. (488)

Whitehead, G.W., On mappings into group-like spaces. Comment. Math. Helv. **28**(1954), 320–328. (303, 347)

Whitehead, G.W., On the homology suspension, Ann. of Math. (2) **62**(1955), 254–268. (297, 301)

Whitehead, G.W., Generalized homology theories, Trans. Amer. Math. Soc. **102**(1962), 227–283. (222, 408, 495)

Whitehead, G.W., Recent advances in homotopy theory, Conference Board of the Mathematical Sciences Regional Conference Series in Mathematics, No. 5. American Mathematical Society, Providence, R.I., 1970. iv+82 pp. (493, 495)

Whitehead, G.W., Elements of Homotopy Theory. Springer-Verlag, New York, (1978). (96, 115, 132, 139, 160, 178, 199, 340, 403)

Whitehead, G.W., Fifty years of homotopy theory, Bull. Amer. Math. Soc (2)**8**(1983), 1–29. (205)

Whitehead, J.H.C., On adding relations to homotopy groups, Ann. of Math. (2) **42**(1941), 409–428. (303)

Whitehead, J.H.C., Combinatorial homotopy I, II, Bull. A.M.S. **55**(1949), 213–245, 453–496. (93, 95)

Whitney, H., Sphere spaces, Proc. Nat. Acad. Sci. U.S.A. **21**(1935), 462–468. (133, 207)

Whitney, H., Differentiable manifolds, Ann. of Math. **37**(1936), 645–680. (409)

Whitney, H., On products in a complex, Annals of Math. **39**(1938), 397–432. (133)

Wolf, J., The cohomology of homogeneous spaces, Amer. J. Math. **99**(1977), 312–340. (277, 294, 296)

Wu, J., On combinatorial descriptions of homotopy groups of certain spaces, to appear in Math. Proc. Cambridge Phil. Soc. (366)

Wu, W.-T., Classes caractéristiques et *i*-carrés d'une variété, C.R. Acad. Sci. Paris **230**(1950), 508–509. (217)

Würgler, U., On products in a family of cohomology theories associated to the invariant prime ideals of $\pi_*(BP)$, Comment. Math. Helv. **52**(1977), 457–481. (482)

Yamaguchi, A., Note on the Eilenberg-Moore spectral sequence, Publ. Res. Inst. Math. Sci. **22**(1986), 889–903. (321)

Yoneda,N., On the homology theory of modules, J. Fac. Sci. Univ. Tokyo. Sect. I. **7**(1954), 193–227. (380, 381)

Zabrodsky, A., Implications in the cohomology of *H*-spaces, Illinois J. Math. **14**(1970), 363–375. (458)

Zabrodsky, A., Secondary operations in the cohomology of *H*-spaces, Illinois J. Math. **15**(1971), 648–655. (476, 477)

Zabrodsky, A., On the construction of new finite CW *H*-spaces, Invent. Math. 16 (1972), 260–266. (476)

Zariski, O., Samuel, P., Commutative algebra. Vol. I and II. The University Series in Higher Mathematics. D. Van Nostrand Co., Inc., Princeton, N. J.-Toronto-London-New York. I. 1958. xi+329 pp.; II. 1960 x+414 pp. (278)

Zeeman, E.C., A proof of the comparison theorem for spectral sequences, Proc. Camb. Philos. Soc. **53**(1957), 57–62. (23)

Zeeman, E.C., A note on a theorem of Armand Borel, Proc. Camb. Philos. Soc. **54**(1958), 396–398. (85, 86)

Zeeman, E. C., Dihomology: I. Relations between homology theories; II. The spectral theories of a map; III. A generalization of the Poincaré duality for manifolds, Proc. London Math. Soc. (3) **12**(1962), 609–638; 639–689; **13**(1963), 155–183. (504)

Zisman, M., Fibre bundles, fibre maps, in History of Topology, ed. James, I.M., North-Holland, Amsterdam, 1999, 605–629. (113, 133)

Symbol Index

Index